Iron and Cobalt Catalysts

Iron and Cobalt Catalysts

Special Issue Editors
Wilson D. Shafer
Gary Jacobs

MDPI • Basel • Beijing • Wuhan • Barcelona • Belgrade • Manchester • Tokyo • Cluj • Tianjin

Special Issue Editors
Wilson D. Shafer
Asbury University
USA

Gary Jacobs
University of Texas at San Antonio
USA

Editorial Office
MDPI
St. Alban-Anlage 66
4052 Basel, Switzerland

This is a reprint of articles from the Special Issue published online in the open access journal *Catalysts* (ISSN 2073-4344) (available at: https://www.mdpi.com/journal/catalysts/special_issues/Fe_Co).

For citation purposes, cite each article independently as indicated on the article page online and as indicated below:

LastName, A.A.; LastName, B.B.; LastName, C.C. Article Title. *Journal Name* **Year**, *Article Number*, Page Range.

ISBN 978-3-03928-388-0 (Pbk)
ISBN 978-3-03928-389-7 (PDF)

Cover image courtesy of Eliza Tan.

© 2020 by the authors. Articles in this book are Open Access and distributed under the Creative Commons Attribution (CC BY) license, which allows users to download, copy and build upon published articles, as long as the author and publisher are properly credited, which ensures maximum dissemination and a wider impact of our publications.

The book as a whole is distributed by MDPI under the terms and conditions of the Creative Commons license CC BY-NC-ND.

Contents

About the Special Issue Editors vii

Wilson D. Shafer and Gary Jacobs
Editorial: Cobalt and Iron Catalysis
Reprinted from: *Catalysts* 2020, *10*, 36, doi:10.3390/catal10010036 1

Neil Simpson, Karin Maaijen, Yfranka Roelofsen and Ronald Hage
The Evolution of Catalysis for Alkyd Coatings: Responding to Impending Cobalt Reclassification with Very Active Iron and Manganese Catalysts, Using Polydentate Nitrogen Donor Ligands
Reprinted from: *Catalysts* 2019, *9*, 825, doi:10.3390/catal9100825 7

Christian A. M. R. van Slagmaat, Khi Chhay Chou, Lukas Morick, Darya Hadavi, Burgert Blom and Stefaan M. A. De Wildeman
Synthesis and Catalytic Application of Knölker-Type Iron Complexes with a Novel Asymmetric Cyclopentadienone Ligand Design
Reprinted from: *Catalysts* 2019, *9*, 790, doi:10.3390/catal9100790 35

Chen-Yu Chou, Jason A. Loiland and Raul F. Lobo
Reverse Water-Gas Shift Iron Catalyst Derived from Magnetite
Reprinted from: *Catalysts* 2019, *9*, 773, doi:10.3390/catal9090773 59

Yizhou Li, Yepeng Yang, Daomei Chen, Zhifang Luo, Wei Wang, Yali Ao, Lin Zhang, Zhiying Yan and Jiaqiang Wang
Liquid-Phase Catalytic Oxidation of Limonene to Carvone over ZIF-67(Co)
Reprinted from: *Catalysts* 2019, *9*, 374, doi:10.3390/catal9040374 77

Bingchuan Yang, Yuanfeng Qi and Rutao Liu
Pilot-Scale Production, Properties and Application of Fe/Cu Catalytic-Ceramic-Filler for Nitrobenzene Compounds Wastewater Treatment
Reprinted from: *Catalysts* 2019, *9*, 11, doi:10.3390/catal9010011 91

Kai Li, Yang Li, Wenchao Peng, Guoliang Zhang, Fengbao Zhang and Xiaobin Fan
Bimetallic Iron–Cobalt Catalysts and Their Applications in Energy-Related Electrochemical Reactions
Reprinted from: *Catalysts* 2019, *9*, 762, doi:10.3390/catal9090762 109

Dmytro S. Nesterov and Oksana V. Nesterova
Polynuclear Cobalt Complexes as Catalysts for Light-Driven Water Oxidation: A Review of Recent Advances
Reprinted from: *Catalysts* 2018, *8*, 602, doi:10.3390/catal8120602 141

Adolph Anga Muleja, Joshua Gorimbo and Cornelius Mduduzi Masuku
Effect of Co-Feeding Inorganic and Organic Molecules in the Fe and Co Catalyzed Fischer–Tropsch Synthesis: A Review
Reprinted from: *Catalysts* 2019, *9*, 746, doi:10.3390/catal9090746 163

Wilson D. Shafer, Muthu Kumaran Gnanamani, Uschi M. Graham, Jia Yang, Cornelius M. Masuku, Gary Jacobs and Burtron H. Davis
Fischer–Tropsch: Product Selectivity–The Fingerprint of Synthetic Fuels
Reprinted from: *Catalysts* 2019, *9*, 259, doi:10.3390/catal9030259 179

Julián López-Tinoco, Rubén Mendoza-Cruz, Lourdes Bazán-Díaz, Sai Charan Karuturi, Michela Martinelli, Donald C. Cronauer, A. Jeremy Kropf, Christopher L. Marshall and Gary Jacobs
The Preparation and Characterization of Co–Ni Nanoparticles and the Testing of a Heterogenized Co–Ni/Alumina Catalyst for CO Hydrogenation
Reprinted from: Catalysts **2020**, *10*, 18, doi:10.3390/catal10010018 237

Yanying Qi, Jia Yang, Anders Holmen and De Chen
Investigation of $C_1 + C_1$ Coupling Reactions in Cobalt-Catalyzed Fischer-Tropsch Synthesis by a Combined DFT and Kinetic Isotope Study
Reprinted from: Catalysts **2019**, *9*, 551, doi:10.3390/catal9060551 267

Anna P. Petersen, Michael Claeys, Patricia J. Kooyman and Eric van Steen
Cobalt-Based Fischer–Tropsch Synthesis: A Kinetic Evaluation of Metal–Support Interactions Using an Inverse Model System
Reprinted from: Catalysts **2019**, *9*, 794, doi:10.3390/catal9100794 279

Ljubiša Gavrilović, Jonas Save and Edd A. Blekkan
The Effect of Potassium on Cobalt-Based Fischer–Tropsch Catalysts with Different Cobalt Particle Sizes
Reprinted from: Catalysts **2019**, *9*, 351, doi:10.3390/catal9040351 293

Pawel Mierczynski, Bartosz Dawid, Karolina Chalupka, Waldemar Maniukiewicz, Izabela Witoska, Krasimir Vasilev and Malgorzta I. Szynkowska
Comparative Studies of Fischer-Tropsch Synthesis on Iron Catalysts Supported on Al_2O_3-Cr_2O_3 (2:1), Multi-Walled Carbon Nanotubes or BEA Zeolite Systems
Reprinted from: Catalysts **2019**, *9*, 605, doi:10.3390/catal9070605 301

Harrison Williams, Muthu K. Gnanamani, Gary Jacobs, Wilson D. Shafer and David Coulliette
Fischer–Tropsch Synthesis: Computational Sensitivity Modeling for Series of Cobalt Catalysts
Reprinted from: Catalysts **2019**, *9*, 857, doi:10.3390/catal9100857 319

Wenping Ma, Gary Jacobs, Wilson D. Shafer, Yaying Ji, Jennifer L. S. Klettlinger, Syed Khalid, Shelley D. Hopps and Burtron H. Davis
Fischer-Tropsch Synthesis: Cd, In and Sn Effects on a 15%Co/Al_2O_3 Catalyst
Reprinted from: Catalysts **2019**, *9*, 862, doi:10.3390/catal9100862 331

Seong Bin Jo, Tae Young Kim, Chul Ho Lee, Jin Hyeok Woo, Ho Jin Chae, Suk-Hwan Kang, Joon Woo Kim, Soo Chool Lee and Jae Chang Kim
Selective CO Hydrogenation Over Bimetallic Co-Fe Catalysts for the Production of Light Paraffin Hydrocarbons (C_2–C_4): Effect of Space Velocity, Reaction Pressure and Temperature
Reprinted from: Catalysts **2019**, *9*, 779, doi:10.3390/catal9090779 347

Rama Achtar Iloy and Kalala Jalama
Effect of Operating Temperature, Pressure and Potassium Loading on the Performance of Silica-Supported Cobalt Catalyst in CO_2 Hydrogenation to Hydrocarbon Fuel
Reprinted from: Catalysts **2019**, *9*, 807, doi:10.3390/catal9100807 359

Nafeezuddin Mohammad, Sujoy Bepari, Shyam Aravamudhan and Debasish Kuila
Kinetics of Fischer–Tropsch Synthesis in a 3-D Printed Stainless Steel Microreactor Using Different Mesoporous Silica Supported Co-Ru Catalysts
Reprinted from: Catalysts **2019**, *9*, 872, doi:10.3390/catal9100872 377

About the Special Issue Editors

Wilson D. Shafer was born in Lexington KY in 1980, attended the University of Kentucky to graduate with a degree in chemistry in 2003. In 2004, Dr. Shafer entered the master's program at Eastern Kentucky University. His research was in collaboration with the USDA in developing a method for the verification and quantification of Ergot alkaloids. He defended his MS in Chemistry in 2007 at EKU. After graduation, he joined the Center for Applied Energy Research (CAER) at the University of Kentucky under Dr. Burtron H. Davis. While working fulltime, Dr. Shafer decided to pursue his Ph.D., under Dr. Davis in 2011. Dr. Shafer, in an untraditional manner, completed the Ph.D. program part-time as a full-time university employee. During his time at CAER, Dr. Shafer became influential in both designing and conducting highly cited research in industrial catalysis. After 10 years at CAER, Dr. Shafer accepted an inorganic chemistry faculty position at Asbury University where he plans to continue his research while continuing to teach.

Gary Jacobs was born in Newmarket, United Kingdom in 1971 He received a B.S. in chemical engineering from the University of Texas in 1993 and a Ph.D. in chemical engineering from the University of Oklahoma in 2000, where he worked with Prof. Daniel E. Resasco on applying synchrotron techniques to the development of dehydrocyclization catalysts. He then joined the Davis group at the University of Kentucky CAER, where he worked for 17 years. He is now an Assistant Professor at the new Department of Biomedical and Chemical Engineering at the University of Texas at San Antonio. He has co-authored over 200 refereed publications and received, with Dr. Davis, four Elsevier top-50 most-cited-author awards in catalysis for articles in cobalt-catalyzed Fischer–Tropsch synthesis and low temperature water-gas-shift for fuel cell reformer applications. In addition to these topics, he has a continuing interest in applying synchrotron methods to the investigation of the local atomic and electronic structure of heterogeneous catalysts pertinent to sustainable energy, fuels, and the environment. He especially enjoys mentoring students and helping them achieve their true potential.

Editorial

Editorial: Cobalt and Iron Catalysis

Wilson D. Shafer [1],* and Gary Jacobs [2],*

[1] Science and Health Department, Asbury University, Lexington, KY 40390, USA
[2] Department of Biomedical Engineering and Chemical Engineering/Department of Mechanical Engineering, University of Texas at San Antonio, One UTSA Circle, San Antonio, TX 78249, USA
* Correspondence: Wilson.Shafer@asbury.edu (W.D.S.); gary.jacobs@utsa.edu (G.J.)

Received: 20 December 2019; Accepted: 20 December 2019; Published: 27 December 2019

Cobalt and iron have long history of importance in the field of catalysis that continues to this day. Currently, both are prevalent metals in a multitude of catalytic processes: the synthesis of ammonia [1], Fischer-Tropsch synthesis (FTS) [2–5], water gas shift [6–9], hydrotreating [10], olefin polymerization [11], hydrotreating [12,13], isomerization [14,15], alcohol synthesis [16–18], and photocatalysis [19–21]. One cannot underestimate the importance of hydrotreating. The U.S. Energy Information Administration (EIA) shows that crude demand for 2019 was ~100 million barrels per day [22]; this crude requires hydrotreating before it can be brought to market [23–25]. FTS is another example, as it is at the core of large-scale industrial processes in coal-to-liquid (CTL) [26–28] and gas-to-liquid (GTL) [23,24,29] processes, as well as futuristic research on biomass-to-liquid (BTL) [30,31] processes. It is responsible for the production of more than 300,000 barrels per day of synthetic fuels [32–35], with some examples being Sasol I and II in South Africa (160,000 BPD), Pearl GTL (140,000 BPD), Oryx GTL (34,000 BPD), Shenhua CTL (20,000 BPD), and Shell Bintulu Malaysia (12,000 BPD). Water gas shift and steam reforming, used as a precursor to many of the aforementioned processes for hydrogen production, is used to produce nearly 70 million metric tons of hydrogen per year (based on 2018 data) [36]. Trajectories of hydrogen are likely to rise provided that the current demand for "clean" energy continues, coupled with the fact that companies are still developing hydrogen-powered cars [36]. It is not clear at this time how such an economy may develop. For example, one possibility is onboard reforming of a chemical carrier of hydrogen (e.g., light alcohols) [37–41], while other options include steam reforming, partial oxidation, and the gasification of various natural resources carried out at centralized or decentralized locations. Furthermore, iron, which is central to the high temperature water–gas shift step in fuel processors for hydrogen production, also serves to catalyze water–gas shift concurrently with FTS when the H_2/CO ratio in syngas is low, as occurs with syngas derived from coal and biomass. In addition, the iron used in FTS helps to boost the selectivities of alcohols and olefins [16,42,43]. In fact, both iron and cobalt can be catalytically tuned to favor products for the oxygenate market or to provide feedstock to the olefin polymerization processes [16,17,44,45]. Plastics produced from the latter process serve as materials for packaging, healthcare, diagnostics, optics, electronics, and a host of other industries. Finally, the topic of CO_2 utilization is becoming increasingly important to alleviate greenhouse gas emissions and combat climate change. One key roadblock is having on-hand a renewable source of hydrogen. If these key metals have the ability to transform CO, there may be a path forward for developing them for CO_2 utilization; research in this area is already under way [46].

However, although many of the catalytic processes mentioned above are run at commercial scales, they are still implemented with a certain degree of "blackbox" thinking. There are good reasons for this; although we understand, in general terms, the catalyst formulations and process conditions involved that will result in a variety of product distributions of interest, we do not, as of yet, have a clear understanding of what precisely constitutes the "active site", which electronic and geometric effects are involved, how promoters and poisons influence the active site, and which elementary steps are involved in the complex surface reaction mechanisms [4,47–52]. A quick search from Google Scholar on the "Fischer Tropsch mechanism" resulted in over 6500 publications published in the past two years on this topic alone. The routes are not fully understood, and much more work is needed to bridge the gap between, on the one hand, density functional theory and microkinetic modeling, and, on the other hand, actual data (i.e., product workup, catalyst characterization) generated from the testing of industrially- and academically-relevant catalyst formulations. A multitude of unanswered questions come to the fore. (1) What is the rate limiting step of the mechanism and how can activity be improved? [53] (2) Which metals and promoters are most suitable for a specific process, and how do they work together to bring about the desired selectivity? [54–56] (3) Which specific reactor configurations and process conditions are superior for targeting specific product ranges? [57–60] (4) Which processes are involved in catalyst deactivation, which ones dominate, and how can they be managed? (5) What are the relationships between homogeneous and heterogeneous catalysts? [61].

Much of the importance of these metals lies in their ability to manage carbon, including the activation of carbon monoxide, the coupling of carbon with carbon, hydrogen, or oxygen [62], and the scission of the same, as well as their ability remove products, once formed [62]. The capacity for iron and cobalt to manage carbon creates an avenue for useful products to be constructed from much simpler molecules. This is a significant advantage over alternative approaches such as direct liquefaction and the upgrading of bio-oil produced from flash pyrolysis. For example, Fischer-Tropsch synthesis produce a plethora of paraffins, 1-olefins, and oxygenates from simply carbon monoxide and hydrogen. These products are then readily hydro-processed to produce alternative fuels (i.e., especially diesel and aviation fuels), as well as lubricants and waxes. The olefins and oxygenates produced can be used as feedstocks for various applications, including the manufacturing of plastics. Thus, deepening our understanding of how the properties of the catalyst and process conditions can be tuned will give rise to greater control of the product distribution [62].

This special series highlights the work of scientists and engineers from all over the world who embody the leading edge of catalyst research on cobalt and iron. This collection provides a broad overview of a multitude of processing routes. In addition to the aforementioned topics, global concerns also drive: environmentally-attractive catalysts through specific homogeneous asymmetric iron complexes [63], the development of consumer products (preservatives, cosmetics, and flavors) [64] and materials (inks, coating, and paints) [65], and effective waste water treatment [66]. Furthermore, environmental concerns necessitate not just the capture of CO_2, but also its utilization as a means of mitigation [46,57,67]. Yet, alternative means of energy, such as hydrogen, may help limit harmful emissions and could also pave the way for change. The importance of cobalt, as a less expensive metal relative to the precious metals, for light-driven water oxidation [68] and electrochemical processes for hydrogen production [69] further highlights the potential of harnessing these metals in a way that benefits society. Regarding current methods that utilize carbon, such as the FTS process, a series of studies included here examine the effect of reaction conditions and promoters [70], investigate the kinetics [53,59,60], and examine the reaction route through computational modeling [71], with the aim of tailoring new formulations in a scientifically-driven manner [55,56,72]. This offers the potential to more effectively utilize the routes that are already available. Our aim is to tune [54] catalysts to efficiently produce fuels and chemicals in a more environmentally-benign manner.

The guest editors wish to thank all of the authors, reviewers, and editorial staff who took the time to contribute to and shape this special issue in a meaningful way. We dedicate this special issue to the memory of our mentor and friend, Professor Burtron H. Davis.

Conflicts of Interest: The authors declare no conflict of interest.

References

1. Van Rooij, A. Engineering Contractors in the Chemical Industry. The Development of Ammonia Processes, 1910–1940. *Hist. Technol.* **2005**, *21*, 345–366. [CrossRef]
2. Schulz, H. Short history and present trends of Fischer–Tropsch synthesis. *Appl. Catal. A Gen.* **1999**, *186*, 3–12. [CrossRef]
3. Khodakov, A.Y. Fischer-Tropsch synthesis: Relations between structure of cobalt catalysts and their catalytic performance. *Catal. Today* **2009**, *144*, 251–257. [CrossRef]
4. Davis, B.H. Fischer–Tropsch Synthesis: Reaction mechanisms for iron catalysts. *Catal. Today* **2009**, *141*, 25–33. [CrossRef]
5. Pirola, C.; Bianchi, C.; Di Michele, A.; Vitali, S.; Ragaini, V. Fischer Tropsch and Water Gas Shift chemical regimes on supported iron-based catalysts at high metal loading. *Catal. Commun.* **2009**, *10*, 823–827. [CrossRef]
6. Ma, W.; Jacobs, G.; Sparks, D.E.; Klettlinger, J.L.; Yen, C.H.; Davis, B.H. Fischer–Tropsch synthesis and water gas shift kinetics for a precipitated iron catalyst. *Catal. Today* **2016**, *275*, 49–58. [CrossRef]
7. Ma, W.; Jacobs, G.; Graham, U.M.; Davis, B.H. Fischer–Tropsch Synthesis: Effect of K Loading on the Water–Gas Shift Reaction and Liquid Hydrocarbon Formation Rate over Precipitated Iron Catalysts. *Top. Catal.* **2013**, *57*, 561–571. [CrossRef]
8. Jacobs, G.; Ricote, S.; Graham, U.M.; Davis, B.H. Low Temperature Water–Gas Shift Reaction: Interactions of Steam and CO with Ceria Treated with Different Oxidizing and Reducing Environments. *Catal. Lett.* **2014**, *145*, 533–540. [CrossRef]
9. Jacobs, G.; Chenu, E.; Patterson, P.M.; Williams, L.; Sparks, D.; Thomas, G.; Davis, B.H. Water-gas shift: Comparative screening of metal promoters for metal/ceria systems and role of the metal. *Appl. Catal. A Gen.* **2004**, *258*, 203–214. [CrossRef]
10. Laurent, E.; Delmon, B. Influence of oxygen-, nitrogen-, and sulfur-containing compounds on the hydrodeoxygenation of phenols over sulfided cobalt-molybdenum/gamma.-alumina and nickel-molybdenum/.gamma.-alumina catalysts. *Ind. Eng. Chem. Res.* **1993**, *32*, 2516–2524. [CrossRef]
11. Stürzel, M.; Mihan, S.; Mülhaupt, R. From Multisite Polymerization Catalysis to Sustainable Materials and All-Polyolefin Composites. *Chem. Rev.* **2016**, *116*, 1398–1433. [CrossRef] [PubMed]
12. Sollner, J.; Gonzalez, D.; Leal, J.; Eubanks, T.; Parsons, J. HDS of dibenzothiophene with $CoMoS_2$ synthesized using elemental sulfur. *Inorg. Chim. Acta* **2017**, *466*, 212–218. [CrossRef]
13. Pelardy, F.; Dos Santos, A.S.; Daudin, A.; Devers, É.; Belin, T.; Brunet, S. Sensitivity of supported MoS_2-based catalysts to carbon monoxide for selective HDS of FCC gasoline: Effect of nickel or cobalt as promoter. *Appl. Catal. B Environ.* **2017**, *206*, 24–34. [CrossRef]
14. Chada, J.P.; Xu, Z.; Zhao, D.; Watson, R.B.; Brammer, M.; Bigi, M.; Rosenfeld, D.C.; Hermans, I.; Huber, G.W. Oligomerization of 1-butene over carbon-supported CoO x and subsequent isomerization/hydroformylation to n-nonanal. *Catal. Commun.* **2018**, *114*, 93–97. [CrossRef]
15. Kostyniuk, A.; Key, D.; Mdleleni, M. 1-hexene isomerization over bimetallic M-Mo-ZSM-5 (M: Fe, Co, Ni) zeolite catalysts: Effects of transition metals addition on the catalytic performance. *J. Energy Inst.* **2019**. [CrossRef]
16. Ao, M.; Pham, G.H.; Sunarso, J.; Tade, M.O.; Liu, S. Active Centers of Catalysts for Higher Alcohol Synthesis from Syngas: A Review. *ACS Catal.* **2018**, *8*, 7025–7050. [CrossRef]
17. Nebel, J.; Schmidt, S.; Pan, Q.; Lotz, K.; Kaluza, S.; Muhler, M. On the role of cobalt carbidization in higher alcohol synthesis over hydrotalcite-based Co-Cu catalysts. *Chin. J. Catal.* **2019**, *40*, 1731–1740. [CrossRef]
18. Pan, W. Direct alcohol synthesis using copper/cobalt catalysts. *J. Catal.* **1988**, *114*, 447–456. [CrossRef]
19. Giannakis, S.; Liu, S.; Carratalà, A.; Rtimi, S.; Amiri, M.T.; Bensimon, M.; Pulgarin, C. Iron oxide-mediated semiconductor photocatalysis vs. heterogeneous photo-Fenton treatment of viruses in wastewater. Impact of the oxide particle size. *J. Hazard. Mater.* **2017**, *339*, 223–231. [CrossRef]
20. Arcanjo, G.S.; Mounteer, A.H.; Bellato, C.R.; Da Silva, L.M.M.; Dias, S.H.B.; Da Silva, P.R. Heterogeneous photocatalysis using TiO_2 modified with hydrotalcite and iron oxide under UV–visible irradiation for color and toxicity reduction in secondary textile mill effluent. *J. Environ. Manag.* **2018**, *211*, 154–163. [CrossRef]

21. Ghosh, K.; Harms, K.; Franconetti, A.; Frontera, A.; Chattopadhyay, S. A triple alkoxo bridged dinuclear cobalt(III) complex mimicking phosphatase and showing ability to degrade organic dye contaminants by photocatalysis. *J. Organomet. Chem.* **2019**, *883*, 52–64. [CrossRef]
22. Annual Energy Outlook 2019 with Projections to 2050. In *E.I. Administration*; U.S. Energy Information Administration Office of Energy Analysis, U.S. Department of Energy: Washington, DC, USA, 2019; p. 161.
23. Panahi, M.; Yasari, E.; Rafiee, A. Multi-objective optimization of a gas-to-liquids (GTL) process with staged Fischer-Tropsch reactor. *Energy Convers. Manag.* **2018**, *163*, 239–249. [CrossRef]
24. Behroozsarand, A.; Zamaniyan, A. Simulation and optimization of an integrated GTL process. *J. Clean. Prod.* **2017**, *142*, 2315–2327. [CrossRef]
25. Ushakov, S.; Halvorsen, N.G.; Valland, H.; Williksen, D.H.; Æsøy, V. Emission characteristics of GTL fuel as an alternative to conventional marine gas oil. *Transp. Res. Part D Transp. Environ.* **2013**, *18*, 31–38. [CrossRef]
26. Kong, Z.; Dong, X.; Jiang, Q. Forecasting the development of China's coal-to-liquid industry under security, economic and environmental constraints. *Energy Econ.* **2019**, *80*, 253–266. [CrossRef]
27. Zhou, X.; Zhang, H.; Qiu, R.; Lv, M.; Xiang, C.; Long, Y.; Liang, Y. A two-stage stochastic programming model for the optimal planning of a coal-to-liquids supply chain under demand uncertainty. *J. Clean. Prod.* **2019**, *228*, 10–28. [CrossRef]
28. Qin, S.; Chang, S.; Yao, Q. Modeling, thermodynamic and techno-economic analysis of coal-to-liquids process with different entrained flow coal gasifiers. *Appl. Energy* **2018**, *229*, 413–432. [CrossRef]
29. Ramos, Á.; García-Contreras, R.; Armas, O. Performance, combustion timing and emissions from a light duty vehicle at different altitudes fueled with animal fat biodiesel, GTL and diesel fuels. *Appl. Energy* **2016**, *182*, 507–517. [CrossRef]
30. Hunpinyo, P.; Narataruksa, P.; Tungkamani, S.; Pana-Suppamassadu, K.; Chollacoop, N.; Sukkathanyawat, H.; Jiamrittiwong, P. A comprehensive small and pilot-scale fixed-bed reactor approach for testing Fischer–Tropsch catalyst activity and performance on a BTL route. *Arab. J. Chem.* **2017**, *10*, S2806–S2828. [CrossRef]
31. Rimkus, A.; Zaglinskis, J.; Rapalis, P.; Skačkauskas, P. Research on the combustion, energy and emission parameters of diesel fuel and a biomass-to-liquid (BTL) fuel blend in a compression-ignition engine. *Energy Convers. Manag.* **2015**, *106*, 1109–1117. [CrossRef]
32. Dimitriou, I.; Goldingay, H.; Bridgwater, A.V. Techno-economic and uncertainty analysis of Biomass to Liquid (BTL) systems for transport fuel production. *Renew. Sustain. Energy Rev.* **2018**, *88*, 160–175. [CrossRef]
33. Trippe, F.; Fröhling, M.; Schultmann, F.; Stahl, R.; Henrich, E. Techno-economic assessment of gasification as a process step within biomass-to-liquid (BtL) fuel and chemicals production. *Fuel Process. Technol.* **2011**, *92*, 2169–2184. [CrossRef]
34. Swain, P.K.; Das, L.; Naik, S. Biomass to liquid: A prospective challenge to research and development in 21st century. *Renew. Sustain. Energy Rev.* **2011**, *15*, 4917–4933. [CrossRef]
35. Vallentin, D. Policy drivers and barriers for coal-to-liquids (CtL) technologies in the United States. *Energy Policy* **2008**, *36*, 3198–3211. [CrossRef]
36. *The Future of Hydrogen, Seizing Today's Opportunities*; International Energy Agency: Paris, France, 2019; p. 203.
37. Palo, D.R.; Dagle, R.A.; Holladay, J.D. Methanol Steam Reforming for Hydrogen Production. *Chem. Rev.* **2007**, *107*, 3992–4021. [CrossRef] [PubMed]
38. Evin, H.N.; Jacobs, G.; Ruiz-Martinez, J.; Graham, U.M.; Dozier, A.; Thomas, G.; Davis, B.H. Low Temperature Water–Gas Shift/Methanol Steam Reforming: Alkali Doping to Facilitate the Scission of Formate and Methoxy C–H Bonds over Pt/ceria Catalyst. *Catal. Lett.* **2007**, *122*, 9–19. [CrossRef]
39. Ribeiro, M.C.; Jacobs, G.; Davis, B.H.; Mattos, L.V.; Noronha, F.B. Ethanol Steam Reforming: Higher Dehydrogenation Selectivities Observed by Tuning Oxygen-Mobility and Acid/Base Properties with Mn in $CeO_2 \cdot MnO_x \cdot SiO_2$ Catalysts. *Top. Catal.* **2013**, *56*, 1634–1643. [CrossRef]
40. De Lima, S.M.; Colman, R.C.; Jacobs, G.; Davis, B.H.; Souza, K.R.; De Lima, A.F.; Appel, L.G.; Mattos, L.V.; Noronha, F.B. Hydrogen production from ethanol for PEM fuel cells. An integrated fuel processor comprising ethanol steam reforming and preferential oxidation of CO. *Catal. Today* **2009**, *146*, 110–123. [CrossRef]
41. Li, Y.; Zhang, Z.; Jia, P.; Dong, D.; Wang, Y.; Hu, S.; Xiang, J.; Liu, Q.; Hu, X. Ethanol steam reforming over cobalt catalysts: Effect of a range of additives on the catalytic behaviors. *J. Energy Inst.* **2019**. [CrossRef]

42. Tian, Z.; Wang, C.; Si, Z.; Ma, L.; Chen, L.; Liu, Q.; Zhanga, Q.; Huanga, H. Fischer-Tropsch synthesis to light olefins over iron-based catalysts supported on KMnO$_4$ modified activated carbon by a facile method. *Appl. Catal., A* **2017**, *541*, 50–59. [CrossRef]
43. Wang, Y.; Davis, B.H. Fischer–Tropsch synthesis. Conversion of alcohols over iron oxide and iron carbide catalysts. *Appl. Catal. A: Gen.* **1999**, *180*, 277–285. [CrossRef]
44. Boahene, P.E.; Dalai, A.K. Higher Alcohols Synthesis: Experimental and Process Parameters Study over a CNH-Supported KCoRhMo Catalyst. *Ind. Eng. Chem. Res.* **2017**, *56*, 13552–13565. [CrossRef]
45. Xiang, Y.; Kruse, N. Tuning the catalytic CO hydrogenation to straight- and long-chain aldehydes/alcohols and olefins/paraffins. *Nat. Commun.* **2016**, *7*, 13058. [CrossRef]
46. Shafer, W.D.; Jacobs, G.; Graham, U.M.; Hamdeh, H.H.; Davis, B.H. Increased CO$_2$ hydrogenation to liquid products using promoted iron catalysts. *J. Catal.* **2019**, *369*, 239–248. [CrossRef]
47. Chen, W.; Filot, I.A.W.; Pestman, R.; Hensen, E.J.M. Mechanism of Cobalt-Catalyzed CO Hydrogenation: 2. Fischer–Tropsch Synthesis. *ACS Catal.* **2017**, *7*, 8061–8071. [CrossRef]
48. Chakrabarti, D.; Gnanamani, M.K.; Shafer, W.D.; Ribeiro, M.C.; Sparks, D.E.; Prasad, V.; De Klerk, A.; Davis, B.H. Fischer–Tropsch Mechanism: 13C18O Tracer Studies on a Ceria–Silica Supported Cobalt Catalyst and a Doubly Promoted Iron Catalyst. *Ind. Eng. Chem. Res.* **2015**, *54*, 6438–6453. [CrossRef]
49. Todic, B.; Ma, W.; Jacobs, G.; Davis, B.H.; Bukur, D.B. CO-insertion mechanism based kinetic model of the Fischer–Tropsch synthesis reaction over Re-promoted Co catalyst. *Catal. Today* **2014**, *228*, 32–39. [CrossRef]
50. Van Santen, R.A.; Markvoort, A.J.; Filot, I.A.W.; Ghouri, M.M.; Hensen, E.J.M. Mechanism and microkinetics of the Fischer–Tropsch reaction. *Phys. Chem. Chem. Phys.* **2013**, *15*, 17038–17063. [CrossRef]
51. Davis, B.H. Fischer–Tropsch synthesis: Current mechanism and futuristic needs. *Fuel Process. Technol.* **2001**, *71*, 157–166. [CrossRef]
52. Muleja, A.A.; Gorimbo, J.; Masuku, C.M. Effect of Co-Feeding Inorganic and Organic Molecules in the Fe and Co Catalyzed Fischer–Tropsch Synthesis: A Review. *Catalysts* **2019**, *9*, 746. [CrossRef]
53. Qi, Y.; Yang, J.; Holmen, A.; Chen, D. Investigation of C1 + C1 Coupling Reactions in Cobalt-Catalyzed Fischer-Tropsch Synthesis by a Combined DFT and Kinetic Isotope Study. *Catalysts* **2019**, *9*, 551. [CrossRef]
54. Shafer, W.D.; Gnanamani, M.K.; Graham, U.M.; Yang, J.; Masuku, C.M.; Jacobs, G.; Davis, B.H. Fischer–Tropsch: Product Selectivity—The Fingerprint of Synthetic Fuels. *Catalysts* **2019**, *9*, 259. [CrossRef]
55. Ma, W.; Jacobs, G.; Shafer, W.D.; Ji, Y.; Klettlinger, J.L.S.; Khalid, S.; Hopps, S.D.; Davis, B.H. Fischer-Tropsch Synthesis: Cd, In and Sn Effects on a 15%Co/Al$_2$O$_3$ Catalyst. *Catalysts* **2019**, *9*, 862. [CrossRef]
56. Mierczynski, P.; Dawid, B.; Chalupka, K.; Maniukiewicz, W.; Witoska, I.; Vasilev, K.; Szynkowska, M.I. Comparative Studies of Fischer-Tropsch Synthesis on Iron Catalysts Supported on Al$_2$O$_3$-Cr$_2$O$_3$ (2:1), Multi-Walled Carbon Nanotubes or BEA Zeolite Systems. *Catalysts* **2019**, *9*, 605. [CrossRef]
57. Iloy, R.A.; Jalama, K. Effect of Operating Temperature, Pressure and Potassium Loading on the Performance of Silica-Supported Cobalt Catalyst in CO$_2$ Hydrogenation to Hydrocarbon Fuel. *Catalysts* **2019**, *9*, 807. [CrossRef]
58. Keyvanloo, K.; Huang, B.; Okeson, T.; Hamdeh, H.H.; Hecker, W.C. Effect of Support Pretreatment Temperature on the Performance of an Iron Fischer–Tropsch Catalyst Supported on Silica-Stabilized Alumina. *Catalysts* **2018**, *8*, 77. [CrossRef]
59. Mohammad, N.; Bepari, S.; Aravamudhan, S.; Kuila, D. Kinetics of Fischer–Tropsch Synthesis in a 3-D Printed Stainless Steel Microreactor Using Different Mesoporous Silica Supported Co-Ru Catalysts. *Catalysts* **2019**, *9*, 872. [CrossRef]
60. Petersen, A.P.; Claeys, M.; Kooyman, P.J.; Van Steen, E. Cobalt-Based Fischer–Tropsch Synthesis: A Kinetic Evaluation of Metal–Support Interactions Using an Inverse Model System. *Catalysts* **2019**, *9*, 794. [CrossRef]
61. Bungane, N.; Welker, C.; Van Steen, E.; Claeys, M. Fischer-Tropsch CO-Hydrogenation on SiO$_2$-supported Osmium Complexes. *Z. Nat. B* **2008**, *63*, 289–292. [CrossRef]
62. Schmal, M. *Heterogeneous Catalysis and its Industrial Applications*; Springer: Berlin/Heidelberg, Germany, 2016.
63. Van Slagmaat, C.A.M.R.; Chou, K.C.; Morick, L.; Hadavi, D.; Blom, B.; De Wildeman, S.M.A. Synthesis and Catalytic Application of Knölker-Type Iron Complexes with a Novel Asymmetric Cyclopentadienone Ligand Design. *Catalysts* **2019**, *9*, 790. [CrossRef]
64. Li, Y.; Yang, Y.; Chen, D.; Luo, Z.; Wang, W.; Ao, Y.; Zhang, L.; Yan, Z.; Wang, J. Liquid-Phase Catalytic Oxidation of Limonene to Carvone over ZIF-67(Co). *Catalysts* **2019**, *9*, 374. [CrossRef]

65. Simpson, N.; Maaijen, K.; Roelofsen, Y.; Hage, R. The Evolution of Catalysis for Alkyd Coatings: Responding to Impending Cobalt Reclassification with Very Active Iron and Manganese Catalysts, Using Polydentate Nitrogen Donor Ligands. *Catalysts* **2019**, *9*, 825. [CrossRef]
66. Yang, B.; Qi, Y.; Liu, R. Pilot-Scale Production, Properties and Application of Fe/Cu Catalytic-Ceramic-Filler for Nitrobenzene Compounds Wastewater Treatment. *Catalysts* **2018**, *9*, 11. [CrossRef]
67. Chou, C.Y.; Loiland, J.A.; Lobo, R.F. Reverse Water-Gas Shift Iron Catalyst Derived from Magnetite. *Catalysts* **2019**, *9*, 773. [CrossRef]
68. Nesterov, D.S.; Nesterova, O.V. Polynuclear Cobalt Complexes as Catalysts for Light-Driven Water Oxidation: A Review of Recent Advances. *Catalysts* **2018**, *8*, 602. [CrossRef]
69. Li, K.; Li, Y.; Peng, W.; Zhang, G.; Zhang, F.; Fan, X. Bimetallic Iron–Cobalt Catalysts and Their Applications in Energy-Related Electrochemical Reactions. *Catalysts* **2019**, *9*, 762. [CrossRef]
70. Bin Jo, S.; Chae, H.J.; Kim, T.Y.; Lee, C.H.; Oh, J.U.; Kang, S.-H.; Kim, J.W.; Jeong, M.; Lee, S.C.; Kim, J.C. Selective CO hydrogenation over bimetallic Co-Fe catalysts for the production of light paraffin hydrocarbons (C2-C4): Effect of H_2/CO ratio and reaction temperature. *Catal. Commun.* **2018**, *117*, 74–78.
71. Williams, H.; Gnanamani, M.K.; Jacobs, G.; Shafer, W.D.; Coulliette, D. Fischer–Tropsch Synthesis: Computational Sensitivity Modeling for Series of Cobalt Catalysts. *Catalysts* **2019**, *9*, 857. [CrossRef]
72. Gavrilović, L.; Blekkan, A.; Save, J. The Effect of Potassium on Cobalt-Based Fischer–Tropsch Catalysts with Different Cobalt Particle Sizes. *Catalysts* **2019**, *9*, 351. [CrossRef]

© 2019 by the authors. Licensee MDPI, Basel, Switzerland. This article is an open access article distributed under the terms and conditions of the Creative Commons Attribution (CC BY) license (http://creativecommons.org/licenses/by/4.0/).

Article

The Evolution of Catalysis for Alkyd Coatings: Responding to Impending Cobalt Reclassification with Very Active Iron and Manganese Catalysts, Using Polydentate Nitrogen Donor Ligands

Neil Simpson [1],*, Karin Maaijen [2], Yfranka Roelofsen [2] and Ronald Hage [2]

1. Borchers, Berghausener Strasse 100, 40764 Langenfeld, Germany
2. Catexel B.V., Galileiweg 8, 2333 BD Leiden, The Netherlands; karin.maaijen@catexel.com (K.M.); yfranka.roelofsen@catexel.com (Y.R.); ronald.hage@catexel.com (R.H.)
* Correspondence: neil.simpson@borchers.com; Tel.: +49-217-33926700

Received: 29 August 2019; Accepted: 27 September 2019; Published: 1 October 2019

Abstract: Autoxidation processes to achieve curing of alkyd resins in paints, inks, and coatings are ubiquitous in many applications. Cobalt soaps have been employed for these applications for many decades and most of the paint and ink alkyd resin formulations have been optimized to achieve optimal benefits of the cobalt soaps. However, cobalt soaps are under increased scrutiny because of likely reclassification as carcinogenic under REACH (Registration, Evaluation, Authorisation, and Restrictions of Chemicals) legislation in Europe. This is critical, since such coatings are available for regular human contact. Alternative manganese- and iron-based siccatives have been developed to address this need for over a decade. They often show very high curing activity depending on the organic ligands bound to the metal centers. Recently, new classes of catalysts and modes of application have been published or patented to create safe paints, whilst delivering performance benefits via their unique reaction mechanisms. Besides the use of well-defined, preformed catalysts, paint formulations have also been developed with mixtures of metal soaps and ligands that form active species in-situ. The change from Co-soaps to Mn- and Fe-based siccatives meant that important coating issues related to radical-based curing, such as skinning, had to be rethought. In this paper we will review the new catalyst technologies and their performance and modes of action, as well as new compounds developed to provide anti-skinning benefits.

Keywords: cobalt carboxylate; coating; autoxidation; alkyd; siccative; polymerization; iron; manganese

1. Introduction

The creation of coatings can be traced to prehistoric cave paintings from around 30,000 years ago, but the advent of lacquers for protection started in China over 2000 years ago and have since developed into a wide range of chemistries, application methods, and formulations to cope with the applications and coating performance criteria [1]. They are ubiquitous, essential, and provide both protection and beauty to all manner of surfaces, for example: Ships, mobile phones, satellites, housing, furniture, light-bulbs, paper, and bridges [2]. The coatings market is significant, worth $178 billion USD in 2018, and growing at a 4.5% compound annual growth rate [3]. A wide range of chemistries have been developed to create thermoplastic or thermosetting chemistries with common examples including (poly)urethane, -acrylic, -epoxy, and -phenolics. Technology continues to progress; in the last decade new functionalities have been developed for coatings including easy-to-clean [4], anti-pollution [5], and self-healing [6].

Recently coatings are moving towards using sustainable, environmentally-friendly raw materials [7,8]. This shift in approaching health, safety, and sustainability concerns does not only originate from change in legislation, but also from agreed restrictions in using less desirable raw materials, including restricting the use of organic solvents. As part of this trend, alkyd resins [2] have found renewed interest as a paint based on natural ingredients [9]—they are essentially polyesters prepared from modified vegetable oils, polyols like glycerol, natural, unsaturated fatty acids, and dibasic acids (see Figure 1 as an example) [10]. They are often used in decorative applications, vehicle coatings, coil coatings, and marine coatings [2].

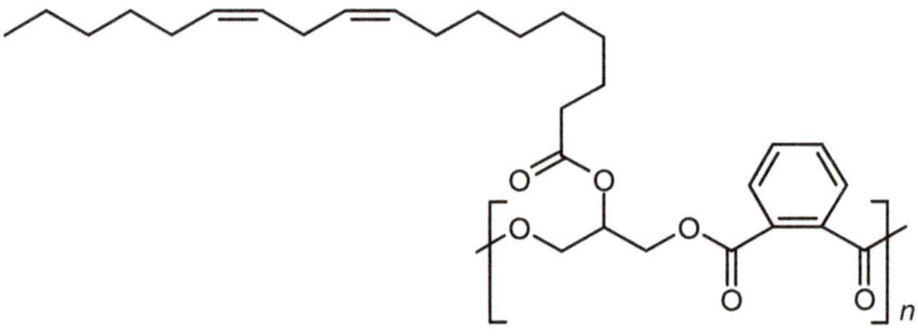

Figure 1. Structure of a typical alkyd resin with linoleic acid as side group.

1.1. Radical Curing Chemistry

Alkyd resins cure via a physical and chemical drying process that requires catalysis [11,12] (Figure 2). A liquid paint dries physically by releasing solvent (water or organic) after application to a surface. The alkyd resin within the paint accounts for most of the formulation and gives the coating its strength and durability. It forms a liquid layer containing the other components, such as pigments and surface-active components. During the physical drying, a tandem process occurs via a chemical process of air-oxidation. This hardens the film beyond a tacky consistency to give the final coating properties.

The chemical air-drying process is facilitated via the available unsaturation for crosslinking. Natural drying oils like linseed oil, tung oil, dehydrated castor oil, tall oil (from wood pulp), and oitica oil can be used as they have large amounts of conjugated double-bonds, essential for better drying properties versus semi-drying oils like sunflower oil or soybean oil. The degree of unsaturation affects the dry time and final resin compatibility and amphiphilicity, important for preparing resins for water- or solvent-borne coatings. Alkyds with 55% *w/w* oil content are long-oil alkyds and mostly suited to solvent-borne formulations and have the slowest dry times. Medium (45–55%) and shorter oils (<45% *w/w* oil) dry faster but give less crosslinking in relation to the oil content. The final alkyd resin can be tuned with additional chemistry, such as the number of hydroxyl groups, degree of branching, reaction with epoxy or urethanes to make polymeric hybrids, or by simply changing the molecular weight, to improve properties such as the time to form a hard film (dry-time) or glass transition state for flexibility versus chemical resistance [10].

The oxidative cure can occur in the presence of oxygen, but it can take days to form a tack-free film and might only be active at the coating surface where the most oxygen is available for curing, leading to a soft coating and a hard surface. Long dry-times and soft coatings are not commercially viable. This has long been solved by using a catalyst, a so-called 'primary drier (siccative)', based on a metal soap carboxylate, that can reduce the dry-time to hours rather than days. The most commonly used are organic-acid salts of cobalt. Coatings formulations often include other (so called 'secondary') metal carboxylate driers (siccatives) that cannot start cross-linking, but facilitate redox processes throughout the coating, examples metals include aluminium, zirconium, and barium. Alkyd coating formulations also contain other additives to prevent adsorption of siccatives onto pigment particles retarding the drying process, such as pigment dispersants or calcium-based soaps, that can block the pigment surface.

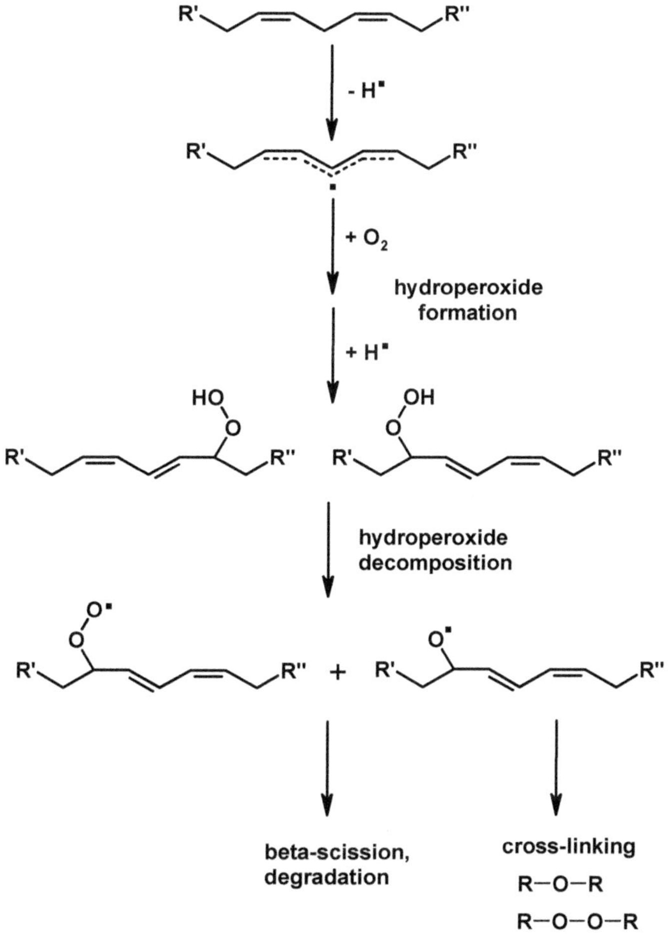

Figure 2. Simplified scheme of mechanisms of alkyd curing via Oxidative Catalysis and O_2.

The oxidative curing process takes place via radical autoxidation reactions (Figure 2). The radical curing reactions are initiated by metal catalysts via a peroxidation reaction involving minute amounts of alkylhydroperoxide (ROOH) present in the resin containing an unsaturated resin yielding an alkoxy radical (RO$^\bullet$) and (formally) a hydroxyl radical (HO$^\bullet$) (reaction (1)). However, as this reaction often occurs via binding of ROOH to the metal ion, an oxidized metal-oxo species is formed (reaction (2)). Both the alkoxy radical and the metal-oxo species react with unsaturated fatty acid moieties (RH), to form the alkyl radical (reactions (3) and (4)), which in turn reacts nearly diffusion limited with dioxygen to form peroxyl radicals (reaction (5)). The peroxyl radicals react efficiently with unsaturated fatty acids, yielding more alkyl radicals. Reactions (1)–(4) are known as initiation reactions and reactions (5) and (6) are propagation reactions.

In some cases, also reactions between metal ions in their lower valence state and dioxygen may take place, leading to an oxidized metal species with a superoxide radical bound (which can then react with an unsaturated fatty acid, such as depicted in reaction (6), where in this case ROO$^\bullet$ should be read as M(OO$^{\bullet-}$)).

$$ROOH \rightarrow RO^\bullet + HO^\bullet \tag{1}$$

$$M^{n+}\text{-OOR} \rightarrow RO^\bullet + O=M^{(n+1)+} \tag{2}$$

$$RO^\bullet + RH \rightarrow ROH + R^\bullet \tag{3}$$

$$O=M^{(n+1)+} + RH \rightarrow M^{n+}\text{-OH} + R^\bullet \tag{4}$$

$$R^\bullet + O_2 \rightarrow ROO^\bullet \tag{5}$$

$$ROO^\bullet + RH \rightarrow ROOH + R^\bullet \tag{6}$$

$$ROO^\bullet + ROO^\bullet \rightarrow RCHO + ROH + O_2 \tag{7}$$

Termination of the radical processes occur by different reactions. First, two peroxyl radicals react to form aldehyde and alcohol terminal products (reaction (7)). Further, importantly for alkyd curing, the radical intermediates ($R^\bullet$, $RO^\bullet$, and $ROO^\bullet$) react further with each other to form species, such as R–R, R–O–R, R–OO–R [13].

The chemistry of alkyd resin curing is rather complicated to study in detail and often a model substrate based on ethyllinoleate has been employed [10,13,14]. Various spectroscopic techniques can be used to study reactions (like Infrared and Raman spectroscopy), and using this substrate can help to understand the mechanisms of the Mn and Fe catalysts on alkyd resin curing as will be highlighted in the subsequent paragraphs.

1.2. Regulatory Changes for Cobalt

Cobalt-based siccatives have faced reclassification as a class 1b carcinogen. The main participants of this reclassification process are the Cobalt REACH consortium [15], a non-profit group tasked with preparing the registration dossiers for cobalt and cobalt compounds and a special consortium of major European producers of metal carboxylates (for siccatives) that was established in 2007 within the CoRC [16].

Recently, five cobalt salts—cobalt sulfate, cobalt dichloride, cobalt dinitrate, cobalt carbonate, and cobalt di(acetate)—have been proposed for restriction by ECHA as Carc. 1B (inhalation), Muta. 2, and classified same as Repr. 1B and skin and respiratory sensitizers. In 2016, RAC (ECHA) agreed that the cobalt salts should be considered as genotoxic carcinogens with a non-threshold mode of action and endorsed a dose–response relationship for these substances. However, in their REACH registration dossiers, industry (CoRC) identified the cobalt salts as non-genotoxic carcinogens with a threshold mode of action. A DNEL value of 40 µg Co/m^3 is used by the registrants in their Chemical Safety Assessments. According to the dose-response relationship derived by RAC, work-life exposure to this DNEL value corresponds to an excess lifetime cancer risk of 4×10^{-2}. The stakeholder engagement window ended June 19, 2019, and a meeting was held between ECHA and the RAC on June 6, 2019—the final opinion on reclassification is expected end of 2019. It is highly likely that the five above-mentioned salts will be classified without threshold resulting in their eventual restriction and subsequent placement on a list of substances requiring authorization before use.

The members of the Cobalt Institute and Cobalt REACH Consortium agreed to self-classify the following substances under the UN Globally Harmonized System for Classification and Labelling of Chemicals (UN GHS) as specific target organ toxicants following repeated exposure (STOT RE 1 (GI Tract); H372).

1. Neodecanoic acid, cobalt salts EC No. 248-373-0. CAS No. 27253-31-2; EC No. 257-798-0; CAS No. 52270-44-7;
2. Cobalt, borate Neodecanoate complexes EC No. 270-601-2; CAS No. 68457-13-6;
3. Stearic acid, cobalt salts EC No. 237-016-4; CAS No. 13586-84-0. EC No. 213-694-7; CAS No. 1002-88-6;
4. Resin and rosin acids, cobalt salts EC No. 273-321-9; CAS No. 68956-82-1;
5. Naphthenic acids, cobalt salts EC No. 263-064-0. CAS No. 61789-51-3; EC No. 285-220-7; CAS No. 85049-49-6;

6. Cobalt bis(2-ethylhexanoate), CAS No. 136-52-7. This substance has been self-classified from Repr. 2 to Repr. 1B.

The industry is therefore increasingly responding to the reality of cobalt and cobalt-compound restrictions. The TSCA Reset in the US has led to Consent Orders (formerly known as SNURs) for new PMNs that restrict the use of cobalt siccatives for consumer use. Cobalt neodecanoate, a classic cobalt-based siccative for alkyds and inks, is a good example of restricted use (to tires, for non-consumer usage only).

It is worth mentioning that the proposed EU Harmonized Classification of cobalt metal as a carcinogen, mutagen, and reproductive toxicant will also have a significant effect on cobalt metal powder users [17]. Further, it should be mentioned that cobalt(II) carboxylates ('soaps') form upon hydrolysis cobalt(II) aqua species, which are the same or very similar to those present in for example cobalt(II)sulfate hexahydrate.

It is likely that cobalt carboxylates will therefore be reclassified as carcinogenic due to the read across strategies used by Competent Authorities (chemical similarity of the salts) and because cobalt itself has been classified.

As a carcinogen, cobalt can be a risk to human health since coatings and inks cured via cobalt-catalysts are often in frequent human contact (especially when applying paints or scratching out old paint layers). The industry recognizes that the continued use of cobalt catalysts risks both human health and the movement towards a sustainable future. As a result, cobalt catalysts have been continuously phased-out of alkyd (and ink) formulations as coating firms prepare for a regulatory-driven shift in the market.

1.3. Cobalt Alternative Technologies

Cobalt reclassification requires forward-thinking firms to develop new strategies and products; this is reinforced by the issue of cobalt pricing affecting manufacturers—cobalt supplies have seen major supply shortages with the advent of increased battery production for electronic cars [18].

Due to the combined pressures of regulatory change and demand leading to volatile pricing, the coatings and ink industries have been looking for alternative technologies. This area of research has been evolving for decades, as cobalt was the replacement for lead and alternatives currently exist from other metal soap carboxylates that are well known in the market, based on manganese, iron, and vanadium. The most credible carboxylate technologies have well-known disadvantages [19]. Manganese siccatives are dark and create discoloration, iron siccatives are too slow at room temperature and vanadium siccatives are toxic [20,21].

The development of cobalt-replacement strategies is discussed below, with a short review of past technologies (1.3.1), followed by a detailed review of new Fe- and Mn-based catalysts studied or used as siccatives for alkyd resin formulation in Section 1.3.2. discussed below.

1.3.1. Traditional Cobalt-Catalyst Replacement Strategies: Manganese and Iron Soaps

The use of manganese salts or soaps as alternatives to cobalt has been an appealing proposition for many years [22]. Especially, manganese(II)(2-ethylhexanoate)$_2$ (abbreviated as Mn^{II}(2-EH)$_2$) has been applied in various paint and ink formulations. However, as the activity towards curing of Mn^{II}(2-EH)$_2$ is significantly lower than that of the cobalt soaps and, therefore, the inclusion level has to be much higher [13,23]. As a consequence, when strongly colored Mn(III) ions are present in the manganese soap formulation, or are formed due to oxidation processes, the color of the manganese in the paint formulation or in the applied coating layer may be apparent and this is undesired in white or light-colored paint formulations. By adding ligands that contain amine donor atoms (either aliphatic amines or aromatic amines or a combination thereof), the activity of the manganese centers can be enhanced and therefore the dose level can be lowered (vide infra).

The use of manganese soaps is limited due to the high dosage needed, which leads to an increased cost in use and often unacceptable coloring of the paint formulation or coating layer. Addition of a bidentate ligand to manganese soap, such as 2,2'-bipyridine (bpy, Figure 3), leads to a significant increase in activity, allowing a reduction of the level of manganese soap at a similar level as what is used for cobalt soap [24]. A benefit is then that the color of manganese is less apparent than without bpy added, but still the coloring issue can prohibit the use in many formulations. Bpy is often used in ink formulations in combination with manganese soaps, because inclusion of a dark colored siccative will not be noticed in these dark formulations. Detailed spectroscopic studies and X-ray analysis revealed that tetranuclear manganese clusters may be formed and involved in the catalysis [25]. However, it should be noted that in general intermediates that are crystallized and analyzed may not be the species involved in the oxidative processes.

The use manganese complexes with acetylacetonate (acac, Figure 3) for paint curing has been investigated [26–28]. When using purified ethyllinoleate that have no hydroperoxide impurities, the initiation reaction is much shorter than when Mn^{II}(2-EH)$_2$ is used. A further improvement has been made by addition of one equivalent of bpy, which has led to the suggestion that coordination of the nitrogen donor leads to a stabilization of the Mn(II) species, thereby increasing the rate of H$^\bullet$ abstraction of the ethyllinoleate by the Mn(III) species. A disadvantage noted was that the Mn(acac)$_2$-bpy leads to more β-scission cleavage of the double bonds leading to formation of volatile aldehydes. As a consequence, the curing of alkyd resins will be less than optimal because the extent of polymerization will be diminished. A benefit noted by using acetylacetonate is that a reduced skinning behavior was observed [28]. As acetylacetone is volatile, it may that when bound to the metal ion it prevents it to react with the alkyd resin (or hydroperoxides present in the alkyd resin), but when the paint formulation is applied on a surface, acetylacetone that is not bound to the metal center may evaporate, allowing the metal ion to react with the peroxides to initiate the curing reactions. This assumes that there is an equilibrium, which is kinetically competent, between acetylacetonate that is bound to the manganese ion and acetylacetone that is in the paint formulation.

Iron soap drying catalysts have also been studied, but their application has been limited due to the low activity. Only in stoving enamels such siccatives have been implemented. Miccichè et al. have shown by X-ray crystallography and Mössbauer analyses that a μ$_3$-oxo carboxylate bridged complex is formed and this may be involved in the alkyd resin oxidation process [29]. Addition of ascorbic acid palmitate enhances the reactivity of the curing process, which has been explained by an increased rate of reduction of one of the iron centers by ascorbic acid, which in turn reacts with the alkyd resin to form the alkyd resin radical species (which reacts with O$_2$ to form peroxyl radicals as explained in Section 1.2) [10].

1.3.2. Contemporary Cobalt-Catalyst Replacement Strategies

Manganese-Based Complexes Using Polydentate Nitrogen Donor Ligands

Various other classes of Mn complexes have been patented as paint siccatives. A very active mononuclear Mn complex with a cross-bridged tetraaza macrocyclic ligand (Bcyclam, Figure 3) has been patented [30]. Model experiments with β-carotene used as a model substrate indicated that radicals are formed when methyllinoleate was added. A very good curing of linseed oil was observed.

More recently, a patent by Dura Chemicals covered the use of Mn porphyrins [31]. Examples using Mn^{III}(5,10,15,20-tetrakisphenylporphyrin) show a good curing activity (Figure 3). As porphyrin complexes are intensely colored, such catalyst can only be used at low concentrations to prevent coloring of the paint layer.

Whilst initially the use of manganese salts or complexes did not seem to be very promising, the finding that [Mn^{IV}_2(μ-O)$_3$(Me$_3$TACN)$_2$](PF$_6$)$_2$ (Me$_3$TACN = 1,4,7-trimethyl-1,4,7-triazacyclononane– Figure 3) was active towards curing opened up new possibilities to formulate paint compositions with alternative siccatives based on manganese [32,33]. Mixtures of Mn^{II} salts and a molar excess of Me$_3$TACN or a molar excess of Mn^{II} salts and Me$_3$TACN ligand have been patented for drying alkyds as an alternative to synthesized catalysts [34,35]. This compound was already known as catalyst to activate hydrogen peroxide for bleaching of stains in laundry and dishwash cleaning applications [36]. One of the features of this catalyst is that the triazacyclononane ligand is tightly bound to the manganese ion, preventing fast formation of manganese hydroxide precipitation commonly observed in the alkaline detergent formulations when manganese salts or complexes are used [37].

When an excess of ligand is used, the level of Mn can be reduced leading to diminished color by Mn in the paint formulation or coating applied [34]. Hardness improvement has been observed when an excess of Mn was used [35]. Furthermore, patents were published wherein mixtures of Mn complexes and Me$_3$TACN were claimed, yielding improved curing activity when [Mn$^{IV}_2$(μ-O)$_3$(Me$_3$TACN)$_2$](PF$_6$)$_2$ was mixed with Me$_3$TACN ligand [38], or when non-Me$_3$TACN Mn complexes were mixed with the Me$_3$TACN ligand [39].

Related to the Mn complexes with Me$_3$TACN is the class of bridging triazacyclononane ligands covered in recent patent filings. The use of 1,2-bis(4,7-dimethyl-1,4,7-triazacyclononan-1-yl)-ethane (Me$_4$DTNE, Figure 3) in alkyd resins (without Mn) has been claimed in a patent along with other classes of ligands, including the Me$_3$TACN ligand [40]. This alkyd resin formulation that does not contain metal ions is not subject to skinning, whilst when Mn soap is added, curing of the alkyd resin can start. In another patent, the use of this ligand with a Mn soap or salt has been claimed [41]. The experiments show good curing activity of such complexes, although the nature of the active species has not been disclosed.

A different class of polydentate nitrogen-based ligands, based on 1,4-diazacycloheptane-6-amines, in conjunction with Mn salts has also been patented (Figure 3) [42]. These ligands contain tridentate aliphatic nitrogen donor ligands as a backbone to which other coordinating donor groups can be attached (such as pyridine). Compared to the triazacyclononane based ligands, the stability constants of nickel complexes are about 10^5 less than for an analogous complex with triazacyclononane [43], indicating that this is also the case when comparing Mn complexes containing this class of ligands with the triazacyclononane ligands. Despite this difference in stability, the experiments disclosed in the patent revealed clearly improved curing activity as compared with Mn^{II}(2-EH)$_2$.

Two classes of tetradentate ligands with Mn salts have been patented recently that are based on tris(pyridin-2-ylmethyl)amine (TPA) and N-methyl-N(pyridin-2-ylmethyl)-bis(pyridin-2-yl)methylamine (MeN3py) (Figure 3) [44]. Despite the distinct different structures and properties of these ligands (3 aromatic N donors + 1 aliphatic ones) vs. the 1,4-diazacycloheptane-6-amines and triazayclononane ligands (3 aliphatic N donors), a clear enhancement in curing has been measured. It would be interesting to understand whether all these different Mn catalysts achieve curing of alkyd resins all following similar pathways or whether there are different reactions catalyzed for the different classes of catalysts (e.g., direct activation of the allylic CH$_2$ moieties, activation of ROOH to yield peroxyl or alkoxy radicals, or interaction with O$_2$ to yield O$_2$ activation).

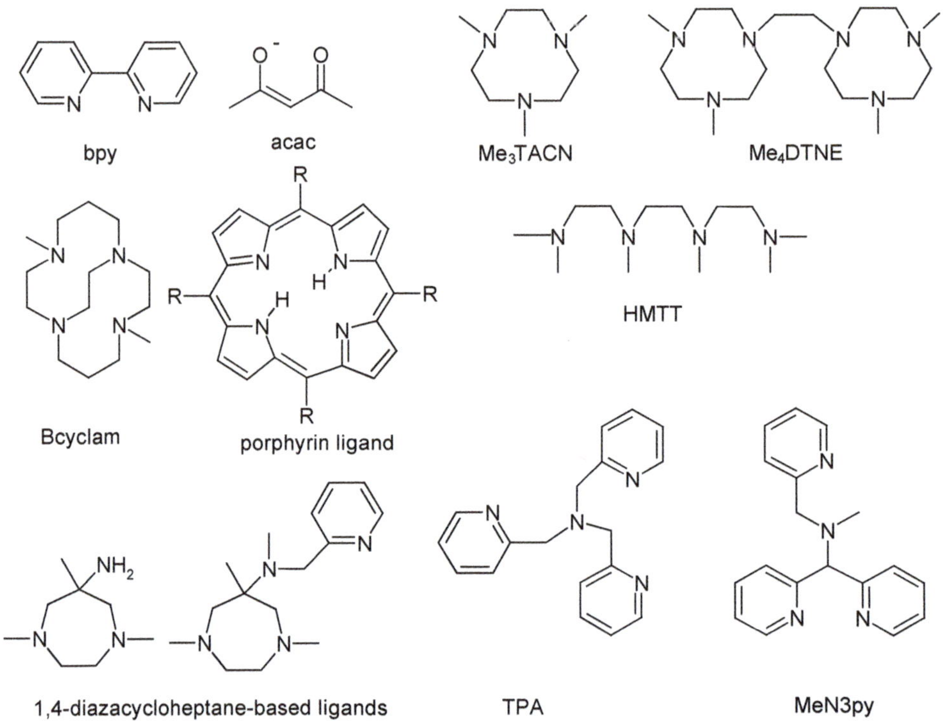

Figure 3. Ligands and additives exhibiting paint drying activity when mixed with Mn salts or as Mn-ligand complexes discussed in the text: 2,2'-bipyridine (bpy), acetylacetonate (acac), 1,4,7-trimethyl-1,4,7-triazacyclononane (Me$_3$TACN), 1,2-bis(4,7-dimethyl-1,4,7-triazacyclononan-1-yl)-ethane (Me$_4$DTNE), 4,11-dimethyl-1,4,8,11-tetraazabicyclo[6.6.2]hexadecane (Bcyclam), 5,10,15,20- tetrakisphenylporphyrin (R = phenyl; porphyrin ligand), N,N,N',N'',N''',N'''-hexamethyltriethylenetetraamine (HMTT), 6-amino-1,4,6-trimethyl-1,4-diazacycloheptane and 1,4,6-trimethyl-6{N-pyridin-2-ylmethyl)-N-methylamino}-1,4-diazacycloheptane (1,4-diazacycloheptane based ligands), tris(pyridin-2-ylmethyl)amine (TPA), and N-methyl-N(pyridin-2-ylmethyl)-bis(pyridin-2- yl)methylamine (MeN3py).

Iron-Based Complexes Using Polydentate Ligands

Iron ferrocene complexes have been studied by Erben and co-workers [45–47]. Whilst initially a good activity of the unsubstituted ferrocene was reported, follow-up studies revealed that most substituted ferrocenes and especially benzoylferrocene may be more suitable to be applied (Figure 4). Interestingly, this class of catalysts appear to be operative via a different mechanism. Ferrocenes are known for the electron transfer capabilities without changing the coordination environment (see for example [48]). Therefore, for this class of siccatives binding of alkylhydroperoxide to the metal center may not be operative. The activity of the benzoylferrocene catalyst is similar to that of Co soap (based on metal level), suggesting that cost of the catalyst may become an issue if it would be applied.

The manganese-based porphyrin claimed for alkyd paint curing has also been covered for Fe, Co and V [31]. Unfortunately, no examples on curing were given for the Fe porphyrins, so its activity compared with the analogous Mn-porphyrin catalyst cannot be done.

1,4-diazacycloheptane-6-amines based ligands that have been described in the previous section, are besides active with Mn salts, some of the derivatives are also active with iron naphthenate salt (Figure 4) [42]. The examples shown in the patent indicated that for this class of ligands, one or more pyridine rings should be present to achieve curing activity. This was also observed for the mixed aliphatic/aromatic N donor ligands based on tris(pyridin-2-ylmethyl)amine (TPA) and N-methyl-N(pyridin-2-ylmethyl)-bis(pyridin-2-yl)methylamine (MeN3py) (Figure 4) that were also shown to give good curing activity with Fe [44].

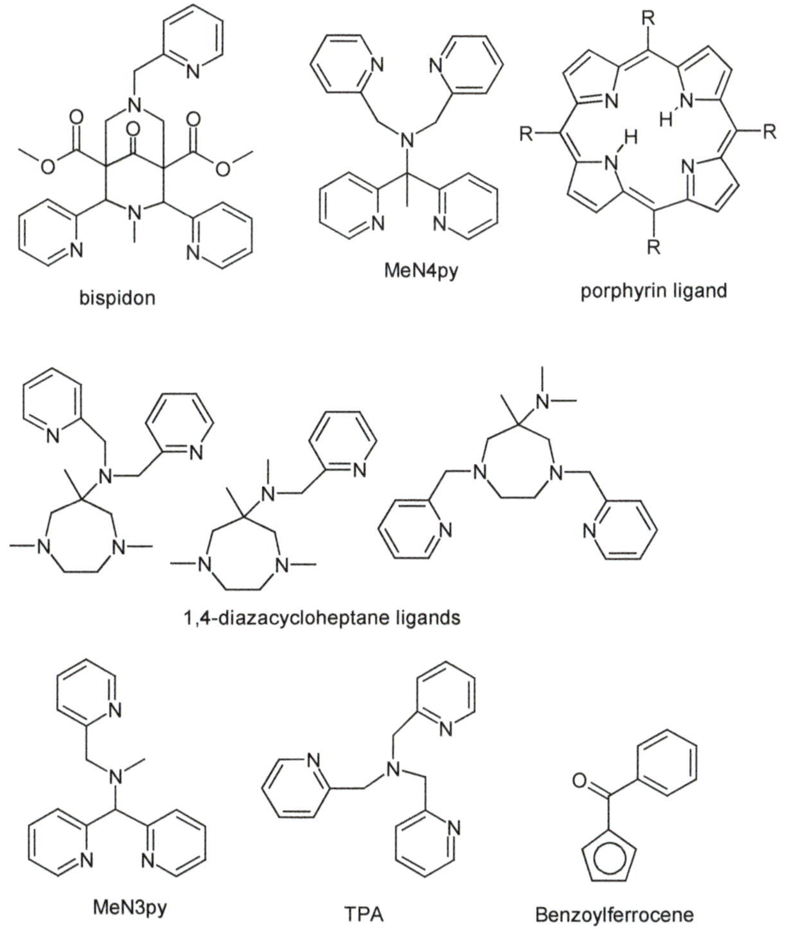

Figure 4. Ligands exhibiting paint drying activity when mixed with Fe salts or as Fe-ligand complexes as Fe-ligand complexes. Bispidon = dimethyl 3-methyl-9-oxo-2,4-di(pyridin-2-yl)-7-(pyridin-2-ylmethyl)-3,7-diazabicyclo[3.3.1]nonane-1,5-dicarboxylate, MeN4Py = N,N-bis(pyridin-2-ylmethyl)-1,1-bis(pyridine-2-yl)-1-amino-ethane, 5,10,15,20-tetrakisphenylporphyrin (R = phenyl; porphyrin ligand), 6-dimethylamino-1,4-bis(pyridine-2-ylmethyl)-6-methyl-1,4- diazacycloheptane, 1,4,6-trimethyl-6{N-(pyridin-2-ylmethyl)-N-methylamino}-1,4-diazacycloheptane, 6,{N,N-bis(pyridin-2-ylmethyl)amino}-1,4,6-trimethyl-1,4-diazacycloheptane (1,4-diazacycloheptane based ligands), tris(pyridin-2-ylmethyl)amine (TPA), and N-methyl-N(pyridin-2-ylmethyl)-bis(pyridin-2-yl)methylamine (MeN3py), benzoylferrocene.

In this research, we will describe our process of catalyst development, and understanding their processes, as supported by applications data. These technologies have been developed for cobalt-free solutions, and via a constant screening and selection process, we have found several candidates that meet the demands of performance in both water and solvent-based alkyds.

2. Results and Discussion

2.1. Iron-Based Technology (Fe-Bispidon Catalyst, Supplied as Borchi® OxyCoat)

Recently, very active iron catalyst based on pentadentate nitrogen bispidon and MeN4py donor ligands class has been described (Figure 4) [30]. The structure of the most active and applied catalyst is [(bispidon)Fe^{II}Cl]Cl, where bispidon is the abbreviation of dimethyl 3-methyl-9-oxo-2,4-di(pyridin-2-yl)-7-(pyridin-2-ylmethyl)-3,7-diazabicyclo[3.3.1]nonane-1,5-dicarboxylate) (Figure 4). This family of alkyd paint siccatives originate from studies for their use in detergent cleaning powders by Unilever. They have been shown to a high activity on food-oil stains in liquid detergents that do not contain bleaching agents such as sodium percarbonate [37,49]. Such stains originate often from unsaturated oils used in the cooking processes, which generates alkylhydroperoxides. In the presence of the Fe-bispidon or Fe-MeN4py catalyst, activation of the peroxide on the stain occurs yielding alkoxy and peroxyl radicals, which in turn react with other unsaturated oil molecules or the colorant, like lycopene found in tomatoes [37,49–51]. As these mechanisms are very similar to the ones described for alkyd resin curing by cobalt soap catalysts (and later-on described also for various manganese and iron catalysts), it is not a surprise that these compounds are amongst the most active ones for alkyd coating curing.

Using a medium-oil solvent borne alkyd resin, the level of Fe to obtain a good curing activity could be reduced from 0.08 wt% of Co (for a standard Co-soap formulation) to 0.0007 wt% of Fe (for the Fe-bispidon siccative) [52]. In a water-based alkyd emulsion, the activity of the Fe-bispidon is even more pronounced, it could be reduced to 0.0002 wt% versus Co-soaps that are used at between 0.005 and 0.02 wt%. Perhaps more importantly, when storing the alkyd resin emulsion, the curing activity of the iron complex remained constant versus the reduced activity of the Co-soap (the latter could be attributed to hydrolysis in the water phase leading to much less active [Co(H$_2$O)$_6$]$^{2+}$ species). As the bispidon catalyst consists of an iron nucleus bound by a pentadentate ligand, the stability in water is very high. Other benefits reported are the low color intensity in the paint formulations thanks to the high dilution of the yellow catalyst and relatively temperature insensitive behavior of curing (whilst Co^{II}(2-EH)$_2$ is clearly less active at low temperatures).

Pirš et al. studied [(bispidon)Fe^{II}Cl]Cl in high-solid (HS) alkyd paints and compared the properties with Co^{II}(2-EH)$_2$ [53]. FTIR (Fourier-transform Infrared spectroscopy) and elasto-viscosity measurements indicated that the cobalt soap siccative induced curing from the outer layer of the film, whilst [(bispidon)Fe^{II}Cl]Cl gave a more uniform curing activity through the whole film. As a consequence, the films cured by the bispidon catalyst are softer than those obtained by curing using cobalt soap. These results are in agreement with the NMR (Nuclear Magnetic Resonance) experiments reported by Erich and co-workers using the same catalyst (as well as two Mn siccatives) [54]. The same bispidon catalyst has also been studied with strontium soap (Sr(EH)$_2$) and compared with cobalt soap by Pirš et al. [55]. It was shown that Sr(EH)$_2$ induces harder film and better curing with especially the cobalt soap. Only at the highest dosage level of Sr(EH)$_2$ a clear increase in hardness when using the bispidon catalyst was observed.

We have developed different formulations of these catalysts to enable improved anti-skinning and rheological properties, for example Borchi® OxyCoat (BOC) supplied as a dilution in propylene glycol, and BOC1101 supplied in water. Improved dry times versus conventional cobalt soaps can be seen in Figure 5, at a fraction of the concentration, and with the added benefit of significantly reduced yellowing (seen visually in Figure 6) and loss of dry after ageing (one week at 40 °C—Figure 5).

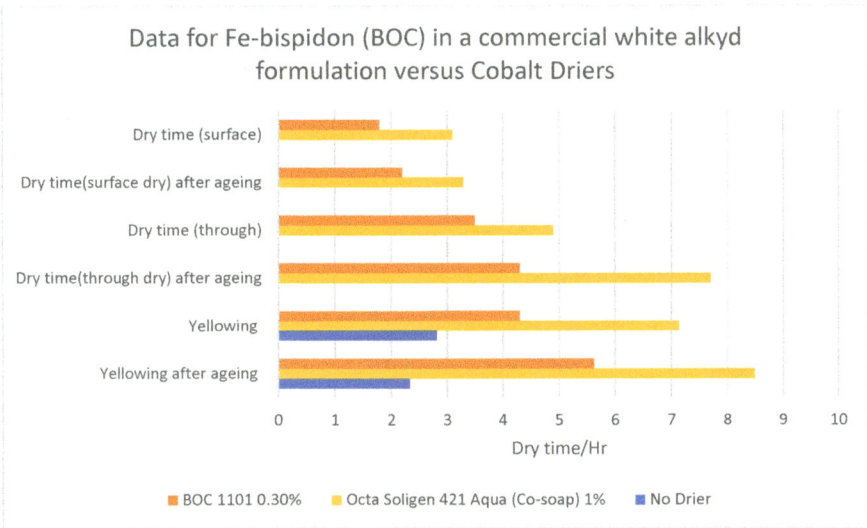

Figure 5. Data from a commercial resin comparing Borchi® OxyCoat (BOC) 1101 with a cobalt-siccative. Note that the resin does not dry without a catalyst (no siccative). BOC 1101 is a 1% solution of Fe-bispidon catalyst in water. Octa Soligen 421 is a mixture of cobalt, zinc, and zirconium carboxylates in water-dispersible oil.

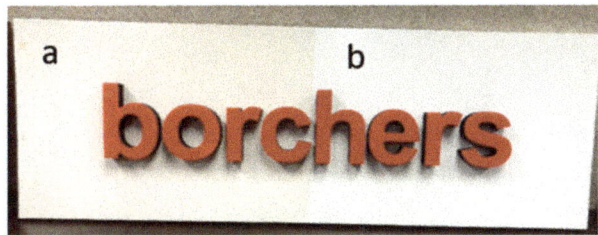

Figure 6. A marketing tool by Borchers showing the real-life effect of yellowing using a white alkyd formulated with a cobalt-soap (**a**) versus BOC (**b**) as the siccative, after storage.

2.2. Developing Alternative Curing Catalysts for Solvent-Borne Alkyds

2.2.1. Generation 1: $[Mn^{IV}_2(\mu\text{-}O)_3(Me_3TACN)_2](CH_3COO)_2$

Our experience has shown that the bispidon-based iron catalyst is not as effective in solvent-based high oil-containing alkyds as it is in water with medium and short-oil alkyds, particularly due to skinning in specific formulations. As a response to this, we examined a series of new ligands and concepts. In the first generation, we used an existing novel catalyst, $[Mn^{IV}_2(\mu\text{-}O)_3(Me_3TACN)_2](CH_3COO)_2$ sold as an aqueous solution with the trade-name DB A350. Improved curing activity using these manganese complexes was claimed in patents from DSM [56–58]. In one of these patents, it was shown that $[Mn^{III}_2(\mu\text{-}O)(\mu\text{-}RCO_2)_2(Me_3TACN)_2]^{2+}$ complexes with RCO_2 = benzoate and 2,6-difluorobenzoate exhibit faster alkyd paint drying activity than $[Mn^{IV}_2(\mu\text{-}O)_3(Me_3TACN)_2](PF_6)_2$ [56]. Even more interestingly, it was shown in another patent with the same priority date that the curing activity of $[Mn^{IV}_2(\mu\text{-}O)_3(Me_3TACN)_2](CH_3COO)_2$ is higher than the analogous compound with PF_6 as counterion [57]. As this complex was supplied as a diluted aqueous solution, most of the water was removed by evaporation at 50 °C under low pressure, it might be that some of the original $[Mn^{IV}_2(\mu\text{-}O)_3(Me_3TACN)_2]^{2+}$ species decomposed, leading to the formation of more reactive species than the original compound. In the last patent of this series, it was shown that the activity of [Mn^{IV}$_2$(µ-O)$_3$(Me$_3$TACN) $_2$](CH$_3$COO)$_2$ is higher when ascorbic acid palmitate is added [58], suggesting that lower-valent manganese (Mn(II) and/or Mn(III)) species with this ligand are involved or closely associated with the curing chemistry of alkyd resins. Whilst the dinuclear Mn^{III} complex has two carboxylate bridges, the dinuclear Mn^{II} complex has either two carboxylate bridges and a hydroxide bridge or three carboxylate bridges, i.e., having structures such as [Mn^{II}$_2$(µ–OH)(µ–RCO$_2$)$_2$(Me$_3$TACN)$_2$]$^+$ or [Mn^{II}$_2$(µ–RCO$_2$)$_3$(Me$_3$TACN)$_2$]$^+$. The structures of both the analogous Mn^{II}$_2$ complexes with acetate bridges have been elucidated by Wieghardt and co-workers [59]. Both species with ethylhexanoate bridges are expected to be more lypophilic than [Mn^{IV}$_2$(µ-O)$_3$(Me$_3$TACN)$_2$](CH$_3$COO)$_2$ due to the combination of the more lypophilic carboxylates in combination with the presence of a mono-charged complex (in contrast to most other dinuclear Mn complexes with Me$_3$TACN which are divalent in nature [59,60]).

In our trials, [Mn^{IV}$_2$(µ-O)$_3$(Me$_3$TACN)$_2$](PF$_6$)$_2$ (Mn-Me$_3$TACN) gave similar or faster drying activity compared to cobalt 2-ethylhexanoate (CoEH$_2$), based on time to fully surface dry (Figure 7); saving 2 h can allow serious cost savings in labour during application.

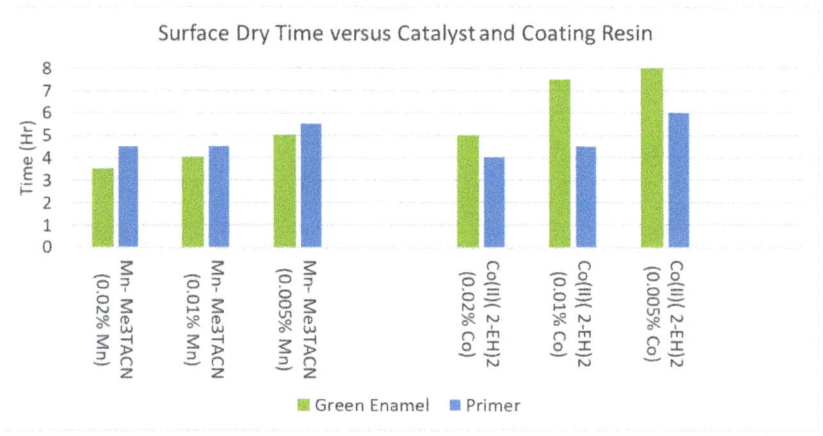

Figure 7. Dry times of [Mn^{IV}$_2$(µ-O)$_3$(Me$_3$TACN)$_2$](PF$_6$)$_2$ (abbreviated as Mn- Me$_3$TACN) versus Co^{II}(2-EH)$_2$ in two formulated, commercial alkyd paints.

[Mn^{IV}$_2$(µ-O)$_3$(L)$_2$](PF$_6$)$_2$ complexes with L = Me$_3$TACN analogues, such as Me$_2$BTACN (1-benzyl-4,7-dimethyl-1,4,7-triazacyclononane) [61] and Et$_3$TACN (1,4,7-triethyl-1,4,7-triazacyclononane) [62] were tested as well for alkyd curing and were found to yield comparable alkyd resin curing activity as [Mn^{IV}$_2$(µ-O)$_3$(Me$_3$TACN)$_2$](PF$_6$)$_2$ (all in the presence of one molar equivalent of ascorbic acid) (Figure 8). Therefore, the presence of a benzyl group (instead of methyl) or three ethyl groups (instead of three methyl groups) gave no further advantage in curing activity and these might only influence ligand compatibility.

We can learn something about the reactivity from the literature, where the [Mn^{IV}$_2$(µ-O)$_3$(Me$_3$TACN)$_2$](PF$_6$)$_2$ catalyst in emulsions towards reactivity of ethyllinoleate with oxygen is higher than that of Mn^{II}(2-EH)$_2$ but less than that of Co^{II}(2-EH)$_2$ [32]. A key problem noted for [Mn^{IV}$_2$(µ-O)$_3$(Me$_3$TACN)$_2$](PF$_6$)$_2$ is that the solubility in various formulations is not high. It might be that by using the complexes with other counterions [58], or by addition of reducing agent in the presence of formulations containing lipophilic soaps [57], species are formed in situ that are much better compatible with the alkyd formulations than the parent [Mn^{IV}$_2$(µ-O)$_3$(Me$_3$TACN)$_2$](PF$_6$)$_2$ compound.

Inclusion of *N,N,N',N'',N''',N'''*-hexamethyltriethylenetetraamine (HMTT–Figure 3) in an emulsion of ethyllinoleate and [Mn^{IV}$_2$(µ-O)$_3$(Me$_3$TACN)$_2$](PF$_6$)$_2$ gave faster oxidations of ethyllinoleate. It is known that polyamines facilitate the decomposition of alkylhydroperoxides. Support of the latter reaction has been found by studying homogeneous solutions of the manganese catalyst, HMTT and ethyllinoleate: The reaction to yield volatile hexanal and heptanal due to β-scission of the substrate is significantly higher in the presence of HMTT [63].

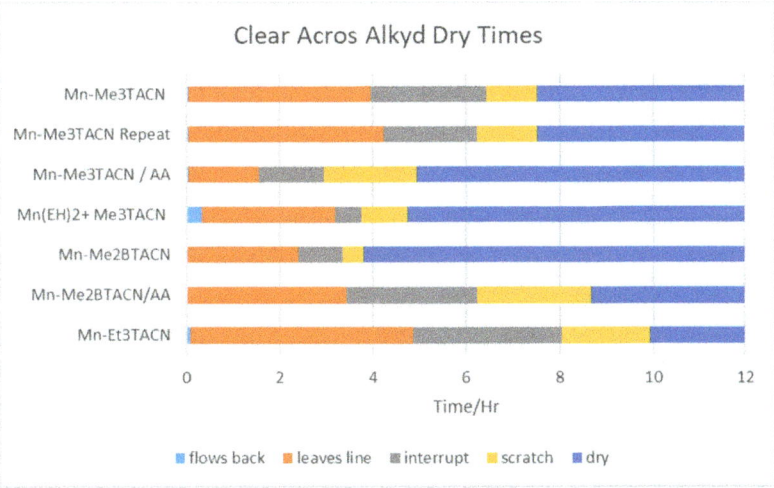

Figure 8. Drying times of alkyd resin in the presence of 0.01 wt% Mn added as [Mn^{IV}$_2$(μ-O)$_3$(Me$_3$TACN)$_2$](PF$_6$)$_2$ (abbr. as Mn-Me$_3$TACN), [Mn^{IV}$_2$(μ-O)$_3$(Me$_3$TACN)$_2$](PF$_6$)$_2$ with one molar equivalent of ascorbic acid (abbr. as Mn-Me$_3$TACN/AA), Mn(EH)$_2$ with one molar equivalent of Me$_3$TACN (abbr. as Mn(EH)$_2$ + Me$_3$TACN), [Mn^{IV}$_2$(μ-O)$_3$(Me$_2$BTACN)$_2$](PF$_6$)$_2$ with one molar equivalent of ascorbic acid abbr. as (Mn-Me$_2$BTACN/AA) and [Mn^{IV}$_2$(μ-O)$_3$(Et$_3$TACN)$_2$](PF$_6$)$_2$ (abbr. as Mn-Et$_3$TACN/AA).

Besides the reported curing activity in water-borne emulsions of this and related catalysts [57,58], this catalyst gives also good activity in solvent borne paints [34], which is of major interest as the Fe-bispidon catalyst exhibits activity at very low levels of catalysts in water-borne emulsions (vide supra). In a short oil formulation, a similar curing time as observed for Co^{II}(2-EH)$_2$ was observed. In a paint formulation containing a long oil alkyd resin the curing activity appeared to be good, with a notable better through-drying activity than Co-soap, albeit yielding a slightly softer film [33]. The observation that C=C bond vibrations at 1650 cm^{-1} as studied by IR are faster depleted at the surface for Co^{II}(2-EH)$_2$ than for [Mn^{IV}$_2$(μ-O)$_3$(Me$_3$TACN)$_2$](PF$_6$)$_2$, suggesting that whilst overall similar mechanisms may be operative, the curing activity of Co^{II}(2-EH)$_2$ on the surface of the film may inhibit O$_2$ to penetrate in the deeper layers, whilst [Mn^{IV}$_2$(μ-O)$_3$(Me$_3$TACN)$_2$](PF$_6$)$_2$ allows O$_2$ to diffuse through the whole paint layer and therefore it can generate an in-depth cured film [33]. This aspect has been discussed much more recently for a mixture of Mn^{II}(2-EH)$_2$ and the Me$_3$TACN ligand (vs. Co^{II}(2-EH)$_2$, a Mn porphyrin catalyst and the Fe bispidon catalyst) using NMR imaging measurements [54].

As shown in Figure 8 and elsewhere [58], adding ascorbic acid has been shown to accelerate drying when using [Mn^{IV}$_2$(μ-O)$_3$(Me$_3$TACN)$_2$](PF$_6$)$_2$. However, at this stage we also saw a significant improvement by moving towards an in-situ formed catalyst by blending Mn^{II}(EH)$_2$ with Me$_3$TACN, almost halving the dry time required (Figure 8). This prompted us to examine alternative catalyst concepts, as discussed in the next sections. The observations that adding a reducing agent to [Mn^{IV}$_2$(μ-O)$_3$(Me$_3$TACN)$_2$]$^{2+}$ and that mixing ligand with Mn^{II}(EH)$_2$ gives increased curing activity suggests that lower-valent Mn species are involved in the catalysis, in disagreement with the proposal put forward by Oyman and co-workers, where [Mn^{IV}$_2$(μ-O)$_3$(Me$_3$TACN)$_2$]$^{2+}$ itself is thought to be involved in the alkylhydroperoxide activation process, via outer-sphere electron transfer (Mn^{IV}Mn^{IV}/Mn^{III}Mn^{IV} couple) [63].

Based on the observations that mixing ligand with $Mn^{II}(EH)_2$ gives good alkyd-curing activity, we developed and patented an innovative way of allowing formulation flexibility and boosting resin performance by preloading the resin with ligand as an 'accelerator'(without metal ions). The paint formulator can then afterwards add the Mn soap into the resin/ligand mixture to form active species [40]. The benefit is that when the ligand is mixed with the resin and no metal ions are present, no reaction with oxygen to obtain peroxide and therefore skinning can occur. Only after addition of the Mn soap, reactivity towards alkyd resin curing can occur. In the same patent, EPR (electron paramagnetic resonance) spectroscopy has been used to obtain more understanding on the various species formed [40]. The ligand in the resin (without $Mn^{II}(2\text{-}EH)_2$ added) did not show any appreciable EPR signal, whilst adding $Mn^{II}(2\text{-}EH)_2$ without ligand furnished an EPR signal of a typical Mn(II) species, as expected. When the ligand is mixed with $Mn^{II}(2\text{-}EH)_2$, signals typical of dinuclear $Mn^{II}{}_2(Me_3TACN)_2$ species were observed [40,60,64]. The dinuclear species may well be the same species formed when ascorbic acid palmitate was added to the $[Mn^{IV}{}_2(\mu\text{-}O)_3(Me_3TACN)_2]^{2+}$ complex [58]. This latter observation suggests that for the activation of dioxygen or hydroperoxide to achieve alkyd curing the dinuclear Mn^{IV} complex is not involved, but that more likely lower-valent Mn species react with oxygen or alkylhydroperoxide to attain alkyd curing. The studies conducted by Oyman and co-workers, indicate that the oxidation process of ethyllinoleate by air and $[Mn^{IV}{}_2(\mu\text{-}O)_3(Me_3TACN)_2](PF_6)_2$, involve alkoxy and peroxyl radicals [32,63], reminiscent to the curing processes ascribed to $Co^{II}(2\text{-}EH)_2$, but no studies have been published using the lower-valent $Mn\text{-}Me_3TACN$ complexes.

2.2.2. Generation 2: A Broad Study of Me_3TACN-Catalyst Analogs

We already saw a benefit of combining Me_3TACN with $Mn^{II}(2\text{-}EH)_2$) vs. $[Mn^{IV}{}_2(\mu\text{-}O)_3(Me_3TACN)_2](PF_6)_2$ (Figure 8), and that $[Mn^{IV}{}_2(\mu\text{-}O)_3(Me_3TACN)_2](PF_6)_2$ compared favorably with $Co^{II}(2\text{-}EH)_2$) (Figure 7) regardless of dose. Although the two derivatives tested (complexes with Me_2BTACN and Et_3TACN) did not yield improved curing activity compared to the parent compound, we decided to investigate a wider range of alternative ligands versus Me_3TACN for creating new complexes that may give improved performance. First a series of tests was conducted using different levels of $Mn^{II}(2\text{-}EH)_2$ vs. $Mn^{II}(2\text{-}EH)_2$ with one molar equivalent of Me_3TACN (between 0.01 and 0.05% Mn) to test curing activity on a formulated alkyd paint, as shown in Figure 9. It was clear that even at 0.05% Mn, only set dry but no surface curing took place, whilst addition of one molar equivalent of Me_3TACN ligand gave already good surface curing activity at 0.01%Mn level, which further improved by increasing the Mn/ Me_3TACN level further (Figure 9).

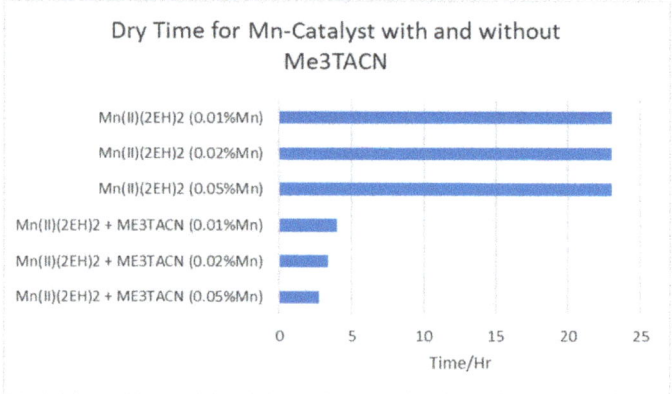

Figure 9. Surface drying activity (denoted in blue) of various level of $Mn^{II}(2\text{-}EH)_2$ and $Mn^{II}(2\text{-}EH)_2$ premixed with molar equivalent of Me_3TACN (1,4,7-trimethyl-1,4,7-triazacyclononane). This data set was obtained from a formulated commercial alkyd.

We therefore screened the ligands shown in Figure 10 in combination with equimolar levels of Mn^{II}(2-EH)$_2$ at 0.01% Mn. In most cases we either blended the ligands with the manganese soap at a 1:1 molar ratio to form the catalyst in-situ, or the manganese complexes have been prepared [61,62] (denoted in Figure 10). For some experiments we increased the ratio of ligand to Mn^{II}(2-EH)$_2$.

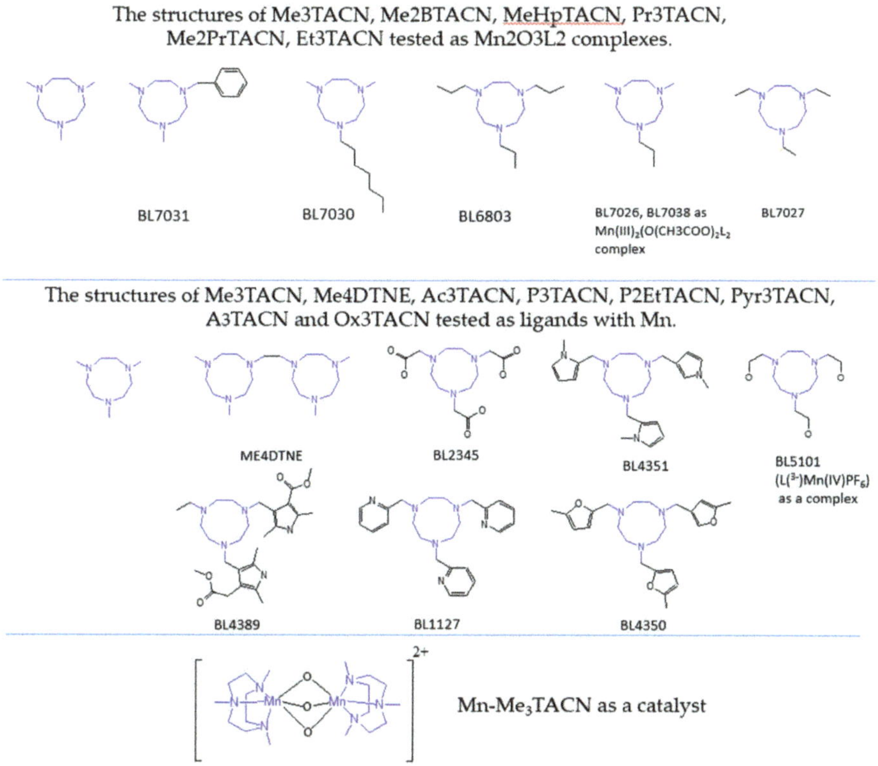

Figure 10. Schematic structures of the ligands tested as [Mn$_2$O$_3$L$_2$]$^{2+}$ complexes (top) or tested as mixtures with Mn^{II}(EH)$_2$ (middle). At the bottom of the figure, the structure of the dinuclear [Mn$_2$O$_3$L$_2$]$^{2+}$ complex with L = Me$_3$TACN is depicted. The codes given are numbers as used for the ligands and complexes and are also used in the results shown in Figures 11–13.

Selected comparative data in an alkyd resin, that does not contain pigments and is additive free, can be seen in Figure 11, where ligands such as Me$_3$TACN, Me$_4$DTNE, and Me$_2$BTACN in combination with Mn^{II}(2-EH)$_2$ yielded the fasted drying activity. Moreover, various preformed dinuclear Mn complexes containing Me$_3$TACN, Me$_2$BTACN, Me$_2$HpTACN, and Me$_2$PTACN ligand in combination with one molar equivalent of ascorbic acid yielded efficient curing. Whilst these complexes were tested as PF$_6$ salt, the manganese complex with Me$_3$TACN gave similar activities as sulfate or acetate salts (all in the presence of ascorbic acid). It should be noted that it has been shown before that the sulfate and acetate salts of the dinuclear Mn-Me$_3$TACN complexes in the absence of ascorbic acid are significantly more active than the analogous complex as PF$_6$ salt [58]. Selected testing showed good reproducibility for the dry times.

Interestingly, the ligands containing three longer chain aliphatic groups (Et$_3$TACN, Pr$_3$TACN) or other groups that may coordinate to Mn (P$_3$TACN, Py$_3$TACN) did not show good curing activity with Mn. More sterically demanding side groups may hinder efficient binding to the metal center. Furthermore, the ligands with the three auxiliary nitrogen donors (P$_3$TACN and Py$_3$TACN) may bind via six nitrogen donors to Mn(II) ions, thereby blocking the coordination sites of the Mn center that otherwise would be involved in activating the alkylhydroperoxide of the alkyd resin as explained in Section 1.2.

We then retested a wide range of the ligands in a white alkyd paint supplied by a commercial supplier. This was an important test, as the paint was ready-formulated, and the data would include (unknown and proprietary) additional surfaces (i.e., pigments) and emulsifiers that could interfere with the drying process. This test would also allow us to compare the effect of yellowing over time by examining the yellowing index. It is well known that alkyds can yellow over time in the dark, caused by a series of oxidative transformations of the unsaturated groups leading to decomposition products such as highly conjugated ketones [65]. It is also known that siccatives can lose performance over time if they are adsorbed onto pigments.

The dry times were very similar between $Me_3TACN/Mn^{II}(2\text{-}EH)_2$, $Me_2BTACN/Mn^{II}(2\text{-}EH)_2$, preformed $[Mn^{IV}_2(\mu\text{-}O)_3(L)_2]^{2+}$ complexes and $Me_4DTNE/Mn^{II}(2\text{-}EH)_2$, all fully drying within 2 h at 0.01% Mn on resin (Figure 11). They also all gave similar hardness values of around 18 s pendulum hardness after initial drying, suggesting little difference in the crosslinking density and indicating similar curing processes. It should be noted that these coatings were particularly soft as they were not optimized via the use of secondary siccatives to prevent additional variables in our comparisons.

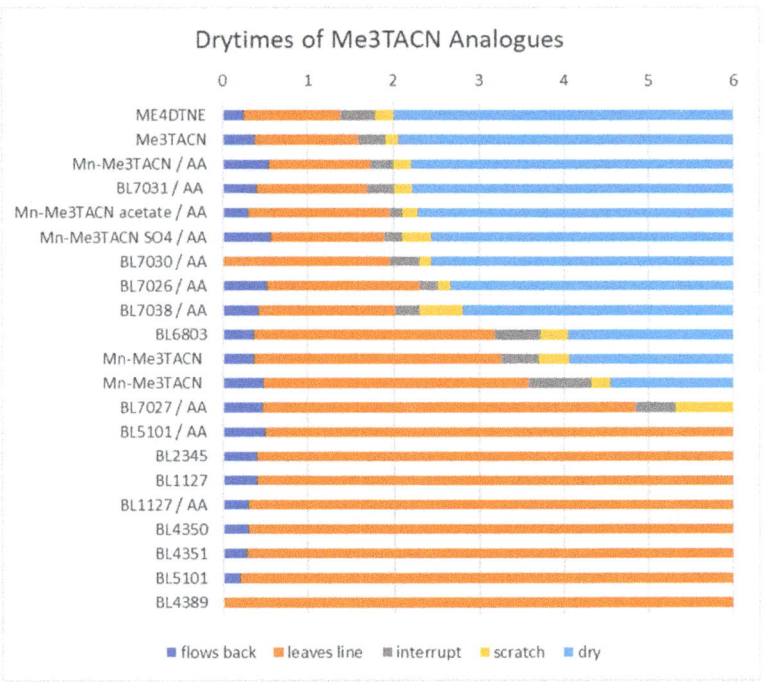

Figure 11. Dry time data from selected catalysts or mixtures of ligands and $Mn^{II}(EH)_2$ based on Me_3TACN analogs shown in Figure 10. In this figure, $Mn\text{-}Me_3TACN$ stands for $[Mn^{IV}_2(\mu\text{-}O)_3(Me_3TACN)_2](PF_6)_2$, $Mn\text{-}Me_3TACN$ acetate stands for $[Mn^{IV}_2(\mu\text{-}O)_3(Me_3TACN)_2](CH_3COO)_2$, and $Mn\text{-}Me_3TACN\ SO_4$ stands for $[Mn^{IV}_2(\mu\text{-}O)_3(Me_3TACN)_2](SO_4)$. When AA is written after the abbreviation, one molar equivalent of ascorbic acid was mixed to the catalyst solution prior application. The other numbers of ligands or complexes are given in Figure 10.

However, we saw very clear differences in both yellowing and loss of dry after four weeks storage for these classes of materials (Figure 12). The coatings made using the $[Mn^{IV}_2(\mu\text{-}O)_3(Me_3TACN)_2]^{2+}$

and Me₃TACN siccatives exhibit significant yellowing after four weeks of storage compared with Me₄DTNE, Me₂BTACN and the other next-generation ligands.

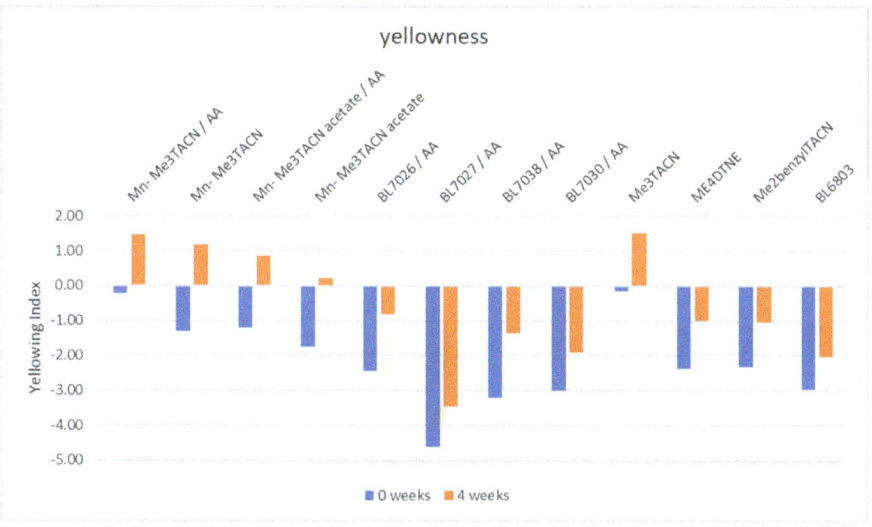

Figure 12. Yellowing Index data from selected catalysts based on Me₃TACN analogues, before and after storage over four weeks. Catalysts tested from Figure 10, some with additional ascorbic acid.

Loss of dry occurs when some (or all) of the catalyst can no longer function, leading to increased dry times. This is typically caused by adsorption to a pigment surface, and can be remedied by formulation, for example by adding a calcium soap or an emulsifier such as a pigment dispersant. We examined the catalysts in the pigmented alkyd, with the best dry times versus yellowing and loss of dry after four weeks of storage (Figure 13). [Mn^{IV}₂(μ-O)₃(Me₃TACN)₂](PF₆)₂ gave the most yellowing, but newer generations such as BL7030/AA, Me2BTACN, BL7038/AA gave a significant improvement without incurring loss of dry. Moreover, they reduced yellowing versus the blended Me₃TACN/Mn^{II}(2-EH)₂ catalyst. BL7027/AA dried faster after storage, perhaps indicating improved solubility over time or slow formation of more active species; we did not investigate further.

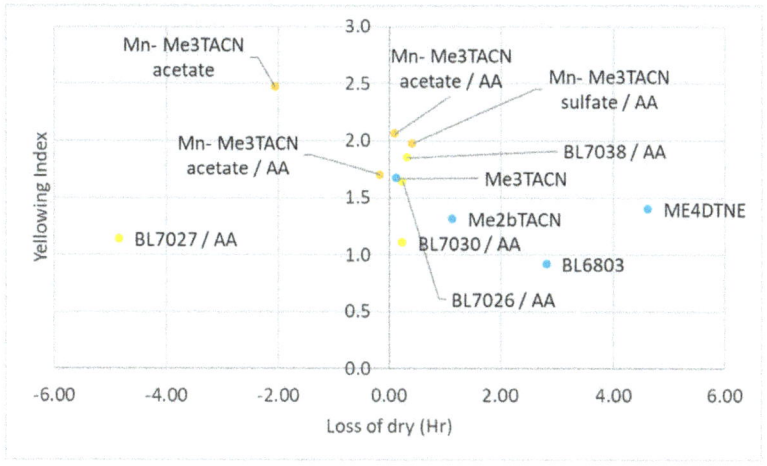

Figure 13. Combined Yellowing index and loss of dry data for selected Me₃TACN analogs.

2.2.3. Generation 3: Non-Me₃TACN-Catalyst Analogs

As mentioned in the introduction, another class of tridentate aliphatic nitrogen donor ligands is based on the 1,4-diazacycloheptane-6-amine backbone (Figure 14). The fact that, in contrast to the triazacyclononane ligands, these ligands do not contain the three nitrogen donors in the macrocyclic ring, the kinetic stability of metal complexes are less than that of analogous complexes with the triazacyclononane ligands [43]. An advantage of this class of ligands is that different modifications can be relatively easily achieved. Methylation of the aliphatic nitrogen atom yielded L2. Attachment of a methylpyridine-2-yl group to L1 yielded L3. In order to prevent the possibility of imine formation of the NH-CH$_2$ unit, methylation of this nitrogen donor yielded L4. Alternatively, another the methylpyridine-2-yl group attached on this nitrogen atom, yielded the pentadentate ligand, L5. L6 and L7 are ligand variants were the nitrogen donors in the cycloheptane ring are functionalized by methylpyridine-2-yl moieties.

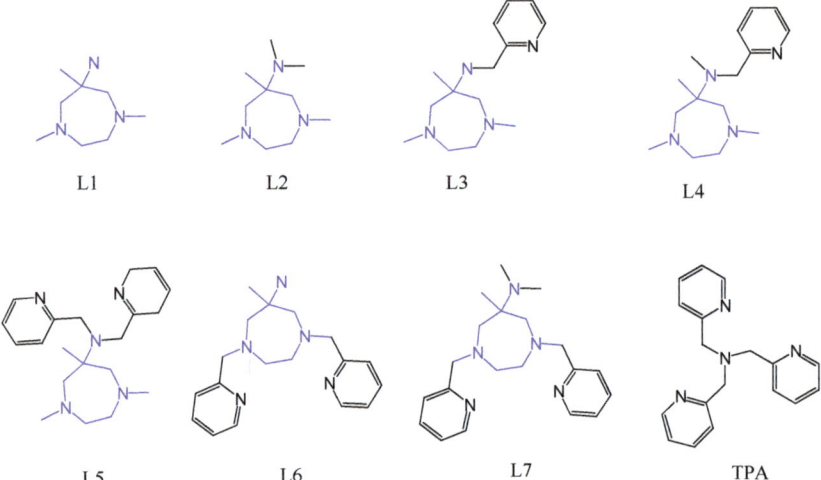

Figure 14. Non-Me₃TACN Ligands tested with Mn and Fe salts towards alkyd resin curing.

As now a wider variety of tridentate, tetradentate, and pentadentate ligands has been prepared, we decided to test these ligands not only with Mn salts, but also with Fe salts (or test the preformed Fe and Mn complexes). Results are given and discussed below.

Ligands L1, L3, L4, and Me₃TACN in combination with Mn^{II}(2-EH)$_2$ at 0.01% Mn level and L5 and L6 with FeCl$_2$ at 0.01% Fe level all showed efficient drying (Figure 15). L2 yielded a lower activity with Mn than L1, L3, or L4. Interestingly, L5 and L6 worked best with the Fe-source, whilst with Mn^{II}(2-EH)$_2$ and L5 no drying activity was observed. For L7 a similar (relatively low) drying activity was found for both Mn and Fe. The ligands L1–L4 did not show activity with FeCl$_2$. L5 and L6 are both pentadentate ligands, like the bispidon and N4py ligands that are known to be very active towards alkyd resin curing when bound to Fe(II) [30]. Probably, ligands with less coordination sites may bind to Fe(II), but will yield less stable iron complexes. On the other hand, tetradentate ligands such as TPA (Figure 14) and N3py (Figure 4) did show activity with the iron source, albeit less efficiently than the pentadentate bispidon ligand. Thus, the choice of the metal centers in combination with the ligand is critical for the alkyd curing activity. In most cases, we found no benefit from using the pre-formed metal complexes vs. mixing the metal-soap/salts with ligands.

As with the other ligands, we generally saw that increasing catalyst concentration increased drying activity as expected. The hardness development of all generation 2 siccatives in a formulated commercial alkyd was in same range, highest hardness for Mn-L2 (Figure 16), although the faster curing L4 gave the softest film, perhaps due to poor through-drying activity.

The increase of yellowness on storage (Figure 17) could be seen for all generation 2 siccatives. In general, Fe based dries showed higher initial yellowness, but Mn siccatives showed slightly more yellowing (increase in yellowness). Based on our research in other resins, we rank the activity thusly: Mn-Me$_3$TACN > Mn-TPA > Mn-L4 > Mn/Fe-L6 > Mn-L1 > Mn-L3 > Fe-L5 > Mn-L2 > Mn/Fe-L7.

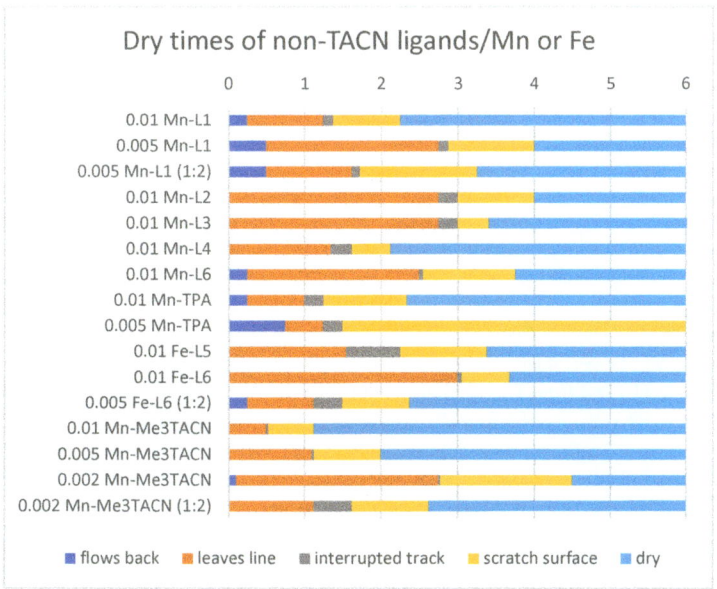

Figure 15. Alkyd resin curing activity of the ligands L1–L6, Me$_3$TACN, and TPA mixed with either Mn^{II}(EH)$_2$ (abbr. as Mn) or FeCl$_2$ (abbr. as Fe). The numbers of the ligands refer to the structures given in Figure 14. Results of Mn-L5 and Mn-L7 were omitted, because no or low curing activity was observed.

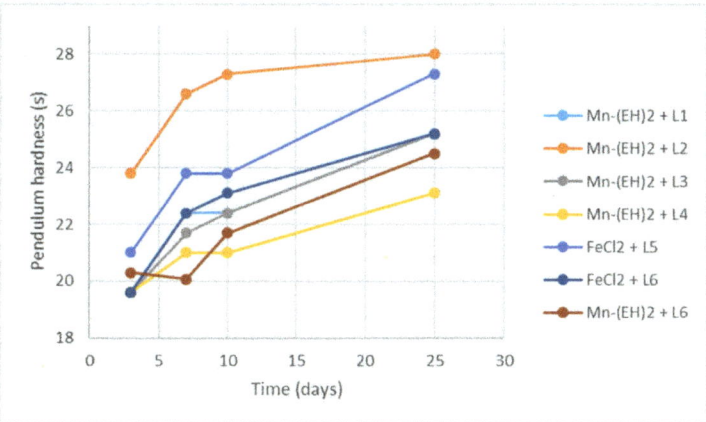

Figure 16. Hardening data in a commercial white alkyd formulation (without optimization using, i.e., secondary and sacrificial siccatives).

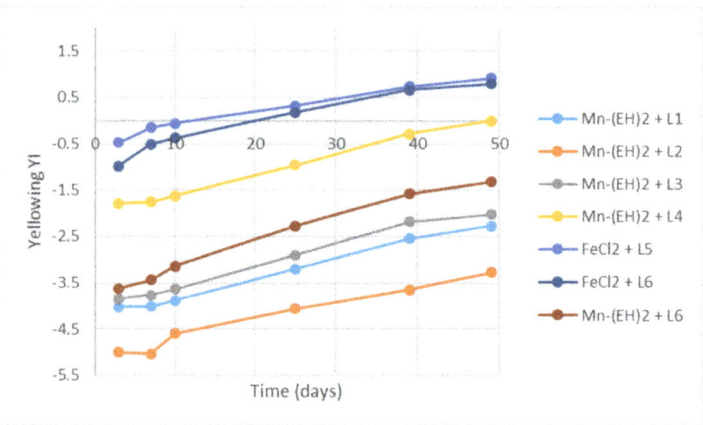

Figure 17. Yellowing data in a commercial white alkyd formulation (without optimization using, i.e., secondary and sacrificial siccatives).

From the tests discussed in Sections 2.2 and 2.3, different interesting metal–ligand complexes have been identified to be of interest for alkyd paint curing. From the TACN family (Sections 2.2.1 and 2.2.2), the fully methylated ligand (Me_3TACN), benzyl-substituted TACN (Me_2BTACN), n-heptyl-substituted TACN (Me_2HpTACN), propyl-substituted TACN (Me_2PrTACN), and bridged TACN ligand (Me_4DTNE) showed all good and similar curing activity in conjunction with a Mn-soap or as a Mn complex (with ascorbic acid added). Similarly, manganese complexes with Mn-TPA, Mn-L4, L6, and Mn-L1, discussed in this section, show a good curing activity and all similar or better than that of Co-soap. Whilst the structures of the active driers are very different, they all have a common feature: The nature of the polydentate nitrogen ligand aids in stabilizing the active metal complexes from hydrolyzing/decomposition into free metal salts, that are much less active towards curing.

However, as shown in subsequent tests, the rate and extent of yellowing of the alkyd resin and the development of the hardness of the cured film are affected to a large extent by the choice of the Mn or Fe drier.

Therefore, whilst it may be tempting to design an optimal alkyd resin curing catalyst based on the curing activity alone, the paint producer will make a choice to replace Co-soaps as driers not only based on the high curing activity. Yellowing, hardness development, but also storage stability will be essential attributes. To a significant extent the properties of the paint formulation can be altered by using different alkyd resins, secondary driers, different pigments, etc. However, as most of the formulations have been previously been optimized for the compatibility with Co-soap, the paint formulations need to be adapted to optimize the properties of the coating (i.e., good curing activity, low yellowing, good storage stability, and appropriate hardness development). At this stage of the development it is not feasible to make a good prediction which is the best metal–ligand combination in conjunction with the optimal paint formulation.

2.3. Anti-Skinning Agents in Combination with Fe Catalysts

Oxygen reacts at the surface of an alkyd paint to begin the crosslinking process, this leads to a skin on the surface of the paint that can account for a 2% loss in production. Therefore, anti-skinning agents are typically added to the alkyd paint formulations that work by retarding oxidation. They typically work by inhibiting free radical-induced oxidations, by reacting alkoxy or peroxyl radicals formed, or by saturating in the head space to reduce the available oxygen to the paint surface or by both mechanisms. Typical examples of anti-skinning agents include amines, phenols, oximes, or the most popular and effective being ketoximes like methyl ethylketoxime (MEKO) [66]. However, ECHA classifies MEKO as toxic and possibly carcinogenic (EC number 202-496-6). As a result, alternatives have been developed such as acetylacetonate [67], isoascorbate [68], ureas [69], and ketoxime-based analogues such as 2-pentanone oxime [70].

MEKO works by suppressing the reaction of radicals on the surface of the alkyd paint layer with oxygen (present in the headspace of the paint can), due to its volatility and capability to react alkoxy radicals [71]. Further, it has weakly coordination capabilities to cobalt soaps [71], thereby blocking the possibility of the cobalt siccative to react with oxygen or hydroperoxyl species to initiate radical-induced polymerization of the paint formulation in the can. When the paint coating is applied on a surface, MEKO can evaporate and the equilibrium shifts to non-coordinated MEKO, allowing the cobalt soap ion to react to induce radical curing of the alkyd resin.

The iron-bispidon ligand catalyst does not work efficiently with conventional anti-skinning agents like MEKO and other ketoximes, which is likely caused by steric hindrance of the bulky pentadentate bispidon ligand preventing co-ordination into the central iron center. Another possibility is that the ketoxime is a poor ligand for the Fe(II) center, and therefore it is much less effective as anti-skinning agent than what has been found for Co soaps. Moreover, for Fe-TPA a poor anti-skinning effect of MEKO was observed, suggesting that for this complex with the less sterically crowded ligand, poor binding of MEKO to the metal center may be the main explanation of these observations.

One solution for the Fe-bispidon catalyst is via the use of water as co-solvent [72]. Although not specified in this patent application, it might be that phase-separation of the hydrophilic Fe-bispidon catalyst in water from the hydrophobic alkyd resin may be the reason why such approach has shown to give benefits.

Typically, volatile anti-skinning agents such Borchers non-ketoxime Borchers® Ascinin 0444 work very effectively with conventional metal carboxylates and iron-bispidon and do not suffer the regulatory issues of MEKO, meaning the combination is a perfect formulation fit for the regulatory environment for water-borne alkyds and inks. Borchers® Ascinin 0444 contains N,N-diethylhydroxylamine, a volatile amine that helps hinder oxidation via saturating the headspace of the paint can and acts as an oxygen scavenger. In this research, we introduce our initial results for developing a bespoke anti-skinning solution for the iron-bispidon catalyst, as part of an ongoing investigation to complement MEKO-free anti-skinning agents.

For some solvent-borne high-oil alkyds, the Fe-bispidon catalyst shows skinning behavior also in the presence of MEKO or Borchers® Ascinin 0444. Raman and FT-IR spectroscopies have been used to study iron-bispidon's reactivity with ethyllinoleate [52]. Like observed for $Co^{II}(2-EH)_2$ and various other Mn and Fe alkyd resin curing catalysts, autoxidation, radical based oxidation processes appear to take place, which is likely also operative in the skinning behavior of Fe-bispidon. However, when using the Fe-bispidon catalyst the oxygen scavenging and headspace process of Borchers® Ascinin 0444 do not have as significant effect on preventing skinning when compared with $Co^{II}(2-EH)_2$.

In our laboratories, through a process of formulating drier combinations, we have discovered that a combination of a ketoxime and cerium octanoate (named LP1810) helped to improve anti-skinning behavior to the same degree as Borchers® Ascinin 0444 (Figure 18). This effect has been observed using different ketoxime anti-skinning agents, including 2-pentanone oxime. We observed a small variation in skinning behavior across seven different resins, of between one–two units, even despite different dose levels. More interestingly, we can see the dry times are significantly impacted by using different anti-skinning agents. As expected, increasing anti-skinning agent dose levels lead to reduced dry time as the oxidative process of crosslinking has been retarded. The best combination of dry time and anti-skinning came from the MEKO and LP1810 agents at low dose, showing we had identified an improved MEKO-free solution.

We are still developing an understanding of why cerium soap and ketoximes furnish a strong retardation effect of the skinning. Even though a full understanding has not been achieved at this stage, highlighting this observation may be useful for developers of new paint formulations which includes the iron bispidon catalyst. We can infer some rational from the literature, for example in a recent paper it has been shown that the Fe-bispidon complex with 2-ethylhexanoate as sixth ligand instead of chloride showed increased solubility in organic solvents and as a consequence shows different activity in the autoxidation processes [73]. Improved catalyst performance has also been seen by adding a potassium salt of a carboxylic acid to an iron-bispidon catalyst [74]—we believe this leads to an ion-exchange of Cl for the carboxylate leading to improved resin compatibility. We believe that with LP1810 the cerium carboxylate works in the same way, whilst the ketoxime saturates the headspace to prevent easy access to oxygen. This combination can be prepared by formulators that wish to use the Fe-bispidon catalyst because of its powerful curing profile and non-yellowing aspect. However, this is not a solution for all alkyds. For that reason, we have also investigated the new ligands made specifically as cobalt-free siccatives for solvent-borne alkyds.

With the new ligands it is expected that less sterically hindered structures might enable a combination of activity retardation through binding to the metal center, as well as headspace protection. This would enable alternative strategies for reducing skin, and this research is currently under investigation.

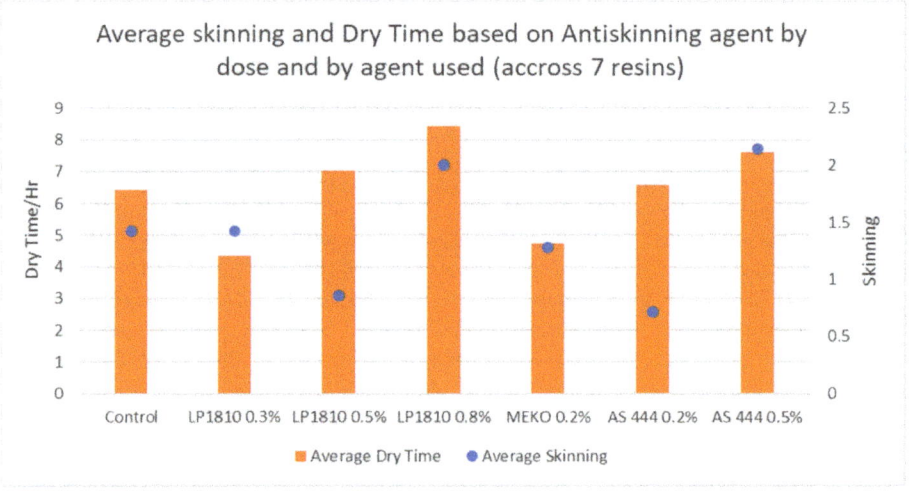

Figure 18. Average skinning (shown as blue dots) and dry times (shown as orange bars) of seven resins using anti-skinning agents with Borchi® Oxy Coat Fe-bispidon catalyst. Seven commercial, high-oil alkyds were used, and an average dry time and skinning was taken for each anti-skinning agent. Formulations were mixed using a speed mixer DAC 150.FVZ at 3500 RPM for 1.5 min and left for 24 h before being applied to glass plates at 100 µM. Dry time conditions: 23 °C at 50% relative humidity. In all cases 0.5% (w/w) of a 1% (w/w) Fe-bispidon catalyst dissolved in propylene glycol was added, together with an anti-skinning agent, LP1810, MEKO, or AS0444 was added in the amounts given in the figure. LP1810 is a mixture of 12% Cerium hex Cem and Borchi® Nox C3, MEKO is methylethylketoxime and AS0444 is Borchers® Ascinin 0444 from Borchers. Skinning is a qualitative scale where 0 = no change, 1 = very thin skin, 2 = thin skin, 3 = medium-thin, 4 = hard skin, 5 = very hard skin.

3. Materials and Methods

3.1. Methods

Coatings were typically applied at a thickness of 75 µM, using a cube applicator, onto 30 × 2.5 cm glass plates. The amount of catalyst added was based on the metal wt% on total resin formulation, since the resin content was not always known.

Yellowing index (YI) was measured according to ASTM method E313, using the equation 8, where X, Y, and Z are the CIE tristimulus values measured using a Minolta CM-3700D spectrophotometer.

$$YI\ E313 = \frac{(100 * (CxX - CzZ))}{Y} \qquad (8)$$

Drying times were measured using a BK-3 drying recorder, where the different stages of drying can be observed by the pattern created by a needle dragged across the coating surface, in Figure 19, starting from (1) wet (reflow of the scratch) to dry (no needle penetration).

Hardness was determined by using a pendulum hardness tester (by TQC Sheen) on a formulated resin applied at a thickness of 75 µM, using a cube applicator, onto glass plates.

Skin formation was monitored by monitoring the flow of a formulated resin (6 g) within a closed vial (20 mL), every working day for 2 weeks and after that on weekly basis till 7 weeks. Once a week, the jar is opened to allow air (oxygen) in/out of the jar. If a skin is formed, the resin did not flow anymore. Skinning is a qualitative scale where:

0 = no change;
1 = very thin skin, can barely be seen when vial is tipped, breaks with gentle force;
2 = thin skin, can be seen when the vial is tipped, breaks with gentle force;
3 = medium-thin, can be seen when tipped, breaks with force;
4 = hard skin, can be seen without tipping vial, breaks with force;
5 = very hard thick skin, can be seen without tipping vial, cannot be broken easily with force.

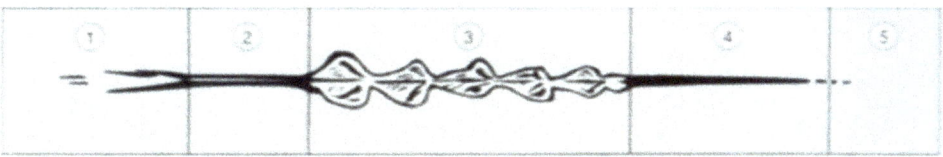

Figure 19. The different stages of dry time as determined visually versus time. Stage 5 is considered dry.

Alkyd Paint and Catalyst Preparation

Table 1 gives an example of how catalysts were prepared as a 1:1 ratio, they were then added into 5 g of the alkyd resin or formulated alkyd paint, mixed, and left for 24 h before application.

Generally, samples were prepared either as 5 or 10 g of resin (or paint) in 20 mL glass vials. As a general rule the siccative was added as 0.2 mL solution per 3 g of resin/formulation. The solvent used to dissolve the ligand/metal salt or complex depended on the solubility in the ligand/metal salt or complex in that solvent and solubility in the alkyd resin formulation. Used solvents were typically propylene glycol, Dowanol PM (1-methoxy-2-propanol), acetonitrile, ethanol, or methanol. For example, if a level of 0.05 wt% Mn (as $Mn(EH)_2$) is included in the alkyd mixture, 25 mg of $Mn(EH)_2$ dissolved in 50 µL ethanol. Separately a solution of the ligand that needs to be mixed with an equimolar amount of Mn in the paint formulation is prepared. Depending on the molecular weight of the ligand, the right amount of ligand is weighed into 50 µL ethanol. Table 1 gives an example of how catalysts were prepared as a 1:1 ratio. Each of the solutions were added to 5 g of the alkyd resin or formulated alkyd paint, mixed via stirring, and then left for 24 h before application.

In case that preformed Mn complexes were tested, these were dissolved generally in ethanol or water to yield 30 mM, which was then added to the paint or alkyd resin formulation to obtain the desired level of Mn (typically between 0.005 and 0.02 wt% Mn). If the influence of ascorbic acid was tested, an equimolar amount was premixed with the above-mentioned Mn complex dissolved in water or ethanol and then added to the paint or alkyd resin formulation.

Table 1. Mn = the preparation of example ligands with $Mn^{II}(2-EH)_2$ (as the Mn) prior to formulation. The ligand numbers refer to those discussed in Figure 14. The dose of these catalysts in a formulation would then be calculated and set as a %Mn content versus formulation mass for comparison purposes.

Sample	Concentration (wt% Mn)	Ligand Amount (mg/50 μL EtOH)	Mn-soap Amount (mg/50 μL EtOH)
Mn + L3 (MW = 248 g/mol)	0.05	11.3	25
Mn + L1 (MW = 157 g/mol)	0.05	7.2	25
Mn + L4 (MW = 262 g/mol)	0.05	11.9	25
Mn + L6 (MW = 311 g/mol)	0.05	14.2	25
Mn + L7 (MW = 339 g/mol)	0.05	15.5	25
Mn + Me$_3$TACN (MW = 171 g/mol))	0.05	8.2	25
Mn only	0.05	-	25

The solutions of Borchi® OxyCoat 1101 (BOC1101) in paint formulations were prepared by adding anything from 0.3 to 1% of BOC1101 by mass on total formulation. As an example, used for data in Figure 5, 0.3% of BOC1101 in 100 g of an alkyd paint would mean adding 0.3 g of BOC1101 (which is 1% w/w) to the formulation, meaning 0.003% (w/w) of BOC actives on total formulation. Due to differences in reactivity, less BOC is required for optimal performance (of anti-skinning and dry time) versus cobalt siccatives, therefore comparisons are based on the optimal does levels, using 1% (w/w) of Octa Soligen 421 cobalt siccative on total formulation. Unlike the Mn materials, the BOC was not calculated as metal on solids due to historical calculation methods for ease of use with unknown customer formulations. For further information, 1% actives BOC contains 8×10^{-4} wt% of iron.

3.2. Materials

Ligands and Mn complexes were supplied by Catexel (Leiden, The Netherlands), the Fe-bispidon catalyst is supplied by Borchers as Borchi® OxyCoat (Langenfeld, Germany), $Mn^{II}(2-EH)_2$ (containing 6 wt% Mn) from Alfa Aesar (Ward Hill, MA, USA), $Mn^{II}(acetate)_2$ from Sigma Aldrich (Saint Louis, MO, USA), $Fe^{II}Cl_2$ from Merck (Kenilworth, NJ, USA). Alkyd resin (70% in white spirits) was supplied by Acros Organics (Geel, Belgium). Commercial formulated resins are confidential, and the information cannot be shared. Dowanol PM is a commercial product from Dow (Midland, MI, USA). All other chemicals, including Dowanol PM, were purchased from VWR International (Amsterdam, The Netherlands).

4. Concluding Remarks

This article covers the history of our development work until present day for iron and manganese catalysts with polydentate nitrogen donor ligands for alkyd coating formulations. The toxic nature of cobalt has presented a difficult future for its use in alkyd resins where they will come into human contact. Examples of new catalyst systems and innovative delivery processes that are meeting the demand and that will continue to deliver the future catalysts for oxidatively cured coatings and inks have been described. Testing of the new ligands with Mn and Fe salts has revealed that even slight modifications of the structure of the ligand can have a profound effect on curing activity, dark yellowing, and hardness development. This may point to potentially different curing mechanisms, or at least different rates of activation of various processes. One aspect that has not been discussed in this paper is the fact that many paint formulations have been optimized for cobalt soaps for many years. It will be likely that the new catalysts will become more efficient when the paint formulations are further optimized, e.g., type of alkyd resin, secondary driers, fillers, or solvents.

We learned that the presence of polydentate nitrogen ligands is important to stabilize active metal–ligand catalyst species, but it is still elusive how each part of the curing/yellowing/hardening mechanisms are affected by each of the catalytically active species. Undoubtedly, new insights by studies both in academia on the fundamentals and industry on applications will enhance this understanding, which in turn will lead to catalysts that meet all requirements in various types of paint formulations.

Author Contributions: Writing—original draft preparation, review and editing, data curation, co-supervision, project administration, and funding, N.S.; investigation, data curation, specific lab applications, and development work on anti-skinning agents and method development, K.M.; method development, dry time evaluations, formal analysis of data, and catalyst preparations, Y.R.; contribution to literature review, co-supervision, review and editing, funding, and catalyst technology development leader, R.H.

Funding: This project was a collaboration between Borchers and Catexel, both sides funded their own research. Some evaluation research of Borchi® Oxycoat catalysts and of anti-skinning agents were performed by Sonja Esser of Borchers under the direction of Neil Simpson.

Acknowledgments: Max Shumba of Borchers, for updated information on the regulatory landscape of cobalt-replacement. We are grateful to Sonja Esser for the studies on the BorchiOxy Coat catalyst and the anti-skinning agents. The authors would also like to acknowledge Hans de Boer for the synthesis of various ligands and complexes described in Section 2.2 and the EPR studies on manganese complexes described in Section 2.2.1.

Conflicts of Interest: The authors declare no conflict of interest.

References

1. Brock, T.; Groteklaes, M.; Mischke, P. *European Coatings Handbook*; Vincentz Network GmbH & Co KG: Hannover, Germany, 2000.
2. Paul, S. Chapter 8—Types of Coatings. In *Surface Coatings Science and Technology*; John Wiley & Sons Ltd.: Chichester, UK, 1996; pp. 654–667.
3. Paints and Coatings Market 2019 Size, Share, Trends, Type, Production Growth, Industry Demand, Global Manufacturers, Equipment, Revenue Analysis and Forecast 2025. Available online: https://www.reuters.com/brandfeatures/venture-capital/article?id=92459 (accessed on 21 August 2019).
4. Ganesh, V.A.; Raut, H.K.; Naira, A.S.; Ramakrishna, S. A review on self-cleaning coatings. *J. Mater. Chem.* **2011**, *21*, 16304–16322. [CrossRef]
5. McInnis, B.M.; Hurt, J.D.; McDaniel, S.; Kemp, L.K.; Hodges, T.W.; Nobles, D.R. Carbon Capture Coatings: Proof of Concept Results and Call to Action. *Coat. World* **2019**, *24*, 82–91.
6. Li, G.; Meng, H. Recent Advances in Smart Self-Healing Polymers and Composites. In *Chapter 8—Self-Healing Coatings*; Hughes, A.E., Ed.; Woodhead Publishing: Cambridge, UK, 2015; pp. 211–241.
7. Challenger, C. An Update on Sustainability in the Coatings Industry. *Coat. Technol.* **2018**, *15*, 22–28.
8. Mash, T. Sustainability in the Coatings Industry. PCI Magazine. Available online: https://www.pcimag.com/articles/100363-sustainability-in-the-coatings-industry (accessed on 1 April 2015).
9. Van Haveren, J.; Oostveen, E.A.; Miccichè, F.; Noordover, B.A.J.; Koning, C.E.; van Benthem, R.A.T.M.; Frissen, A.E.; Weijnen, J.G.J. Resins and additives for powder coatings and alkyd paints, based on renewable sources. *J. Coat. Technol. Res.* **2007**, *4*, 177–186. [CrossRef]
10. Honzíček, J. Curing of Air-Drying Paints: A critical Review. *Ind. Eng. Chem. Res.* **2019**, *58*, 12485–12505. [CrossRef]
11. Tuck, N. Waterborne and Solvent Based Alkyds and their End User Application. In *Surface Coating Technology—Volume VI*; SITA Technology; Wiley: London, UK, 2000.
12. Hubert, J.C.; Venderbosch, R.A.M.; Muizebelt, W.J.; Klaassen, R.P.; Zabel, K.H. Singlet Oxygen Drying of Alkyd Resins and Model Compounds. *J. Coat. Technol.* **1997**, *69*, 59–64. [CrossRef]
13. Bieleman, J.H. Driers. *Chimia* **2002**, *56*, 184–190. [CrossRef]
14. Miccichè, F.; van Haveren, J.; Oostveen, E.; Ming, W.; van der Linde, R. Oxidation and oligomerization of ethyl linoleate under influence of the combination of ascorbic acid 6 palmitate/iron-2-ethylhexanoate. *Appl. Catal. A* **2006**, *297*, 174–181. [CrossRef]
15. Cobalt REACH Consortium. Cobalt REACH Consortium (CoRC). Available online: http://cobaltreachconsortium.org/ (accessed on 7 August 2019).
16. Reach Metal Consortium. Welcome to the REACH Metal Carboxylates (Driers & Catalysts) Consortium Ltd. (RMC). Retrieved from Reach Metal (Driers and Catalysts) Consortium. Available online: https://www.metal-carboxylates.org/ (accessed on 7 August 2019).

17. Powder Metallurgy Review. Proposed Cobalt Reclassification Raises Concerns for Metal Powder Users. Powder Metallurgy Review. Available online: https://www.pm-review.com/proposed-cobalt-reclassification-raises-concerns-metal-powder-users/ (accessed on 18 October 2017).
18. Sanderson, H. Cobalt hits 2-year low as DRC ramps up supply. *Financial Times*, 6 February 2019.
19. Steinert, A. Effective drying without cobalt. *Eur. Coat. J.* **2005**, *3*, 84–88.
20. Ha, D.; Joo, H.; Ahn, G.; Kim, M.J.; Bing, S.J.; An, S.; Kim, H.; Kang, K.G.; Lim, Y.K.; Jee, Y. Jeju ground water containing vanadium induced immune activation on splenocytes of low dose γ-rays-irradiated mice. *Food Chem. Toxicol.* **2012**, *50*, 2097–2105. [CrossRef]
21. Srivastava, A.K. Anti-diabetic and toxic effects of vanadium compounds. *Mol. Cell. Biochem.* **2000**, *206*, 177–182. [CrossRef] [PubMed]
22. Soucek, M.D.; Khattab, T.; Wu, J. Review of autoxidation and driers. *Prog. Org. Coat.* **2012**, *73*, 435–454. [CrossRef]
23. Bieleman, J.H. Progress in the Development of Cobalt-free Drier Systems. *Macromol. Symp.* **2002**, *187*, 811–821. [CrossRef]
24. Van Gorkum, R.; Bouwman, E. The oxidative drying of alkyd paint catalysed by metal complexes. *Coord. Chem. Rev.* **2005**, *249*, 1709–1728. [CrossRef]
25. Warzeska, S.T.; Zonneveld, M.; van Gorkum, R.; Muizebelt, W.J.; Bouwman, E.; Reedijk, J. The influence of bipyridine on the drying of alkyd paints. A model study. *Prog. Org. Coat.* **2002**, *44*, 243–248. [CrossRef]
26. van Gorkum, R.; Bouwman, E. Fast Autoxidation of ethyl linoleate catalyzed by [Mn(acac)$_3$] and bipyridine: A possible drying catalyst for alkyd paints. *Inorg. Chem.* **2004**, *43*, 2456–2458. [CrossRef]
27. Oyman, Z.O.; Ming, W.; van der Linde, R.; Gorkum, R.; Bouwman, E. Effect of [Mn(acac)$_3$] and its combination with 2,2'-bipyridine on the autoxidation and oligomerisation of ethyl linoleate. *Polymer* **2005**, *46*, 1731–1738. [CrossRef]
28. Van Gorkum, R.; Bouwman, E.; Reedijk, J. Drier for Alkyd-Based Coating. Patent EP 1382648 A, 21 January 2004.
29. Miccichè, F.; Long, G.J.; Shahin, A.M.; Grandjean, F.; Ming, W.; van Haveren, J.; van der Linde, R. The combination of ascorbic acid 6-palmitate and [Fe(III)(μ_3-O)]$^{7+}$ as a catalyst for the oxidation of unsaturated lipids. *Inorg. Chim. Acta* **2007**, *360*, 535–545. [CrossRef]
30. Hage, R.; Wesenhagen, P.V. Unilever, Liquid Hardening. Patent WO2008003652, 10 January 2008.
31. Santhanam, R. Additives for Curable Liquid Compositions. Patent WO2012092034, 5 July 2012.
32. Oyman, Z.O.; Ming, W.; Micciche, F.; Oostveen, E.; van Haveren, J.; van der Linde, R. A promising environmentally friendly manganese-based catalyst for alkyd emulsion coatings. *Polymer* **2004**, *45*, 7431–7436. [CrossRef]
33. Oyman, Z.O.; Ming, W.; van der Linde, R.; ter Borg, J.; Schut, A.; Bieleman, J.H. Oxidative drying of alkyd paints catalysed by a dinuclear manganese complex (MnMeTACN). *Surf. Coat. Int. Part B* **2005**, *88*, 269–275. [CrossRef]
34. Meijer, M.D.; van Weelde, E.; van Dijk, J.T.M.; Flapper, J. Drier for Auto-Oxidisable Coating Compositions. Patent EP2794786B, 27 July 2013.
35. Meijer, M.D.; van Weelde, E.; van Dijk, J.T.M.; Flapper, J. Drier for Auto-Oxidisable Coating Compositions. Patent EP2794787B, 27 July 2013.
36. Hage, R.; Iburg, J.E.; Kerschner, J.; Koek, J.H.; Lempers, E.L.M.; Martens, R.J.; Racherla, U.S.; Russell, S.W.; Swarthoff, T.; van Vliet, M.R.P.; et al. Efficient manganese catalysts for low-temperature bleaching. *Nature* **1994**, *369*, 637–639. [CrossRef]
37. Hage, R.; Lienke, A. Applications of Transition-Metal Catalysts to Textile and Wood-Pulp Bleaching. *Angew. Chem. Int. Ed. Engl.* **2006**, *45*, 206–222. [CrossRef] [PubMed]
38. Meijer, M.D.; Flapper, J. Drier for Auto-Oxidisable Coating Compositions. Patent EP2935435B, 26 June 2014.
39. Maaijen, K.; Hage, R. Method of Preparing an Oxidatively Curable Coating Formulation. Patent WO2017/134463, 10 August 2017.
40. de Boer, J.W.; Hage, R.; Maaijen, K. Drier for Alkyd-Based Coating. Patent WO2014/122433, 31 January 2014.
41. de Boer, J.W.; Hage, R.; Maaijen, K. Oxidative Curable Coating Composition. Patent WO2014/122432, 31 January 2014.
42. Hage, R.; de Boer, J.W.; Maaijen, K. Oxidative Curable Coating Composition. Patent WO2014/122434, 31 January 2014.

43. Peralta, R.A.; Neves, A.; Bortoluzzi, A.J.; Casellato, A.; dos Anjos, A.; Greatti, A.; Xavier, F.R.; Szpoganicz, B. First transition-metal complexes containing the ligands 6-amino-6-methylperhydro-1,4-diazepine (AAZ) and a new functionalized derivative: Can AAZ act as a mimic for 1,4,7-triazacyclononane? *Inorg. Chem.* **2005**, *44*, 7690–7692. [CrossRef]
44. Maaijen, K.; Hage, R. Oxidatively Curable Coating Composition. Patent WO2017103620, 22 June 2017.
45. Kalenda, P.; Holeček, J.; Veselý, D.; Erben, M. Influence of methyl groups on ferrocene on rate of drying of oxidizable paints by using model compounds. *Prog. Org. Coat.* **2006**, *56*, 111–113. [CrossRef]
46. Stava, V.; Erben, M.; Veselý, D.; Kalenda, P. Properties of metallocene complexes during the oxidative crosslinking of air drying coatings. *J. Phys. Chem. Solids* **2007**, *68*, 799–802. [CrossRef]
47. Erben, M.; Veselý, D.; Vinklárek, J.; Honzíček, J. Acyl-substituted ferrocenes as driers for solvent-borne alkyd paints. *J. Mol. Catal. A* **2012**, *353–354*, 13–21. [CrossRef]
48. Dong, S.; Wang, B.; Lui, B. Amperometric glucose sensor with ferrocene as an electron transfer mediator. *Biosens. Bioelectron.* **1992**, *7*, 215–222. [CrossRef]
49. Hage, R.; de Boer, J.W.; Gaulard, F.; Maaijen, K. Manganese and iron bleaching and oxidation catalysts. *Adv. Inorg. Chem.* **2013**, *65*, 85–116.
50. Appel, R.; Hage, R.; van der Hoeven, P.C.; Lienke, J.; Smith, R.G. Enhancement of Air Bleaching Catalysts. Patent EP1368450, 27 June 2002.
51. Appel, R.; Hage, R.; Lienke, J. Device Suitable for Analysing Edible Oil Quality. Patent EP1326074, 9 July 2003.
52. de Boer, J.W.; Wesenhagen, P.V.; Wenker, E.C.M.; Maaijen, K.; Gol, F.; Gibbs, H.; Hage, R. The quest for cobalt-free alkyd paint driers. *Eur. J. Inorg. Chem.* **2013**, *21*, 3581–3591. [CrossRef]
53. Pirš, B.; Bogdan, Z.; Skale, S.; Zabret, J.; Godnjavec, J.; Venturini, P. Iron as an alternative drier for high-solid alkyd coatings. *J. Coat. Technol. Res.* **2015**, *12*, 965–974. [CrossRef]
54. Gezici-Koc, O.; Thomas, C.A.M.M.; Michel, M.E.B.; Erich, S.J.F.; Huinink, H.P.; Flapper, J.; Duivenvoorde, F.L.; van der Ven, L.G.J.; Adan, O.C.G. In-depth study of drying solvent-borne alkyd coatings in the presence of Mn- and Fe-based catalysts as Co alternatives. *Mater. Today Commun.* **2016**, *7*, 22–31. [CrossRef]
55. Pirš, B.; Bogdan, Z.; Skale, S.; Zabret, J.; Godnjavec, J.; Berce, P.; Venturini, P. The influence of Co/Sr and Fe/Sr driers on film formation of high-solid alkyd coatings. *Acta Chim. Slov.* **2015**, *62*, 52–59. [CrossRef] [PubMed]
56. Jansen, J.; Kleuskens, E.C.; van Summeren, R.; Alsters, P.L. Manganese Complex Drier for Coating Compositions. Patent EP2534215B, 18 August 2011.
57. Jansen, J.; Bergman, F.A.C.; Kleuskens, E.C. Manganese Salt Complex as Drier for Coating Compositions. Patent EP2534216B, 18 August 2011.
58. Jansen, J.; Bergman, F.A.C.; Kleuskens, E.C.; Hage, R. Manganese Salt Complex as Drier for Coating Compositions. Patent EP2534217B, 18 August 2011.
59. Wieghardt, K.; Bossek, U.; Nuber, B.; Weiss, J.; Bonvoisin, J.; Corbella, M.; Vitols, S.E.; Girerd, J.J. Synthesis, Crystal Structures, Reactivity, and Magnetochemistry of a series of binuclear complexes of Manganese(II), -(III), and -(IV) of Biological Relevance. The crystal structure of [L'Mn^{IV}(μ-O)$_3$Mn^{IV}L'] (PF$_6$)$_2$. H$_2$O Containing an Unprecedented Short Mn ... Mn distance of 2.296 Å. *J. Am. Chem. Soc.* **1988**, *110*, 7398–7411.
60. De Boer, J.W.; Browne, W.R.; Brinksma, J.; Alsters, P.L.; Hage, R.; Feringa, B.L. Mechanism of cis-dihydroxylation and epoxidation of alkenes by highly efficient manganese catalysts. *Inorg. Chem.* **2007**, *46*, 6353–6372. [CrossRef] [PubMed]
61. Koek, J.H.; Kohlen, E.W.M.J.; Russell, S.W.; van der Wolf, L.; ter Steeg, P.F.; Hellemons, J.C. Synthesis and properties of hydrophobic [Mn^{IV}(μ-O)$_3$(L)$_2$]$^{2+}$ complexes, derived from alkyl substituted 1,4,7-triazacyclononane ligands. *Inorg. Chim. Acta* **1999**, *295*, 189–199. [CrossRef]
62. Koek, J.H.; Russell, S.W.; van der Wolf, L.; Hage, R.; Warnaar, J.B.; Spek, A.L.; Kerschner, J.; DelPizzo, L. Improved synthesis, structures, spectral and electrochemical properties of and properties of [Mn^{III}$_2$(μ-O) (μ-O$_2$CMe) L)$_2$]$^{2+}$ and [Mn^{IV}$_2$(μ-O)$_3$(L)$_2$]$^{2+}$ complexes. Two homologous series derived from eight N-substituted 1,4,7-triazacyclononanes. *J. Chem. Soc. Dalton Trans.* **1996**, *3*, 353–362. [CrossRef]
63. Oyman, Z.O.; Ming, W.; van der Linde, R. Catalytic activity of a dinuclear manganese complex (MnMeTACN) on the oxidation of ethyllinoleate. *Appl. Catal. A* **2007**, *316*, 191–196. [CrossRef]
64. Golombek, A.P.; Hendrich, M.P. Quantitative analysis of dinuclear manganese(II) EPR spectra. *J. Magn. Res.* **2003**, *165*, 33–48. [CrossRef]
65. Kumarathasan, R.; Rajkumar, A.B.; Hunter, N.R.; Gesser, H.D. Autoxidation and yellowing of methyl linolenate. *Prog. Lipid Res.* **1992**, *31*, 109–126. [CrossRef]

66. NIIR Board. *Modern Technology of Paints*, 2nd ed.; ASIA PACIFIC BUSINESS PRESS Inc.: Delhi, India, 2007; p. 49.
67. Figge, H.J.; Weber, B. Anti-Skinning Agent for Coating Composition. Patent WO00/11090, 2 March 2000.
68. Martyak, N.; Alford, D. Antiskinning Compound and Compositions Containing them. Patent WO 2007/024592, 1 March 2007.
69. Mason, A. Composition Containing an Oxime-Free Anti-Skinning Agent and Methods for Making and Using the Same. Patent WO 2016/100527A1, 23 June 2016.
70. Asirvatham, E.; Amerson, E.J.; Militch, E.D. Coating Composition Including Alkyl Oximes. U.S. Patent U.S. 2016/0304732A1, 20 October 2016.
71. Tanase, S.; Hierso, J.C.; Bouwman, E.; Reedijk, J.; ter Borg, J.; Bieleman, J.H.; Schut, A. New insights on the anti-skinning effect of methyl ethyl ketoxime in alkyd paints. *New J. Chem.* **2003**, *27*, 854–859. [CrossRef]
72. Hage, R.; Gol, F.; Gibbs, H.W.; Maaijen, K. Antiskinning Compositions. Patent WO 2012093250, 12 July 2012.
73. Krizan, M.; Vinklárek, J.; Erben, M.; Cisarova, I.; Honzíček, J. Autoxidation of alkyd resins catalysed by iron(II) bispidine complex: Drying performance and in-depth infrared study. *Prog. Org. Coat.* **2017**, *111*, 361–370. [CrossRef]
74. Weijnen, J.; Bloem, M.; Klomp, D. Drier Composition and Use Thereof. U.S. Patent U.S. 20130274386A1, 17 October 2013.

© 2019 by the authors. Licensee MDPI, Basel, Switzerland. This article is an open access article distributed under the terms and conditions of the Creative Commons Attribution (CC BY) license (http://creativecommons.org/licenses/by/4.0/).

Article

Synthesis and Catalytic Application of Knölker-Type Iron Complexes with a Novel Asymmetric Cyclopentadienone Ligand Design

Christian A. M. R. van Slagmaat [1], Khi Chhay Chou [1], Lukas Morick [1], Darya Hadavi [2], Burgert Blom [1] and Stefaan M. A. De Wildeman [1,*]

[1] Faculty of Science and Engineering, Maastricht University, 6167 RD Geleen, The Netherlands; c.vanslagmaat@maastrichtuniversity.nl (C.A.M.R.v.S.); k.chou@student.maastrichtuniversity.nl (K.C.C.); lmorick@gmail.com (L.M.); burgert.blom@maastrichtuniversity.nl (B.B.)
[2] Maastricht MultiModal Molecular Imaging Institute (M4I), Maastricht University, 6229 ER Maastricht, The Netherlands; d.hadavi@maastrichtuniversity.nl
* Correspondence: s.dewildeman@maastrichtuniversity.nl

Received: 31 August 2019; Accepted: 17 September 2019; Published: 22 September 2019

Abstract: Asymmetric catalysis is an essential tool in modern chemistry, but increasing environmental concerns demand the development of new catalysts based on cheap, abundant, and less toxic iron. As a result, Knölker-type catalysts have emerged as a promising class of iron catalysts for various chemical transformations, notably the hydrogenation of carbonyls and imines, while asymmetric versions are still under exploration to achieve optimal enantio-selectivities. In this work, we report a novel asymmetric design of a Knölker-type catalyst, in which the C_2-rotational symmetric cyclopentadienone ligand possesses chiral substituents on the 2- and 5-positions near the active site. Four examples of the highly modular catalyst design were synthesized via standard organic procedures, and their structures were confirmed with NMR, IR, MS, and polarimetry analysis. Density functional theory (DFT) calculations were conducted to elucidate the spatial conformation of the catalysts, and therewith to rationalize the influence of structural alterations. Transfer- and H_2-mediated hydrogenations were successfully established, leading to appreciable enantiomeric excesses (ee) values up to 70%. Amongst all reported Knölker-type catalysts, our catalyst design achieves one of the highest ee values for hydrogenation of acetophenone and related compounds.

Keywords: asymmetric hydrogenation; homogeneous catalysis; iron; structural design; conformational analysis; NMR spectroscopy; DFT

---

1. Introduction

The importance of asymmetric catalysis is emphasized by the Nobel Prize of 2001 awarded to Noyori and Knowles, for their contributions to asymmetric hydrogenations [1,2]. Their efforts harvested pivotal discoveries and developments, such as the enantioselective synthesis of the anti-Parkinson drug Levodopamine [3]. Catalysts for these transformations generally rely on precious metals such as ruthenium or rhodium, but the ever-increasing scale of global application in combination with environmental concerns demands the development of new catalysts based on first-row transition metals; preferably cheap, abundant, and less toxic iron [4–6].

Fortunately, many developments of iron-based catalysts have been conducted meanwhile by the groups of Gao [7], Morris [8], Chirik [9] and Milstein [10], to name a few, and their sophisticated ligand structures often show highly enhanced catalytic properties compared to simple iron salts or iron pentacarbonyl [11]. However, perhaps the most versatile class of iron catalysts is derived from η4-(cyclopentadienone) iron tricarbonyl complexes, which are often called "Knölker-type complexes".

Although iron compounds of such kind were synthesized for the first time in 1953 by Reppe and Vetter [12], and revisited by several groups subsequently [13], it was the pioneering work of Knölker et al. upon the delicate chemistry of η_5-(1,3-bis(trimethylsilyl)-4,5,6,7-tetrahydro-2H-inden-2-one)iron tricarbonyl (**1a**) [14–19], for which its trivial name "Knölker's complex" became generally accepted in literature. Ever since the discovery of Knölker's complex possessing catalytic activity for the hydrogenation of polarized double bonds by Casey and Guan in 2007 [20], this bifunctional iron catalyst has received tremendous attention in modern research [13].

The modes of interaction between a Knölker-type catalyst and a substrate usually rely on an 'outer-sphere' coordination mechanism, as shown in Scheme 1 for hydrogenation [21–23]. When starting from a Knölker-type complex (**1**), it first undergoes mono-decarbonylation by treatment with a base, Me$_3$NO, or UV-light [24], to afford an unsaturated 16e species with a vacant site (**2**). Via the heterolytic addition of (a) hydrogen (donor), followed by heterolytic cleavage, the metal center is oxidized from Fe(0) to Fe(II), while the η_4-cyclopentadienone ligand turns into an aromatic η_5-hydroxycyclopentadiene moiety. As such, the bifunctional active species (**3**) is formed, featuring protic binding site on the ligand, and a hydridic site on the iron. A polarized unsaturated substrate (e.g., a ketone) can then associate onto these hydrogen atoms, initiating hydrogen transfer to furnish reduction product (e.g., alcohol), while **2** is formed back.

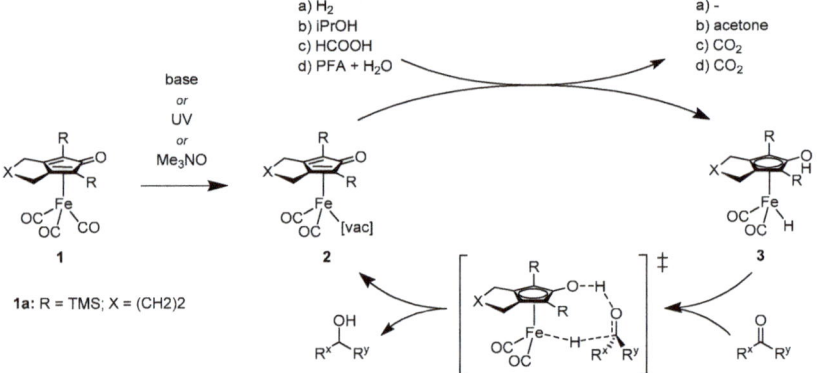

Scheme 1. Activation procedures and the catalytic cycle of a Knölker-type catalyst for the hydrogenation of ketones, featuring the "concerted outer-sphere mechanism" for hydrogen transfer. Suitable hydrogen donors can be dihydrogen, isopropyl alcohol, formic acid, and paraformaldehyde with water.

Following associative interactions conform this mechanism; the catalyst was found applicable in both transfer and pressure hydrogenation [20,25,26], Oppenauer oxidation [27], reductive amination [28], water–gas-shift-reactions [29], electro-reductions [30], alkylations [31], photocatalysis [32], and enantioselective dual catalysis typically for allylic alcohols [33–38] with formidable results. The capabilities of Knölker-type catalysts in asymmetric catalysis for normal ketones and imines have also been explored [39], however, their performance regarding enantioselectivity are generally found mediocre, as well for the corresponding ruthenium-based complexes (see Figure 1). Hence, in order to make Knölker-type catalysts industrially more competitive for the production of asymmetric fine chemicals, we considered further development a necessity.

To our knowledge, the first asymmetric catalyst of such kind is reported by Yamamoto et al. [40], who designed ruthenium complexes derived from spirocyclic C-arylribosides. Shortly after, cyclopentadienone metal complexes with different substituents on the 2,5-positions, and with an enantiopure carbon center in the cyclic backbone of the ligand were developed by the group of Wills [25,41]. In these early examples the ligand design did not allow complete blockage of one face of the cyclopentadienone moiety, leading to diastereomeric mixtures of the complexes, and therefore to rather low enantio-selectivities in catalysis.

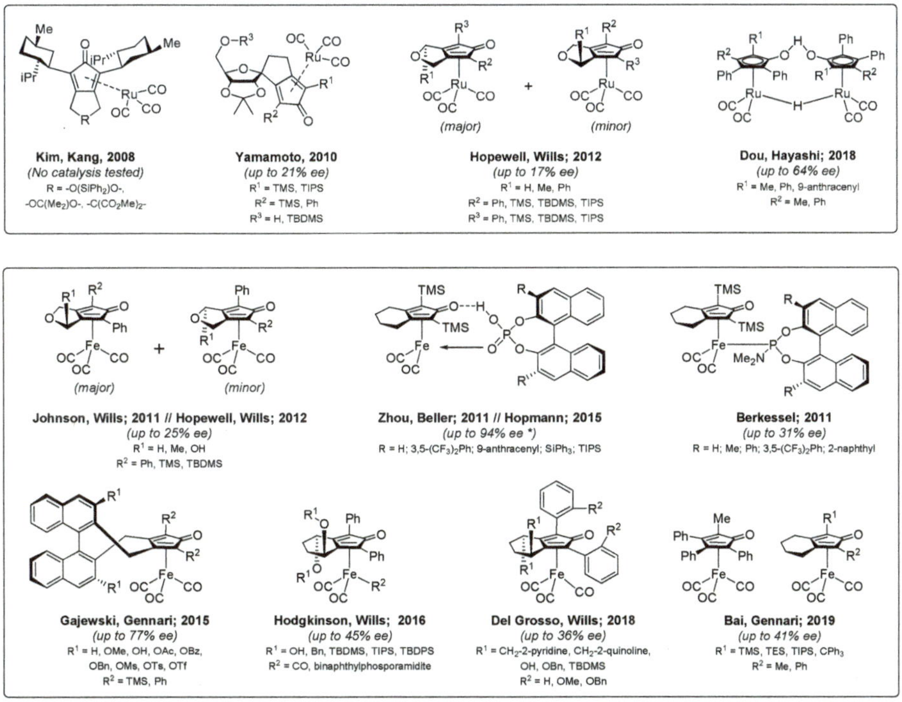

Figure 1. Overview of reported structures of asymmetric cyclopentadienone-ligated metal tricarbonyl complexes, featuring ruthenium (top panel) and iron (bottom panel), and their best enantioselectivity in asymmetric hydrogenation. * This result concerns the hydrogenation of imines.

In later years, this problem was circumvented with the use of ligands featuring a C_2-rotational symmetry axis along the catalytically functional carbonyl bond, because for such a design coordination of the metal to either face of the ligand will lead to the same complex. As such, higher catalytic enantioselectivities could be achieved [42,43], with the best example featuring BINOL-derived cyclopentadienone ligand provided by Gajewski et al. [44–46].

Alternatively, racemic mixtures of asymmetric ruthenium [47] and iron catalysts [48] with different 2,5-substituents were separated using preparative chiral HPLC chromatography, yielding batches of both pure enantiomers. These catalyst versions achieved appreciable enantioselectivities for certain substrates, however, they was not high enough to make them industrially relevant.

Meanwhile, different concepts had also emerged which derived their chirality from a different chemical source to induce enantio-selectivity using a symmetric Knölker-type complex. Binaphthyl phosphoric acids were employed as co-catalysts by the Beller group, and they achieved outstanding enantiomeric excesses (ee) values over 90%. However, their system is only applicable to imines [49–52], quinoxalines and benzoxazines [53]. Binaphthyl phosphoramides were coordinated to a symmetric Knölker catalyst [24], as well as asymmetric iron complexes [42]. Furthermore, enzymatic approaches have also been explored, for example iron-catalyzed hydrogenation in combination with dynamic kinetic resolution using *Candida antarctica* Lipase B [54,55], but also covalent bonding of Knölker-type catalysts into a *Streptadivin* enzyme [56]. However, the latter example rendered low to moderate enantioselectivities.

By in-depth study of all available literature on asymmetric (cyclopentadienone)metal catalysts, we deduced that desirable structural aspects are C_2-rotary symmetry of the ligand, and having the chirality-inducing moieties closer located to the active site of the catalyst, in order to envision improved enantioselectivity by Knölker-type catalysts. Therefore, we hypothesize that a cyclopentadienone ligand, bearing two identical enantiopure substituents on the 2- and 5-position with *R,R* or *S,S* orientations only, could be a suitable design for this goal. To our knowledge, only one example of such a design is reported for a ruthenium complex by Kim et al. [57], which possesses enantiopure menthyl substituents. Surprisingly, the complexes in this work were not tested in any catalytic reaction, and successive research upon ligand designs of this kind has been dormant for over a decade until now.

2. Results and Discussion

2.1. Design, Synthesis and Characterization of the Pre-Ligands and Catalysts

In our pursuit of designing a novel asymmetric Knölker-type catalyst, bearing a C_2- rotational symmetric cyclopentadienone ligand with chiral centers on the 2,5-positions (**4**), we considered two conventional synthesis strategies by applying retrosynthetic analysis, as shown in Scheme 2. Herein, Route A describes the formation of the desired iron complex from an iron carbonyl precursor and a 3,4-aryl appended cyclopentadienone (**5**) [58], which could be obtained from the double aldol condensation [59] of an *R,R* or *S,S* diastereopure β,β'-substituted acetone (**6**), and an aril (**7**). Alternatively, via Route B the iron complex with the desired structural features (**4**) can also be derived from the 2 + 2 + 1 cycloaddition of an iron carbonyl precursor with a tethered dialkyne [60], containing identically enantiopure groups on the termini (**8**). Such dialkyne pre-ligands can be obtained via coupling reactions of enantiopure terminal alkynes (**9**) with a dihalide (**10**) [61–64].

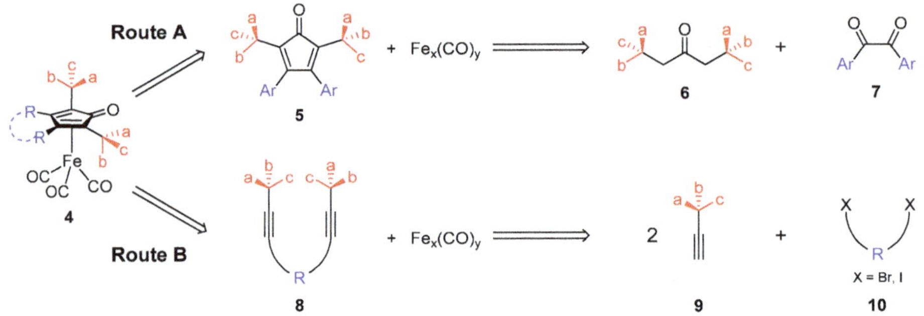

Scheme 2. Retro-syntheses of an iron cyclopentadienone complex, with optically active substituents on the 2,5-positions of the ligand.

Although syntheses of compounds in accordance with the structure of **6** have been reported [65,66], the acquisition of such diastereopure product in decent yields remains a significant challenge. On the contrary, enantiopure terminal alkynes are easier to synthesize, because they possess only one chiral center, and are readily available from commercial sources. For these reasons we selected Route B to target the first examples of our asymmetric catalyst design. Herein, we considered enantiopure 1-substituted prop-2-yn-1-ols as the ideal synthon, because the possible combinations of 1-substituents and applicable alcohol protection groups offer a broad variety of structures to tune the envisioned catalyst. In addition, the dihalide synthon allows structural modularity in the backbone of the cyclopentadienone ligand.

Considering the nature of the reaction steps in the total synthesis, a proper protection group for the alcohol of the starting material is required to ensure chemical orthogonality against the strong base required in the next step for alkynic proton abstraction, while a successful double C–C coupling with the dihalide is a requisite to furnish the pre-ligand. In addition, a microwave-assisted procedure for the complexation onto iron pentacarbonyl giving high yields was recently published [67]. These facts led to the establishment of the total synthesis described in Scheme 3.

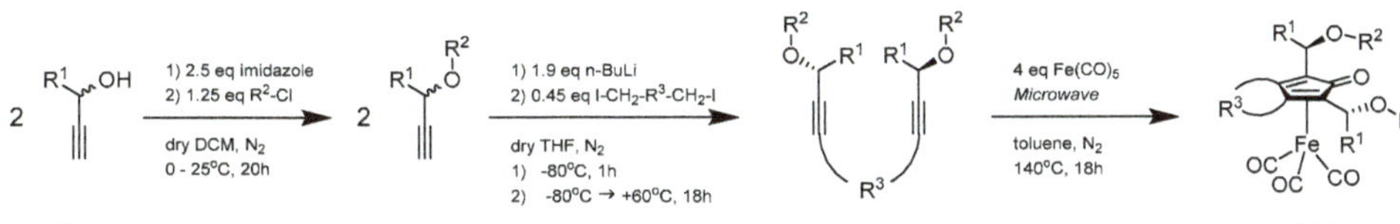

Scheme 3. Total synthesis of a novel asymmetric iron cyclopentadienone complex series from enantio-pure 1-substituted prop-2-ynols. R^1 = Me, Ph; R^2 = TIPS, TBDPS; R^3 = $(CH_2)_4$, $(CH_2)_3$.

As the first example, R-but-3-yn-2-ol (**11a**) was protected with a triisopropylsylil (TIPS) group using Corey's conditions [68], affording the desired silyl ether product (**12a**) in 93% yield, and complete preservation of the enantiopurity was confirmed by chiral GC-fid and polarimetry (Table 1).

However, the subsequent coupling reaction to produce the pre-ligand **13a** proved to be far more challenging. Although complete and selective abstraction of the alkynic proton be established in THF using nBuLi, we experienced through numerous attempts that only iodides are sufficient leaving groups for the alkyl dihalide reagent to achieve appreciable conversion. A strict temperature program featuring the addition of nBuLi at −80 °C to react for 1 h allows complete deprotonation, followed by the addition of 1,4-diiodobutane still at −80 °C, after which the resulting mixture is kept at 60 °C overnight to drive the reaction. Moreover, delicate excess amounts of reagents (i.e., **12a** > nBuLi >> 1,4-diiodobutane) were essential to acquire bis-coupled product, and to minimize the remainder of mono-coupled intermediate. Although having trace amounts of mono-coupled species in the crude product was practically inevitable, separation over a long silica column ultimately provided chemo-pure **13a** in a commendable 89% yield. Polarimetry indicated enantiomeric excess of the R,R-diastereomer being present in **13a**, but no absolute diastereo-purity could be ensured, since the pre-ligand is too heavy for GC-fid, and diastereomers of **13a** could not be distinguished in NMR.

Table 1. Obtained yields and enantiomeric properties of the starting materials, silylated building blocks, pre-ligands, and catalysts.

Compound	R^1	R^2	R^3	Isolated Yield (%)	Enantiomeric Configuration	Specific Rotation
11a	Me	-	-	n/a	R	n/d
11b	Ph	-	-	n/a	R	n/d
11c	Me	-	-	n/a	rac	0
12a	Me	TIPS	-	92.7	R	+96.1
12b	Me	TBDPS	-	91.3	R	+288.3
12c	Ph	TIPS	-	85.5	R	−14.8
12d	Me	TIPS	-	76.3	rac	0
13a	Me	TIPS	$(CH_2)_4$	89.1	R,R	+122.6
13b	Me	TBDPS	$(CH_2)_4$	38.7	R,R	+364.1
13c	Ph	TIPS	$(CH_2)_4$	60.8	R,R	−37.0
13d	Me	TIPS	$(CH_2)_3$	70.3	R,R	+131.9
13e	Me	TIPS	$(CH_2)_4$	66.8	rac	0
14a	Me	TIPS	$(CH_2)_4$	51.4	R,R	+16.5
14b	Me	TBDPS	$(CH_2)_4$	33.8	R,R	+62.5
14c	Ph	TIPS	$(CH_2)_4$	18.2	R,R	−107.5
14d	Me	TIPS	$(CH_2)_3$	51.9	R,R	+8.4
14e [1]	Me	TIPS	$(CH_2)_4$	48.2	rac	0

[1] **14e** serves as a collective abbreviation for all obtained diastereo-isomers in this reaction.

The complexation reaction of **13a** with Fe(CO)$_5$ in a closed system at 140 °C by microwave irradiation was carried on nevertheless, and the expected increase of pressure by CO release was indeed observed within few hours. From the retrieved dark suspension a bright yellow solid was isolated through column chromatography, and identified as a Knölker-type complex by the characteristic IR absorptions in the region of 1980–2070 cm^{-1} for the iron-bound carbonyl vibration modes, plus the vibration of 1630 cm^{-1} for the ligand's ketone. In addition, detected masses of 675.31648 Da and 697.29835 Da, corresponding to the protonated species and sodium ion, respectively, exhibited the isotopic patterns of a mono-iron species. Altogether, with an optical rotation in polarimetry observed, and with thorough ^{1}H- and ^{13}C-NMR analysis (vide infra), the successful synthesis of enantiopure complex **14a** with 51% yield was confirmed (see Appendix A).

The optimal reaction conditions and purification procedures found for all three synthetic steps towards **14a** were also applicable for certain structural variations of the building blocks and reagents (Table 1). Protection of **11a** with tert-butyldiphenyl silyl (TBDPS) group, followed by dicoupling onto 1,4-diiodobutane and complexation with iron pentacarbonyl afforded the more bulky complex **14b**. 1-phenylprop-2-yn-1-ol (**11b**) was demonstrated as well to be an eligible building block for silylation, and subsequent nBuLi-mediated coupling, which ultimately led to sterically- and electronically altered complex **14c**. Silyl ether **12a** was also coupled to 1,3-diiodopropane, to finally afford complex **14d**, which possesses a smaller five-membered ring as backbone of its ligand. Furthermore, the use of racemic but-3-yn-2-ol **11c** in the total synthesis led to the diastereomeric mixture **14e**. In total, the syntheses of complex series **14a–d** (Figure 2) give rise to a structural variation in each of R^1, R^2, and R^3, respectively, enabling the study of the influence of each separate R-group on the structural conformation of the ligand, as well as enantioselectivity in catalysis (vide infra).

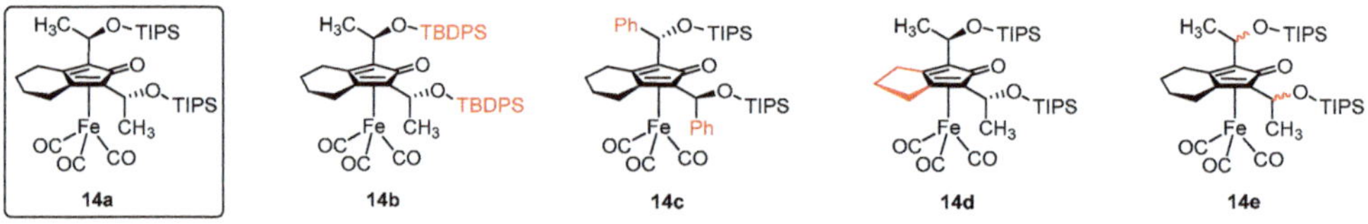

Figure 2. Structures of the asymmetric Knolker-type complexes synthesized in this work. **14a** serves as the base case complex, while the fragments of **14b–e** marked in red are structural alterations upon **14a**.

NMR analysis was conducted as the main technique to verify the success of each step in the total catalyst syntheses. In ^{1}H-NMR (Figure 3, top), upon silylation of **11a–c** the disappearance of the alcohol signal, and the rise of silyl-appended hydrocarbon signals around 1 ppm were observed for **12a–**textbfd. The introduction of the alkyl tether via the coupling reaction then provided two characteristic peaks between 1.5 and 2.5 ppm for **13a–**textbfc (different signals for **13d**). Finally, in ^{13}C-NMR the characteristic signals of Knölker-type complexes were observed for iron-bound carbonyls between 200–210 ppm, and the cyclopentadienone carbonyl in the proximity of 170 ppm (Figure 3, bottom).

However, the most significant effects were observed upon changing the molecular architecture of the linear dialkyne into a bicyclic ring structure via the 2 + 2 + 1 cyclo-addition with iron pentacarbonyl. As such, a spatial confinement of the bulky chiral substituents was created, while the C$_2$-rotary symmetry of the newly formed cyclopentadienone ligand is broken by coordination of iron to one plane. This inflicted an intriguing splitting phenomenon with remarkably high differences in chemical shift (Δδ) for most of the signals in both ^{1}H- and ^{13}C-NMR. For the molecular fragments being closer to the centers of asymmetry, notably R_1, the methine groups of the 2,5-substituents, and the cyclopentadienone carbons, Δδ generally appeared to be larger (see the section 'Materials and Methods', Appendix A, and the Supplementary Materials for in-depth details).

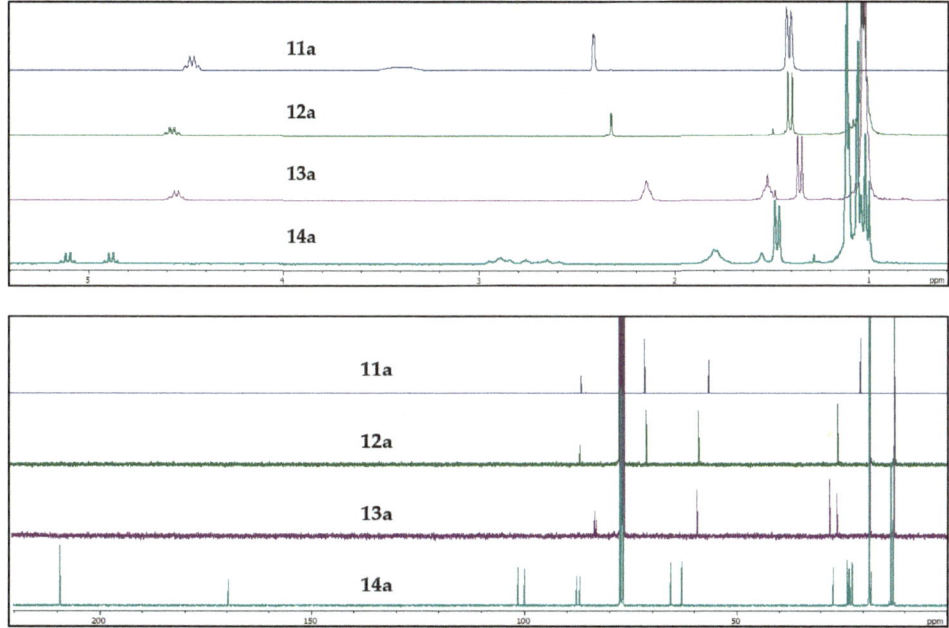

Figure 3. ^{1}H-NMR (top) and ^{13}C-NMR (bottom) spectra of compounds **11a**, **12a**, **13a**, and **14a** in CDCl$_3$.

Similar signal patterns in the NMR spectra for complexes **14b–d** were observed as well, but with different chemical shifts and Δδ splits, of which the most noteworthy ones are listed in Tables 2 and 3. Although these spectroscopic values are to an extent subjected to the electronic nature of the molecular surroundings of their corresponding atoms, we believe that they are also affected by the dihedral rotation of the chiral 2,5-substituents, as a result from structural variations in R^1, R^2, and R^3. Therefore, the comparison of complexes **14b–d** with **14a** could provide useful insights for each separate structural deviation.

In general, a higher Δδ for a signal in NMR is in accordance with a slower and less facile spatial movement of the corresponding molecular group [69]. Although a clear trend for such peak segregation is not easily recognized by taking all represented split signals into account, some observed NMR signals may still be linked to possible structural properties. The heavier and sterically more encumbered 2,5-substituents of **14c** are expected to exhibit a more restricted dihedral rotation, which is in agreement with the increased values for Δδ in ^{13}C-NMR.

On the other hand, **14d** possesses a fused five-membered ring as backbone of the ligand, which is narrower than the fused six-membered ring of **14a**, while R^1 and R^2 are identical for these complexes. Remarkably, all Δδ values (except for the methyl group in ^{1}H-NMR) are significantly smaller than the values observed for **14a**. This result suggests a considerably more facile rotation of the 2,5-substituents for **14d**, which may be enabled by the narrower backbone of the ligand, and highlights the unexpectedly significant effect of the R^3-group.

Regarding R^1, no statement on the effect of certain silyl groups can be declared yet, whereas the observed differences in Δδ values for **14b** are not understood at this moment, and the steric difference (e.g., cone angles) between TIPS and TBDPS are rather similar [70]. For a better understanding, a larger scope of alcohol protection groups for R^1 should be included in a future study.

Table 2. Observed chemical shifts and values of Δδ of characteristic signal splitting in ^{1}H-NMR spectroscopy for the synthesized iron complexes in CDCl$_3$.

Complex	^{1}H-NMR			
	Methine		Methyl	
	δ (ppm)	Δδ (Hz)	δ (ppm)	Δδ (Hz)
14a	4.99	66.3	1.47	6.4
14b	4.82	85.7	1.32	6.4
14c	5.94	52.7	-	-
14d	4.89	28.1	1.47	18.8

Table 3. Observed chemical shifts and values of Δδ of characteristic signal splitting in ^{13}C-NMR spectroscopy for the synthesized iron complexes in CDCl$_3$.

Complex	^{13}C-NMR							
	3,4-Cp		2,5-Cp		Methine		Methyl	
	δ (ppm)	Δδ (Hz)	δ (ppm)	Δδ (Hz)	δ (ppm)	Δδ (Hz)	δ (ppm)	Δδ (Hz)
14a	100.5	113.9	87.2	61.2	64.1	194.6	25.2	261.2
14b	100.6	74.5	86.0	110.4	64.3	98.0	25.1	237.1
14c	99.8	156.4	87.7	248.0	69.2	223.6	-	-
14d	105.5	60.6	88.4	28.0	64.5	52.6	27.6	5.2

The occurrence of such notable signal splitting in NMR spectra, as a result from the structural asymmetry in these complexes, prompted us to investigate the diastereo-isomeric mixture **14e**, which was obtained from racemic building block **11c** through the total iron complex synthesis (Scheme 3). Herein, **12d** and **13e** were analyzed and compared with their structural contenders from enantio-pure sources, to confirm that their spectroscopic properties and masses are identical, but no optical rotation in polarimetry can be detected (Table 1).

However, upon complexation of **13e** with Fe(CO)$_5$ a statistical mixture all possible diastereo-isomers is expected, comprising the asymmetric R,R-complex **14a** and its enantiomer S,S-complex **14a'**, as well as two meso-isomers: R,S-complex **14f** and S,R-complex **14g**, which are not enantiomers of each other (Scheme 4). Because enantiomers **14a** and **14a'** exhibit identical molecular environments and conformations, though mirrored, they cannot be distinguished from each other by spectroscopy. In contrast, the pairs of chiral centers of symmetric compounds **14f** and **14g** have opposite optical configurations pertaining to the iron core, which is expected to give different dihedral rotations of their 2,5-substituents. Therefore, the diastereomerically racemic mixture **14e** should display three sets of signals in NMR spectroscopy, in theory.

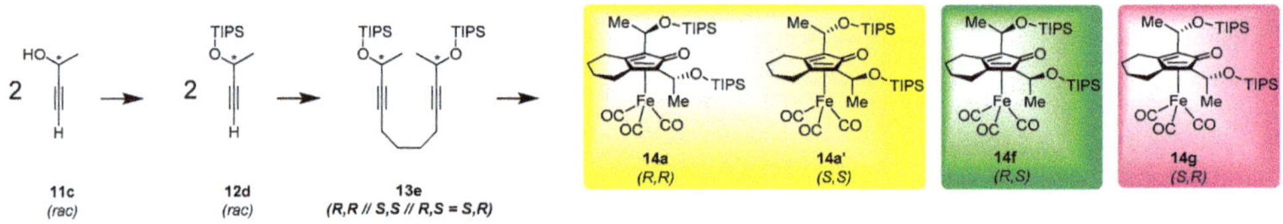

Scheme 4. Total synthesis of a diastereomeric mixture of (cyclopentadienone)iron complexes derived from racemic but-3-yn-2-ol. Each colored box indicates a (group of) diastereo-isomer(s) that can be distinguished by spectroscopic analysis. The collective mixture of **14a** + **14a'** + **14f** (+ **14g**) is denoted as **14e**.

The actual NMR spectra of **14e**, however, only revealed two sets of signals, of which one set belongs to both **14a** and **14a'**, which is in accordance with the absence of optical rotation in polarimetry. The other set comprises non-split signals that are characteristic for a symmetric Knölker-type complex (see Supplementary Materials). In ^{13}C-NMR, the new peaks at 209.35 ppm and 169.58 ppm were assigned to the iron-bound carbonyls and the cyclopentadienone carbonyl, respectively. Clear singular peaks at 146.87 ppm, 101.39 ppm, and 87.76 ppm were also observed for the other pair of cyclopentadienone carbons and the methine carbons, respectively (Figure 4, bottom). The ^{1}H-NMR spectrum shows through the doublet of quartets for the methine protons of **14a** and **14a'** in the region of 5.0–5.2 ppm another singular quartet at 5.08 ppm, in agreement with a symmetric species within **14e** (Figure 4, top). These observations evidence that only three out of four expected diastereo-isomers are formed.

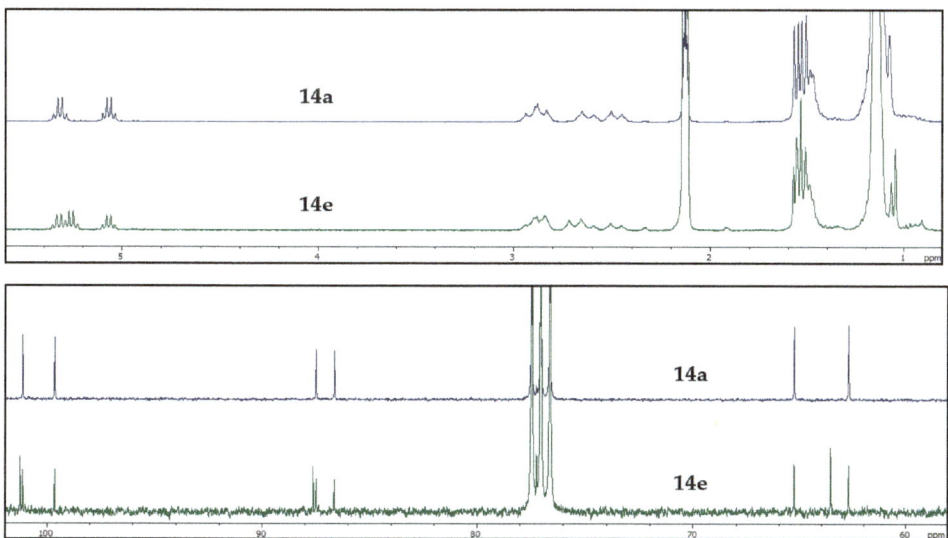

Figure 4. ^{1}H-NMR spectrum of **14a** and **14e** in d8-toluene (top), and a zoom of the ^{13}C-NMR spectrum of compounds **14a**, and **14e** in CDCl$_3$ (bottom).

The quartet signal for the symmetric diastereo-isomer rendered an integral of 1.3H compared to 2H for **14a** + **14a'**, which is significantly higher than the statistically expected integral of 1.0H. Since all synthesized complexes **14a–e** were found stable in air at room temperature over the course of several months, as well as during column chromatography over silica, this unusual diastereomeric product distribution is not likely a result from decomposition during these stages. However, partial decomposition during the synthetic reaction cannot be excluded, as incomplete mass balances of the iron complexes but no unreacted pre-ligand were obtained.

Alternatively, we propose the possibility that one of the symmetric diastereomers, supposedly **14g**, is sterically constrained to such an extent that ring closure of the alkynes cannot be achieved. Steric restrictions of such kind have been reported for the formation of a similar cyclopentadienone ruthenium complex as well [71]. In the case of **14g** the bulky silyl groups are oriented towards the iron core (deduced from its DFT structure, vide infra) inhibiting ring closure. However, rearrangement of the alkynes onto the iron by dissociation, displacement, and re-coordination of one the alkyne moieties may lead to the formation of **14f** in extra quantities (Scheme 5), as observed in NMR.

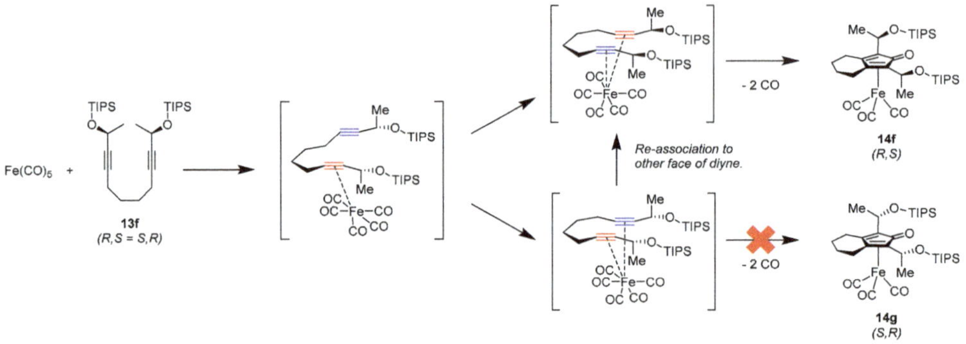

Scheme 5. Proposed complexation mechanism of mesomeric pre-ligand **13f** with Fe(CO)$_5$, yielding complex **14f** only. Herein, formation of complex **14g** is excluded by steric aspects, and rearrangement of the coordinated alkyne moieties of intermediate species leads to **14f** as well.

Another important implication is that potential racemization during any stage of the total synthesis would result in elimination of optical rotation, and the occurrence of more than one set of signals in NMR spectroscopy, as partial racemization of **14a** into **14f** cannot be excluded in this scenario. This is fortunately not the case for **14a–d**, which evidences preservation of chiral information through their total synthesis, thus rendering these complexes enantio-pure.

2.2. Computational Structure Assessment of the Novel Catalyst Design

In order to assess the spatial conformation of complexes **14a–e** a series of computational structure optimizations were conducted in lieu of a crystal structure determination, since the acquisition of suitable crystals of **14a–d** proved very difficult. These calculations were conducted on DFT level using they hybrid B3LYP function [72,73], and the LANL2DZ core potential [74,75] was applied on all atoms.

The optimized structure of **14a** exhibits the typical "piano-stool" complex, featuring η$_4$-coordination of the cyclopentadienone ligand onto iron, as expected (Figure 5, left). The enantio-pure 2,5-substituents are rotated in different dihedral angles, in which the bulky TIPS-groups are directed away from the core of the molecule, and the least sterically demanding methine hydrogens point towards the backbone of the ligand. As such, one TIPS group is oriented *endo* with respect to the iron, while the other is oriented *exo*, establishing the visibly asymmetric nature of the complex.

Assessment of the symmetric **14f** and **14g** diastereo-isomers reveals a conformational change regarding the TIPS groups, as they are oriented *exo–exo*, and *endo–endo*, respectively. These conformations bear different degrees of steric clashing, as the difference in Gibbs free energy (ΔG) between **14f** and **14g** is 5.29 kcal·mol^{-1}, and plausibly induces the formation of **14f** only from the mesomeric diastereo-isomer of **13e**.

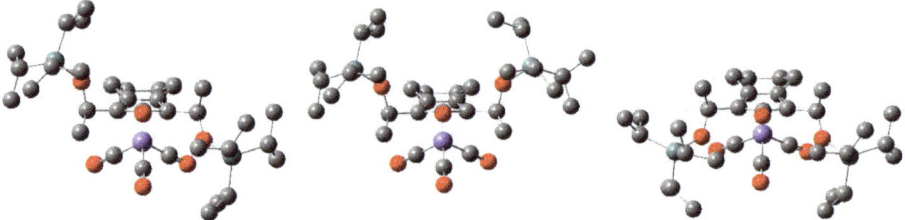

Figure 5. Optimized structures of complexes **14a** (left), **14f**, (middle), and **14g** (right)—which are each other's diastereo-isomers—derived from DFT calculations. Hydrogens are omitted for clarity.

The optimized structures for **14b–d** reveal a similar conformation compared to **14a**, regarding the orientation of the 2,5-substituents (Figure 6). Since the 2,5-substituents do possess some degree of rotational freedom, which is assumed different for each complex, we defined the corresponding dihedral angles θ and φ for comparative reasons, as indicated in Figure 7.

Figure 6. Optimized structures of complexes **14b** (left), **14c′**, (middle), and **14d** (right) derived from DFT calculations. Hydrogens are omitted for clarity. The *S,S*-enantiomer of **14c** was chosen to be calculated for comparative reasons.

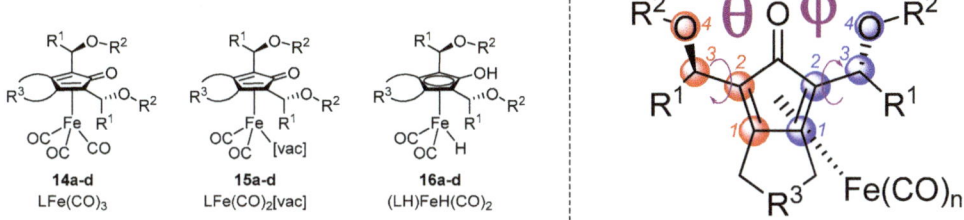

Figure 7. General structures for complexes **14**, decarbonylated complexes **15**, and active catalysts **16**, in which angle θ describes the dihedral angle along labeled atoms 1–4 for the left-hand chiral substituent (as seen from top view), while φ describes the dihedral angle for the right-hand chiral substituent.

In addition, the structures for the (cyclopentadienone)iron dicarbonyl (**15a–d**) and (hydroxycyclopentadiene)iron(II) dicarbonyl hydride (**16a–d**) were optimized as well, because they are more representative for the iron species within the catalytic cycle. Their values for θ and φ, and the absolute difference between θ and φ, are listed in Table 4. From the overview it can be seen that for complexes **14** and **15** the *endo*-oriented substituent is affected the most upon structural variation, as indicated by θ. Moreover, the rotations of θ and φ, and the increased angle of |θ − φ| for **14c′** and **15c′** deviate a lot from those of the other complexes. These observations indicate that **14c** is significantly more constrained in rotational freedom than the other complexes, which is complementary to the deductions from NMR-analysis and enantioselectivity in catalysis (vide infra), concerning **14c**.

Table 4. DFT-computed dihedral angles of the 2,5-substituents of the cyclopentadienone ligand.

Entry	LFe(CO)$_3$ (14)			LFe(CO)$_2$[vac] (15)			(LH)FeH(CO)$_2$ (16)		
	θ (°)	φ (°)	|θ − φ|	θ (°)	φ (°)	|θ − φ|	θ (°)	φ (°)	|θ − φ|
a	−133.2	−76.6	56.6	−135.6	−80.7	54.9	−168.7	−80.8	87.9
b	−142.5	−78.1	64.3	−127.6	−84.6	43.0	−169.2	−86.6	82.6
c′	−159.2	−90.8	68.4	−155.0	−92.8	62.3	−169.0	−96.5	72.5
d	−132.5	−77.5	54.9	−129.8	−80.0	49.8	−170.1	−82.4	87.7
f	75.4	−75.4	0.0	n/a			n/a		
g	−135.2	134.9	0.3						

Therefore, DFT may serve as a preliminary tool for asymmetric Knölker complexes of this kind, to predict the enantioselectivity of one complex relative to another. However **16a–d** all exhibit a very similar rotation for θ of about 170°. This is attributed to hydrogen bonding between the hydroxy group and the oxygen of the left-hand chiral substituent, and may qualify structures **16a–d** unsuitable to base predictions upon.

2.3. Performance of the Catalysts in Asymmetric Hydrogenation

To probe the catalytic abilities of the synthesized complexes, the reduction of the representative substrate acetophenone was targeted. Initial experimentation conducted using the base case complex **14a** in transfer hydrogenation with isopropyl alcohol (iPrOH) as hydrogen donor.

Since this reaction is subject to the Meerwein–Ponndorf–Verley (MPV) equilibrium, a substantial amount of excess iPrOH is required to drive the reaction forward [76]. Several amounts of iPrOH were tested at 80 °C (Table 5, Entries 1–4), which afforded modest conversions up to 24% for the most concentrated reaction mixture. To our delight, enantiomeric excesses (ee) of 40% in average favoring S-1-phenylethanol production were achieved, with slightly better enantioselectivities for more dilute systems. To find a suitable mitigation between conversion and enantioselectivity, 10 equivalents of iPrOH (10% substrate dilution) was selected for further experiments. Herein, addition of solvent without hydrogen-donative properties showed even further inhibition of conversion (Entries 5 and 6), while variation of the temperature indicated slightly better enantioselectivity at lower temperatures (Entries 7 and 8). Moreover, the advanced conversion upon higher temperature suggests that the MPV equilibrium was not reached within 24 h, thus the reaction proceeds rather slowly.

Table 5. Asymmetric transfer hydrogenation of acetophenone catalyzed by complex **14a**.

Entry	Solvent	equiv. iPrOH	Temperature (°C)	Conversion (%)	Ee (%)	Enantiomeric Configuration
1 [1]	none	2.5	80	24	36	S
2 [1]	none	10	80	15	41	S
3 [1]	none	17.5	80	9	42	S
4 [1]	none	25	80	8	42	S
5 [2]	toluene	10	80	7	39	S
6 [2]	tBuOH	10	80	3	42	S
7	none	10	100	25	40	S
8	none	10	60	11	45	S

[1] With different amounts of iPrOH under solvent-free conditions, these concentrations vary accordingly. [2] Delicate solvent volumes were added to afford the same concentration as in Entry 4.

In order to boost the reaction rates, and to eliminate the MPV equilibrium for allowing higher conversions, pressure hydrogenation using 50 bar H_2 was conducted subsequently. Although near-complete conversion was readily achieved, the ee was decreased to 33% (Table 6, Entry 1). Variation of the solvent did not alter the enantioselectivity significantly, which appears a common factor for Knölker-type catalysts with the origin of asymmetry in the cyclopentadienone ligand [39], while the enantioselectivity for other types of catalysts is sometimes more susceptible to solvent effects [77]. However, significantly lower conversions were observed in methanol and 2-methyltetrahydrofuran (Entries 2–5).

Comparative screening of **14a–e** at 80 °C rendered 99% conversion for all catalysts, while ee values varied from 21% to 41% ee in the order **14c** > **14a** > **14b** > **14d** (Entries 6–8). Upon decreasing the temperature to 60 °C, lower conversions are attained in the order of **14d** > **14a** > **14b** > **14c**. The corresponding ee values still showed the same trend with respect to the catalyst structures, but rendered significantly higher with a commendable peak of 62% for **14c**. (Entries 9–12).

Table 6. Iron-catalyzed asymmetric pressure hydrogenation of acetophenone.

Entry	Catalyst	Solvent	Temperature (°C)	H$_2$ Pressure (bar)	Conversion (%)	Ee (%)	Enantiomeric Configuration
1	14a	iPrOH	80	50	99	33	S
2	14a	MeOH	80	50	5	30	S
3	14a	EtOH	80	50	99	33	S
4	14a	toluene	80	50	99	33	S
5	14a	2-MeTHF	80	50	65	30	S
6	14b	iPrOH	80	50	99	21	S
7	14c	iPrOH	80	50	99	41	R
8	14d	iPrOH	80	50	99	21	S
9	14e	iPrOH	80	50	99	0	-
10	14a	iPrOH	60	50	74	40	S
11	14b	iPrOH	60	50	29	36	S
12	14c	iPrOH	60	50	14	62	R
13	14d	iPrOH	60	50	97	34	S
14 [1]	14a	iPrOH	60	1	45	44	S
15 [1]	14a	iPrOH	22	1	4	60	S

[1] The reaction was carried out in a Schlenk flask, equipped with a H$_2$-filled balloon.

Moreover, the outcome of these catalyst-screening experiments suggest few structural trends: (1) an increased bulkiness of the asymmetric 2,5-substituents seems to enhance enantioselectivity, but at the cost of conversion, notably for **14c**; (2) alteration of the ligand's backbone (i.e., R^3) again proves to be a more dominating factor than expected, as seen for **14d** versus **14a**. Furthermore, balloon experiments indicate that even milder conditions (e.g., H$_2$ pressure, temperature) can already achieve a decent mitigation between conversion and ee (Entries 14–15).

With the influences of the investigated reaction conditions upon the yield and enantioselectivity for acetophenone hydrogenation using **14a–d** now known, the most representative circumstance was selected to assess the catalytic performance for hydrogenation of different substrates. Several substituted acetophenone derivatives with varying electronic properties were screened (Table 7, Entries 1–5), as well as differently substituted ketones (Entries 6–9), and bio-derived methyl levulinate (Entry 10).

Decent to complete conversions were obtained using catalyst **14a**, and enantio-selectivity favoring the *S*-product was always observed. Compared to the acetophenone hydrogenations, lower ee values were generally observed for the substituted analogues of this substrate, however, hydrogenation of 2,6-dihydroxyacetophenone and propiophenone rendered 47% ee. Low enantio-selectivity was observed towards non-aromatic substrates such as *tert*-butyl methyl ketone and methyl levulinate. Some substrates were subjected to catalysts **14c** and **14d** additionally. These examples followed a general trend, that **14c** is a more selective but less active catalyst, and **14d** is slightly less selective than **14a**. Herein, the **14c**-catalyzed hydrogenation of propiophenone yielded the highest enantioselectivity obtained in this work, rendering 70% ee.

Table 7. Substrate scope for the iron-catalyzed asymmetric pressure hydrogenation.

Entry	Substrate	Catalyst	Conversion (%)	Ee (%)	Enantiomeric Configuration
1	4-MeO-acetophenone	14a	99	29	S
2	4-O₂N-acetophenone	14a	66	26	S
3	4-Br-acetophenone	14a	97	29	S
4	3,5-(CF₃)₂-acetophenone	14a	97	38	S
		14c	93	56	R
		14d	72	34	S
5	2,6-(OH)₂-acetophenone	14a	99	47	S
6	1-acetonaphthone	14a	69	26	S
		14c	32	63	R
		14d	95	23	S
7	2-acetyl-5-methylfuran	14a	100	14	S
		14c	41	28	R
8	pinacolone	14a	100	3	S
		14c	2	22	R
9	acetophenone	14a	98	47	S
		14c	53	70	R
		14d	68	38	S
10 [1]	methyl levulinate	14a	50	1	S
		14c	32	4	R

[1] The reaction temperature was 80 °C.

3. Materials and Methods

3.1. General Statements

(R)-but-3-yn-2-ol (99% pure, >99.5% ee); (R)-1-phenylprop-2-yn-1-ol (99% pure, >99.5% ee); (rac)-but-3-yn-2-ol (99% pure); triisopropylsilyl chloride (97% pure); *tert*-butyldiphenylsilyl chloride (98% pure); imidazole (99% pure); n-butyllithium (1.6 M in hexanes); 1,4-diiodobutane (98% pure); 1,3-diiodopropane (98% pure); iron pentacarbonyl (99% pure); diiron nonacarbonyl (99% pure); trimethylamine-N-oxide (99% pure); acetophenone (≥99% pure) were purchased from Sigma Aldrich (St. Louis, MO, USA) verified by NMR (Bruker, Billerica, MA, USA) and chiral GC-fid (Hewlett Packard, Palo Alto, CA, USA), and used without further purification. Ketones in the substrate scope were obtained from miscellaneous sources. Dry DCM was purchased from Sigma Aldrich (St. Louis, MO, USA); dry THF and dry toluene were drawn from a SPS MBraun solvent dispenser (Garching,

Germany), and were always verified to contain <100 ppm water by titration using a Mettler Toledo C30S Compact Karl Fischer Coulometer (Columbus, OH, USA). All other solvents were HPLC-grade pure, and were purchased from Biosolve (Valkenswaard, The Netherlands).

All organic reactions were performed in oven-dried glassware under a N_2 atmosphere using Schlenk techniques. Complexation reactions were performed in pressure tubes equipped with a magnetic stirring bar and a crimp-cap septum, using a Biotage Initiator + microwave.

Thin layer chromatography was conducted using aluminum TLC plates, coated with μm 60 mesh normal phase silica and fluorescent indicator F254. Column chromatography was conducted manually using glass columns and 80–200 μm mesh silica.

NMR spectra were recorded on a Bruker AVANCE-300 Ultra Shield spectrometer (Billerica, MA, USA), 300 MHz for ^{1}H-NMR and 75 MHz for ^{13}C-NMR at 25 °C. FT-IR were recorded on a Shimadzu Miracle 10 FT-IR spectrometer (Kyoto, Japan) in the range of 400–4000 cm^{-1}. Mass spectrometry was conducted using a Bruker solariX XR FT-ICR-MS mass-spectrometer (Billerica, MA, USA) with an ultra-high resolution over 10^5. The applied MS parameters in positive mode were as following. Capillary: 4.4 kV; end plate offset: −800 V; nebulizer: 1 bar, dry gas 4 L/min, dry temperature 200 °C; mass range: 50–1000 Da. Samples were prepared by dissolving the compounds in CH_2Cl_2–CH_3CN–CH_3COOH (49.95–49.95–0.1 vol%).

Chiral organic products from the hydrogenation reactions were analyzed using a Hewlett Packard 5890 Series II gas chromatograph (Palo Alto, CA, USA), equipped with a flame-ionization detector, and a CP-Chirasil-Dex CB capillary column (length = 25 m; internal diameter = 0.25 mm; film thickness = 0.25 mm). The heating program was 5 min at 40 °C, a ramp of 5 °C/min up to 200 °C and a final 8 min at 200 °C. The (direction of) optical rotations of synthetic products and hydrogenation reaction mixtures were measured using a Bellingham + Stanley ADP410 polarimeter (Kent, UK), equipped with a 5.0 cm pathway sample chamber.

Computational calculations for all chemical geometries were performed by using the Gaussian software package (Version 09, Gaussian Inc., Wallingford, CT, USA, 2015). Optimizations were performed at the level of DFT by means of the hybrid B3LYP [72,73] functional and the basis set LANL2DZ [74,75] was employed for all elements. All calculations were performed without freezing any atom. Frequency calculations were performed for all stationary points at the same level to identify the minima (zero imaginary frequencies) and transition states (TS, only one imaginary frequency) and to provide free energies at 298.15 K and 1 atm.

3.2. General Procedure for the Synthesis of Alkynic Silyl Ethers 12a–c

In oven-dried glassware under a nitrogen atmosphere, alkynol 11 (12.4 mmol, 1.0 eq), and imidazole (25 mmol, 2.0 eq) were dissolved together in 30 mL of dry DCM, and silyl chloride (14.5 mmol, 1.2 eq) was dissolved separately in 15 mL of dry DCM. The alkynol/imidazole solution was cooled to 0 °C, and the silyl chloride solution was added dropwise under vigorous stirring, upon which the formation of imidazolyl chloride salt was observed quite readily. Nevertheless, the mixture was stirred overnight at room temperature to ensure complete conversion. For workup, 20 mL of demineralized water was added for quenching. The biphasic mixture was separated, and the aqueous layer was back-extracted using 3 × 15 mL ethyl acetate. The combined organic phases were subsequently extracted using 3 × 30 mL of a saturated aqueous NH_4Cl solution and dried over $MgSO_4$. Upon concentration, chromatography over a silica column (±30 cm × 1 cm) using pure hexane as eluent, and thorough rotary vaporization the pure product was obtained. Staining with $KMnO_4$/alkaline solution is required to visualize the product in TLC.

3.3. General Procedure for the Synthesis of Pre-Ligands 13a–e

NOTE: For a successful reproduction, it is essential to ensure absolutely dry conditions, and to apply the stoichiometry of reagents and the temperature programming exactly as described below!

In oven-dried glassware under a nitrogen atmosphere, alkynic silyl ether **13** (6.0 mmol, 4.0 eq) was dissolved in 20 mL of dry THF, and diiodo alkane (1.5 mmol, 1.0 eq) was dissolved separately in 10 mL of dry THF. The alkynic silyl ether solution was cooled to −80 °C using an isopropanol/liquid N_2 bath, and a commercial solution of n-butyllithium 1.6 M in hexanes (9.1 mL, 5.7 mmol, 3.8 eq) was added dropwise over a course of 10 min. The reaction was kept at −80 °C for 60 min to achieve complete alkyne deprotonation. Then, the diiodo alkane solution was added dropwise at −80 °C under inert conditions, and the resulting reaction mixture was slowly heated to 60 °C subsequently, and allowed to react for 18 h. For workup, the reaction mixture was quenched with 5 mL of a saturated aqueous NH_4Cl solution and 10 mL ethyl acetate was added to promote better phase separation. The biphasic mixture was separated, and the aqueous layer was back-extracted with 5 mL ethyl acetate, and using 15 mL of a saturated aqueous NH_4Cl solution and dried over $MgSO_4$. The final organic layer was dried using $MgSO_4$ and the solution was concentrated. The crude product was purified by chromatography over a silica column (±80 cm × 2 cm) using 0–3% (slow increment) ethyl acetate/hexane eluent. Rotary vaporization of the carefully selected chromatography fractions yielded the pure dialkyne product, and the excess of alkynic silyl ether **12** could also be recovered for recycling. Staining with $KMnO_4$/alkaline solution is required to visualize the product in TLC.

3.4. General Procedure for the Synthesis of Iron Complexes **14a–e**

CAUTION: This reaction builds up a significant pressure (±7 bar) of CO gas, and a very thorough consideration on how to handle and neutralize this pressurized lethal gas is essential to warrant safety! We used dedicated microwave equipment from Biotage®, which firmly clamps the crimp-cap onto the vial, monitors the pressure in real-time, and has a steel-cage chamber equipped with a sponge to absorb any potential leakage or explosion. We neutralized the CO-pressurized vials via: (1) ensuring that the reaction mixture was cooled to room temperature; (2) applying a 10 mL syringe equipped with a Luer lock and a thin needle to release the CO pressure manually, while keeping all materials deep inside a well-ventilated fume hood. (Hold the syringe firmly with your thumb on the plunger, pierce the septum carefully with the needle and collect the CO gas in the syringe in a controlled manner. Then, withdraw the needle and release the CO gas from the syringe deep and high inside the fume hood (the septum from Biotage® will close and withstand the remaining pressure). Repeat the manual CO extractions with the syringe, until all pressure is released.); (3) Purging the headspace of the microwave vial via needles using a balloon of nitrogen inside a well-ventilated fume hood. Then finally, the crimp-cap can be safely removed from the vial.

A 30-mL glass pressure vial from with a stirring magnet was mounted inside a large Schlenk flask, and the system was purged under a nitrogen atmosphere. Under outflow of a nitrogen stream, dialkyne **6** (1.2 mmol, 1.0 eq) and 15 mL of dry toluene, and finally $Fe(CO)_5$ (5 mmol, 4–5 eq) were added into the pressure vial, and the vial was sealed using a dedicated crimp-cap septum. The mixture was reacted by microwave irradiation to a constant temperature of 140 °C for 18 h, during which the formation of CO pressure was observed to build up to 10 bar over the course of 4–8 h. After cooling down to room temperature, the vial was still pressurized with 7 bar CO gas, which was very carefully released as described above in the red 'caution' section. The neutralized reaction mixture was passed through a Celite column (±30 cm × 1 cm) using 100 mL ethyl acetate to remove solid iron carbonyl particles, and resulting solution was pushed through a millipore filter subsequently to remove paramagnetic iron nanoparticles. After removal of the ethyl acetate by rotary vaporization, the concentrated crude product was purified by chromatography over a silica column (±50 cm × 2 cm) using 0–5% ethyl acetate/hexane eluent. Rotary vaporization of the product fractions yielded the pure Knölker-type iron complex. Visualization of the product in TLC is possible under UV-light, but also by staining with $KMnO_4$/alkaline solution.

3.5. General Procedure for Transfer–Hydrogenation

In oven-dried Schlenk flasks under a nitrogen atmosphere, separate stock solutions of pre-catalyst **14a** (0.052 M), Me$_3$NO (0.156 M), in degassed isopropanol were prepared. Generally, 20.0 µL of these stock solutions, 117 µL of acetophenone, aliquots of isopropanol, and optionally amounts of extra solvent were added into oven-dried 5-mL Schlenk flasks under a nitrogen atmosphere to obtain the ratios as described in Table 5. The reaction mixtures were heated in pre-heated oil baths at the desired temperature for 24 h. Samples for GC-analysis were prepared by dissolving delicate aliquots of reaction mixture in 1.00 mL of a 0.1 vol% solution of hexadecane in DCM.

3.6. General Procedure for Pressure–Hydrogenation

Hydrogenation reactions were performed in 3-mL glass vials equipped with a magnetic stirring bar. First, solid reagents (i.e., catalyst (0.010 mmol), Me$_3$NO (0.03 mmol), and certain substrates (1.00 mmol)) were loaded. Each vial was then inserted in a Schenk flask and purged under N$_2$ by applying three vacuum/N$_2$ cycles, and isopropanol (1.00 mL) and liquid substrates (1.00 mmol). A disposable snap-cap was fitted on each vial under outflow of N$_2$. Four of such reaction vials at a time were mounted inside a 5500 HP compact 100 mL autoclave from Parr Instrument Company, and the snap-caps were pierced once with a thick needle to allow gas exchange. The reactor was sealed, purged with 5 × 2.5 bar N$_2$ and 3 × 10 bar H$_2$, and finally charged with 50 bar H$_2$ pressure. Subsequently, the system was heated to the desired temperature using 'mode 1' (i.e., using 40% electric power) to avoid a temperature overshoot, and the reaction mixtures were magnetically stirred at 300 rpm using a stirring plate placed underneath the autoclave. After 24 h, the reactions were stopped by allowing the reactor to cool down below 40 °C within 15 min, after which the reactor was purged with 3 × 2.5 bar N$_2$ before opening.

Samples for GC-analysis were prepared by dissolving 50.0 µL reaction mixture in 1.00 mL of a 0.1 vol% solution of hexadecane in DCM.

4. Conclusions

We have successfully synthesized a new series of asymmetric Knölker-type iron catalysts, which exhibit the centers of chirality at the front side of the catalyst near the catalytically active site. Amongst the four asymmetric complexes a structural alteration in each variable fragment of the cyclopentadienone ligand (i.e., R^1, R^2, and R^3) compared to base–case complex **14a** was made. NMR analysis was particularly useful to study the asymmetric nature of the complexes, and it could even be used as a tool to confirm preservation of enantio-purity throughout the total synthesis of the complexes. In addition, DFT calculations provided useful insights on the structural conformation of the chiral substituents on the ligand, and the role of the ligand's backbone herein.

All synthesized iron complexes exhibited catalytic activity in the pressure hydrogenation of ketones, while transfer hydrogenation was also established using **14a**. Complexes **14a–d** rendered ee values in the range of 21–62% for the hydrogenation of acetophenone, and 70% as highest ee in the substrate screening catalyzed by **14c**, which ranks our structural design in the top three of enantioselective Knölker-type catalysts.

Future research will focus on extending the structural examples within the series of our catalyst design, aiming for a superior catalytic performance to make environmentally considerate iron more attractive in applied and industrial chemistry.

Supplementary Materials: The following are available online at http://www.mdpi.com/2073-4344/9/10/790/s1.

Author Contributions: Conceptualization, C.A.M.R.v.S.; methodology, C.A.M.R.v.S., L.M., K.C.C., D.H., and B.B.; software, C.A.M.R.v.S.; validation, C.A.M.R.v.S., B.B., and S.M.A.D.W.; formal analysis, C.A.M.R.v.S., L.M., K.C.C. and D.H.; investigation, C.A.M.R.v.S., L.M., and K.C.C.; resources, S.M.A.D.W.; data curation, C.A.M.R.v.S., L.M., K.C.C. and D.H.; writing—original draft preparation, C.A.M.R.v.S.; writing—review and editing, B.B. and S.M.A.D.W.; visualization, C.A.M.R.v.S.; supervision, S.M.A.D.W.; project administration, C.A.M.R.v.S., and S.M.A.D.W.; funding acquisition, S.M.A.D.W.

Funding: This research and the APC were funded by the framework of the public–private knowledge institute of *Chemelot InSciTe*.

Acknowledgments: We would like to thank Maarten Honing for offering his mass spectrometry equipment for our use; Andrij Pich, Ayse Deniz, and Wenjing Xu for arranging the elemental analysis; and Gerard Verzijl for his inspirational support to the general catalyst structure. Special appreciation to Martien Brouwers, Gerard Verzijl, and Jeroen Welzen for the donation and installation of their GC-fid equipment.

Conflicts of Interest: The authors declare no conflict of interest.

Appendix A

(R)-(but-3-yn-2-yloxy)triisopropylsilane (12a)

Appearance: colorless liquid. Yield (isolated): 92.7%. ^{1}H-NMR (300 MHz), 25 °C, CDCl$_3$ (7.26 ppm): δ = 4.60 (qd, J_1 = 6.5 Hz, J_2 = 2.0 Hz, 1H, CHO), 2.37 (d, J = 2.0, 1H, C≡CH), 1.46 (d, J = 6.5 Hz, 3H, Me), 1.34–0.83 (m, 6H, Si(CHMe$_2$)$_3$), 1.09 (d, J = 3.5 Hz, 36H, Si(CHMe$_2$)$_3$) ppm. ^{13}C-NMR (75 MHz), 25 °C, CDCl$_3$ (77.16 ppm): δ = 86.80, 71.18, 58.94, 25.72, 18.11, 18.09, 12.30 ppm. FT-IR: 629, 654, 679, 752, 835, 881, 918, 974, 991, 1014, 1059, 1101, 1121, 1250, 1313, 1337, 1370, 1384, 1464, 2866, 2891, 2943, 3312 cm^{-1}. $[\alpha]_D^{25}$ = +96.1 (c = 2.44 in DCM).

(R)-(but-3-yn-2-yloxy)(tert-butyl)diphenylsilane (12b)

Appearance: colorless liquid. Yield (isolated): 91.3%. ^{1}H-NMR (300 MHz), 25 °C, CDCl$_3$ (7.26 ppm): δ = 7.91–7.65 (m, 4H, C$_6$H$_5$), 7.53–7.37 (m, 6H, C$_6$H$_5$), 4.52 (qd, J_1 = 6.5 Hz, J_2 = 2.0 Hz, 1H, CHO), 2.37 (d, J = 2.1 Hz, 1H, C≡CH), 1.45 (d, J = 6.5 Hz, 3H, Me), 1.15 (s, 9H, tBu) ppm. ^{13}C-NMR (75 MHz), 25 °C, CDCl$_3$ (77.16 ppm): δ = 136.09, 135.90, 133.84, 133.55, 129.90, 129.86, 127.77, 127.66, 86.20, 71.72, 59.93, 27.00, 25.30, 19.33 ppm. FT-IR: 544, 611, 657, 700, 739, 762, 822, 841, 938, 974, 1057, 1098, 1427, 1471, 2859, 2889, 2932, 2959, 3070, 3306 cm^{-1}. $[\alpha]_D^{24}$ = +288.3 (c = 2.23 in DCM).

(R)- triisopropyl((1-phenylprop-2-yn-1-yl)oxy)silane (12c)

Appearance = viscous yellow liquid. Yield (isolated): 85.5%. ^{1}H-NMR (300 MHz), 25 °C, CDCl$_3$ (7.26 ppm): δ = 7.56 (d, J = 7.3, 2H, C$_6$H$_5$), 7.48–7.29 (m, 3H, C$_6$H$_5$), 5.62 (d, J = 2.0 Hz, 1H, CHO), 2.58 (d, J = 2.2 Hz, 3H, Me), 1.36–1.21 (m, 2H), 1.22–1.10 (m, 18H, Si(CHMe$_2$)$_3$) ppm. ^{13}C-NMR (75 MHz), 25 °C, CDCl$_3$ (77.16 ppm): δ = 142.04, 128.49, 127.88, 125.98, 85.32, 77.58, 77.16, 76.74, 73.53, 64.83, 18.13, 12.39 ppm. FT-IR: 575, 656, 681, 695, 733, 824, 834, 881, 918, 961, 997, 1015, 1028, 1063, 1092, 1192, 1265, 1317, 1341, 1366, 1385, 1462, 1493, 2866, 2891, 2943, 3308 cm^{-1}. $[\alpha]_D^{26}$ = −14.8 (c = 2.16 in DCM).

(5R,14R)-3,3,16,16-tetraisopropyl-2,5,14,17-tetramethyl-4,15-dioxa-3,16-disilaoctadeca-6,12-diyne (13a)

Appearance: pale yellow liquid. Yield (isolated): 89.1%. ^{1}H-NMR (300 MHz), 25 °C, CDCl$_3$ (7.26 ppm): δ = 4.57 (q, J = 6.4 Hz, 2H, CHO), 2.19 (m, 4H, C≡C–CH$_2$), 1.58 (m, 4H, CH$_2$), 1.41 (d, J = 6.4 Hz, 6H, Me), 1.22–0.93 (m, 6H, Si(CHCH$_3$)$_3$), 1.08 (d, J = 3.6 Hz, 36H, Si(CHCH$_3$)$_3$) ppm. ^{13}C-NMR (75 MHz), 25 °C, CDCl$_3$ (77.16 ppm): δ = 83.48, 83.08, 77.58, 77.36, 77.16, 76.74, 59.25, 27.81, 26.11, 18.35, 18.15, 18.13, 12.35 ppm. FT-IR: 666, 677, 756, 881, 920, 949, 976, 1013, 1030, 1069, 1099, 1159, 1248, 1316, 1335, 1368, 1383, 1462, 2236, 2864, 2892, 2941 cm^{-1}. ESI-MS (in CH$_2$Cl$_2$/CH$_3$CN 1:1 v/v with CH$_3$COOH for M = C$_{30}$H$_{58}$Si$_2$O$_2$): m/z = 529.38515 {[M + Na]$^+$, calcd 693.356006. $[\alpha]_D^{22}$ = +122.6 (c = 2.01 in DCM).

(5R,14R)-2,2,5,14,17,17-hexamethyl-3,3,16,16-tetraphenyl-4,15-dioxa-3,16-disilaoctadeca-6,12-diyne (13b)

Appearance: viscous colorless liquid. Yield (isolated): 38.7%. ^{1}H-NMR (300 MHz), 25 °C, CDCl$_3$ (7.26 ppm): δ = 7.73 (dd, J_1 = 20.6 Hz, J_2 = 6.1 Hz, 8H, C$_6$H$_5$), 7.37 (d, J = 7.4 Hz, 12H, C$_6$H$_5$), 4.47 (q, J = 6.3 Hz, 2H, CHO), 2.06 (m, 4H, C≡C–CH$_2$), 1.40 (m, 4H, CH$_2$), 1.38 (d, J = 6.4 Hz, 6H, Me), 1.07 (s, 18H, tBu) ppm. ^{13}C-NMR (75 MHz), 25 °C, CDCl$_3$ (77.16 ppm): δ = 136.10, 135.91, 134.14, 134.10, 129.74, 129.63, 127.66, 127.50, 83.88, 83.00, 60.31, 27.61, 27.02, 25.69, 19.33, 18.29 ppm. FT-IR: 611, 698, 738, 822, 951, 974, 998, 1028, 1080, 1098, 1161, 1341, 1368, 1390, 1427, 1472, 1589, 2857, 2891,

2930, 3070 cm^{-1}. ESI-MS (in CH$_2$Cl$_2$/CH$_3$CN 1:1 v/v with CH$_3$COOH for M = C$_{44}$H$_{54}$Si$_2$O$_2$): m/z = 693.35186 {[M + Na]$^+$, calcd 529.387306}. [α]$_D^{26}$ = +364.1 (c = 2.17 in DCM).

(5R,14R)-3,3,16,16-tetraisopropyl-2,17-dimethyl-5,14-diphenyl-4,15-dioxa-3,16-disilaoctadeca-6,12-diyne (13c)

Appearance: viscous orange liquid. Yield (isolated): 60.8%. ^{1}H-NMR (300 MHz), 25 °C, CDCl$_3$ (7.26 ppm): δ = 7.52–7.17 (m, 4H, C$_6$H$_5$), 7.14 (m, J = 6.8, 6H, C$_6$H$_5$), 5.46 (s, 2H, CHO), 2.10 (b, 4H, C≡C–CH$_2$), 1.48 (b, 4H, CH$_2$), 1.08 (m, 6H), 1.00 (dd, J$_1$ = 12.3, J$_2$ = 6.0, 36H) ppm. ^{13}C-NMR (75 MHz), 25 °C, CDCl$_3$ (77.16 ppm): δ = 143.00, 128.26, 127.44, 125.94, 85.47, 82.10, 65.07, 27.67, 18.44, 18.17, 12.44 ppm. FT-IR: 581, 629, 659, 681, 694, 721, 820, 881, 918, 996, 1013, 1026, 1057, 1083, 1134, 1194, 1273, 1329, 1367, 1383, 1462, 1492, 2864, 2891, 2941 cm^{-1}. ESI-MS (in CH$_2$Cl$_2$/CH$_3$CN 1:1 v/v with CH$_3$COOH for M = C$_{44}$H$_{54}$Si$_2$O$_2$): m/z = 653.41845 {[M + Na]$^+$, calcd 653.418606}; 669.41330 {[M + K]$^+$, calcd 669.392544}. [α]$_D^{26}$ = −37.0 (c = 2.00 in DCM).

(5R,13R)-3,3,15,15-tetraisopropyl-2,5,13,16-tetramethyl-4,14-dioxa-3,15-disilaheptadeca-6,11-diyne (13d)

Appearance: pale yellow liquid. Yield (isolated): 70.3%. ^{1}H-NMR (300 MHz), 25 °C, CDCl$_3$ (7.26 ppm): δ = 4.57 (q, J = 6.4 Hz, 2H, CHO), 2.28 (td, J$_1$ = 7.0 Hz, J$_2$ = 1.6 Hz, 4H, C≡C–CH$_2$), 1.66 (p, J = 7.1 Hz, 2H, CH$_2$), 1.41 (d, J = 6.4 Hz, 6H, Me), 1.22–0.93 (m, 6H, Si(C$\underline{H}$Me$_2$)$_3$), 1.08 (d, J = 3.7 Hz, 36H, Si(CHMe$_2$)$_3$)) ppm. ^{13}C-NMR (75 MHz), 25 °C, CDCl$_3$ (77.16 ppm): δ = 83.76, 82.55, 77.58, 77.16, 76.74, 59.23, 27.91, 26.08, 18.15, 18.12, 12.35 ppm. FT-IR: 657, 679, 756, 881, 920, 952, 973, 996, 1014, 1028, 1069, 1099, 1157, 1248, 1316, 1337, 1368, 1383, 1464, 2866, 2891, 2941 cm^{-1}. [α]$_D^{26}$ = +131.9 (c = 2.28 in DCM).

η^4-[1,3-bis((R)-1-((triisopropylsilyl)oxy)ethyl)-4,5,6,7-tetrahydro-2H-inden-2-one]iron tricarbonyl (14a)

Appearance: yellow solid. Yield (isolated): 51.4%. ^{1}H-NMR (300 MHz), 25 °C, CDCl$_3$ (7.26 ppm): δ = 4.99 (dq, J$_1$ = 66.3 Hz, J$_2$ = 6.4 Hz, 2H, CHO), 3.01–2.42 (m, 4H, (CH$_2$)$_2$), 1.80 (b, 4H, Cp–CH$_2$), 1.47 (d, J = 6.4 Hz, 6H, Me), 1.19–0.93 (m, 42H, TIPS) ppm. ^{1}H-NMR (300 MHz), 25 °C, toluene-d8 (2.08 ppm): δ 5.14 (dq, J$_1$ = 75.3, J$_1$ = 6.5, 2H, CHO), 2.84–2.30 (m, 4H, Cp–CH$_2$), 1.63–1.34 (m, 4H), 1.49 (dd, J$_1$ = 11.7, J$_2$ = 6.5, 6H, Me), 1.25–0.8 (m, 42H, TIPS). ^{13}C-NMR (75 MHz), 25 °C, CDCl$_3$ (77.16 ppm): δ = 209.20, 169.47, 101.26, 99.75, 87.60, 86.79, 65.40, 62.82, 26.92, 23.46, 23.08, 22.90, 22.37, 22.24, 18.42, 18.29, 18.27, 17.84, 13.16, 12.56, 12.43 ppm. FT-IR: 575, 594, 628, 653, 678, 737, 756, 770, 820, 881, 928, 993, 1012, 1065, 1130, 1256, 1387, 1464, 1630, 1987, 2000, 2060, 2864, 2943 cm^{-1}. ESI-MS (in CH$_2$Cl$_2$/CH$_3$CN 1:1 v/v with CH$_3$COOH for M = C$_{34}$H$_{58}$Si$_2$O$_6$Fe): m/z = 675.31648 {[M + H]$^+$, calcd 675.319960}; 697.29835 {[M + Na]$^+$, calcd 697.301905}. [α]$_D^{22}$ = +16.5 (c = 2.06 in DCM).

η^4-[1,3-bis((R)-1-((tert-butyldiphenylsilyl)oxy)ethyl)-4,5,6,7-tetrahydro-2H-inden-2-one]iron tricarbonyl (14b)

Appearance: yellow solid. Yield (isolated): 33.8%. ^{1}H-NMR (300 MHz), 25 °C, CDCl$_3$ (7.26 ppm): δ = 7.88–7.62 (m, 8H, C$_6$H$_5$), 7.52–7.28 (m, 12H, C$_6$H$_5$), 4.81 (dq, J$_1$ = 85.7 Hz, J$_2$ = 6.4 Hz, 2H, CHO), 2.90–2.35 (m, 4H, (CH$_2$)$_2$), 1.74 (m, 4H, Cp–CH$_2$), 1.32 (dd, J$_1$ = 6.4 Hz, J$_2$ = 3.1 Hz, 6H, Me), 1.08 (d, J = 5.0 Hz, 18H, tBu) ppm. ^{13}C-NMR (75 MHz), 25 °C, CDCl$_3$ (77.16 ppm): δ = 209.22, 169.74, 136.12, 136.09, 135.92, 135.78, 134.92, 134.39, 134.34, 133.72, 133.34, 129.85, 129.81, 129.80, 129.65, 127.76, 127.73, 127.61, 101.11, 100.12, 87.20, 85.74, 77.58, 77.16, 76.74, 65.98, 64.68, 27.22, 27.20, 26.68, 26.34, 23.54, 23.17, 22.50, 22.15, 22.02, 19.45, 19.36 ppm. FT-IR: 574, 594, 604, 628, 700, 739, 822, 927, 991, 1078, 1105, 1186, 1260, 1371, 1427, 1472, 1645, 1983, 2060, 2361, 2857, 2930, 2956, 3046, 3071 cm^{-1}. ESI-MS (in CH$_2$Cl$_2$/CH$_3$CN 1:1 v/v with CH$_3$COOH for M = C$_{48}$H$_{54}$Si$_2$O$_6$Fe): m/z = 839.28749 {[M + H]$^+$, calcd 839.288660}; 861.270605 {[M + Na]$^+$, calcd 861.26817}. Elemental analysis for C$_{48}$H$_{54}$Si$_2$O$_6$Fe: calcd: C 68.72%, H 6.49%; found, C 68.32%, H 6.352%. [α]$_D^{22}$ = +62.5 (c = 2.59 in DCM).

η⁴-[1,3-bis((R)-phenyl((triisopropylsilyl)oxy)methyl)-4,5,6,7-tetrahydro-2H-inden-2-one]iron tricarbonyl (14c)

Appearance: orange solid. Yield (isolated): 18.2%. ^{1}H-NMR (300 MHz), 25 °C, CDCl$_3$ (7.26 ppm): δ = 7.48 (dd, J_1 = 45.1 Hz, J_2 = 7.3 Hz, 4H, C$_6$H$_5$), 7.16 (m, 6H, C$_6$H$_5$), 5.94 (d, J = 52.7 Hz, 2H, CHO), 2.84 (dd, J_1 = 53.3 Hz, J_2 = 17.0 Hz, 2H, Cp–CH$_2$), 2.51–2.20 (m, 2H, Cp–CH$_2$), 1.76–1.42 (m, 4H, (CH$_2$)$_2$), 1.05–0.66 (m, 42H, TIPS) ppm. ^{13}C-NMR (75 MHz), 25 °C, CDCl$_3$ (77.16 ppm): δ = 208.11, 168.62, 145.90, 143.69, 127.95, 127.87, 127.48, 127.16, 126.50, 126.40, 100.88, 98.81, 89.38, 86.10, 77.58, 77.16, 76.74, 70.69, 67.72, 23.43, 22.05, 22.01, 21.95, 18.37, 18.28, 18.15, 17.96, 12.82, 12.20 ppm. FT-IR: 571, 598, 613, 647, 704, 731, 752, 806, 833, 881, 918, 972, 1015, 1055, 1084, 1104, 1173, 1260, 1288, 1366, 1423, 1452, 1492, 1, 1630, 1993, 2062, 2359, 2866, 2891, 2943, 3030, 3062 cm^{-1}. ESI-MS (in CH$_2$Cl$_2$/CH$_3$CN 1:1 v/v with CH$_3$COOH for M = C$_{44}$H$_{62}$Si$_2$O$_6$Fe): m/z = 799.35063 {[M + H]$^+$, calcd 799.351260}; 821.33173 {[M + Na]$^+$, calcd 821.333205}. $[\alpha]_D^{26}$ = +107.5 (c = 2.14 in DCM).

η⁴-[1,3-bis((R)-1-((triisopropylsilyl)oxy)ethyl)-5,6-dihydropentalen-2(4H)-one]iron tricarbonyl (14d)

Appearance: yellow solid. Yield (isolated): 51.9%. ^{1}H-NMR (300 MHz), 25 °C, CDCl$_3$ (7.26 ppm): δ = 4.89 (dq, J_1 = 28.1 Hz, J_2 = 6.2 Hz, 2H, CHO), 2.88–2.70 (m, 2H, Cp–CH$_2$), 2.69–2.45 (m, 2H, Cp–CH$_2$), 2.43–2.21 (m, 1H, CH$_2$), 1.98–1.77 (m, 1H, CH$_2$), 1.47 (dd, J_1 = 18.8 Hz, J_2 = 6.2 Hz, 6H, Me), 1.08 (m, 42H, TIPS) ppm. ^{13}C-NMR (75 MHz), 25 °C, CDCl$_3$ (77.16 ppm): δ = 209.15, 171.58, 105.90, 105.09, 88.58, 88.21, 64.83, 64.13, 27.61, 27.54, 26.52, 26.48, 24.26, 18.41, 18.30, 18.24, 13.17, 12.90 ppm. FT-IR: 576, 590, 605, 618, 635, 675, 737, 799, 841, 880, 918, 953, 1000, 1013, 1030, 1065, 1094, 1125, 1169, 1206, 1258, 1364, 1384, 1438, 1464, 1622, 1996, 2062, 2864, 2891, 2941, 2967 cm^{-1}. ESI-MS (in CH$_2$Cl$_2$/CH$_3$CN 1:1 v/v with CH$_3$COOH for M = C$_{33}$H$_{56}$Si$_2$O$_6$Fe): m/z = 661.30282 {[M + H]$^+$, calcd 661.304310}; 683.28437 {[M + Na]$^+$, calcd 683.286255}; 699.25943 {[M + K]$^+$, calcd 699.260193}. Elemental analysis for C$_{33}$H$_{56}$Si$_2$O$_6$Fe: calcd: C 59.98%, H 8.54%; found: C 59.55%, H 8.59%. $[\alpha]_D^{26}$ = +8.4 (c = 2.15 in DCM).

14f (R,S-complex extrapolated from racemic mixture 14e; signals of 14a also present in spectrum)

^{1}H-NMR (300 MHz), 25 °C, CDCl$_3$ (7.26 ppm): δ = 5,08 (q, J = 6.4 Hz, 2H, CHO), 3.01–2.42 (m, 4H, (CH$_2$)$_2$), 1.80 (b, 4H, Cp–CH$_2$), 1.47 (d, J = 6.4 Hz, 6H, Me), 1.19–0.93 (m, 42H, TIPS) ppm. ^{1}H-NMR (300 MHz), 25 °C, toluene-d8 (2.08 ppm): δ = 5,21 (q, J = 6.3 Hz, 2H, CHO), 3.01–2.42 (m, 4H, (CH$_2$)$_2$), 1.80 (b, 4H, Cp–CH$_2$), 1.47 (d, J = 6.4 Hz, 6H, Me), 1.19–0.93 (m, 42H, TIPS) ppm. ^{13}C-NMR (75 MHz), 25 °C, CDCl$_3$ (77.16 ppm): δ = 209.35, 169.58, 146.87, 101.39, 87.76, 63.69, 34.66, 27.65, 23.47, 22.33, 12.75 ppm. $[\alpha]_D^{25}$ = 0.00 (c = 2.02 in DCM).

References

1. Knowles, W.S. Asymmetric Hydrogenations (Nobel Lecture). *Angew. Chem. Int. Ed.* **2002**, *41*, 1998–2007. [CrossRef]
2. Noyori, R. Asymmetric Catalysis: Science and Opportunities (Nobel Lecture). *Angew. Chem. Int. Ed.* **2002**, *41*, 2008–2022. [CrossRef]
3. Knowles, W.S. Application of Organometallic Catalysis to the Commercial Production of L-DOPA. *J. Chem. Ed.* **1986**, *63*, 222–225. [CrossRef]
4. Enthaler, S.; Junge, K.; Beller, M. Sustainable Metal Catalysis with Iron: From Rust to a Rising Star? *Angew. Chem. Int. Ed.* **2008**, *47*, 3317–3321. [CrossRef] [PubMed]
5. Bauer, G.; Kirchner, K.A. Well-Defined Bifunctional Iron Catalysts for the Hydrogenation of Ketones: Iron, the New Ruthenium. *Angew. Chem. Int. Ed.* **2011**, *50*, 5798–5800. [CrossRef] [PubMed]
6. Darwish, M.; Wills, M. Asymmetric catalysis using iron complexes—'Ruthenium Lite'? *Catal. Schi. Technol.* **2012**, *2*, 243–255. [CrossRef]
7. Li, Y.; Yu, S.; Wu, X.; Xiao, J.; Shen, W.; Dong, Z.; Gao, J. Iron Catalyzed Asymmetric Hydrogenation of Ketones. *J. Am. Chem. Soc.* **2014**, *136*, 4031–4039. [CrossRef]
8. Sues, P.E.; Demmans, K.Z.; Morris, R.H. Rational development of iron catalysts for asymmetric transfer hydrogenation. *Dalton Trans.* **2014**, *43*, 7650–7667. [CrossRef]

9. Obligacion, J.V.; Chirik, P.J. Earth-abundant transition metal catalysts for alkene hydrosilylation and hydroboration. *Nat. Rev. Chem.* **2018**, *2*, 15–34. [CrossRef]
10. Zell, T.; Milstein, D. Hydrogenation and Dehydrogenation Iron Pincer Catalysts Capable of Metal-Ligand Cooperation by Aromatization/Dearomatization. *Acc. Chem. Res.* **2015**, *48*, 1979–1994. [CrossRef]
11. Bauer, I.; Knölker, H.-J. Iron Catalysis in Organic Synthesis. *Chem. Rev.* **2015**, *115*, 3170–3387. [CrossRef] [PubMed]
12. Reppe, W.; Vetter, H. Carbonylierung VI. Synthesen mit Metallcarbonylwasserstoffen. *Liebigs Ann. Chem.* **1953**, *582*, 133–161. [CrossRef]
13. Quintard, A.; Rodriquez, J. Iron Cyclopentadienone Complexes: Discovery, Properties, and Catalytic Reactivity. *Angew. Chem. Int. Ed.* **2014**, *53*, 4044–4055. [CrossRef] [PubMed]
14. Knölker, H.-J.; Heber, J.; Mahler, C.H. Transition Metal-Diene Complexes in Organic Synthesis, Part 14. Regioselective Iron-Mediated [2 + 2 + 1] Cycloadditions of Alkynes and Carbon Monoxide: Synthesis of Substituted Cyclopentadienones. *Synlett* **1992**, *12*, 1002–1004. [CrossRef]
15. Knölker, H.-J.; Heber, J. Transition Metal-Diene Complexes in Organic Synthesis, Part 18. Iron-Mediated [2 + 2 + 1] Cycloadditions of Diynes and Carbon Monoxide: Selective Demetalation Reactions. *Synlett* **1993**, *12*, 924–926. [CrossRef]
16. Knölker, H.-J.; Baum, E.; Klauss, R. Transition Metal-Diene Complexes in Organic Synthesis, Part 25. Cycloadditions of Annulated 2,5-Bis(trimethylsilyl)cyclopentadienones. *Tetrahedron Lett.* **1995**, *36*, 7647–7650. [CrossRef]
17. Knölker, H.-J.; Baum, E.; Goesmann, R.; Klauss, R. A Novel Method for the Demetalation of Tricarbonyliron–Diene Complexes by a Photolytically Induced Ligand Exchange Reaction with Acetonitrile. *Angew. Chem. Int. Ed.* **1999**, *38*, 702–705. [CrossRef]
18. Knölker, H.-J.; Goesmann, R.; Klauss, R. Demetalation of Tricarbonyl(cyclopentadienone)iron Complexes Initiated by a Ligand Exchange Reaction with NaOH—X-Ray Analysis of a Complex with Nearly Square-Planar Coordinated Sodium. *Angew. Chem. Int. Ed.* **1999**, *38*, 2064–2066. [CrossRef]
19. Knölker, H.-J.; Braier, A.; Bröcher, D.J.; Cämmerer, S.; Fröhner, W.; Gonser, P.; Hermann, H.; Herzberg, D.; Reddy, K.R.; Rohde, G. Recent applications of tricarbonyliron-diene complexes to organic synthesis. *Pure Appl. Chem.* **2001**, *73*, 1075–1086. [CrossRef]
20. Casey, C.P.; Guan, H. An Efficient and Chemoselective Iron Catalyst for the Hydrogenation of Ketones. *J. Am. Chem. Soc.* **2007**, *129*, 5816–5817. [CrossRef]
21. Casey, C.P.; Guan, H. Cyclopentadienone Iron Alcohol Complexes: Synthesis, Reactivity, and Implications for the Mechanism of Iron-Catalyzed Hydrogenation of Aldehydes. *J. Am. Chem. Soc.* **2009**, *131*, 2499–2507. [CrossRef] [PubMed]
22. von der Höh, A.; Berkessel, A. Insights into the Mechanism of Dihydrogen-Heterolysis at Cyclopentadienone Iron Complexes and Subsequent C=X Hydrogenation. *ChemCatChem* **2011**, *3*, 861–867. [CrossRef]
23. Lu, X.; Zhang, Y.; Yun, P.; Zhang, M.; Li, T. The mechanism for the hydrogenation of ketones catalyzed by Knölker's iron-catalyst. *Org. Biomol. Chem.* **2013**, *11*, 5264–5277. [CrossRef] [PubMed]
24. Berkessel, A.; Reichau, S.; von der Höh, A.; Leconte, N.; Neudörfl, J.-M. Light-Induced Enantioselective Hydrogenation Using Chiral Derivatives of Casey's Iron–Cyclopentadienone Catalyst. *Organometallics* **2011**, *30*, 3880–3887. [CrossRef]
25. Hopewell, J.P.; Martins, J.E.D.; Johnson, T.C.; Godfrey, J.; Wills, M. Developing asymmetric iron and ruthenium-based cyclone complexes; complex factors influence the asymmetric induction in the transfer hydrogenation of ketones. *Org. Biomol. Chem.* **2012**, *10*, 134–145. [CrossRef] [PubMed]
26. Natte, K.; Li, W.; Zhou, S.; Neumann, H.; Wu, X.-F. Iron-catalyzed reduction of aromatic aldehydes with paraformaldehyde and H_2O as the hydrogen source. *Tetrahedron Lett.* **2015**, *56*, 1118–1121. [CrossRef]
27. Coleman, M.G.; Brown, A.N.; Bolton, B.A.; Guan, H. Iron-Catalyzed Oppenauer-Type Oxidation of Alcohols. *Adv. Synth. Catal.* **2010**, *352*, 967–970. [CrossRef]
28. Pagnoux-Ozherelyeva, A.; Pannetier, N.; Mbaye, M.D.; Gaillard, S.; Renaud, J.-L. Knölker's Iron Complex: An Efficient in Situ Generated Catalyst for Reductive Amination of Alkyl Aldehydes and Amines. *Angew. Chem. Int. Ed.* **2012**, *124*, 5060–5064. [CrossRef]
29. Tlili, A.; Schrank, J.; Neumann, H.; Beller, M. Discrete Iron Complexes for the Selective Catalytic Reduction of Aromatic, Aliphatic, and α, β-Unsaturated Aldehydes under Water-Gas Shift Conditions. *Chem. Eur. J.* **2012**, *18*, 15935–15939. [CrossRef]

30. Rosas-Hernández, A.; Junge, H.; Beller, M.; Roemelt, M.; Francke, R. Cyclopentadienone iron complexes as efficient and selective catalysts for the electroreduction of CO_2 to CO. *Catal. Sci. Technol.* **2017**, *7*, 459–465. [CrossRef]

31. Elangovan, S.; Sortais, J.-B.; Beller, M.; Darcel, C. Iron-Catalyzed α-Alkylation of Ketones with Alcohols. *Angew. Chem. Int. Ed.* **2015**, *54*, 14483–14486. [CrossRef] [PubMed]

32. Sun, Y.-Y.; Wang, H.; Chen, N.-Y.; Lennox, A.J.J.; Friedrich, A.; Xia, L.-M.; Lochbrunner, S.; Junge, H.; Beller, M.; Zhou, S.; et al. Efficient Photocatalytic Water Reduction Using in Situ Generated Knölker's Iron Complexes. *ChemCatChem* **2016**, *8*, 2340–2344. [CrossRef]

33. Quintard, A.; Constantieux, T.; Rodriquez, J. An Iron/Amine-catalyzed Cascade Process for the Enantioselective Functionalization of Allylic Alcohols. *Angew. Chem.* **2013**, *125*, 13121–13125. [CrossRef]

34. Roudier, M.; Constantieux, T.; Quintard, A.; Rodriquez, J. Enantioselective Cascade Formal Reductive Insertion of Allylic Alcohols into the C(O)-C Bond of 1,3-Diketones: Ready Access to Synthetically Valuable 3-Alkylpentanol Units. *Org. Lett.* **2014**, *16*, 2802–2805. [CrossRef] [PubMed]

35. Roudier, M.; Constantieux, T.; Quintard, A.; Rodriquez, J. Triple Iron/Copper/Iminium Activation for the Efficient Redox Neutral Catalytic Enantioselective Functionalization of Allylic Alcohols. *ACS Catal.* **2016**, *6*, 5236–5244. [CrossRef]

36. Quintard, A.; Rodriquez, J. A Step into an eco-Compatible Future: Iron- and Cobalt-catalyzed Borrowing Hydrogen Transformation. *ChemSusChem* **2016**, *9*, 28–30. [CrossRef]

37. Quintard, A.; Rodriquez, J. Catalytic enantioselective OFF↔ON activation processes initiated by hydrogen transfer: Concepts and challenges. *Chem. Commun.* **2016**, *52*, 10456–10473. [CrossRef]

38. Rodriquez, J.; Quintard, A. Discovery of Eco-compatible Synthetic Paths by a Multi-catalysis Approach. *Chimia* **2018**, *72*, 580–583. [CrossRef]

39. Piarulli, U.; Vailati Fachini, S.; Pignataro, L. Enantioselective Reductions Promoted by (Cyclopentadienone)iron Complexes. *Chimia* **2017**, *71*, 580–585. [CrossRef]

40. Yamamoto, Y.; Akagi, M.; Shimanuki, K.; Kuwahara, S.; Watanabe, M.; Harada, N. A general method for the synthesis of enantiopure aliphatic chain alcohols with established absolute configurations. Part 1. Application of the MαNP acid method to acetylene alcohols. *Tetrahedron Asym.* **2010**, *25*, 1456–1465. [CrossRef]

41. Johnson, T.C.; Clarkson, G.J.; Wills, M. (Cyclopentadienone)iron Shvo Complexes: Synthesis and Application to Hydrogen Transfer Reactions. *Organometallics* **2011**, *30*, 1850–1868. [CrossRef]

42. Hodgkinson, R.; Del Grosso, A.; Clarkson, G.; Wills, M. Iron cyclopentadienone complexes derived from C_2-symmetric bis-propargylic alcohols; preparation and application to catalysis. *Dalton Trans.* **2016**, *45*, 3992–4005. [CrossRef] [PubMed]

43. Del Grosso, A.; Chamberlain, A.E.; Clarkson, G.J.; Wills, M. Synthesis and applications to catalysis of novel cyclopentadienone iron tricarbonyl complexes. *Dalton Trans.* **2018**, *47*, 1451–1470. [CrossRef] [PubMed]

44. Gajewski, P.; Renom-Carrasco, M.; Vailati Facchini, S.; Pignataro, L.; Lefort, L.; de Vries, J.G.; Ferraccioli, R.; Forni, A.; Piarulli, U.; Gennari, C. Chiral (Cyclopentadienone)iron Complexes for the Catalytic Asymmetric Hydrogenation of Ketones. *Eur. J. Org. Chem.* **2015**, *9*, 1887–1893. [CrossRef]

45. Gajewski, P.; Renom-Carrasco, M.; Vailati Facchini, S.; Pignataro, L.; Lefort, L.; de Vries, J.G.; Ferraccioli, R.; Piarulli, U.; Gennari, C. Synthesis of (R)-BINOL-Derived (Cyclopentadienone)iron Complexes and Their Application in the Catalytic Asymmetric Hydrogenation of Ketones. *Eur. J. Org. Chem.* **2015**, *25*, 5526–5536. [CrossRef]

46. Cettolin, M.; Bai, X.; Lübken, D.; Gatti, M.; Vailati Facchini, S.; Piarulli, U.; Pignataro, L.; Gennari, C. Improving C = N Bond Reductions with (Cyclopentadienone)iron Complexes: Scope and Limitations. *Eur. J. Org. Chem.* **2019**, *4*, 647–654. [CrossRef]

47. Dou, X.; Hayashi, T. Synthesis of Planar Chiral Shvo Catalysts for Asymmetric Transfer Hydrogenation. *Adv. Synth. Catal.* **2016**, *358*, 1054–1058. [CrossRef]

48. Bai, X.; Cettolin, M.; Mazzoccanti, G.; Pierini, M.; Piarulli, U.; Colombo, V.; Dal Corso, A.; Pignataro, L.; Gennari, C. Chiral (cyclopentadienone)iron complexes with a stereogenic plane as pre-catalysts for the asymmetric hydrogenation of polar double bonds. *Tetrahedron* **2019**, *75*, 1415–1424. [CrossRef]

49. Zhou, S.; Fleischer, S.; Junge, K.; Beller, M. Cooperative Transition-Metal and Chiral Brønsted Acid Catalysis: Enantioselective Hydrogenation of Imines to Form Amines. *Angew. Chem.* **2011**, *50*, 5120–5124. [CrossRef]

50. Fleischer, S.; Werkmeister, S.; Zhou, S.; Junge, K.; Beller, M. Consecutive Intermolecular Reductive Hydroamination: Cooperative Transition-Metal and Chiral Brønsted Acid Catalysis. *Chem. Eur. J.* **2012**, *18*, 9005–9010. [CrossRef]
51. Zhou, S.; Fleischer, S.; Jiao, H.; Junge, K.; Beller, M. Cooperative Catalysis with Iron and a Chiral Brønsted Acid for Asymmetric Reductive Amination of Ketones. *Adv. Synth. Catal.* **2014**, *356*, 3451–3455. [CrossRef]
52. Hopmann, K. Iron/Brønsted Acid Catalyzed Asymmetric Hydrogenation: Mechanism and Selectivity-Determining Interactions. *Chem. Eur. J.* **2015**, *21*, 10020–10030. [CrossRef] [PubMed]
53. Fleischer, S.; Zhou, S.; Werkmeister, S.; Junge, K.; Beller, M. Cooperative Iron-Brønsted Acid Catalysis: Enantioselective Hydrogenation of Quinoxalines and 2*H*-1,4-Benzoxazines. *Chem. Eur. J.* **2013**, *19*, 4997–5003. [CrossRef] [PubMed]
54. El-Sepelgy, O.; Brzozowska, A.; Rueping, M. Asymmetric Chemoenzymatic Reductive Acylation of Ketones by a Combined Iron-Catalyzed Hydrogenation-Racemization and Enzymatic Resolution Cascade. *ChemSusChem* **2017**, *10*, 1664–1668. [CrossRef] [PubMed]
55. Gustafson, K.P.J.; Gudmundsson, A.; Lewis, K.; Bäckvall, J.-E. Chemoenzymatic Dynamic Kinetic Resolution of Secondary Alcohols Using and Air- and Moisture-Stable Iron Racemization Catalyst. *Chem. Eur. J.* **2017**, *23*, 1048–1051. [CrossRef] [PubMed]
56. Mérel, D.S.; Gaillard, S.; Ward, T.R.; Renaud, J.-L. Achiral Cyclopentadienone Iron Tricarbonyl Complexes Embedded in Streptadivin: An Access to Artificial Iron Hydrogenases and Application in Asymmetric Hydrogenation. *Catal. Lett.* **2016**, *146*, 564–569. [CrossRef]
57. Kim, M.; Lee, J.W.; Lee, J.E.; Kang, J. Synthesis of Enantiopure Ruthenium Tricarbonyl Complexes of a Bicyclic Cyclopentadienone Derivative. *Eur. J. Inorg. Chem.* **2008**, *16*, 2510–2513. [CrossRef]
58. Funk, T.W.; Mahoney, A.R.; Sponenburg, R.A.; Zimmerman, K.P.; Kim, D.K.; Harrison, E.E. Synthesis and Catalytic Activity of (3,4-Diphenylcyclopentadienone)Iron Tricarbonyl Compounds in Transfer Hydrogenations and Dehydrogenations. *Organometallics* **2018**, *37*, 1133–1140. [CrossRef]
59. Cesari, C.; Sambri, L.; Zacchini, S.; Zanotti, V.; Mazzoni, R. Microwave-Assisted Synthesis of Functionalized Shvo-Type Complexes. *Organometallics* **2014**, *33*, 2814–2819. [CrossRef]
60. Moulin, S.; Dentel, H.; Pagnoux-Ozherelyeva, A.; Gaillard, S.; Poater, A.; Cavallo, L.; Lohier, J.-F.; Renaud, J.-L. Bifunctional (Cyclopentadienone)Iron-Tricarbonyl Complexes: Synthesis, Computational Studies and Application in Reductive Amination. *Chem. Eur. J.* **2013**, *19*, 17881–17890. [CrossRef]
61. Boss, C.; Keese, R. Synthesis of Cycloalkadiynes of Various Ring Size. *Tetrahedron* **1997**, *53*, 3111–3122. [CrossRef]
62. Weaving, R.; Roulland, E.; Monneret, C.; Florent, J.-C. A rapid acces to chiral alkylidene cyclopentenone prostaglandins involving ring-closing methathesis reaction. *Tetrahedron Lett.* **2003**, *44*, 2579–2581. [CrossRef]
63. Melikyan, G.G.; Voorhees, E.; Wild, C.; Spencer, R.; Molnar, J. Carbon tether rigidity as a stereochemical tool directing intramolecular cyclizations. *Tetrahedron Lett.* **2010**, *51*, 2287–2290. [CrossRef]
64. Matsuya, Y.; Ihara, D.; Fukuchi, M.; Honma, D.; Itoh, K.; Tabuchi, A.; Nemoto, H.; Tsuda, M. Synthesis and biological evaluation of pyrethroid insecticide-derivatives as a chemical inducer for Bdnf mRNA expression in neurons. *Bioorg. Med. Chem.* **2012**, *20*, 2564–2571. [CrossRef] [PubMed]
65. Šebesta, R.; Pizzuti, M.G.; Minnaard, A.J.; Feringa, B.L. Copper-Catalyzed Enantioselective Conjugate Addition of Organometallic Reagents to Acyclic Dienones. *Adv. Synth. Catal.* **2007**, *349*, 1931–1937. [CrossRef]
66. Pan, L.; Yang, K.; Li, G.; Ge, H. Palladium-catalyzed site-selective arylation of aliphatic ketones enabled by a transient ligand. *Chem. Commun.* **2018**, *54*, 2759–2762. [CrossRef] [PubMed]
67. Richard, C.J.; Macmillan, D.; Hogarth, G. Microwave-assisted synthesis of cyclopentadienone iron tricarbonyl complexes: Molecular structures of [{η^4-C$_4$R$_2$C(O)C$_4$H$_8$}Fe(CO)$_3$] (R = Ph, 2,4-F$_2$C$_6$H$_3$, 4-MeOC$_6$H$_4$) and attempts to prepare Fe(II) hydroxycyclopentadienyl-hydride complexes. *Transit. Met. Chem.* **2018**, *43*, 421–430. [CrossRef]
68. Corey, E.J.; Venkateswarlu, A. Protection of hydroxyl groups as tert-butyldimethylsilyl derivatives. *J. Am. Chem. Soc.* **1972**, *94*, 6190–6191. [CrossRef]
69. Johnson, C.S.; Mottley, C. Theory of the temperature dependence of the NMR spectra of tunneling methyl groups. *J. Phys. C Solid State Phys.* **1976**, *9*, 2789–2795. [CrossRef]

70. Sobieski, J.W. Assessing Steric Bulk of Protecting Groups Via A Computational Determination of Exact Cone Angle (θ°) and Exact Solid Cone Angle (Θ°). Ph.D. Thesis, Kent State University Honors College, Kent, OH, USA, May 2018.
71. Xu, L.; Li, S.; Jiang, L.; Zhang, G.; Zhang, W. Electronic and steric effects of substituents in 1,3-diphenylprop-2-yn-1-one during ist reaction with $Ru_3(CO)_{12}$. *RSC Adv.* **2018**, *8*, 4354–4361. [CrossRef]
72. Stevens, P.J.; Devlin, F.J.; Chablowski, C.F.; Frisch, M.J. Ab Initio Calculation of Vibrational Absorption and Circular Dichroism Spectra Using Density Functional Force Fields. *J. Phys. Chem.* **1994**, *98*, 11623–11627. [CrossRef]
73. Becke, A.D. A new mixing of Hartree-Fock and local density-functional theories. *J. Chem. Phys.* **1993**, *98*, 5648. [CrossRef]
74. Hay, P.J.; Wadt, W.R. Ab initio effective core potentials for molecular calculations. Potentials for K to Au including the outermost core orbitals. *J. Chem. Phys.* **1985**, *82*, 299–310. [CrossRef]
75. Huzinaga, S.; Anzelm, J.; Klobukowski, M.; Radzio-Andzelm, E.; Sakai, Y.; Tatewaki, H. *Gaussian Basis Sets for Molecular Calculations*; Elsevier: Amsterdam, The Netherlands, 1984.
76. Hach, V. Meerwein-Ponndorf-Verley Reduction of Mono- and Bicyclic Ketones. Rate of Reaction. *J. Org. Chem.* **1973**, *38*, 293–299. [CrossRef]
77. Tukacs, J.M.; Fridrich, B.; Dibó, G.; Székely, E.; Mika, L.T. Direct asymmetric reduction of levulinic acid to gamma-valerolactone: Synthesis of a chiral platform molecule. *Green Chem.* **2015**, *17*, 5189–5195. [CrossRef]

© 2019 by the authors. Licensee MDPI, Basel, Switzerland. This article is an open access article distributed under the terms and conditions of the Creative Commons Attribution (CC BY) license (http://creativecommons.org/licenses/by/4.0/).

Article

Reverse Water-Gas Shift Iron Catalyst Derived from Magnetite

Chen-Yu Chou, Jason A. Loiland and Raul F. Lobo *

Center for Catalytic Science and Technology, Department of Chemical and Biomolecular Engineering, University of Delaware, Newark, DE 19716, USA; cychou@udel.edu (C.-Y.C.); jasonloiland@gmail.com (J.A.L.)
* Correspondence: lobo@udel.edu; Tel.: +1-302-831-1261

Received: 19 August 2019; Accepted: 30 August 2019; Published: 14 September 2019

Abstract: The catalytic properties of unsupported iron oxides, specifically magnetite (Fe_3O_4), were investigated for the reverse water-gas shift (RWGS) reaction at temperatures between 723 K and 773 K and atmospheric pressure. This catalyst exhibited a fast catalytic CO formation rate (35.1 mmol h^{-1} $g_{cat.}^{-1}$), high turnover frequency (0.180 s^{-1}), high CO selectivity (>99%), and high stability (753 K, 45000 $cm^3 h^{-1} g_{cat.}^{-1}$) under a 1:1 H_2 to CO_2 ratio. Reaction rates over the Fe_3O_4 catalyst displayed a strong dependence on H_2 partial pressure (reaction order of ~0.8) and a weaker dependence on CO_2 partial pressure (reaction order of 0.33) under an equimolar flow of both reactants. X-ray powder diffraction patterns and XPS spectra reveal that the bulk composition and structure of the post-reaction catalyst was formed mostly of metallic Fe and Fe_3C, while the surface contained Fe^{2+}, Fe^{3+}, metallic Fe and Fe_3C. Catalyst tests on pure Fe_3C (iron carbide) suggest that Fe_3C is not an effective catalyst for this reaction at the conditions investigated. Gas-switching experiments (CO_2 or H_2) indicated that a redox mechanism is the predominant reaction pathway.

Keywords: RWGS; iron oxides; CO_2 conversion; gas-switching

1. Introduction

Today's anthropogenic emissions of carbon dioxide to the atmosphere amount to about 35,000 Tg per year, and the greenhouse effect of these accumulated emissions has been recognized as an alarming hazard to the well-being of modern societies. Although multiple approaches have been considered to mitigate these emissions, recently more emphasis has been placed on the potential synergy of carbon capture and the utilization of these large outflows of carbon dioxide. Among several possibilities, a recent National Academies of Science and Engineering report [1] on CO_2 waste gas utilization highlights the conversion of CO_2 by hydrogenation into CO and water—the reverse water-gas shift (RWGS) reaction, Equation (1)— as critical, and points to the need for improved catalysts with high stability and durability.

The RWGS reaction can be part of a two-step hydrogenation process for the conversion of CO_2 to valuable products. First, CO_2 is reduced to CO via the RWGS reaction, and second, CO can be converted to either hydrocarbons via the Fischer-Tropsch (FT) process or methanol via CAMERE (CArbon dioxide hydrogenation to form MEthanol via a Reverse WGS reaction) process [2]. The RWGS reaction is an endothermic reaction ($\Delta H°_{298\,K}$ = 41.2 kJ mol^{-1}), and thus is thermodynamically favorable at higher temperatures.

$$CO_2 + H_2 \leftrightarrow CO + H_2O \tag{1}$$

Noble metals such as Pt [3–5] and Pd [6,7], and other various metals such as Cu [8–11], Ni [12,13] and Fe [14] supported on oxides were reported to be active for the production of CO. Among them, Cu-based materials have been widely studied, and thus have also been investigated in many instances for the RWGS reaction. For example, Cu-Ni/Al_2O_3 [15], Cu/ZnO [16], Cu-Zn/Al_2O_3 [16] and Cu/SiO_2

promoted with potassium [17] have all shown good RWGS activity. However, Cu materials tend to deactivate by sintering (frittage) at high temperatures (T > 773 K), which are required for high RWGS activity. For high temperature applications, iron can be added as a thermal stabilizer: Chen et al. [10] showed that adding small amounts of iron to 10% Cu/SiO$_2$ resulted in stable RWGS activity for 120 h at 873 K and atmospheric pressure, while non-promoted 10% Cu/SiO$_2$ deactivated rapidly.

Iron oxides (Fe$_x$O$_y$) are often used industrially for FT synthesis (473 K–623 K, 1 MPa) [18,19] and the high-temperature (623 K–723 K) WGS reaction [20–22]. In FT synthesis, alkalized, iron-based catalysts are combined with Cu for reduction promotion. Schulz et al. [18,23] showed that the working FT catalysts contain several iron carbide phases and elemental carbon formed after the hydrogen reduction period. Iron oxides and the metallic iron are still present in the catalyst composition, but iron carbide phases are identified as active sites [23]. The RWGS and WGS reactions are often carried out in conjunction with FT synthesis at a higher temperature regime on iron catalysts, and iron oxide or oxidic amorphous iron phases are known as the active phases for WGS and RWGS [24–26].

Extensive research on iron-based catalysts has been reported mainly on the WGS reaction over decades [20]. Chromium is a structural promoter that helps prevent the iron from sintering at high temperature. A more recent survey on Cr-free, Fe-based WGS catalysts shows the current strong interest in this topic [27]. However, the studies on RWGS reactions over iron-based catalysts are much less frequent. Fishman et al. [28] synthesized hematite nano-sheets to obtain a 28% CO$_2$ conversion at 783 K, and hematite nanowires to obtain a 50% CO$_2$ conversion at a very high temperature of 1023 K. Hematite was reduced to magnetite during the reaction. The catalytic behavior over time on stream and the stability of the CO production were, however, not investigated. Fe nanoparticles have also shown good stability and activity (35% CO$_2$ conversion, >85% CO selectivity) in RWGS by Kim et al. [29], yet no kinetic parameters were determined, and the mechanism was not discussed.

Two principal mechanisms of the WGS (or RWGS) reaction have been investigated extensively: The "redox mechanism" and the "associative" mechanism [30,31]. Different catalysts may lead to a different reaction pathway. The redox mechanism was suggested to be active for the WGS reaction over iron catalysts promoted with chromium [32]. A distinguishing feature of the redox mechanism is that products can be generated in the absence of both reactants. The catalyst is first reduced by the adsorbed H$_2$, and is subsequently oxidized by CO$_2$ (in RWGS) or H$_2$O (in WGS). The associative mechanism was proposed to be dominant in the WGS reaction over iron oxide catalysts [33]. In this mechanism, both reactants are adsorbed on the catalyst surface at the same time to create products. Several carbon-containing intermediates, including formate, carbonate, carbonyl and carboxyl species, have been proposed. In a previous report in alumina-supported iron catalysts [14], we showed that the redox mechanism is the only pathway for RWGS over Fe/γ-Al$_2$O$_3$, and the predominant pathway over Fe-K/γ-Al$_2$O$_3$. The addition of the potassium promoter activates a secondary pathway for CO formation, which is probably the associative pathway.

In the present report, unsupported Fe$_3$O$_4$-derived catalyst is identified as a highly active, selective and stable catalyst for the reverse water-gas shift reaction at temperatures between 723 K and 753 K. The characterization of surface composition, bulk properties, and the evaluation of the CO production specific rate showed that the working catalyst is constructed during the H$_2$-activation and the period of reaction conditions. Quantitative gas-switching experiments in combination with isotopic switching experiments allowed the redox and associative reaction pathway to be differentiated. The catalysts appear to be highly stable under the reaction conditions investigated.

2. Results and Discussion

The catalytic CO formation rates on the Fe$_3$O$_4$ catalyst with various H$_2$ to CO$_2$ ratios (Figure 1) show that after an induction period of ~120 min, the catalyst produced CO at a steady rate of 35.1 mmol h^{-1} g$_{cat.}$$^{-1}$ at 753 K with 12.5% CO$_2$ conversion. The selectivity to CO was greater than 99% under equimolar CO$_2$ and H$_2$. After 950 min, the partial pressure of CO$_2$ was raised to 60 kPa, while the partial pressure of H$_2$ was kept constant. The rate increased to 54.6 mmol h^{-1} g$_{cat.}$$^{-1}$ with

4.4% CO_2 conversion. Deactivation also occurred during this period: Starting at a deactivation rate of 3.71 mmol h^{-1} $g_{cat.}^{-1}$ per h, this rate gradually decreased to 0.23 mmol h^{-1} $g_{cat.}^{-1}$ per h. When the partial pressure of H_2 was 60 kPa, and the CO_2 was switched back to 15 kPa, the CO formation rate increased first to 91.3 mmol h^{-1} $g_{cat.}^{-1}$ and gradually stabilized to a value of 95.3 mmol h^{-1} $g_{cat.}^{-1}$ with 33.7% CO_2 conversion. The final rate of the catalyst reactivation was about 0.50 mmol h^{-1} $g_{cat.}^{-1}$ per h. It should be noted that the differential condition (see Equation (8)) was used to determine reaction rates, and preferred for the investigation of the kinetic properties of our materials. Under higher concentration of H_2 (15 kPa CO_2 + 60 kPa H_2), high CO_2 conversion (>12%) could lead to small errors in the estimation of the reaction rate and the reaction order. The trend observed in Figure 1, however, should not be affected by this approximation.

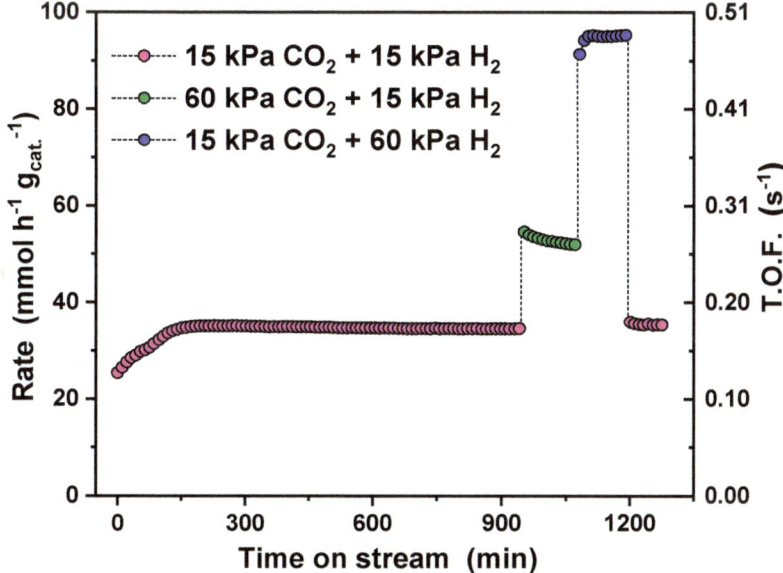

Figure 1. CO formation rates and their turnover frequencies (T.O.F.) on Fe_3O_4 at partial pressures of CO_2 and H_2 indicated in the legend. Other reaction conditions: P_{tot} = 1 bar, T = 753 K, $F_{tot.}$ = 75 sccm, GHSV = 4.5×10^4 cm^3 h^{-1} $g_{cat.}^{-1}$.

With 60 kPa of H_2 and 15 kPa of CO_2 in the feed, a small amount of CH_4—the only side-product—was produced at the rate of 1.35 mmol h^{-1} $g_{cat.}^{-1}$, reducing the CO selectivity from near 100% to 98.6%. Methane production implies that C–H bond formation is facilitated at higher partial pressure of H_2. There was no further C–C chain growth under this reaction condition, indicating that the FT synthesis was not active over the working catalyst. The effect of H_2 partial pressure on the CO formation rate was much higher than the effect of CO_2, implying a higher reaction order on H_2 than CO_2. The catalyst showed overall high stability in 1300 min, and the final reactivation rate in excess H_2 was higher than the deactivation rate under the excess CO_2 condition. This indicates that this Fe_3O_4-derived catalyst is easy to regenerate in a very short period (<100 min), making the catalyst attractive in industrial use for long-term application.

The activation of the catalyst was carried out by a pretreatment under reducing conditions (H_2 gas). The bulk structure of the catalyst after the pretreatment can be identified. Based on the body-centered cubic (BCC) structure of α-Fe (JCPDS PDF 00-006-0696) after the pretreatment (Figure S1), the surface density of Fe atoms on the Fe(110) surface can be calculated as 1.297×10^{19} Fe atoms m^{-2}. Assuming all Fe atoms on Fe(110) were active sites, the observed CO formation rates can be converted to turnover frequency (TOF). This is reported in Figure 1 based on the measured CO formation rates, atomic surface density, and BET surface area (2.52 m^2/g for Fe_3O_4) of the pristine catalyst. The turnover frequency of this catalyst under the equimolar condition was as high as 0.18 s^{-1} (P_{tot} = 1 bar, T = 753 K, $F_{tot.}$ = 75 sccm, GHSV = 4.5 × 10^4 cm^3 h^{-1} g$_{cat.}$$^{-1}$).

The stable reaction rates observed after the initial break-in period at 753 K allow for the determination of kinetic parameters without having to model deactivation profiles. The kinetic parameters, including reaction orders with respect to CO_2 and H_2, and measured activation energies (E_{meas}) over Fe_3O_4 under near equimolar CO_2 and H_2 (~1:1), and in H_2 excess (2:1, 4:1, and 9:1)—see Table 1— indicate that CO formation rates have a higher dependence on H_2 partial pressure (order of ~0.8) than CO_2 partial pressure (order of ~0.33) under equimolar composition. In general, rate orders depend on reaction conditions: Increasing the H_2 partial pressure increases the order on CO_2 to 0.39 and decreases the rate order on H_2 to 0.72. At a ratio of H_2:CO_2 near 4:1, the reaction orders still show the same trend: An increasing dependence on CO_2 (order of 0.43) and decreasing dependence on H_2 (order of 0.31). In a high excess of H_2 (H_2:CO_2 = ~9:1), the reaction rate over Fe_3O_4 was of the order 1.30 with respect to CO_2, and was independent of H_2 pressure. The activation energies (E_{meas}) also depend on the H_2:CO_2 partial pressure ratios; that is, different reaction pathways may occur under these conditions. This behavior is not unique to iron catalysts. Similar reaction orders were also observed by Ginés et al. [34] in the same regime of P_{H2}/P_{CO2} < 3 (CO_2 order ≈ 0.3, H_2 order ≈ 0.8) on the $CuO/ZnO/Al_2O_3$ catalyst, and by Kim et al. [3] on Pt/Al_2O_3 catalysts (CO_2 order = 0.32, H_2 order = 0.70). It was also suggested by Ginés et al. [34] that different reaction pathways should be existed for P_{H2}/P_{CO2} < 3 and P_{H2}/P_{CO2} > 3 regions.

Table 1. Measured reaction orders with respect to CO_2 and H_2, and measured activation energies (E_{meas}) over Fe_3O_4. Reaction conditions: 100 mg Fe_3O_4, F_{tot} = 75 sccm, T = 723 K.

P_{H2} (kPa)	P_{CO2} (kPa)	P_{H2}: P_{CO2}	Reaction Order in CO_2	Reaction Order in H_2	E_{meas} (kJ/mol)
15	10–20	~1:1	0.33	-	28.9 ± 0.9
10–20	15		-	0.79	
30	10–20	~2:1	0.39	-	27.1 ± 0.5
25–35	15		-	0.72	
40	5–12.5	~4:1	0.43	-	34.2 ± 1.9
35–45	10		-	0.31	
85	5–12.5	~9:1	1.30	-	39.0 ± 3.4
70–90	10		-	0	

Figure 2 presents the temperature-programmed reduction (TPR) profiles of (a) the fresh Fe_3O_4 sample and (b) the post-reaction Fe_3O_4 sample. In Figure 2a, the fresh Fe_3O_4 was heated to 753 K in a hydrogen atmosphere, kept for 2 h at these conditions, and then heated to 1,073 K at the rate of 5 K/min. The small peak observed in the TPR trace at about 563 K is assigned to an impurity of hematite present in the initial sample of magnetite (Fe_3O_4), but not detected in the XRD pattern, as shown in the report by Jozwiak et al. [35]. The following broader and asymmetric peak suggests a two-step reduction process that has been previously postulated in literature [36], as the following: (1) $Fe_3O_4 \xrightarrow{H_2} FeO$ and (2) $FeO \xrightarrow{H_2} Fe^0$. These two steps can be deconvoluted into two peaks located at ~ 688 K and 773 K in the TPR traces. After the 2 h reduction period at 773 K, there was no further H_2 consumption at higher temperatures. That is, the sample, after the reduction pretreatment used in our activation protocol, has been converted into metallic iron. This result is also consistent with the XRD pattern in Figure S1, which shows that α-Fe was the crystal formed after the reduction pretreatment of the Fe_3O_4 sample in the microreactor.

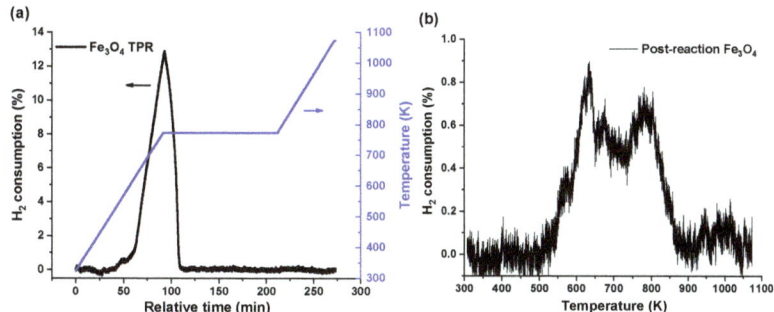

Figure 2. (a) H_2 consumption (%) of the reduction period during the pretreatment following the temperature-programmed reduction (TPR) of 100 mg Fe_3O_4. (b) TPR curve of the post-reaction Fe_3O_4 sample.

As soon as CO_2 was fed into the reactor, the surface of the catalyst was partially oxidized. This is known from the results of Figure 2b that illustrates the presence of two significant peaks in the TPR profile for the post-reaction catalysts (at 630 K and 780 K, respectively). The location of these two predominant peaks shows good agreement with the results of Figure 2a and results reported elsewhere [35,37]. There is at least a two-step reduction at ~630 K and 780 K, implying the coexistence of different oxidation states of iron on the post-reaction sample due to the partial oxidation from CO_2. For this post-reaction sample, the H_2 concentration in the effluent stream decreased by only a small amount (less than 1%), suggesting that the consumption of the H_2 feed would not affect the reduction rate during the TPR reaction.

Figure 3 displays the diffraction pattern of fresh Fe_3O_4 and the catalyst after the RWGS reaction (post-reaction Fe_3O_4). The 2θ degree peak positions in fresh Fe_3O_4 were 30.15°, 35.45°, 37.15°, 43.15°, 53.50°, 56.95° and 62.55°, which are all consistent with magnetite (JCPDS PDF 01-071-6336). The post-reaction Fe_3O_4 shows a very different XRD pattern: This pattern was composed of metallic iron (α-Fe, 44.67°, shown in the inset of Figure 3), iron carbide (Fe_3C), and a small peak of FeO_X (35.47°).

Iron oxides can be converted directly into carbides in a reducing and carburizing atmosphere [38], and the carbon source of the Fe_3C production can be either from impurities in the fresh Fe_3O_4 sample or due to reaction with the product CO, after the RWGS reaction as indicated by Equation (2) [39]:

$$3Fe + 2CO \leftrightarrow Fe_3C + CO_2 \qquad (2)$$

The bulk composition of the catalyst after the reaction is also consistent with the TPR results in Figure 2a,b. During TPR, the amount of H_2 consumption of magnetite relative to the amount of H_2 consumption after the reaction was 12:1, therefore the iron species in the catalyst has been changed into a more reduced chemical state after the pretreatment and RWGS reaction. The reduction was mainly caused by the pretreatment, while the following CO_2/H_2 reaction shifted the metallic iron back to a slightly more oxidized state; a combination of iron carbide, metallic iron and some iron oxides.

Besides bulk property information obtained from XRD, XPS analyses were conducted to characterize the surface composition of the initial fresh Fe_3O_4 and the change of the catalyst after the RWGS reaction. Figure 4 shows the XPS spectra, peak deconvolutions and the fitting envelopes for the Fe $2p_{3/2}$ spectra of Fe_3O_4 and post-reaction Fe_3O_4.

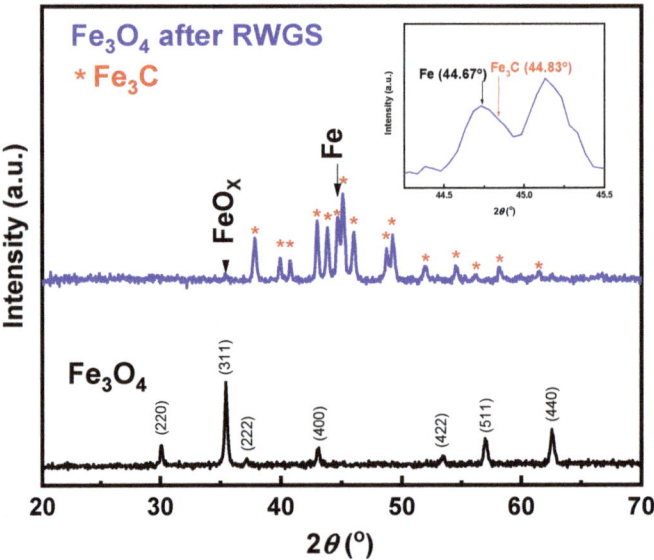

Figure 3. XRD patterns of fresh Fe_3O_4 (down) and post-reaction Fe_3O_4 (up); inset is the magnification of post-reaction Fe_3O_4 from 44° to 45.5°.

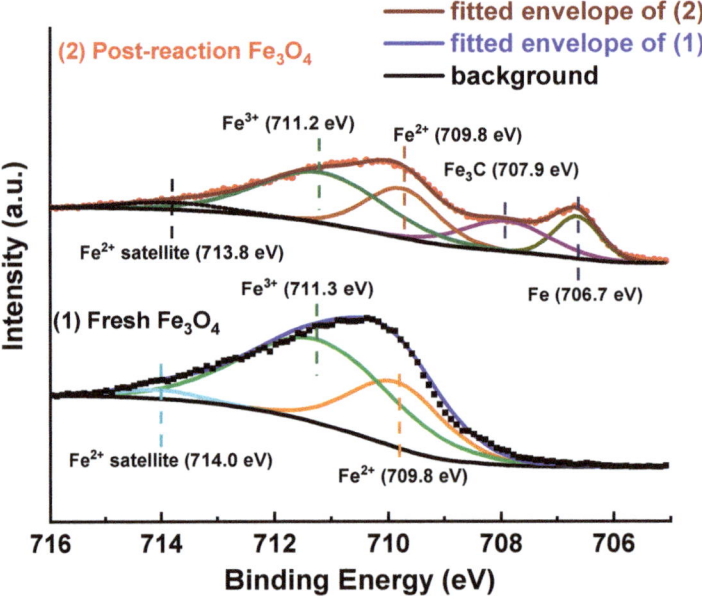

Figure 4. XPS Fe $2p_{3/2}$ spectra of Fe_3O_4 and post-reaction Fe_3O_4. The curves under the fitted envelope and above the background are contributions of estimated components from peak fitting.

Atomic percent contributions are calculated from the fitted peaks of Fe $2p_{3/2}$ due to its larger intensity (Area of Fe $2p_{3/2}$:Fe $2p_{1/2}$ = 2:1). The Fe $2p_{3/2}$ spectra were fitted over the range of 705–722 eV. The spectra between 716–722 eV were not shown in the figure for clarity. In this range, there were only small Fe^{3+} satellite peaks located at 719.2 eV and 719.4 eV for both the fresh and post-reaction Fe_3O_4 samples, respectively, although the area of the satellite peaks was still included in the corresponding components when calculating the relative atomic percentage. The fits, including the binding energy, full width at half maximum (FWHM), and the relative iron composition, are summarized in Table 2. The fitted XPS spectrum of fresh Fe_3O_4 was composed of doublets for Fe^{2+} at 709.8 eV and Fe^{3+} at 711.3 eV. Fe_3O_4 has an inverse spinel structure which can be written as $Fe^{3+}_{TET}[Fe^{2+}Fe^{3+}]_{OCT}O_4$, with one Fe^{3+} on a tetrahedral site, and Fe^{2+} and the other Fe^{3+} distributed on octahedral sites. Therefore, the theoretical relative composition of Fe^{2+}/Fe^{3+} is 0.5, which is close to the area fitted and the calculated relative composition in our fresh Fe_3O_4 ($Fe^{2+}:Fe^{3+}$ = 34.8/65.2 = 0.53). The Fe^{3+} peak has a larger FWHM than Fe^{2+}. This is as expected because the electronic configuration of Fe^{2+} is $3d^6$, while that of Fe^{3+} is $3d^5$, that is, Fe^{2+} will have a longer life time compared to Fe^{3+}; and therefore the FWHM of the Fe^{2+} peak should be smaller than the Fe^{3+} peak [40]. Additionally, the Fe^{3+} peaks can be attributed to two different structures, octahedral Fe^{3+} and tetrahedral Fe^{3+}, a factor that will also lead to broader Fe^{3+} peaks.

Table 2. XPS peak fittings for Fe $2p_{3/2}$ spectra for Fe_3O_4 and post-reaction Fe_3O_4.

Sample	Peak Deconvolution	Binding Energy (eV)	FWHM (eV)	Atomic %
Fe_3O_4	Fe^{3+}	711.3	3.08	65.2
	Fe^{2+}	709.8	1.95	34.8
post-reaction Fe_3O_4	Fe^{3+}	711.2	2.58	43.9
	Fe^{2+}	709.8	1.51	27.5
	Fe_3C	707.9	1.78	16.4
	Metallic Fe	706.7	0.94	12.2

After the RWGS reaction, the spectrum was fitted using four different components corresponding to metallic Fe (706.7 eV), Fe_3C (707.9 eV), Fe^{2+} (709.8 eV) and Fe^{3+} (711.2 eV). The peak locations of Fe^{2+} and Fe^{3+} were the same or very close to the fresh sample, indicating that there was only a small surface charging effect with the flood gun on. The binding energies of the components are in agreement with literature results [38,40–42]. In terms of the atomic percentages, the overall peak area was re-allocated to a more reduced regime after the RWGS reaction. The Fe^{2+} decreased from 34.8% to 27.5%, the Fe^{3+} decreased from 65.2% to 43.9%, while there were two components formed: Fe (12.2%) and Fe_3C (16.4%). The shift of the spectra was due to the H_2 reduction pretreatment before operating the RWGS reaction, and the flow of H_2/CO_2 reactants through the system would balance each other to make the catalyst partially oxidized or reduced. Though the sample surface could be oxidized by the air during the transportation from the reactor to the XPS analysis chamber, the result from XPS still can confirm the reduction of the surface during the reaction, since new crystal structures such as Fe and Fe_3C are detected by XRD.

The XPS analyses indicate that the active catalyst consisted of a mixture of metallic Fe, Fe_3C, Fe^{2+} and Fe^{3+}; however, it cannot establish the relative contributions of these components to the observed rate of the RWGS reaction. To determine if the iron carbide formed in our reaction can catalyze the RWGS reaction, reaction rates over pure Fe_3C were measured (Figure 5). In Figure 5a, Fe_3C showed an initial CO formation rate of 26.0 mmol h^{-1} g^{-1}, but dropped 48% to near 13.5 mmol h^{-1} g^{-1} in 10 min, and then further down to 5.7 mmol h^{-1} g^{-1} in 160 min at 753 K. This is clearly different from the properties of the Fe_3O_4-derived catalyst in Figure 1, since the Fe_3O_4-derived catalyst displayed high stability for at least 1300 min. To evaluate the fast deactivation shown in Figure 5a and remove the initial reduction effect during the ramping by hydrogen, another measurement was carried out with respect to temperature, (see Figure 5b). This experiment was conducted flowing a CO_2/H_2/He gas mixture in the same relative concentration used in the standard activity test during the ramping procedure from room temperature to 773 K. At 573 K, the material did not catalyze the formation of CO; however, when the temperature was increased to 623 K, the catalyst immediately showed catalytic rates in the range of 2.60 mmol h^{-1} g^{-1} to 2.99 mmol h^{-1} g^{-1}. After 1 h of the reaction, the rate remained at quasi-steady state at this low temperature (623 K).

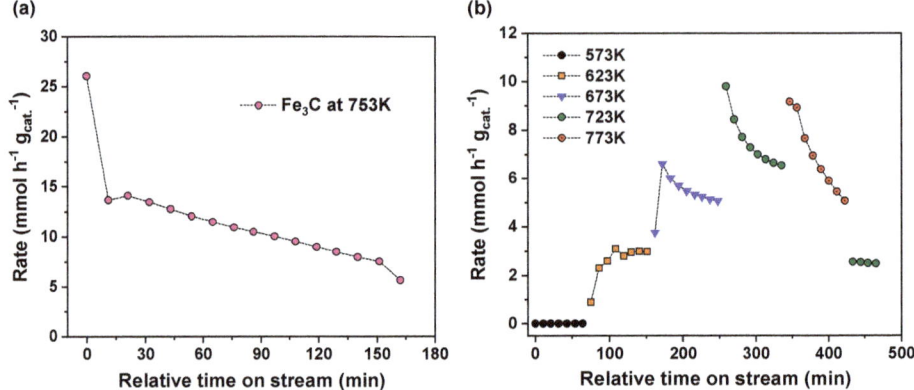

Figure 5. (a) CO formation rate on 100 mg Fe$_3$C. Reaction conditions: F$_{tot}$ = 75 sccm; the reactor was ramped to 753 K with P$_{H2}$ = 15 kPa and He as remainder. No further reduction was applied after the ramping. During the reaction, T = 753 K, P$_{tot}$ = 1 bar, P$_{H2}$ = P$_{CO2}$ = 15 kPa. (b) CO formation rate on 100 mg Fe$_3$C. The temperature was directly ramped from room temperature to 773 K as shown in the figure with P$_{tot}$ = 1 bar, P$_{H2}$ = P$_{CO2}$ = 15 kPa, and the remainder is He.

The CO formation rates were 6.60 mmol h^{-1} g^{-1}, 9.80 mmol h^{-1} g^{-1}, and 9.16 mmol h^{-1} g^{-1}, as the temperature was increased to 673 K, 723 K and 773 K, respectively. The formation rate at 773 K was not higher than that at the lower temperature because of rapid deactivation at this temperature. Faster deactivation rates were observed at higher temperatures: The average deactivation rates were 1.36 mmol h^{-1} g^{-1} per h, 2.57 mmol h^{-1} g^{-1} per h and 3.50 mmol h^{-1} g^{-1} per h.

After reacting at 773 K, the temperature was reduced to 723 K to monitor the reaction rate and compare to the previous value. Much lower rates (2.55 mmol h^{-1} g^{-1}) were observed than in the previous measurement at the same temperature (723 K), indicating that there was an irreversible change in the catalyst structure or composition. At higher temperatures, the reverse reaction of Equation (2), where Fe$_3$C reacts with CO$_2$ and forms Fe and CO, is more favorable than the Fe$_3$C formation [39]. Therefore, the initial CO formation rate was probably due to the formation of CO from decomposition, but quickly dropped since it is harder to convert the metallic iron back to Fe$_3$C at higher temperatures. The iron carbide catalyst only showed steady CO production at 623 K. This explains why the deactivation at 723 K over Fe$_3$C is very different from the steady-state magnetite catalyst reported in Figure 1, which showed a very stable CO formation rate at 753 K. This observation suggests that the operating temperature of RWGS in this study was not an environment conducive for a stable iron carbide for CO production. In addition, the iron carbide (Fe$_5$C$_2$ or Fe$_3$C) is normally considered to be the active phase of iron for hydrocarbon production [43,44], and iron oxide is the active phase for WGS and RWGS [25]. Several reports have suggested that the stability of the iron catalyst in either FT synthesis [45] or RWGS [29] can be related to an iron carbide layer. Davis [45] suggested that catalyst composition and reaction condition will define the existence of the pseudo-equilibrium layer of iron carbide to ensure a very slow deactivation condition. Kim et al. [29] concluded that the stability of the catalyst could have originated from migration of C and O into the catalyst bulk, forming iron oxide and iron carbide, which likely prevented the nanoparticles on the surface from agglomerating. Based on the XPS results and Fe$_3$C catalytic tests, the iron carbide of the working catalyst is less likely to be the main active site for CO production, but is an important species to provide stability in the overall catalytic performance.

Gas-switching experiments, in which H_2 and CO_2 are flown on and off, were used to distinguish and quantify contributions from redox and associative reaction pathways [3,14,46]. In the simplest form of the redox mechanism, gas-phase CO_2 adsorbs on a reduced site to form CO and an oxidized site (Equation (3)), which can then be reduced by gas phase H_2 to reform the reduced site (Equation (4)). The simplest redox cycle can be described as follows:

$$CO_2(g) + s_{red.} \rightarrow CO(g) + O \cdot s \quad (3)$$

$$H_2 + O \cdot s \rightarrow H_2O(g) + s_{red.} \quad (4)$$

A simplified associative pathway can be described generally by Equation (5). CO_2 and H_2 adsorb on the catalyst surface to form a carbon-containing intermediate (i.e. formate, carbonate, or bicarbonate), which then decomposes in the presence of H_2 to form CO and H_2O.

$$CO_2(g) + H_2(g) \rightarrow COOH \cdot s + H \cdot s \rightarrow CO \cdot s + H_2O \cdot s \quad (5)$$

CO and H_2O were the main products formed during gas-switching experiment (Figure 6). In the first three cycles, CO was formed when switching from H_2 to CO_2, and a very small amount of CO was formed when switching from CO_2 to H_2 at 30 min, 70 min and 110 min, respectively. When the catalyst was purged 20 min with helium before switching from CO_2 to H_2, CO was not formed, and H_2O was produced at 195 min. Water was formed when switching from H_2 to CO_2 and when switching from CO_2 to H_2. After flowing H_2 and purging the reactor with He for 20 min, only a negligible amount of H_2O was formed upon the admission of CO_2 (at 150 min).

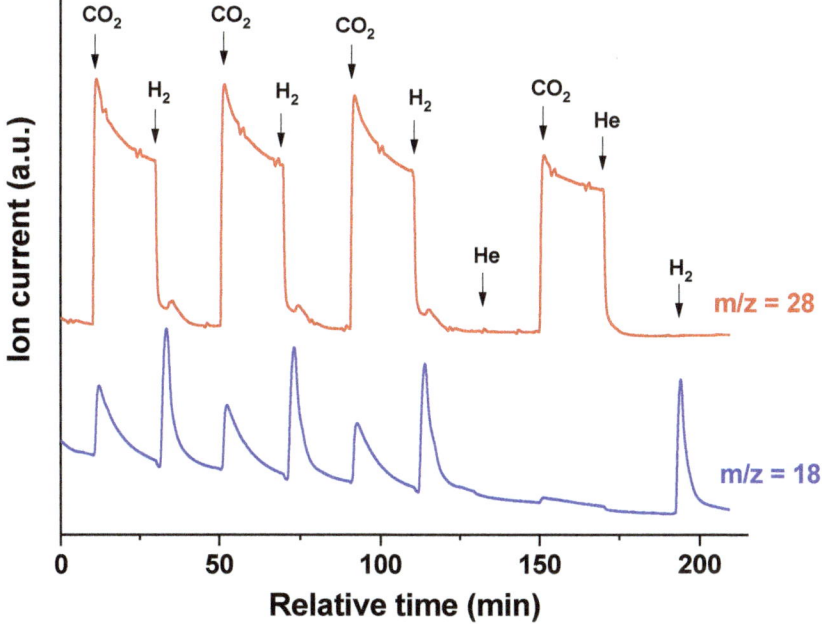

Figure 6. Ion current at m/z = 18 (H_2O) and 28 (CO) during H_2/CO_2 switching experiments on Fe_3O_4. Arrows with a label indicate a change in gas composition to the indicated gas. The catalyst first was reduced in flowing H_2 for 2 h, following the reaction in H_2 and CO_2 for 2 h, and H_2/He in 20 min before the first admission of CO_2. Reaction conditions: T = 773 K, F_{He} = 36 sccm, F_{H2} or F_{CO2} = 4 sccm.

In Figure 6, the fact that CO was formed when the reduced form of the Fe_3O_4 catalyst was contacted with CO_2, (even after the purge with He to decrease the concentration of any surface H_2), is evidence of a redox pathway. During the first 125 min of gas-switching experiments, H_2O was produced during flows of only CO_2 or only H_2. This differs from what is expected in the traditional redox cycle, in which H_2O is only produced during the H_2 feeding period (see Equation (4)). However, after flowing H_2 and purging the reactor with He for 20 min, the admission of CO_2 only produced negligible amounts of water. Table 3 summarizes the estimated initial rates of CO production on the Fe_3O_4 catalyst during each segment of the gas-switching experiments. The CO production rates were calculated from the initial slopes of the concentration vs. time data in Figure 6, essentially modeling the system as a batch reactor (Equation (6)).

$$\frac{dC_A}{dt} = r_A \qquad (6)$$

Table 3. Estimated initial rates of CO production after gas switches from H_2 to CO_2 and from CO_2 to H_2 during gas-switching experiment on Fe_3O_4 in Figure 6.

Period	Rate after H_2 to CO_2 Gas Switch (μmol L^{-1} s^{-1} $g_{cat.}^{-1}$)	Rate after CO_2 to H_2 Gas Switch (μmol L^{-1} s^{-1} $g_{cat.}^{-1}$)	(H_2 to CO_2 Rate)/(CO_2 to H_2 Rate) Ratio
1st CO_2	2.78	1.64	1.70
2nd CO_2	2.98	1.00	2.98
3rd CO_2	2.74	0.91	3.01
4th CO_2 (after He purge)	1.94	0	-

It is observed (Table 3) that the rate after switch from H_2 to CO_2 fluctuated between 2.74 μmol L^{-1} s^{-1} $g_{cat.}^{-1}$ and 2.98 μmol L^{-1} s^{-1} $g_{cat.}^{-1}$ in the first three periods of CO_2 admission, and it decreased after the He purge. The rate after switching from CO_2 to H_2 decreased in the first three periods, and it was zero (with no CO produced) during the last admission of CO_2 after the He purge. The decrease of the CO initial rate after switching from CO_2 to H_2, especially when equal to zero after the purge, raises doubts about the existence of residual CO_2 during the first three admissions of H_2 in the switching experiment. As a control, when the gas was switched from CO_2 to H_2, the CO_2 gas did not exit from the surface very quickly (see Figure S2). Therefore, a small amount of CO can be produced by the residual CO_2 with the available reduced sites; evidence of this interpretation in the detection of very small peaks after the H_2 admissions (Figure 6). The negligible CO production (relative time = 34 min, 74 min, and 114 min, respectively) after the H_2 admissions should not be considered evidence of the associative mechanism. In summary, the CO formation upon switching from H_2 to CO_2 is evidence consistent with the redox mechanism, while the small contribution of CO production upon switching from CO_2 to H_2 was suppressed by the confirmation of the helium purge. Thus, from the gas-switching experiment, only the redox pathway is active on our Fe_3O_4-derived catalyst.

Isotopic experiments were conducted to gain insight into the mechanism of the reaction. The isotopic $C^{18}O_2$ to the CO_2 switching experiment is shown in Figure 7. Here it can be seen that $C^{18}O$ (m/z = 30) formed and CO (m/z = 28) decreased when the gas (CO_2/H_2) was switched to $C^{18}O_2/H_2$. CO can be only formed from CO_2, and not from the lattice oxygen.

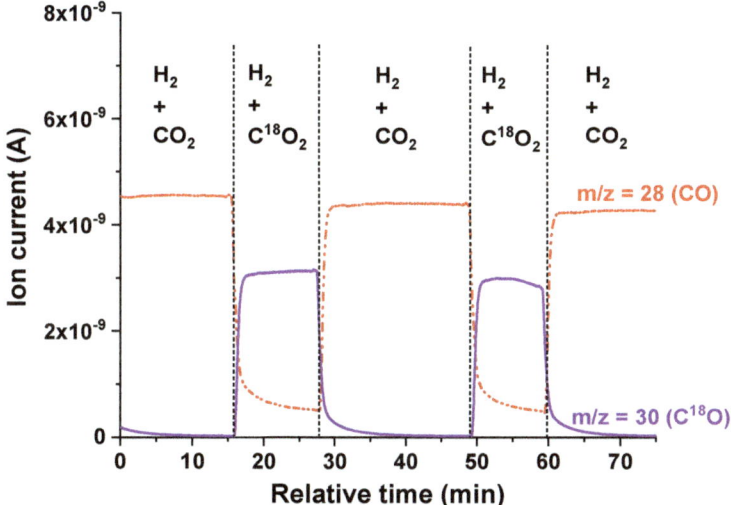

Figure 7. Ion current at m/z = 28 (CO) and 30 ($C^{18}O$) during $CO_2/C^{18}O_2$ switching experiments on Fe_3O_4 (100 mg). Labels in each region indicate a change in gas composition. A standard pretreatment (reduced in flowing H_2 for 2 h) and RWGS reaction (>2 h) had been done before the switching experiment. Switching experiment conditions: T = 753 K, F_{tot} = 40 sccm, F_{H2} = 4 sccm, F_{CO2} or F_{C18O2} = 4 sccm.

The gas-switching experiments with CO_2 and H_2 led us to conclude that a redox pathway is active on the Fe_3O_4-derived catalyst. A model that can present a redox reaction pathway for this catalyst is given in Scheme 1. It includes the adsorption of both of the reactants, CO_2 and H_2. In the surface redox mechanism, the dissociation of CO_2 at the catalyst surface (step 2 in Scheme 1) is known to be the RDS [8,30,34]. Evidence for H_2 dissociation (step 5 in Scheme 1) was observed when H_2/D_2 mixtures were fed to the catalyst in the presence of CO_2 (see Figure S3).

$$1)\ CO_{2\,(g)} + s \rightleftharpoons CO_2 \cdot s$$

$$2)\ CO_2 \cdot s + s \longrightarrow CO \cdot s + O \cdot s$$

$$3)\ CO \cdot s \rightleftharpoons CO_{(g)} + s$$

$$4)\ H_{2\,(g)} + s \rightleftharpoons H_2 \cdot s$$

$$5)\ H_2 \cdot s + s \rightleftharpoons 2\,H \cdot s$$

$$6)\ 2\,H \cdot s + O \cdot s \rightleftharpoons H_2O \cdot s$$

$$7)\ H_2O \cdot s \rightleftharpoons H_2O_{(g)} + s$$

Scheme 1. Redox reaction pathway for CO formation.

HD formation was observed to occur quickly, since the amount of CO_2 to CO conversion decreased on the same time scale when switching the concentration from H_2/D_2 (7.5 kPa/7.5 kPa) to H_2 (7.5 kPa), indicating that H_2 dissociation is reversible and not rate limiting.

An additional gas switching experiment was conducted to monitor the H_2O production (Figure S4). The amount of H_2O produced during H_2 flow periods was consistent between each cycle (Figure S4a), that is, the adsorbed O·s species formed upon CO_2 reduction are stable at these reaction conditions. However, the amount of H_2O produced during the period of CO_2 flow decreased as the purge time in helium increased. After only a five min purge, the amount of H_2O produced during the period of CO_2 flow was much greater than that produced following a 20 min purge in helium, and the rate fitted from the initial slope of this region dropped dramatically (see Table S1). This suggests that H* atoms from the catalyst surface appeared to desorb (as H_2) during the He purge. This was a slow process, because even following a 20 min purge, there were enough H* atoms on the sample to form small amounts of H_2O when CO_2 was administered. Both gas-switching experiments (Figure 6 and Figure S4) suggest that the redox mechanism should be the dominant reaction pathway for this Fe_3O_4-derived catalyst during the CO_2 hydrogenation at our reaction conditions (753 K, 1 atm).

3. Materials and Methods

3.1. Materials

Magnetite (Fe_3O_4, 99.99%, 50–100 nm particle size, Sigma-Aldrich, St. Louis, MO, USA) powder was pressed by hydraulic press and sieved (mesh 40 to mesh 60) to obtain particle sizes within the range of 250–425 μm before the reaction. Iron carbide (Fe_3C, 99.5%, American Element, Los Angeles, CA, USA) granules were crushed to a powder using a ball-mill, pressed using a hydraulic press, and sieved within the range of 250–425 μm. The gases used were CO_2 (Grade 5.0, Keen, Wilmington, DE, USA), H_2 (UHP, Matheson, Basking Ridge, NJ, USA), Helium (Grade 5.0, Keen, Wilmington, DE, USA), $C^{18}O_2$ (95 atom% ^{18}O, Sigma-Aldrich, St. Louis, MO, USA), and D_2 (99.6% gas purity, 99.8% isotope purity, Cambridge Isotopes, Tewksbury, MA, USA).

3.2. Reactor Setup for Kinetics and Gas-Switching Experiments

The reaction rates and other kinetic parameters were measured using a packed-bed microreactor operated in a down-flow mode at atmospheric pressure. The catalyst particles were supported on a plug of quartz wool within a 7 mm I.D. quartz tube reactor. The quartz tube was positioned inside an Omega CRFC-26/120-A ceramic radiant full cylinder heater. The temperature was controlled by an Omega CN/74000 temperature controller using the input from a K-type, 1/16 in. diameter thermocouple (Omega Engineering, Inc., Norwalk, CT, USA) placed around the outside of the quartz tube at the center of the catalyst bed. Gas flows were controlled by mass flow controllers (Brooks Instruments) through the reactor or the other instruments.

Gas transfer lines were heated to a temperature above 373 K at all times to avoid water condensation. The composition of the effluent stream was analyzed online by a gas chromatograph (GC, 7890A, Agilent, Santa Clara, CA, USA) during continuous flow experiments or by a mass spectrometer (MS, GSD320, Pfeiffer Vacuum Technology AG, Aßlar, Germany) during gas-switching experiments or isotopic experiments. The GC was equipped with both a thermal conductivity detector (TCD) and a flame-ionization detector (FID). The TCD was used to quantify CO_2, CO and H_2 concentrations, while the FID was used to quantify hydrocarbon concentrations. An Agilent 2 mm ID × 12 ft Hayesep Q column was used in the GC to separate products quantified with the TCD, and an Agilent 0.32 mm ID × 30 m HP-Plot Q column was used to separate products quantified with the FID.

3.3. Catalyst Characterization

Temperature-programmed reduction (TPR) was performed by using a MS including the reduction period (773 K, P_{tot} = 1 bar, P_{H2} = 15 kPa in He for 2 h), and then the temperature was ramped up to 1073 K. The ramping rate of each step was 5 K min^{-1}. Another TPR experiment was performed after the reverse water-gas shift (RWGS) reaction (553K, P_{tot} = 1 bar, GHSV = 4.5 × 10^4 cm^3 h^{-1} g$_{cat.}^{-1}$, P_{CO2} = P_{H2} = 15 kPa with He as an inert balance gas). The post-reaction sample was cooled down to room temperature in He flow inside the microreactor. The TPR was then continued in 15 kPa H_2 and 86.3 kPa He from room temperature to 1073 K at the rate of 5 K min^{-1}. The H_2 consumption was converted from the ion current (m/z = 2) after the calibration of the H_2 signal each time before the experiment using a mass spectrometer.

X-Ray Diffraction (XRD) patterns of catalyst powders were collected at room temperature on a Bruker diffractometer using Cu Kα radiation (λ = 1.5418 Å). Measurements were taken over the range of 5° < 2θ < 70°, with a step size of 0.02° before and after the RWGS reaction. X-ray photoelectron spectroscopy (XPS) measurements were performed on a K-alpha Thermo Fisher Scientific spectrometer using a monochromated Al Kα X-ray source. The measurements of iron oxide samples (pre- and post-reaction) were done with a spot size of 50 μm at ambient temperature and a chamber pressure of ~10^{-7} mbar. A flood gun was used for charge compensation. All the spectra measured were calibrated by setting the reference binding energy of carbon 1s at 284.8 eV. The spectra were analyzed by Thermo Fisher Scientific (Waltham, MA, USA) Avantage® commercial software (v5.986). For the fitting, each component consists of a linear combination of Gaussian and Lorentzian product functions, and the full width at half maximum (FWHM) and differences in binding energy of the same species between the Fe2p$_{3/2}$ and Fe2p$_{1/2}$ scan were kept constant. SMART background in Avantage® was used over the region to define the peaks.

3.4. Measurement of Product Formation Rates and Reaction Rates with Gas-switching or Isotopic Experiments

Most of the procedures in the measurement of the reaction rates, gas-switching, or isotopic experiments are identical to the ones described in our previous report [14]. Fe_3O_4 samples were pretreated before all experiments by increasing the reactor temperature at a rate of 5 K min^{-1} to 773 K under a gas flow of 15 kPa H_2. After the pretreatment at 773 K for 2 h, the temperature was lowered to the initial reaction temperature of 753 K. During the measurement of the reaction rates, the partial pressures of the reactants were P_{CO2} = P_{H2} = 15 kPa. A constant total flow rate of 75 cm^3 min^{-1} (sccm) was maintained with He as an inert balance gas.

Gas hourly space velocity (GHSV) in cm^3 h^{-1} g$_{cat.}^{-1}$ was calculated under STP condition (273 K and 1 atm) according to the equation below:

$$GHSV = \frac{F_{tot}}{m_{cat.}} \quad (7)$$

where $m_{cat.}$ is the mass of the catalyst and F_{tot} is the total flow rate.

CO_2 and H_2 reaction orders were measured by independently varying the inlet CO_2 and H_2 partial pressures. The total pressure remained at 1 bar. The activation energy was estimated by using the Arrhenius plot with the temperature varied between 723 K and 773 K while monitoring the CO formation rate.

Rates of CO formation were calculated assuming differential reactor operation according to Equation (8):

$$r_{CO} = \frac{\dot{V} \Delta C_{CO}}{m_{cat.}} \quad (8)$$

where $\dot{V}$ is the total volumetric flow rate, ΔC_{CO} is the change in CO concentration. Measured reaction rates are the net rate of the forward and reverse reactions; therefore, the observed rate must be transformed into the reaction rate for the forward reaction by using Equations (9)–(11). The equilibrium constant (K_C) is low (<1) for the RWGS at the temperatures investigated, although the reverse reaction had a negligible contribution to the observed rates because of the low conversion (<12%) under conditions at which the reactor was operated. Note that C_o (Equation (11)) represents the standard state (1 mol L^{-1}) and equals 1, since the reaction is equimolar.

$$r_{obs.} = r_+ - r_- = r_+(1-\eta) \quad (9)$$

$$\eta = \frac{[CO][H_2O]}{K_C[CO_2][H_2]} \quad (10)$$

$$K_C = \left(\prod_i C_{i_{eq.}}^{\gamma}\right)\frac{1}{C_o} \tag{11}$$

where $r_{obs.}$ is the observed rate, r_+ and r_- are the rate of forward and reverse reactions, η is the ratio of the rate of the reverse and forward reactions and K_C is the equilibrium constant.

Experiments were also conducted to *(i)* determine reaction rates in excess (i.e. non-equimolar) CO_2 or H_2, and *(ii)* to determine apparent kinetic parameters. In the first case, CO_2 and H_2 were fed with the catalyst—Fe_3O_4 (100 mg)—held at a temperature of 753 K. The initial partial pressure of both CO_2 and H_2 was 15 kPa. After a period of 16 h, the partial pressure of CO_2 was increased to 60 kPa, while the partial pressure of H_2 was held at 15 kPa. After another period of 2 h, the partial pressure of CO_2 was decreased to 15 kPa and the partial pressure of H_2 was increased to 60 kPa. Finally, both partial pressures were returned to 15 kPa. CO_2 conversion was quantified under the same conditions.

For the second case, apparent kinetic parameters (activation energy and reaction orders) were determined with near equimolar concentrations of CO_2 and H_2 on Fe_3O_4, and under large H_2 excess on Fe_3O_4 as well. With near equimolar concentrations of CO_2 and H_2, the reaction was first performed for 15–16 h at a temperature of 753 K with reactant partial pressures of 15 kPa. The temperature was then lowered in 10 K increments to 723 K, with 5–6 GC injections (a period of about 60 min) taken at each temperature. After the period at 723 K, the CO_2 partial pressure was reduced to 10 kPa and increased in 2.5 kPa increments to a final partial pressure of 20 kPa. Finally, the CO_2 partial pressure was returned to 15 kPa and the H_2 partial pressure was lowered to 10 kPa and increased in 2.5 kPa increments. The basic outline of experiments conducted with excess H_2 was the same as that used for near equimolar reactant concentrations (see also our previous report) [14]. Reactant partial pressures during the initial period were 90 kPa H_2 and 10 kPa CO_2. During the variable CO_2 partial pressure period, the H_2 partial pressure was 85 kPa, and the CO_2 partial pressure was varied between 5 and 12.5 kPa in 2.5 kPa increments. To investigate the effect of H_2 partial pressure, the CO_2 partial pressure was kept at 10 kPa and the H_2 partial pressure was varied between 70–90 kPa in 5 kPa increments.

The measurement of the CO formation rate over Fe_3C has been performed in two different manners. First, the temperature was ramped under a flow of H_2 (15 kPa) and He at the rate of 5 K min^{-1} up to 753 K. 15 kPa CO_2 was then added into the flow while monitoring the CO formation rate. The second part of the Fe_3C activity test was done by ramping the temperature to 573 K, 623 K, 673 K, 723 K and then cooling back to 673 K again at the rate of 5 K min^{-1} under the flow of H_2 (15 kPa), CO_2 (15 kPa), and He (remainder). Each temperature was held for 80 min respectively for GC injection.

Gas-switching experiments were done by measuring CO formation rates while alternating between CO_2 and H_2 gas flows. Catalysts were pretreated as described above, and after pretreatment, CO_2 was added into the reactor to allow the RWGS reaction to proceed for 2 h.

After the reaction, the gas flow rates were changed to 36 sccm helium and 4 sccm H_2. After 20 min, H_2 flow was stopped and was replaced by 4 sccm of CO_2. After 20 min, CO_2 in the gas stream was replaced by H_2. This $CO_2 \rightarrow H_2$ sequence was repeated three times. The reactor was then purged with helium for 20 min before CO_2 was readmitted into the gas stream. After 20 min, the reactor was again purged with helium before H_2 was readmitted to the gas stream. All sequences with a given gas composition lasted for 20 min, and the temperature of the reactor was 773 K throughout the duration of the gas switching portion of the experiment.

Additional gas-switching experiments involving purge times of varying length with an inert gas were carried out at 753 K. Following the same pretreatment and the reaction in 15 kPa H_2 and 15 kPa for 2 h at 773 K, 15 kPa H_2 was admitted to the reactor. After 15 min, H_2 was replaced by 15 kPa CO_2 for 15 min, and CO_2 was then replaced by H_2 for another 15 min. Then the reactor was purged with helium for 5 min. This sequence ($CO_2 \rightarrow H_2 \rightarrow He$) was repeated several times, but each time the length of the inert purge was increased by 5 min.

An isotopic experiment for the CO formation rate was monitored by MS on Fe_3O_4 (100 mg) while alternating between CO_2 (4 sccm) and $C^{18}O_2$ (4 sccm) after the RWGS reaction for 2 h. The temperature was kept constant at 753 K. The total flow rate was 40 sccm with H_2 maintained at 4 sccm.

The kinetic isotope effect (KIE) of H_2/D_2 was investigated on Fe_3O_4 (100 mg) for various $H_2:CO_2$ ratios. After pretreatment, the reaction began at a temperature of 753 K with CO_2 and H_2 partial pressures of 15 kPa. After 16 h, the temperature was lowered to 723 K, and after 1.5 h, H_2 in the feed was replaced by D_2.

4. Conclusions

An unsupported Fe_3O_4-derived catalyst showed very promising activity toward CO formation via CO_2 hydrogenation. The high selectivity (~100% under $H_2:CO_2$ = 1:1) and great stability make the catalyst feasible to consider in extensive use. The catalyst exhibited only slight deactivation under conditions of excess CO_2, but it can be quickly regenerated under excess H_2. Reaction rates depended more strongly on H_2 (0.8 in reaction order) compared to CO_2 (0.33 in reaction order) under a near equimolar gas-phase composition. The post-reaction analyses of the catalyst indicated that the catalyst was reduced to metallic iron first in the pretreatment of H_2, but the working catalyst remained partially oxidized with the composition of Fe^{2+}, Fe^{3+}, Fe^0 and Fe_3C. The main active sites are believed to be the combination of the above species, except that Fe_3C is unlikely to directly contribute to the very steady CO formation at our reaction conditions (1 atm, 723–773 K). Gas-switching experiments revealed that CO was formed only when switching from H_2 to CO_2, and H_2O was formed when switching from CO_2 to H_2, but not when switching from H_2 to CO_2 if purging of helium was in between with the gas admission. The redox mechanism is identified as the dominant reaction pathway for the unsupported iron catalyst.

Supplementary Materials: The following are available online at http://www.mdpi.com/2073-4344/9/9/773/s1, Figure S1. XRD pattern of Fe_3O_4 after the reduction in H_2, Figure S2. Ion current at m/z = 18 (H_2O), 28 (CO), and 44 (CO_2) during H_2/CO_2 switching experiments on Fe_3O_4. Arrows with a label indicate a change in gas composition to the indicated gas. Reaction conditions: T = 773 K, F_{He} = 36 sccm, F_{H2} or F_{CO2} = 4 sccm. The figure is a modification of Figure 6, Figure S3. Ion current at m/z = 2 (H_2), 3 (HD), 4 (D_2), and 28 (CO) during flow of 7.5 kPa H_2 + 7.5 kPa D_2 +15 kPa CO_2 and 7.5 kPa H_2 +15 kPa CO_2 on Fe_3O_4. Reaction conditions: T = 753 K, F_{tot} = 75 sccm, Figure S4. (a) Ion current at m/z = 28 (CO) and m/z = 18 (H_2O) during H_2/CO_2 switching experiments on Fe_3O_4. Arrows with a label indicate a change in gas composition to the indicated gas. The catalysts were in flowing H_2 for 2 h followed by the reaction in CO_2+H_2 for 2 h before the first admission of H_2 (relative time: 31 min) and CO_2 (relative time: 46 min). Reaction conditions: T = 753 K, F_{tot} = 75 sccm, P_{H2} or P_{CO2} = 15 kPa. (b) is the modification of (a), Figure S5. XPS O1s spectra of Fe_3O_4 and post-reaction Fe_3O_4. The curves under the fitted envelope and above the background are contributions of estimated components from peak fitting. The peak deconvolution and their atomic % are listed on the right of the spectra for each sample, Table S1 Fitted initial slopes and area of H_2O in H_2/CO_2 switching experiment with different He purging time in Figure S4

Author Contributions: Conceptualization, C.-Y.C. and R.F.L.; methodology, C.-Y.C., J.A.L., and R.F.L.; investigation, C.-Y.C. and R.F.L.; resources, R.F.L.; data curation, C.-Y.C.; writing—original draft preparation, C.-Y.C.; writing—review and editing, C.-Y.C., J.A.L., and R.F.L.; supervision, R.F.L.; project administration, R.F.L.; funding acquisition, R.F.L."

Funding: This research was funded by U.S. Army, grant number GTS-S-17-013.

Acknowledgments: The authors would also like to acknowledge Terry Dubois for the help and suggestions.

Conflicts of Interest: The authors declare no conflict of interest.

References

1. National Academies of Sciences, Engineering, and Medicine. *Gaseous Carbon Waste Streams Utilization: Status and Research Needs*; The National Academies Press: Washington, DC, USA, 2019.
2. Joo, O.S.; Jung, K.D.; Moon, I.; Rozovskii, A.Y.; Lin, G.I.; Han, S.H.; Uhm, S.J. Carbon dioxide hydrogenation to form methanol via a reverse-water-gas-shift reaction (the CAMERE process). *Ind. Eng. Chem. Res.* **1999**, *38*, 1808–1812. [CrossRef]
3. Kim, S.S.; Lee, H.H.; Hong, S.C. A study on the effect of support's reducibility on the reverse water-gas shift reaction over Pt catalysts. *Appl. Catal. A Gen.* **2012**, *423–424*, 100–107. [CrossRef]
4. Kim, S.S.; Lee, H.H.; Hong, S.C. The effect of the morphological characteristics of TiO_2 supports on the reverse water-gas shift reaction over Pt/TiO_2 catalysts. *Appl. Catal. B Environ.* **2012**, *119–120*, 100–108. [CrossRef]

5. Goguet, A.; Shekhtman, S.O.; Burch, R.; Hardacre, C.; Meunier, F.C.; Yablonsky, G.S. Pulse-response TAP studies of the reverse water-gas shift reaction over a Pt/CeO$_2$ catalyst. *J. Catal.* **2006**, *237*, 102–110. [CrossRef]
6. Pettigrew, D.J.; Cant, N.W. The effects of rare earth oxides on the reverse water-gas shift reaction on palladium/alumina. *Catal. Lett.* **1994**, *28*, 313–319. [CrossRef]
7. Park, J.N.; McFarland, E.W. A highly dispersed Pd-Mg/SiO$_2$ catalyst active for methanation of CO$_2$. *J. Catal.* **2009**, *266*, 92–97. [CrossRef]
8. Wang, G.C.; Nakamura, J. Structure sensitivity for forward and reverse water-gas shift reactions on copper surfaces: A DFT study. *J. Phys. Chem. Lett.* **2010**, *1*, 3053–3057. [CrossRef]
9. Wang, G.-C.; Jiang, L.; Pang, X.-Y.; Cai, Z.-S.; Pan, Y.-M.; Zhao, X.-Z.; Morikawa, Y.; Nakamura, J. A theoretical study of surface-structural sensitivity of the reverse water gas shift reaction over Cu(hkl) surfaces. *Surf. Sci.* **2003**, *543*, 118–130. [CrossRef]
10. Chen, C.S.; Cheng, W.H.; Lin, S.S. Enhanced activity and stability of a Cu/SiO$_2$ catalyst for the reverse water gas shift reaction by an iron promoter. *Chem. Commun.* **2001**, *1*, 1770–1771. [CrossRef]
11. Fujitani, T.; Nakamura, J. The effect of ZnO in methanol synthesis catalysts on Cu dispersion and the specific activity. *Catal. Lett.* **1998**, *56*, 119–124. [CrossRef]
12. Sun, F.M.; Yan, C.F.; Wang, Z.D.; Guo, C.Q.; Huang, S.L. Ni/Ce-Zr-O catalyst for high CO$_2$ conversion during reverse water gas shift reaction (RWGS). *Int. J. Hydrogen Energy* **2015**, *40*, 15985–15993. [CrossRef]
13. Wolf, A.; Jess, A.; Kern, C. Syngas Production via Reverse Water-Gas Shift Reaction over a Ni-Al$_2$O$_3$ Catalyst: Catalyst Stability, Reaction Kinetics, and Modeling. *Chem. Eng. Technol.* **2016**, *39*, 1040–1048. [CrossRef]
14. Loiland, J.A.; Wulfers, M.J.; Marinkovic, N.S.; Lobo, R.F. Fe/γ-Al$_2$O$_3$ and Fe–K/γ-Al$_2$O$_3$ as reverse water-gas shift catalysts. *Catal. Sci. Technol.* **2016**, *6*, 5267–5279. [CrossRef]
15. Liu, Y.; Liu, D. Study of bimetallic Cu–Ni/γ-Al$_2$O$_3$ catalysts for carbon dioxide hydrogenation. *Int. J. Hydrogen Energy* **1999**, *24*, 351–354. [CrossRef]
16. Stone, F.S.; Waller, D. Cu–ZnO and Cu–ZnO/Al$_2$O$_3$ catalysts for the reverse water-gas shift reaction. The effect of the Cu/Zn ratio on precursor characteristics and on the activity of the derived catalysts. *Top. Catal.* **2003**, *22*, 305–318. [CrossRef]
17. Chen, C.S.; Cheng, W.H.; Lin, S.S. Study of reverse water gas shift reaction by TPD, TPR and CO$_2$ hydrogenation over potassium-promoted Cu/SiO$_2$ catalyst. *Appl. Catal. A Gen.* **2002**, *238*, 55–67. [CrossRef]
18. Schulz, H. Short history and present trends of Fischer-Tropsch synthesis. *Appl. Catal. A Gen.* **1999**, *186*, 3–12. [CrossRef]
19. Van Der Laan, G.P.; Beenackers, A.A.C.M. Kinetics and Selectivity of the Fischer-Tropsch Synthesis: A Literature Review. *Catal. Rev. Sci. Eng.* **1999**, *41*, 255–318. [CrossRef]
20. Zhu, M.; Wachs, I.E. Iron-Based Catalysts for the High-Temperature Water–Gas Shift (HT-WGS) Reaction: A Review. *ACS Catal.* **2016**, *6*, 722–732. [CrossRef]
21. Newsome, D.S. The Water-Gas Shift Reaction. *Catal. Rev.* **1980**, *21*, 275–318. [CrossRef]
22. Reddy, G.K.; Gunasekara, K.; Boolchand, P.; Smirniotis, P.G. Cr- and Ce-doped ferrite catalysts for the high temperature water-gas shift reaction: TPR and mossbauer spectroscopic study. *J. Phys. Chem. C* **2011**, *115*, 920–930. [CrossRef]
23. Schulz, H.; Riedel, T.; Schaub, G. Fischer-Tropsch principles of co-hydrogenation on iron catalysts. *Top. Catal.* **2005**, *32*, 117–124. [CrossRef]
24. Riedel, T.; Schulz, H.; Schaub, G.; Jun, K.W.; Hwang, J.S.; Lee, K.W. Fischer-Tropsch on iron with H$_2$/CO and H$_2$/CO$_2$ as synthesis gases: The episodes of formation of the Fischer-Tropsch regime and construction of the catalyst. *Top. Catal.* **2003**, *26*, 41–54. [CrossRef]
25. Ratnasamy, C.; Wagner, J.P. Water Gas Shift Catalysis. *Catal. Rev.* **2009**, *51*, 325–440. [CrossRef]
26. Visconti, C.G.; Martinelli, M.; Falbo, L.; Infantes-Molina, A.; Lietti, L.; Forzatti, P.; Iaquaniello, G.; Palo, E.; Picutti, B.; Brignoli, F. CO$_2$ hydrogenation to lower olefins on a high surface area K-promoted bulk Fe-catalyst. *Appl. Catal. B Environ.* **2017**, *200*, 530–542. [CrossRef]
27. Lee, D.W.; Lee, M.S.; Lee, J.Y.; Kim, S.; Eom, H.J.; Moon, D.J.; Lee, K.Y. The review of Cr-free Fe-based catalysts for high-temperature water-gas shift reactions. *Catal. Today* **2013**, *210*, 2–9. [CrossRef]

28. Fishman, Z.S.; He, Y.; Yang, K.R.; Lounsbury, A.W.; Zhu, J.; Tran, T.M.; Zimmerman, J.B.; Batista, V.S.; Pfefferle, L.D. Hard templating ultrathin polycrystalline hematite nanosheets: Effect of nano-dimension on CO_2 to CO conversion: Via the reverse water-gas shift reaction. *Nanoscale* **2017**, *9*, 12984–12995. [CrossRef] [PubMed]
29. Kim, D.H.; Han, S.W.; Yoon, H.S.; Kim, Y.D. Reverse water gas shift reaction catalyzed by Fe nanoparticles with high catalytic activity and stability. *J. Ind. Eng. Chem.* **2014**, *23*, 67–71. [CrossRef]
30. Kunkes, E.L.; Studt, F.; Abild-Pedersen, F.; Schlögl, R.; Behrens, M. Hydrogenation of CO_2 to methanol and CO on $Cu/ZnO/Al_2O_3$: Is there a common intermediate or not? *J. Catal.* **2015**, *328*, 43–48. [CrossRef]
31. Kalamaras, C.M.; Americanou, S.; Efstathiou, A.M. "Redox" vs "associative formate with -OH group regeneration" WGS reaction mechanism on Pt/CeO_2: Effect of platinum particle size. *J. Catal.* **2011**, *279*, 287–300. [CrossRef]
32. Temkin, M.I. The Kinetics of Some Industrial Heterogeneous Catalytic Reactions. *Adv. Catal.* **1979**, *28*, 173–291.
33. Rhodes, C.; Hutchings, G.J.; Ward, A.M. Water-gas shift reaction: Finding the mechanistic boundary. *Catal. Today* **1995**, *23*, 43–58. [CrossRef]
34. Ginés, M.J.L.; Marchi, A.J.; Apesteguía, C.R. Kinetic study of the reverse water-gas shift reaction over $CuO/ZnO/Al_2O_3$ catalysts. *Appl. Catal. A Gen.* **1997**, *154*, 155–171. [CrossRef]
35. Jozwiak, W.K.; Kaczmarek, E.; Maniecki, T.P.; Ignaczak, W.; Maniukiewicz, W. Reduction behavior of iron oxides in hydrogen and carbon monoxide atmospheres. *Appl. Catal. A Gen.* **2007**, *326*, 17–27. [CrossRef]
36. Pineau, A.; Kanari, N.; Gaballah, I. Kinetics of reduction of iron oxides by H2. Part II. Low temperature reduction of magnetite. *Thermochim. Acta* **2007**, *456*, 75–88. [CrossRef]
37. Zhang, C.L.; Li, S.; Wang, L.J.; Wu, T.H.; Peng, S.Y. Studies on the decomposing carbon dioxide into carbon with oxygen-deficient magnetite. II. The effects of properties of magnetite on activity of decomposition CO_2 and mechanism of the reaction. *Mater. Chem. Phys.* **2000**, *62*, 52–61. [CrossRef]
38. Bonnet, F.; Ropital, F.; Lecour, P.; Espinat, D.; Huiban, Y.; Gengembre, L.; Berthier, Y.; Marcus, P. Study of the oxide/carbide transition on iron surfaces during catalytic coke formation. *Surf. Interface Anal.* **2002**, *34*, 418–422. [CrossRef]
39. Mondal, K.; Lorethova, H.; Hippo, E.; Wiltowski, T.; Lalvani, S.B. Reduction of iron oxide in carbon monoxide atmosphere—Reaction controlled kinetics. *Fuel Process. Technol.* **2004**, *86*, 33–47. [CrossRef]
40. Yamashita, T.; Hayes, P. Analysis of XPS spectra of Fe^{2+} and Fe^{3+} ions in oxide materials. *Appl. Surf. Sci.* **2008**, *254*, 2441–2449. [CrossRef]
41. Biesinger, M.C.; Payne, B.P.; Grosvenor, A.P.; Lau, L.W.M.; Gerson, A.R.; Smart, R.S.C. Resolving surface chemical states in XPS analysis of first row transition metals, oxides and hydroxides: Cr, Mn, Fe, Co and Ni. *Appl. Surf. Sci.* **2011**, *257*, 2717–2730. [CrossRef]
42. McIntyre, N.S.; Zetaruk, D.G. X-ray photoelectron spectroscopic studies of iron oxides—Analytical Chemistry (ACS Publications). *Anal. Chem.* **1977**, *49*, 1521–1529. [CrossRef]
43. Shroff, M.D.; Kalakkad, D.S.; Coulter, K.E.; Kohler, S.D.; Harrington, M.S.; Jackson, N.B.; Sault, A.G.; Datye, A.K. Activation of Precipitated Iron Fischer-Tropsch Synthesis Catalysts. *J. Catal.* **1995**, *156*, 185–207. [CrossRef]
44. De Smit, E.; Cinquini, F.; Beale, A.M.; Safonova, O.V.; Van Beek, W.; Sautet, P.; Weckhuysen, B.M.; Supe, N.; Horowitz, R.J.; Cedex, F.-G. Stability and Reactivity of E-X-θ Iron Carbide Catalyst Phases in Fischer—Tropsch Synthesis: Controlling μC. *J. Am. Chem. Soc.* **2010**, *132*, 14928–14941. [CrossRef] [PubMed]
45. Davis, B.H. Fischer-Tropsch Synthesis: Reaction mechanisms for iron catalysts. *Catal. Today* **2009**, *141*, 25–33. [CrossRef]
46. Fujita, S.I.; Usui, M.; Takezawa, N. Mechanism of the reverse water gas shift reaction over Cu/ZnO catalyst. *J. Catal.* **1992**, *134*, 220–225. [CrossRef]

© 2019 by the authors. Licensee MDPI, Basel, Switzerland. This article is an open access article distributed under the terms and conditions of the Creative Commons Attribution (CC BY) license (http://creativecommons.org/licenses/by/4.0/).

Article

Liquid-Phase Catalytic Oxidation of Limonene to Carvone over ZIF-67(Co)

Yizhou Li, Yepeng Yang, Daomei Chen, Zhifang Luo, Wei Wang, Yali Ao, Lin Zhang, Zhiying Yan * and Jiaqiang Wang *

National Center for International Research on Photoelectric and Energy Materials, Yunnan Provincial Collaborative Innovation Center of Green Chemistry for Lignite Energy, Yunnan Province Engineering Research Center of Photocatalytic Treatment of Industrial Wastewater, The Universities' Center for Photocatalytic Treatment of Pollutants in Yunnan Province, School of Chemical Sciences & Technology, Yunnan University, Kunming 650091, China; zh111111ou@163.com (Y.L.); mondaysunday1234@163.com (Y.Y.); dmchen@ynu.edu.cn (D.C.); zhifangluo@126.com (Z.L.); wangwei2@ynu.edu.cn (W.W.); yl_a2019@163.com (Y.A.); echolanchen@163.com (L.Z.)
* Correspondence: zhyyan@ynu.edu.cn (Z.Y.); jqwang@ynu.edu.cn (J.W.);
  Tel.: +86-871-6503-1567 (Z.Y.); +86-871-6503-1567 (J.W.)

Received: 15 March 2019; Accepted: 16 April 2019; Published: 21 April 2019

Abstract: Liquid-phase catalytic oxidation of limonene was carried out under mild conditions, and carvone was produced in the presence of ZIF-67(Co), cobalt based zeolitic imidazolate framework, as catalyst, using t-butyl hydroperoxide (t-BHP) as oxidant and benzene as solvent. As a heterogeneous catalyst, the zeolitic imidazolate framework ZIF-67(Co) exhibited reasonable substrate–product selectivity (55.4%) and conversion (29.8%). Finally, the X-ray diffraction patterns of the catalyst before and after proved that ZIF-67(Co) acted as a heterogeneous catalyst, and can be reused without losing its activity to a great extent.

Keywords: liquid-phase catalytic oxidation; limonene; carvone; zeolitic imidazolate frameworks

---

1. Introduction

As we know, the reaction about the allylic oxidation plays a critical role in developing fine chemicals with high additional value from biomass, and has great value in the synthesis of unsaturated aldehydes and ketones [1,2]. Carvone is a main ingredient, derived from plant essential oils, used for cosmetics and food flavors, and is also used in the preservation of meat, fruits, and vegetables because it has good antioxidant activity, analgesic effects, and antibacterial effects [3,4]. In addition, carvone is an important intermediate in industrial chemistry, where it can further be used to synthesize carvone thioether and cyanoacetone [5]. The applicability of traditional methods of transforming limonene to carvone through catalytic oxidation, such as epoxidation and nitrosochlorination, is limited; largely because environmentally unfriendly reagents are often used and poisonous secondary products are produced [6]. Compared with these methods, the use of heterogeneous catalysts has broader applicability prospects in the allylic oxidation of cycloolefins, due to their remarkable advantages in catalyst recovery and stability.

However, current research results show that the selectivity of the catalytic oxidation of limonene to carvone by heterogeneous reaction has not been satisfactory. For instance, 36% substrate conversion and 25% selectivity for carvone were obtained by using chromium-containing mesoporous molecular sieves MCM-41 [7]. Lower than 5% selectivity for carvone was obtained by using the [Fe^{III} (BPMP)Cl(μ-O)Fe^{III}Cl$_3$] complex as catalyst [8]. A 21% conversion and 20% selectivity for carvone were obtained in a reaction using Fe/EuroPh catalysts [9]. In another study, high conversion up to 93% of limonene was attained, while the selectivity for carvone was less than 7% [10]. In our previous study, cobalt-doped mesoporous silica templated by reed leaves exhibited high substrate conversion (100%) and relatively good product (carvone) selectivity (40.2%) for the allylic oxidation of limonene to carvone [11].

Metal–organic frameworks (MOFs) have attracted considerable attention in catalysis and adsorption in recent years, owing to their advantages, such as high surface areas, well-defined structures, special metal centers, their ease of processing, and their structural diversity [12–16]. Thus, the performance of MOF materials in this reaction has also been reported [17,18], but their application is limited by their thermal and chemical stability [19]. However we are sure that it is valuable to use some MOFs materials in this reaction, and this research has aroused our great interest.

Zeolitic imidazolate frameworks (ZIFs), which are constructed from tetrahedrally coordinated divalent cations (Zn^{2+} or Co^{2+}) linked by the uninegative imidazolate ligands, are a new class of porous metal–organic framework (MOF) [20]. They are widely used in organic synthesis [21–25], CO_2 capture of [26], and olefin/paraffin separation [27] because of their excellent thermal and chemical stability. They benefit from strong interactions between imidazolium salts and metal ions, which make the frameworks of zeolitic imidazolates maintain their structural integrity even in water, and this is difficult for other MOFs [28]. Additionally, ZIF-67 has a high-porosity zeolite structure, which is conducive to catalytic reactions. According to our surveys, ZIFs have been widely used to catalyze different types of organic synthesis reactions in recent years, such as cyclohexene hydrogenation [29], glycerol esterification [30], and transesterification reactions [31].To this end, the zeolitic imidazole framework ZIF-67(Co) was selected because ZIF-67 is easy to prepare at large scale in water under ambient conditions without using toxic solvents, and the preparation of ZIF-67(Co) has good reproducibility.

In this continuation of our work, we successfully synthesized ZIF-67(Co) by hydrothermal synthesis, and it displayed good catalytic properties in the catalytic oxidation of limonene. Furthermore, the application of tert-butyl hydroperoxide (t-BHP) to the allylic oxidation of cycloolefins offers an alternative to traditional unfriendly oxidants. In this work, the effects of solvents, oxidants, and reaction temperature on the productivity and reaction time were investigated. Under the optimal conditions, although the conversion of limonene was only 29.8%, we increased the selectivity for carvone to 55.4%, which is more than that in all the literature we have been able to refer to. For comparison, the catalytic activities of more MOFs were also investigated. This is the first report about the oxidation of limonene to carvone over ZIF-67(Co).

2. Results and Discussion

2.1. Characterization of ZIF-67(Co)

Powder XRD patterns of the ZIF-67(Co) are shown in Figure 1. The diffraction peaks of the 2θ values at 7.48°, 10.42°, 12.74°, and 18.06°, were assigned to the (011), (002), (112), and (222) planes of ZIF-67, which has been reported by other researchers [20,21,29]. Figure 1 also reflects the fact that synthetic ZIF-67(Co) is a phase-pure product. Besides, in order to compare the different catalysts' ability about catalyzed the limonene to carvone, some other MOFs have been synthesized successfully and their XRD patterns have been revealed in supplementary material (Figures S1–S5)

The N_2 sorption experiments for the ZIF-67(Co) yielded the typical type I isotherms in Figure 2. The Brunauer–Emmett–Teller (S_{BET}) and Langmuir surface areas of ZIF-67 were 1378 and 1805 m^2/g, and the pore volume was 0.62 cm^3/g, which are near to the previously reported values [32,33]. At lower relative pressures, the micropore in the material makes the adsorption capacity increase rapidly, while at higher relative pressures, due to the existence of spin/large porosity nanoparticles structure, the enhancement of adsorption capacity is lower. The pore size distribution curve was calculated by the Horvath–Kawazoe (HK) method, which clearly showed that there were two kinds of micropores in ZIF-67(Co). This result is similar to an earlier work [22].

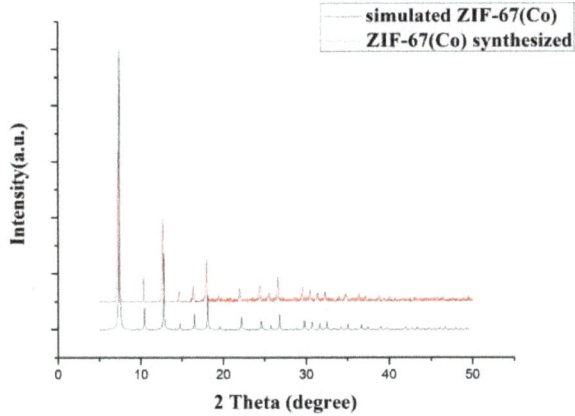

Figure 1. XRD patterns of the zeolitic imidazolate framework ZIF-67(Co).

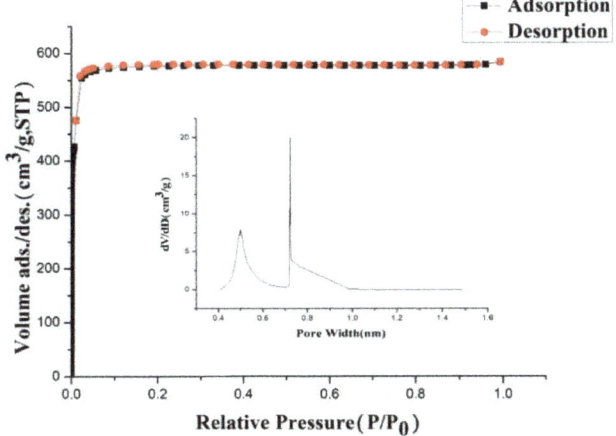

Figure 2. N_2 adsorption/desorption isotherms of ZIF-67(Co) samples at 77 K and Horvath–Kawazoe (HK) pore size distribution curve.

The morphology and particle size of the samples revealed by scanning electron microscopy (SEM) are shown in Figure 3. The SEM picture reveals that the particles were submicroscopic crystals with polyhedral shapes, each surface was a homogeneous quadrilateral, and the particle sizes ranged from 200 to 500 nm. This also proves that the as-synthesized ZIF-67(Co) was composed of submicroscopic crystals.

FT-IR and TGA were employed to obtain further structural information about the ZIF-67(Co). Figure 4 presents FT-IR spectra of ZIF-67(Co) between 400 and 4000 cm^{-1}. The peak at 3420 cm^{-1} is a contribution of the asymmetric characteristic absorption peaks of O–H bonds in hydroxyl groups adsorbed on the surface of materials. There were also some peaks between 500 and 1500 cm^{-1} caused by plane-bending vibrations and stretching vibrations of the imidazole ring. Meanwhile, the two small peaks at 758 and 534 cm^{-1} were attributed to the stretching vibration peak of the Co–N bond. In addition, there were two bands at 3134 and 2931 cm^{-1} caused by the asymmetric absorption vibrations of the C–H bond in the methyl groups.

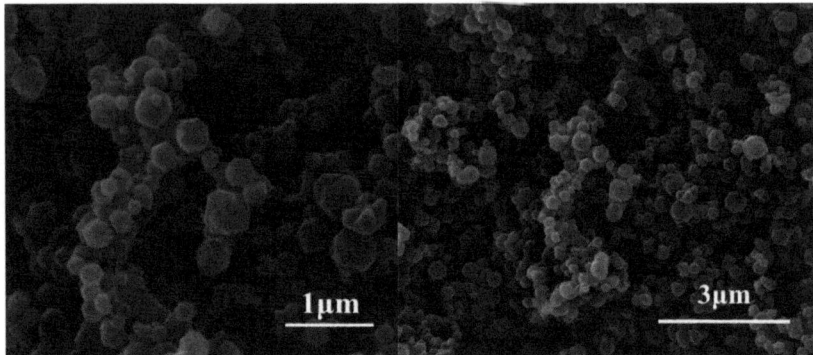

Figure 3. SEM pictures of ZIF-67(Co).

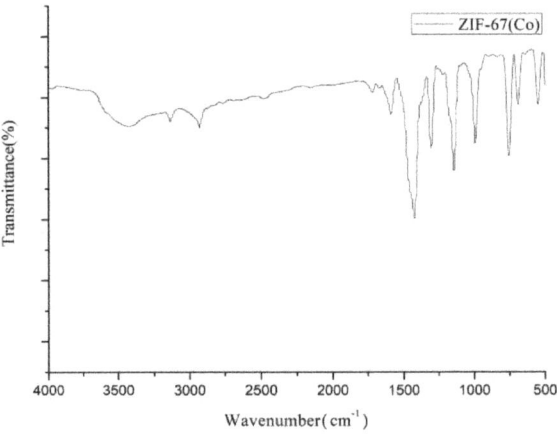

Figure 4. FT-IR spectra of ZIF-67(Co).

The synthesized ZIF-67(Co) submicroscopic crystals had high thermostability, which was proved by TGA and differential scanning calorimetry (DSC) as shown in Figure 5. The test increased temperature to 873 K from room temperature at 5 K/min. When calcined below 373 K, the total weight loss of the sample was very small (about 8%), relating to the removal of guest molecules such as water. Between 373 and 510 K, the curve remained smooth overall, indicating that the skeleton of ZIF-67(Co) has good thermal stability below 510 K. When the temperature was greater than 520 K, the mass loss began at the stage of skeletal decomposition. This was attributed to the collapse of the ligands, which was caused by the decomposition of 2-methylimidazole. The weight-loss process was complete at 620 K, and the residue was Co_3O_4.

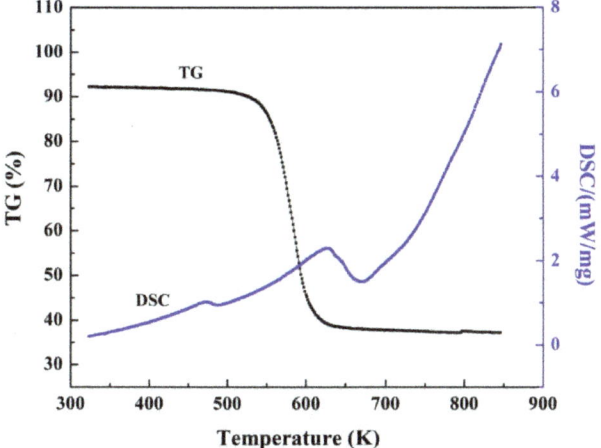

Figure 5. TGA and differential scanning calorimetry (DSC) curves of ZIF-67(Co) submicroscopic crystals.

2.2. Catalytic Experiments

2.2.1. The Effect of Various Catalysts on the Reaction

The comparison of the catalytic activities of different catalysts for the oxidation of limonene is summarized in Table 1. Using t-BHP as oxidant yielded carvone as the main product, and the reaction time was 8 h. Besides, the synthetic method and XRD patterns about these MOFs have been revealed in supplementary material (Figures S1–S5)

Table 1. Comparison of the catalytic activities of various metal–organic frameworks (MOFs) for the oxidation of limonene.

Catalysts	Conversion (%)	Selectivity (%)
MIL-101(Fe)	50.8	20.3
ZIF-8(Zn)	16.5	9.2
ZIF-67(Co)	35.7	32.6
MIL-101(Cr)	44.1	26.6
HKUST-1(Cu)	20.4	7.9
MIL-125(Ti)	18.2	10.1
No catalyst	26.7	0.05

Conditions: limonene: 2 mL, catalyst: 100 mg, solvent: 15 mL acetic acid, oxidant: 15 mL t-BHP (tert-butyl hydroperoxide), temperature: 85 °C, reaction time: 8 h.

First, the reactivity of several common catalysts was compared without any condition optimization. The activity of MIL-101(Fe)(Materials of Institute Lavoisier Frameworks) was the most outstanding for the oxidation of limonene—the conversion rate was 50.8% after 8 h; followed by MIL-101(Cr), whose conversion rate was 44.1%. ZIF-8 showed the worst activity, with a conversion rate of only 16.5%. However, after the reaction, the filtrates from HKUST-1(Cu)([$Cu_3(BTC)_2(H_2O)_3$]$_n$), MIL-101(Fe), and MIL-101(Cr) were light blue, dark red, and yellow-green, respectively. This indicates that metal ions, as metal centers, leaked during the reactions with HKUST-1(Cu), MIL-101(Fe), and MIL-101(Cr). On the other hand, ZIF-67 showed the highest selectivity, and the selectivity for carvone reached 36.6% after 8 h of reaction. The high selectivity of ZIF-67(Co) may be attributed to its unique ion centers. Heterogeneous catalysis with MOF materials is one of their extensively investigated applications [21].

2.2.2. The Effect of the Catalyst Dosage on the Reaction

The catalyst dosage had a significant effect on the oxidation of limonene. Five different dosages of ZIF-67(Co)—35, 60, 85, 110, and 135 mg—were used while keeping all the other reaction parameters fixed, and the experimental results are shown in Figure 6. Conversions of 4.75%, 13.2%, 28.1%, 30.3%, and 33.5%, corresponding to the 35, 60, 85, 110, and 135 mg catalyst loadings, respectively, were obtained in the experiment. With increasing ZIF-67(Co) dosage, the conversion of limonene increased. When the amount of catalyst was 135 mg, the conversion of limonene was 33.5%. Interestingly, the selectivity for carvone increased at low catalyst contents and then decreased, relative to increasing content of the catalyst. When the dosage of the catalyst was 85 mg, the selectivity for carvone reached the maximum value (36.7%). After that, the yield of carvone decreased slightly as the amount of catalyst continued to increase, but the rate of change was small.

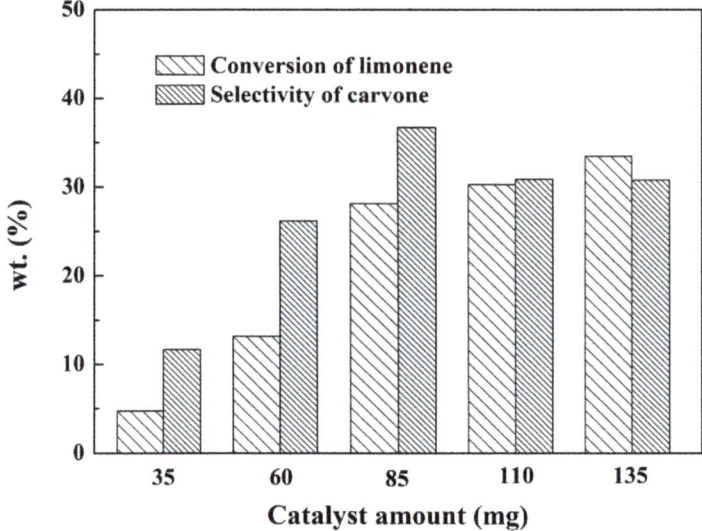

Figure 6. The effect of the catalyst dosage on the reaction. Conditions: limonene: 2 mL, catalyst: ZIF-67(Co), solvent: 15 mL acetic acid, oxidant: 15 mL t-BHP, temperature: 85 °C, reaction time: 8 h.

2.2.3. The Effects of Various Oxidizing Agents on the Reaction

Choosing a suitable oxidant is very important for improving the conversion and selectivity of catalytic oxidation, because oxidants are one of the key factors in catalytic oxidation. We tried to get air in the reaction to use oxygen as a reagent. In addition, hydrogen peroxide and t-BHP were used as oxidants. The results of these experiments are recorded in Table 2. When t-BHP (70%) was added to the system, the conversion of limonene increased to 28.1%, and the selectivity for carvone was 36.7%. However, if hydrogen peroxide (30%) was used as oxidant, the conversion of limonene and the selectivity decreased significantly.

Table 2. The effect of various oxidizing agents on the reaction.

Oxidant	Conversion (%)	Selectivity (%)
Air	11.3	30.9
30% H_2O_2	6.8	9.9
70% t-BHP	28.1	36.7

Conditions: limonene: 2 mL, catalyst: 85 mg of ZIF-67(Co), solvent: 15 mL acetic acid, oxidant: 5 mL H_2O_2 (30%) or 5 mL t-BHP, temperature: 85 °C, reaction time: 8 h.

2.2.4. The Effect of the Dosage of the Oxidant on the Reaction

Experimental results on the effect of oxidant dosage on the reaction were shown in Figure 7. We noticed that the conversion of limonene increased with increasing amounts of t-BHP. When the amount of t-BHP was 7 mL, the conversion of limonene was 33.1%. Meanwhile, the selectivity for carvone increased and then decreased with the increase in t-BHP. We also found that the selectivity for carvone reached its maximum (43.5%) when the dosage of oxidant was 3 mL. After that, increasing the amount of oxidant increased the degree of oxidation, resulting in a decrease of selectivity and yield.

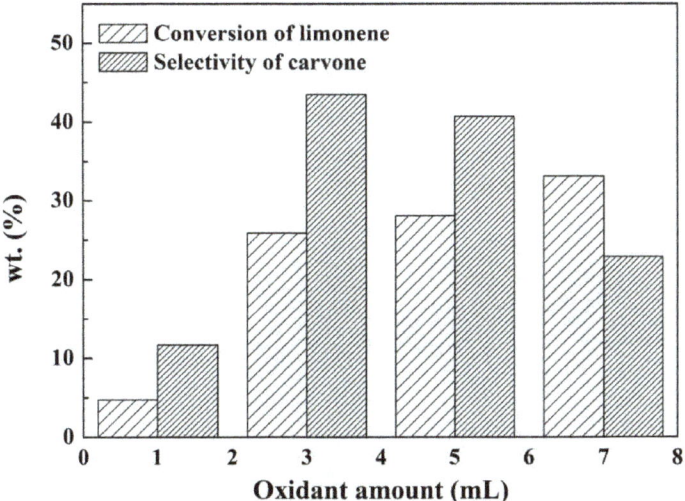

Figure 7. The effect of the oxidant dosage on the reaction. Conditions: limonene: 2 mL, catalyst: ZIF-67(Co), solvent: 15 mL acetic acid, oxidant: t-BHP, temperature: 85 °C, reaction time: 8 h.

2.2.5. The Effects of Different Solvents on the Reaction

The solvent is an important factor affecting heterogeneous catalytic reaction systems. The polarity, acidity, viscosity, and volatility of solvents have a great influence on the performance and reactivity of the catalyst. Some experiments were carried out to study the effects of different solvents on the catalytic oxidation of limonene by ZIF-67(Co), and the results are shown in Table 3. We found that limonene conversion and carvone selectivity were higher when benzene was used as solvent than with acetic acid, acetic anhydride, or ethyl acetate.

Table 3. The effect of different solvents on the reaction.

Solvent	Conversion (%)	Selectivity (%)
Acetic acid	25.9	33.3
Acetic anhydride	38.1	15.9
Ethyl acetate	42.6	23.1
Benzene	20.3	49.4

Conditions: limonene: 2 mL, catalyst: 85 mg of ZIF-67(Co), solvent: 15 mL, oxidant: 3 mL t-BHP, temperature: 85 °C, reaction time: 8 h.

2.2.6. The Effect of the Temperature on the Reaction

The oxidation of limonene was examined over the temperature range of 55–95 °C, and the results are shown in Figure 8. The conversion of limonene increased quickly from 6.5% to 30.5% for a reaction time of 8 h, when the temperature was increased from 55 to 65 °C. Thereafter, the trend of conversion had a slow decline. Interestingly enough, the selectivity for carvone reached its maximum (48.7%) at 85 °C. This might be because the activity of the catalyst definitely increased at the beginning of the temperature rise, so the conversion of limonene increased rapidly. However, after the temperature reached a certain value, the catalytic activity of the material did not increase and the byproduct formation did increase, so the selectivity for carvone decreased.

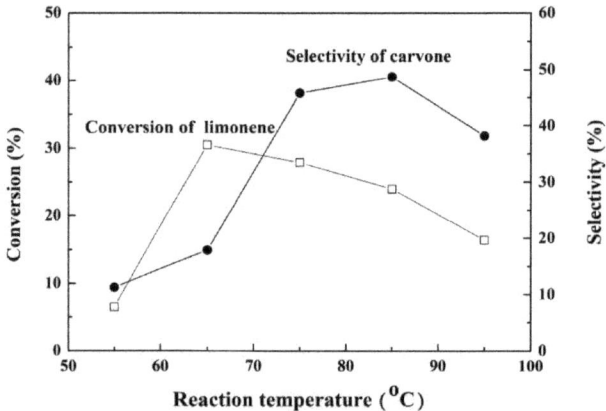

Figure 8. The effect of temperature on the reaction. Conditions: limonene: 2 mL, catalyst: 85 mg ZIF-67(Co), solvent: 12 mL benzene, oxidant: 3 mL t-BHP, reaction time: 8 h.

2.2.7. The Effect of the Reaction Time on the Reaction

The catalytic oxidation of limonene, using the above conditions for different reaction times, was studied. As can be seen in Figure 9, the conversion of limonene increased with increasing reaction time, and the conversion was 37.8% at 12 h. Meanwhile, the selectivity for carvone increased first, reached a maximum at 6 h (57.7%), and then decreased. Owing to the fact that t-BHP is not completely decomposed over short times, the oxidation efficiency was undesirable, and with increasing reaction time, there were corresponding increases in the oxidants' oxidation capacity as well as the conversion rate. However, selectivity of carvone decreased with reaction time increased because of the depth oxidation of carvone and by-product increased.

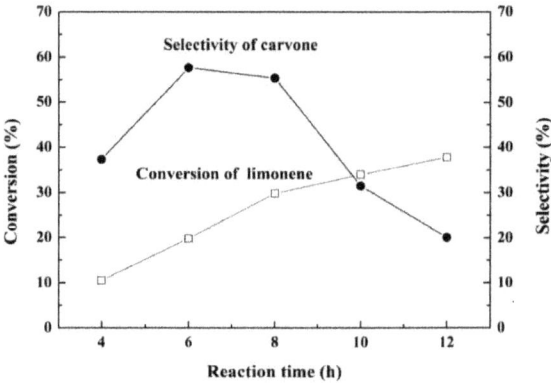

Figure 9. The effect of the reaction time on the reaction. Conditions: limonene: 2 mL, catalyst: 85 mg ZIF-67(Co), solvent: 12 mL benzene, oxidant: 3 mL t-BHP, temperature: 75 °C.

2.2.8. Study of the Catalyst Stability

Stability is an important parameter of catalyst performance. In order to investigate the stability of ZIF-67(Co), XRD spectra of ZIF-67(Co) before and after the first reaction round were obtained and are presented here. As shown in Figure 10, the characteristic peak was still present, but its strength decreased slightly. So, we can be sure that the structure of ZIF-67 still existed.

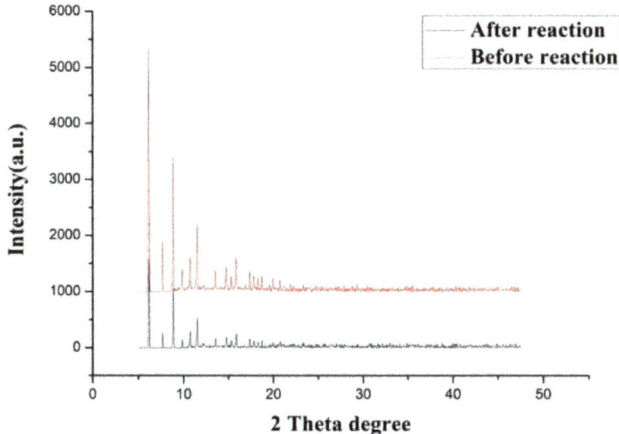

Figure 10. XRD spectra of ZIF-67(Co) as synthesized and after the first recycle.

Concurrently, the specific data about repetitive experiments are revealed in Table 4.

Table 4. The data from reuse experiments.

Times	Conversion (%)	Selectivity (%)	TON
First round	29.8	55.4	18.1
Second round	33.5	41.6	13
Third round	32.5	28.0	8.7

Conditions: limonene: 2 mL, catalyst: 85 mg of ZIF-67(Co), solvent: 12 mL benzene, oxidant: 3 mL t-BHP, temperature: 75 °C, reaction time: 8 h. Turnover number (TON) = carvone (yield) mmol/total Co site mmol.

The data in Table 4 reveal the repeatability of ZIF-67(Co) in the catalytic oxidation of limonene. As can be seen from the data in Table 4, we found that the selectivity for carvone declined slightly in the second round. At the same time, the conversion of limonene declined as well. We conclude that this may be related to a partial collapse of the basic structure of ZIF-67(Co).

3. Materials and Methods

3.1. Materials and Solvents

Cobaltous nitrate hexahydrate was supplied by Xilong Chemical Co., Ltd. (Shantou, China) and 2-methylimidazole was obtained from Sinopharm Chemical Reagent Co., Ltd. (Shanghai, China). All other reactants used in the synthesis of the ZIF-67(Co) catalyst samples were supplied by Xilong Chemical Co., Ltd. Methanol and acetic acid were purchased from Xilong Chemical Co., Ltd. The tert-butyl hydroperoxide and limonene were obtained from Adamas Reagent Co., Ltd. (Shanghai, China).

3.2. Catalyst Preparation

In a typical preparation process [33], 0.45 g cobaltous nitrate hexahydrate was dissolved in 3 mL of distilled water, and then 5.5 g of 2-methylimidazole was dissolved in 100 mL of distilled water. Then the two solutions were mixed evenly and stirred at room temperature for 6 h. The products obtained were washed with distilled water and methanol, and then dried in at 80 °C for 24 h.

3.3. Catalyst Characterization

ZIF-67(Co) was characterized through a series of techniques. The diffraction patterns of the materials and the residual catalyst were obtained by powder X-ray diffraction (XRD, Rigaku Co., Tokyo, Japan) experiments using Cu Kα band radiation on a D/max-3B spectrometer. Scans were made over the 2θ range of 10–90° with a scan rate of 10°/min (wide-angle diffraction). Pore size distributions, pore volumes, and specific surface area were measured by nitrogen adsorption/desorption using a Micromeritics ASAP 2460(Micromeritics, Norcross, GA, USA) at 77 K. The specific surface areas were calculated by the Brunauer–Emmett–Teller (S_{BET}) and Langmuir methods, and the pore volumes and pore sizes were calculated by the BJH (Barrett–Joyner–Halenda) method. Thermogravimetric analysis (TGA) curves and differential scanning calorimetry (DSC) determinations were carried out on a NETZSCH STA 449 F3 synchronous TG-DSC thermal analyzer (Nuremberg, Germany) with a scanning rate of 5 K/min in a dry nitrogen atmosphere and heating from room temperature to 1073 K. The FT-IR(Fourier transform infrared spectroscopy) measurements were performed on a Thermo Nicolet 8700 instrument(Wilmington, MA, USA). Potassium bromide pellets were used in FT-IR experiments at a spectral resolution of 4 cm^{-1}. Scanning electron microscopy (SEM) images were taken on a Quanta 200FEG microscope (FEI, Hillsboro, OR, USA) at an accelerating voltage of 15 kV with the pressure in the sample chamber set to 2.7×10^{-5} Pa. Double-sided adhesive tape was bonded on the carrier disc, then a small amount of powder sample was placed near the center of the carrier disc on the adhesive tape. Then, a rubber ball was used to blow the sample outward along the radial direction of the load plate, so that the powder was evenly distributed on the tape. The tape was then coated with conductive silver paste to connect the sample to the carrier plate. After the silver paste was dried, the final gold steaming process could be carried out.

3.4. Catalytic Performance

In a typical reaction, 2 mL limonene with varying amounts of catalyst and solvents were used as received, without further purification, in a 25 mL three-neck flask. Each set of experiments was performed at a corresponding temperature, under atmospheric pressure and magnetic stirring, using an appropriate amount of solvent, e.g., acetic acid, acetic anhydride, ethyl acetate, benzene, etc. Afterwards, the catalyst was separated by filtration, and the sample solution was dried with anhydrous magnesium sulfate. A final filtration was carried out, and the samples were qualitatively and quantitatively analyzed by gas chromatography (GC-9560, HUAAI Chromatography, Shanghai, China), with a capillary column AE.OV-624, 30 m in length, 0.25 mm i.d., 0.5 μm film thickness, and an FID (Flame ionization detector).

In order to assess the stability of the ZIF-67(Co), some sets of a repeatability experiment were performed. After the first reaction, the catalyst was centrifuged and separated from the reaction system, and after repeated washing with distilled water and anhydrous ethanol, the catalyst was reactivated at 90 °C in the vacuum drying chamber for 12 h. Thereafter, the reactor was supplied with the recovered dry catalyst, new reagents, and solvent for the second reaction. We repeated the above process three times.

4. Conclusions

In conclusion, we proved that the zeolitic imidazolate framework ZIF-67(Co) was an efficient catalyst for the liquid-phase catalytic oxidation of limonene, under relatively mild reaction conditions, to carvone. Based on current research, the catalytic oxidation of limonene comprises two parallel reactions, namely a synergistic reaction and a free-radical reaction. Carvone is the product of the free-radical reaction of limonene at the allyl group. The synergistic reaction produced limonene-1,2-oxides and limonene-1,2-diols. Meanwhile, under acidic conditions, the 1,2-oxide of limonene could be hydrolyzed to produce the 1,2-diol of limonene, and the 1,2-diol of limonene could be further rearranged and converted to carvone.

This is the first report on catalyzing the oxidation of limonene using ZIF-67(Co) as catalyst. It was revealed that ZIF-67(Co) was an efficient catalyst, with selectivity of 55.4% for carvone in the catalytic oxidation of limonene, and could be reused with similar activity.

Supplementary Materials: The following are available online at http://www.mdpi.com/2073-4344/9/4/374/s1, Figure S1: XRD pattern of MIL-101(Cr), Figure S2: XRD pattern of MIL-125(Ti), Figure S3: XRD pattern of HKUST-1(Cu), Figure S4: XRD pattern of ZIF-8(Zn), Figure S5: XRD pattern of MIL-101(Fe).

Author Contributions: Conceptualization, J.W. and Y.L.; Data curation, Z.Y. and Y.L.; Formal analysis, Y.L. and W.W.; Investigation, Y.L., Y.Y., D.C., and Z.L.; Methodology, J.W.; Project administration, J.W. and Z.Y.; Software, Z.L., Y.A., and L.Z.; Supervision, J.W. and Z.Y.; Writing—original draft, Y.L.; Writing—review and editing, J.W. and Z.Y.

Funding: This research was funded by National Natural Science Foundation of China (Project 21573193 and 21464016). The authors also thank the Key Projects for Research and Development of Yunnan Province (2018BA065), the Scientific Research Fund of Department of Yunnan Education (Project 2016CYH04 and 2017ZZX223) for financial support.

Acknowledgments: The work was supported by National Natural Science Foundation of China (Project 21573193 and 21464016). The authors also thank the Key Projects for Research and Development of Yunnan Province (2018BA065), the Scientific Research Fund of Department of Yunnan Education (Project 2016CYH04 and 2017ZZX223), and the Program for Innovation Team of Yunnan Province and Key Laboratory of Advanced Materials for Wastewater Treatment of Kunming for financial support.

Conflicts of Interest: The authors declare no conflict of interest.

References

1. Murphy, E.F.; Mallat, T.; Baiker, A. Allylic oxofunctionalization of cyclic olefins with homogeneous and heterogeneous catalysts. *Catal. Today* **2000**, *31*, 115–126. [CrossRef]
2. Sakthivel, A.; Dapurkar, S.E.; Selvam, P. Allylic oxidation of cyclohexene over chromium containing mesoporous molecular sieves. *Appl. Catal. A-Gen.* **2003**, *246*, 283–293. [CrossRef]
3. Goncalves, J.C.; Oliveira, F.S.; Benedito, R.B.; Sousa, D.P.; de Almeida, R.N.; de Araújo, D.A. Antinociceptive Activity of (-)-Carvone: Evidence of Association with Decreased Peripheral Nerve Excitability. *Biol. Pharm. Bull.* **2008**, *31*, 1017–1020. [CrossRef] [PubMed]
4. de Carvalho, C.C.C.R.; da Fonseca, M.M.R. Carvone: Why and how should one bother to produce this terpene. *Food Chem.* **2006**, *95*, 413–422.
5. Li, N.; Wang, F.; Li, M. The Chemical Modification of L(-)-Carvone and its Biological Activity to Three Fungal Pathogens of Plant. *J. Mt. Agric. Biol.* **2010**. [CrossRef]
6. Linder, S.M.; Greenspan, F.P. Reactions of Limonene Monoxide. The Synthesis of Carvone. *J. Org. Chem.* **1957**, *22*, 949–951. [CrossRef]
7. Robles-Dutenhefner, P.A.; Brandão, B.B.N.S.; de Sousa, L.F.; Gusevskaya, E.V. Solvent-free chromium catalyzed aerobic oxidation of biomass-based alkenes as a route to valuable fragrance compounds. *Appl. Catal. A Gen.* **2011**, *399*, 172–178. [CrossRef]
8. Caovilla, M.; Caovilla, A.; Pergher, S.B.C.; Esmelindro, M.C.; Fernandes, C.; Dariva, C.; Bernardo-Gusmão, K.; Oestreicher, E.G.; Antunes, O.A.C. Catalytic oxidation of limonene, α-pinene and β-pinene by the complex [Fe^{III}(BPMP)Cl(μ-O)Fe^{III}Cl$_3$] biomimetic to MMO enzyme. *Catal. Today* **2008**, *133*, 695–698.
9. Młodzik, J.; Wróblewska, A.; Makuch, E.; Wróbel, R.J.; Michalkiewicz, B. Fe/EuroPh catalysts for limonene oxidation to 1,2-epoxylimonene, its diol, carveol, carvone and perillyl alcohol. *Catal. Today* **2015**, *268*, 111–120. [CrossRef]
10. Wróblewska, A.; Makuch, E.; Miądlicki, P. The oxidation of limonene at raised pressure and over the various titanium-silicate catalysts. *Pol. J. Chem. Technol.* **2015**, *17*, 82–87. [CrossRef]

11. Li, J.; Li, Z.; Zi, G.; Yao, Z.; Luo, Z.; Wang, Y.; Xue, D.; Wang, B.; Wang, J. Synthesis, characterizations and catalytic allylic oxidation of limonene to carvone of cobalt doped mesoporous silica templated by reed leaves. *Catal. Commun.* **2015**, *59*, 233–237. [CrossRef]
12. Drake, T.; Ji, P.; Lin, W. Site Isolation in Metal–Organic Frameworks Enables Novel Transition Metal Catalysis. *Acc. Chem. Res.* **2018**, *51*, 2129–2138. [CrossRef]
13. Dhakshinamoorthy, A.; Li, Z.; Garcia, H. Catalysis and photocatalysis by metal organic frameworks. *Chem. Soc. Rev.* **2018**, *47*, 8134–8172. [CrossRef]
14. Mukherjee, S.; Desai, A.V.; Ghosh, S.K. Potential of metal–organic frameworks for adsorptive separation of industrially and environmentally relevant liquid mixtures. *Coord. Chem. Rev.* **2018**, *367*, 82–126. [CrossRef]
15. Li, H.; Wang, K.; Sun, Y.; Lollar, C.T.; Li, J.; Zhou, H.-C. Recent advances in gas storage and separation using metal–organic frameworks. *Mater. Today* **2018**, *21*, 108–121. [CrossRef]
16. Li, J.; Wang, X.; Zhao, G.; Chen, C.; Chai, Z.; Alsaedi, A.; Hayat, T.; Wang, X. Metal–organic framework-based materials: superior adsorbents for the capture of toxic and radioactive metal ions. *Chem. Soc. Rev.* **2018**, *47*, 2322–2356. [CrossRef] [PubMed]
17. Lima, I.F.; Corraza, M.L.; Cardozo-Filho, L.; Márquez-Alvarez, H.; Antunes, O.A.C. Oxidation of limonene catalyzed by Metal(Salen) complexes. *Braz. J. Chem. Eng.* **2006**, *23*. [CrossRef]
18. Raj, N.K.K.; Puranik, V.G.; Gopinathan, C.; Ramaswarmy, A.V. Selective oxidation of limonene over sodium salt of cobalt containing sandwich-type polyoxotungstate [WCo$_3$(H$_2$O)$_2${W$_9$CoO$_{34}$}$_2$]$^{10-}$. *Appl. Catal. A-Gen.* **2003**, *256*, 265–273.
19. Jiang, D.; Mallat, T.; Meier, D.M.; Urakawa, A.; Baiker, A. Copper metal–organic framework: Structure and activity in the allylic oxidation of cyclohexene with molecular oxygen. *J. Catal.* **2010**, *270*, 26–33. [CrossRef]
20. Yang, H.; He, X.-W.; Wang, F.; Kang, Y.; Zhang, J. Doping copper into ZIF-67 for enhancing gas uptake capacity and visible-light-driven photocatalytic degradation of organic dye. *J. Mater Chem.* **2012**, *22*, 21849–21851. [CrossRef]
21. Kuruppathparambil, R.R.; Jose, T.; Babu, R.; Hwang, G.-Y.; Kathalikkattil, A.M.; Kim, D.-W.; Park, D.-W. A room temperature synthesizable and environmental friendly heterogeneous ZIF-67 catalyst for the solvent less and co-catalyst free synthesis of cyclic carbonates. *Appl. Catal. B-Environ.* **2016**, *182*, 562–569. [CrossRef]
22. Yang, L.; Yu, L.; Sun, M.; Gao, C. Zeolitic imidazole framework-67 as an efficient heterogeneous catalyst for the synthesis of ethyl methyl carbonate. *Catal. Commun.* **2014**, *54*, 86–90. [CrossRef]
23. Zhu, M.; Srinivas, D.; Bhogeswararao, S.; Ratnasamy, P.; Carreon, M.A. Catalytic activity of ZIF-8 in the synthesis of styrene carbonate from CO$_2$ and styrene oxide. *Catal. Commun.* **2013**, *32*, 36–40. [CrossRef]
24. Yang, L.; Yu, L.; Diao, G.; Sun, M.; Cheng, G.; Chen, S. Zeolitic imidazolate framework-68 as an efficient heterogeneous catalyst for chemical fixation of carbon dioxide. *J. Mol. Catal. A-Chem.* **2014**, *392*, 278–283. [CrossRef]
25. Jose, T.; Hwang, Y.; Kim, D.-W.; Kim, M.-I.; Park, D.-W. Functionalized zeolitic imidazolate framework F-ZIF-90 as efficient catalyst for the cycloaddition of carbon dioxide to allyl glycidyl ether. *Catal. Today* **2015**, *245*, 61–67. [CrossRef]
26. Banerjee, R.; Phan, A.; Wang, B.; Knobler, C.; Furukawa, H.; O'Keeffe, M.; Yaghi, O.M. High-throughput synthesis of zeolitic imidazolate frameworks and application to CO$_2$ capture. *Science* **2008**, *319*, 939–943. [CrossRef] [PubMed]
27. Li, K.H.; Olson, D.H.; Seidel, J.; Emge, T.J.; Gong, H.W.; Zeng, H.P.; Li, J. Zeolitic imidazolate frameworks for kinetic separation of propane and propene. *J. Am. Chem. Soc.* **2009**, *131*, 10368–10369. [CrossRef] [PubMed]
28. Yang, J.; Zhang, F.; Lu, H.; Hong, X.; Jiang, H.; Wu, Y.; Li, Y. Hollow Zn/Co ZIF Particles Derived from Core-Shell ZIF-67@ZIF-8 as Selective Catalyst for the Semi-Hydrogenation of Acetylene. *Angew. Chem. Int. Ed. Engl.* **2015**, *127*, 11039–11043. [CrossRef]
29. Kuo, C.-H.; Tang, Y.; Chou, L.-Y.; Sneed, B.T.; Brodsky, C.N.; Zhao, Z.; Tsung, C.K. Yolk–shell nanocrystal@ZIF-8 nanostructures for gas-phase heterogeneous catalysis with selectivity control. *J. Am. Chem. Soc.* **2012**, *134*, 14345–14348. [CrossRef]
30. Wee, L.H.; Lescouet, T.; Ethiraj, J.; Bonino, F.; Vidruk, R.; Garrier, E.; Packet, D.; Bordiga, S.; Farrusseng, D.; Herskowitz, M.; et al. Hierarchical zeolitic imidazolate framework-8 catalyst for monoglyceride synthesis. *ChemCatChem* **2013**, *5*, 3562–3566. [CrossRef]

31. Chizallet, C.; Lazare, S.; Bazer-Bachi, D.; Bonnier, F.; Lecocq, V.; Soyer, E.; Quoineaud, A.-A.; Bats, N. Catalysis of transesterification by a nonfunctionalized metal– organic framework: acido-basicity at the external surface of ZIF-8 probed by FTIR and ab initio calculations. *J. Am. Chem. Soc.* **2010**, *132*, 12365–12377. [CrossRef] [PubMed]
32. Shi, Q.; Chen, Z.; Song, Z.; Li, J.; Dong, J. Synthesis of ZIF-8 and ZIF-67 by Steam-Assisted Conversion and an Investigation of Their Tribological Behaviors. *Angew. Chem. Int. Ed. Engl.* **2011**, *50*, 672–675. [CrossRef] [PubMed]
33. Qian, J.; Sun, F.; Qin, L.; Li, J.; Dong, J. Hydrothermal synthesis of zeolitic imidazolate framework-67 (ZIF-67) nanocrystals. *Mater. Lett.* **2012**, *82*, 220–223. [CrossRef]

© 2019 by the authors. Licensee MDPI, Basel, Switzerland. This article is an open access article distributed under the terms and conditions of the Creative Commons Attribution (CC BY) license (http://creativecommons.org/licenses/by/4.0/).

Article

Pilot-Scale Production, Properties and Application of Fe/Cu Catalytic-Ceramic-Filler for Nitrobenzene Compounds Wastewater Treatment

Bingchuan Yang [1,2], Yuanfeng Qi [2,3,*] and Rutao Liu [2,*]

1. School of Chemistry and Chemical Engineering, Liaocheng University, Liaocheng 252059, China; yangbingchuan@lcu.edu.cn
2. School of Environmental Science and Engineering, Shandong University, Jinan 250100, China
3. School of Environmental Science and Municipal Engineering, Qingdao University of Technology, Qingdao 266000, China
* Correspondence: lewis19886@163.com (Y.Q.); rutaoliu@sdu.edu.cn (R.L.)

Received: 4 December 2018; Accepted: 19 December 2018; Published: 25 December 2018

Abstract: Iron powder, Kaolin powder and $CuSO_4 \cdot 5H_2O$ were employed as the main materials for the pilot-scale production of Fe/Cu catalytic- ceramic-filler (CCF) by way of wet type replacement-thermo-solidification. The physical properties, half-life, microstructure, removal rate of nitrobenzene compounds and the biodegradability-improvement of military chemical factory comprehensive wastewater were tested in comparison with commercial Fe/C ceramic-filler (CF). Catalytic micro-electrolysis bed reactors (CBRs) designed as pretreatment process and BAFs (Biological Aerated Filters) were utilized in a 90 days field pilot-scale test at last. The results showed the characteristics of optimum CCF were: 1150 kg/m^3 of bulk density, 1700 kg/m^3 of grain density, lower than 3.5% of shrinking ratio, 3.5% of 24 h water absorption, 6.0 Mpa of numerical tube pressure, 0.99 acid-resistance softening co-efficiency and 893.55 days of half-life. 25% addition of Fe with 1% of copper plating rate was efficient for the removal of nitrobenzene compounds and significant in promoting the biodegradability of military chemical factory comprehensive wastewater. The two-stage design of CBRs and BAFs showed high dependability and stability for the practical engineering application.

Keywords: Fe/Cu catalytic-ceramic-filler; nitrobenzene compounds wastewater; pilot-scale test; biodegradability-improvement

---

1. Introduction

As the chemical industry developed rapidly in China, the consumption of nitrobenzene compounds which was widely utilized as the raw materials especially in pharmaceutical, perfume, military and dye industries dramatically increased. As a result, the discharge amount of refractory wastewater which abounds of nitrobenzene compounds grows sharply as well [1–3]. Nitrobenzene compounds wastewater can make great damage to the environment [4] which is stable and hard to be decomposed under national conditions. The toxicity of the nitrobenzene compounds in wastewater lower down the biodegradability, and harmful to the survival of creatures or microorganism. For example, as shown by [5–7], fishes will die when the concentration of trinitrotoluene is higher than 1 mg/L, and aerobic microbes in active sludge will be inhibited when the concentration of tri-nitrobenzene compounds over 5–10 mg/L.

Biological methods are widely used in wastewater treatment which cost less than the other treatments [8]. But biological treatment is not suitable for nitrobenzene compounds wastewater degradation (BOD$_5$/CODcr = 0.09–0.2). In order to improve the biodegradability (BOD/CODcr ratio), a large amount of glucose and methanol must be added into the wastewater [9–11]. Other ways to improve the biodegradability are mostly chemical oxidation method, such as Fenton, [12,13] ozone oxidation [14–16] and supercritical water oxidation [17]. All the mentioned methods have common disadvantages which with high operating cost and secondary pollutions [18] and diseconomy for the practical engineering application.

Micro-electrolysis methods are proved to be efficient for certain nitrobenzene compounds wastewater treatment [19–22]. Traditional Fe/C micro-electrolysis [23,24] are usually subject to the short-circuiting and clogging [25,26] during the actual operation, resulting in passivation and inefficiency. It is well known that Fe/Cu bimetal-system may accelerate the corrosion of iron, hence improve the micro-electrolysis ability (could be called as catalytic micro-electrolysis) and improve performance of the traditional Fe/C micro-electrolysis. Lab-scale Fe/Cu catalytic micro-electrolysis material which was applied in the TNT (Trinitrotoluene) wastewater treatment in our previous studies [27,28] were proved more effective than that of traditional Fe/C micro-electrolysis filler. But the Fe/Cu catalytic micro-electrolysis filler was prepared regardless the cost and was lack of basic properties test

As a follow-up, there were three the goals in this study: first, pilot-scale of Fe/Cu catalytic ceramic filler of cost-effective and high-efficiency was produced; second, the basic properties for practical engineering application of Fe/Cu catalytic ceramic filler compared with commercial Fe/C ceramic filler was test; third, pilot-scale application in military chemical factory comprehensive wastewater were performed and parameters for practical project design were verified.

2. Result and Discussion

2.1. Basic Properties of CCF

The basic properties such as bulk density, grain density, 24 h water absorption, shrinking ratio, numerical tube pressure and acid-resistance softening co-efficiency were tested during the pilot-scale production of CCF. And effect of additions of iron and copper plating rate of iron were shown in Figures 1 and 2, respectively.

From Figures 1 and 2, as the addition of iron raised, the bulk density and grain density raised as well (density of iron powder is much higher than Kaolin powder). The shrinking ratio increased slowly (lower than 2%) before the addition of iron lower than 25%, but obviously increased when the addition of iron become more than 25% (the copper plating rate affect the shrinking ratio seriously as well). During the preparation of the raw material mixture, 1% of $NaHCO_3$ was added as swelling agent. $NaHCO_3$ decomposes and releases CO_2 and H_2O at 600 °C, and the generated gases would be captured by the melting Kaolin powder which prevented the shrink of raw pellets before the addition of iron lower than 25%. As the addition of iron increased over than 30% (meanwhile the percentage of Kaolin powder lower than 70%), the generated gases cannot be captured by the melting Kaolin powder and released out of the pellets bodies, which leaded the seriously increase of shrinking ratio [29,30]. 24 h water absorption decreased when the iron content increased. To sum up, high percentage of iron affects the density, water absorption and shrinking ratio a lot.

As for the numerical tube pressure and the acid-resistance softening co-efficiency, all the materials showed good performances than CF. The acid-resistance softening co-efficiency grown rapidly when the percentage of iron increased (before 25%), after that, the pressure drops off. This phenomenon occurred because of the high percentage of iron which can react with acid and release hydrogen, [31] then the surface of the catalyst would be destroyed. According to the acid-resistance softening co-efficiency, 25% of iron content was the best option for the preparation of CCF applied in acidic environment.

Compared with CF (shown in Table 1) and CCF prepared at 25% of iron content, CF has lower bulk density (1085 kg/m^3 to 1150 kg/m^3) and grain density (1548 kg/m^3 to 1700 kg/m^3) but higher 24 h water absorption (3.8% to 3.5%), which contributed to the addition of carbon during the manufacture of CF. CCF has higher numerical tube pressure (6.0 Mpa) and the acid-resistance softening co-efficiency (about 99%) than that of CF (4.2 Mpa, 92.5%), which indicated CCF showed better compression performance and more suitable applied as filler in fixed bed reactor operated in acidic environment.

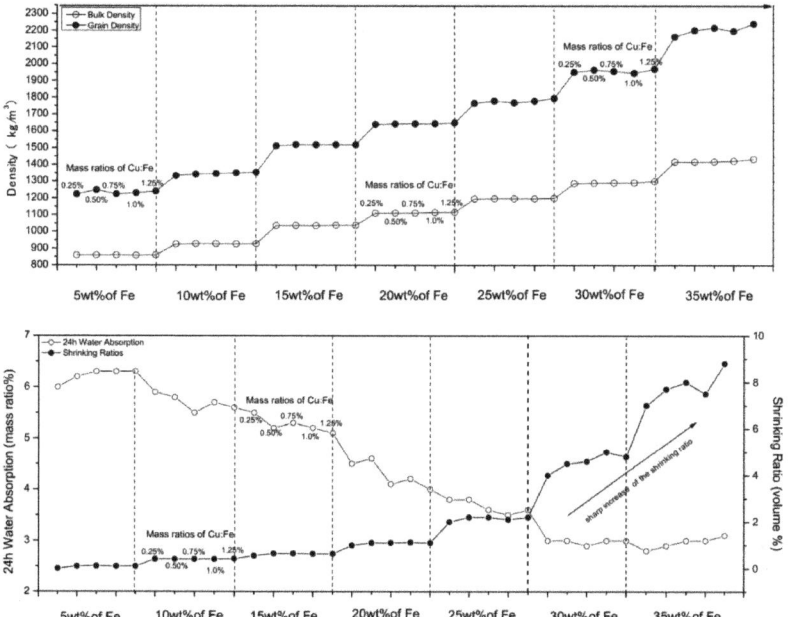

Figure 1. Bulk density, grain density, 24 h water absorption and shrinking ratio test of CCF (Copper Carbon Fe Catalyst).

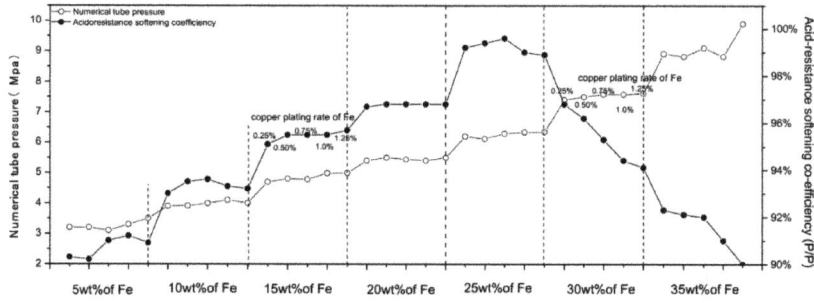

Figure 2. Numerical tube pressure and Acid-resistance softening co-efficiency test of CCF.

To verify the optimum copper plating rate, 5 species of CCF with 25% of iron content but different copper plating rate were manufactured and 6 groups of sequence nitrobenzene compounds wastewater treatment test was performed (No.1 group to No.5 group for CCFs and No.0 group for CF as control).

2.2. Result of the Sequence Wastewater Treatment Test

2.2.1. Effect of Iron Content and Copper Plating Rate for the Removal of TNCs

Normally and traditionally, ratios of BOD_5 to COD_{cr} (BOD_5/COD_{cr} or B/C) were employed as the parameters for the evaluation of biodegradability in the practical engineering design, and the B/C ratio of 0.3 was the basic requirement for the survival of biofilm attached to LCF (which applied as filler in BAFs). But in the practical project operation, B/C was restricted as the daily monitoring index mainly related to the relatively long test time (5 days for the BOD_5 at least).

The nitrobenzene compounds wastewater is not suitable for biological treatment because of the bio-toxicity of nitrobenzene compounds which lower down the biodegradability. And as in this study, the main substances reduced the biodegradability of wastewater were the mono-nitrobenzene, di-nitrobenzene and tri-nitrobenzene compounds. As a result, the concentration of TNCs could be considered as substitute parameter instead of B/C for the evaluation of biodegradability in comprehensive nitrobenzene compounds wastewater treatment during the field pilot-scale test or further applications.

Result of 6 groups of sequence wastewater treatment test performed at aerobic/anaerobic conditions were shown in Figure 3A,B respectively.

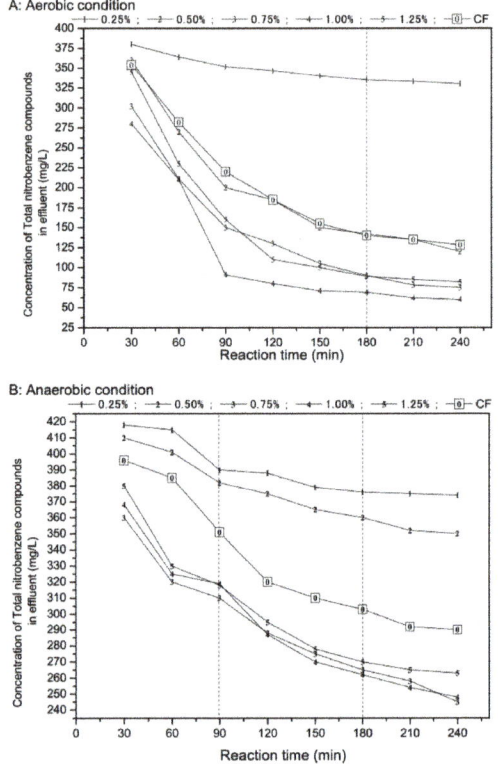

Figure 3. Effect of CCF and CF (Carbon Fe Catalyst) performed at aerobic/anaerobic conditions. (CCF was prepared with 25% of iron, and copper plating rate for iron range from 0.25% to 1.25%).

As shown in Figure 3, all of the degradation procedure can be divided into three phases. Phase 1 fast degradation period (0–90 min), phase 2 slow degradation periods (90–180 min), phase 3 stable degradation periods (180–240 min), and the TNCs could be degraded by CCF and CF, obviously. When CCF and CF were settled under anaerobic condition and acid environment, Zero-valent iron would release electron (e^-_{aq}) by electrical catalysis of copper/carbon, and the followed free radical reaction would occur [32]:

$$Fe \xrightarrow{Cu/C} Fe^{2+} + 2e^-_{aq} \qquad (1)$$

$$e^-_{aq} + H^+ \rightarrow H\cdot \text{ or } e^-_{aq} + H_2O \rightarrow H\cdot + OH^- \qquad (2)$$

$$H\cdot + H_2O \rightarrow H_2 + OH \tag{3}$$

The reducibility/oxidizability of H radical/OH would destroy the nitro and benzene ring of nitrobenzene compounds [28].

When the 6 groups of sequence wastewater treatment test performed at aerobic conditions, concentrations of TNCs in effluent dropped faster than at anaerobic conditions, which attributed to the dissolved oxygen produced more oxidative radicals (such as $\cdot O_2^-$; OH·; HO_2^-; H_2O_2) and enhanced the degradation of nitrobenzene compounds [22,24]. The pathway of related oxidative radicals generated was shown as followed:

$$e_{aq}^- + O_2 \rightarrow \cdot O_2^-; \tag{4}$$

$$e_{aq}^- + \cdot O_2^- + H_2O \rightarrow HO_2^- + OH^- \tag{5}$$

$$H^+ + HO_2^- \rightarrow H_2O_2 \tag{6}$$

$$H_2O_2 \xrightarrow{Fe^{2+}} OH\cdot \text{ or } H_2O_2 + e_{aq}^- \rightarrow OH\cdot + OH^- \tag{7}$$

The core of the reaction (2)~(7) was the reaction rate of (1) and the amount of generated free electrons. As it was shown in Figure 3, when the copper plating rate over 0.75%, the removal rate of TNCs for CCF were higher than that for CF, which indicated a higher efficiency of copper promoted the corrosion of iron than that of carbon, obviously. When the copper plating rate was 1.25% (No.5 group CCF), the removal rate of TNCs was lower than that of 1.00% (No.4 group CCF). This phenomenon occurred mainly attributed to the over coverage of copper which isolated the oxygen to iron [24]. However, this is the first study to examine the degradation of TNCs by Fe/Cu ceramic filler; it was not possible to predict the behavior of those compounds in any detail. Addition studies will have to be conducted to follow the destruction of individual nitrobenzene compounds and is was beyond the scope of the present study.

As a result, 25% of the iron with 1.00% copper plating rate was optimum content for the preparation of CCF, and aeration were benefit for the removal of TNCs.

2.2.2. Effect of CFF and CF for the Biodegradability of Wastewater

The concentrations test of CODcr, BOD_5 and the TNCs in effluent for No.5 group (CCF with 25% of iron content and 1.00% of copper plating rate) and No.0 group (CF) which both operated at aerobic conditions (Figure 3A) were shown in Figure 4, and the corresponding biodegradability of the effluent was calculated by BOD_5/CODcr and shown as well.

In Figure 4, CODcr, and TNCs dropped, and BOD_5 increased before 120 min then fall off. The BOD_5/CODcr ratio for CCF over 0.3 began from 90 min and the peak value was 0.38 (appeared at 120–150 min). The BOD_5/CODcr ratio kept lower than 0.3 until 210 min. When the BOD_5/CODcr ratio approached to 0.3, the TNCs in effluent was 91 mg/L (CCF, Figure 4A) and 128 mg/L (CF, Figure 4B), respectively, but the concentration ratio of CODcr to TNCs (COD/TNC) of CCF and CF was 18.92 and 11.71, respectively. We inferred the different reaction process and pathways between CCF and CF caused the same BOD_5/CODcr but different COD/TNC. The minimum efficient hydraulic retention time (HRT) for CCF and CF was 90 min and 210 min, respectively. The optimum HRT for CCF was 150 min but more than 240 min for CF. The concentration of nitrobenzene compounds (91 mg/L for CCF and 128 mg/L for CF) in effluent was suitable for the biological treatment process and feasible as critical value for the pilot-scale test and/or practical engineering application. In addition, 30 min of extra reaction time was added in the followed test and enhanced the removal rate of TNCs which guaranteed the BOD_5/CODcr ratio over 0.3.

Both the degradation of CODcr and nitrobenzene compounds of CCF and CF were fitted the first order reaction kinetics. The reaction kinetic equation fitting for the removal of CODcr and nitrobenzene compounds were respectively shown in Table 1: and the corresponding reaction rate was calculated as well. CCF had a higher efficiency both in the removal of CODcr (K_{COD} = 6.45/min) and the degradation of TNCs (K_{NC} = 4.15/min) than that of CF (K_{COD} = 5.52/min and K_{NC} = 2.18/min, respectively).

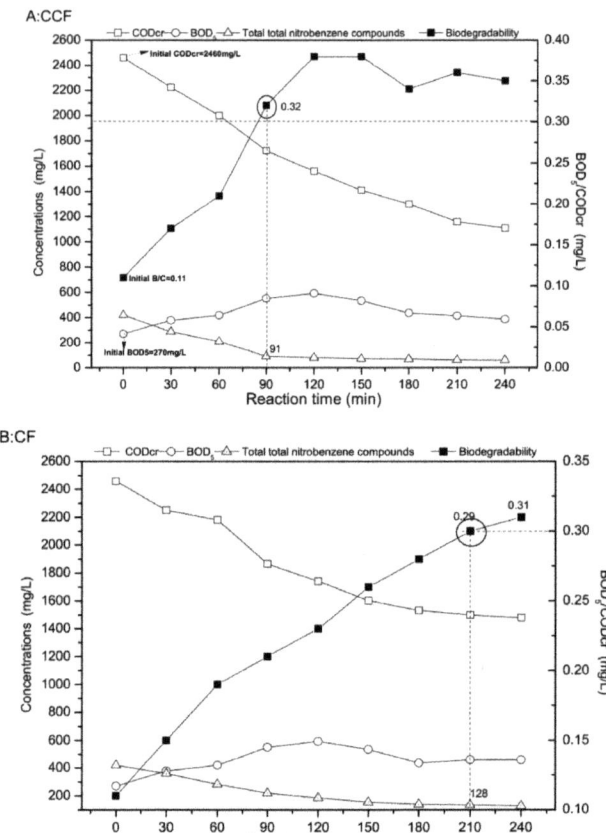

Figure 4. Effect of CCF and CF applied in comprehensive nitrobenzene compounds wastewater treatment.

Table 1. Reaction kinetic fitting for the degradation of CODcr and TNCs.

Degradation of CODcr	Equation: $Y(COD_{cr}) = A_{1-COD} \times \exp^{\frac{-x}{t_{1COD}}} + Y_0(CODcr)$				
	A_{1-COD}	t_{1COD}	$Y_0(CODcr)$	R^2	K_{COD} [1]
CCF	1964.4 mg/L	192.44 min	519.4 mg/L	0.9973	6.45/min
CF	1344.97 mg/L	154.01 min	1147.93 mg/L	0.9777	5.52/min
Degradation of TNCs	Equation: $Y(NC) = A_{1-NC} \times \exp^{\frac{-x}{t_{1NC}}} + Y_0(NC)$				
	A_{1-NC}	t_{1NC}	$Y_0(NC)$	R^2	K_{NC} [2]
CCF	383.27 mg/L	58.32 min	45.13 mg/L	0.9798	4.15/min
CF	341.78 mg/L	99.13 min	89.11 mg/L	0.9913	2.18/min

Note: 1, 2: the reaction rate.

Generally, as compared with CCF and CF applied in the sequence test, CCF performed more efficient in the degradation of CODcr and TNCs, which benefit for the improvement of biodegradability (the B/C increased from 0.11 to 0.32–0.38).

2.3. Results of Backwash Frequency and the Volume Half-Life Test

2.3.1. Result of Backwash Frequency Test for CCF and CF

Two evaluation reactors (filled with CCF and CF, respectively) were operated at aerobic conditions (aeration intensity was set up to 50 L/min) in sequence operating model. HRT for CCF and CF was 180 min and 270 min, and the he backwash process was not operated until the concentration of nitrobenzene compounds in the effluent over 91 mg/L and 128 mg/L, respectively. Result of backwash frequency test was shown in Figure 5. The surface appearance of CCF and CF after/before applied was shown in Figure 6.

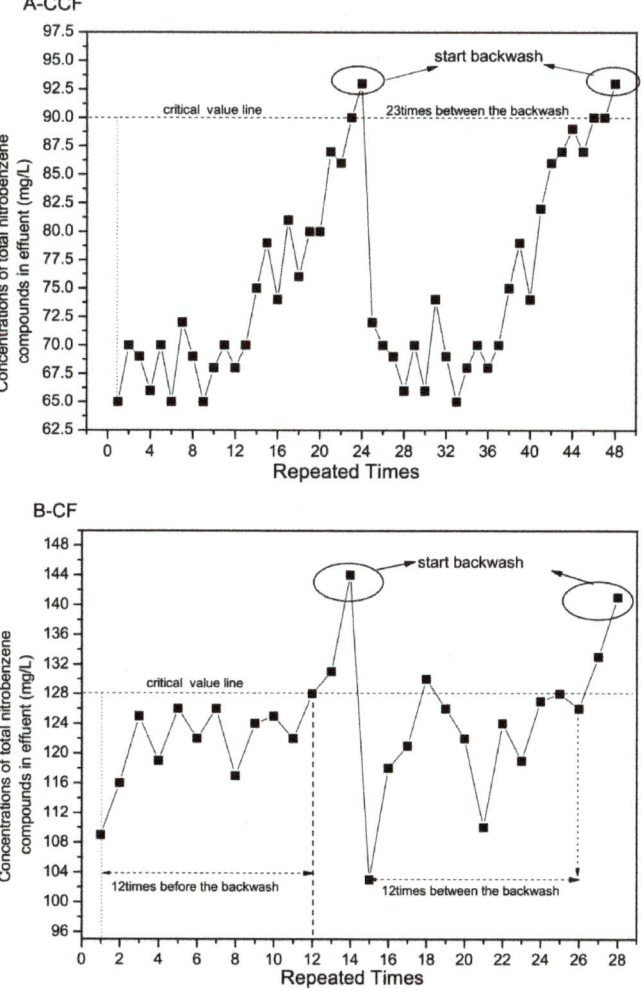

Figure 5. The backwash frequency of the optimum iron-based catalyst.

As it was shown in Figure 5, concentration of TNCs in effluent from the reactor filled with CCF or CF appeared a rising trend before or after backwash procedure operated. The backwash procedure not operated until the repeated times of 24 and 48 for CCF (the effective repeated times was 23) but 14 and 28 for CF (the effective repeated times was 12). The effective time of a single cycle (time of reaction and backwash procedure) for CCF and CF was 4160 min (69.3 h) and 3260 min (54.3 h), respectively. CCF had a longer single cycle running time than that of CF. After the operation of backwash procedure, the activity of CCF and CF were resumed to the initial competence. Therefore, the backwash procedure designed as regeneration process was essential and efficient for the application of CCF or CF in practice.

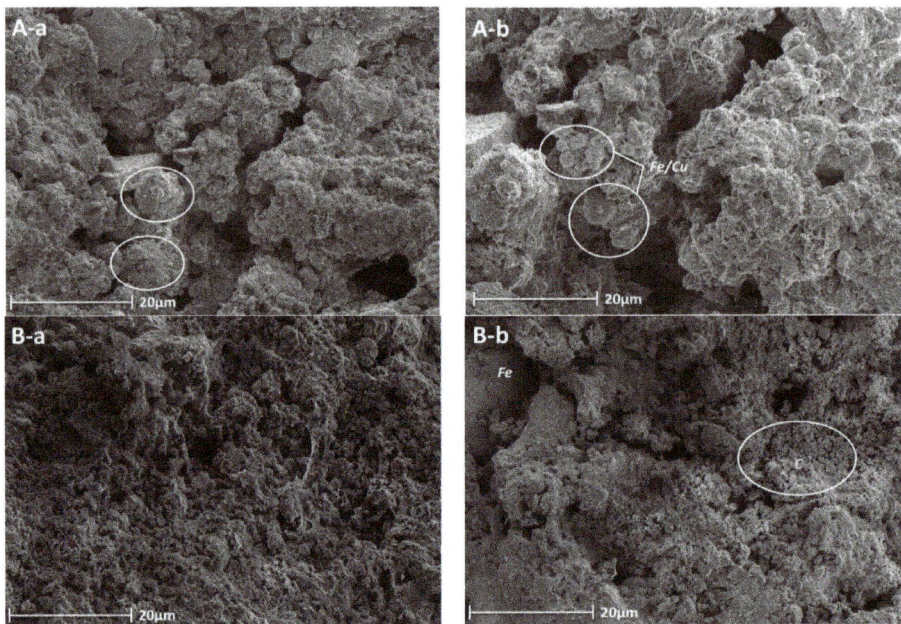

Figure 6. The surface appearance of CCF and CF ((**A**)-CCF, (**B**)-CF; (**a**)-after applied, (**b**)-before applied).

From Figures 5 and 6: Before the application of CFF (Figure 6A-b) and CF (Figure 6B-b), Fe/Cu and Fe/C compounds could be detected from the image, obviously. After the application in a single cycle, both the Fe/Cu and Fe/C compounds were totally covered by the by-products (FeOOH$_2$ or Fe(OH)$_3$) [23] generated from the complicated electrochemical reaction of iron, H$_2$O and oxygen. The increase of TNCs in effluent might be caused as follow steps: firstly, the by-products generated and partially covered the surface of iron which isolated the oxygen attached to iron, gradually; secondly, the corrosion of iron was prevented by the covering, and fewer free electrons and radicals were generated, then the destruction rate of TNCs dropped; CCF and CF were further covered by the by-products and deactivation appeared on the surface, and concentration of TNCs over the critical value at last. (Repeated times was 24 for CCF (Figure 5A) and 13 for CF (Figure 5B), respectively).

2.3.2. Result of Volume Half-Life Test for CCF and CF

The operation time of backwash procedure and remaining height of CCF/CF were shown in Figure 7.

The initial packing volume and height of CCF and CF filled in each evaluation reactor was 50 L and 500 mm, respectively. As it was shown in Figure 7, when the loss of height for CCF/CF was 20%, 40% and 50%, the backwash operating time was 48.3 h/53.2 h, 85 h/108.3 and 103.6/127.8, respectively. The backwash operating time and remaining height of CCF/CF approximately fitted linear relation.

As mentioned, the time operated for backwash procedure was 20 min. When the initial volume reduced to half, repeated times of backwash procedure for CCF and CF were 310.8 and 383.4, respectively. The effective time of a single cycle (time of reaction and backwash procedure) for CCF and CF was 4160 min (69.3 h) and 3260 min (54.3 h), respectively. Therefore, the half-life of CCF and CF could be calculated by the reduced times and effective time of a single cycle: about 21,538.4 h (897.4 days) of volume half-life for CCF and 20,818.6 h (867.4 days) for CF, respectively.

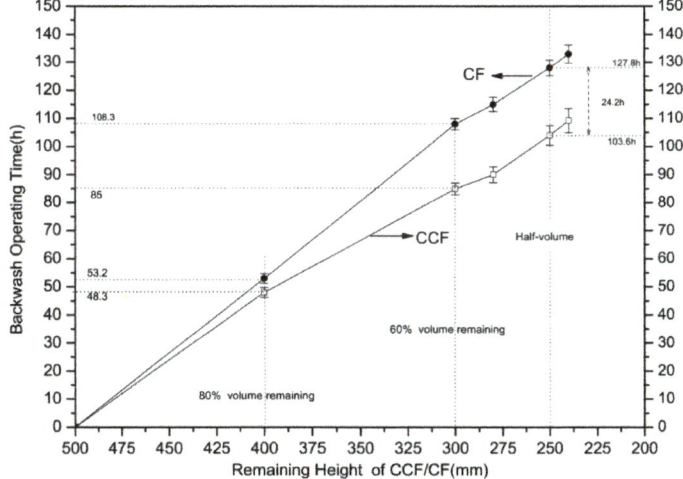

Figure 7. The half-volume test of CCF and CF.

In general, when the BOD_5/COD_{cr} ratio higher than 0.3 and the concentration of TNCs in effluent was not exceed the critical value as well, the operating conditions for CCF or CF were gathered and shown as follows: the effective time of a single cycle (time of reaction and backwash procedure) was 69.3 h or 54.3 h, the backwash procedure operated every 69 h or 54 h, and the volume half-life was 897.4 days or 867.4 days, respectively. CCF showed more excellent efficiency in the degradation of COD_{cr} and TNCs, and in the improvement of biodegradability than that of CF.

2.4. Results of Field Pilot-Scale Test

2.4.1. The Design and Operating Method of the Two-Stage Wastewater Treatment System

The maximum wastewater treatment capacity of the pilot-scale test was 2 m^3/d. Because of the low pH of wastewater, PP (polypropylene) which has excellent corrosion resistance property was employed as the main material for the manufacture of CBRs and BAFs. The effective HRT for each CBR and BAF was 180 min (3 h) and 8 h, respectively. The Process Flow Diagram of the two-stage wastewater treatment system was shown in Figure 8, scene images of the running CBRs and BAFs were shown in Figure 9, and initial wastewater, the effluent of CBRs and BAFs of 90th day were shown in Figure 10.

The PFD of wastewater treatment could be divided into three portions:

Part 1: Preliminary treatment portion: the initial wastewater was stored at regulation pool (effective volume was 2 m^3). And at the bottom of the pool, ball valves and vent pipe were designed for the discharge of sediment and sludge. HCl which stored in the dosing tank was added into the regulation pool by automatic dosing unit (adjust the pH of initial wastewater lower than 3.00).

Part 2: The catalytic-biological treatment portion: the catalytic treatment process (stage1) and biological treatment process (stage2), which was considered as the core of the system, both shown in Figure 9. Wastewater was pump by lifting pump of regulation pool into the bottom of CBR-1, and the effluent of CBR-1 flow automatically through the outlet pipe and stored in Reaction sedimentation tank 1, then the lifting pump of CBR-2 pumped the wastewater into the bottom of CBR-2 and the effluent was stored at Reaction sedimentation tank 2. The same pathway was implemented in the BAF-1 and BAF-2 treatment process.

Part 3: Backwash portion: both the CBRs and BAFs are performed the same backwash method which has been mentioned above. Each CBRs was separately backwash every 3 days. The backwash procedure for BAFs was not start until the system operated for 50 days, and from 52–90 day each BAFs was separately backwash every 7 days. In addition, the lifting pump was applied as backwash pump at backwash process.

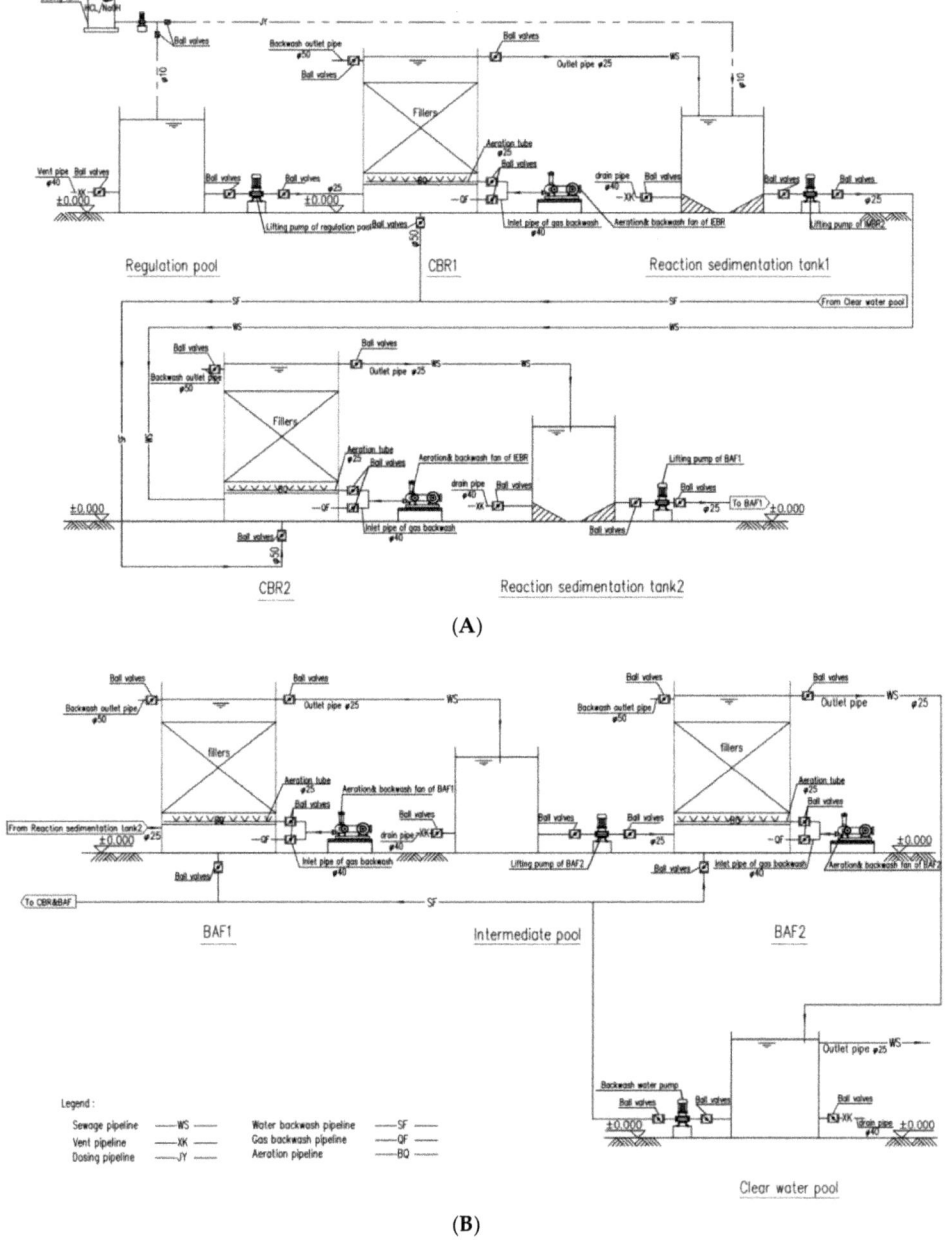

Figure 8. The Process Flow Diagram (PFD) of the two-stage wastewater treatment system.

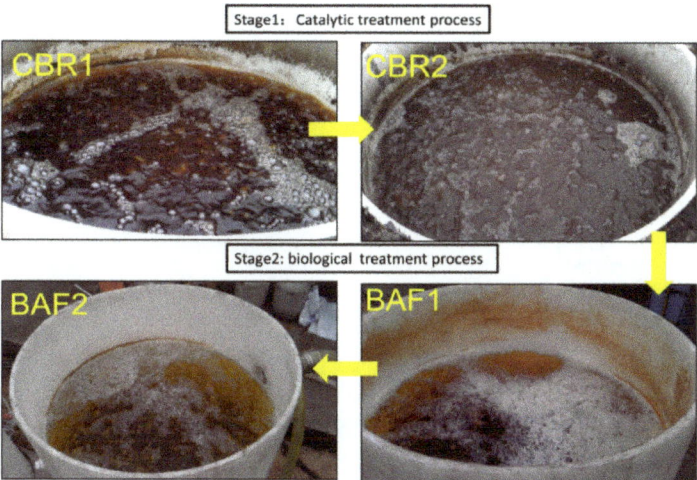

Figure 9. The scene images of the catalytic-biological treatment portion (CBRs and BAFs).

Figure 10. Effluent of catalytic-biological treatment portion (at 90th day). (Left to right: pure water/initial water/CBR-1/CBR-2/BAF-1/BAF-2).

Site commissioning and operation of the test divided into three portions:

1. 1st day to 30th day, quantity of influent was 2 m³ per day (Q = 2 m³/d) and no other preliminary process operated except for regulation of pH.
2. 31th day to 60th day, PAM was extra added into the regulation pool for the removal of the suspended solid (SS). And the quantity of influent was improved to 2.5 m³/d in order to verify the resistance of the system.
3. 61th day to 90th day, the quantity of influent was 2 m³ per day (Q = 2 m³/d) in order to verify the resumption performance of the system. The scene images of effluent at 90th day were shown in Figure 10.

2.4.2. The TNCs and CODcr Removal by System

The results for the removal of TNCs and CODcr by the system were shown in Figures 11 and 12, respectively.

Catalysts **2019**, *9*, 11

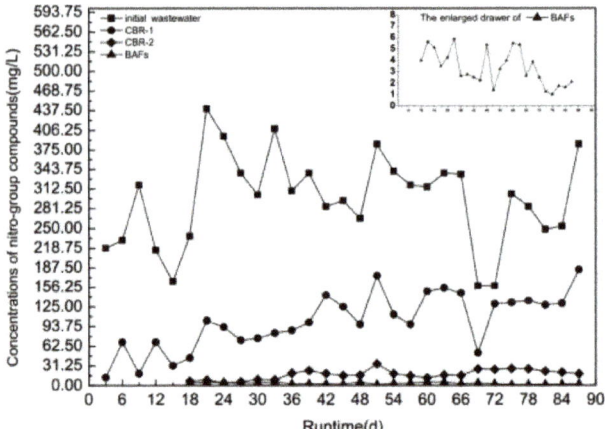

Figure 11. The concentration of nitrobenzene compounds in effluent of stage 1.

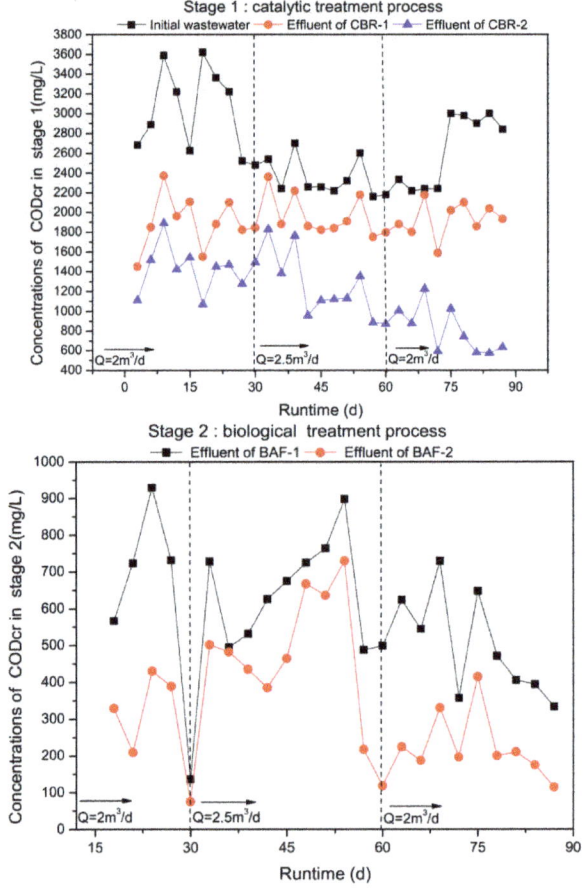

Figure 12. The concentration of CODcr in effluent in each stage. (stage 1: CBR-1 and CBR-2; stage 2: BAF-1 and BAF-2).

As shown in Figure 11: the concentration of nitrobenzene compounds existed in the initial wastewater ranged from 156.25 mg/L to 437.5 mg/L (mean concentration was 312 mg/L). Concentrations of nitrobenzene compounds in effluent of CBR-1 had a slight upward tendency before 42nd day (from 15.6 mg/L to 140 mg/L), then kept small fluctuation stability ranged from 42nd day to 90th 42–90. For CBR-2, the concentration of nitrobenzene compounds in effluent stayed lower than 31.25 mg/L. The average removal rate of nitrobenzene compounds in CBRs was higher than 90% which showed obviously dependability in dispose the nitro-group compounds.

As shown in Figure 12, combined with the site commissioning and operation situation:

① The removal of CODcr kept excellent stability in CBR-1 when the quantity of influent and SS changed (which was removed by PAM from 31st day to 90th day); And the removal of CODcr was slightly affected by the quantity of influent in CBR-2.

② For the BAFs, the removal of CODcr was obviously affected by the quantity of influent. And the removal rate of CODcr of BAFs was 20–30% from 31st day to 60th day (the average volume load of biological process was 1.875 kg m^{-3} d^{-1}), but 65–75% at 1st day to 30th day 30 and 61st day to 90th day (the average volume load of biological process was 1.5 kg m^{-3} d^{-1}). Compared with our previous studies [24] (CF and lab-scale produced CCF were utilized as pretreatment for the TNT wastewater treatment): when the effective HRT of two BAFs were 16 h and the BOD/CODcr was over 0.3 after the catalytic process, the CODcr and total nitrobenzene compound in comprehensive wastewater were more difficult to be removed than TNT wastewater (only the tri-nitrobenzene compounds existed). This probably attributed to the lower growth velocity of biomass and content of biomass for BAFs which operated in extreme conditions. And it can be further inferred that the membrane biological reactor (MBR), which had longer sludge retention time and higher content of biomass, would be more suitable as the biological process stage for the nitrobenzene compounds wastewater treatment during the application of practical project.

To sum up, the backwash process enabled the EBRs had anti-clogging ability and Fe/Cu ceramic-catalyst showed excellent influent impact resistance and nitro-group compounds impact resistance. BAFs showed high removal rate of CODcr with the average volume load was 1.5 kg m^{-3} d^{-1}.

3. Materials and Methods

3.1. Pilot-Scale Production and Basic Property Test of Fe/Cu Catalytic-Ceramic-Filler

3.1.1. Raw Materials

Iron powder and Kaolin powder ($Al_2O_3 \cdot 2SiO_2 \cdot 2H_2O$) were purchased from Zibo city (Shandong province, China) and crushed in a ball mill and dried at 110 °C, respectively. PAM (polyacrylamide), $NaHCO_3$ and $CuSO_4 \cdot 5H_2O$ were obtained from Alfa-Aesar. Commercial Fe/C ceramic-filler (CF, employed as comparison materials) and commercial lightweight-ceramic-filler (LCF, applied as filler for BAFs in pilot-scale test) which had been frequently utilized in our previous studies [24–26] were shown in Table 2.

Table 2. Commercial Fe-C ceramic filler (CF) and commercial lightweight ceramic filler (LCF).

Materials	BD [4]/kg m^{-3}	GD [5]/kg m^{-3}	24 WA [6]/%	NTP [7]/Mpa	ARSC [8]/%	Diameter/mm	Iron/%	Carbon %
CF	1085	1548	3.8	4.2	92.5%	4–6	26.5%	5.2%
LCF	981	1238	3.2	8.6	100%	3–6	~	~

Note: 4~6-BD, GD and 24 WA were the bulk density, grain density and 24 water absorption. 7 numerical tube pressure, 8-acid-resistance softening co-efficiency.

3.1.2. Pilot-Scale Production Process Flow of Fe/Cu Catalytic-Ceramic-Filler

Fe/Cu catalytic-ceramic-filler (CCF) was prepared by six steps. (1) Prepare the raw material mixture: Iron powder and Kaolin powder were mixed in different mass ratios, then about 1% of $NaHCO_3$ was added; (2) Prepare the binder: 2% $CuSO_4 \cdot 5H_2O$ was dissolved in PAM solution

(1000 mg/L); (3) Mix the ceramic powder and the binder by 10:1 (w/w); (4) Pour the mixture into a pelletizer to produce pellets, then we got raw pellets (the diameters were 4.0 mm to 5.0 mm); (5) The raw pellets were dried in dry oven under N_2 circumstance for 24 h; (6) the raw pellets were heated in a muffle with N_2 at 600 °C for 1.0 h, then the sintered pellets were sealed in a steel drum cool down to room temperature.

3.1.3. Basic Property Test of CCF/CF

$$\rho_{bd} = \frac{M_c}{V_{bd}} kg \cdot m^{-3} \qquad (8)$$

$$\rho_{gd} = \frac{M_c}{V_{cd}} kg \cdot m^{-3} \qquad (9)$$

$$\alpha_{24h} = \frac{M_{c+w} - M_c}{M_c} \times 100\% \qquad (10)$$

$$\eta_{es} = \frac{V_{rcd} - V_{cd}}{V_{rcd}} \times 100\% \qquad (11)$$

The physical properties of CCF or CF were calculated by the formulas above. P_{bd} stands for the bulk density; ρ_{gd} stands for the grain density; $\alpha_{24\,h}$ is the water absorption in 24 h; η_{es} is the shrinking rate; Mc stands for the mass of dry ceramic bodies; M_{c+w} is the 24 h saturated mass of ceramic bodies; V_{bd} stands for the accumulation volume; V_{cd} is the real volume; V_{rcd} stands for the real volume of raw pellets.

Numerical tube pressure and acid-resistance softening co-efficiency test were proceeded according to GB/T17431-1998 [21]. Firstly, 20 L of CCF was random sampled and divided into group A and B. After dried at 110 °C, CCF in group A was filled into a stainless steel tank and flatted (the effective volume/height were 1 L/100 mm, respectively). Secondly, completely covered the stainless steel tank, and pressure was exerted (with accelerated pressure was 0.01 Mpa.s^{-2}) from the top of the stainless steel tank until the pressed depth was 20 mm. The final pressure was the numerical tube pressure of CCF. CCF in group B was settled in 0.1 mol/L of hydrochloric acid for 24 h then dried at 110 °C and tested the numerical tube pressure. Ratios of numerical tube pressure of CCF in group B to that of group A was acid-resistance softening co-efficiency for CCF. According to CCF, the numerical tube pressure and acid-resistance softening co-efficiency of CF were tested as well.

3.2. Performance Test of CCF/CF

3.2.1. Half-Life and Effectiveness Evaluation Reactor and Wastewater

Half-life and Effectiveness evaluation reactor, which reference for fixed bed reactor widely utilized in practical engineering, was designed and self-prepared (Figure 13). The total volume and height of reactor was 100 L and 1000 mm, respectively. And the equipment was made of PP (polypropylene) material and automatic controlled by PLC (Programmable Logic Controller, Siemens 300, Beijing, China).

Nitrobenzene compounds wastewater for the performance test was obtained from a comprehensive military special-chemicals factory (Liaoning Provence, China) which the main products were civil and military explosive (2,4,6-trinitrotoluene-TNT), dye and pharmaceutical nitrobenzene-intermediates. The components of total nitrobenzene compounds (TNCs)in comprehensive wastewater were mono-nitrobenzene/di-nitrobenzene groups and tri-nitrobenzene groups. The main features of wastewater are shown in Table 3.

Concentrations of mono-nitrobenzene and di-nitrobenzene compounds were tested by reduction-azo spectrophotometric method, and tri-nitrobenzene compounds in wastewater were tested by n-cetyl pridinium chloride- sodium sulfite spectrophotometric method, respectively [19]. The TNCs were calculated by the concentrations of mono-nitrobenzene, di-nitrobenzene and tri-nitrobenzene.

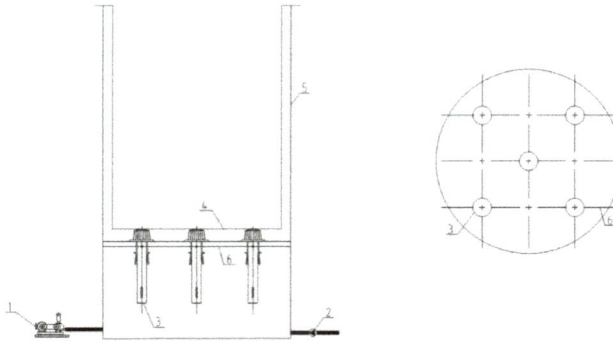

Figure 13. Catalyst life and Effectiveness evaluation reactor. (Left: front view; Right: top view). 1—frequency conversion fan, 2—plunger metering pump, 3—long handle filter nob, 4—Stainless steel tube (bottom mesh 40, DN350), 5—tank body, 6—filter board.

Table 3. Characteristics of nitrobenzene compounds wastewater.

Materials	CODcr/ mg·L^{-1}	BOD$_5$/ mg·L^{-1}	Biodegradability BOD$_5$/ CODcr	mono-NCs/ di-NCs [1]/mg·L^{-1}	tri-NCs [2]/ mg·L^{-1}	TNCs [3]/ mg·L^{-1}	pH
Wastewater	2460	270	0.11	297	123	420	2.2

Note: 1-mono-nitrobenzene/di-nitrobenzene compounds, 2-tri-nitrobenzene compounds, 3-total nitrobenzene compounds.

3.2.2. The Sequence Wastewater Treatment Test of CCF and CF

50 L of CCF/CF was soaked in the nitrobenzene compounds wastewater for 24 h to avoid the adsorption factor, respectively. Then the CCF/CF were separately added into the evaluation reactor shown in Figure 13. After that, the nitrobenzene compounds wastewater was poured into the unit, the test was operated in aerobic condition (50 L/min) and anaerobic condition, respectively [18]. The optimum operating conditions of CCF and CF were gathered from the test of TNCs, CODcr and BOD$_5$, respectively.

3.2.3. The Backwash Frequency and the Volume Half-Life Test of CCF and CF

The backwash frequency test: the evaluation reactor was operated with CCF (CF) and nitrobenzene compounds wastewater. The backwash procedure was not operated until the BOD$_5$/CODcr $\leq$ 0.3. The initial surface appearance of CCF and CF were examined by scanning electron microscopy (Sirion200, Au coated, Fei, Beijing, China), respectively. And both the images of CCF and CF before the backwash procedure were examined as well.

Air-water combination backwash method which reference to the practical engineering was utilized for the backwash procedure of CCF or CF: firstly, air backwash was set for 5 min (Air backwash intensity was 15 L m^{-2} s^{-1}); then air and water combination backwash 5 min (Air and water backwash intensity was 15 L m^{-2} s^{-1} and 5 m^{-2} s^{-1}, respectively); finally, water backwash last for 5 min (water backwash intensity was 10 m^{-2} s^{-1}). The air, air-water and water backwash combined as a complete backwash cycle, and the time of a cycle was 20 min (4 cycles per hour). The backwash process was controlled by PLC (Siemens 300).

The volume half-life test: Evaluation reactors (CCF and CF were filled, respectively) were operated at in air, air-water and water backwash procedure model. The volume and height of both CCF and CF were 50 L and 500 mm, respectively. Each volume half-life test for CCF and CF test was conducted in five replicates.

3.3. The Field Pilot-Scale Test

For the feasibility of CCF utilized in practical project, a 90 days of field pilot-scale test which formed by two-stage catalytic-biological system was designed and performed in the military chemical factory comprehensive wastewater treatment. The catalytic portion of the system was designed according to the result of performance test, and biological portion reference to our previous studies [24]. CBR-1 and CBR-2, both filled with CCF, was connected in series as the catalytic portion; BAF-1 (biological aerated filter) and BAF-2, both filled with LCC, was connected in series as the biological portion. In addition, the backwash procedure for CBRs and BAFs were all controlled by PLC (Siemens 300).

4. Conclusions

The addition of iron was 25% with the copper plating rate was 1% were benefit for the pilot-scale production of CCF. The characteristics of optimum CCF were: 1150 kg/m^3 of bulk density, 1700 kg/m^3 of grain density, lower than 3.5% of shrinking ratio, 3.5% of 24 h water absorption, 6.0 Mpa of numerical tube pressure and 0.99 acid-resistance softening co-efficiency.

When the BOD_5/CODcr ratio higher than 0.3 and the concentration of TNCs in effluent was not exceed the critical value as well, the operating conditions for CCF or CF were: the effective time of a single cycle was 69.3 h or 54.3 h, the backwash procedure operated every 69 h or 54 h, and the volume half-life was 897.4 days or 867.4 days, respectively. Both the removal of CODcr and TNCs fitted the first order reaction kinetics for CCF and CF. Compared with CCF and CF, CCF showed better performances and more efficiency applied in the improvement of the biodegradability.

From the pilot-scale test: the CBRs had excellent influent impact resistance and total nitro-group compounds impact resistance. BAFs showed high stability at the average volume load was 1.5 kg m^{-3} d^{-1}. As a prediction, MBR might be better than BAFs in practical engineering application. Our study offers a new solution to the highly toxic organic wastewater treatment. The catalysts are easy to be manufactured. The application condition in industrial area has been mature. It is hopeful that we could get more achievements in the industrial wastewater treatment.

Author Contributions: Investigation, B.Y.; Methodology, B.Y.; Project administration, R.L.; Supervision, Y.Q.; Writing-original draft, B.Y.

Funding: We are grateful to National Science Foundation of China (No.51378306, 21402079), Primary Research and Developement Plan of Shandong Province (2015GGH317001) and Liaocheng University fund (318051515, 201610447013) for financial support of this research.

Conflicts of Interest: The authors declare no conflict of interest.

References

1. Zhao, Q.; Ye, Z.; Wang, Z.; Zhang, M. Progress on the treatment of TNT waste water. *Environ. Chem.* **2010**, *5*, 796–801.
2. Nguyen, T.D.; Le, T.V.; Show, P.L.; Nguyen, T.T.; Tran, M.H.; Tran, T.N.; Lee, S.Y. Bioflocculation formation of microalgae-bacteria in enhancing microalgae harvesting and nutrient removal from wastewater effluent. *Bioresour. Technol.* **2019**, *272*, 34–39. [CrossRef] [PubMed]
3. National Standard. *Discharge Standard for Water Pollutions from Ordnance Industry Powder and Explosive*; GB 14470.1-2002. 2002. Available online: http://www.gbstandards.org/China_standards/GB/GB%2014470.1-2002.htm (accessed on 18 November 2002).
4. Ma, C.; Peng, Y. *Treatment and Control for High-Concentration Recalcitrant Organic Wastewater*, 2nd ed.; Chemical Industry Press: Beijing, China, 2010; pp. 266–288.
5. Ryosuke, S.; Toshinari, M.; Yoichiro, H.; Nobuo, N.; Hiroaki, O. Biological treatment of harmful Nitrobenzene compounds wastewater containing a high concentration of nitrogen compounds by waste activated sludge. *J. Biotechnol.* **2010**, *150*, 226–227. [CrossRef]
6. Shi, J.; Han, Y.; Xu, C.; Han, H. Biological coupling process for treatment of toxic and refractory compounds in coal gasification wastewater. *Rev. Environ. Sci. Bio/Technol.* **2018**, *17*, 765–790. [CrossRef]

7. Maloney, S.W.; Adrian, N.R.; Hickey, R.F.; Heine, R.L. Anaerobic treatment of pinkwater in a fluidized bed reactor containing GAC. *J. Hazard. Mater.* **2002**, *92*, 77–88. [CrossRef]
8. Ye, Z.; Zhao, Q.; Zhang, M. Acute toxicity evaluation of explosive wastewater by bacterial bioluminescence assays using a freshwater luminescent bacterium, *Vibrio qinghaiensis* sp. Nov. *J. Hazard. Mater.* **2011**, *186*, 1351–1354. [CrossRef] [PubMed]
9. He, Y.; Bai, H. Progress in Biodegradation of Nitrobenzene compounds wastewater. *Chem. Intem.* **2011**, *10*, 10–12.
10. Lin, H.; Lin, Y.; Wen, Y.; Gan, L. Degradation of TNT in Aqueous Solution by Uncultureed Soil Bacterium Clone UD3. *Chin. J. Energ. Mater.* **2009**, *17*, 630–634.
11. Payan, A.; Fattahi, M.; Jorfi, S.; Roozbehani, B.; Payan, S. Synthesis and characterization of titanate nanotube/single-walled carbon nanotube (TNT/SWCNT) porous nanocomposite and its photocatalytic activity on 4-chlorophenol degradation under UV and solar irradiation. *Appl. Surf. Sci.* **2018**, *434*, 336–350. [CrossRef]
12. Matta, R.; Hanna, K.; Kone, T.; Chiron, S. Oxidation of 2,4,6-trinitrotoluene in the presence of different iron-bearing minerals atneutral pH. *Chem. Eng. J.* **2008**, *144*, 453–458. [CrossRef]
13. Yu, Y.; Guoji, D.; Yongqiang, Z.; Deyong, L.; Zheng, J. Study on treatment of the midcourse wastewater from alkali straw pulp by using a new biological film integrative reactor. *Ind. Water Treat.* **2010**, *30*, 42–45.
14. Wu, Y.G.; Zhao, D.W. Experiment studies on the Degradation of TNT-containing Wastewater by Ozone Oxidization. *Chin. J. Energy Mater.* **2003**, *11*, 201–204.
15. Wu, Y.; Zhao, C.; Wang, Q.; Ding, K. Integrated effects of selected ions on 2,4,6-trinitrotoluene removal by O_3/H_2O_2. *J. Hazard. Mater.* **2006**, *132*, 232–236. [CrossRef]
16. Diao, J.; Liu, Y.; Wang, H.; Li, P.; Kang, R. O_3/H_2O_2 Oxidative treatment of TNT Red-Water in a Rotating Packed Bed. *Chin. J. Energy Mater.* **2007**, *15*, 281–284.
17. Chang, S.; Liu, C. Treatment of Nitrobenzene compounds wastewater by supercritical water oxidation. *Chin. J. Energy Mater.* **2007**, *25*, 285–288.
18. Li, G.X.; Huaug, Y.H.; Chen, T.C.; Shih, Y.J.; Zhang, H. Reduction and Immobilization of Potassium Permanganate on Iron Oxide Catalyst by Fluidized-Bed Crystallization Technology. *Appl. Sci.* **2012**, *2*, 166–174. [CrossRef]
19. Hernandez, R.; Zappi, M.; Kuo, C.H. Chloride effect on TNT degradation by zerovalent iron or zinc duringwater treatment. *Environ. Sci. Technol.* **2004**, *38*, 5157–5163. [CrossRef] [PubMed]
20. Barreto-Rodrigues, M.; Silva, F.T.; Paiva, T.C. Optimization of brazilian TNT industry wastewater treatment using combined zero-valent iron and fenton processes. *J. Hazard. Mater.* **2009**, *168*, 1065–1069. [CrossRef] [PubMed]
21. Ma, L. *Catalytic Reduction Treatment for Wastewater—Mechanism and Application*; Beijing Science Press: Beijing, China, 2008; Volume 169–179, pp. 260–279.
22. Wang, G.; Zhang, J.; Liu, L.; Zhou, J.Z.; Liu, Q.; Qian, G.; Xu, Z.P.; Richards, R.M. Novel multi-metal containing MnCr catalyst made from manganese slag and chromium wastewater for effective selective catalytic reduction of nitric oxide at low temperature. *J. Clean. Prod.* **2018**, *183*, 917–924. [CrossRef]
23. Li, G. *Analysis and Test Method for Water and Wastewater*; Chemical Industry Press: Beijing, China, 2012; Volume 65–70, pp. 201–230.
24. China Association for Engineering Construction Standardization. *Technical Specification for Biological Aerated Filter Engineering CECS 265-2009*; China Planning Press: Beijing, China, 2009; pp. 15–16.
25. National Standard. *Lightweight Aggregates and Its Test Methods*. GB/T 17431-1998. 1998. Available online: http://218.196.240.38/root/eWebEditor/uploadfile/20170503133740612.pdf (accessed on 15 July 1998).
26. Lemos, B.R.; Teixeira, A.P.; Ardisson, J.D.; Macedo, W.A.; Fernandez-Outon, L.E.; Amorim, C.C.; Moura, F.C.; Lago, R.M. Magnetic Amphiphilic Composites Applied for the Treatment of Biodiesel Wastewaters. *Appl. Sci.* **2012**, *2*, 513–524. [CrossRef]
27. Barreto-Rodrigues, M.; Silva, F.T.; Paiva, T.C. Combined zero-valent iron and fenton processes for the treatment of brazilian TNT industry wastewater. *J. Hazard. Mater.* **2009**, *165*, 1224–1228. [CrossRef] [PubMed]
28. Wu, S.; Qi, Y.; He, S.; Fan, C.; Dai, B.; Zhou, W.; Gao, L.; Huang, J. Preparation and application of novel catalytic-ceramic-filler in a coupled system for TNT manufacturing wastewater treatment. *Chem. Eng. J.* **2015**, *280*, 417–425. [CrossRef]

29. Qi, Y.; Dai, B.; He, S.; Wu, S.; Huang, J.; Xi, F.; Ma, Y.; Meng, M. Effect of chemical constituents of oxytetracycline mycelia residue and dredged sediments on characteristics of ultra-lightweight ceramsite. *J. Taiwan Inst. Chem. E* **2016**, *65*, 225–232. [CrossRef]
30. Qi, Y.F.; He, S.B.; Wu, S.Q.; Dai, B.B.; Hu, C.H. Utilization of micro-electrolysis, up-flow anaerobic sludge bed, anoxic/oxic-activated sludge process, and biological aerated filter in penicillin G wastewater treatment. *Desalin. Water Treat.* **2015**, *55*, 1480–1487. [CrossRef]
31. Qi, Y.; Yue, Q.; Han, S.; Yue, M.; Gao, B.; Yu, H.; Shao, T. Preparation and mechanism of ultra-lightweight ceramics produced from sewage sludge. *J. Hazard. Mater.* **2010**, *176*, 76–84. [CrossRef]
32. López Peñalver, J.J.; Gómez Pacheco, C.V.; Sánchez Polo, M.; Rivera Utrilla, J. Degradation of tetracyclines in different water matrices by advanced oxidation/reduction processes based on gamma radiation. *J. Chem. Technol. Biotechnol.* **2013**, *88*, 1096–1108. [CrossRef]

 © 2018 by the authors. Licensee MDPI, Basel, Switzerland. This article is an open access article distributed under the terms and conditions of the Creative Commons Attribution (CC BY) license (http://creativecommons.org/licenses/by/4.0/).

Review

Bimetallic Iron–Cobalt Catalysts and Their Applications in Energy-Related Electrochemical Reactions

Kai Li [†], Yang Li [†], Wenchao Peng, Guoliang Zhang, Fengbao Zhang and Xiaobin Fan *

Lab of Advanced Nano Structures & Transfer Processes, Department of Chemical Engineering,
Tianjin University, Tianjin 300354, China; tju_likai@163.com (K.L.); liyang1895@tju.edu.cn (Y.L.); wenchao.peng@tju.edu.cn (W.P.); zhangguoliang@tju.edu.cn (G.Z.); fbzhang@tju.edu.cn (F.Z.)
* Correspondence: xiaobinfan@tju.edu.cn; Tel.: +86-022-85356119
† These authors contributed equally to this work.

Received: 19 August 2019; Accepted: 5 September 2019; Published: 11 September 2019

Abstract: Since the persistently increasing trend of energy consumption, technologies for renewable energy production and conversion have drawn great attention worldwide. The performance and the cost of electrocatalysts play two crucial roles in the globalization of advanced energy conversion devices. Among the developed technics involving metal catalysts, transition-metal catalysts (TMC) are recognized as the most promising materials due to the excellent properties and stability. Particularly, the iron–cobalt bimetal catalysts exhibit exciting electrochemical properties because of the interior cooperative effects. Herein, we summarize recent advances in iron–cobalt bimetal catalysts for electrochemical applications, especially hydrogen evolution reaction (HER), oxygen evolution reaction (OER), and oxygen reduction reaction (ORR). Moreover, the components and synergetic effects of the composites and catalytic mechanism during reaction processes are highlighted. On the basis of extant catalysts and mechanism, the current issues and prospective outlook of the field are also discussed.

Keywords: energy conversion; iron–cobalt bimetal catalysts; electrochemical application; hydrogen evolution; oxygen evolution; oxygen reduction

1. Introduction

Growing depletion of fossil fuels and rapid increase of pollution pose new challenges to the environment and ecosystem. It is urgent to explore renewable and sustainable energy as a substitute to traditional energy sources to balance the economic and ecological development. Among the related technologies, energy storage and conversion devices are needed and widely concentrated. In the past decades, a series of potential electrochemical energy storage and conversion facilities such as fuel cells, water splitting technologies, and metal–air batteries have been deeply investigated, which can significantly decrease the reliance on traditional fossil fuels and promote energy conversion efficiency [1]. Electrocatalytic reactions, like hydrogen evolution reaction (HER), oxygen evolution reaction (OER), and oxygen reduction reaction (ORR), play vital roles in these electrochemical techniques. For example, water splitting demands high HER and OER activities, in which the cathodic reaction is HER ($4H^+ + 4e^- \rightarrow 2H_2$ in acid, $4H_2O + 4e^- \rightarrow 2H_2 + 4OH^-$ in alkali) while the anodic reaction is OER ($2H_2O \rightarrow O_2 + 4H^+ + 4e^-$ in acid, $4OH^- \rightarrow O_2 + 2H_2O + 4e^-$ in alkali). Meanwhile, rechargeable metal–air batteries require OER in the charging process and ORR ($O_2 + 4H^+ + 4e^- \rightarrow 2H_2O$, $4e^-$ pathway, $O_2 + 2H^+ + 2e^- \rightarrow H_2O_2$) in the discharging process. Developing electrocatalysts of high efficiency and low cost is the main obstruction to break through the limit of the sluggish dynamics of such reactions [2]. Up to now, the noble metals, like platinum (Pt), iridium (Ir), and ruthenium (Ru) and their derivatives have presented favorable catalytic activities, where Pt-based materials show excellent activity for both HER and ORR, and Ir-based and Ru-based catalysts are more active for OER. However, the scarcity and high cost of the noble metals severely limit their large-scale applications and commercial promotions. In this regard, it is necessary to develop nonprecious metals and their derivatives with high efficiency and long-term stability as promising candidates for electrocatalysis.

Currently, the 3d transition nonnoble metals, especially Fe, Co, and Ni and their derivatives, have been regarded as the most promising substitute to the noble metals like Pt because of their high catalytic activities to the energy conversion reactions [3]. However, bare 3d transition metal materials are not active or even stable enough to catalyze the electrochemical reactions for long-term operation under acid or alkaline condition. On the contrary, the bimetallic catalysts can make use of the synergistic effects between the metal contents themselves and the supports, thus exhibiting favorable reaction kinetics with high activity and efficiency. Theoretically, the related calculation results, such as density functional theory calculation (DFT), have also proved that the bimetallic electrocatalysts perform better than the monometallic ones. The calculated ΔG of the single 3d transition metallic materials to develop intermediates during the reactions is much further from the optimal value than the precious metals. Therefore, designing bimetallic catalysts can modify the ΔG of the single metallic catalysts and improve the catalytic efficiency. Recently, the bimetallic materials combining Fe and Co or their derivatives and their electrochemical applications have been investigated by researchers. Numerous studies have been devoted to the bimetallic electrocatalysts, such as alloys [4], oxides [5], hydroxides [6], sulfides [7], phosphides [8], and hierarchical structures. The strategies to construct the materials and the electronic modulations inside have been also deeply researched. By both calculative and experimental approaches, the kinetics and mechanisms during the catalytic processes are also explored.

In this review, we summarized recent development of the Fe–Co bimetal electrocatalysts and their electrochemical applications, with the focus on both the constructive strategies and the catalytic mechanisms. The synthesized materials are categorized by the structural characteristics and the preparation process, followed by an inductive section of the characterization methods. The electrochemical applications are then discussed, which mainly concentrated on electrochemical HER, OER, and ORR. The relation between the structures and the catalytic properties is highlighted and elaborated emphatically. By this work, we aim to give a summarization of the bimetallic electrocatalysts containing Fe and Co, so as to update the conception of the electrocatalytic kinetics for the aforementioned electrochemical reactions.

2. Synthesis Methods for Fe/Co Bimetallic Catalysts

The catalyst supports in the hybrid materials affect the electronic and surface structure, shape, size, and decide the accessible active sites and the mass transfer rate during the reaction process and the accessible part of the active sites; hence, they largely control the electrocatalytic activity. Such a strategy is found to be prospective for the three essential requirements for an electrocatalysts' activity, as it can simultaneously ensure abundant active site densities, controllable performance characteristics, and reactant accessibility by narrowing down the overpotential and enhancing the current density. We present the synthesis process of the Fe–Co bimetal hybrids into three categories based on the catalyst supports, as bimetallic alloy-based hybrid, carbon materials supported bimetallic hybrid, and metal–organic frameworks (MOF)-based bimetallic hybrid catalysts.

2.1. Bimetallic Alloy-Based Hybrid Catalysts

Fe–Co bimetallic alloys [9] exhibit multifarious structural, mechanical, optical, electronic, and electrocatalytic properties. Spherical, uniform, and highly monodisperse Fe–Co nanoparticles with different metal content ratios were synthesized through the microemulsion method [4]. The microemulsions of metal salt precursors, reducing agent, and surfactants were annealed in H_2 atmosphere at 700 °C for 6 h and formed monophasic bimetallic nanoparticles. The average sizes of the obtained nanoparticles vary with the initial loaded stoichiometries of Fe and Co metal precursors.

Meanwhile, the oxidation currents obtained by cyclic voltammetry are a function of both the surface area and the composition of the bimetallic nanoparticles, illustrating that the $Fe_{33}Co_{67}$ nanoparticles are of superior hydrogen and oxygen evolution activity.

The positively charged layered hydroxide sheets formed by tilted edge-sharing MO_6 octahedra and negatively charged counterions in the interlayers are namely layered double hydroxides (LDHs), with intrinsic OER activity according to the abundance tetrahedral Co^{2+}, octahedral Co^{3+}, and oxygen vacancies on the high surface sheet of two-dimensional materials [10,11]. A platelet-like-shaped Fe substituted α-Co(OH)$_2$ was synthesized under a N_2 atmosphere at room temperature [12]. To reach an equivalent number of anions, Fe (III) ions replaced some Co (II) ions in α-Co(OH)$_2$ by the interaction within the edge sharing MO_6 (M = Co or Fe) octahedral layers, resulting in the formation of CoFe LDH structure. Similar CoFe- and CoAl-based LDH catalysts with various Fe and Al contents were also fabricated by an efficient coprecipitation method [13,14]. The Fe impurity makes OH^- intercalation easier in facilitating $Co(OH)_2 + OH^- \rightarrow CoOOH + H_2O + e^-$. Herein, the incorporation of Fe into the layered (oxy)hydroxide structure may accelerate interlayer spacing of the sheets, as well as the defect or edge sites on the Co oxyhydroxide structure. The appearance of the low-angle reflection for the Co–Fe phase in the X-ray diffraction (XRD) pattern is the evidence for such incorporation [14]. Considering the strong electronic coactions between Co and Fe, the alternation of Co/Fe ratio during the CoFe-based LDHs synthesis process may change the situation of Fe^{3+} cations substituted into the lattice of α-Co(OH)$_2$, modifying the electronic structure of the catalyst and the formation of LDH structure [15,16]. The facile synthesized process takes advantage of the chemically stable and areal layered structure, good conductivity of α-Co(OH)$_2$, as well as the rich redox properties and earth abundance of Fe, substantially facilitating low-cost clean energy production.

Metal borides are known as the newly developed OER catalysts, as the boron can drop off the M–M bonds and cut down the thermodynamic and kinetic barrier in the rate-limiting step during OER [17]. A bimetallic Co_x–Fe–B catalyst, with the average particle size of about 30 nm, was synthesized by the reduction of sodium borohydride and sodium hydroxide solution under whisking at room temperature [18]. Through this chemical reduction method, Fe has been successfully incorporated into the Co–B, resulting in the stabilization of Co at a higher oxidation level and the generation of OOH-like species.

The influence of non-metal elements doping in the transition metal-based materials are also investigated through bifunctional electrocatalytic application of phosphorus-doped Co–Fe–B material [19]. Co and Fe chlorides with different molar ratios are solved in polyvinyl pyrrolidone (PVP) and $NaBH_4$, which serve as surfactant and reducing agent. One-dimensional phosphorous-doped Co–Fe–B (Co–Fe–B–P) nanochains were synthesized via a low-temperature phosphorization procedure of annealing in Ar atmosphere to 300 °C for 2 h (Figure 1). The XRD pattern of the obtained Co–Fe–B–P materials illustrates the amorphous nature, which exhibit higher electrochemical activities compared with the corresponding crystalline material [20]. The structure comparation of bimetal borides and mono-metal counterparts verifies that Co plays a major role in morphology control. The synergistic effect of different metals and nonmetal components incorporation are characterized by XPS and electrochemical application, which is beneficial for lowering the kinetic energy barriers of electrocatalysis and improving the activity. The one-dimensional chain-like nanostructures possessing high aspect ratio and easy surface atoms contribute to the high OER electrocatalytic activity [21].

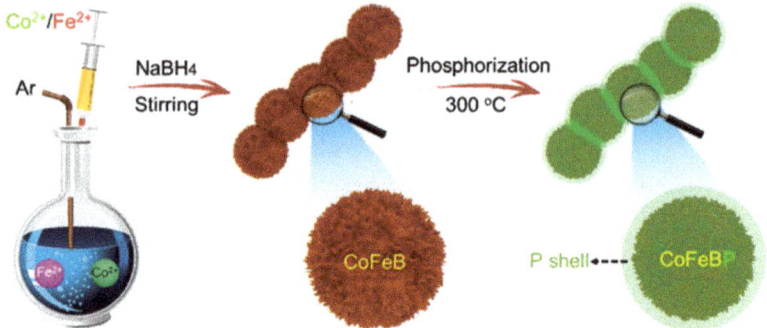

Figure 1. Schematic synthesis process of the Co_1–Fe_1–B–P nanochains, reproduced with permission from [19]. Copyright Royal Society of Chemistry, 2019.

2.2. Carbon Materials Supported Bimetallic Hybrid Catalysts

Dispersing metal particles on appropriate supports is an efficient strategy to improve the catalytic performance and further reduce the production cost. Large-area conductive substrates, carbon-based materials, such as graphene, are favorable for ameliorating the materials' electronic conductivity and dispersion in order to promote the electrocatalytic activity. Such carbon-based materials take on unique expediency for assigned catalysis owing to their accessibility, tunable structures, and good resilience in acidic and basic environments.

2.2.1. Graphene-Based Bimetallic Hybrid Catalysts

The two-dimensional material functionalized as the catalyst support can not only increase the quantity of accessible active sites by minimizing the particle sizes, but also manipulate the electronic properties of the obtained metal particles, thus leading to the amelioration in instinct catalytic activity of the hybrid materials. On the other hand, the mesoporous structure and large specific surface area in the graphene-based materials are in favor of the high exposure of active sites, fasten the mass transport of the reactants, as well as fabricate conducting networks for fast electron transfer during the whole reaction process.

Hydrothermal and solvothermal treatment of the metal precursor and graphene oxide are the general approaches to synthesize metal nanoparticles embedded graphene framework [22,23]. Wu et al. [7] decorated cobalt sulfide (Co_9S_8) nanoparticles grown in situ on the reduced graphene oxide surface with Fe_3O_4 nanoparticles in two solvothermal steps. In the first step, Co_9S_8 nanoparticles are formed onto reduced graphene oxide nanosheets. Then, iron ions are selectively adsorbed on the surface of Co_9S_8 by forming the strong bonding of sulfur species, and subsequently reduced to Fe_3O_4 at 600 °C. The interface–orientation relationship determined by HRTEM illustrates the loading of Fe_3O_4 on the surface of Co_9S_8 and may induce a relatively high stability of the composite. It has been demonstrated that the breaking of the Co–O bond in the stable configure ration (Co–O–O superoxo group) to free the O_2 molecules may represent a rate-limiting step [24]. Electron transfer behavior from Fe species to Co_9S_8, causing the down-shift in the electron binding energy of Co $2p_{3/2}$, may induce a relatively lower oxidation state for cobalt ions, which would promote the Co–O bond break and the O_2 release, and improve the catalytic activity.

Considering the possible structural damage during the high-temperature treatment, a conversion tailoring strategy was designed to assemble nanometer-sized Fe-modulated CoOOH nanoparticles (Fe–CoOOH) on 2D graphene in a mild synthesis process [25]. The CoFeAl-layered double hydroxide (CoFeAl-LDH) sheets in situ grow on the GO surface by the electrostatic interactions with metal ions. The Fe components of ultrasensitive triggered behavior would restrict the process for the structural rearrangement toward $Co(OH)_2$ and oxidize to CoOOH and yield the Fe–CoOOH nanoparticles.

Such a process indicates the chemical selective accelerative conversion led by Fe. According to DFT calculation, the adsorption energies of OH, OOH, and O species are remarkably enhanced at the Fe sites and decreased at the Co sites of Fe–CoOOH, facilitating the whole redox reactions and enhancing the catalytic activity of the hybrids.

The introduction of chemical functionalities through chemical functionalization and heteroatom (non-metal elements such as N, S, B, F, and P) may improve not only the immobilization of different species, but also the electrocatalytic activity by tuning the band gap and electronic structure (for instance, charge and/or spin density redistribution) of carbon to increase the active sites density towards favorable water splitting reactions. The synergistic effect between the transition metal/metal oxide and heteroatom-doped carbon along with the enhanced electronic conductivity may largely improve the ORR and OER kinetics [26]. Wang et al. [27] reported the method of combining ball milling and pyrolysis to fabricate FeCo nanoparticles/N-doped carbon with core–shell structure spheres supported on N-doped graphene sheets (Figure 2). The ball-milling process was carried out with carbon nitride (C_3N_4) and metal acetylacetonates, followed by the pyrolysis step at 700 °C. The acetylacetonates were decomposed to acetone and carbon dioxide and adsorbed on C_3N_4, then transferred to graphene as the supporter. The change of pyrolysis temperature led to the structure alteration of layered graphitic carbon encapsulated in the FeCo nanoparticles and the growth of carbon nanotubes on graphene surface [28]. Pyridinic-N and graphitic-N in carbon produced during pyrolysis can decline the adsorption energy of O_2 owing to the N doping-induced charge redistribution and have been pointed out to be the efficient active sites for ORR. The synergy of N-doped carbon shell-covered FeCo bimetallic nanoparticles, the suitable pore structure, along with the graphene supporter facilitate the reactant transport and promote charge transfer during the reaction.

Figure 2. Schematic illustration synthesis process of FeCo/NC catalysts from Fe(acac)$_2$, Co(acac)$_2$, and bulk C_3N_4, reproduced with permission from [27]. Copyright Springer nature, 2017.

Quantum confinement effect along with edge effect are other efficient methods in controlling the band gap of graphene, which may make graphene materials compatible for direct exploitation in nanoelectronics [29]. The introduction of nanoholes on graphene sheet may increase the defects and dangling bonds of edge sites as the appropriate active sites for the heteroatom doping [30], as well as promote the reactant ions diffusion to the active sites to enhance the catalytic activity in the reaction. Herein, nano-porous graphene has been applied as the support for metal carbides to fabricate Fe/CoN-doped porous graphene by annealing at 900 °C (Figure 3) [31]. A chemically assisted oxidative treatment of graphene is used to break small pieces through epoxide formation and prepare porous graphene with more edged sites [32]. The TEM and XRD analyze that the pyrolysis of the M-phenanthroline (M–Fe and Co) complexes with graphene at 900 °C leads to the generation of the M–N–C active sites along with the metallic or carbide phases. The obtained hybrids have homogeneously distributed Fe/Co–N-doped active centers, which confirms that the uniform distribution of nanoholes plays a vital role in enriching the coordinated active sites on the graphene surface, enhancing mass transfer and modulating reaction kinetics. Such high activity of the obtained catalyst highlights the important role of the active M–N–C coordinations conceived in the hybrids as the potential ORR active centers. By comparing the ORR activity with its single metallic counterpart, the Fe/CoN-doped porous graphene show higher catalytic activity by taking advantage of the existence of the bimetallic coordinated sites in the company of the existence of the nanopores on graphene surface. The electrocatalytic test results likewise illustrate that the pyrolysis of the mixture containing the metal macrocyclic complexes and suitable carbon nanomaterials could be the appropriate method for preparing effective M–N–C electrocatalysts for ORR rather than developing harmonious cooperation of the metal ions and doped nitrogen in the carbon matrix.

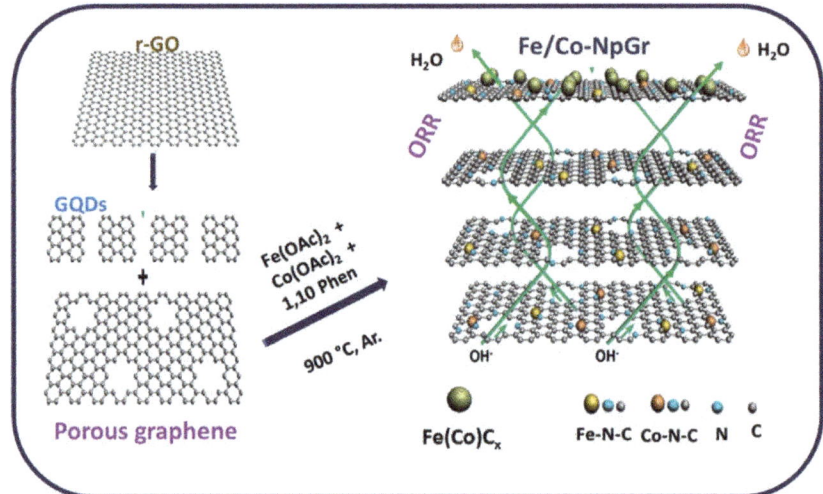

Figure 3. The scheme of the preparation of active Fe/Co-NpGr representing facile ion transport attainable within the hybrids through the holes in the porous graphene, which may enhance the accessibility of the active sites and lead to improved oxygen reduction reaction (ORR) activity, reproduced with permission from [31]. Copyright Wiley, 2016.

Aforementioned graphene-based electrocatalysts are of a high surface area to volume ratio. Indeed, only a small fraction of metal nanoparticles is exposed to reactants in such materials. Herein, promoting the metal atom utilization of a catalyst is an incredibly effective way to further improve the catalytic activity. 100% atom utilization efficiency by featured atomically dispersed metal atoms as robust active centers is of long-standing interest in fabricating catalysts. The latest development of fabricating single-atom catalysts (SACs) [33–35] with atomic-scale metal catalytic center, making individual metal atoms accessible and active, is a possible method for maximizing the atom efficiency. Ball milling is one of the low-cost and high-efficiency exfoliation and functionalization methods for establishing graphene-based SACs. Wu et al. [36] synthesized highly dispersed Fe/Co/N on graphene through a one-step ball milling protocol in 12 h. Many defective sites may appear during the ball milling process and facilitate the heteroatom doping. The existence of metal-bonded N/O, which are taken for the efficient active sites for OER [37], are confirmed by XPS characterization. The electrocatalytic activity results also show that the electrochemical surface area (ESCA) and valid active sites of the obtained hybrids are substantially augmented after ball milling and heteroatom doping, resulting in greatly ameliorating the OER performance.

2.2.2. Other Carbon Materials Supported Bimetallic Hybrid Catalysts

Raj et al. [38] used a single-step thermal annealing approach of potassium cobalt hexacyanoferrate (KCoHCF) to synthesize FeCo bimetal nanoparticle decorating on the nitrogen-doped reduced graphene oxide (N–rGO–CoFe) and graphitic carbon (N–C–CoFe), and discuss the effect of catalyst support on the overall water splitting performance. Compared with that thermal annealing with graphene oxide, the N–C–CoFe hybrids have the structure of a carbon shell strongly coupled with bimetal nanoparticles and high graphitization degree, which ensure the durability of the obtained catalyst. The electrocatalytic activities of the catalysts on different supports confirm that the chemical nature of nitrogen and degree of graphitization of the catalyst largely support control of the overall performance of the catalyst. The higher graphitization of the carbon (sp^2 C–C) support containing a larger amount of graphitic nitrogen, the better performance the catalyst has for OER and ORR.

One-dimensional mesoporous Fe/Co–N–C nanofibers with depositing FeCo nanoparticles (denoted as FeCo@MNC) were synthesized through an electrospinning process [39] of Fe and Co 2, 2-bipyridine chelates, PAN and PVP polymers, following stabilization at 280 °C in air and carbonization at 900 °C under N_2 [40]. Porous structures of FeCo@MNC with bunch-like nanoparticles embedded on the nanofibers are determined by TEM. Fe and Co appeared both inside and on the surface of the whole nanofibers, implying the simultaneous existing structures of FeCo and Fe/Co–N–C in the carbon skeleton. The formation process may be described firstly as formed Fe (Co)–N chelates electro-spin with polymers and compose Fe (Co)–N and N–C structures after carbonization with a high proportion of pyridinic-N structures. Polymers serving as a carbon precursor supplied the perfect template for the fabrication of one-dimensional hollow carbon fibers. The carbon fibers supporting Fe, Co, and N show a larger aspect ratio and excellent electrical conductivity, likewise benefitting from active sites exposure and rapid transportation [41].

Another N-doped graphitic carbon nanotube decorated with bimetal FeCo nanoparticles (denoted as N-GCNT/FeCo) were prepared through temperature-programmed carbonization strategy by pyrolyzing the mixture of metal salt precursors, glutamic acid, and melamine [42] (Figure 4). After the pyrolysis temperature increase, glutamic acid carbonization derives many defects in the graphitic carbon framework, as well as reinforcing N atoms into the carbon skeleton [41]. The nanotubes with cavities generated through the temperature-controlled carbonization are illustrated by TEM. On account of the strong metal–support interaction, the bimetal alloy nanoparticles catalyze the N-doped graphitic carbon species render into N-GCNT and grow on the inner wall of N-GCNT. The obtained multiphase materials also possess a large specific surface area, with potentially favorable property in the mass transfer during the electrocatalytic reactions.

2.3. MOF-Based Bimetallic Hybrid Catalysts

Carbon-based materials encapsulated with transition metals and their alloys have been emerging as promising candidates for water splitting catalysts. Some of the synthesis methods consume complex and expensive multi-step synthesis strategies, delaying the large-scale commercialization. On the other hand, heteroatom doping as mentioned above is an efficient method to accelerate the electrocatalytic activity of the carbon-based materials. Therefore, metal–organic frameworks (MOFs), composed of designable metal ion centers and organic ligands, are the promising precursors for the facile synthesis of metal/metallic alloy@carbon composites via pyrolysis. The easy control of composition, morphology, and construction also endow MOFs with multiple functions and broad applications. Chen et al. [43] carefully designed MOF precursors of desirable metal ion centers and organic ligands with different dopants, to fabricate carbon materials doped by nitrogen cooperating with non-precious metals through one-step annealing (Figure 5). Fe and Co atoms decomposed from the Prussian blue analogues (PBAs, $Fe_3[Co(CN)_6]_2$) precursor form FeCo alloy nanoparticles during the annealing process under N_2. Meanwhile, the CN^- groups, as double effect carbon and nitrogen sources, form nitrogen-doped graphene layers that cover the FeCo alloy particles. The obtained materials have the FeCo alloys encapsulated in a nitrogen-doped graphene layers structure. The nitrogen content in the material can be altered by different annealing temperatures, which may also affect the catalytic activity. The catalyst obtained by annealing at 600 °C with the highest nitrogen content (8.2 atom%) exhibits the best HER catalytic activity. Such results can be explained by the DFT calculation, as the increase of nitrogen content may accelerate the adsorption sites for H* and cut down the ΔG_{H^*} for H adsorption at the same time. Peng et al. [44] also chose PBAs as the structural template and fabricated Co–Fe phosphides by etching with urea and phosphorization. The obtained hybrid materials are used as the bifunctional electrocatalyst materials for overall water splitting.

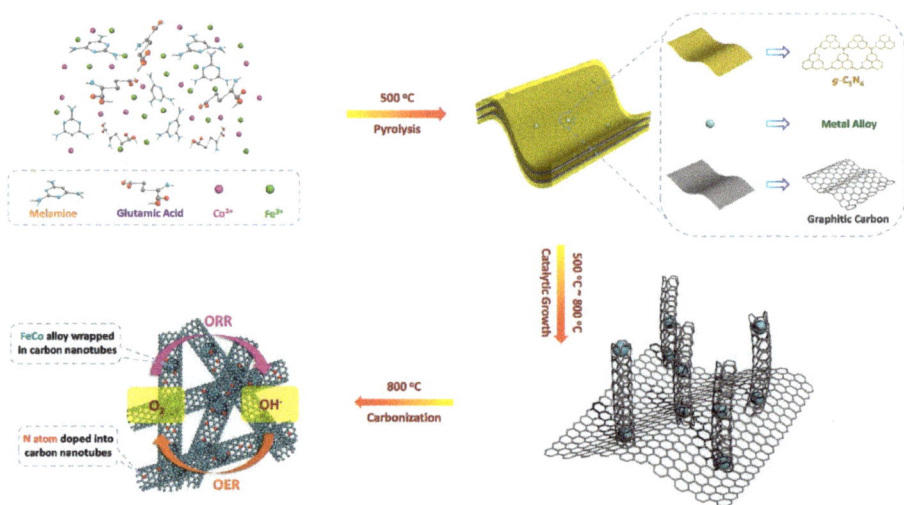

Figure 4. Illustration for the fabrication of N-doped graphitic carbon nanotube (N-GCNT)/FeCo bifunctional hybrids through the temperature-programed carbonization process, reproduced with permission from [42]. Copyright Wiley, 2017.

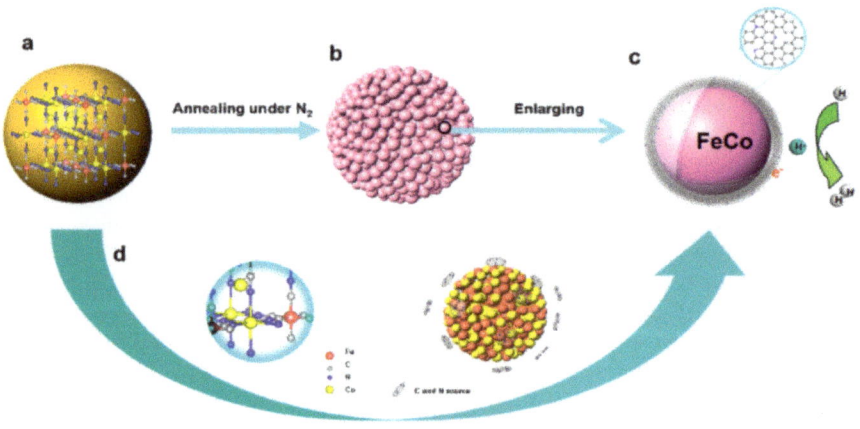

Figure 5. The illustration of synthetic process and model of the FeCo alloys decorated on nitrogen doped graphene layers by annealing corresponding metal–organic frameworks (MOFs), reproduced with permission from [43]. Copyright Royal Society of Chemistry, 2015.

Metallic/non-metallic heteroatom doping of the hybrids is one of the most efficient approaches to tune the catalyst activity. PBAs precursors were converted to a (Fe–Co)Se$_2$ composite by self-assembled strategy and, subsequently, a post-selenization method [45]. K$_3$[Co(CN)$_6$] and Fe(NO$_3$)$_3$ are dissolved in the polyvineypirrolydone (PVP)-HCl solution at 80 °C to assemble the Fe–Co PBAs nanocubes, and are then converted to MOFs-derived bimetallic selenides composites by annealing with selenium powder. TEM images confirm the conversion of the uniform small cubes on the stack nanostructure to stack nanospheres during the post-selenization process. MOFs-derived bimetallic selenides composites with the synthetic bimetal-selenide [46] are more favorable for electrochemical application with the advantages of rapid electron and proton transfer along with dioxygen molecules delivery. Not only does a loose and porous structure offer larger specific surface area and more active sites, but it also expedites the transfer of the electrolyte and the emission of gas bubbles, leading to superior catalytic performance. Co–Fe–P nanotubes were synthesized through a calcination process in air with the Co incorporating with a MIL-88B MOFs (MIL, Materials of Institute Lavoisier) template [47] (Figure 6). The TEM characterizations confirm the hollow structure of Co–Fe–P nanostructure, as well as the fact that the morphology of MOFs-derived materials is strongly connected to the molar ratio of the metal salt precursors following the heterogeneous coordination process in this synthetic case. The positive effect of Co dopant could increase the density of states (DOS) and electrical conductivity and fasten the charge transfer kinetics [48]. The one-dimensional structure of the hollow nanotubes also helps to enhance the structural stability of Co–Fe–P, showing excellent durability and long-term stability.

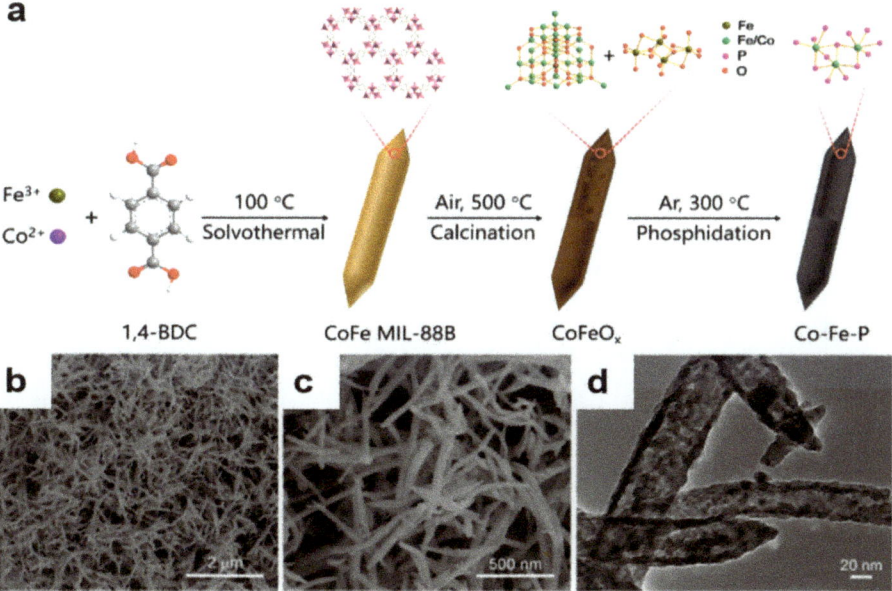

Figure 6. (**a**) Schematic illustration of the synthesis of Co–Fe–P nanotubes; (**b**, **c**) SEM and (**d**) TEM images of Co–Fe–P nanotubes, reproduced with permission from [47]. Copyright Elsevier, 2019.

Besides using single composition precursors, controllable conversion of two or more MOFs to synthesize desirable functional materials favors the construction of complex MOF-derived nanomaterials from the point of view of both architecture and chemical composition. Lou et al. [49] reported the synthesis process with zeolitic imidazolate framework-67 (ZIF-67, a Co-based MOF) and Co–Fe Prussian blue analogue (PBA) (Figure 7a). ZIF-67 nanocube precursors and [Fe(CN)$_6$]$^{3-}$ ions proceed a facile anion-exchange reaction with an obvious reaction solution color change from purple to brick red at room temperature and transform into ZIF-67/Co–Fe PBA yolk–shell nanocubes (YSNCs).

Through a subsequent annealing treatment in air, the YSNCs then further convert to Co_3O_4/Co–Fe oxide double-shelled nanoboxes (DSNBs), of slightly concave Co–Fe PBA shell and the inner sharp corner ZIF-67 in each particle (Figure 7b). Constructing nanostructures with specific morphology including solid or hollow nanoparticles, nanowires, nanosheets, and nanotubes, is another approach to change the catalytic properties of the hybrids. The extra active sites embedded on the inner Co_3O_4 shells enhance the ECSA of Co_3O_4/Co–Fe oxide DSNBs. The structure of PBA nanoshells can be changed by increasing the water fraction and altering the ion exchange reaction rate, which may also enhance the conductivity of the hybrids by the attendance of Fe element in cobalt oxide [50]. Co–Fe PBA hollow structures are also fabricated by employing Co-glycerate nanospheres as the template and precursor synthesis to validate the generality of anion-exchange reaction.

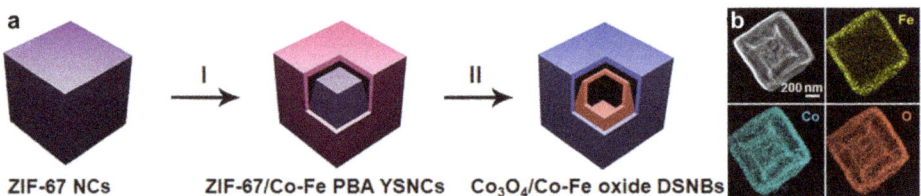

Figure 7. (a) The formation process of Co_3O_4/Co–Fe oxide double-shelled nanoboxes (DSNBs): (I) Ion-exchange reaction between ZIF-67 NCs and $[Fe(CN)_6]^{3-}$ ions to fabricate ZIF-67/Co–Fe PBA yolk–shell nanocubes (YSNCs) and (II) subsequent conversed to the Co_3O_4/Co–Fe oxide DSNBs through thermal annealing; (b) TEM and mapping images of the Co_3O_4/Co–Fe -Fe oxide DSNBs, reproduced with permission from [49]. Copyright Wiley, 2018.

The latest development of single-atom electrocatalysts possess optimal activity, stability, and selectivity through the tune strategy of well-defined active centers [51]. MOFs are also used as the substrate to confine Fe–Co dual atom scale sites. The MOF-derived SACs catalysts were synthesized through the pyrolysis of encapsulating Fe^{3+} in a type of ZIF-8 structure Zn/Co bimetallic MOF [52] (Figure 8). The existence of Fe species may help to modulate the geometric constructions of carbon support, generate N-doped graphene fragments and defects during the pyrolysis, as well as accelerate disintegration of metal–imidazolate–metal linkages. The interior cavities and size enlargement construction through decomposition and graphitization coincide with the Kirkendall effect [28]. All these factors positively impact the mass transport and electronic transfer in ORR.

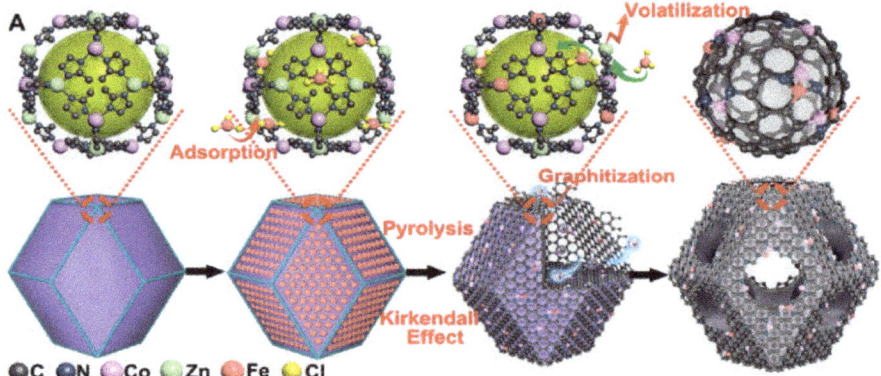

Figure 8. Preparation route of (Fe,Co)/N–C, reproduced with permission from [52]. Copyright American Chemical Society, 2017.

3. Applications of Fe/Co Bimetallic Electrocatalysts in Electrochemical Reactions

Due to the synergistic effects between Fe and Co in electrochemical reactions, effective electrocatalysts based on that have been developed for solving energy and environmental crisis. Among the reactions, hydrogen evolution reaction (HER), oxygen evolution reaction (OER), and oxygen reduction reaction (ORR) are most widely concentrated on by researchers, owing to their efficient applications in the energy transfer and conversion process. In the following part, we will summarize the detailed applications and mechanisms of the Fe/Co bimetallic catalysts.

3.1. HER

Hydrogen is a promising substitute for traditional fossil fuels because of its low pollution and high energy density, while generating hydrogen by electrochemical methods has been thought of as an effective and low-cost way. During this process, developing high-efficiency catalysts for HER is the pivot to decrease the whole cost and enhance the economic value. Electrocatalysts with excellent HER performance can sharply reduce the electronic energy, thus limiting the cost of hydrogen production. The HER process occurs via either Volmer–Heyrovsky mechanism or Volmer–Tafel mechanism depending on different pH values (Equations (1)–(5)). As H* is the only intermediate involved in the catalytic process in both acid and alkaline electrolytes, the hydrogen adsorption free energy (ΔG_H) is the key indicator for HER catalysts. In the typical volcano plot [53] (Figure 9a), Pt locates near the top, behaving with the best HER activity. However, the non-precious metals Fe and Co and their derivatives also perform well, even exceeding Pt in activity and reduction of cost.

$$H^+ + e^- \rightarrow H^* \text{ (Volmer reaction in acid solution)} \quad (1)$$

$$H_2O + e^- \rightarrow H^* + OH^- \text{ (Volmer reaction in alkaline solution)} \quad (2)$$

$$H^* + H^* \rightarrow H_2 \text{ (Tafel reaction)} \quad (3)$$

$$H^+ + H^* + e^- \rightarrow H_2 \text{ (Heyrovsky reaction in acid solution)} \quad (4)$$

$$H_2O + H^* + e^- \rightarrow H_2 + OH^- \text{(Heyrovsky reaction in alkaline solution)} \quad (5)$$

3.1.1. Bimetallic Alloy-Based Electrocatalysts

Among the composites involving Fe and Co, alloys can reflect the synergistic effects of the two elements most directly. However, the early binary Fe–Co alloy nanoparticles exhibit poor HER activity. Ahmed et al. [4] microemulsion-based synthesized spherical, uniform, and highly monodisperse nanoparticles, including $Fe_{75}Co_{25}$, $Fe_{67}Co_{33}$, $Fe_{50}Co_{50}$, and $Fe_{33}Co_{67}$ with average sizes of 20, 25, 10, and 40 nm, respectively. Electrocatalytic tests indicate that $Fe_{33}Co_{67}$ nanoparticles show better performance compared with $Fe_{67}Co_{33}$ and $Fe_{75}Co_{25}$ nanoparticles, whereas the HER performance is still unsatisfactory. To improve the HER activities of Fe–Co alloy-based materials in wide pH ranges, an N-doped carbon shell is introduced to the catalysts as a substrate. An N-doped carbon shell can not only provide abundant adsorption sites for H*, but also prevents the aggregation and combination of the alloy nanoparticles and exposes a larger active area. For HER in acid solution, Yang et al. [43] reported a facile process to fabricate Fe–Co alloy nanoparticles that are encapsulated in high-level nitrogen content-doped (8.2 atom%) graphene layers by annealing MOF precursors at 600 °C in N_2 directly. The HER performance of the material is highly improved compared with the former $Fe_{33}Co_{67}$, showing a low onset potential (88 mV) and achieving 10 mA cm^{-2} at 262 mV in 0.5 M H_2SO_4 (Figure 9b). This design employs Prussian blue analogues (PBAs) and transition metals as precursors, obtaining an ideal electrocatalyst with outstanding stability for HER owing to protection of the N-doped graphene shell. According to the density functional theory calculation (DFT) (Figure 9c), the active sites for HER are still mainly the nitrogen atoms doped in carbon matrix as the metallic alloy is almost encapsulated in graphene layers entirely. Nitrogen doping can improve the activity not only by providing adsorption sites for H*, but also by decreasing ΔG_H for HER in acid. The alloy core and the nitrogen-doped graphitic shell can generate the synergism of metal components and carbon support, which could facilitate HER. ΔG_H of pure metal models (Co_4 and Co_2Fe_2) are also calculated in this work. Both models show relatively negative ΔG_H, which is unfavorable for the desorption step. However, ΔG_H of the Co_2Fe_2 is closer to zero than pure Co, indicating the alloy performs better than the pure metal, showing the synergistic effect between the metals. Similar alloy materials are also reported and applied in alkaline solution for HER. Li et al. [37] showed a novel 3D hierarchically porous flower-like structure of tiny FeCo@NC ultrathin with N-doped carbon nanosheets core-shell groups dispersed on the surfaces. The reported favorable $Fe_{0.5}Co_{0.5}$@NC/NCNS-800 sample shows a high activity for catalyzing HER in 1.0 M potassium hydroxide (KOH), with an onset potential of −63 mV and a low overpotential of 150 mV to achieve 10 mA cm^{-2}. In order to test where the activity for HER originates from, pure $Fe_{0.5}Co_{0.5}$ (reduced by NH_3) and acid-$Fe_{0.5}Co_{0.5}$@NC/NCNS-800 are synthesized as comparisons. Catalytic activities of both the comparisons obviously decay, demonstrating that only perfect FeCo@NC core–shell structure could effectively catalyze the HER process in alkaline electrolyte.

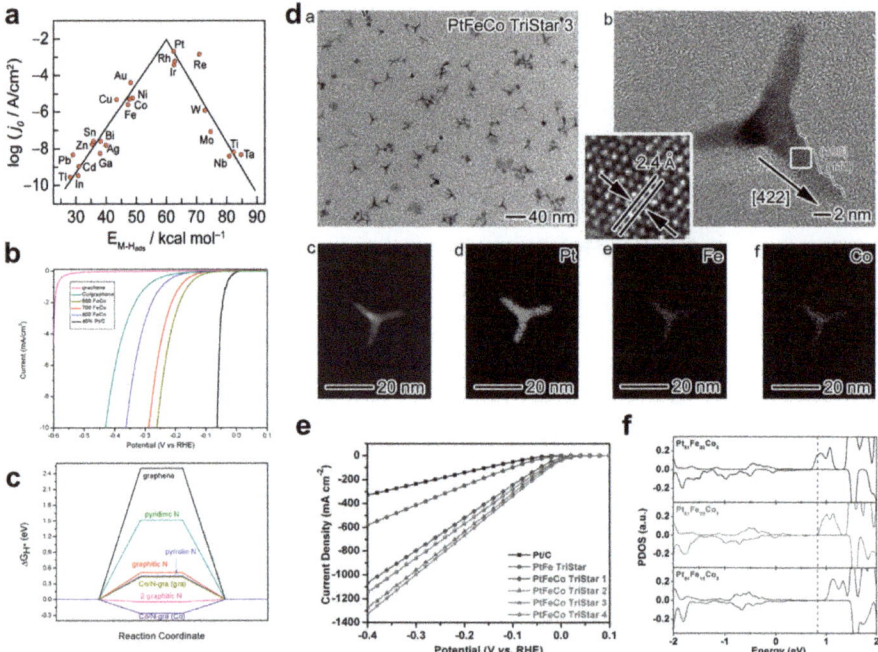

Figure 9. (a) The typical volcano curve for the hydrogen evolution reaction (HER) performance of the metal materials in acid solution, reproduced with permission from [53]. Copyright American Chemical Society, 2010. (b) The polarization curves of FeCo alloy samples of different temperatures; (c) calculated ΔG_H* diagram of some models; the words in the brackets differ from the H* adsorbed on the graphene side or Co_4 side, reproduced with permission from [43]. Copyright Royal Society of Chemistry, 2015. (d) TEM, HRTEM, and STEM mapping images of $Pt_{81}Fe_{28}Co_{10}$ nanostructures in a TriStar shape; (e) Polarization curves for Pt-based alloy TriStar nanostructures in different chemical compositions compared with Pt/C at the same total metal loading weights in 0.5 M H_2SO_4 (aq) electrolyte; (f) calculated projected density of states (PDOS) diagrams for the d orbitals of Pt atoms in three types of PtFeCo lattices in neutral condition, reproduced with permission from [54]. Copyright Wiley, 2016.

Apart from coating the alloy with carbon shells, there are other methods to decorate the alloy, thereby exhibiting favorable activities in different pH. By combining non-noble metals like Fe and Co with noble metal like Pt, trimetallic tristar nanostructures were reported by Du et al. [54], and they tuned electronic and surface structures for enhanced electrocatalytic hydrogen evolution in acid. Different from the former ones, the activity for HER of the PtFeCo comes mainly from metal Pt. However, each type of transition metal would have a specific function in influencing the Pt-based alloy material behaviors for HER. Fe can induce more intense charge polarization on Pt compared with Co, which coincides with the experimental results of PtFe NP exhibiting higher HER activity. However, when an Fe atom in the $Pt_{81}Fe_{27}$ lattice is replaced by Co, the charge accumulation on the Pt atoms close to the Co would increase, generating highly active sites. As a result, it is necessary to incorporate abundant Co atoms into the PtFe lattice to form trimetallic alloy. In addition, by calculating the projected density of states (PDOS) of Pt d orbitals, the PDOS distribution gradually shifts along with the increase of Co content. By controlling the Co:Fe ratio, the d-band center is adjusted to an optimal distance to the Fermi level, obtaining the best HER activity (Figure 9d–f). Electrochemical methods are also widely applied to alloy materials for catalyzing HER in alkaline condition. Müller et al. [55] prepared $Fe_{60}Co_{20}Si_{10}B_{10}$ alloy for HER in a rapid solidification process using a melt spinning device. The alloy with a composition of $Fe_{0.75}Co_{0.25}$ is suggested to be the active surface species. Liu et al. [56] carried out a one-step electro-reductive deposition method to form a Fe–Co composite film above carbon fiber paper in solution. The obtained Fe–Co composite films exhibit efficient electrocatalytic performance and durability for HER, reaching 10 mA cm^{-2} at a low overpotential of 163 mV in 1.0 M KOH.

3.1.2. Bimetal Phosphide Electrocatalysts

Since the simple alloys lack sufficient active sites for HER and show poor performance, several classes of transition metal derivatives have been developed as candidates to replace the rare Pt. For HER, transition metal-based phosphides (TMPs) have exhibited activities superior to other types of compounds. By both experimental and theoretical methods for HER in acid environment, Kibsgaard et al. [8] designed an enhanced transition metal phosphide electrocatalyst. By using DFT calculation, the ΔG_H is consistent with a volcano trend and is located within a narrow range at the top of the volcano plot (Figure 10a,b). The results show that CoP and FeP exhibit hydrogen adsorption free energies in each side of the optimum ΔG_H value. A $Fe_{0.5}Co_{0.5}P$ with a ΔG_H value that is closer to the optimum value than either the value of CoP or FeP is designed. The high efficiency of such a structure is also confirmed by experimental methods. Regarding applications of TMPs in wider pH ranges, Zhang et al. [57] found that the different doping level of Fe in CoP has various effects on catalytic properties in different conditions. By adjusting the ratio of precursors, hybrids denoted as CoP/CNT or $Co_{1-x}Fe_xP$/CNT (with x = 0.1, 0.2, 0.3, 0.4, and 0.5) are synthesized by a simple two-step device. The catalytic activity of $Co_{1-x}Fe_xP$/CNT for HER improves following the increase of Fe mixing ratio from x = 0 to the optimal x = 0.4. The $Co_{0.6}Fe_{0.4}P$/CNT electrocatalyst can afford a current density of 10 mA cm^{-2} at an ultra-low overpotential of 67 mV. In alkaline electrolyte, the $Co_{1-x}Fe_xP$/CNT hybrid also shows excellent HER activities; however, the trend of performance catalyzing HER along with the Fe mixing ratio is different from that in acidic solution. With higher content of Fe, the hybrids exhibit worse catalytic performance. In acid electrolyte, the best activity is denoted by Fe-free CoP/CNT electrode with only 76 mV to afford 10 mA cm^{-2}. Alloyed Fe–Co mono-phosphide exhibits a medial ΔG_H between that of CoP and FeP, leading to an optimized hydrogen binding. Therefore, the better HER performance of the obtained $Co_{1-x}Fe_xP$/CNT catalysts with increased Fe content can be mainly ascribed to the better hydrogen adsorption thermodynamic on the surface of $Co_{1-x}Fe_xP$ catalyst, which is induced by Fe-doping. However, in an alkaline condition, the negatively charged P sites play the role of the proton acceptor centers to facilitate the generation of metal hydrides on the proximate metal centers. Since the electron negativity of Fe atom is lower than that of Co, more positive charges on metal sites are expected with increased Fe incorporation. The larger positive charges will induce stronger adsorption of hydroxyl on metal centers, resulting in less accessibility for proton adsorption to form metal hydrides. These factors can explain the opposite trend in alkaline electrolyte induced by Fe-doping.

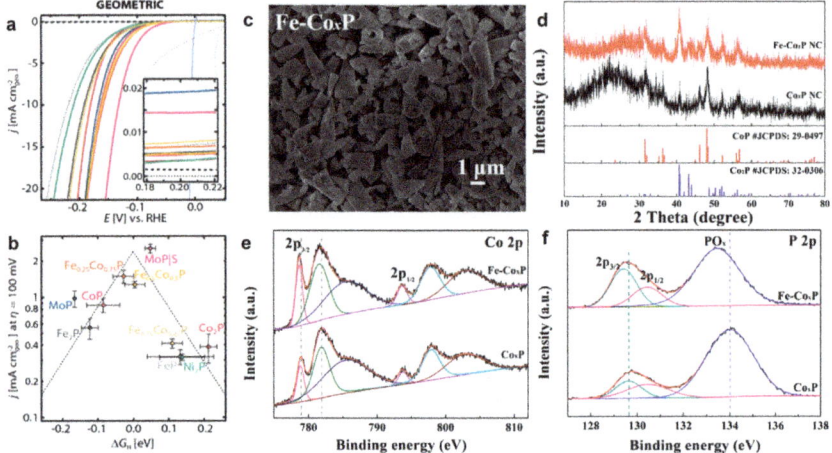

Figure 10. (**a**) The polarization curves per geometric area for transition metal phosphide electrodes. The inset figure is a zoom-in on a potential region. The polarization curve of Pt nanoparticles (NPs) is displayed for comparison; (**b**) activity volcano plot for HER performance, showing the geometric current density from (**a**) at η = 100 mV as a function of hydrogen adsorption free energy (ΔG_H), reproduced with permission from [8]. Copyright Royal Society of Chemistry, 2015. (**c**) SEM image of Fe-doped Co_xP nanocones; (**d**) XRD patterns of Co_xP and Fe-doped Co_xP nanocones; XPS spectra of (**e**) Co 2p and (**f**) P 2p for Co_xP and Fe-doped Co_xP nanocones, reproduced with permission from [58]. Copyright American Chemical Society, 2018.

Compared with FeP, the CoP crystals are easier to form when post-processing the bimetal precursors like MOF, Prussian blue analogues, and metal hydroxides. The obtained catalysts show outstanding HER activities in various electrolytes. For example, Guo et al. [58] reported PBAs nanocones and their transition into Fe-doped Co_xP nanocones by intercalation synthesis. The Fe–Co PBAs nanocones are prepared by an intercalation strategy, employing layer structured α-$Co(OH)_2$ NCs as self-sacrificed templates. Electrochemical analysis indicates that incorporating Fe can effectively improve the HER activities in acid solution because of the increase of electrochemical surface area (ECSA) and more suitable free energy of hydrogen adsorption on Co-centered sites (Figure 10c–f). For the alkaline electrolyte, Li et al. [59] used cobalt- and iron-containing metal–organic frameworks (MOF, ZIF-67@MIL-88B) as precursors followed by a phosphorization process, thus obtaining Fe-doped porous cobalt phosphide polyhedron ($Co_{0.68}Fe_{0.32}P$). The morphology of the polyhedron structure remains after the phosphorization process but exhibits high porosity. As-prepared porous $Co_{0.68}Fe_{0.32}P$ shows a small onset overpotential of 84 mV in 1.0 M KOH and an η_{10} (the overpotential to achieve the current density of 10 mA cm^{-2}) of 116 mV. When doping Fe into the CoP, the HER performance will increase significantly. With its weaker electronegativity, Fe doping results in much weaker binding between Co and P, leading to the formation of low oxidation state Co and elemental Co. The unsaturated Co will further change the charge-density distribution on the surface of the electrocatalyst, giving rise to different H adsorption energies. Prussian blue analogues are also a serious of precursors for Fe–CoP hybrids. Similarly, Cao et al. [60] studied an easy and practical approach to obtain a bimetal catalyst by directly synthesizing a Co–Fe Prussian blue analogue on a 3D porous conductive support (nickel foam), which is then phosphorized into Fe-doped CoP. The Fe-CoP/NF electrocatalyst exhibits efficient electrocatalytic HER activities with a very low overpotential of 78 mV to achieve the current density of 10 mA cm^{-2} in 1.0 M KOH.

On the contrary, Co incorporating FeP nanotubes by a metal–organic frameworks templating strategy was also proposed by Chen et al. [47] for efficient hydrogen evolution. The Co–Fe–P nanotubes achieve excellent HER catalytic performance within a wide pH range, reaching 10 mA cm^{-2} at low overpotentials of 86, 138, and 66 mV in 1 M KOH, 1 M phosphate buffer solution (PBS), and 0.5 M H_2SO_4, respectively. Based on DFT calculations, Co incorporation significantly increases the density of states (DOS) for d-orbital near Fermi level, resulting in higher electrocatalytic activity. Different from the discussed situations, Kim et al. [61] fabricated a 3D porous Co–Fe–P framework utilizing an electrodeposition method. The optimal electrodeposited layers have an amorphous structure while owning excellent HER performance (73 mV at 10 mA cm^{-2}) in alkaline solution. Except for the synthesis of metal phosphides, P was also applied for modification. Kuo et al. [62] improved HER activity of cobalt-doped FeS_2 electrocatalysts by surface phosphorization. The P/Co–FeS_2 nanocomposites grown on the carbon fiber paper substrate achieve geometric current densities of 10 mA cm^{-2} at much lower voltage of –63 mV compared to that of Co–FeS_2 on the substrate (–102 mV).

3.2. OER

Oxygen evolution reaction (OER) is a crucial part in water splitting and recharging metal–air batteries, whereas it is limited by sluggish kinetics. The mechanisms and pathways for OER are more complicated than those for HER. Up to this point, Ir/Ru-based metal materials have been regarded as state-of-the-art electrocatalysts to burst through the kinetic barriers, but their high expense and rare resources hinder their wider applications. To overcome the difficulties, transition metals and their derivatives have been developed as candidates to replace the noble metals. It is generally acknowledged that the OER includes the following steps, where the symbol '*' represents the active sites.

$$* + OH^- \rightarrow HO* + e^- \tag{6}$$

$$HO* + OH^- \rightarrow H_2O + O* + e^- \tag{7}$$

$$O* + OH^- \rightarrow HOO* + e^- \tag{8}$$

$$HOO* + OH^- \rightarrow O_2* + H_2O + e^- \tag{9}$$

$$O_2* \rightarrow O_2 + * \tag{10}$$

3.2.1. Bimetallic Alloy-Based Electrocatalysts

Fe–Co bimetal electrocatalysts exhibit outstanding properties for OER even exceeding Ir/Ru-based materials due to their synergistic effects. To investigate the intrinsic mechanisms, the Fe–Co alloy materials are firstly summarized and analyzed. As we know, the nitrogen doped with metal center is generally regarded as active for OER. For instance, Wu et al. [27] reported a facile preparation of FeCo@NC core–shell nanospheres with graphene substrate displaying outstanding OER performance (Figure 11a,b). The process for fabricating the FeCo/NC electrocatalysts is a facile one-step pyrolysis of graphitic carbon nitride and acetylacetonates. Calculated by the first-principles calculation, the nitrogen doping can activate the adjacent carbon for the electro reactions by modifying the spin status, charge density, and energy bandgap. The encapsulated bimetal alloy nanoparticles in the material will more deeply modulate the electronic environment of the M–N–C sites and enhance the free-energy of adsorption of the oxygen-containing reactants onto the surface of the catalyst. When pyrolyzing carbon source and Co containing materials, carbon nanotubes are easy to generate due to the catalytic effect of Co. As a result, the hierarchical structures consisting of alloy nanoparticles, carbon nanotubes, and carbon sheets are widely synthesized. Liu et al. [63] also reported a two-step fabrication of CoFe alloy nanoparticles supported by nitrogen-doped carbon nanosheets/carbon nanotubes for OER. The CoFe–N–CN/CNTs hybrid material achieves 10 mA cm^{-2} at a small overpotential of 285 mV and shows a low Tafel slope of 51.09 mV dec^{-1}. It is hypothesized that the synergistic effect from the bimetal alloy and nitrogen dopant may be due to the formation of M–N_x active sites. To further prove the mechanism, the OER linear sweep voltammetry (LSV) curves are obtained in alkaline solution with and without SCN$^-$. As a probe, SCN$^-$ can poison M–N$_x$ sites through combining strongly with the M–N$_x$ active sites, thus it can be utilized to investigate the function of the metal-centered sites. The OER performance of the CoFe–N–CN/CNTs exhibits a negative shift after the introduction of SCN$^-$ ions. The decreased activity can be ascribed to the fact that the CoFe–N–CN/CNTs containing M–N$_x$ sites play a critical role in the improvement of the OER activity.

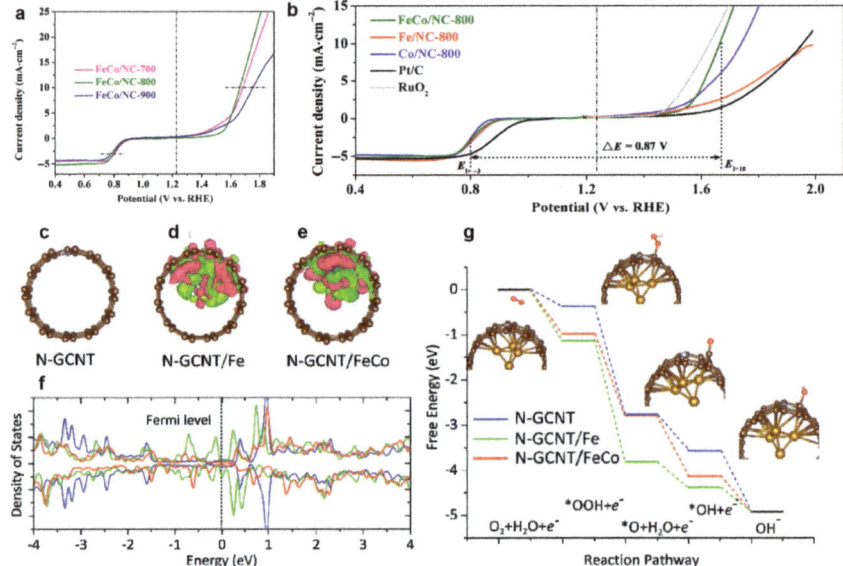

Figure 11. (**a**) OER and ORR activities of FeCo/NC-700, FeCo/NC-800, and FeCo/NC-900 in 0.1 M KOH solution from 0.4 to 1.9 V; (**b**) LSV curves of Fe/NC-800, Co/NC-800, FeCo/NC-800, Pt/C, and RuO$_2$ samples from 0.4 to 2.0 V, reproduced with permission from [27]. Copyright Springer Nature, 2017. (**c**) The optimized morphology of bare N-GCNT. (**d**) The isosurface (0.1e Å^{-3}) of charge density variation in N-GCNT/Fe. Green area represents decrease in charge density (Δρ < 0) and purple area represents increase in charge density (Δρ > 0). (**e**) The isosurface of charge density variation in N-GCNT/FeCo. (**f**) The simulated density of states for the p orbitals of the active C atoms in N-GCNT, N-GCNT/Fe, and N-GCNT/FeCo. (**g**) The free-energy plot of the *OOH, *O, *OH intermediates in the typical 4e$^-$ reaction-pathway for ORR catalyzed by N-GCNT, N-GCNT/Fe, and N-GCNT/FeCo, reproduced with permission from [42]. Copyright Wiley, 2017.

Similar testimony exists in another work by Liu et al. [64], who developed a bamboo-like hierarchical structure consisting of CoFe alloy nanoparticles embedded in N-doped carbon nanotubes supported by reduced graphene oxide, presenting favorable ORR/OER activity and promising application in Zn–air batteries. In this work, in situ X-ray adsorption spectroscopy is applied to reveal the detailed origin of the CoFe alloy. In particular, the K-edge X-ray absorption near edge structure (XANES) spectrum of Fe and Co shows minor variation with larger potential, suggesting the valence states of Fe and Co do not change, evidently. The corresponding FT-EXAFS data demonstrate that the length of Fe–Fe bonds experience a variation of extension and then shorten with the increase of potential, suggesting that the ORR process at lower potentials primarily happens on the Fe species. On the contrary, the Co–Co bond length sees a process of first decreasing and then increasing with the increase of potential, indicating the OER at higher potentials mainly occurs on the Co surface. Moreover, the obvious variation of Fe–Fe and Co–Co bond lengths is found in the generation process of OOH* and O* species, respectively, which further implies that the active positions of the ORR and OER are mainly provided by the Fe and Co species. The investigations prove that the existence of FeCo could give more active sites for OER and ORR. A similar structure of alloy nanoparticles wrapped by nanotubes was proposed using an atomic modulation strategy for FeCo–nitrogen–carbon electrodes by Su et al. [42]. By investigating by both experimental and simulation devices, an efficient synergetic couple between FeCo alloy and N-doped carbon nanotubes is reported as an appreciable coordination structure and electronic environment. Due to the effects provided, the catalyst exhibits a small operating potential of 1.73V to reach 10 mA cm^{-2} for OER in 0.1 M KOH. As we know, single metal-based Fe–N–C materials often lack efficient OER catalytic activity. According to this work, the poor OER activities of Fe–N–C-based materials are mainly because of the semi-conductive FeOOH species formed in OER process. The weak FeOOH bonding strength also poisons the Fe sites and consequently falls inactive. As a result, heteroatom doping is a facile strategy to improve the activities, where the highly OER active Co is thought of as a suitable candidate. This work prepares the efficient N-GCNT/FeCo with bamboo-like structures by a one-pot temperature-programmed calcination process. Investigated by XPS, it is found that the bimetal alloy with appropriate Co incorporation content in N-GCNT/FeCo can dominantly promote the generation of pyridinic N. As widely acknowledged electro-active sites, the electron-accepting pyridinic N species can give a quite large positive charge density to the adjacent sp^2-bonded C atoms, accelerate adsorption of oxygen-containing reactant such as OH$^-$ and O$_2$, and facilitate the electron transfer between the surface of the electrocatalyst and intermediates, thus efficiently improving OER activities with fast reaction kinetics. These mechanisms are also proved by DFT calculations shown in Figure 11c–g. For OER, the rate-determining step is the second electron transfer step for the prepared three electrocatalysts, with the values of U_L(OER), 1.16 V for N-GCNT, 1.45 V for N-GCNT/Fe, and 0.58 V for N-GCNT/FeCo. Obviously, N-GCNT/FeCo exhibits the most efficient catalytic activity for OER. Therefore, DFT calculations are highly consistent with the experimental data, shedding light on the synergetic effects between FeCo alloy nanoparticles and N-GCNT in forming considerable surface electronic modulation and thereby promoting the OER performance.

MOFs are also acknowledged as efficient precursors for alloy-based catalysts. Feng et al. [65] prepared a series of CoM (M = Fe, Cu, Ni)-embedded nitrogen porous carbon frameworks by a facile thermal conversion method with metal-doped zeolitic imidazolate frameworks (ZIF). The optimized $Co_{0.75}Fe_{0.25}$ nitrogen-doped carbon catalyst shows a low overpotential of only 303 mV to reach the current density of 10 mA cm^{-2}. For these samples, the incorporation of the second metal into Co–NC not only provides more metal-centered active sites, but also changes the electronic environment of the centers in Co–NC, improving the catalytic activity. The metal alloy and nitrogen could also trigger the OER activity of the inactive carbon atoms. The enhanced outer graphitic layers would act as active sites for OER. By now, it has been clarified that the metal sites of CoFe alloy could catalyze the OER.

Although the N-doped carbon-based FeCo bimetal alloy materials display excellent activities towards OER, hybrid catalysts combining alloy nanoparticles with other metals or derivatives have also been developed by researchers. By electrospinning Fe/Co–N compounds with polyvinylpyrrolidone (PVP) and polyacrylonitrile (PAN), hybrid mesoporous Fe/Co–N–C nanofibers with embedded FeCo alloy nanoparticles were fabricated by Li et al. [40] The FeCo alloy nanoparticles distribute on the surface of nanofibers uniformly. With increased Fe and Co precursors, the resulting OER activities are improved, indicating FeCo nanoparticles are active towards OER. Zhu et al. [66] used an in-situ coupling approach to prepare FeCo alloys and Co_4N hybrid for catalyzing OER. The active hybrid is obtained through a facile pyrolysis of bimetallic porous phthalocyanine-based network. The simultaneous formation of multiple diverse FeCo alloys and Co_4N nanoparticles plays a crucial role for the excellent OER performance. The new catalyst shows an outstanding OER performance at low overpotential of 280 mV to achieve 10 mA cm^{-2} and high durability in an alkaline solution. Co_3O_4-hydroxides (LDHs) were used as precursors to fabricate Co_3O_4-doped Co/CoFe nanoparticles by thermal decomposition by Li et al. [67] The prepared core–shell structure performs excellent OER activities.

3.2.2. Bimetal Oxides and Hydroxides Electrocatalysts

The development of OER electrocatalysts has been impeded by identifying the catalytic active sites accurately. However, it is acknowledged that the oxides and (oxy)hydroxides for transition metals, especially Fe, Co, and Ni, are active for OER.

As for Fe–Co bimetal oxides, there have been conflicting reports about whether the incorporation of Fe into CoO_x could benefit the OER activity, due to the difficulty of quantifying the number of active sites. A binary transition-metal oxide complex was synthesized by Peng et al. [5] with morphology of hollow nanoparticles. Utilizing the Kirkendall effect, the hollow polycrystalline and highly disordered nanoparticles are formed with OER activity in alkali media. Hollow Fe–Co oxide NPs with different ratios of Fe/Co were prepared and their electrocatalytic performance were tested. The optimal Fe_xCo_{100-x} oxides NPs results in an excellent performance, and a current density of 10 mA cm^{-2} is achieved at an overpotential of ~0.30 V. This work shows that the Co-rich surfaces are active while the presence of surface iron is also critical. However, the origin of the outstanding OER performance is still blurry. To further clarify the intrinsic electrocatalytic activities of Fe-doped cobalt oxides for OER, theoretical studies were performed by Kim et al. [68] using first-principle calculations. The OER mechanisms are analyzed for each site containing the terminal and bridge oxygen sites neighboring the dopant or Co cations in the molecules. The oxide cluster model shown in Figure 12a,b illustrates the dopant terminal site, bridge site, and cobalt terminal site. The mechanisms and the free-energy diagrams for OER of each site of the model were investigated by DFT and given in this work. As a result, the main efficiency of the enhanced OER activity of Fe-containing cobalt oxides are supplied by new active bridge sites, rather than the improvement of the terminal sites. This work helps to deeply demonstrate the OER activity of the Fe-doped Co-oxide-based materials.

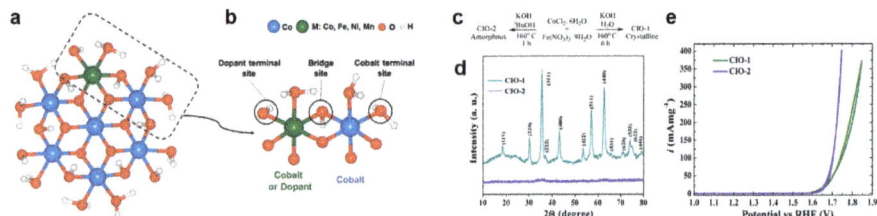

Figure 12. (a) Original or Fe-doped Co oxide cluster model with the composition $Co_6FeO_{24}H_{27}$. (b) The part of the cluster containing the oxygen sites regarded as the OER mechanism simulation. The cobalt terminal site represents the terminal oxygen connected to the Co ion, located at the second coordination range of the dopant. The dopant terminal site represents the terminal oxygen connected to the doped Fe. The bridge site represents the µ2-O bonding the doped Fe and Co ion, reproduced with permission from [68]. Copyright CCC Republication, 2018. (c) Synthetic process for the fabrication of the electrocatalysts CIO-1 and CIO-2; (d) powder XRD patterns of CIO-1 and CIO-2. The peaks of CIO-1 correspond to $CoFe_2O_4$ spinel whereas no apparent peaks are observed for CIO-2; (e) cyclic voltammograms (CV) of CIO-1 and CIO-2 in 0.1 M KOH with a scan rate of 6 mV s^{-1}. Reproduced with permission from [69]. Copyright American Chemical Society, 2014.

Except for the discussions for the original mechanisms, there are also various reports about the materials themselves. For instance, Indra et al. [69] made a comparison between the amorphous and crystalline cobalt iron oxides. By differing solvents and reaction time (Figure 12c) in a simple solvothermal process, a highly crystalline (CIO-1: $CoFe_2O_4$) and an amorphous cobalt iron oxide (CIO-2: $CoFe_2O_n$, n = ~3.66) were prepared. As demonstrated in Figure 12e, CIO-2 exhibits higher OER activity than CIO-1 in alkaline condition, indicating better OER performance. Compared with CIO-1, CIO-2 includes a larger content of low spin Co^{3+} species, evidencing that the Co^{3+} in the octahedral structure improves OER, while the Co^{2+} in the tetrahedral sites is inactive. In addition, from the XPS spectra for

Co 2p edges of the samples before and after electrochemical measurement, it can be observed that the peak broadens and shifts to higher energy for both CIO-1 and CIO-2, proving the generation of Co^{3+} during the OER. Similarly, the same results are found in Fe XPS 2p peaks. As a result, electrocatalytic active Co^{3+} and Fe^{3+} are generated through the OER process, leading to higher OER activities.

Cobalt iron hydroxide is also regarded as a precious OER electrocatalyst as it can easily combine with oxygen-containing groups and then change into OER reaction intermediates. Babar et al. [6] reported a convenient and scalable approach to obtain a 3D self-supported NF electrode substrate integrated with Co–Fe hydroxide. The obtained CoFe/NF electrode performs excellent OER activity with a small potential of 1.45 V to reach 10 mA cm^{-2}. Synergistic effects between the metallic characters of Fe and Co are the main reason for the excellent electrocatalytic activity. The introduction of Fe could significantly promote the OER performance of $Co(OH)_2$ catalyst because easier OH^- intercalation can be achieved by increasing disorder/porosity as well as more defect or edge sites on $Co(OH)_2$ surface. The CoFe/NF facilitates the proton-coupled electron transfer steps by forming oxo species at lower potentials, due to the synergistic couple of Co and Fe. Jin et al. [12] also presented an economic and highly efficient catalyst for OER, which was denoted as α-$Co_{1-m}Fe_m(OH)_2$. The obtained layered α-$Co_4Fe(OH)_x$ nanosheets can catalyze OER at a low potential of 1.525 V to achieve 10 mA cm^{-2} and a low Tafel slope of 52 mV dec^{-1}. During the OER test, the first CV cycle of α-$Co_4Fe(OH)_x$ shows a much higher anodic peak than the second CV cycle without reduction peak (Figure 13a,b), which coincides with the previously published result that $Co(OH)_2$ could be irreversibly changed into CoOOH. It is noticeable that with the increased Fe, the anodic wave of $Co^{2+/3+}$ shifts positively. This indicates that an evident electronic function between Co and Fe could modify the electronic environment of α-$Co(OH)_2$ and make Co^{2+} oxidation more difficult. Besides, the Fe enlarges the spacing between the $Co(OH)_2$ sheets and increases the electrochemical surface area. The DFT calculations also demonstrate that the introduction of Fe influences the adsorption energy of O and OH, leading to a higher $\Delta G^0_{O*} - \Delta G^0_{HO*}$ and lower overpotential for OER (Figure 13c–g).

Burke et al. [14] systematically investigated the function of Fe in Co-based materials and the intrinsic OER activity of $Co_{1-x}Fe_x(OOH)$. The intrinsic OER p of $Co_{1-x}Fe_x(OOH)$ is demonstrated by turnover frequency (TOF) calculation, which is defined as the number of O_2 molecules produced per second per active site. As Figure 13h,i shows, the intrinsic activity of $Co_{1-x}Fe_x(OOH)$ is ~100-fold higher for $x \approx 0.6$–0.7 than for $x = 0$ on a per-metal TOF basis. The role of Fe in $Co_{1-x}Fe_x(OOH)$ for OER could be divided into three aspects. For the structure and electrolyte accessibility, Fe incorporation would allow for easier OH^- intercalation and facilitate the transformation from $Co(OH)_2$ into CoOOH evidently, due to the increased disordered structure. For the effective electrical conductivity, the FeOOH films exhibit apparently poorer conductivity than those with Co. However, the addition of Fe into CoOOH does not enhance the conductivity of $Co_{1-x}Fe_x(OOH)$ compared with CoOOH; there is only a shift towards more anodic potentials for the beginning of the $Co^{2+/3+}$ oxidation peak after Fe incorporation. For the active sites, the results indicate that the intrinsic activity of the iron-based sites is ~130-fold higher than that of the cobalt-based sites. The element Fe provides the primary OER activity while CoOOH provides an inherently large surface-area and superior conductivity.

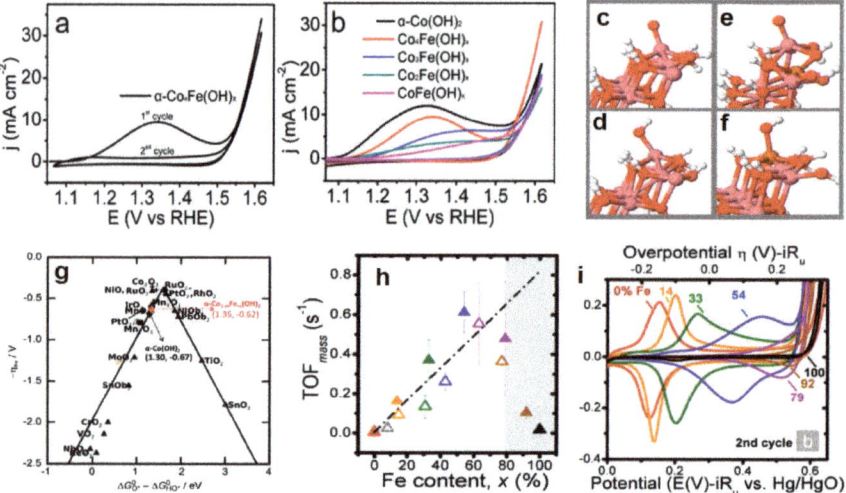

Figure 13. (**a**) Voltammetry of α-Co$_4$Fe(OH)$_x$ in purified KOH showing the difference between the first and second CV cycles; (**b**) the first voltammetry of different Co/Fe ratio catalysts showing a systematic anodic shift of the (nominally) Co$^{2+/3+}$ wave with the increasing Fe content, O binding on (**c**) Co(OH)$_2$ and (**d**) α-Co$_{1-m}$Fe$_m$(OH)$_2$, OH binding on (**e**) Co(OH)$_2$ and (**f**) α-Co$_{1-m}$Fe$_m$(OH)$_2$ (red: Oxygen; pink: Cobalt; brown: Ferrum; white: Hydrogen); (**g**) the positions of the pristine and Fe-doped α-Co(OH)$_2$ on the volcano curve, reproduced with permission from [12]. Copyright Royal Society of Chemistry, 2017. (**h**) Turnover frequency (TOF) data depicted based on the total film mass and composition assuming all metal sites are available for catalysis; (**i**) voltammetry of Co$_{1-x}$Fe$_x$(OOH) showing systematic anodic shift of the (nominally) Co$^{2+/3+}$ wave with increasing Fe content, reproduced with permission from [14]. Copyright American Chemical Society, 2015.

Although bimetal oxide and hydroxide have been regarded as outstanding catalysts for OER, researchers still make efforts to combine them with other components to improve the electrocatalytic activity. Chen et al. [70] synthesized electrochemically tuned cobalt–nickel–iron oxides by an in situ electrochemical oxidation approach from their corresponding sulfides. An electrochemical deposition-sulfurization-tuning strategy was applied for nano-porous transition metal oxides with significantly enhanced OER activity. The prepared Co–Ni–Fe oxides in situ fabricated on the 3D carbon-fiber electrode exhibits a low potential of 1.462 V at 10 mA cm^{-2} and a small Tafel slope of 37.6 mV dec^{-1}. By this approach, the OER activity is enormously enhanced owing to an enlargement of specific surface area and active catalytic sites caused by the in situ electrochemical tuning. Zhang et al. [71] developed cobalt–iron phytate (Co–Fe–phy) nanoparticles as OER catalyst, which has a big degree of amorphization, nano-porous geometry, and large electrochemically active surface area. The hybrid exhibits a low overpotential of 278 mV at 10 mA cm^{-2} and a small Tafel slope of 34 mV dec^{-1} in 1.0 M KOH. As previously reported, the bulky phosphate could induce the crystal lattice of the metal to become disordered. Hence, enormous phytate groups may play a similar function in generation of the Co–Fe–phy, resulting in larger surface area and sufficient active sites. According to the role of Fe, it is observed that the oxidation peak of Co–Fe–phy shifts positively and becomes larger compared to Co–phy. The phenomenon could be explained as the doping iron tuning the electronic states of cobalt-based materials. Similar discussion was given by Liu et al. [72] who reported a facile approach to a porous iron–cobalt phosphide (Fe–Co–P) alloy structure with good conductivity employing an Fe–Co bimetal MOF as a precursor. The obtained complex possesses outstanding OER performance with the typical 10 mA cm^{-2} being reached at 1.482 V. Density of states (DOS) values of the Fe–Co–P and the compared FeP and CoP are calculated to inspect the electronic structures.

The DOS value of Fe–Co–P has a bigger ratio of states located near the Feimi level than that of FeP and CoP, indicating higher electronic conductivity. The local surface changes of the material during OER are analyzed by XPS. After electrochemical test, an additional evident peak is observed in the Fe $2p_{3/2}$ spectrum corresponded to the generated FeOOH during the electrochemical process. The Co $2p_{3/2}$ peaks after OER show a similar valence status with those of fresh Fe–Co–P sample, suggesting the Co element is stable during the reaction. As a result, the Fe is regarded as being beneficial for stabilizing the Co-centered species in a low state, resulting in the simultaneous improvement of activity and durability for OER. Former experimental data and DFT results have demonstrated that the generation of high-oxidation state M–OOH intermediates is the rate-determining step for OER. Introducing heteroatoms is thought of as an efficient way to induce easier transformation toward M–OOH. The boron could debilitate the M–M bonds, thus decreasing the thermodynamic and kinetic barriers for M–OOH generation. Chen et al. [18] reported an amorphous Co_x–Fe–B compound with a low overpotential of 0.298 V at 10 mA cm^{-2} for OER. From the experimental results, the function mechanism of the synergistic effect is demonstrated in Figure 14a; the incorporation of Fe enhances the third step in the OER process, increasing the quantity of OOH species, while enhancing the conductivity. The existence of B content decreases the energy barrier for the generation of the OOH.

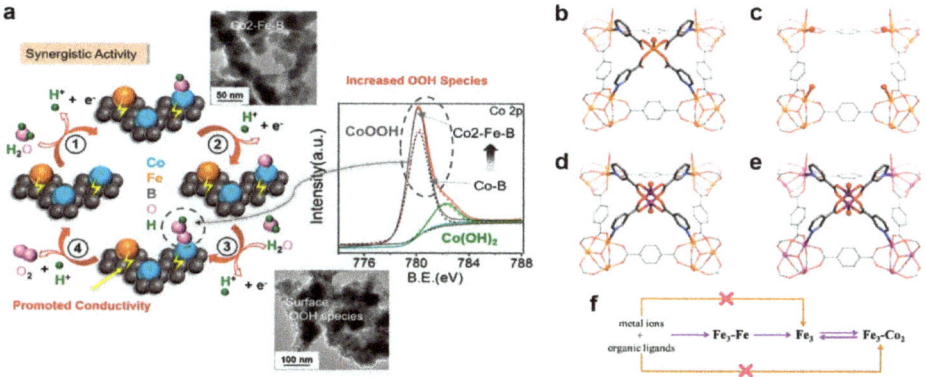

Figure 14. (a) Illustration of the OER mechanism of the synergistic couple of Co_2–Fe–B, reproduced with permission from [18]. Copyright American Chemical Society, 2017. Key partial structures of (b) Fe_3–Fe, (c) Fe_3, (d) Fe_3–Co_2, and (e) Co_3–Co_2; (f) synthesis demonstration of Fe_3–Co_2, reproduced with permission from [73]. Copyright American Chemical Society, 2018.

3.2.3. Dual-Metal M–N–C Site-Based Electrocatalysts

As a facile method to design dual-metal sites, the construction of metal–organic frameworks has emerged as a series of potential electrocatalysts, as the unsaturated metal centers tend to be active catalytic sites. An unusual host/guest geometry and modular synthetic strategy was provided by Shen et al. [73] through a two-step, single-crystal to single-crystal, post-synthetic modifications, a thermal-, water-, and alkaline-stable MOF [{$Fe_3(\mu_3$-O)(bdc)$_3$}$_4$-{Co_2(na)$_4$(L^T)$_2$}$_3$] (H_2bdc = 1,4-benzenedicaboxylic acid, Hna = nicotinic acid, L^T = terminal ligand)-containing dicobalt cluster. The hybrid MOF processes high electrocatalytic oxygen evolution activity with a low overpotential of 225 mV at 10 mA cm^{-2} in aqueous solution at pH = 13. By constructing the dual-metal structure, the paddle-wheel type dinuclear metal carboxylate cluster M_2(RCOO)$_4$(L^T)$_2$ could be stabilized in water since the specific coordination geometries give them high adsorption affinities and low chemical stabilities. The chemical durability of the Fe_3–Co_2, Fe_3, Fe_3–Fe, and Co_3–Co_2 (Figure 14b–f) are obtained in aqueous solution at different pH values. The Fe_3–Co_2, Fe_3, and Fe_3–Fe can remain undamaged at pH = 13, whereas Co_3–Co_2 collapses even at pH = 7. In the OER electrocatalytic investigations, the Fe_3–Co_2 sample gives much higher activity than the single metal samples, including lower overpotential and smaller Tafel slope, indicating the synergistic couple between Fe and Co. The intrinsic synergistic effect has been proved by dual-metal single-atom catalysts. Due to the structural homogeneity, single-atom catalysts can facilitate precise identification and characterization of catalytic active sites. The high atom economy of the single-atom catalysts also coincides with the request of sustainable development [74]. Using two types of templates concurrently, Li et al. [75] reported a hierarchical meso/microporous FeCo–N_x–C nanosheets with high OER activity. The synergistic application of two kinds of templates, an active metal salt and silica nanoparticles, achieves a reversible catalyst of efficient OER–ORR performance. Compared with single-metal Fe–N_x–CN electrocatalyst, the obtained bimetal FeCo–N_x–NC possesses lower onset potential and smaller Tafel slope, suggesting the coupling of Fe and Co with doped nitrogen can tune the electronic environment and surface polarities and improve the OER performance.

3.3. ORR

As the reverse reaction of OER, ORR occurs in the cathode for recharging process of the metal–air batteries and fuel cells. Platinum-based materials are thought of as the most efficient catalysts for proton exchange membrane fuel cells (PEMFCs) and direct methanol fuel cells (DMFCs). However, due to the high cost and scarcity of Pt, it is essential to develop low-cost materials with high ORR efficiency. The charge transfer between different metal sites modulates the electronic structures, enhancing the ORR activity by adjusting the adsorption energy for the intermediates. Among the non-noble metal catalysts, 3d transition metal-based materials are most promising for their unique catalytic properties.

3.3.1. Bimetallic Alloy-Based Electrocatalysts

For ORR, Fe–Co alloy-based doped carbon hybrid materials are highly desirable due to the various active sites. Sultan et al. [76] reported a precisely tuned atomic ratio of Fe and Co embedded in nitrogen-doped graphitic tube (NCT). The optimal $Co_{1.08}Fe_{3.34}$ complex with M–N bonds exhibits remarkable ORR activities of an onset potential of 1.03 V and half-wave potential of 0.94 V, surpassing Pt/C in 0.1 M KOH. The excellent ORR performance is ascribed to the coexistence of Co–N, Fe–N, and sufficient metallic FeCo alloys, which are inherently better at interacting with reactants and favor faster electron movement due to the presence of lone-pair electrons. The Fourier transform extended X-ray absorption fine structure (FT-EXAFS) analysis of cobalt and iron reveals the coexistence of peak corresponding to Co(Fe)–N and Co(Fe)–Co(Fe). Alloying Fe and Co induces favorable changes in intrinsic properties by altering the DOS at the Fermi level of the metallic sites, which enhance their catalytic activity to a large extent. Therefore, not only the acknowledged M–N_x, but also FeCo alloy provides numerous active sites for ORR, resulting in high electrocatalytic activity for ORR. Moreover, the high activity and stability also originate from nitriding and encapsulation in the graphitic shell, which improves the exposed surface of active sites, electronic, and mass transfer. For ORR, a specific class of metal/N-doped carbon matrix materials, which include N-coordinated sites with iron or cobalt embedded, have been regarded as promising electrocatalysts. Similar to the mechanisms of nitrogen doping, some nitrogenous functional groups like 2-pyrridone could also facilitate the complexation of metal irons, generating applicable precursors of ORR electrocatalysts. For example, Lim et al. [77] gave a highly efficient non-noble metal catalyst based on poly(vinylpyrrolidone)-wrapped carbon nanotubes combined with Fe–Co metal ions. First, the nanotubes and PVP are mixed in water with ultrasonic treatment to obtain a durable dispersion of MWCNT-PVP. A metal loading is then added to the dispersion and stirred. After pyrolysis in NH_3, the hybrid material with Fe and FeCo alloy nanoparticles wrapped in N-doped carbon nanotubes is prepared. The total metal loading of the non-noble metal in the FeCo–CNT electrocatalysts is about 1 wt%. During the process, the metal ions are chemisorbed onto the 2-pyttidone groups, because of the high polarity of the near-planar lactam ring. It is notable that according to XPS and XRD results, the nature of metallic nanoparticles has an effect on the formation of nitrogen. Fe nanoparticles induce the generation of graphitic-carbon, while Co nanoparticles promote the formation of pyridinic-nitrogen and protonated pyridyl moiety.

Therefore, the alloy particles have an effect on the ORR performance by both taking part in the reaction and determining the type of the N species. The MWCNT-FeCo catalyst exhibits an enormous improvement in the ORR performance with an onset potential and half-wave potential of 0.82 V and 0.73 V compared with the MWCNT-Fe and MWCNT-Co. The improved activities originate from the active sites provided by FeCo alloy and metal-N_4 cores. The Co–N_4 and Fe–N_4 cores are thought to generate a face-to-face bimetal structure with a specified stereo distance, which could induce easy reduction of oxygen to water and facilitate ORR process.

As a typical type of porous materials, MOF materials have been considered as suitable precursors to derive catalysts with the M–N_x active sites for ORR. Guan et al. [78] developed the "MOF-in-MOF hybrid" confined pyrolysis method to prepare porous Fe–Co alloy/N-doped carbon cages. The obtained the obtained $Fe_{0.3}Co_{0.7}$/BC with porous structures shows superior ORR activity with a half-wave potential of 0.88 V, which is higher than the commercial Pt/C electrocatalysts (0.855 V). The specific strategy from MOF provides sufficient N-coordinated bimetallic active sites together with graphitic N-doped carbon matrix, rendering high activity and selectivity, while the porous structure also improves the mass and charge transport.

The roles of the alloy nanoparticles and M–N_x sites for ORR were further provided by Tan et al. [79]. A nanofiber network of bimetal/nitrogen co-doped carbon electrocatalyst for ORR shows excellent activity with the peak potential (E_{peak}) at 0.85 V. From the TEM and XRD patterns, both cubic $CoFe_2O_4$ and cubic FeCo alloy are encapsulated in the N-doped carbon shells. XPS spectrum also shows Fe and Co are co-doped into carbon. The introduction of metals into a N-doped carbon system can form different intrinsic active sites such as M–N_x, which gives an indispensable effect on the substantial improvement of the ORR activity. After the acid-leached dispose for the catalyst, a significant inferior ORR activity is observed with a potential shift by 51 mV. It can be concluded that upgradation originates from the activity provided by a bimetallic center ($CoFe_2O_4$/FeCo). Regarding the role of the metallic Co and Fe, the Nyquist plots also indicate that the Fc–F (ferrocenoyl–phenylalanine)/Co@N-C800 shows the lowest charge-transfer resistance, which is possibly due to the Fe and Co inducing the graphitization of carbon during the pyrolysis process. Higher graphitization degree can improve the conductivity and stability of the electrocatalyst, resulting in higher activity. In summary, the excellent performance of the catalyst can be attributed to the coexistence of FeCo alloy, $CoFe_2O_4$, and Fe(Co)–N_x sites, which together provide various active catalytic sites for ORR. The activity from the Fe–Co alloy has been discussed in this paragraph, while the roles of Co–Fe bimetal derivatives and metal–N_x sites will be mentioned in the following parts.

3.3.2. Bimetal Derivatives Electrocatalysts

By different precursors and strategies, researchers have developed various bimetal derivatives. The bimetal oxides exhibit excellent activities for ORR. Typically, a template- and surfactant-free strategy was provided by Wang et al. [80] to synthesize a hollow Co_2FeO_4/MWCNT hybrid. The hollow structure is transformed from the alloy Co_2Fe nanoparticles via the Kirkendall effect, which is based on the different diffusion speeds of different atoms under certain conditions. The specific hollow morphology with a rough surface can enlarge the active area and improve the mass transfer of the electrolyte, thus enhancing the electrocatalytic activity. In O_2-saturated 0.1 M KOH solution, the oxygen reduction begins at 0.91 V and the half-wave potential is 0.73 V. The ORR performance of Co_2FeO_4/MWCNT is obviously higher than that of Co_3O_4/MWCNT and Fe_2O_3/MWCNT samples, suggesting the synergistic effect from the Co and Fe. A similar trend due to the synergistic effect can also be observed in hybrids of metallic nanoparticles with metal oxides. Kim et al. [81] rationally designed a cost-effective and efficient ORR catalyst. Well-distributed nanopolyhedron Co_3O_4 grows on nitrogenated graphitic porous two-dimensional layers (C_2N), which encapsulates Fe nanoparticles. The obtained NP Co_3O_4/Fe@C_2N presents outstanding ORR activity with the onset and half-wave potentials comparable to those of Pt/C. The excellent ORR performance comes from various reasons, as follows. Firstly, the Fe@C_2N catalyst performs superior oxygen transfer and electron tunneling with the polar C_2N matrices. The iron cores wrapped by the nitrogenated carbon periphery provide ORR activity and outstanding stability. Then, the increased chemically effective surface area by introducing Co_3O_4 is owing to both the improved conductive properties and the induced junction sites due to the strong function between Co_3O_4 and carbon matrix. Thirdly, the relative area ratio of Co^{3+} to Co^{2+} in the prepared NP $Co_3O_4/Fe@C_2N$ is higher than that of the hybrid $Co_3O_4+Fe@C_2N$. This change gives larger number of donor-acceptor reduction sites, which can also improve ORR activity. Finally, the introduction of Co^{3+} induces an increase in oxygen vacancies and results in a lower ratio of $O_{lattice}/O_{ad}$, which coincides with the XPS analysis. The lower ratio of $O_{lattice}/O_{ad}$ indicates the strong interaction between the NP $Co_3O_4/Fe@C_2N$ and surface oxygen-adsorbed species, thus enhancing the electrocatalytic activity. Furthermore, the decreased $O_{lattice}$ binding energy is proposed to increase the electron donor capacity related to the improved ORR activity, resulting from the synergistic effect between Co_3O_4 and C_2N network. Wang et al. [82] presented cobalt–iron (II, III) oxide (Co–Fe_3O_4) hybrid nanoparticles supported by carbon as an efficient catalyst of ORR in alkaline media. The Co–Fe_3O_4 hybrid nanoparticles embedded on carbon substrate (Co–Fe_3O_4/C) are synthesized by a two-step strategy. Co nanoparticles are prepared first and Fe_3O_4 nanoparticles are then fabricated onto them. The Fe_3O_4 possesses a cubic inverse spinel structure as well as a close-packed, face-centered cubic configuration of O^{2-} ions, where each Fe^{2+} ion occupies half of the octahedral structures while the Fe^{3+} ions are distributed uniformly between the left octahedral structures and the tetrahedral structures. Based on the square wave voltammetry (SMV) results (Figure 15), it can be concluded that the Fe^{2+}/Fe^{3+} transition of Co–Fe_3O_4/C apparently has higher potential than Fe_3O_4/C. The anodic shift of Co–Fe_3O_4/C will lead to higher ORR activity. From the EXAFS spectrum of Co K-edge and Fe K-edge, the specific peak is at 3.21 Å, which corresponds to the substitution of Co^{2+} ion by Fe^{2+} ion to form Co–O–Fe in the hybrid. This replacement may cause the change of the cationic distribution of the catalytic surface, improving the ORR performance. XAS results indicate that the distance of Fe–O and the Fe–Fe in Fe_3O_4 differ throughout the structure, forming lattice strain in Co–Fe_3O_4/C. The desorption and adsorption of the oxygen-containing O and OH intermediates are the two determining steps of ORR, in which the lattice strain would have an effect on the oxygen binding energy and change the ORR efficiency.

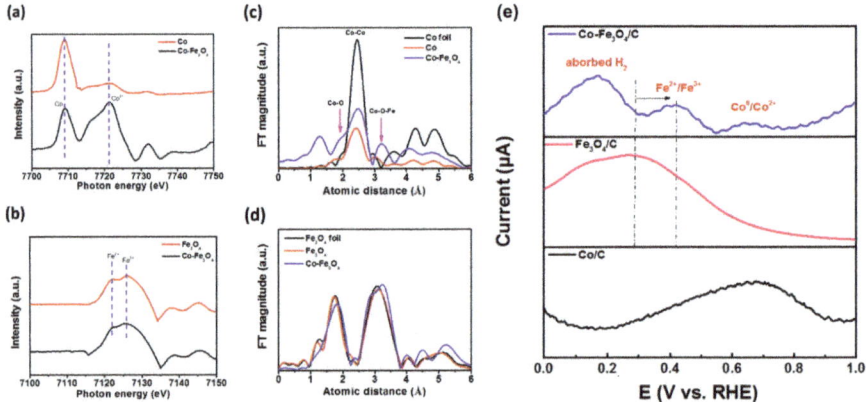

Figure 15. XANES spectra for (**a**) Co K-edge and (**b**) Fe K-edge; extended X-ray absorption fine structure (EXAFS) spectra for (**c**) Co K-edge and (**d**) Fe K-edge, (**e**) square wave voltammetry (SWV) curves of Co–Fe_3O_4/C, Fe_3O_4/C, and Co/C tested in 0.1 M KOH, reproduced with permission from [82]. Copyright Elsevier, 2015.

Apart from metal oxides, there are other classes of derivatives synthesized by scientists. For instance, Jin et al. [83] reported a carbon iron 3D porous carbon-supported carbonate hydroxide hydrate as efficient and stable ORR electrocatalysts. This experiment innovatively applies super absorbent polymer (SAP) to provide carbon source, which is the important constituent of baby diapers with efficient adsorption of water and ions. The obtained cobalt iron carbonate hydroxide hydrate in situ loaded on 3D porous carbon (CICHH@C) is prepared by a two-step strategy (Figure 16a–d). Firstly, the mixture is freeze-dried. Then, the resulting mixture is calcined and cooled down to environment temperature. During the preparation, the addition of potassium hydroxide is the key step because it could both accelerate the hydrolysis reactions to form colloid intermediate and induce the graphitization by consuming the amorphous carbon. This CICHH@C exhibits fast ORR kinetics and excellent catalytic activity with a small half-wave potential (0.780 V) and a low Tafel slope (73 mV dec^{-1}). As a Co–Fe binary material, CICHH is a novel low-cost material for ORR catalysis, of intrinsic superior activity. In addition to SAP, biomass has also been used as precursors for bimetal ORR electrocatalysts. As reported by Sun et al. [84], a three-dimensional hierarchically porous carbon framework embedded with cobalt–iron–phosphide nanodots nanocomposite was synthesized through a lyophilization–pyrolysis–phosphorization process, in which egg white is chosen as the precursor of the N-doped carbon matrix due to its high nitrogen content. The egg white dispersion containing metal ions is freeze-dried and then pyrolyzed in Ar atmosphere to form a 3D honeycomb-like morphology. After a phosphorization process, CoFeP nanoparticles anchored on the conductive substitute are obtained. The XRD patterns clearly shows that once Co and Fe are doped simultaneously, a synergetic effect would happen between Fe and Co, leading to the different structures of the phosphides. XPS results also prove the existence of interactions between Fe and Co, achieving superior catalyst performance. The ORR performance of the materials in 0.1 M KOH is measured by the rotating disk electrode method. The onset potential and half-wave potential of the CoFeP/EWC sample is 0.94 V and 0.83 V respectively, much higher than those of Co$_2$P/EWC and Fe$_2$P/EWC.

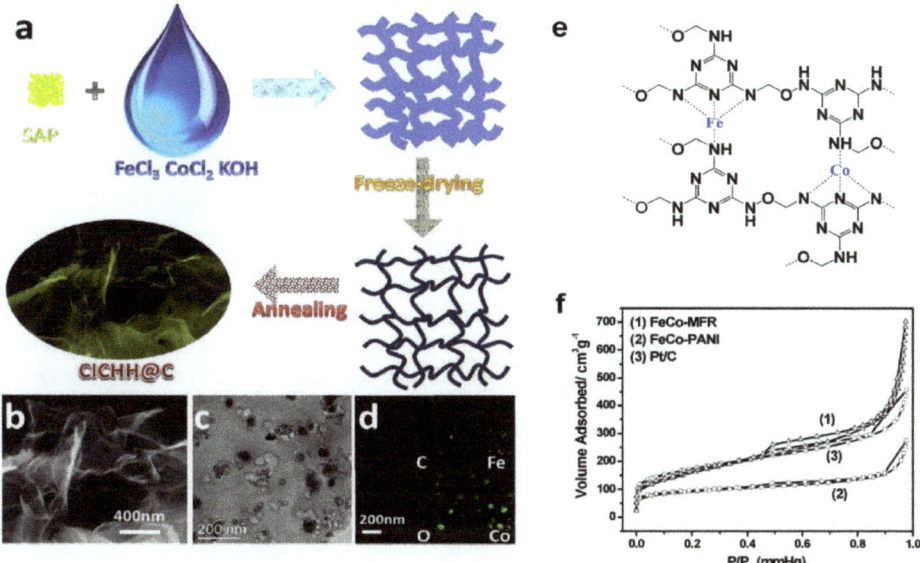

Figure 16. (a) The synthetic scheme of CICHH@C, (b) SEM image of CICHH@C, (c) TEM image of CICHH@C. (d) the C, O, Fe, and Co element mapping of CICHH@C, reproduced with permission from [83]. Copyright Elsevier, 2018. (e) A hypothetical chemical structure of Fe- and Co-coordinating melamine-formaldehyde resin (FeCo-MFR), (f) steady-state polarization curves of ORR for the FeCo-MFR (1), FeCoPANI (2), and Pt/C (3) electrodes in the phosphate buffer solution (PBS) solution, reproduced with permission from [85]. Copyright Royal Society of Chemistry, 2013.

3.3.3. Dual Metal M–N–C Sites-Based Electrocatalysts

To investigate the active M–N_x sites of ORR, a series of macrocyclic organic compounds containing both Co and Fe are fabricated as precursors of the electrocatalysts. Zhao et al. [85] synthesized a stable Fe/Co/C/N nano-porous structure using melamine network polymers as carbon and nitrogen source. The predicted chemical structure of the Fe- and Co-coordinating melamine formaldehyde is shown in Figure 16e. The procedure of the synthesis includes hydroxylation of melamine with formaldehyde, polymerization with carbon supports, coordination of transition metals, and pyrolysis. By this method, the Fe/Co/C/N catalyst behaves with an onset potential of 0.88 V and a half-wave potential of 0.78 V in neutral media (Figure 16f). Corresponding to the XPS analysis, pyridinic-N is suggested to constitute the main transition metal-binding site in the M–CN material. Protons can be trapped by free pyridinic-N and subsequently transferred to active sites for ORR. To arrange the Co and Fe species with more precise control, Lin et al. [86] altered monomeric iron and cobalt metalloporphyrins to fabricate heterometal-embedded organic conjugate frameworks (N–FeCo–C). The ordered distribution of iron and cobalt represents that the innovative method realizes the uniform of active components. The pore volume can also be tuned by utilizing diverse types of monomers with different geometries. In both acid and alkaline electrolyte, the heterometallic product shows excellent ORR activity compared to the pure-Co(Fe) containing samples. In 0.1 M KOH, the PCN-FeCo/C gives the onset potential of 0.90 V and the peak potential of 0.88 V, while in 0.1 M $HClO_4$, the onset and half-wave potentials value 0.90 V and 0.76 V, respectively. From XPS measurements, the peaks at 710.8 eV and 780.6 eV are assigned to Fe and Co respectively, which prove the existence of Fe–N_x and Co–N_x sites for providing high ORR performance. The durability may derive from the high graphitization of carbon structures, which is partly induced by Co. Similar active sites were also mentioned by Zhang et al. [87] by reporting a facile approach to Fe–Co and nitrogen co-doped 3D porous graphitic carbon networks. A facile and rational two-step route is provided to prepare the efficient bimetal ORR catalyst, which exhibits a smaller onset potential (1.05 V) than the single-metal comparisons and a low Tafel slope (65 mV dec^{-1}) in 0.1 M KOH. The observed excellent ORR performance is owing to its hierarchical porosity and the synergistic effect of Fe and Co co-doping, which improve the reactant and electrolyte and provide accessible active sites including pyridinic N and metal-bonded N. The development of single-atom catalysts provides a more detailed and intrinsic origin for the M–N_x sites. Wang et al. [52] designed a stable and efficient non-platinum catalyst for ORR based on N-coordinated dual-metal sites. A host–guest strategy is developed to construct the material within the confined space of MOFs. A type of Co/Zn bimetallic MOF served as a host to absorb $FeCl_3$ molecules with cavities. During the preparation, a double solvents method is applied to remit the diffusion resistance due to the narrow aperture. The X-ray adsorption fine structure spectroscopy and Mössbauer spectroscopic both prove the single-atom structure of the Fe and Co. From theoretical simulation and EXAFS results, the coordination structure is confirmed and shown in Figure 17a–f. The catalyst displays remarkable ORR activity, with the onset potential of 1.06 V and half-wave potential of 0.863 V in acid electrolyte. The ORR mechanism is simulated by employing DFT calculation and utilizing a simulated model with Fe–Co dual sites deposited on N-doped graphene, which is depicted in Figure 17g. The calculation indicates the break of the O–O bond easily occurs on Fe–Co dual site, the reason for which is owing to strong bond of O_2 molecule on the dual site. The Fe–Co dual site provides two anchored position for the free O atoms, resulting in a stable output and exothermic reaction heat. Based on the calculation results, the rate-determining step of ORR is the hydrogenation step of changing adsorbed OH to H_2O with an energy barrier of 0.26 eV, which is much lower than that of the rate-determining O–O bond break on Fe/N–C (0.65 V) and Pt-based materials (~0.80 eV). In summary, the (Fe, Co)/N–C dual site favors breaking of the O–O bond, leading to a higher ORR selectivity and efficiency to the four-electron path.

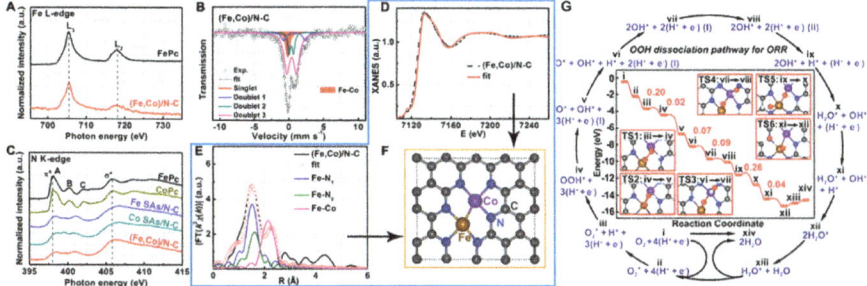

Figure 17. (**a**) XANES spectra of Fe L-edge of the prepared (Fe,Co)/N–C and FePc. (**b**) Fe Mössbauer transmission spectra measured at 298 K for (Fe,Co)/N–C, and fittings with spectral components. (**c**) N K-edge XAS spectra of Fe SAs/N–C, Co SAs/N–C, and (Fe,Co)/N–C. (**d**) Comparison between K-edge XANES experimental spectrum of (Fe,Co)/N–C (black dashed line) and simulated spectrum obtained with depicted structure (solid red line). (**e**) Corresponding Fe K-edge EXAFS fittings of (Fe,Co)/N–C. (**f**) Hypothetical structures of Fe–Co dual sites. (**g**) Energies of intermediates and transition states in mechanism of ORR at (Fe,Co)/N–C from density functional theory calculation (DFT), reproduced with permission [52]. Copyright American Chemical Society, 2017.

4. Conclusions

In spite of the enormous progress already achieved in fabricating bimetallic electrocatalysts with various methodologies, challenging opportunities for optimized and novel synthesis process, which deeply influence the catalytic activity, still remain to be grasped for electrocatalytic applications. In this review, we firstly categorize catalyst supports in the synthesis progresses, including carbon and MOF-based supports, which affect the surface and electronic structure, and further determine the mass transfer and electronic transport in the reactions. Methods such as hydrothermal and solvothermal treatment, electrospinning method, ball milling, and pyrolysis, etc., have been utilized in the construction of highly efficient Fe/Co bimetallic nanoparticles or even atoms decorated with electrocatalysts. The introduction of the catalysts supports great improvements in the electrical conductivity and the texture structure of the catalytic materials, confines the nanoparticle size uniform distribution, and benefits the exposure of the active sites in the reaction process. Furthermore, we reviewed the outcomes of Fe, Co, and their derivatives, combined with hybrid materials for energy-related electrochemical reaction, including HER, ORR, and OER, emphasizing the synergistic effects between the bimetal content and their supports, and the correlation of the material structure with the electronic structure for enhanced electrocatalytic performances. The specific active structures and the mechanisms for the energy-related reactions are characterized and simulated. The synergistic coupling effect of substrates and the active metal phase further facilitates better charge transfer, and thus enhances the activity more forward. Incorporation of heteroatoms (N, P, B) into the lattice of carbon material and MOFs, altering the electronegativity of the support, is speculated to greatly ameliorate the electrocatalytic water splitting performance via modulating the electronic structure.

Therefore, the concepts in catalyst design, advanced approaches in determining active sites and intermediates, are imperative for the electrocatalysts. The molecule design, as well as the combination of computational and experimental methods, have significantly influenced understanding of the reaction mechanisms and led to the development of advanced electrocatalysts.

It may help to investigate the hybrid synthesis progress and material structure effect on the particle size and dispersion situation, metal–support interactions, charge transfer, and adsorbing energy of certain intermediates. However, the chemical modification of the catalyst with multi-atom doping and multilevel structure make the materials more complicated and difficult to identify actual active sites. Specifically, the single-atom catalysts combining both metals, especially those containing dual-metal sites as catalytic active positions, are still urgently needed. The theoretical significance of accurately identifying the catalytic active sites of the single-atom catalysts is promising and notable. However, to design and construct catalysts that could ideally link theoretical calculation and intrinsic properties of the material is still a tough challenge. Herein, it is critical to fill the gap between the theoretical models and experimental methods. Furthermore, considering the intrinsic properties of the transition bimetallic catalysts, more energy-related applications remain to be explored in width and depth.

Author Contributions: Conceptualization, K.L. and Y.L.; validation, W.P., G.Z. and F.Z.; resources, X.F.; writing—original draft preparation, K.L.; writing—review and editing, Y.L.; visualization, X.F.; funding acquisition, X.F. and Y.L.

Funding: This research was funded by the National Natural Science Funds, grant number 21676198, the Specialized Research Funds for the National Natural Science Foundation of China, grant number 21506157, and the Program of Introducing Talents of Discipline to Universities, grant number B06006.

Conflicts of Interest: The authors declare no conflict of interest.

References

1. Tao, L.; Wang, Y.; Zou, Y.; Zhang, N.; Zhang, Y.; Wu, Y.; Wang, Y.; Chen, R.; Wang, S. Charge Transfer Modulated Activity of Carbon-Based Electrocatalysts. *Adv. Energy Mater.* **2019**, 1901227. [CrossRef]
2. Dong, D.; Wu, Z.; Wang, J.; Fu, G.; Tang, Y. Recent progress in Co_9S_8-based materials for hydrogen and oxygen electrocatalysis. *J. Mater. Chem. A* **2019**, *7*, 16068–16088. [CrossRef]
3. Yan, Y.; Xia, B.Y.; Zhao, B.; Wang, X. A review on noble-metal-free bifunctional heterogeneous catalysts for overall electrochemical water splitting. *J. Mater. Chem. A* **2016**, *4*, 17587–17603. [CrossRef]
4. Ahmed, J.; Kumar, B.; Mugweru, A.M.; Trinh, P.; Ramanujachary, K.V.; Lofland, S.E.; Govind; Ganguli, A.K. Binary Fe-Co Alloy Nanoparticles Showing Significant Enhancement in Electrocatalytic Activity Compared with Bulk Alloys. *J. Phys. Chem. C* **2010**, *114*, 18779–18784. [CrossRef]
5. Peng, P.; Lin, X.-M.; Liu, Y.; Filatov, A.S.; Li, D.; Stamenkovic, V.R.; Yang, D.; Prakapenka, V.B.; Lei, A.; Shevchenko, E.V. Binary Transition-Metal Oxide Hollow Nanoparticles for Oxygen Evolution Reaction. *ACS Appl. Mater. Interfaces* **2018**, *10*, 24715–24724. [CrossRef]
6. Babar, P.; Lokhande, A.; Shin, H.H.; Pawar, B.; Gang, M.G.; Pawar, S.; Kim, J.H. Cobalt Iron Hydroxide as a Precious Metal-Free Bifunctional Electrocatalyst for Efficient Overall Water Splitting. *Small* **2018**, *14*, 1702568. [CrossRef]
7. Yang, J.; Zhu, G.; Liu, Y.; Xia, J.; Ji, Z.; Shen, X.; Wu, S. Fe_3O_4-Decorated Co_9S_8 Nanoparticles in Situ Grown on Reduced Graphene Oxide: A New and Efficient Electrocatalyst for Oxygen Evolution Reaction. *Adv. Funct. Mater.* **2016**, *26*, 4712–4721. [CrossRef]
8. Kibsgaard, J.; Tsai, C.; Chan, K.; Benck, J.D.; Norskov, J.K.; Abild-Pedersen, F.; Jaramillo, T.F. Designing an improved transition metal phosphide catalyst for hydrogen evolution using experimental and theoretical trends. *Energy Environ. Sci.* **2015**, *8*, 3022–3029. [CrossRef]
9. Jo, C.; Lee, J.I.; Jang, Y. Electronic and Magnetic Properties of Ultrathin Fe–Co Alloy Nanowires. *Chem. Mater.* **2005**, *17*, 2667–2671. [CrossRef]
10. Bergmann, A.; Martinez-Moreno, E.; Teschner, D.; Chernev, P.; Gliech, M.; de Araujo, J.F.; Reier, T.; Dau, H.; Strasser, P. Reversible amorphization and the catalytically active state of crystalline Co_3O_4 during oxygen evolution. *Nat. Commun.* **2015**, *6*, 8625. [CrossRef]
11. Song, F.; Hu, X. Exfoliation of layered double hydroxides for enhanced oxygen evolution catalysis. *Nat. Commun.* **2014**, *5*, 4477. [CrossRef] [PubMed]
12. Jin, H.; Mao, S.; Zhan, G.; Xu, F.; Bao, X.; Wang, Y. Fe incorporated α-$Co(OH)_2$ nanosheets with remarkably improved activity towards the oxygen evolution reaction. *J. Mater. Chem. A* **2017**, *5*, 1078–1084. [CrossRef]
13. Yang, F.; Sliozberg, K.; Sinev, I.; Antoni, H.; Baehr, A.; Ollegott, K.; Xia, W.; Masa, J.; Gruenert, W.; Roldan Cuenya, B.; et al. Synergistic Effect of Cobalt and Iron in Layered Double Hydroxide Catalysts for the Oxygen Evolution Reaction. *Chemsuschem* **2017**, *10*, 156–165. [CrossRef] [PubMed]
14. Burke, M.S.; Kast, M.G.; Trotochaud, L.; Smith, A.M.; Boettcher, S.W. Cobalt-Iron (Oxy)hydroxide Oxygen Evolution Electrocatalysts: The Role of Structure and Composition on Activity, Stability, and Mechanism. *J. Am. Chem. Soc.* **2015**, *137*, 3638–3648. [CrossRef] [PubMed]

15. Griboval-Constant, A.; Butel, A.; Ordomsy, V.V.; Chernavskii, P.A.; Khodakova, A.Y. Cobalt and iron species in alumina supported bimetallic catalysts for Fischer-Tropsch reaction. *Appl. Catal. A* **2014**, *481*, 116–126. [CrossRef]
16. Antonio Diaz, J.; Akhavan, H.; Romero, A.; Maria Garcia-Minguillan, A.; Romero, R.; Giroir-Fendler, A.; Luis Valverde, J. Cobalt and iron supported on carbon nanofibers as catalysts for Fischer-Tropsch synthesis. *Fuel Process. Technol.* **2014**, *128*, 417–424. [CrossRef]
17. Masa, J.; Weide, P.; Peeters, D.; Sinev, I.; Xia, W.; Sun, Z.; Somsen, C.; Muhler, M.; Schuhmann, W. Amorphous Cobalt Boride (Co_2B) as a Highly Efficient Nonprecious Catalyst for Electrochemical Water Splitting: Oxygen and Hydrogen Evolution. *Adv. Energy Mater.* **2016**, *6*, 1502313. [CrossRef]
18. Chen, H.; Ouyang, S.; Zhao, M.; Li, Y.; Ye, J. Synergistic Activity of Co and Fe in Amorphous Co_x-Fe-B Catalyst for Efficient Oxygen Evolution Reaction. *ACS Appl. Mater. Interfaces* **2017**, *9*, 40333–40343. [CrossRef]
19. Wu, Z.; Nie, D.; Song, M.; Jiao, T.; Fu, G.; Liu, X. Facile synthesis of Co-Fe-B-P nanochains as an efficient bifunctional electrocatalyst for overall water-splitting. *Nanoscale* **2019**, *11*, 7506–7512. [CrossRef]
20. Yang, L.; Guo, Z.; Huang, J.; Xi, Y.; Gao, R.; Su, G.; Wang, W.; Cao, L.; Dong, B. Vertical Growth of 2D Amorphous $FePO_4$ Nanosheet on Ni Foam: Outer and Inner Structural Design for Superior Water Splitting. *Adv. Mater.* **2017**, *29*, 1704574. [CrossRef]
21. Yang, Y.; Zhuang, L.; Lin, R.; Li, M.; Xu, X.; Rufford, T.E.; Zhu, Z. A facile method to synthesize boron-doped Ni/Fe alloy nano-chains as electrocatalyst for water oxidation. *J. Power Sources* **2017**, *349*, 68–74. [CrossRef]
22. Li, Y.; Fan, X.; Qi, J.; Ji, J.; Wang, S.; Zhang, G.; Zhang, F. Palladium nanoparticle-graphene hybrids as active catalysts for the Suzuki reaction. *Nano Res.* **2010**, *3*, 429–437. [CrossRef]
23. Zhang, J.; Ma, J.; Fan, X.; Peng, W.; Zhang, G.; Zhang, F.; Li, Y. Graphene supported Au-Pd-Fe_3O_4 alloy trimetallic nanoparticles with peroxidase-like activities as mimic enzyme. *Catal. Commun.* **2017**, *89*, 148–151. [CrossRef]
24. Mattioli, G.; Giannozzi, P.; Bonapasta, A.A.; Guidonili, L. Reaction Pathways for Oxygen Evolution Promoted by Cobalt Catalyst. *J. Am. Chem. Soc.* **2013**, *135*, 15353–15363. [CrossRef] [PubMed]
25. Han, X.; Yu, C.; Zhou, S.; Zhao, C.; Huang, H.; Yang, J.; Liu, Z.; Zhao, J.; Qiu, J. Ultrasensitive Iron-Triggered Nanosized Fe–CoOOH Integrated with Graphene for Highly Efficient Oxygen Evolution. *Adv. Energy Mater.* **2017**, *7*, 1602148. [CrossRef]
26. Jiao, Y.; Zheng, Y.; Jaroniec, M.; Qiao, S.Z. Origin of the Electrocatalytic Oxygen Reduction Activity of Graphene-Based Catalysts: A Roadnnap to Achieve the Best Performance. *J. Am. Chem. Soc.* **2014**, *136*, 4394–4403. [CrossRef] [PubMed]
27. Wu, N.; Lei, Y.; Wang, Q.; Wang, B.; Han, C.; Wang, Y. Facile synthesis of FeCo@NC core–shell nanospheres supported on graphene as an efficient bifunctional oxygen electrocatalyst. *Nano Res.* **2017**, *10*, 2332–2343. [CrossRef]
28. Seo, W.S.; Lee, J.H.; Sun, X.; Suzuki, Y.; Mann, D.; Liu, Z.; Terashima, M.; Yang, P.C.; McConnell, M.V.; Nishimura, D.G.; et al. FeCo/graphitic-shell nanocrystals as advanced magnetic-resonance-imaging and near-infrared agents. *Nat. Mater.* **2006**, *5*, 971–976. [CrossRef]
29. Zhu, S.J.; Zhang, J.H.; Qiao, C.Y.; Tang, S.J.; Li, Y.F.; Yuan, W.J.; Li, B.; Tian, L.; Liu, F.; Hu, R.; et al. Strongly green-photoluminescent graphene quantum dots for bioimaging applications. *Chem. Commun.* **2011**, *47*, 6858–6860. [CrossRef]
30. Navalon, S.; Dhakshinamoorthy, A.; Alvaro, M.; Garcia, H. Carbocatalysis by Graphene-Based Materials. *Chem. Rev.* **2014**, *114*, 6179–6212. [CrossRef]
31. Palaniselvam, T.; Kashyap, V.; Bhange, S.N.; Baek, J.-B.; Kurungot, S. Nanoporous Graphene Enriched with Fe/Co-N Active Sites as a Promising Oxygen Reduction Electrocatalyst for Anion Exchange Membrane Fuel Cells. *Adv. Funct. Mater.* **2016**, *26*, 2150–2162. [CrossRef]
32. Palaniselvam, T.; Valappil, M.O.; Illathvalappil, R.; Kurungot, S. Nanoporous graphene by quantum dots removal from graphene and its conversion to a potential oxygen reduction electrocatalyst via nitrogen doping. *Energy Environ. Sci.* **2014**, *7*, 1059–1067. [CrossRef]
33. Yang, X.-F.; Wang, A.; Qiao, B.; Li, J.; Liu, J.; Zhang, T. Single-Atom Catalysts: A New Frontier in Heterogeneous Catalysis. *Acc. Chem. Res.* **2013**, *46*, 1740–1748. [CrossRef] [PubMed]
34. Qiao, B.T.; Wang, A.Q.; Yang, X.F.; Allard, L.F.; Jiang, Z.; Cui, Y.T.; Liu, J.Y.; Li, J.; Zhang, T. Single-atom catalysis of CO oxidation using Pt-1/FeO_x. *Nat. Chem.* **2011**, *3*, 634–641. [CrossRef] [PubMed]

35. Bakandritsos, A.; Kadam, R.G.; Kumar, P.; Zoppellaro, G.; Medved, M.; Tucek, J.; Montini, T.; Tomanec, O.; Andryskova, P.; Drahos, B.; et al. Mixed-Valence Single-Atom Catalyst Derived from Functionalized Graphene. *Adv. Mater.* **2019**, *31*, e1900323. [CrossRef] [PubMed]
36. Wang, W.; Babu, D.D.; Huang, Y.; Lv, J.; Wang, Y.; Wu, M. Atomic dispersion of Fe/Co/N on graphene by ball-milling for efficient oxygen evolution reaction. *Int. J. Hydrog. Energy* **2018**, *43*, 10351–11035. [CrossRef]
37. Li, M.; Liu, T.; Bo, X.; Zhou, M.; Guo, L. A novel flower-like architecture of FeCo@NC-functionalized ultra-thin carbon nanosheets as a highly efficient 3D bifunctional electrocatalyst for full water splitting. *J. Mater. Chem. A* **2017**, *5*, 5413–5425. [CrossRef]
38. Samanta, A.; Raj, C.R. Catalyst Support in Oxygen Electrocatalysis: A Case Study with CoFe Alloy Electrocatalyst. *J. Phys. Chem. C* **2018**, *122*, 15843–15852. [CrossRef]
39. Liu, Q.; Cao, S.; Qiu, Y.; Zhao, L. Bimetallic Fe-Co promoting one-step growth of hierarchical nitrogen-doped carbon nanotubes/nanofibers for highly efficient oxygen reduction reaction. *Mater. Sci. Eng. B* **2017**, *223*, 159–166. [CrossRef]
40. Li, C.; Wu, M.; Liu, R. High-performance bifunctional oxygen electrocatalysts for zinc-air batteries over mesoporous Fe/Co-N-C nanofibers with embedding FeCo alloy nanoparticles. *Appl. Catal. B* **2019**, *244*, 150–158. [CrossRef]
41. Yan, Y.; Miao, J.; Yang, Z.; Xiao, F.-X.; Yang, H.B.; Liu, B.; Yang, Y. Carbon nanotube catalysts: Recent advances in synthesis, characterization and applications. *Chem. Soc. Rev.* **2015**, *44*, 3295–3346. [CrossRef] [PubMed]
42. Su, C.-Y.; Cheng, H.; Li, W.; Liu, Z.-Q.; Li, N.; Hou, Z.; Bai, F.-Q.; Zhang, H.-X.; Ma, T.-Y. Atomic Modulation of FeCo-Nitrogen-Carbon Bifunctional Oxygen Electrodes for Rechargeable and Flexible All-Solid-State Zinc-Air Battery. *Adv. Energy Mater.* **2017**, *7*, 1602420. [CrossRef]
43. Yang, Y.; Lun, Z.; Xia, G.; Zheng, F.; He, M.; Chen, Q. Non-precious alloy encapsulated in nitrogen-doped graphene layers derived from MOFs as an active and durable hydrogen evolution reaction catalyst. *Energy Environ. Sci.* **2015**, *8*, 3563–3571. [CrossRef]
44. Lian, Y.; Sun, H.; Wang, X.; Qi, P.; Mu, Q.; Chen, Y.; Ye, J.; Zhao, X.; Deng, Z.; Peng, Y. Carved nanoframes of cobalt-iron bimetal phosphide as a bifunctional electrocatalyst for efficient overall water splitting. *Chem. Sci.* **2019**, *10*, 464–474. [CrossRef] [PubMed]
45. Zhang, W.; Zhang, H.; Luo, R.; Zhang, M.; Yan, X.; Sun, X.; Shen, J.; Han, W.; Wang, L.; Li, J. Prussian blue analogues-derived bimetallic iron-cobalt selenides for efficient overall water splitting. *J. Colloid Interface Sci.* **2019**, *548*, 48–55. [CrossRef] [PubMed]
46. Hu, C.; Zhang, L.; Zhao, Z.-J.; Li, A.; Chang, X.; Gong, J. Synergism of Geometric Construction and Electronic Regulation: 3D Se-(NiCo)S$_x$/(OH)$_x$ Nanosheets for Highly Efficient Overall Water Splitting. *Adv. Mater.* **2018**, *30*, 1705538. [CrossRef] [PubMed]
47. Chen, J.; Liu, J.; Xie, J.-Q.; Ye, H.; Fu, X.-Z.; Sun, R.; Wong, C.-P. Co-Fe-P nanotubes electrocatalysts derived from metal-organic frameworks for efficient hydrogen evolution reaction under wide pH range. *Nano Energy* **2019**, *56*, 225–233. [CrossRef]
48. Wang, J.; Xu, F.; Jin, H.; Chen, Y.; Wang, Y. Non-Noble Metal-based Carbon Composites in Hydrogen Evolution Reaction: Fundamentals to Applications. *Adv. Mater.* **2017**, *29*, 1605838. [CrossRef]
49. Wang, X.; Yu, L.; Guan, B.Y.; Song, S.; Lou, X.W. Metal–Organic Framework Hybrid-Assisted Formation of Co$_3$O$_4$/Co-Fe Oxide Double-Shelled Nanoboxes for Enhanced Oxygen Evolution. *Adv. Mater.* **2018**, *30*, 1801211. [CrossRef] [PubMed]
50. Trotochaud, L.; Young, S.L.; Ranney, J.K.; Boettcher, S.W. Nickel–Iron Oxyhydroxide Oxygen-Evolution Electrocatalysts: The Role of Intentional and Incidental Iron Incorporation. *J. Am. Chem. Soc.* **2014**, *136*, 6744–6753. [CrossRef]
51. Chen, W.X.; Pei, J.J.; He, C.T.; Wan, J.W.; Ren, H.L.; Zhu, Y.Q.; Wang, Y.; Dong, J.C.; Tian, S.B.; Cheong, W.C.; et al. Rational Design of Single Molybdenum Atoms Anchored on N-Doped Carbon for Effective Hydrogen Evolution Reaction. *Angew. Chem. Int. Edit.* **2017**, *56*, 16086–16090. [CrossRef] [PubMed]
52. Wang, J.; Huang, Z.; Liu, W.; Chang, C.; Tang, H.; Li, Z.; Chen, W.; Jia, C.; Yao, T.; Wei, S.; et al. Design of N-Coordinated Dual-Metal Sites: A Stable and Active Pt-Free Catalyst for Acidic Oxygen Reduction Reaction. *J. Am. Chem. Soc.* **2017**, *139*, 17281–17284. [CrossRef] [PubMed]
53. Cook, T.R.; Dogutan, D.K.; Reece, S.Y.; Surendranath, Y.; Teets, T.S.; Nocera, D.G. Solar energy supply and storage for the legacy and nonlegacy worlds. *Chem. Rev.* **2010**, *110*, 6474–6502. [CrossRef] [PubMed]

54. Du, N.; Wang, C.; Wang, X.; Lin, Y.; Jiang, J.; Xiong, Y. Trimetallic TriStar Nanostructures: Tuning Electronic and Surface Structures for Enhanced Electrocatalytic Hydrogen Evolution. *Adv. Mater.* **2016**, *28*, 2077–2084. [CrossRef] [PubMed]
55. Müller, C.I.; Sellschopp, K.; Tegel, M.; Rauscher, T.; Kieback, B.; Röntzsch, L. The activity of nanocrystalline Fe-based alloys as electrode materials for the hydrogen evolution reaction. *J. Power Sources* **2016**, *304*, 196–206. [CrossRef]
56. Liu, W.; Du, K.; Liu, L.; Zhang, J.; Zhu, Z.; Shao, Y.; Li, M. One-step electroreductively deposited iron-cobalt composite films as efficient bifunctional electrocatalysts for overall water splitting. *Nano Energy* **2017**, *38*, 576–584. [CrossRef]
57. Zhang, X.; Zhang, X.; Xu, H.; Wu, Z.; Wang, H.; Liang, Y. Iron-Doped Cobalt Monophosphide Nanosheet/Carbon Nanotube Hybrids as Active and Stable Electrocatalysts for Water Splitting. *Adv. Funct. Mater.* **2017**, *27*, 1606635. [CrossRef]
58. Guo, X.; Yu, X.; Feng, Z.; Liang, J.; Li, Q.; Lv, Z.; Liu, B.; Hao, C.; Li, G. Intercalation Synthesis of Prussian Blue Analogue Nanocone and Their Conversion into Fe-Doped Co_xP Nanocone for Enhanced Hydrogen Evolution. *ACS Sustain. Chem. Eng.* **2018**, *6*, 8150–8158. [CrossRef]
59. Li, F.; Bu, Y.; Lv, Z.; Mahmood, J.; Han, G.-F.; Ahmad, I.; Kim, G.; Zhong, Q.; Baek, J.-B. Porous Cobalt Phosphide Polyhedrons with Iron Doping as an Efficient Bifunctional Electrocatalyst. *Small* **2017**, *13*, 1701167. [CrossRef] [PubMed]
60. Cao, L.M.; Hu, Y.W.; Tang, S.F.; Iljin, A.; Wang, J.W.; Zhang, Z.M.; Lu, T.B. Fe-CoP Electrocatalyst Derived from a Bimetallic Prussian Blue Analogue for Large-Current-Density Oxygen Evolution and Overall Water Splitting. *Adv. Sci.* **2018**, *5*, 1800949. [CrossRef] [PubMed]
61. Kim, H.; Oh, S.; Cho, E.; Kwon, H. 3D Porous Cobalt–Iron–Phosphorus Bifunctional Electrocatalyst for the Oxygen and Hydrogen Evolution Reactions. *ACS Sustain. Chem. Eng.* **2018**, *6*, 6305–6311. [CrossRef]
62. Kuo, T.-R.; Chen, W.-T.; Liao, H.-J.; Yang, Y.-H.; Yen, H.-C.; Liao, T.-W.; Wen, C.-Y.; Lee, Y.-C.; Chen, C.-C.; Wang, D.-Y. Improving Hydrogen Evolution Activity of Earth-Abundant Cobalt-Doped Iron Pyrite Catalysts by Surface Modification with Phosphide. *Small* **2017**, *13*, 1603356. [CrossRef] [PubMed]
63. Liu, Y.; Li, F.; Yang, H.; Li, J.; Ma, P.; Zhu, Y.; Ma, J. Two-Step Synthesis of Cobalt Iron Alloy Nanoparticles Embedded in Nitrogen-Doped Carbon Nanosheets/Carbon Nanotubes for the Oxygen Evolution Reaction. *ChemSusChem* **2018**, *11*, 2358–2366. [CrossRef] [PubMed]
64. Liu, X.; Wang, L.; Yu, P.; Tian, C.; Sun, F.; Ma, J.; Li, W.; Fu, H. A Stable Bifunctional Catalyst for Rechargeable Zinc-Air Batteries: Iron-Cobalt Nanoparticles Embedded in a Nitrogen-Doped 3D Carbon Matrix. *Angew. Chem. Int. Ed. Engl.* **2018**, *57*, 16166–16170. [CrossRef] [PubMed]
65. Feng, X.; Bo, X.; Guo, L. CoM(M = Fe,Cu,Ni)-embedded nitrogen-enriched porous carbon framework for efficient oxygen and hydrogen evolution reactions. *J. Power Sources* **2018**, *389*, 249–259. [CrossRef]
66. Zhu, X.; Jin, T.; Tian, C.; Lu, C.; Liu, X.; Zeng, M.; Zhuang, X.; Yang, S.; He, L.; Liu, H.; et al. In Situ Coupling Strategy for the Preparation of FeCo Alloys and Co_4N Hybrid for Highly Efficient Oxygen Evolution. *Adv. Mater.* **2017**, *29*, 1704091. [CrossRef]
67. Li, T.; Lu, Y.; Zhao, S.; Gao, Z.-D.; Song, Y.-Y. Co_3O_4-doped Co/CoFe nanoparticles encapsulated in carbon shells as bifunctional electrocatalysts for rechargeable Zn–Air batteries. *J. Mater. Chem. A* **2018**, *6*, 3730–3737. [CrossRef]
68. Kim, B.; Park, I.; Yoon, G.; Kim, J.S.; Kim, H.; Kang, K. Atomistic Investigation of Doping Effects on Electrocatalytic Properties of Cobalt Oxides for Water Oxidation. *Adv. Sci.* **2018**, *5*, 1801632. [CrossRef]
69. Indra, A.; Menezes, P.W.; Sahraie, N.R.; Bergmann, A.; Das, C.; Tallarida, M.; Schmeisser, D.; Strasser, P.; Driess, M. Unification of Catalytic Water Oxidation and Oxygen Reduction Reactions: Amorphous Beat Crystalline Cobalt Iron Oxides. *J. Am. Chem. Soc.* **2014**, *136*, 17530–17536. [CrossRef]
70. Chen, W.; Wang, H.; Li, Y.; Liu, Y.; Sun, J.; Lee, S.; Lee, J.-S.; Cui, Y. In Situ Electrochemical Oxidation Tuning of Transition Metal Disulfides to Oxides for Enhanced Water Oxidation. *ACS Cent. Sci.* **2015**, *1*, 244–251. [CrossRef]
71. Zhang, Y.; Gao, T.; Jin, Z.; Chen, X.; Xiao, D. A robust water oxidation electrocatalyst from amorphous cobalt-iron bimetallic phytate nanostructures. *J. Mater. Chem. A* **2016**, *4*, 15888–15895. [CrossRef]
72. Liu, K.; Zhang, C.; Sun, Y.; Zhang, G.; Shen, X.; Zou, F.; Zhang, H.; Wu, Z.; Wegener, E.C.; Taubert, C.J.; et al. High-Performance Transition Metal Phosphide Alloy Catalyst for Oxygen Evolution Reaction. *ACS Nano* **2018**, *12*, 158–167. [CrossRef]

73. Shen, J.-Q.; Liao, P.-Q.; Zhou, D.-D.; He, C.-T.; Wu, J.-X.; Zhang, W.-X.; Zhang, J.-P.; Chen, X.-M. Modular and Stepwise Synthesis of a Hybrid Metal–Organic Framework for Efficient Electrocatalytic Oxygen Evolution. *J. Am. Chem. Soc.* **2017**, *139*, 1778–1781. [CrossRef]
74. Chen, Y.; Ji, S.; Chen, C.; Peng, Q.; Wang, D.; Li, Y. Single-Atom Catalysts: Synthetic Strategies and Electrochemical Applications. *Joule* **2018**, *2*, 1242–1264. [CrossRef]
75. Li, S.; Cheng, C.; Zhao, X.; Schmidt, J.; Thomas, A. Active Salt/Silica-Templated 2D Mesoporous FeCo-N_x-Carbon as Bifunctional Oxygen Electrodes for Zinc-Air Batteries. *Angew. Chem. Int. Ed. Engl.* **2018**, *57*, 1856–1862. [CrossRef]
76. Sultan, S.; Tiwari, J.N.; Jang, J.-H.; Harzandi, A.M.; Salehnia, F.; Yoo, S.J.; Kim, K.S. Highly Efficient Oxygen Reduction Reaction Activity of Graphitic Tube Encapsulating Nitrided Co_xFe_y Alloy. *Adv. Energy Mater.* **2018**, *8*, 1801002. [CrossRef]
77. Lim, S.H.; Li, Z.; Poh, C.K.; Lai, L.; Lin, J. Highly active non-precious metal catalyst based on poly(vinylpyrrolidone)–wrapped carbon nanotubes complexed with iron–cobalt metal ions for oxygen reduction reaction. *J. Power Sources* **2012**, *214*, 15–20. [CrossRef]
78. Guan, B.Y.; Lu, Y.; Wang, Y.; Wu, M.; Lou, X.W.D. Porous Iron-Cobalt Alloy/Nitrogen-Doped Carbon Cages Synthesized via Pyrolysis of Complex Metal-Organic Framework Hybrids for Oxygen Reduction. *Adv. Funct. Mater.* **2018**, *28*, 1706738. [CrossRef]
79. Tan, M.; He, T.; Liu, J.; Wu, H.; Li, Q.; Zheng, J.; Wang, Y.; Sun, Z.; Wang, S.; Zhang, Y. Supramolecular bimetallogels: A nanofiber network for bimetal/nitrogen co-doped carbon electrocatalysts. *J. Mater. Chem. A* **2018**, *6*, 8227–8232. [CrossRef]
80. Wang, J.; Xin, H.L.; Zhu, J.; Liu, S.; Wu, Z.; Wang, D. 3D hollow structured Co2FeO4/MWCNT as an efficient non-precious metal electrocatalyst for oxygen reduction reaction. *J. Mater. Chem. A* **2015**, *3*, 1601–1608. [CrossRef]
81. Kim, J.; Gwon, O.; Kwon, O.; Mahmood, J.; Kim, C.; Yang, Y.; Lee, H.; Lee, J.H.; Jeong, H.Y.; Baek, J.B.; et al. Synergistic Coupling Derived Cobalt Oxide with Nitrogenated Holey Two-Dimensional Matrix as an Efficient Bifunctional Catalyst for Metal-Air Batteries. *ACS Nano* **2019**, *13*, 5502–5512. [CrossRef]
82. Wang, C.-H.; Yang, C.-W.; Lin, Y.-C.; Chang, S.-T.; Chang, S.L.Y. Cobalt–iron (II, III) oxide hybrid catalysis with enhanced catalytic activities for oxygen reduction in anion exchange membrane fuel cell. *J. Power Sources* **2015**, *277*, 147–154. [CrossRef]
83. Jin, Y.Q.; Lin, Z.; Zhong, R.; Huang, J.; Liang, G.; Li, J.; Jin, Y.; Meng, H. Cobalt iron carbonate hydroxide hydrate on 3D porous carbon as active and stable bifunctional oxygen electrode for Zn–air battery. *J. Power Sources* **2018**, *402*, 388–393. [CrossRef]
84. Sun, K.; Li, J.; Huang, L.; Ji, S.; Kannan, P.; Li, D.; Liu, L.; Liao, S. Biomass-derived 3D hierarchical N-doped porous carbon anchoring cobalt-iron phosphide nanodots as bifunctional electrocatalysts for Li O_2 batteries. *J. Power Sources* **2019**, *412*, 433–441. [CrossRef]
85. Zhao, Y.; Watanabe, K.; Hashimoto, K. Efficient oxygen reduction by a Fe/Co/C/N nano-porous catalyst in neutral media. *J. Mater. Chem. A* **2013**, *1*, 1450–1456. [CrossRef]
86. Lin, Q.; Bu, X.; Kong, A.; Mao, C.; Bu, F.; Feng, P. Heterometal-Embedded Organic Conjugate Frameworks from Alternating Monomeric Iron and Cobalt Metalloporphyrins and Their Application in Design of Porous Carbon Catalysts. *Adv. Mater.* **2015**, *27*, 3431–3436. [CrossRef]
87. Zhang, Z.; Dou, M.; Liu, H.; Dai, L.; Wang, F. A Facile Route to Bimetal and Nitrogen-Codoped 3D Porous Graphitic Carbon Networks for Efficient Oxygen Reduction. *Small* **2016**, *12*, 4193–4199. [CrossRef]

© 2019 by the authors. Licensee MDPI, Basel, Switzerland. This article is an open access article distributed under the terms and conditions of the Creative Commons Attribution (CC BY) license (http://creativecommons.org/licenses/by/4.0/).

Review

Polynuclear Cobalt Complexes as Catalysts for Light-Driven Water Oxidation: A Review of Recent Advances

Dmytro S. Nesterov * and Oksana V. Nesterova *

Centro de Química Estrutural, Instituto Superior Técnico, Universidade de Lisboa, Av. Rovisco Pais, 1049-001 Lisboa, Portugal
* Correspondence: dmytro.nesterov@tecnico.ulisboa.pt (D.S.N.); oksana.nesterova@tecnico.ulisboa.pt (O.V.N.); Tel.: +351-218-419-241 (O.V.N.)

Received: 5 November 2018; Accepted: 27 November 2018; Published: 2 December 2018

Abstract: Photochemical water oxidation, as a half-reaction of water splitting, represents a great challenge towards the construction of artificial photosynthetic systems. Complexes of first-row transition metals have attracted great attention in the last decade due to their pronounced catalytic efficiency in water oxidation, comparable to that exhibited by classical platinum-group metal complexes. Cobalt, being an abundant and relatively cheap metal, has rich coordination chemistry allowing construction of a wide range of polynuclear architectures for the catalytic purposes. This review covers recent advances in application of cobalt complexes as (pre)catalysts for water oxidation in the model catalytic system comprising $[Ru(bpy)_3]^{2+}$ as a photosensitizer and $S_2O_8^{2-}$ as a sacrificial electron acceptor. The catalytic parameters are summarized and discussed in view of the structures of the catalysts. Special attention is paid to the degradation of molecular catalysts under catalytic conditions and the experimental methods and techniques used to control their degradation as well as the leaching of cobalt ions.

Keywords: polynuclear cobalt complexes; water oxidation; artificial photosynthesis

---

1. Introduction

Currently, worldwide energy demand is satisfied mostly with fossil fuels, such as gas, oil, and coal [1,2]. Although these energy resources are rather easy to use, they contain a few principal drawbacks: limited stock, high consumption of O_2 as well as high CO_2 emission and, finally, high risk of environmental pollution. It is for these reasons that the search for highly efficient energy sources has been of high interest during the last decades. The use of sunlight, as an inexhaustible energy source with the ground level flux of ca. 1360 W m^{-2} [3], is an attractive alternative to fossil fuels. The chemical approach to sunlight energy utilization is to develop photo-driven processes able to replace the existing environmentally non-friendly ones. The photosynthesis of carbohydrates from CO_2 and water, occurring in natural biosystems, represents a perfect model to be studied [4]. The energy for this process is provided by photosystem (PSI and PSII) protein complexes, and through the formation of ATP and NADPH under visible light exposure. In brief, (for recent reviews see [5–10]), upon irradiation, the photosensitizer (chlorophyll) in the PSII abstracts electrons from a water-oxidizing complex (WOC). The latter oxidizes water and releases O_2 to close the catalytic cycle within the PSII system. The four-electron process of water oxidation, which also produces four protons, consumes significant energy and, at the same time, is potentially attractive for practical applications when protons are reduced to H_2 [10–12].

While the existence of biological photosynthesis has been known for a long time [13], the principal details about the structure of the PSII and WOC were obtained only in 21st century [14,15]. The first single crystal X-ray structure analysis of the catalytically active PSII from cyanobacterium at 3.8 Å resolution was accomplished in 2001, but the structure of the WOC was not resolved in detail [16]. Just few years later (in 2004 and 2005) a better set of data, collected at 3.5 and 3.0 Å resolutions, allowed researchers to propose a cubane-like Mn_3CaO_4 structure of the WOC [17,18]. In 2011 the experimental results were improved by collecting the data at 1.9 Å [19], revealing that the structure of the WOC contains four manganese atoms, where cubane fragment Mn_3CaO_4 is accompanied with the fourth Mn atom bridged to cubane with two oxygen atoms (Figure 1). The oxygen-evolving mechanism presumes stepwise oxidation of manganese atoms to Mn^{IV} and/or Mn^V with formation of high-valent metal-oxo species, which oxygen atoms "join" to release O_2 molecules [5,20–22]. The role of redox-inactive calcium is still not fully clear [5,23]. Most probably, the calcium center acts as a Lewis acid and coordinates water molecules, thus assisting the mechanism [23,24]. Calcium could influence the redox potentials of the manganese cluster, as evidenced by the model studies [25–27]. Also, it was demonstrated that Ca^{2+} is required for the assembly of the oxygen-evolving center (OEC) [23,28]. The mechanism of action of the PSII has been the subject of numerous experimental and theoretical investigations (which are far from the final conclusion) and we would like to refer readers to the recent reviews on this topic [5,6,8,9,15].

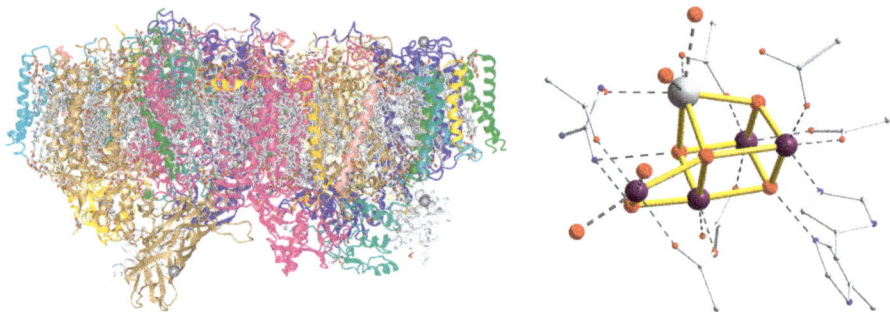

Figure 1. Overall structure (left) and Mn_4CaO_4 site (right) in PSII (PDB code 3WU2). Colour scheme: Mn, violet; Ca, grey; O, red; N, blue; C, grey (small balls). M–O–M bonds are highlighted by yellow.

From the point of view of coordination chemistry, the complex bearing Mn_3CaO_4 or Mn_4CaO_4 core looks relatively simple. Cubane, $M_4(\mu_3\text{-O})_4$, is the most widespread molecular structure type for tetranuclear coordination compounds [29] with more than 1000 examples in the Cambridge Structural Database (CSD) [30]. Discovery of the WOC structure of the PSII, inspired chemists to synthesize model coordination compounds having related polynuclear assemblies (biomimetic approach) and redox potentials [25–27]. The principal idea of this approach was to design well-defined molecular systems that allow for the studying of the mechanisms of action. An artificial water oxidation system comprises the same principal components as the natural one: photosensitizer, catalyst, and electron acceptor [10,31]. One of the most simple, efficient, and studied photosensitizers is $[Ru(bpy)_3]^{2+}$ (bpy = 2,2'-bipyridine) [10,32]. Absorption of a visible light photon with λ = 450 nm transfers this complex to an exited state which easily loses electron to produce $[Ru(bpy)_3]^{3+}$, a strong oxidant. Sacrificial electron acceptor is required for this scheme of action. The most common known is persulfate, $S_2O_8^{2-}$ [10,33]. Finally, a buffer is required for keeping the pH value at the proper level. The overall catalytic system for the screening of the WOC activity is depicted in Scheme 1. A large series of metal complexes have been tested as WOCs using this scheme, mainly compounds of ruthenium and iridium [10,31,34]. However, the first-row transition metals, as the more abundant and less expensive ones, also represent a great interest form a practical point of view [35–38]. In this review we focus on the cobalt polynuclear complexes as the catalysts for light-driven water oxidation, with special attention to their stability under the studied catalytic conditions.

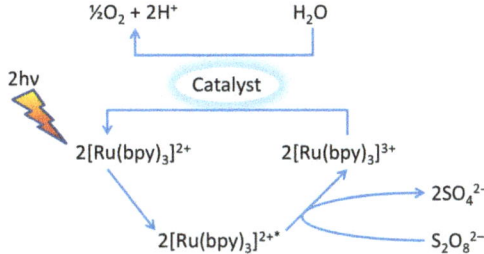

Scheme 1. General functioning mechanism of the light driven water oxidation system [Ru(bpy)$_3$]$^{2+}$/Catalyst/S$_2$O$_8$$^{2-}$.

Despite the impressive number of H$_2$-evolving water splitting (reduction) cobalt catalysts [39], the field of cobalt-catalysed water oxidation has risen only during the last decade. The activity of simple cobalt compounds (salts and oxides) in the catalytic chemical water oxidation with [Ru(bpy)]$^{3+}$ or other strong oxidants has been known for some time [10], but did not attract much attention until 2008, where Kanan and Nocera presented the Co-P electrocatalytic film for water oxidation formed in situ from aqueous Co^{2+} in phosphate buffer at neutral pH [40]. In 2009, Jiao and Frei reported the light driven system with the Co$_3$O$_4$ nanoparticles as a WOC, operating at mild conditions (room temperature and pH 5.8) [41]. The system [Ru(bpy)$_3$]Cl$_2$/Co$_3$O$_4$/Na$_2$S$_2$O$_8$ shows the TOF (turnover frequency, moles of product produced per mol of catalyst per a certain period of time) of more than 1000 s^{-1} per Co$_3$O$_4$ nanoparticle. For comparison, the TOF of the natural PSII is estimated at 100 to 400 s^{-1} level in live cells and 1000 s^{-1} in vitro [42]. Just one year after, in 2010, a study was published that highlighted a stable carbon-free WOC based on tetranuclear cobalt moiety [Co$_4$(H$_2$O)$_2$(PW$_9$O$_{34}$)$_2$]$^{10-}$ for chemical water oxidation with [Ru(bpy)]$^{3+}$, allowing the TOF of 5 s^{-1} and TON (turnover number, moles of product per mol of catalyst) of 78, and then (in 2011) for light-driven oxidation to reach the TON of > 220 [43,44]. Remarkably, in the same year, 2011, Stracke and Finke reported that under the conditions of electrocatalytic water oxidation, this compound releases Co^{2+} ions to form CoO$_x$ as an active catalyst [45]. These reports caused considerable interest in the field but also demonstrated the complexity of the problem of clear identification of the true WOC.

2. Mono- and Binuclear Cobalt Complexes

The complex [Co(L^1)] (**1**) (Figure 2) with Schiff base ligand H$_2$L^1 (N,N'-bis(salicylidene) ethylenediamine, known also as "salen"), shows high TON and TOF values, as well as the quantum yield of O$_2$, using [Ru(bpy)$_3$](ClO$_4$)$_2$ as a photosensitizer (Table 1) [46]. When the ClO$_4$$^-$ anion was replaced with Cl$^-$ or SO$_4$$^{2-}$ one the TON drops to ca. 780 and 190, respectively. Complex **1** retains its activity when [Ru(bpy)$_3$](ClO$_4$)$_3$ was used as a terminal chemical oxidant, although showing lower catalytic parameters (Table 1). Further investigations disclosed that **1** was a pre-catalyst and its catalytic activity was associated to cobalt nanoparticles formed during the photocatalysis. According to ESI-MS (electrospray mass spectrometry) and DLS (dynamic light scattering) measurements, complex **1** was shown to be stable in the presence of the main components of the catalytic system, in the absence of light. However, after illumination the presence of nanoparticles of various sizes (1–1000 nm) were detected by the DLS method. Moreover, no signals from salen ligand were observed by ^{1}H NMR, suggesting its complete decomposition. The precipitate formed after the illumination was recovered and used again as a catalyst, showing somewhat lower water oxidation activity than **1** (40.9 and 54.6% of the yield of O$_2$ based on S$_2$O$_4$$^{2-}$, respectively). The cobalt nitrate was tested as WOC (**31a**, Table 1) under the same conditions as above, showing nearly the same TON as **1** as well as similar O$_2$ accumulation kinetics [46]. Cobalt salts form CoO$_x$ nanoparticles in the [Ru(bpy)$_3$]$^{2+}$/S$_2$O$_8$$^{2-}$ water oxidation systems (**31–33**, Table 1), providing a non-direct proof for the instability of **1**, which appears to be just a pre-catalyst.

Table 1. Selected catalytic parameters for the catalysts 1–33 in oxidation of water using $[Ru(bpy)_3]^{2+}$ as a photosensitizer and $S_2O_8^{2-}$ as a sacrificial electron acceptor.

N	Core	(Pre)catalyst	O_2 Evolution, µmol/µM	TON[1]	TOF[2], s^{-1}	QE[3], %	λ[4], nm	Irradiation Source	pH	NP[5]	Anion[6]	[cat]$_0$, µM	Ref.
1a	Co$_1$	[Co^{II}(L^1)]	13.7/1366	854	6.4	38.6	>420	LED, 5.1 mW/cm^2	9	+	ClO$_4$	1.6	[46]
1b	Co$_1$	[Co^{II}(L^1)]	n.d.[7]	194	2.0	-	-	Chemical oxidation with $[Ru(bpy)_3]^{3+}$	9	n.d.[7]	ClO$_4$	n.d.	[46]
2	Co$_1$	[Co^{II}(TPPS)]	12.2/1220	122	0.17	n.d.	400–800	Xe lamp, 300 W	11	-	NO$_3$	10	[47]
3	Co$_1$	[Co^{II}(L^2)(H$_2$O)$_2$](ClO$_4$)$_2$	0.55/67	335	n.d.	n.d.	475	Hg arc lamp, 500 W	8	-	Cl	0.2	[48]
4	Co$_1$	[Co^{II}(bpy)$_2$(H$_2$O)](ClO$_4$)$_2$	0.34/41	206	n.d.	n.d.	475	Hg arc lamp, 500 W	8	-	Cl	0.2	[48]
5a	Co$_1$	[Co^{II}(L^3)(H$_2$O)](ClO$_4$)$_2$	5.4/2700	54	n.d.	32	>420	Xe lamp, 500 W	8	+	ClO$_4$	50	[49]
5b	Co$_1$	[Co^{II}(L^3)(H$_2$O)](ClO$_4$)$_2$	0.9/450	9	n.d.	n.d.	>420	Xe lamp, 500 W	8	-	Cl	50	[49]
6a	Co$_2$	[Co$^{III}_2$(µ-OH)$_2$(TPA)$_2$](ClO$_4$)$_4$	9.3/1855	742	n.d.	44	420	Xe lamp	9.3	-	ClO$_4$	2.5	[50]
6b	Co$_2$	[Co$^{III}_2$(µ-OH)$_2$(TPA)$_2$](ClO$_4$)$_4$	5.4/1080	2.7	n.d.	n.d.	420	Xe lamp	9.3	-	ClO$_4$	400	[50]
7a	Co$_2$	[(TPA)Co^{III}(µ-OH)$_2$(µ-O)Co^{III}(TPA)](ClO$_4$)$_3$	0.1/99	58	1.4	n.d.	470	LED, 820 µE/s	8	-/+	ClO$_4$	5	[51]
7b	Co$_2$	[(TPA)Co^{III}(µ-OH)$_2$(µ-O$_2$)Co^{III}(TPA)](ClO$_4$)$_3$ + 1 eq. bpy	0/0	0	0	0	470	LED, 150 mW/cm^2	8	n.d.	ClO$_4$	5	[52]
8a	Co$_2$	[(L^4)Co^{III}(µ-OH)(µ-O$_2$)Co^{III}(L^4)](ClO$_4$)$_3$	35/2333	233	0	>420	470	LED, 5.0 mW/cm^2	8	n.d.	Cl	10	[53]
8b	Co$_2$	[(L^4)Co^{III}(µ-OH)(µ-O$_2$)Co^{III}(L^4)](ClO$_4$)$_3$ + 5 eq. bpy	0/0	0	0	0	>420	LED, 5.0 mW/cm^2	9	n.d.	Cl	10	[53]
9	Co$_4$	[Co$^{III}_4$O$_4$(OAc)$_4$(py)$_4$]	16/1143	>40	0.02	n.d.	>395	Arc lamp, 250W, 2 mW/cm^2	7	n.d.	Cl	330	[54]
10	Co$_4$	[Co$^{III}_4$O$_4$(OAc)$_4$(p-py-OCH$_3$)$_4$]	5.0/2520	140	−0.04	80	>400	Halogen lamp	8	n.d.	Cl	18	[55]
11a	Co$_4$	[Co$^{II}_4$(L^5)$_4$(OAc)$_4$(H$_2$O)$_2$]	15.5/1940	20	1.8	n.d.	470	LED, 26.1 mW/cm^2	7	-	Cl	97	[56]
11b	Co$_4$	[Co$^{II}_4$(L^5)$_4$(OAc)$_4$(H$_2$O)$_2$]	14.4/1800	35	4.4	n.d.	470	LED, 26.1 mW/cm^2	8	-	Cl	60	[56]
11c	Co$_4$	[Co$^{II}_4$(L^5)$_4$(OAc)$_4$(H$_2$O)$_2$]	13.4/1680	28	7	n.d.	470	LED, 26.1 mW/cm^2	9	-	Cl	60	[56]
12	Co$_4$	[Co$^{III}_4$O$_4$(OAc)$_2$(bpy)$_4$](ClO$_4$)$_2$	n.d.	n.d.	0.02	n.d.	>380	LED	8	n.d.	Cl	0.2	[57]
13a	Co$_4$	[Co$^{III}_4$O$_4$(OAc)$_4$(py)$_4$], crude sample	0.3/167	0.5	n.d.	n.d.	>400	Hg/Xe arc lamp, 1000 W	7	n.d.	Cl	330	[58]
13b	Co$_4$	[Co$^{III}_4$O$_4$(OAc)$_4$(py)$_4$] after purification	0.06/31	0.09	2.3 × 10^{-4}	n.d.	>400	Hg/Xe arc lamp, 1000 W	7	n.d.	Cl	330	[58]
14a	Co$_4$	[Co$^{II}_4$(L^5)$_4$(OAc)$_2$(H$_2$O)$_2$](ClO$_4$)$_2$	9.6/1200	96	1.2	n.d.	470	LED, 26.1 mW/cm^2	8.5	-	Cl	12.5	[59]
14b	Co$_4$	[Co$^{II}_4$(L^5)$_4$(OAc)$_2$(H$_2$O)$_2$](ClO$_4$)$_2$	16.0/2000	20	0.24	n.d.	470	LED, 26.1 mW/cm^2	8.5	-	Cl	100	[59]
15	Co$_{4-x}$Ni$_x$	[Co$^{II}_{1.15}$Ni$^{II}_{2.85}$(L^6)$_4$(OAc)$_2$(H$_2$O)$_2$](ClO$_4$)$_2$	5.5/690	6.9	0.1	n.d.	450	LED, 26.1 mW/cm^2	8.5	-	Cl	100	[59]
16	Co$_3$Ho	[Co$^{II}_3$Ho(L^5)$_4$(OAc)$_5$(H$_2$O)]	13.0/1630	163	5.8	n.d.	450	LED, 26.1 mW/cm^2	8	-	Cl	10	[60]
17	Co$_3$Er	[Co$^{II}_3$Er(L^5)$_4$(OAc)$_5$(H$_2$O)]	16.9/2110	211	5.7	n.d.	450	LED, 26.1 mW/cm^2	8	-	Cl	10	[60]
18	Co$_3$Tm	[Co$^{II}_3$Tm(L^5)$_4$(OAc)$_5$(H$_2$O)]	7.4/920	92	5.3	n.d.	450	LED, 26.1 mW/cm^2	8	-	Cl	10	[60]
19	Co$_3$Yb	[Co$^{II}_3$Yb(L^5)$_4$(OAc)$_5$(H$_2$O)]	12.8/1600	160	6.8	n.d.	450	LED, 26.1 mW/cm^2	8	-	Cl	10	[60]
20a	Co$_4$	[Co$^{III}_4$O$_4$(OAc)$_4$(py)$_4$](L^{7}Rubpy)$_2$]	0.15/75	5	7 × 10^{-3}	n.d.	>400	Xe lamp, 300 W	7	n.d.	Cl	15	[61]
20b	Co$_4$	[Co$^{III}_4$O$_4$(OAc)$_4$(py)$_4$] + 1 eq. [Ru(bpy)$_3$]Cl$_2$	n.d.	2	5 × 10^{-3}	n.d.	>400	Xe lamp, 300 W	7	n.d.	Cl	15	[61]
21	Co$_4$	[Co$^{III}_4$O$_4$(OAc)$_2$(py)$_4$]$_2$[(L^8)Rubpy)$_2$]$_2$	0.74/360	24	0.02	n.d.	>400	Xe lamp, 300 W	7	n.d.	Cl	15	[61]
22a	Co$_4$	Na$_{10}$[Co$_4$(H$_2$O)$_2$(VW$_9$O$_{34}$)$_2$]	n.d./0.15	300	>1600	-	-	Chemical oxidation with $[Ru(bipy)_3]^{3+}$	9	-	ClO$_4$	0.5	[62]
22b	Co$_4$	Na$_{10}$[Co$_4$(H$_2$O)$_2$(VW$_9$O$_{34}$)$_2$]	3.0/1484	742	4	61	455	LED, 135 mW/cm^2	9	-	Cl	2.0	[62]
22c	Co$_4$	Na$_{10}$[Co$_4$(H$_2$O)$_2$(VW$_9$O$_{34}$)$_2$]	1.7/842	4210	n.d.	48	455	LED, 135 mW/cm^2	9	-	Cl	0.2	[62]
23	Co$_4$	[Co$_4$(H$_2$O)$_4$(HL^9)$_2$(L^9)$_2$]	13.2/1324	662	0.03	n.d.	>420	Xe lamp, 300 W, 26.4 mW/cm^2	9	-	ClO$_4$	2	[63]

Table 1. Cont.

N	Core	(Pre)catalyst	O_2 Evolution, μmol/μM	TON [1]	TOF [2], s^{-1}	QE [3], %	λ [4], nm	Irradiation Source	pH	NP [5]	Anion [6]	[cat]₀, μM	Ref.
24a	Co_6	$K_{12}[CoL^9]_6$	1.1/113	1125	35.3	5.9	460	LED, 33.8 mW/cm²	9	n.d.	Cl	0.1	[64]
24b	Co_6	$K_{12}[CoL^9]_6$	4.4/435	4350	162.6	27.1	460	LED, 33.8 mW/cm²	9	n.d.	ClO_4	0.1	[64]
24c	Co_6	$K_{12}[CoL^9]_6$	11.7/1169	167	4.4	51.5	460	LED, 33.8 mW/cm²	9	n.d.	Cl	7	[64]
25	Co_7	$Na_{12}[Co^{II}_7As_6O_9(OH)_6\{SiW_9O_{34}\}_2]$	2.3/115	115.2	0.14	n.d.	>420	Xe lamp, 300W	8	–	Cl	1	[65]
26a	Co_7	$[Co^{II}_5Co^{III}_2(mdea)_4(N_3)_2(CH_3CN)_6(OH)_2(H_2O)_2][ClO_4]_4$	10.7/1070	43	n.d.	n.d.	450	LED	9	n.d.	Cl	25	[66]
26b	Co_7	$[Co^{II}_5Co^{III}_2(mdea)_4(N_3)_2(CH_3CN)_6(OH)_2(H_2O)_2][ClO_4]_4$	22/2200	88	n.d.	n.d.	450	LED	9	n.d.	ClO_4	25	[66]
26c	Co_7	$[Co^{II}_5Co^{III}_2(mdea)_4(N_3)_2(CH_3CN)_6(OH)_2(H_2O)_2][ClO_4]_4$	10.5/1050	210	n.d.	n.d.	450	LED	9	n.d.	Cl	5	[66]
27a	Co_8	$K_8Na_8[(SiW_9O_{34})_2Co_8(OH)_6(H_2O)_2(CO_3)_3]$	16.4/1090	545	3.1	35.8	>420	LED, 5.1 mW/cm²	9	n.d.	Cl	2	[67]
27b	Co_8	$K_8Na_8[(SiW_9O_{34})_2Co_8(OH)_6(H_2O)_2(CO_3)_3]$	10.8/718	1436	10	28.8	>420	LED, 5.1 mW/cm²	9	n.d.	Cl	0.5	[67]
28	Co_9	$K_{16}[Co_9(H_2O)_6(OH)_3(PW_9O_{34})_3]$	1.0/67	10	n.d.	n.d.	>375	Tungsten lamp, 150 W, 90 mW/cm²	8	n.d.	n.d.	6.6	[68]
29a	Co_{15}	$Na_5[Co_6(H_2O)_{30}\{Co_9Cl_2(OH)_3(H_2O)_9(SiW_8O_{31})_3\}]$	3.1/207	53	21×10^{-3}	n.d.	>375	Tungsten lamp, 150 W, 90 mW/cm²	8	n.d.	n.d.	3.4	[68]
29b	Co_{15}	$Na_5[Co_6(H_2O)_{30}\{Co_9Cl_2(OH)_3(H_2O)_9(SiW_8O_{31})_3\}]$	4.1/273	28	19×10^{-3}	5.5	450	LED, 7 mW	8	n.d.	n.d.	9.8	[68]
30	Co_{16}	$Na_{22}Rb_6[\{Co_4(OH)_3PO_4\}_4(PW_9O_{34})_4]$	2.0/133	37	24×10^{-3}	n.d.	>375	Tungsten lamp, 150 W, 90 mW/cm²	8	n.d.	n.d.	3.6	[68]
31a	Co_1	$Co(NO_3)_2$	12.2/1219	762	n.d.	n.d.	>420	LED, 5.1 mW/cm²	9	+	ClO_4	1.6	[46]
31b	Co_1	$Co(NO_3)_2$	5.2/2600	52	n.d.	n.d.	>420	Xe lamp, 500 W	8	+	ClO_4	50	[49]
31c	Co_1	$Co(NO_3)_2$	4.6/310	4.3	1.92×10^{-3}	11	450	LED, 42 mW	8	–	Cl	72	[69]
31d	Co_1	$Co(NO_3)_2$	11.1/1112	139	n.d.	n.d.	>420	Xe lamp, 300 W, 26.4 mW/cm²	8	+	ClO_4	8	[63]
31e	Co_1	$Co(NO_3)_2$	19.5/1952	244	n.d.	n.d.	>420	Xe lamp, 300 W, 26.4 mW/cm²	9	+	ClO_4	8	[63]
31f	Co_1	$Co(NO_3)_2$	3/300	150	n.d.	n.d.	>420	Xe lamp, 300 W, 26.4 mW/cm²	8	+	ClO_4	2	[63]
32a	Co_1	$Co(ClO_4)_2$	0.4/400	570	19	n.d.	470	LED, 150 mW/cm²	8	+	ClO_4	0.7	[52]
32b	Co_1	$Co(ClO_4)_2$	0.25/250	500	16	n.d.	470	LED, 150 mW/cm²	8	+	ClO_4	0.5	[52]
32c	Co_1	$Co(ClO_4)_2$ + 8 eq. bpy	0/0	0	0	n.d.	470	LED, 150 mW/cm²	8	+	ClO_4	0.5	[52]
33a	Co_1	$Co(OAc)_2$	5.4/680	7	n.d.	n.d.	470	LED, 26.1 mW/cm²	7	+	Cl	250	[56]
33b	Co_1	$Co(OAc)_2$	9.2/924	132	3.1	36.1	460	LED, 33.8 mW/cm²	9	n.d.	ClO_4	7	[64]

[1] Moles of product per mol of catalyst; [2] Moles of product produced per mol of catalyst per a certain period of time; [3] Quantum efficiency, moles of O_2 produced per moles of absorbed photons; [4] Irradiation wavelength, as reported; [5] Formation of nanoparticles; [6] anion of the $[Ru(bpy)_3]^{2+}$ photosensitizer; [7] no data.

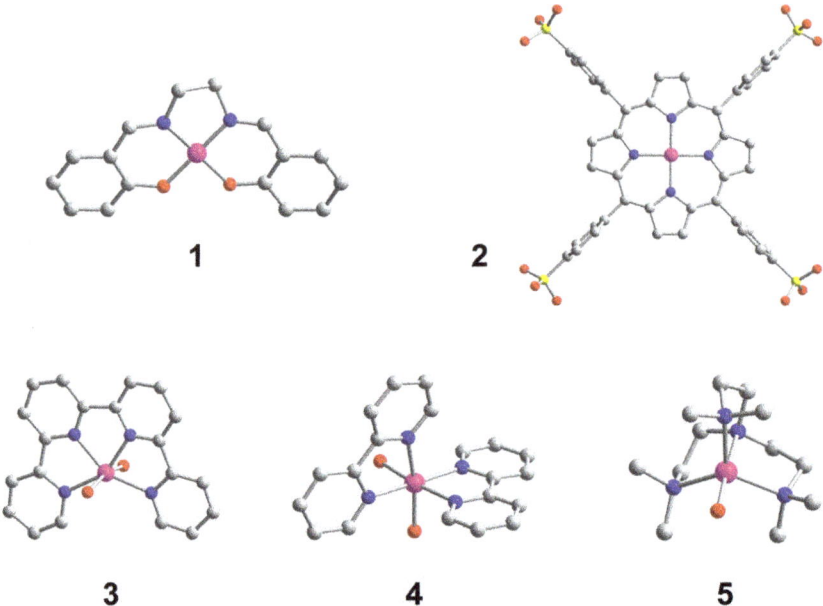

Figure 2. Structures of complexes **1–5**. Color scheme: Co, pink; O, red; N, blue; S, yellow; C, grey.

Complexes of porphyrin ligands are known to catalyze a wide range of reactions due to the pronounced stability of porphyrins, their redox activity, and ability to stabilize high-valent metal-oxo species of iron, manganese, ruthenium, and other metals [70–72]. A series of the water-soluble porphyrin complexes of cobalt have been tested as WOCs. The best results were obtained using the catalyst [Co(TPPS)] (**2**) with porphyrin ligand TPPS (meso-tetrakis(4-sulfophenyl)porphyrin) (Figure 2), reaching the TOF value of $0.17\ s^{-1}$ at pH 11 [47]. At lower or higher pH, TOF did not exceed $0.05\ s^{-1}$. From the second order dependence of the initial rate of O_2 formation on the $[\mathbf{2}]_0$, as well as from the DFT calculations, it was proposed that two molecules of **2** are required for completion of the water oxidation catalytic cycle. The proposed catalytic cycle presumes coupling of the Co–O• radicals as the principal mechanistic step, corresponding to the so-called I2M type of mechanisms (Scheme 2) [7]. One may notice, however, that for the electrocatalytic water oxidation catalyzed by cobalt porphyrins, the water nucleophilic attack (WNA) mechanism was proposed [73]. No nanoparticles formation was observed by DLS measurements. Appearance of the Soret band of porphyrin after mixing of $[Ru(bpy)_3]^{3+}$ with **2** also suggests that **2** is responsible for the catalytic activity. An unusual feature of this catalytic system is the use of nitrate salt of the $[Ru(bpy)_3]^{2+}$ photosensitizer (Table 1) instead of common chloride or perchlorate ones.

Two complexes, **3** and **4**, with polypyridine ligands (Figure 2) were tested as WOCs and revealed similar activities (Table 1) [48]. Chemical oxidation with $[Ru(bpy)]^{3+}$ was performed resulting on TON values of 160 and 70 for **3** and **4**, respectively. No formation of cobalt oxide nanoparticles or leaching of the cobalt ions was observed under catalytic conditions (both in light-driven and chemical oxidations), as evidenced by ESI-MS and DLS tests. The authors proposed the existence of two active intermediates, which ratio is defined by relative stability of the molecular intermediate [48]. One may notice that very low concentrations of the catalysts in the case of **3** and **4** (0.2 µM) may prevent correct determination of the nanoparticles or other decomposition products by conventional methods.

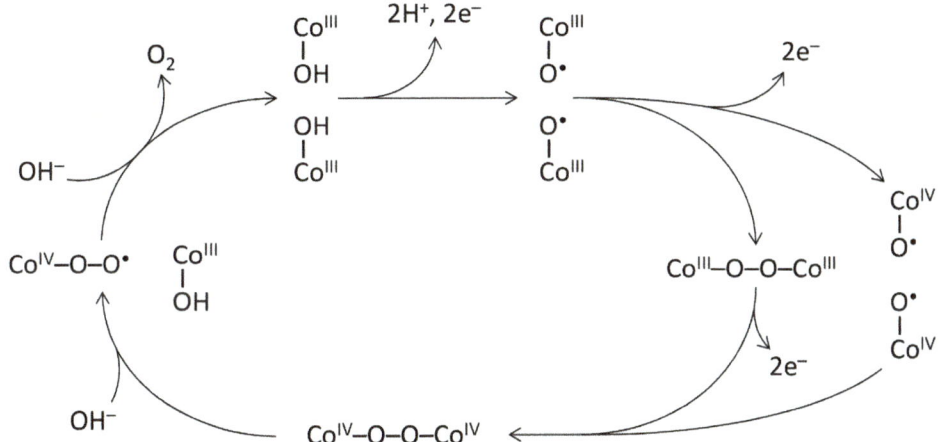

Scheme 2. Proposed pathways for the O$_2$ formation in the water oxidation catalyzed by **2**.

A series of mononuclear cobalt complexes with polydentate N-donor ligands was prepared and studied as catalysts in the light-driven water oxidation. The most active complex [Co(L^3)] (**5**) having tris(N,N'-(dimethylamino)ethyl)amine ligand (Figure 2) showed the TON values up to 54 and quantum yield of O$_2$ up to 32% (**5a**, Table 1) [49]. Very similar results were obtained for Co(NO$_3$)$_2$ as pre-catalyst (**31b**, Table 1), suggesting decomposition of complex **5** under catalytic conditions. Formation of nanoparticles with sizes ranging from 10 to 50 nm was confirmed by DLS and TEM (transmission electron microscopy) for the case of **5**. The size of particles and their aggregation behavior were different for **5** and Co(NO$_3$)$_2$. Moreover, the dependence of the yield of O$_2$ on the pre-catalyst concentration showed no changes at [Co(NO$_3$)$_2$]$_0$ > 1 mM, while for [**5**]$_0$ > 0.1 mM the catalytic activity showed rapid decay until no evolution of O$_2$ at [**5**]$_0$ = 2.5 mM. Formation of CO$_2$ in the catalytic system [Ru(bpy)$_3$]$^{2+}$/**5**/S$_2$O$_8$$^{2-}$, with elevated concertation of persulfate (50 mM), unambiguously pointed out the oxidation of organic ligand of **5**. The authors concluded that although complex **5** (and other complexes of the series) acts as a pre-catalyst, the nature of the ligand influences the structure and composition of the oxide nanoparticles formed and therefore the catalytic activity as well [49]. As a final remark, one may notice a strong influence of the nature of anion of [Ru(bpy)$_3$]$^{2+}$ photosensitizer, where perchlorate system (**5a**) is six times more active than the chloride one (**5b**).

The cobalt complex [Co$_2$(μ-OH)$_2$(TPA)$_2$](ClO$_4$)$_4$ (**6**) with tris(2-pyridylmethyl)amine (TPA) was synthesized by dimerization of the mononuclear complexes [Co^{III}(TPA)Cl$_2$](ClO$_4$) in the presence of AgClO$_4$ as a halogen acceptor [50]. The crystal structure of **6**, features a binuclear core where metal atoms are mediated by the bridging oxido ligands (Figure 3). The highest TON of 742 was obtained for low concentration of the catalyst (**6a**, Table 1). Increase of [**6**]$_0$ led to lower evolution of O$_2$ (**6b**). Chemical water oxidation with [Ru(bpy)]$^{3+}$ showed the TON of 4.3. According to DFT calculations, the active species are formed from the two-electron oxidation of **6** (Scheme 3). As evidenced by the ^{1}H NMR and DLS experiments performed for [**6**]$_0$ = 0.4 mM, complex **6** is stable during the photocatalytic reaction. Also, no CO$_2$ evolution, attributable to TPA ligand oxidation, was detected. These results are in accord with the ones observed for the TPA family of ligands in various oxidative processes. For instance, iron complexes with TPA are known to stabilize Fe^{IV} and Fe^V species capable of abstracting H atoms from sp^3 C–H bonds [74,75].

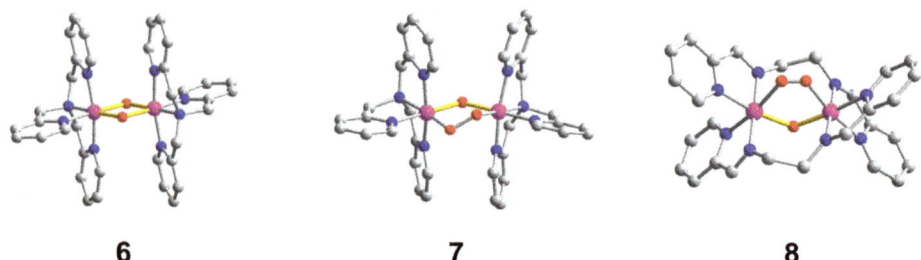

Figure 3. Structures of complexes **6–8**. Color scheme: Co, pink; O, red; N, blue; S, yellow; C, grey. Co–O–Co bonds are highlighted by yellow.

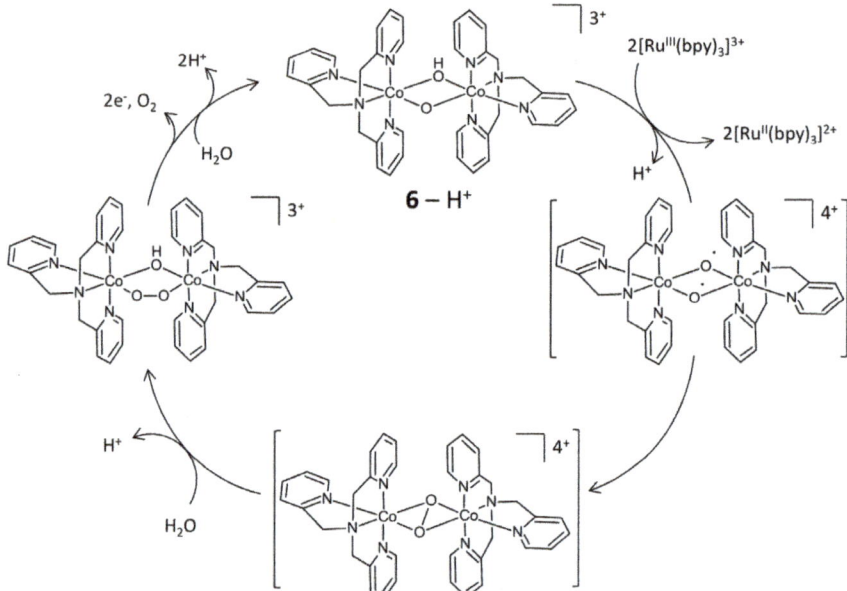

Scheme 3. Proposed pathways for the O_2 formation in the water oxidation catalyzed by **6**.

In contrast to **6**, obtained from the Co^{III} starting material, dimerization of $[Co^{II}(TPA)Cl]Cl$ complexes in the presence of $LiClO_4$ in open air, led to in situ oxidation of cobalt to Co^{3+} and formation of the Co–O–O–Co bridge in the complex $[(TPA)Co(\mu\text{-OH})(\mu\text{-}O_2)Co(TPA)](ClO_4)_3$ (**7**) (Figure 3) [51]. The compound was found to be stable at ambient conditions and showed moderate activity as a WOC at pH 8 with TON = 58. The authors performed DLS measurements and concluded that no CoO_x nanoparticles are formed during the photocatalytic reaction, while the control DLS experiment using $Co(ClO_4)_2$ as a catalyst revealed nanoparticles with ca. 100 nm size after irradiation. However, just two years later these results were reinvestigated by a different group of authors [52]. The use of an extended set of methods disclosed that the catalytic system based on complex **7** is in fact heterogeneous and contains CoO_x nanoparticles. Specifically, the water oxidation activity using the pre-catalyst **7** was completely suppressed when adding chelating ligands such as EDTA (ethylenediaminetetraacetic acid) or bpy (Figure 4a). The presence of just 0.1 mM of EDTA led to zero yields of O_2 for 0.5 mM of **7**. The test with bipyridine requires its larger amounts, but one equivalent of bpy (5 μM of bpy per 5 μM of **7**) is enough for quenching the catalytic activity of **7** [52]. Such behavior is consistent with that observed for $Co(ClO_4)_2$ salt as a catalyst, which activity at 0.7 μM concentration (**32a**, Table 1; Figure 4b) is completely quenched with eight equivalents of bpy (**32b**, Table 1). A question appears, namely, why earlier DSL measurements did not detect nanoparticles presumably formed upon degradation of complex **7**? A general explanation could be the different experimental conditions in these cases. It was mentioned that the DLS method might not detect small particles (at low amounts) formed in the course of reaction. Also, as can be seen from the systems **31** and **32** (Table 1), cobalt cations act as very efficient pre-catalysts at micromolar concentrations. Thus, even a small leaching of cobalt cations (captured by EDTA and bpy) from the initial coordination compound may be responsible for the pronounced catalytic activity, initially associated to the complex. In any case, these studies clearly demonstrate that the conclusion about the catalyst stability should not be made based on a single experimental method, such as DLS.

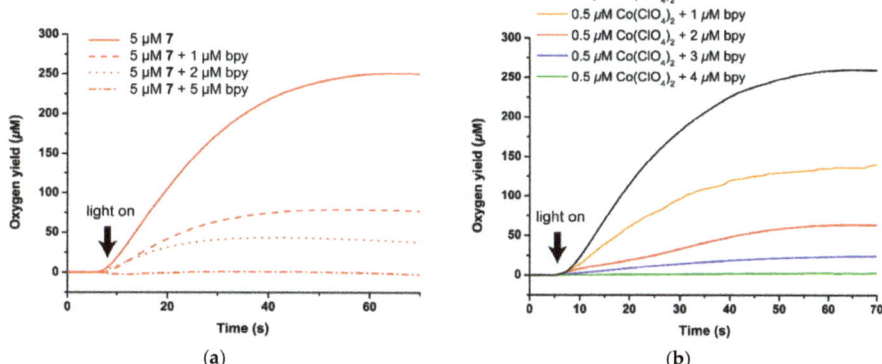

Figure 4. Accumulations of the oxygen in the course of water oxidation catalyzed by complex **7** (**a**) or $Co(ClO_4)_2$ (**b**), in the presence of various amounts of bipyridine (bpy) chelating agent. Adapted with permission from [52], American Chemical Society, 2016.

By reacting of the Schiff base ligand, N,N-bis(pyridin-2-ylmethylene)ethane-1,2-diimine (L^4), and $Co(ClO_4)_2$ in excess of pyridine the binuclear complex $[(L^4)Co^{III}(\mu\text{-}OH)(\mu\text{-}O_2)Co^{III}(L^4)](ClO_4)_3$ (**8**) has been prepared [53]. The structure of its core is similar to that observed for **7** (Figure 3). The light-driven water oxidation activity was studied by the independent group of authors, revealing TON values and oxygen concentrations much higher than those for **7** (Table 1). However, the tests performed in the presence of bpy ligand showed complete suppression of the catalytic activity with five equivalents of bpy (**7b**, Table 1), indicating that the observed activity could be associated to the free cobalt cations, rather than complex **7**.

3. Tetranuclear Cubanes Co_4O_4 or $Co_{4-x}M_xO_4$

In 2011 McCool at al. reported the water oxidation catalytic activity of the tetranuclear complex $[Co^{III}_4O_4(Ac)_4(py)_4]$ (**9**) [54]. This coordination compound was known earlier, and was also studied as a catalyst for the alcohols oxidation [76]. Its structure (Figure 5) represents a cubane core $Co_4(\mu_3\text{-}O)_4$, resembling that found in the PSII protein ($CaMn_3O_4$). The TOF of 0.02 s^{-1} exhibited by **9** for the light-driven oxidation was among, as of 2011, the highest known, resulting in high interest in the tetranuclear cobalt complexes as WOCs. The stability of the catalyst under the conditions of experiment was confirmed by the ^{1}H NMR (constant signals from acetate and pyridine ligands). Furthermore, the authors studied dependence of the lag time on the catalyst concentration. Lag period is expected when a catalyst undergoes transformations to active species (such as CoO_x nanoparticles) after irradiation begins, thus a lag time smaller than that for cobalt salt suggests the stability of the catalyst. The catalytic system based on complex **9** showed small lag time (less than 20 s for $[\mathbf{9}]_0 \sim 4$ mM), and this was associated by the authors to low oxidant/catalyst ratio, while the cobalt salt showed a lag time of almost 200 s under the same conditions [54].

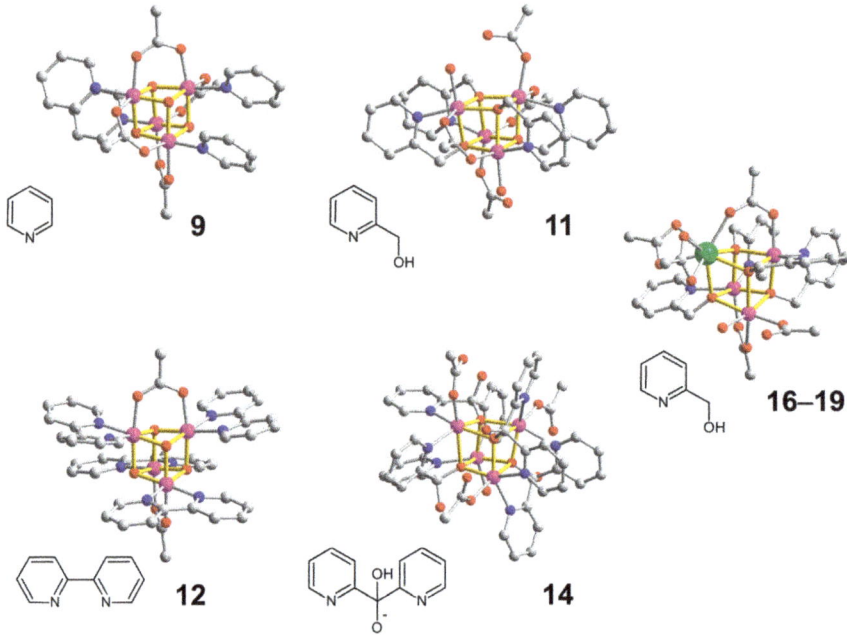

Figure 5. Structures of complexes **9**, **11**, **12**, **14**, and **16–19** with the schemes of respective N-donor ligands. Color scheme: Co, pink; Ln, green; O, red; N, blue; C, grey. M–O–M bonds are highlighted by yellow.

The water oxidation activity of a series of related complexes [Co$^{III}_4$O$_4$(OAc)$_4$(p-py-X)$_4$] having para-substituted pyridine (X = H, Me, t-Bu, OCH$_3$, Br, COOCH$_3$, CN) ligands has been reported in 2012 [55]. The nature of the substituent had strong influence on the catalytic activity, with the quantum yields (yield of oxygen based on the photon flux) ranging from 10% (for t-Bu) to 80% (for OCH$_3$; **10**, Table 1), suggesting the molecular nature of catalytically active species. In contrast to most of the other studies, 1:1 mixture of water:acetonitrile was used as a solvent.

Replacement of pyridine with 2-(hydroxymethyl)pyridine ligand (HL^5), which provides oxygen donor atoms, has led to the formation of cubane complex [Co$^{II}_4$(L^5)$_4$(OAc)$_4$(H$_2$O)$_2$] (**11**) where cobalt has +2 oxidation state (Figure 5), as reported one year after, in 2013 [56]. Another difference from complexes **9** and **10** is the presence of water molecules coordinated to cobalt centers. Water ligands take an important place in the structure of the active centers of PSII (Figure 1), thus the authors indicate that the structure of complex **11** is close to that of the PSII Mn$_4$CaO$_4$ cubane. Studies of methanol and water solutions of **11** by ESI-MS and pH-dependent UV-Vis titration disclosed that water ligands remain intact at neutral pH, becoming labile in the alkali media. The presence of labile water represents the main feature of **11** comparing to **9** and **10**, where, in the absence of coordinated water molecules, the protonation–deprotonation may occur at the Co$_4$O$_4$ core. Dependences of the TON and TOF in the course of catalytic water oxidation on pH were studied, disclosing that the highest TON of 35 is achieved at pH 8 (**11b**, Table 1) and the highest TOF of 7.0 s^{-1} at pH 9 (**11c**, Table 1). Stability of the core of **11**, under catalytic conditions, was investigated by the cyclic voltammetry (CV) experiments, suggesting that the structure of the compound does not undergo degradation. No nanoparticles in the catalytic solutions after 60 min of visible light irradiation were detected by the DLS method. In the comparative test with Co(OAc)$_2$, formation of the two fractions of nanoparticles (of ca. 5 and 120 nm diameter) under the same conditions was observed. Furthermore, significant difference in the lag times between **11** and Co(OAc)$_2$ catalysts has been observed: While cobalt acetate shows a lag period of 25 s, complex **11** exhibits a much shorter time (ca. 5 s), which does not depend on the catalyst concentration.

A series of bi-, tri-, and tetranuclear complexes with pyridine and bipyridine ligands have been tested as the WOCs by Smith et al. in 2014 [57]. While the complexes containing two or three metal centers were found to be catalytically inactive, the tetranuclear compound bearing cubane core [Co$^{III}_4$O$_4$(OAc)$_2$(bpy)$_4$](ClO$_4$)$_2$ (**12**) (Figure 5) showed the activity with maximum TOF of 0.05 s^{-1}, achieved at [**11**]$_0$ = 0.02 µM. At higher concentrations of the catalyst, the TOF drops until 0.02 s^{-1} (Table 1). The catalyst **9** was also tested under the same conditions, showing a bit higher activity, with a maximum TOF of 0.09 s^{-1} [57]. The catalytic mixture containing **11** was analyzed by means of the ESI-MS spectroscopy, revealing the presence of the unaltered complex **11** at 521 m/z. The authors interpreted this observation, with the support of 1H NMR data, as evidence that **11** is a truly molecular WOC. No DLS experiments data were reported.

In 2014, the group of Nocera reinvestigated the water oxidation properties of complexes **9** and **10** [58]. It was demonstrated that the catalytic activity previously associated to the cubane complexes is, in fact, exhibited by the small amounts of Co^{2+} impurities present in the samples. The ^{1}H NMR spectra of the crude samples of **9** and **10** showed many small peaks. Furthermore, thin layer chromatography disclosed the presence of a few bands in addition to that of the complexes. Purification of the samples using the column chromatography afforded the products with clear ^{1}H NMR spectra. The light-driven water oxidation tests showed that purified samples of **9** possesses ca. five times lower activity than the crude ones (entries **13a** and **13b**, Table 1). The exact nature of the Co^{2+} impurities was not identified. It was mentioned that they do not elute on silica and act as a precursor for a heterogeneous oxygen evolving catalyst. Notably, even after purification, complex **9** reveals detectable water oxidation activity (entry **13b**, Table 1).

A series of cubane cobalt Co$_4$O$_4$ (**14**) and mixed-metal Co$_{4-x}$Ni$_x$O$_4$ (**15**) complexes having bipyridine-like ligand was reported in 2017 [59]. The base complex [Co$^{II}_4$(L^6)$_4$(OAc)$_2$(H$_2$O)$_2$](ClO$_4$)$_2$ (**14**) was prepared starting from the pro-ligand di(2-pyridyl)ketone, which is hydrolyzed to produce the ligand L^6 (Figure 5). Complex **14** revealed moderate activity in the water oxidation with maximum TON of 96 (Table 1) at pH 8.5. In general, the catalytic parameters exhibited by **14** are close to other cobalt cubane complexes (**9–13**). No nanoparticles formation was observed according to the DLS experiments. The possibility of significant cobalt leaching was rejected by applying the Chelex® resin able to capture free metal ions: only a negligible drop of catalytic activity was observed in the case of **14**, while a simple cobalt salt showed 66% suppression caused by the resin trap. The presence of cobalt in the +2 oxidation state allowed isomorphic substitution of cobalt ions with nickel ones to obtain a series of Co$_{4-x}$Ni$_x$O$_4$ cubane complexes with maximum x = 2.85 (not including pure nickel compound with x = 4), bearing the same structures. The catalytic activity in the water oxidation gradually dropped with the increase of nickel portion, showing the TON and TOF ca. three times lower than for pure cobalt complex. The complex Ni$_4$O$_4$ did not show any notable catalytic activity under the same conditions [59].

Substitution of one of the metal positions with lanthanide led to a series of Co$_3$LnO$_4$ cubane complexes [Co$^{II}_3$Ln(L^5)$_4$(OAc)$_5$(H$_2$O)] (**16–19**) with 2-(hydroxymethyl)pyridine) ligand (HL^5) [60]. Compounds of this series were firstly synthesized by mixing cobalt and lanthanide acetates in stoichiometric ratio of 3:1 and characterized by X-ray diffraction [77]. The modified synthetic protocol using Co:Ln ratio of 2.3:1 afforded Co$_3$Ln compounds, where Ln = Ho, Er, Tm, and Yb [60]. All complexes show the best performance at pH = 8, being slightly less active at pH = 9 and considerably less active at pH = 7. Erbium complex is the most active with TON > 200, while the compound containing thulium shows the weakest results (Table 1). The authors followed a special workflow designed for clear identification of the complexes **16–19** as the water oxidation catalysts. The most important evidence was obtained from the DLS tests (no nanoparticles formation was detected) and trapping of "free" Co^{2+} ions with EDTA or Chelex® resin (no influence on the catalytic parameters observed). Stability of the tetranuclear cores in solution was confirmed by EXAFS, XANES and ESI-MS tests. The coordinated water molecule undergoes rapid exchange with solvent water, as evidenced by the tests in CD_3CN and CD_3CN/D_2O mixture, monitored by FT-IR. The authors account that lanthanide Ln^{3+} serves as a catalytic promoter, possibly working in analogy to Ca^{2+} in the natural $CaMn_3O_4$ OEC [60].

Typically, the components of the model catalytic system $[Ru(bpy)]^{2+}/catalyst/S_2O_8^{2-}$ are dissolved separately and do not form aggregates in a solution. In such system the electron transfer between the components is strongly influenced by their concentrations. This obstacle could be overcome by joining the photosensitizer, catalyst, and electron acceptor into one module, as it is found in a natural PSII complex. Such an approach is exemplified by the supramolecular assemblies $\{Co_4O_4(OAc)_3(Py)_4\}\{(L^7)Ru(bpy)_2\}$ and $\{Co_4O_4(OAc)_2(Py)_4\}_2\{(L^8)Ru(bpy)_2\}_2$ (**20** and **21**, respectively, Table 1) (L^7 = bpy-4-CH_3,4'-COOH; L^8 = bpy-4-COOH, 4'-COOH) [61], which include two well-studied components: Ruthenium photosensitizer and Co_4O_4 cubane WOC (similar to **9**). In these compounds, chemical linkage is realized through the modification of bipyridine ligand, attaching to them one or two carboxylic groups (Scheme 4). The integrity of the structures of **20** and **21** is supported by ESI-MS and ^1H-^1H COSY NMR spectroscopes. Assemblies **20** and **21** have been tested in the light-driven water oxidation (Figure 6) using a rather low concentration of the catalyst of 15 µM (calculated per Co_4O_4 cubane), showing the TON value of 25 in case of compound **21** (Table 1). The results exhibited by **20** were more modest, with TON of 5. The authors verified that the catalytic activity of the non-assembled components at the same concentrations is at a negligible level with TON of ca. 2 (**20b**, Table 1). Furthermore, the dependence of oxygen evolution rate on the catalyst concentration was studied for **20** and **20b** (the assembly and equimolar mixture, respectively). While the multicomponent mixture **20b** exhibits saturation at $[20b]_0 > 60$ µM (with the oxygen evolution rate of 1 µM s^{-1} and TOF of 0.02 s^{-1} at $[20b]_0 = 60$ µM), the assembly **20** shows linear increase of the reaction rate in the $10 < [20]_0 < 100$ µM concentrations rate, exhibiting a constant TOF of 7×10^{-3} s^{-1}. Monitoring of the reaction mixtures by UV-Vis spectroscopy disclosed a gradual decay of the absorption at 465 nm, attributed to the metal to ligand charge transfer (MLCT) band [10,32] of $[Ru(bpy)_3]^{2+}$, dropping ca. twice after 60 min. However, the authors note that degradation of the photosensitizer is not a limiting factor because the oxygen evolution stops much earlier, within 30 min [61]. Since the decoordination of the carboxylic group was suggested to be a part of the water oxidation mechanism, the inactivation pathway may involve irreversible decoordination of the ligands L^7 and L^8 with following degradation of the supramolecular assembly.

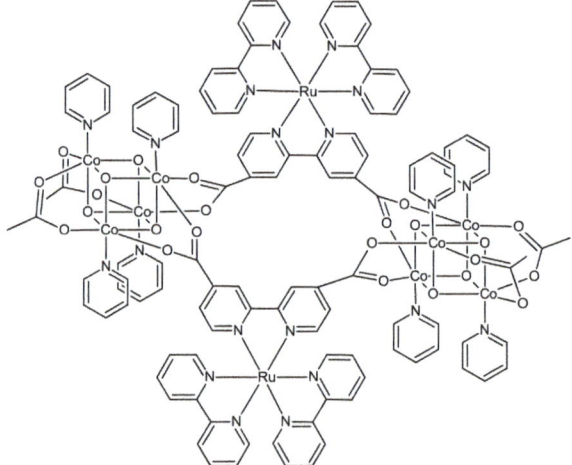

Scheme 4. Schematic representation of assembly **21**.

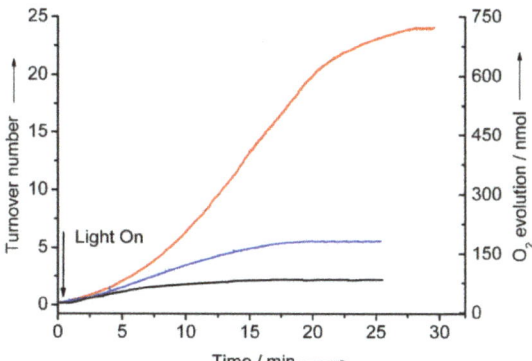

Figure 6. Accumulations of the oxygen in the course of water oxidation catalyzed by assemblies {Co$_4$O$_4$(OAc)$_3$(Py)$_4$}{(L^7)Ru(bpy)$_2$} (**20a**, blue line) and {Co$_4$O$_4$(OAc)$_2$(Py)$_4$}$_2${(L^8)Ru(bpy)$_2$}$_2$ (**21**, red line), as well as the equimolar mixture of [Co$_4$O$_4$(OAc)$_3$(Py)$_4$] and [Ru(bpy)$_3$]Cl$_2$ (**20b**, black line). Complex [Co$_4$O$_4$(OAc)$_3$(Py)$_4$] is equivalent to **9**. Adapted with permission from [61],Wiley, 2014.

4. Other Tetranuclear Complexes and Complexes of Higher Nuclearity

The polynuclear all-inorganic complex Na$_{10}$[Co$_4$(H$_2$O)$_2$(VW$_9$O$_{34}$)$_2$] (**22**) bearing polyoxometalate ligand (Figure 7) was tested as a catalyst for the water oxidation under chemical oxidation with [Ru(bpy)$_3$]$^{3+}$ as well as the light-driven conditions [62]. Chemical oxidation revealed exceptionally high TOF values ranging from 1600 to 2200 s^{-1} (adjusted for four cobalt centers in **22**), supported by TONs up to 75 (**22a**, Table 1). With such a high TOF the reaction completes within 1 s. In contrast, photocatalytic oxidation with in situ generated [Ru(bpy)$_3$]$^{3+}$ revealed the lower TOFs of ca. 4 s^{-1} (**22b**), but higher TON of 742 (or even up to 4210 for the lowest [**22**]$_0$, Table 1). The quantum efficiency of the catalytic system is at rather high level, up to 68% (for [**22**]$_0$ = 6 μM). The integrity of the catalyst during the catalysis was monitored by means of UV-Vis and ^{51}V NMR spectroscopes, which suggested no changes in the structure of **22**. The DLS experiments showed no presence of nanoparticles after both chemical and light-driven oxidations catalyzed by **22**. The comparative test using Co(NO$_3$)$_2$ as catalyst revealed formation of CoO$_x$ nanoparticles with the radius of 220 nm [62].

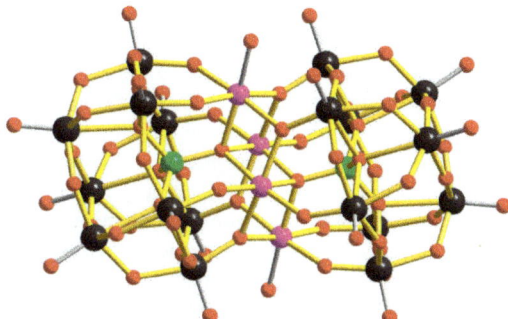

Figure 7. Structure of the anion of the compound **22**. Color scheme: Co, pink; V, green; W, black; O, red; M–O–M bonds are highlighted by yellow.

Two coordination compounds [Co$_4$(H$_2$O)$_4$(HL^9)$_2$(L^9)$_2$] and K$_{12}$[Co(L^9)]$_6$ (**23** and **24**, respectively) bearing tetra- and hexanuclear structures were prepared starting from N-(phosphonomethyl)iminodiacetic acid (H$_4$L^9) as pro-ligand (Figure 8) and cobalt chloride or acetate as a metal source, respectively [63,64]. Complex **23** disclosed TONs up to 662 (Table 1) in the light-driven water oxidation. The standard set of experiments (UV-Vis and comparative DLS tests of **23** and cobalt nitrate pre-catalyst) were applied to confirm that no nanoparticles were formed after light-driven water oxidation using catalyst **23** [63]. Moreover, a set of extractions with following ICP–MS (inductively coupled plasma mass spectrometry) and CAdSV (catalytic adsorptive stripping voltammetry) experiments were performed, indicating that less than 0.35% of catalyst **23** released free Co^{2+} ions in the buffer solution. One may notice that cobalt nitrate, as pre-catalyst under the same conditions, showed TON of 150 (for $[Co^{2+}]_0 = 2$ μM). This value is even higher than that exhibited by catalyst **23** if adjusted for six cobalt atoms in its molecule (TON = 110).

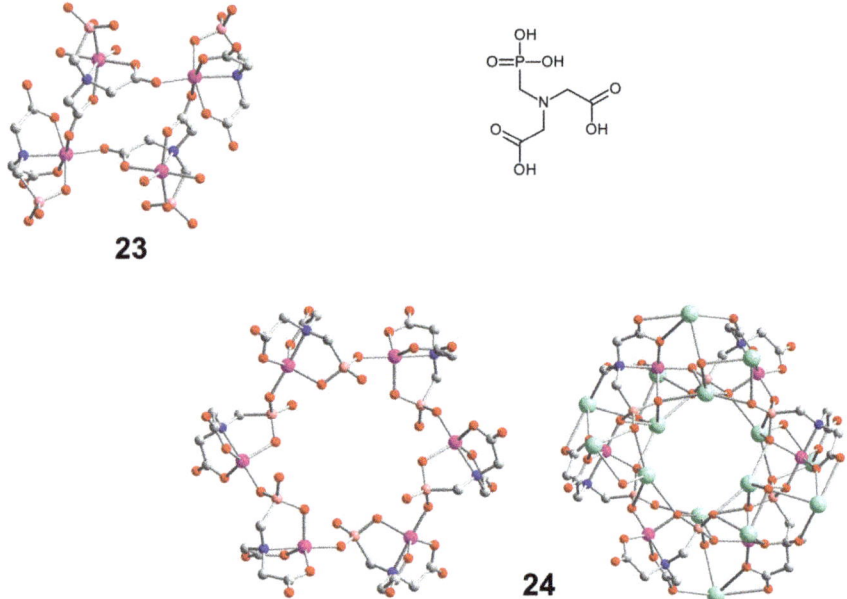

Figure 8. Structure of complexes **23** and **24**. Color scheme: Co, pink; K, pale green; O, red; N, blue; P, pale pink. Structure **24** is presented in two views, without (left) and with (right) nearest potassium atoms.

The hexanuclear complex **24** features amongst the highest values of the TOF reported to date (162.6 s^{-1} per cobalt atom) for the light-driven water oxidation (Table 1) [64]. The use of perchlorate salt of the ruthenium photosensitizer is crucial since the chloride salt shows a ca. 4.7 times lower reaction rate (**24a**, Table 1). The maximum quantum efficiency was found to be 62.6%. Considering such a high parameters, it was important to establish the origin of the catalytically active species. The comparative test with Co(OAc)$_2$ revealed that the latter shows large lag period of 30 s at 5 μM, while catalyst **24** shows no or negligible lag time. Furthermore, the chelation experiment using three equivalents of EDTA revealed no influence on the activity of **24**, while the oxygen evolution catalyzed by cobalt acetate was ca. 50% suppressed (both catalysts were at 7 μM initial concentration). The precipitate recovered after the reaction was studied by X-ray photoelectron spectroscopy (XPS) and was recognized as the authentic complex **24**. The evidence reported by the authors looks convincing, although absence of the DLS experiment is an essential drawback of the investigation. The proposed mechanism of water oxidation involves participation of a single cobalt center only, what is in agreement with the first-order dependence of the reaction rate on the concentration of **24** (for $[24]_0 < 5$ μM) [64]. Kinetic isotope effect (KIE) k_H/k_D between parallel reactions in normal and deuterated water was found to be 2.53, suggesting that catalyst **24** acts via the WNA mechanism, where a dioxygen molecule is constructed by the oxygen atoms from metal-oxo species and water molecule [7]. This value of KIE is very close to that reported (2.45) for ruthenium complex [Ru(bda)bpb]$_3^+$ having cyclic structure (H$_2$bda = 2,2′-bipyridine-6,6′-dicarboxylic acid; bpb = 1,4-Bis(pyrid-3-yl)benzene) believed to operate via the WNA pathway [78]. Unfortunately, in the case of **24**, no ^{18}O-labeling experiments were performed to confirm the attribution of the mechanism type.

The authors presented the structure of **24** as the open Co$_6$ ring (Figure 8, bottom-left) [62]. Such architecture could be rather labile in solution, especially under conditions of oxidative catalysis. Thus, the following questions appear: (1) Why is this compound is stable? (2) Why does it show such a high catalytic parameters? For instance, complex **23** is built using the same ligand (Figure 8) and does not possess water oxidation properties drastically different from those already reported (Table 1). The answer to both these questions could be found in the supramolecular structure of **24** (Figure 8, botom-right). The six-membered Co$_6$ ring is strengthened by potassium atoms, coordinating free oxygen atoms of the ligand, as well as numerous water molecules. Here, one may see two opportunities for the enhancement of the water oxidation reactions. At first, potassium may influence the redox potential of the cobalt centers, as it happens in the Ca$_3$Mn$_4$O$_4$ center of PSII and in the respective model complexes. Furthermore, recent experiments demonstrate that hydrogen bonding around the OEC in PSII is crucial for the water oxidation activity. The complex network of hydrogen bonds, formed by the water molecules coordinated by potassium atoms in the structure of **24**, may exist in a solution and be of great importance for successful water splitting process.

Another example of an all-inorganic compound of cobalt is the heptanuclear complex Na$_{12}$[{Co$^{II}_7$As$_6$O$_9$(OH)$_6$}(SiW$_9$O$_{34}$)$_2$] (**25**) [65]. The structure of **25** (Figure 9) resembles that of **22**. In general, the catalytic activity of **25** in the water photocatalytic oxidation is in the range typically exhibited by the other cobalt complexes of lower nuclearity (Table 1). DLS measurements showed no nanoparticles after the light-driven reaction, while in comparative test with Co(NO$_3$)$_2$ as catalyst, the presence of nanoparticles of ca. 190 nm in diameter were reported.

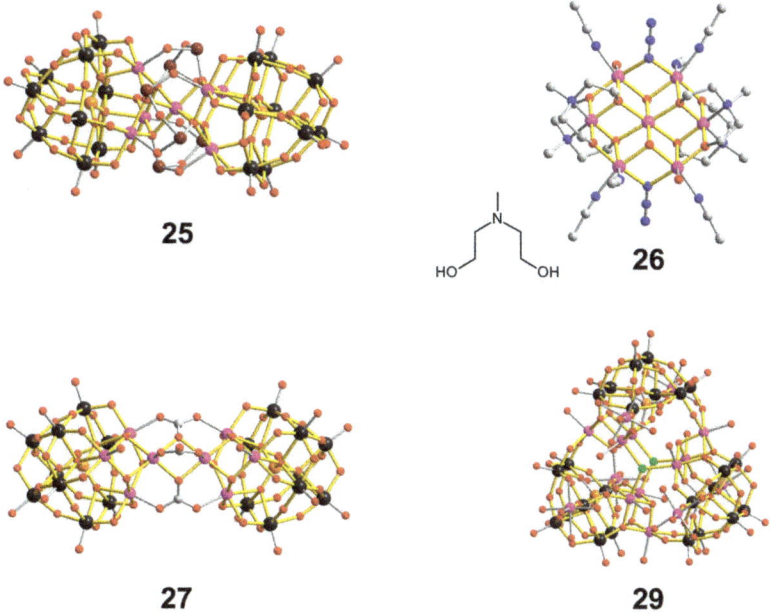

Figure 9. Structures of complexes **25–27**, and **29** (for **25**, **27**, and **29** cationic counterions are omitted). Color scheme: Co, pink; W, black; As, dark red; Si, orange; Cl, green; O, red; N, blue.

The structure of the heptanuclear mixed-valence complex $[Co^{II}{}_5Co^{III}{}_2(mdea)_4(N_3)_2(CH_3CN)_6$ $(OH)_2(H_2O)_2](ClO_4)_4$ (**26**) (H_2mdea = N-methyldiethanolamine) is based on the so-called Anderson structure type (Figure 9) [66]. Oxygen evolution was studied for chloride and perchlorate salts of ruthenium photosensitizer, showing that the latter produces higher TON values (**26b** and **26c**, respectively; Table 1). Using the chelating agents (bpy and EDTA), the decay of the oxygen evolution was observed when taken in equimolar ratio with the pre-catalyst **26**. The ESI-MS spectra indicated that **26** undergoes fragmentation in water solution since no peak attributable to Co_7 cluster was detected.

The octanuclear cobalt complex $K_8Na_8[Co_8(OH)_6(H_2O)_2(CO_3)_3SiW_9O_{34})_2]$ (**27**) containing bulky polyoxometalate ligand (the same as in **25**) has been described [79] and studied [67] as catalyst for water oxidation. An interesting feature of the structure of **27** is the presence of carbonate ligands (Figure 9). Complex **27** acts as a catalyst for water oxidation, showing the highest efficiency at pH 9 with the highest TON of 1435 and highest TOF of 10 s^{-1} (**27b**, Table 1). According to the DLS experiments, no nanoparticles are formed from **27** during the catalytic reaction. The comparative test with 2 µM of $Co(NO_3)_2$ revealed nanoparticles with diameter of 78 nm. FT-IR and UV-Vis spectroscopes also evidenced that the complex keeps its integrity after the catalysis. ESI-MS data are also provided, but the spectra appear to be too noisy (low overall intensity) to use them to make definite conclusions.

A series of high-nuclear cobalt complexes with polyoxometalate capping ligands $K_{16}[Co_9(H_2O)_6$ $(OH)_3(PW_9O_{34})_3]$ (**28**) [80], $Na_5[Co_6(H_2O)_{30}\{Co_9Cl_2(OH)_3(H_2O)_9(SiW_8O_{31})_3\}]$ (**29**) [81] and $Na_{22}Rb_6[\{Co_4(OH)_3PO_4\}_4(PW_9O_{34})_4]$ (**30**) [82] have been recently tested as catalysts for light-driven water oxidation [68]. Compounds **28–30** obey rather complicated architectures (complex **29** is depicted at Figure 9 as an example). The electron transfer between complexes **28–30** and photogenerated $[Ru(bpy)_3]^{3+}$ has been studied in detail. No lag period has been detected in all cases. EXAFS and XANES experiments suggested the existence of water exchange processes with small alterations of the complexes cores, without their degradation. The catalytic parameters exhibited by **28–30** are modest with TONs not exceeding 105 (for $[\mathbf{29}]_0 = 1.27$ µM). In general, Complex **29** shows better performance among this series. The authors note that the structure of **29**, containing 39 coordinated terminal water molecules, may facilitate its WOC activity [68].

5. Concluding Remarks

The field of the cobalt catalyzed water oxidation under visible light irradiation has received much attention in recent years. Pronounced activity of many molecular catalysts demonstrates great potential towards the construction of modular artificial water splitting systems. The TOF values exhibited by some of the catalytic systems approach those exhibited by the PSII system. However, one should always keep in mind that PSII and artificial systems operate at quite different conditions, thus any direct comparison of their parameters should be done with care. Furthermore, depending on the rate-limiting step of the catalytic reaction, the TOF may characterize, for instance, charge transfer between photosensitizer and catalyst, but not the rate of the dioxygen formation by the catalyst.

Discrimination of the true molecular (homogeneous) and heterogeneous systems should be always conducted to ensure that the molecular catalyst keeps its integrity and/or does not undergo heterogenization during the catalysis. The simplest methods involve comparative tests using cobalt salts with the same concentrations as that of the catalyst studied. Chelation experiments employing EDTA or bipyridine additives can provide fast evidence for the metal ions leaching. Purification of the sample from the metal impurities can also play an important role, as demonstrated by the examples of Co_4O_4 cubane complexes (**9–13**, Table 1). The typical instrumental method for nanoparticles detection is the dynamic light scattering (DLS), which is routinely applied in most of the literature reports discussed in the present review. However, one should remember that DLS has its own limitations and should not be used as a single method (see samples **7a** and **7b** as a representative example). Even the design of the experimental setup could have strong influence on the DLS results; for example, the use of magnetic stirring bars may complicate detection of nanoparticles possessing ferromagnetic properties.

Catalytic activity depends on many parameters, often implicit ones. For instance, the catalytic parameters exhibited by the systems containing [Ru(bpy)$_3$](ClO$_4$)$_2$ as photosensitizer are systematically higher than those obtained for using [Ru(bpy)$_3$]Cl$_2$. Thus, careful documentation of the experimental conditions is of exceptional significance for further analysis.

In spite of great efforts towards the study of the catalytic activity and verification of the catalyst stability, the reported reaction mechanisms are typically limited to general descriptions of charge transfers between photosensitizer and sacrificial electron acceptor, while attempts to study the mechanism of action of the WOC itself are scarce.

As a final remark, the use of cheap and abundant metals, such as cobalt, allows for the construction of highly efficient water oxidation catalysts. A special interest is in regards the bioinspired heterometallic complexes involving polynuclear cobalt core with attached redox-inactive metals. The exact mechanisms of many known artificial water oxidation systems are still unexplored and, beyond any doubt, further synthetic and catalytic efforts are required to establish structure-properties correlations towards practically feasible artificial photosynthesis.

Author Contributions: N.D.S. and O.V.N. contributed equally.

Funding: This work was supported by the Foundation for Science and Technology (FCT), Portugal (projects UID/QUI/00100/2013 and PTDC/QEQ-QIN/3967/2014, fellowships SFRH/BPD/99533/2014 and SFRH/BPD/63710/2009).

Conflicts of Interest: The authors declare no conflicts.

References

1. Chu, S.; Majumdar, A. Opportunities and challenges for a sustainable energy future. *Nature* **2012**, *488*, 294–303. [CrossRef] [PubMed]
2. Lewis, N.S.; Nocera, D.G. Powering the planet: Chemical challenges in solar energy utilization. *Proc. Natl. Acad. Sci. USA* **2006**, *103*, 15729–15735. [CrossRef] [PubMed]
3. Kopp, G.; Lean, J.L. A new, lower value of total solar irradiance: Evidence and climate significance. *Geophys. Res. Lett.* **2011**, *38*. [CrossRef]
4. Berardi, S.; Drouet, S.; Francas, L.; Gimbert-Surinach, C.; Guttentag, M.; Richmond, C.; Stoll, T.; Llobet, A. Molecular artificial photosynthesis. *Chem. Soc. Rev.* **2014**, *43*, 7501–7519. [CrossRef] [PubMed]
5. Pantazis, D.A. Missing Pieces in the Puzzle of Biological Water Oxidation. *ACS Catal.* **2018**, *8*, 9477–9507. [CrossRef]
6. Shamsipur, M.; Pashabadi, A. Latest advances in PSII features and mechanism of water oxidation. *Coord. Chem. Rev.* **2018**, *374*, 153–172. [CrossRef]
7. Garrido-Barros, P.; Gimbert-Surinach, C.; Matheu, R.; Sala, X.; Llobet, A. How to make an efficient and robust molecular catalyst for water oxidation. *Chem. Soc. Rev.* **2017**, *46*, 6088–6098. [CrossRef]
8. Najafpour, M.M.; Renger, G.; Holynska, M.; Moghaddam, A.N.; Aro, E.-M.; Carpentier, R.; Nishihara, H.; Eaton-Rye, J.J.; Shen, J.-R.; Allakhverdiev, S.I. Manganese Compounds as Water-Oxidizing Catalysts: From the Natural Water-Oxidizing Complex to Nanosized Manganese Oxide Structures. *Chem. Rev.* **2016**, *116*, 2886–2936. [CrossRef]
9. Yano, J.; Yachandra, V. Mn$_4$Ca Cluster in Photosynthesis: Where and How Water is Oxidized to Dioxygen. *Chem. Rev.* **2014**, *114*, 4175–4205. [CrossRef]
10. Karkas, M.D.; Verho, O.; Johnston, E.V.; Akermark, B. Artificial Photosynthesis: Molecular Systems for Catalytic Water Oxidation. *Chem. Rev.* **2014**, *114*, 11863–12001. [CrossRef]
11. Miyoshi, A.; Nishioka, S.; Maeda, K. Water Splitting on Rutile TiO$_2$-Based Photocatalysts. *Chem. Eur. J.* **2018**. [CrossRef] [PubMed]
12. Brennaman, M.K.; Dillon, R.J.; Alibabaei, L.; Gish, M.K.; Dares, C.J.; Ashford, D.L.; House, R.L.; Meyer, G.J.; Papanikolas, J.M.; Meyer, T.J. Finding the Way to Solar Fuels with Dye-Sensitized Photoelectrosynthesis Cells. *J. Am. Chem. Soc.* **2016**, *138*, 13085–13102. [CrossRef] [PubMed]
13. Sauer, K. A Role for Manganese in Oxygen Evolution in Photosynthesis. *Acc. Chem. Res.* **1980**, *13*, 249–256. [CrossRef]

14. Zhang, M.; Bommer, M.; Chatterjee, R.; Hussein, R.; Yano, J.; Dau, H.; Kern, J.; Dobbek, H.; Zouni, A. Structural insights into the light-driven auto-assembly process of the water oxidizing Mn_4CaO_5-cluster in photosystem II. *eLife* **2017**, *6*, e26933. [CrossRef] [PubMed]
15. Shen, J.-R. The Structure of Photosystem II and the Mechanism of Water Oxidation in Photosynthesis. *Ann. Rev. Plant Biol.* **2015**, *66*, 23–48. [CrossRef] [PubMed]
16. Zouni, A.; Witt, H.T.; Kern, J.; Fromme, P.; Krauss, N.; Saenger, W.; Orth, P. Crystal structure of photosystem II from Synechococcus elongatus at 3.8 angstrom resolution. *Nature* **2001**, *409*, 739–743. [CrossRef] [PubMed]
17. Ferreira, K.N.; Iverson, T.M.; Maghlaoui, K.; Barber, J.; Iwata, S. Architecture of the photosynthetic oxygen-evolving center. *Science* **2004**, *303*, 1831–1838. [CrossRef]
18. Loll, B.; Kern, J.; Saenger, W.; Zouni, A.; Biesiadka, J. Towards complete cofactor arrangement in the 3.0 angstrom resolution structure of photosystem II. *Nature* **2005**, *438*, 1040–1044. [CrossRef]
19. Umena, Y.; Kawakami, K.; Shen, J.R.; Kamiya, N. Crystal structure of oxygen-evolving photosystem II at a resolution of 1.9 angstrom. *Nature* **2011**, *473*, 55–61. [CrossRef]
20. Suga, M.; Akita, F.; Sugahara, M.; Kubo, M.; Nakajima, Y.; Nakane, T.; Yamashita, K.; Umena, Y.; Nakabayashi, M.; Yamane, T.; et al. Light-induced structural changes and the site of O=O bond formation in PSII caught by XFEL. *Nature* **2017**, *543*, 131. [CrossRef]
21. Retegan, M.; Krewald, V.; Mamedov, F.; Neese, F.; Lubitz, W.; Cox, N.; Pantazis, D.A. A five-coordinate Mn(IV) intermediate in biological water oxidation: Spectroscopic signature and a pivot mechanism for water binding. *Chem. Sci.* **2016**, *7*, 72–84. [CrossRef] [PubMed]
22. Young, I.D.; Ibrahim, M.; Chatterjee, R.; Gul, S.; Fuller, F.D.; Koroidov, S.; Brewster, A.S.; Tran, R.; Alonso-Mori, R.; Kroll, T.; et al. Structure of photosystem II and substrate binding at room temperature. *Nature* **2016**, *540*, 453. [CrossRef] [PubMed]
23. Bao, H.; Burnap, R.L. Photoactivation: The Light-Driven Assembly of the Water Oxidation Complex of Photosystem II. *Front. Plant Sci.* **2016**, *7*. [CrossRef] [PubMed]
24. Yocum, C.F. The calcium and chloride requirements of the O_2 evolving complex. *Coordi.Chem. Rev.* **2008**, *252*, 296–305. [CrossRef]
25. Zhang, C.; Chen, C.; Dong, H.; Shen, J.-R.; Dau, H.; Zhao, J. A synthetic Mn_4Ca-cluster mimicking the oxygen-evolving center of photosynthesis. *Science* **2015**, *348*, 690–693. [CrossRef] [PubMed]
26. Kanady, J.S.; Tsui, E.Y.; Day, M.W.; Agapie, T. A Synthetic Model of the Mn_3Ca Subsite of the Oxygen-Evolving Complex in Photosystem II. *Science* **2011**, *333*, 733–736. [CrossRef]
27. Tsui, E.Y.; Tran, R.; Yano, J.; Agapie, T. Redox-inactive metals modulate the reduction potential in heterometallic manganese-oxido clusters. *Nat. Chem.* **2013**, *5*, 293–299. [CrossRef]
28. Chen, C.G.; Kazimir, J.; Cheniae, G.M. Calcium Modulates the Photoassembly of Photosystem-II Mn_4-Clusters by Preventing Ligation of Nonfunctional High-Valency States of Manganese. *Biochemistry* **1995**, *34*, 13511–13526. [CrossRef]
29. Buvaylo, E.A.; Nesterova, O.V.; Kokozay, V.N.; Vassilyeva, O.Y.; Skelton, B.W.; Boca, R.; Nesterov, D.S. Discussion of Planarity of Molecular Structures Using Novel Pentanuclear Cu/Ni Complexes as an Example. *Cryst. Grow. Des.* **2012**, *12*, 3200–3208. [CrossRef]
30. Groom, C.R.; Bruno, I.J.; Lightfoot, M.P.; Ward, S.C. The Cambridge Structural Database. *Acta Cryst. B* **2016**, *72*, 171–179. [CrossRef]
31. Puntoriero, F.; Sartorel, A.; Orlandi, M.; La Ganga, G.; Serroni, S.; Bonchio, M.; Scandola, F.; Campagna, S. Photoinduced water oxidation using dendrimeric Ru(II) complexes as photosensitizers. *Coord. Chem. Rev.* **2011**, *255*, 2594–2601. [CrossRef]
32. Prier, C.K.; Rankic, D.A.; MacMillan, D.W.C. Visible Light Photoredox Catalysis with Transition Metal Complexes: Applications in Organic Synthesis. *Chem. Rev.* **2013**, *113*, 5322–5363. [CrossRef] [PubMed]
33. Bolletta, F.; Juris, A.; Maestri, M.; Sandrini, D. Quantum Yield of Formation of the Lowest Excited-State of $Ru(bpy)_3^{2+}$ and $Ru(phen)_3^{2+}$. *Inorg. Chim. Acta Lett.* **1980**, *44*, L175–L176. [CrossRef]
34. Blakemore, J.D.; Crabtree, R.H.; Brudvig, G.W. Molecular Catalysts for Water Oxidation. *Chem. Rev.* **2015**, *115*, 12974–13005. [CrossRef] [PubMed]
35. Karkas, M.D.; Akermark, B. Water oxidation using earth-abundant transition metal catalysts: Opportunities and challenges. *Dalton Trans.* **2016**, *45*, 14421–14461. [CrossRef] [PubMed]
36. Singh, A.; Spiccia, L. Water oxidation catalysts based on abundant 1st row transition metals. *Coord. Chem. Rev.* **2013**, *257*, 2607–2622. [CrossRef]

37. Han, Q.; Ding, Y. Recent advances in the field of light-driven water oxidation catalyzed by transition-metal substituted polyoxometalates. *Dalton Trans.* **2018**, *47*, 8180–8188. [CrossRef]
38. Wenger, O.S. Photoactive Complexes with Earth-Abundant Metals. *J. Am. Chem. Soc.* **2018**. [CrossRef]
39. Artero, V.; Chavarot-Kerlidou, M.; Fontecave, M. Splitting Water with Cobalt. *Angew. Chem. Int. Ed.* **2011**, *50*, 7238–7266. [CrossRef]
40. Kanan, M.W.; Nocera, D.G. In situ formation of an oxygen-evolving catalyst in neutral water containing phosphate and Co^{2+}. *Science* **2008**, *321*, 1072–1075. [CrossRef]
41. Jiao, F.; Frei, H. Nanostructured Cobalt Oxide Clusters in Mesoporous Silica as Efficient Oxygen-Evolving Catalysts. *Angew. Chem. Int. Ed.* **2009**, *48*, 1841–1844. [CrossRef] [PubMed]
42. Dismukes, G.C.; Brimblecombe, R.; Felton, G.A.N.; Pryadun, R.S.; Sheats, J.E.; Spiccia, L.; Swiegers, G.F. Development of Bioinspired Mn_4O_4-Cubane Water Oxidation Catalysts: Lessons from Photosynthesis. *Acc. Chem. Res.* **2009**, *42*, 1935–1943. [CrossRef] [PubMed]
43. Yin, Q.; Tan, J.M.; Besson, C.; Geletii, Y.V.; Musaev, D.G.; Kuznetsov, A.E.; Luo, Z.; Hardcastle, K.I.; Hill, C.L. A Fast Soluble Carbon-Free Molecular Water Oxidation Catalyst Based on Abundant Metals. *Science* **2010**, *328*, 342–345. [CrossRef] [PubMed]
44. Huang, Z.; Luo, Z.; Geletii, Y.V.; Vickers, J.W.; Yin, Q.; Wu, D.; Hou, Y.; Ding, Y.; Song, J.; Musaev, D.G.; et al. Efficient Light-Driven Carbon-Free Cobalt-Based Molecular Catalyst for Water Oxidation. *J. Am. Chem. Soc.* **2011**, *133*, 2068–2071. [CrossRef] [PubMed]
45. Stracke, J.J.; Finke, R.G. Electrocatalytic Water Oxidation Beginning with the Cobalt Polyoxometalate $Co_4(H_2O)_2(PW_9O_{34})_2^{10-}$: Identification of Heterogeneous CoO_x as the Dominant Catalyst. *J. Am. Chem. Soc.* **2011**, *133*, 14872–14875. [CrossRef] [PubMed]
46. Fu, S.; Liu, Y.; Ding, Y.; Du, X.; Song, F.; Xiang, R.; Ma, B. A mononuclear cobalt complex with an organic ligand acting as a precatalyst for efficient visible light-driven water oxidation. *Chem. Commun.* **2014**, *50*, 2167–2169. [CrossRef] [PubMed]
47. Nakazono, T.; Parent, A.R.; Sakai, K. Cobalt porphyrins as homogeneous catalysts for water oxidation. *Chem. Commun.* **2013**, *49*, 6325–6327. [CrossRef]
48. Leung, C.-F.; Ng, S.-M.; Ko, C.-C.; Man, W.-L.; Wu, J.; Chen, L.; Lau, T.-C. A cobalt(II) quaterpyridine complex as a visible light-driven catalyst for both water oxidation and reduction. *Energ. Environ. Sci.* **2012**, *5*, 7903–7907. [CrossRef]
49. Hong, D.; Jung, J.; Park, J.; Yamada, Y.; Suenobu, T.; Lee, Y.M.; Nam, W.; Fukuzumi, S. Water-soluble mononuclear cobalt complexes with organic ligands acting as precatalysts for efficient photocatalytic water oxidation. *Energ. Environ. Sci.* **2012**, *5*, 7606–7616. [CrossRef]
50. Ishizuka, T.; Watanabe, A.; Kotani, H.; Hong, D.; Satonaka, K.; Wada, T.; Shiota, Y.; Yoshizawa, K.; Ohara, K.; Yamaguchi, K.; et al. Homogeneous Photocatalytic Water Oxidation with a Dinuclear Co^{III} Pyridylmethylamine Complex. *Inorg. Chem.* **2016**, *55*, 1154–1164. [CrossRef]
51. Wang, H.-Y.; Mijangos, E.; Ott, S.; Thapper, A. Water Oxidation Catalyzed by a Dinuclear Cobalt-Polypyridine Complex. *Angew. Chem. Int. Ed.* **2014**, *53*, 14499–14502. [CrossRef] [PubMed]
52. Wang, J.-W.; Sahoo, P.; Lu, T.-B. Reinvestigation of Water Oxidation Catalyzed by a Dinuclear Cobalt Polypyridine Complex: Identification of CoO_x as a Real Heterogeneous Catalyst. *ACS Catal.* **2016**, *6*, 5062–5068. [CrossRef]
53. Lin, J.; Ma, B.; Chen, M.; Ding, Y. Water oxidation catalytic ability of polypyridine complex containing a μ-OH, μ-O_2 dicobalt(III) core. *Chin. J. Catal.* **2018**, *39*, 463–471. [CrossRef]
54. McCool, N.S.; Robinson, D.M.; Sheats, J.E.; Dismukes, G.C. A Co_4O_4 "Cubane" Water Oxidation Catalyst Inspired by Photosynthesis. *J. Am. Chem. Soc.* **2011**, *133*, 11446–11449. [CrossRef] [PubMed]
55. Berardi, S.; La Ganga, G.; Natali, M.; Bazzan, I.; Puntoriero, F.; Sartorel, A.; Scandola, F.; Campagna, S.; Bonchio, M. Photocatalytic Water Oxidation: Tuning Light-Induced Electron Transfer by Molecular Co_4O_4 Cores. *J. Am. Chem. Soc.* **2012**, *134*, 11104–11107. [CrossRef] [PubMed]
56. Evangelisti, F.; Guettinger, R.; More, R.; Luber, S.; Patzke, G.R. Closer to Photosystem II: A Co_4O_4 Cubane Catalyst with Flexible Ligand Architecture. *J. Am. Chem. Soc.* **2013**, *135*, 18734–18737. [CrossRef] [PubMed]
57. Smith, P.F.; Kaplan, C.; Sheats, J.E.; Robinson, D.M.; McCool, N.S.; Mezle, N.; Dismukes, G.C. What Determines Catalyst Functionality in Molecular Water Oxidation? Dependence on Ligands and Metal Nuclearity in Cobalt Clusters. *Inorg. Chem.* **2014**, *53*, 2113–2121. [CrossRef]

58. Ullman, A.M.; Liu, Y.; Huynh, M.; Bediako, D.K.; Wang, H.; Anderson, B.L.; Powers, D.C.; Breen, J.J.; Abruna, H.D.; Nocera, D.G. Water Oxidation Catalysis by Co(II) Impurities in $Co^{III}_4O_4$ Cubanes. *J. Am. Chem. Soc.* **2014**, *136*, 17681–17688. [CrossRef]
59. Song, F.Y.; More, R.; Schilling, M.; Smolentsev, G.; Azzaroli, N.; Fox, T.; Luber, S.; Patzke, G.R. {Co_4O_4} and {$Co_xNi_{4-x}O_4$} Cubane Water Oxidation Catalysts as Surface Cut-Outs of Cobalt Oxides. *J. Am. Chem. Soc.* **2017**, *139*, 14198–14208. [CrossRef]
60. Evangelisti, F.; More, R.; Hodel, F.; Luber, S.; Patzke, G.R. 3d-4f {$Co^{II}_3Ln(OR)_4$} Cubanes as Bio-Inspired Water Oxidation Catalysts. *J. Am. Chem. Soc.* **2015**, *137*, 11076–11084. [CrossRef]
61. Zhou, X.; Li, F.; Li, H.; Zhang, B.; Yu, F.; Sun, L. Photocatalytic Water Oxidation by Molecular Assemblies Based on Cobalt Catalysts. *ChemSusChem* **2014**, *7*, 2453–2456. [CrossRef] [PubMed]
62. Lv, H.; Song, J.; Geletii, Y.V.; Vickers, J.W.; Sumliner, J.M.; Musaev, D.G.; Koegerler, P.; Zhuk, P.F.; Bacsa, J.; Zhu, G.; et al. An Exceptionally Fast Homogeneous Carbon-Free Cobalt-Based Water Oxidation Catalyst. *J. Am. Chem. Soc.* **2014**, *136*, 9268–9271. [CrossRef] [PubMed]
63. Xu, Q.; Li, H.; Chi, L.; Zhang, L.; Wan, Z.; Ding, Y.; Wang, J. Identification of homogeneous $Co_4(H_2O)_4(HPMIDA)_2(PMIDA)_2^{6-}$ as an effective molecular-light-driven water oxidation catalyst. *Appl. Catal. B Environ.* **2017**, *202*, 397–403. [CrossRef]
64. Lin, J.; Meng, X.; Zheng, M.; Ma, B.; Ding, Y. Insight into a hexanuclear cobalt complex: Strategy to construct efficient catalysts for visible light-driven water oxidation. *Appl. Catal. B Environ.* **2019**, *241*, 351–358. [CrossRef]
65. Chen, W.-C.; Wang, X.-L.; Qin, C.; Shao, K.-Z.; Su, Z.-M.; Wang, E.-B. A carbon-free polyoxometalate molecular catalyst with a cobalt-arsenic core for visible light-driven water oxidation. *Chem. Commun.* **2016**, *52*, 9514–9517. [CrossRef] [PubMed]
66. Xu, J.H.; Guo, L.-Y.; Su, H.-F.; Gao, X.; Wu, X.-F.; Wang, W.-G.; Tung, C.-H.; Sun, D. Heptanuclear $Co^{II}_5Co^{III}_2$ Cluster as Efficient Water Oxidation Catalyst. *Inorg. Chem.* **2017**, *56*, 1591–1598. [CrossRef] [PubMed]
67. Wei, J.; Feng, Y.; Zhou, P.; Liu, Y.; Xu, J.; Xiang, R.; Ding, Y.; Zhao, C.; Fan, L.; Hu, C. A Bioinspired Molecular Polyoxometalate Catalyst with Two Cobalt(II) Oxide Cores for Photocatalytic Water Oxidation. *ChemSusChem* **2015**, *8*, 2630–2634. [CrossRef]
68. Natali, M.; Bazzan, I.; Goberna-Ferron, S.; Al-Oweini, R.; Ibrahim, M.; Bassil, B.S.; Dau, H.; Scandola, F.; Galan-Mascaros, J.R.; Kortz, U.; et al. Photo-assisted water oxidation by high-nuclearity cobalt-oxo cores: Tracing the catalyst fate during oxygen evolution turnover. *Green Chem.* **2017**, *19*, 2416–2426. [CrossRef]
69. Genoni, A.; La Ganga, G.; Volpe, A.; Puntoriero, F.; Di Valentin, M.; Bonchio, M.; Natali, M.; Sartorel, A. Water oxidation catalysis upon evolution of molecular Co(III) cubanes in aqueous media. *Faraday Discuss.* **2015**, *185*, 121–141. [CrossRef]
70. Costas, M. Selective C-H oxidation catalyzed by metalloporphyrins. *Coord. Chem. Rev.* **2011**, *255*, 2912–2932. [CrossRef]
71. Barona-Castano, J.C.; Carmona-Vargas, C.C.; Brocksom, T.J.; de Oliveira, K.T. Porphyrins as Catalysts in Scalable Organic Reactions. *Molecules* **2016**, *21*, 310. [CrossRef] [PubMed]
72. Gao, W.Y.; Chrzanowski, M.; Ma, S.Q. Metal-metalloporphyrin frameworks: A resurging class of functional materials. *Chem. Soc. Rev.* **2014**, *43*, 5841–5866. [CrossRef]
73. Wang, D.; Groves, J.T. Efficient water oxidation catalyzed by homogeneous cationic cobalt porphyrins with critical roles for the buffer base. *Proc. Natl. Acad. Sci. USA* **2013**, *110*, 15579–15584. [CrossRef] [PubMed]
74. Bryliakov, K.P.; Talsi, E.P. Active sites and mechanisms of bioinspired oxidation with H_2O_2, catalyzed by non-heme Fe and related Mn complexes. *Coord. Chem. Rev.* **2014**, *276*, 73–96. [CrossRef]
75. Olivo, G.; Cusso, O.; Borrell, M.; Costas, M. Oxidation of alkane and alkene moieties with biologically inspired nonheme iron catalysts and hydrogen peroxide: From free radicals to stereoselective transformations. *J. Biol. Inorg. Chem.* **2017**, *22*, 425–452. [CrossRef] [PubMed]
76. Chakrabarty, R.; Sarmah, P.; Saha, B.; Chakravorty, S.; Das, B.K. Catalytic Properties of Cobalt(III)-Oxo Cubanes in the TBHP Oxidation of Benzylic Alcohols. *Inorg. Chem.* **2009**, *48*, 6371–6379. [CrossRef] [PubMed]
77. Wang, P.; Shannigrahi, S.; Yakovlev, N.L.; Hor, T.S.A. General One-Step Self-Assembly of Isostructural Intermetallic $Co^{II}_3Ln^{III}$ Cubane Aggregates. *Inorg. Chem.* **2012**, *51*, 12059–12061. [CrossRef] [PubMed]

78. Schulze, M.; Kunz, V.; Frischmann, P.D.; Wuerthner, F. A supramolecular ruthenium macrocycle with high catalytic activity for water oxidation that mechanistically mimics photosystem II. *Nat. Chem.* **2016**, *8*, 577–584. [CrossRef]
79. Lisnard, L.; Mialane, P.; Dolbecq, A.; Marrot, J.; Clemente-Juan, J.M.; Coronado, E.; Keita, B.; de Oliveira, P.; Nadjo, L.; Secheresse, F. Effect of cyanato, azido, carboxylato, and carbonato ligands on the formation of cobalt(II) polyoxometalates: Characterization, magnetic, and electrochemical studies of multinuclear cobalt clusters. *Chem. Eur. J.* **2007**, *13*, 3525–3536. [CrossRef]
80. Galanmascaros, J.R.; Gomezgarcia, C.J.; Borrasalmenar, J.J.; Coronado, E. High-Nuclearity Magnetic Clusters–Magnetic-Properties of a 9 Cobalt Cluster Encapsulated in a Polyoxometalate, $Co_9(OH)_3(H_2O)_6(HPO_4)_2(PW_9O_{34})_3^{16-}$. *Adv. Mater.* **1994**, *6*, 221–223. [CrossRef]
81. Bassil, B.S.; Nellutla, S.; Kortz, U.; Stowe, A.C.; van Tol, J.; Dalal, N.S.; Keita, B.; Nadjo, L. The satellite-shaped Co_{15} polyoxotungstate, $Co_6(H_2O)_{30}\{Co_9Cl_2(OH)_3(H_2O)_9(SiW_8O_{31})_3\}^{5-}$. *Inorg. Chem.* **2005**, *44*, 2659–2665. [CrossRef] [PubMed]
82. Ibrahim, M.; Lan, Y.H.; Bassil, B.S.; Xiang, Y.X.; Suchopar, A.; Powell, A.K.; Kortz, U. Hexadecacobalt(II)-Containing Polyoxometalate-Based Single-Molecule Magnet. *Angew. Chem. Int. Ed.* **2011**, *50*, 4708–4711. [CrossRef] [PubMed]

© 2018 by the authors. Licensee MDPI, Basel, Switzerland. This article is an open access article distributed under the terms and conditions of the Creative Commons Attribution (CC BY) license (http://creativecommons.org/licenses/by/4.0/).

Review

Effect of Co-Feeding Inorganic and Organic Molecules in the Fe and Co Catalyzed Fischer–Tropsch Synthesis: A Review

Adolph Anga Muleja [1], Joshua Gorimbo [2] and Cornelius Mduduzi Masuku [3,*]

[1] Nanotechnology and Water Sustainability (NanoWS) Research Unit, University of South Africa, Private Bag X6, Florida 1710, South Africa
[2] Institute for the Development of Energy for African Sustainability (IDEAS), University of South Africa (UNISA), Florida Campus, Private Bag X6, Johannesburg 1710, South Africa
[3] Department of Civil and Chemical Engineering, University of South Africa, Private Bag X6, Florida 1710, South Africa
* Correspondence: masukcm@unisa.ac.za; Tel.: +27-11-471-2343

Received: 15 August 2019; Accepted: 31 August 2019; Published: 4 September 2019

Abstract: This short review makes it clear that after 90 years, the Fischer–Tropsch synthesis (FTS) process is still not well understood. While it is agreed that it is primarily a polymerization process, giving rise to a distribution of mainly olefins and paraffins; the mechanism by which this occurs on catalysts is still a subject of much debate. Many of the FT features, such as deactivation, product distributions, kinetics and mechanism, and equilibrium aspects of the FT processes are still subjects of controversy, regardless of the progress that has been made so far. The effect of molecules co-feeding in FTS on these features is the main focus of this study. This review looks at some of these areas and tries to throw some light on aspects of FTS since the inception of the idea to date with emphasis and recommendation made based on nitrogen, water, ammonia, and olefins co-feeding case studies.

Keywords: Fischer–Tropsch product distribution; reaction mechanism; catalysis; process synthesis and design; catalyst deactivation

---

1. Introduction

The addition of molecules other than syngas in the reactor during Fischer–Tropsch synthesis (FTS) is considered co-feeding of that molecule to the FTS. These molecules, usually with lower molecular weights, could be water, organic or inorganic additives. For instance, there have been controversies about the co-feeding of water to the FT reactor. Several researchers have investigated the addition of water to elucidate the water effect on both activity and selectivity [1–4] on the catalyst deactivation [5–7], the kinetics and mechanism [8–10], and product distribution [3]. Due to the different views on the effect of water co-feeding in syngas during FTS, some authors have focused their studies on shedding light on the subject of positive effects of water co-feeding on the FT rate, which was observed for some, but not all, Co-based catalysts [9]. Reports on the negative effects of water on FT processes are also available [11]. It is worth noting that the FTS mechanism is not entirely clear, and the distribution of products does not typically follow a typical Anderson–Schulz–Flory (ASF) distribution. As such, various molecules have been used in the FT reactions in an attempt to better the understanding of the reaction mechanism and explain the deviations observed from the ASF distribution [12–14]. In particular, the additives that have been co-fed to syngas to investigate their effects on the mechanism and influence on product distribution in FTS include light olefins [14–17], alcohols [18,19], water [1,20–26], CO_2 [27–31], and many more, as illustrated in Figure 1.

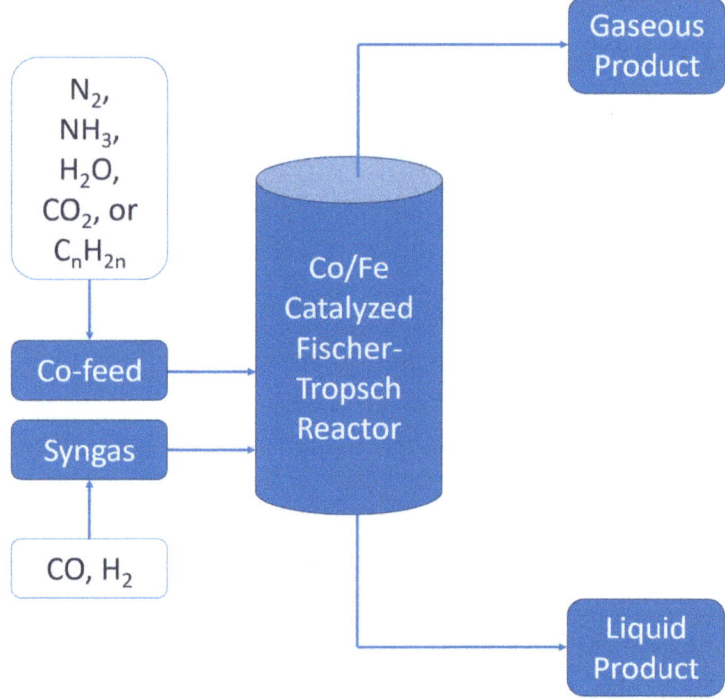

Figure 1. Illustration of co-feeding components considered in this review.

Several researchers [32–34] have reported the effect of ammonia in syngas on the FT reaction. The authors studied the effect to explore possible phase changes of the Fe catalyst with change in time on the stream. What they also assessed was the deactivation mechanism of the Fe catalyst by poisoning during the FT reaction [32]. The effect of co-feeding ammonia on the conversion of syngas using traditional cobalt catalyst was investigated to probe the catalyst deactivation and ability to affect the selectivity of methane and C_{5+} hydrocarbon formation [33]. Sango et al. [34] also reported the co-feeding of ammonia in different concentrations to a slurry phase, iron-based FT synthesis to study catalyst deactivation, methane selectivity, the chain growth probability, the olefin selectivity, and the FT reactions mechanisms. In a separate experimental study, controlled and verified quantities of hydrogen sulfide and ammonia were added into the feed gas. Hydrogen sulfide was employed to probe the impact of irreversible catalyst poisoning relative to the intrinsic deactivation profile, whereas ammonia co-feeding was done to quantify the impact of regenerable catalyst poisoning. The combined effects provided insight regarding catalyst deactivation and possibly prediction of time dependant catalyst performance during FTS [35].

In recent studies, the synthesis gas mainly obtained from the reforming of coal, biomass, and natural gas contains significant amounts of carbon dioxide (CO_2) [36]. CO_2 removal from the synthesis gas is quite complex and expensive, hence, consequently, takes part in the FT catalytic activity [36]. A series of FTS experiments, which entail the co-feeding of CO_2 were conducted in a fixed bed reactor over a cobalt-based catalyst to investigate the catalyst deactivation, catalytic activity, and product selectivity and formation rates [28]. The influence of CO_2 content in the feed stream was shown to improve the product distribution toward valuable hydrocarbons and conversion rates [36]. CO_2 was added to the FT feed to investigate the CO conversion, chain growth probability, the products distribution, and FT reaction mechanisms [37,38]. Further investigations of the CO_2 influence in the feed gas relate to the economy and sustainability developments of FT technology [29,39,40]. Still, on the cost-effectiveness of the FT system, it has been reported that the use of nitrogen-rich syngas could be a better approach than the classical processes where syngas is free of nitrogen [39,40]. Inert gases, such as nitrogen, have also been co-fed to the FT reactor. The effects of co-feeding nitrogen (N_2) in the FTS in a fixed bed reactor over a Co/TiO_2 catalyst was investigated by Muleja et al. (2016) to elucidate the effect on the catalytic activity, selectivity, and the implication on the FTS kinetics [41]. Nitrogen-rich syngas was used to gain an understanding of its behavior in thermochemical catalytic conversion to gasoline range hydrocarbon [42]. As such, researchers [43] have recently proposed that when operating a cobalt-based FTS process, the olefin and paraffin formation rates should be considered separately. The reason being that the rates before, during, and after deactivation during FTS are comparatively inconclusive, for instance, in the studies by [41,43] the olefin formation rate was fairly constant over time, and in some cases even increased, while the paraffin formation rate dropped, indicating that the deactivation mainly affects the paraffin formation rate. This review is undertaken to summarize and consolidate the co-feeding work done so far and to try explaining findings in terms of controlling the catalytic activity, the selectivity of desired products, products distributions, understanding the mechanism, kinetics, and/or thermodynamics of the Fischer–Tropsch (FT) processes.

2. Co-Feeding in Fischer–Tropsch Synthesis

Syngas is a mixture of carbon monoxide (CO) and hydrogen(H_2) often used as the feed to the FT process. As the feed in FTS is central to the process, a plethora of studies has been conducted on the feed composition and co-feeding of different molecules with the syngas for the FT reactions [39,44,45]. In terms of co-feeding, extensive investigations have been conducted, and some of the molecules that have been co-fed include water [1,11,45–48] additives, such as CO_2 [28,30,31,36,49,50], and hydrocarbons, such as olefins [51–56]. Although researchers [32–34,57] have published findings on nitrogen-containing compounds including ammonia, inorganic compounds, such as nitrogen gas, have not been reported on intensively. But to show the importance of co-feeding in FT reaction, already in the 1950s, scientists had investigated FT processes with both inorganic and organic compounds, to be exact CO_2, CH_4, and N_2, and they have shown that the addition of these diluents poses a negative effect on FT reaction rate [58]. Some major co-feeding effect studies are summarized in Table 1.

The co-feeding of molecules to the syngas during FTS is important. However, each of these molecules has advantages and inconveniences. As such, the choice of one molecule over another additive is dictated by the aim of the study. For example, water is an inherently oxygen-containing by-product in the FTS. Oxygen atoms furnished by the CO in the feed are mainly removed as water in FT reactions. The water may negatively affect syngas conversion as it contributes to catalyst deactivation by oxidation. In addition, the products selectivity product distribution, and secondary reactions and catalyst longevity is also affected. [46]. It is, therefore, necessary to identify the aspect of the FT process that requires improvement to devise an appropriate experiment for the co-feeding of water into the syngas for FT reactions.

In FTS, high methane selectivities are generally not preferred as it is comparatively a low-value product than higher olefins and paraffins. Recycling the formed methane back to syngas would be complex and uneconomical, therefore, suppressing its formation would be key [46]. Most studies should, therefore, focus on reducing the production of methane while increasing heavy hydrocarbons formation.

The literature findings indicate that the effect of the co-feeds, whether organic or inorganic, are almost the same regardless of the operating condition or the catalyst used (See Table 1).

Inorganic co-feeds are often incorporated in FTS to enhance selectivity to higher hydrocarbons. In the case of N_2, however, peer-reviewed publications indicate negative effects of the addition, as in the case of H_2O and CO_2. The inorganic feed H_2O source affected catalyst durability; however, overall co-feed type differences in effects were significantly different with the likelihood of affecting the FTS performance upon feeding.

Table 1. Summary of some effects of the co-feeds used in Fischer–Tropsch synthesis (FTS). The Co-feeds are grouped as organic or inorganic.

Nature of Co-Feed	Co-Feed	Catalyst Used	Reaction Conditions	Effect in FTS	Amount Used	Ref
Inorganic Co-feeds	Water	Pt (0.5%)–15%Co/Al$_2$O$_3$	2.93 MPa, and H$_2$/CO was 2.0.	Decreased CO conversion, permanent deactivation of the catalyst.	3–25 vol%	[6]
	Water	0.27%Ru–25%Co/Al$_2$O$_3$ 25%Co/γ-Al$_2$O$_3$	205–230 °C, 1.4–2.5 MPa, H$_2$/CO = 1.0–2.5, and 3–16	Reduction in the CH4 rate by 12% catalyst deactivation observed during the addition of water.	10 vol%	[10]
	CO$_2$	Co/γ-Al2O3	(H$_2$/CO = 2), fixed-bed reactor; 220 °C, 20 bar, and SV (L/kg cat/h) = 2000.	Decrease in CO conversion and C5+ selectivity partial oxidation of surface cobalt metal	20 vol%	[30]
	Nitrogen	Co/TiO$_2$	220 °C; 60 (NTP)mL/min to 75 (NTP) mL/min; 20.85 bar abs 25.85 bar abs	Reduced selectivity to the undesired light hydrocarbons (mainly CH$_4$)	28% N2	[41]
	Ammonia	100Fe/5.1Si/2.0Cu/3.0K	220–270 °C and 1.3 MPa using a 1-L slurry phase reactor.	Rapidly deactivation of catalyst simultaneously changed the product selectivities.	0.1–400 ppm	[32]
	Ammonia	iron-catalyst	250 °C, 75 mL NTP/min, H$_2$:CO = 2 and 5 bar, respectively.	The selectivities toward nitrogen-containing compounds enhanced with increasing NH$_3$ content. Rates of formation of alcohols, aldehydes, and organic acids decreased	0–10 vol%	[59]
	Ethene	62 wt% cobalt oxide and was supported on kieselguhr	473 K and 110 kPa pressure.	The selectivity of the higher hydrocarbons was improved.	1% to 2%	[16]
Organic co-feeds	Ethanol addition	10% Co/TiO$_2$ catalyst	T = 220 °C, P = 8 bar, H$_2$/CO = 2)	The selectivity to light products increased, as well as the olefin to paraffin ratio. A significant decrease in the catalyst activity.	2 vol% and 6 vol%)	[19]
	Small oxygenates	Iron catalysts. gas with a mol ratio H2:CO = 0.5	2.0 MPa; 543 K flow (VHSV = 1000)	Aldehydes suppress and entirely change normal synthesis behavior.	10 mol% dimethyl ether (DME). Diethyl ether (DEE) is 3.3 mol%. Acetaldehyde is 3 mol%	[44]
	1-olefins as additives	Co/ZrO$_2$–SiO$_2$ bimodal catalyst	513 K, 1.0 MPa, W/F Syngas = 10 g – cat h/mol.	Resulted in an anti-Anderson–Schulz–Flory (anti-ASF) product distribution. 1-decene and 1-tetradecene mixed with the volume ratio of 1:1, showed the highest selectivity to jet-fuel-like hydrocarbons. Formation rates of CH$_4$ and CO$_2$, as well as light hydrocarbons (C2–C4), suppressed	20 mol%	[60]

3. Water Co-Feeding in FTS

Water is of particular interest because the FTS produces a significant volume of water [61]. Conventional and green processes rely on accurate accounts of products and by-products [62]. An account of all large quantities of produced material should be thoroughly conducted to ensure the integration of various scenarios happening at different levels [63]. This is crucial not only for the FT reaction but also for downstream processes [64–66]. Okoye-Chine et al. [67] recently proposed that phase of H_2O impacts FT reaction in their report about the effect of water co-feeding on a cobalt-based FT catalyst. They have also proposed that the relationship between water affinity and/or resistance behavior of the supported catalyst affect the activity and selectivity, which could assist the explanation of the kinetic influence of H2O in FT reactions. Fratalocchi et al. [3] found that added water reduced the selectivity to CH_4 and alcohol and increased C_{25+} hydrocarbon, olefin, and CO_2 selectivities, which they ascribed to the suppression of hydrogenation reactions by water. The decrease of the selectivity to methane was also observed with in situ experiments designed for laboratory X-ray diffractometers in the laboratory [68]. The ability to influence FT selectivity could be useful in the design of intensified processes, such as FT reactive distillation (RD) systems [69], which would enable streams to be directed to particular trays in an RD to achieve the desired product specification on that tray.

Iglesia et al. [70] have reported that water influences chain growth, selectivity, and depending on the catalyst, the co-feeding of water also affects the rate of the initial CO hydrogenation reaction. When the pressures of the reactants or the reactants conversions are low, the addition of water improves the FTS reaction productivity (rates and selectivity) of olefins and C5+ hydrocarbons. Furthermore, at low CO conversion, the effects of water co-feeding in FTS are influenced by the type of supports and lead to support effects. The pressure is one of the most critical operating conditions of the FT process. The feed partial pressure is equally important. The effect of water co-feeding stands to affect the performance of catalysts. Scientists [1] have studied the effect of water while changing the conversion and thereby, raising the partial pressure of water. They have also investigated the influence with the co-feeding of water to the feed gas. Upon increasing the water partial pressures, a shift from the ASF distribution was observed with the FTS total product distribution, the olefin reinsertion caused typical deviations, to a much narrower distribution. The sole C_1-wise chain growth process cannot be explained by such mechanisms. As such, another product formation route which takes into consideration the combination of adjacent alkyl chains leading to paraffin ("reverse hydrogenolysis"), has, therefore, been suggested [2]. Certain features (physico–chemical properties) of the catalysts affect the catalytic activity behavior in an FT reactor. The reduction in the reaction rates was found when the water was co-fed with an inlet partial pressure ratio (P_{H2O}/P_{H2} = 0.4) for the narrow-pore catalysts. Whereas, the reaction rates increased for the larger pores catalysts with the same water pressure. Overall, when the quantity of water added was equal to P_{H2O}/P_{H2} = 0.7 at the reactor inlet, the reaction rates were suppressed, resulting in permanent deactivation. Furthermore, the addition of water improved the selectivity to C_{5+} and decreased the selectivity to CH_4 selectivity for all type of catalysts. Hence, the pore characteristics appear to define the impact of water on the rates [23]. Bertole et al. [47] added water at partial pressures amounting to 8 bar to an FT functioning unsupported cobalt catalyst and managed to increase the reactivity of CO which was adsorbed on the surface without affecting the reactivity of the active surface of carbon intermediate. An increase in surface concentration was obtained for the monomeric carbon precursors leading to hydrocarbon formation [47]. The interaction of water with the catalysts differs in many ways. The effect of water with Co particle sizes, adsorbed species on the catalyst surface, in terms of secondary reactions, and diffusion in liquid-filled is reported in the literature [1]. The complexity of the effect of water on the product distribution has also been reported [3]. It has been revealed that water has both irreversible and reversible effects. The former becomes evident upon raising the concentration of co-fed water, and the latter is clearly visible, starting from small concentrations of co-fed water. These two effects result in the decline of the selectivity to the undesired CH4 and alcohol while the selectivity to C_{25+} hydrocarbon, olefin, and those for CO_2 increase. The authors concluded that these findings can be described by an assumption that water suppresses hydrogenation reactions [3]. Another negative effect of co-feeding water to syngas is the deactivation of the catalysts. At high partial pressures, the presence of water deactivates unsupported cobalt catalysts. The deactivation of the catalyst is noticeable through the reduction of site activity of the catalyst and/or lower CO surface inventory. The treatment of the catalyst with hydrogen generally recovers site activity, and it does not affect surface inventory. The latter observation indicates that cobalt surface loss is due to sintering. This sintering process is prominently facilitated by high water partial pressure, such as >4 bar [47]. Claeys and van Steen [8] reported a noticeable increase in the formation rates of the product and significant changes in the selectivity to the FT product, in particular, lower methane selectivity and enhanced chain growth ensuing from the effect of water. The flow rate of the main reactants (CO and H_2) is another FT operating process parameter that affects the performance of the FT processes. Li et al. [71,72] investigated the effects of adding water to the feed gas by changing the values of the space velocity. They found that the addition of water did not affect the CO conversion significantly at higher syngas space velocity values. At lower space velocity, the water co-feeding decreased the CO conversion; yet, the negative effect was reversible with the catalyst quickly recovering the activity obtained before the feeding of water. When the space velocity is low, the CO conversion is high, and researchers have found that the addition of water permanently deactivates the catalyst [71,72]. The quantity of water added to the syngas feed also plays a role in the catalytic activity. For example, it was found that co-feeding of small amounts of water slightly affects the CO conversion, but interestingly the effect was reversible. However, a large amount of added water which could equal the partial pressures $P_{H2O}/P_{CO} \sim 1$ in the feed permanently deactivated the catalyst. When the selectivity to CO_2 is higher, it indicated that cobalt oxide or another catalytic form of cobalt, such as cobalt aluminate, was formed in the presence of higher water partial pressure [11]. Dalai and Davis [46] reviewed the influence of water on the performances of numerous cobalt catalysts for FTS and confirmed that the effect of co-feeding water into an FT reactor is quite complicated. They found that the influence of water addition depends on various aspects, including the catalyst support and its nature, the Co metal loading, the promoters used with noble metals, and also the preparation procedure.

It is, therefore, generally accepted by researchers [1,9,11,45–48] that the effects of water on FTS are quite complex. Dalai and Davis [46] summarised the effects of water in FTS with three scenarios:

(i) An oxidation process for the cobalt supported catalyst with the extent of oxidation being a function of features of the cobalt namely the crystallite size but also the ratio values of the reactor partial pressures of hydrogen and water (P_{H2O}/P_{H2}).
(ii) The average support pore diameter influences the water-co-feeding.
(iii) The effects could be kinetic in particular the CO dissociation by direct interaction with co-adsorbed CO can be lowered with water co-feeding while the secondary hydrogenation of olefin products can be inhibited as a result of competitive adsorption of water [46].

On the other hand, Yan et al. used Fe-Pd/ZSM-5 catalyst (a bi-functional catalyst) which yielded relatively high activity and selectivity in producing liquid hydrocarbons when the FT reaction was carried out with nitrogen-rich syngas [73]. Conversely, Visconti and Mascellaro [74] co-fed nitrogen to the reactor, increasing the nitrogen content (23.5% to 45.1%) and kept the total pressure unchanged and found that the CO conversion dropped. It is most likely that the CO conversion decreased because the reactant partial pressures values declined when adjustment was made to maintain the total pressure while increasing the nitrogen content. Nevertheless, researchers who investigated the effect FTS with nitrogen-rich syngas have reported that such FT process is feasible because it could potentially be cost-effective [39,75–78]. The addition to the FTS of molecules, namely olefins, alcohols, carbon dioxide, water, and isotope markers, has contributed to a better understanding of the reaction mechanism. These same molecules were also used by researchers who were studying the deviations observed from the ASF distribution [12].

4. Organic Co-Feeds

Small hydrocarbons are an important organic additive in FTS. The most commonly used co-feeds include acetylene (C_2H_2), ethylene (C_2H_4), ethane (C_2H_6), propyne (C_3H_4), propene (C_3H_6), and propane (C_3H_8). These hydrocarbons are needed for elucidating reaction mechanisms. These molecules exist in the gas-phase during the reaction and through a series of temperatures used in FTS.

For instance, the influence of low molecular weight olefins on FTS has been examined with the aim of either understanding the mechanism of chain growth or influencing the distribution of products formed [14]. Experiments conducted with co-feeding of ethene, 1-alkenes, and diazomethane as a source of surface methylene have been undertaken to strongly support the hypothesis of two independent mechanisms during FTS with the methylene insertion mechanism as one of them [13]. Furthermore, FT experiments with 1-alkene and ethane co-feeding in a reactor loaded with cobalt catalysts and iron catalysts were carried out to study the product distributions, chain growth probabilities, and different chain growth mechanisms [16,79,80]. In brief, olefins, alcohols or CO_2 additives have all been co-fed to investigate the chain growth and mechanism during FTS [12,14,18,81], the ASF distribution [13], and the role of secondary reactions of olefins [15,17]. Specifically, olefins have been used as co-feeds in FT which resulted in enhancing the selectivity to hydrocarbons (C_8–C_{16}), while at the same time, the formation of light hydrocarbons, such as CH_4 and CO_2 was suppressed [60].

On the other hand, when additives, such as small 1-alkenes, alcohols, or CO_2, were co-fed into the FT reactor, they acted as chain initiators for the FTS. The probe-initiated and conventional FTS progress concurrently, with their separate products overlapping. It has been shown, overwhelmingly, that co-feeding additives do not contribute to chain growth, although when olefins and alcohols were added, their incorporation into FTS products was evident usually through chain initiation [12]. However, up to now, the effects and influences of the additives reported [12] on the overall FT product distributions and on the FT catalysts activity are yet to convince the majority of researchers universally. However, Sage and Burke [12] reported in their review that olefins adsorb on the catalyst surface influenced the overall FTS product distribution. As such, it could be assumed that co-feeding olefins might be useful for FTS to modify the catalyst surface. For example, the catalyst surface modification could be achieved by selectively binding to and/or inhibiting certain active sites favorable to chain growth of higher hydrocarbons, or reactions, such as secondary hydrogenation, to improve the overall FT efficiency and selectivity of the process [12].

Alkane, such as n-hexane, has also been co-fed into the FT reaction. The addition of this molecule to the FTS feed under realistic conditions will not result in supercritical FT conditions. However, such an addition will change the composition of the liquid phase in the FT reactor leading in a higher liquid flow rate and a greater diffusivity of reactants (H_2 and CO). Therefore, the effectiveness of the process was increased if the FTS is conducted under internal mass transport limiting conditions [82]. Patzlaff et al. [79] conducted experiments with 1-alkene and ethane co-feeding using cobalt and iron catalysts and revealed superimposed distributions with different chain growth probabilities resulted from different chain growth mechanisms. When alcohols were co-fed, the results showed reliance on the ratio of the two distributions on the reactant pressures while the promoter effect on iron catalysts also supported the hypothesis of the two mechanisms. They also suggested the CO insertion mechanism as the second approach that is exhibited by the higher growth probability of the resulting ASF distribution [13].

It is known that the FT synthesis is highly exothermic, which makes the heat removal and control of the reaction temperature critical steps since the damage of the catalyst at high temperatures decreases the conversion rate. Product selectivity also shifts with increasing temperature towards the production of more unwanted short-chain hydrocarbons, including methane [75]. FT processes are also categorized based on the temperature at which the operation is being carried out, and the products obtained differ. For example, the syncrude produced during high temperature (300–350 °C) Fischer–Tropsch (HTFT) synthesis has more light hydrocarbons, hence, is gaseous at reaction conditions; more than one product phase is formed on cooling. Low temperature (200–260 °C) Fischer–Tropsch (LTFT) syncrude is acknowledged to be a two-phase mixture at reaction conditions, but at ambient conditions, the syncrude from LTFT synthesis consists of four different product phases: gaseous, organic liquid, organic solid (wax), and aqueous [83]. Medium temperature Fischer–Tropsch (MTFT) ranges from 270 to 300 °C, the product phases are similar to those of LTFT. Although FTS products are useful, it is usually more profitable to obtain fewer light hydrocarbons, such as methane and oxygenate products, which are often unwanted. It has been reported that the commercialization of the FT technology suffers from two of major limitations. These are limited selectivity for the main products since it produces a broad spectrum of products and catalyst deactivation, which leads to frequent replacement of the catalyst [39]. Cobalt-based catalysts are only used in LTFT processes as at the higher temperatures excess CH_4 is produced [84].

5. Effect of Nitrogen as a Co-Feed

In addition to the contribution of cost reduction during syngas preparation, the presence of nitrogen in the syngas is significant for safety purpose, mass balance determination, and controlling the heat in the FT system. Jess and co-workers [77] suggested a process with nitrogen co-feeding to the syngas. They reported that the proposed process with syngas co-fed does not utilize a recycle loop, hence, skipping any build-up of nitrogen in the system. This configuration is probably cost-effective because a recycle compressor is not required. The presence of the syngas with nitrogen plays a significant role which removes substantial amounts of heat generated during the FT reaction [85]. Furthermore, researchers have reported that α-olefins formation was enhanced by the two processes between FTS and N_2 purging. The improved result was achieved, when the liquid filled in pores of the catalyst purged with nitrogen, prompting the release of olefins from liquid to gas phase and minimizing its secondary reaction [86]. On the other hand, Jess et al. [75,77] and Xu et al. [78] separately conducted studies on the effect of nitrogen co-feeding in FT reactor and found that nitrogen only dilutes syngas. The presence of nitrogen in the reactor has little influence on the kinetics if the reactants (carbon monoxide and hydrogen) partial pressures are kept constant. However, Lu et al. [42] used feed composed of nitrogen-rich syngas in FTS and obtained more C_{10} hydrocarbons and smaller amounts of C_8 hydrocarbons which were more when compared to FTS carried out with pure syngas. These results are different from the findings of Jess et al. [75] and Xu et al. [78]. Lu et al. [42] findings are partially in agreement with conclusions from Muleja et al. [41] studies. Muleja et al. [41] investigated the effect of nitrogen co-feeding to an FT reactor while maintaining the partial pressures of reactants and concluded that the selectivity to all light hydrocarbons decreased while the selectivity to C_{5+} in mainly C_5–C_{19} fraction is enhanced. They [41], therefore, suggested such findings could not be explained by kinetics alone but compared the vapour–liquid equilibrium (VLE) and boiling effect and drew attention to the boiling phenomenon during FTS which is depicted in Figure 2.

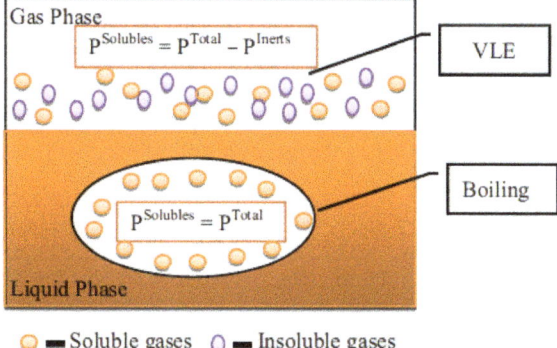

Figure 2. A schematic diagram for the comparison between the pressure in vapour–liquid equilibrium and boiling [41]. P^{Total} is the total pressure of the system, P^{Inerts} are the total partial pressures of all the virtually insoluble gases, and $P^{Solubles}$ is the total partial pressure of the soluble gases.

However, it is relatively agreed that FTS is a polymerization process leading to a distribution of mainly olefins and paraffins, although the mechanism by which this occurs on catalysts is still a subject of much debate. Todic et al. [87] reported well detailed schematic reactional representation of the FTS process as depicted in Figure 3.

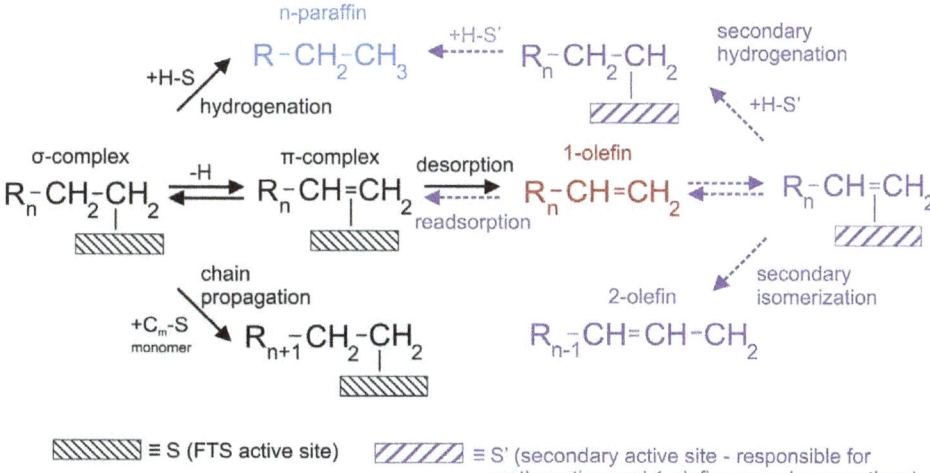

Figure 3. Schematic representation of Fischer–Tropsch synthesis (FTS) (chain propagation, hydrogenation to n-paraffins, and desorption to 1-olefins) and 1-olefin secondary reaction (hydrogenation, isomerization, and readsorption) [87].

On the other hand, van Steen and Schulz [88] have simplified the scheme representing (Figure 4) the formation of chain initiators and insertion into the developing chains. This is what is used to show the growth of rate equations using the polymerization principle.

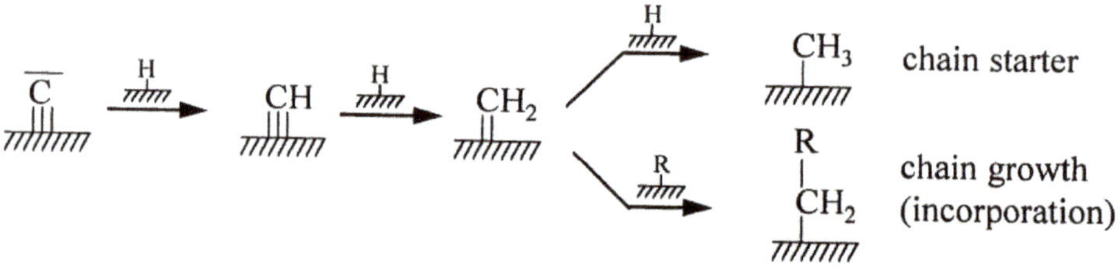

Figure 4. Simplified kinetic scheme of the successive hydrogenation of surface carbon yielding chain starters and incorporation into growing chains [88].

Furthermore, Muleja et al. [89] attempted to explain the effect of nitrogen co-feeding into the FT reactor with a reactive distillation mapping leading to the suggestion that further to kinetics, thermodynamic equilibrium and VLE should potentially be considered for FT processes modeling.

6. Concluding Remarks and Recommendations

Maximizing the yield of high-value product is a critical factor for the commercialization and successful implementation of the FT process. Coal or biomass gasification and natural gas reforming generally produce syngas that contains some short-chain hydrocarbons and impurities [90]. Some feed streams have contaminants that affect the FT reactor in various ways. The ability to integrate some contaminants, such as CO_2, into the process design might be crucial to the success of a new design. Taking advantage of N_2 that is inherent in air could save capital installation costs in eliminating the need for cryogenic air separation units. The by-product water could be used to absorb the heat of reaction upon vaporization in an intensified unit, such as an FT reactive distillation system [91]. Since olefin readsorption has been demonstrated in an FT reactor [17,92–94], the light gas stream could be recycled to various trays in a reactive distillation column to influence product selectivity [95]. Insights gained from FT co-feeding studies are not only applicable to FTS but could also be extended to methanol-to-hydrocarbons synthesis and other hydrocarbons processing systems [96]. There are some ongoing ethylene co-feeding studies to try to understand the impact on olefin selectivity over Co-based catalysts [97]. Thus, insight from co-feeding studies leads to process synthesis and integration opportunities that were not envisaged by the researchers probing various phenomena.

Author Contributions: This manuscript went through various iterations. All the authors collected articles on the topic, reviewed them, and edited each other's contributions on the manuscript.

Funding: This research was partially funded by the National Research Foundation of South Africa, Grant Numbers: 113652 and 120270.

Acknowledgments: We gratefully acknowledge the partial financial support from the University of South Africa and the National Research Foundation of South Africa Grant Numbers: 113652 and 120270. The opinions, findings, and conclusions or recommendations expressed in this publication are those of the authors. The funding bodies accept no liability whatsoever in this regard.

Conflicts of Interest: The authors declare no conflict of interest.

References

1. Storsæter, S.; Borg, Ø.; Blekkan, E.A.; Holmen, A. Study of the effect of water on Fischer–Tropsch synthesis over supported cobalt catalysts. *J. Catal.* **2005**, *231*, 405–419. [CrossRef]
2. Claeys, M.; van Steen, E. On the effect of water during Fischer–Tropsch synthesis with a ruthenium catalyst. *Catal. Today* **2002**, *71*, 419–427. [CrossRef]
3. Fratalocchi, L.; Visconti, C.G.; Lietti, L.; Groppi, G.; Tronconi, E.; Roccaro, E.; Zennaro, R. On the performance of a Co-based catalyst supported on modified γ-Al 2 O 3 during Fischer–Tropsch synthesis in the presence of co-fed water. *Catal. Sci. Technol.* **2016**, *6*, 6431–6440. [CrossRef]

4. Krishnamoorthy, S.; Tu, M.; Ojeda, M.P.; Pinna, D.; Iglesia, E. An investigation of the effects of water on rate and selectivity for the Fischer–Tropsch synthesis on cobalt-based catalysts. *J. Catal.* **2002**, *211*, 422–433. [CrossRef]
5. Sadeqzadeh, M.; Chambrey, S.; Piché, S.; Fongarland, P.; Luck, F.; Curulla-Ferré, D.; Khodakov, A.Y. Deactivation of a Co/Al$_2$O$_3$ Fischer–Tropsch catalyst by water-induced sintering in slurry reactor: Modeling and experimental investigations. *Catal. Today* **2013**, *215*, 52–59. [CrossRef]
6. Li, J.; Davis, B.H. Effect of water on the catalytic properties of supported cobalt Fischer-Tropsch catalysts. In *Studies in Surface Science and Catalysis*; Elsevier: Amsterdam, The Netherlands, 2004; Volume 147, pp. 307–312.
7. Ma, W.; Jacobs, G.; Sparks, D.E.; Spicer, R.L.; Davis, B.H.; Klettlinger, J.L.; Yen, C.H. Fischer–Tropsch synthesis: Kinetics and water effect study over 25% Co/Al$_2$O$_3$ catalysts. *Catal. Today* **2014**, *228*, 158–166. [CrossRef]
8. Rytter, E.; Borg, Ø.; Enger, B.C.; Holmen, A. α-alumina as catalyst support in Co Fischer-Tropsch synthesis and the effect of added water; encompassing transient effects. *J. Catal.* **2019**, *373*, 13–24. [CrossRef]
9. Lögdberg, S.; Boutonnet, M.; Walmsley, J.C.; Järås, S.; Holmen, A.; Blekkan, E.A. Effect of water on the space-time yield of different supported cobalt catalysts during Fischer–Tropsch synthesis. *Appl. Catal. A Gen.* **2011**, *393*, 109–121. [CrossRef]
10. Ma, W.; Jacobs, G.; Das, T.K.; Masuku, C.M.; Kang, J.; Pendyala, V.R.R.; Yen, C.H. Fischer–Tropsch Synthesis: Kinetics and Water Effect on Methane Formation over 25% Co/γ-Al$_2$O$_3$ Catalyst. *Ind. Eng. Chem. Res.* **2014**, *53*, 2157–2166. [CrossRef]
11. Li, J.; Zhan, X.; Zhang, Y.; Jacobs, G.; Das, T.; Davis, B.H. Fischer–Tropsch synthesis: Effect of water on the deactivation of Pt promoted Co/Al2O3 catalysts. *Appl. Catal. A Gen.* **2002**, *228*, 203–212. [CrossRef]
12. Sage, V.; Burke, N. Use of probe molecules for Fischer-Tropsch mechanistic investigations: A short review. *Catal. Today* **2011**, *178*, 137–141. [CrossRef]
13. Gaube, J.; Klein, H.F. Studies on the reaction mechanism of the Fischer-Tropsch synthesis on iron and cobalt. *J. Mol. Catal. A Chem.* **2008**, *283*, 60–68. [CrossRef]
14. Jordan, D.S.; Bell, A.T. Influence of ethylene on the hydrogenation of carbon monoxide over ruthenium. *J. Phys. Chem.* **1986**, *90*, 4797–4805. [CrossRef]
15. Boelee, J.H.; Cüsters, J.M.G.; Van Der Wiele, K. Influence of reaction conditions on the effect of Co-feeding ethene in the Fischer-Tropsch synthesis on a fused-iron catalyst in the liquid phase. *Appl. Catal.* **1989**, *53*, 1–13. [CrossRef]
16. Adesina, A.A.; Hudgins, R.R.; Silveston, P.L. Effect of ethene addition during the Fischer-Tropsch reaction. *Appl. Catal.* **1990**, *62*, 295–308. [CrossRef]
17. Schulz, H.; Claeys, M. Reactions of α-olefins of different chain length added during Fischer–Tropsch synthesis on a cobalt catalyst in a slurry reactor. *Appl. Catal. A Gen.* **1999**, *186*, 71–90. [CrossRef]
18. Kummer, J.T.; Emmett, P.H. Fischer—Tropsch synthesis mechanism studies. The addition of radioactive alcohols to the synthesis gas. *J. Am. Chem. Soc.* **1953**, *75*, 5177–5183. [CrossRef]
19. Jalama, K.; Coville, N.J.; Hildebrandt, D.; Glasser, D.; Jewell, L.L. Fischer–Tröpsch synthesis over Co/TiO2: Effect of ethanol addition. *Fuel* **2007**, *86*, 73–80. [CrossRef]
20. Jacobs, G.; Patterson, P.M.; Das, T.K.; Luo, M.; Davis, B.H. Fischer-Tropsch synthesis: Effect of water on Co/Al2O3 catalysts and XAFS characterization of reoxidation phenomena. *Appl. Catal. A Gen.* **2004**, *270*, 65–76. [CrossRef]
21. Rytter, E.; Borg, Ø.; Tsakoumis, N.E.; Holmen, A. Water as key to activity and selectivity in Co Fischer-Tropsch synthesis: γ-alumina based structure-performance relationships. *J. Catal.* **2018**, *365*, 334–343. [CrossRef]
22. Rytter, E.; Holmen, A. Perspectives on the effect of water in cobalt Fischer-Tropsch synthesis. *ACS Catal.* **2017**, *7*, 5321–5328. [CrossRef]
23. Borg, Ø.; Storsæter, S.; Eri, S.; Wigum, H.; Rytter, E.; Holmen, A. The effect of water on the activity and selectivity for γ-alumina supported cobalt Fischer-Tropsch catalysts with different pore sizes. *Catal. Lett.* **2006**, *107*, 95–102. [CrossRef]
24. Hibbitts, D.D.; Loveless, B.T.; Neurock, M.; Iglesia, E. Mechanistic role of water on the rate and selectivity of Fischer–Tropsch synthesis on ruthenium catalysts. *Angew. Chem. Int. Ed.* **2013**, *52*, 12273–12278. [CrossRef] [PubMed]
25. Yang, J.; Ma, W.; Chen, D.; Holmen, A.; Davis, B.H. Fischer-Tropsch synthesis: A review of the effect of CO conversion on methane selectivity. *Appl. Catal. A Gen.* **2014**, *470*, 250–260. [CrossRef]

26. Borg, Ø.; Yu, Z.; Chen, D.; Blekkan, E.A.; Rytter, E.; Holmen, A. The effect of water on the activity and selectivity for carbon nanofiber supported cobalt Fischer–tropsch catalysts. *Top. Catal.* **2014**, *57*, 491–499. [CrossRef]

27. Sirikulbodee, P.; Ratana, T.; Sornchamni, T.; Phongaksorn, M.; Tungkamani, S. Catalytic performance of Iron-based catalyst in Fischer–Tropsch synthesis using CO_2 containing syngas. *Energy Procedia* **2017**, *138*, 998–1003. [CrossRef]

28. Yao, Y.; Liu, X.; Hildebrandt, D.; Glasser, D. The effect of CO2 on a cobalt-based catalyst for low temperature Fischer–Tropsch synthesis. *Chem. Eng. J.* **2012**, *193*, 318–327. [CrossRef]

29. James, O.O.; Mesubi, A.M.; Ako, T.C.; Maity, S. Increasing carbon utilization in Fischer–Tropsch synthesis using H2-deficient or CO2-rich syngas feeds. *Fuel Process. Technol.* **2010**, *91*, 136–144. [CrossRef]

30. Kim, S.M.; Bae, J.W.; Lee, Y.J.; Jun, K.W. Effect of CO2 in the feed stream on the deactivation of Co/γ-Al2O3 Fischer–Tropsch catalyst. *Catal. Commun.* **2008**, *9*, 2269–2273. [CrossRef]

31. Yao, Y.; Hildebrandt, D.; Glasser, D.; Liu, X. Fischer– Tropsch synthesis using H2/CO/CO2 syngas mixtures over a cobalt catalyst. *Ind. Eng. Chem. Res.* **2010**, *49*, 11061–11066. [CrossRef]

32. Ma, W.; Jacobs, G.; Sparks, D.E.; Pendyala, V.R.R.; Hopps, S.G.; Thomas, G.A.; Davis, B.H. Fischer–Tropsch synthesis: Effect of ammonia in syngas on the Fischer–Tropsch synthesis performance of a precipitated iron catalyst. *J. Catal.* **2015**, *326*, 149–160. [CrossRef]

33. Pendyala, V.R.R.; Gnanamani, M.K.; Jacobs, G.; Ma, W.; Shafer, W.D.; Davis, B.H. Fischer–Tropsch synthesis: Effect of ammonia impurities in syngas feed over a cobalt/alumina catalyst. *Appl. Catal. A Gen.* **2013**, *468*, 38–43. [CrossRef]

34. Sango, T.; Fischer, N.; Henkel, R.; Roessner, F.; van Steen, E.; Claeys, M. Formation of nitrogen containing compounds from ammonia co-fed to the Fischer–Tropsch synthesis. *Appl. Catal. A Gen.* **2015**, *502*, 150–156. [CrossRef]

35. Steynberg, A.P.; Deshmukh, S.R.; Robota, H.J. Fischer-Tropsch catalyst deactivation in commercial microchannel reactor operation. *Catal. Today* **2018**, *299*, 10–13. [CrossRef]

36. Díaz, J.A.; de la Osa, A.R.; Sánchez, P.; Romero, A.; Valverde, J.L. Influence of CO2 co-feeding on Fischer–Tropsch fuels production over carbon nanofibers supported cobalt catalyst. *Catal. Commun.* **2014**, *44*, 57–61. [CrossRef]

37. Liao, P.Y.; Zhang, C.; Zhang, L.J.; Yang, Y.Z.; Zhong, L.S.; Wang, H.; Sun, Y.H. Effect of promoter and CO2 content in the feed on the performance of CuFeZr catalyst in the synthesis of higher alcohol from syngas. *J. Fuel Chem. Technol.* **2017**, *45*, 547–555. [CrossRef]

38. Liu, Y.; Zhang, C.H.; Wang, Y.; Li, Y.; Hao, X.; Bai, L.; Li, Y.W. Effect of co-feeding carbon dioxide on Fischer–Tropsch synthesis over an iron–manganese catalyst in a spinning basket reactor. *Fuel Process. Technol.* **2008**, *89*, 234–241. [CrossRef]

39. Lu, Y.; Lee, T. Influence of the feed gas composition on the Fischer-Tropsch synthesis in commercial operations. *J. Nat. Gas Chem.* **2007**, *16*, 329–341. [CrossRef]

40. Daramola, M.O.; Matamela, K.; Sadare, O.O. Effect of CO2 co-feeding on the conversion of syngas derived from waste to liquid fuel over a bi-functional Co/H-ZSM-5 catalyst. *J. Environ. Chem. Eng.* **2017**, *5*, 54–62. [CrossRef]

41. Muleja, A.A.; Yao, Y.; Glasser, D.; Hildebrandt, D. Effect of feeding nitrogen to a fixed bed Fischer–Tropsch reactor while keeping the partial pressures of reactants the same. *Chem. Eng. J.* **2016**, *293*, 151–160. [CrossRef]

42. Lu, Y.; Hu, J.; Han, J.; Yu, F. Synthesis of gasoline-range hydrocarbons from nitrogen-rich syngas over a Mo/HZSM-5 bi-functional catalyst. *J. Energy Inst.* **2016**, *89*, 782–792. [CrossRef]

43. Muleja, A.A.; Yao, Y.; Glasser, D.; Hildebrandt, D. Variation of the short-chain paraffin and olefin formation rates with time for a cobalt Fischer–Tropsch catalyst. *Ind. Eng. Chem. Res.* **2017**, *56*, 469–478. [CrossRef]

44. Snel, R.; Espinoza, R.L. Fischer-tropsch synthesis on iron-based catalysts: The effect of co-feeding small oxygenates. *J. Mol. Catal.* **1989**, *54*, 213–223. [CrossRef]

45. Sarup, B.; Wojciechowski, B.W. The effect of time on stream and feed composition on the selectivity of a cobalt fischer—Tropsch catalyst. *Can. J. Chem. Eng.* **1984**, *62*, 249–256. [CrossRef]

46. Dalai, A.K.; Davis, B.H. Fischer–Tropsch synthesis: A review of water effects on the performances of unsupported and supported Co catalysts. *Appl. Catal. A Gen.* **2008**, *348*, 1–15. [CrossRef]

47. Bertole, C.J.; Mims, C.A.; Kiss, G. The effect of water on the cobalt-catalyzed Fischer–Tropsch synthesis. *J. Catal.* **2002**, *210*, 84–96. [CrossRef]

48. Rothaemel, M.; Hanssen, K.F.; Blekkan, E.A.; Schanke, D.; Holmen, A. The effect of water on cobalt Fischer-Tropsch catalysts studied by steady-state isotopic transient kinetic analysis (SSITKA). *Catal. Today* **1997**, *38*, 79–84. [CrossRef]
49. Gnanamani, M.K.; Shafer, W.D.; Sparks, D.E.; Davis, B.H. Fischer–Tropsch synthesis: Effect of CO_2 containing syngas over Pt promoted Co/γ-Al_2O_3 and K-promoted Fe catalysts. *Catal. Commun.* **2011**, *12*, 936–939. [CrossRef]
50. Yao, Y.; Liu, X.; Hildebrandt, D.; Glasser, D. Fischer–Tropsch synthesis using $H_2/CO/CO_2$ syngas mixtures: A comparison of paraffin to olefin ratios for iron and cobalt based catalysts. *Appl. Catal. A Gen.* **2012**, *433*, 58–68. [CrossRef]
51. Zhang, R.; Hao, X.; Li, Y. Investigation of acetylene addition to Fischer–Tropsch Synthesis. *Catal. Commun.* **2011**, *12*, 1146–1148. [CrossRef]
52. Kuipers, E.W.; Vinkenburg, I.H.; Oosterbeek, H. Chain length dependence of α-olefin readsorption in Fischer-Tropsch synthesis. *J. Catal.* **1995**, *152*, 137–146. [CrossRef]
53. Iglesia, E.; Reyes, S.C.; Madon, R.J. Transport-enhanced α-olefin readsorption pathways in Ru-catalyzed hydrocarbon synthesis. *J. Catal.* **1991**, *129*, 238–256. [CrossRef]
54. Snel, R.; Espinoza, R.L. Secondary reactions of primary products of the Fischer-Tropsch synthesis: Part 1. The role of ethene. *J. Mol. Catal.* **1987**, *43*, 237–247. [CrossRef]
55. Fan, L.; Fujimoto, K. Fischer–Tropsch synthesis in supercritical fluid: Characteristics and application. *Appl. Catal. A Gen.* **1999**, *186*, 343–354. [CrossRef]
56. Snel, R.; Espinoza, R.L. Secondary reactions of primary products of the fischer-tropsch synthesis: Part II. The role of propene. *J. Mol. Catal.* **1989**, *54*, 103–117. [CrossRef]
57. Henkel, R. The Influence of Ammonia on Fischer-Tropsch Synthesis and Formation of N-Containing Compounds. Ph.D. Thesis, Universität Oldenburg, Oldenburg, Germany, 2013.
58. Gibson, E.J.; Hall, C.C. Fischer-tropsch synthesis with cobalt catalysts. II. The effect of nitrogen, carbon dioxide and methane in the synthesis gas. *J. Appl. Chem.* **1954**, *4*, 464–468. [CrossRef]
59. Ordomsky, V.V.; Carvalho, A.; Legras, B.; Paul, S.; Virginie, M.; Sushkevich, V.L.; Khodakov, A.Y. Effects of co-feeding with nitrogen-containing compounds on the performance of supported cobalt and iron catalysts in Fischer–Tropsch synthesis. *Catal. Today* **2016**, *275*, 84–93. [CrossRef]
60. Li, J.; Yang, G.; Yoneyama, Y.; Vitidsant, T.; Tsubaki, N. Jet fuel synthesis via Fischer–Tropsch synthesis with varied 1-olefins as additives using Co/ZrO_2–SiO_2 bimodal catalyst. *Fuel* **2016**, *171*, 159–166. [CrossRef]
61. Lu, X.; Zhu, X.; Masuku, C.M.; Hildebrandt, D.; Glasser, D. A study of the Fischer–Tropsch synthesis in a batch reactor: Rate, phase of water, and catalyst oxidation. *Energy Fuels* **2017**, *31*, 7405–7412. [CrossRef]
62. Seadira, T.; Sadanandam, G.; Ntho, T.A.; Lu, X.; Masuku, C.M.; Scurrell, M. Hydrogen production from glycerol reforming: Conventional and green production. *Rev. Chem. Eng.* **2018**, *34*, 695–726. [CrossRef]
63. Masuku, C.M.; Biegler, L.T. Recent advances in gas-to-liquids process intensification with emphasis on reactive distillation. *Curr. Opin. Chem. Eng.* **2019**. [CrossRef]
64. Eze, P.C.; Masuku, C.M. Vapour–liquid equilibrium prediction for synthesis gas conversion using artificial neural networks. *S. Afr. J. Chem. Eng.* **2018**, *26*, 80–85. [CrossRef]
65. Eze, P.C.; Masuku, C.M. Supporting plots and tables on vapour–liquid equilibrium prediction for synthesis gas conversion using artificial neural networks. *Data Brief* **2018**, *21*, 1435–1444. [CrossRef] [PubMed]
66. Zhang, Y.; Masuku, C.M.; Biegler, L.T. Equation-oriented framework for optimal synthesis of integrated reactive distillation systems for Fischer–Tropsch processes. *Energy Fuels* **2018**, *32*, 7199–7209. [CrossRef]
67. Okoye-Chine, C.G.; Moyo, M.; Liu, X.; Hildebrandt, D. A critical review of the impact of water on cobalt-based catalysts in Fischer-Tropsch synthesis. *Fuel Process. Technol.* **2019**, *192*, 105–129. [CrossRef]
68. Fischer, N.; Clapham, B.; Feltes, T.; Claeys, M. Cobalt-based Fischer–Tropsch activity and selectivity as a function of crystallite size and water partial pressure. *ACS Catal.* **2014**, *5*, 113–121. [CrossRef]
69. Masuku, C.M.; Lu, X.; Hildebrandt, D.; Glasser, D. Reactive distillation in conventional Fischer–Tropsch reactors. *Fuel Process. Technol.* **2015**, *130*, 54–61. [CrossRef]
70. Iglesia, E. Design, synthesis, and use of cobalt-based Fischer-Tropsch synthesis catalysts. *Appl. Catal. A Gen.* **1997**, *161*, 59–78. [CrossRef]
71. Li, J.; Jacobs, G.; Das, T.; Davis, B.H. Fischer–Tropsch synthesis: Effect of water on the catalytic properties of a ruthenium promoted Co/TiO_2 catalyst. *Appl. Catal. A Gen.* **2002**, *233*, 255–262. [CrossRef]

72. Li, J.; Jacobs, G.; Das, T.; Zhang, Y.; Davis, B. Fischer–Tropsch synthesis: Effect of water on the catalytic properties of a Co/SiO2 catalyst. *Appl. Catal. A Gen.* **2002**, *236*, 67–76. [CrossRef]
73. Yan, Q.; Yu, F.; Cai, Z.; Zhang, J. Catalytic upgrading nitrogen-riched wood syngas to liquid hydrocarbon mixture over a Fe–Pd/ZSM-5 catalyst. *Biomass Bioenergy* **2012**, *47*, 469–473. [CrossRef]
74. Visconti, C.G.; Mascellaro, M. Calculating the product yields and the vapor–liquid equilibrium in the low-temperature Fischer–Tropsch synthesis. *Catal. Today* **2013**, *214*, 61–73. [CrossRef]
75. Jess, A.; Popp, R.; Hedden, K. Fischer–Tropsch-synthesis with nitrogen-rich syngas: fundamentals and reactor design aspects. *Appl. Catal. A Gen.* **1999**, *186*, 321–342. [CrossRef]
76. Dai, X.; Yu, C. Characterization and catalytic performance of CeO2-Co/SiO2 catalyst for Fischer-Tropsch synthesis using nitrogen-diluted synthesis gas over a laboratory scale fixed-bed reactor. *J. Nat. Gas Chem.* **2008**, *17*, 17–23. [CrossRef]
77. Jess, A.; Hedden, K.; Popp, R. Diesel Oil from Natural Gas by Fischer-Tropsch Synthesis Using Nitrogen-Rich Syngas. *Chem. Eng. Technol. Ind. Chem. Plant Equip. Process Eng. Biotechnol.* **2001**, *24*, 27–31. [CrossRef]
78. Xu, D.; Duan, H.; Li, W.; Xu, H. Investigation on the Fischer–Tropsch Synthesis with nitrogen-containing syngas over CoPtZrO$_2$/Al$_2$O$_3$ catalyst. *Energy Fuels* **2006**, *20*, 955–958. [CrossRef]
79. Patzlaff, J.; Liu, Y.; Graffmann, C.; Gaube, J. Studies on product distributions of iron and cobalt catalyzed Fischer–Tropsch synthesis. *Appl. Catal. A Gen.* **1999**, *186*, 109–119. [CrossRef]
80. Patzlaff, J.; Liu, Y.; Graffmann, C.; Gaube, J. Interpretation and kinetic modeling of product distributions of cobalt catalyzed Fischer–Tropsch synthesis. *Catal. Today* **2002**, *71*, 381–394. [CrossRef]
81. Hanlon, R.T.; Satterfield, C.N. Reactions of selected 1-olefins and ethanol added during the Fischer-Tropsch synthesis. *Energy Fuels* **1988**, *2*, 196–204. [CrossRef]
82. Biquiza, L.D.; Claeys, M.; Van Steen, E. Thermodynamic and experimental aspects of 'supercritical' Fischer–Tropsch synthesis. *Fuel Process. Technol.* **2010**, *91*, 1250–1255. [CrossRef]
83. De Klerk, A. Fischer–Tropsch refining: Technology selection to match molecules. *Green Chem.* **2008**, *10*, 1249–1279. [CrossRef]
84. Dry, M.E. The fischer–tropsch process: 1950–2000. *Catal. Today* **2002**, *71*, 227–241. [CrossRef]
85. Jess, A.; Popp, R.; Hedden, K. From natural gas to liquid hydrocarbons. Pt. 4. Production of diesel oil and wax by Fischer-Tropsch-Synthesis using a nitrogen-rich synthesis gas-investigations on a semi-technical scale. *Erdoel Erdgas Kohle* **1997**, *113*, 531–540.
86. Shi, H.X.; Li, Z.K.; Liu, K.F.; Xiao, H.C.; Kong, F.H.; Zhang, J.; Chen, J.G. Enhanced formation of α-olefins by the pulse process between Fischer-Tropsch synthesis and N2 purging. *J. Fuel Chem. Technol.* **2016**, *44*, 822–829. [CrossRef]
87. Todic, B.; Ma, W.; Jacobs, G.; Davis, B.H.; Bukur, D.B. Effect of process conditions on the product distribution of Fischer–Tropsch synthesis over a Re-promoted cobalt-alumina catalyst using a stirred tank slurry reactor. *J. Catal.* **2014**, *311*, 325–338. [CrossRef]
88. Van Steen, E.; Schulz, H. Polymerisation kinetics of the Fischer–Tropsch CO hydrogenation using iron and cobalt based catalysts. *Appl. Catal. A Gen.* **1999**, *186*, 309–320. [CrossRef]
89. Muleja, A.A.; Yao, Y.; Glasser, D.; Hildebrandt, D. A study of Fischer-Tropsch synthesis: Product distribution of the light hydrocarbons. *Appl. Catal. A Gen.* **2016**, *517*, 217–226. [CrossRef]
90. Kapfunde, N.; Masuku, C.M.; Hildebrandt, D. Optimization of the thermal efficiency of a fixed-bed gasifier using computational fluid dynamics. In *Computer Aided Chemical Engineering*; Elsevier: Amsterdam, The Netherlands, 2018; Volume 44, pp. 1747–1752.
91. He, N.; Hu, Y.; Masuku, C.M.; Biegler, L.T. 110th Anniversary: Fischer–Tropsch Synthesis for Multiphase Product Recovery through Reactive Distillation. *Ind. Eng. Chem. Res.* **2019**, *58*, 13249–13259. [CrossRef]
92. Masuku, C.M.; Hildebrandt, D.; Glasser, D.; Davis, B.H. Steady-state attainment period for Fischer–Tropsch products. *Top. Catal.* **2014**, *57*, 582–587. [CrossRef]
93. McNab, A.I.; McCue, A.J.; Dionisi, D.; Anderson, J.A. Combined quantitative FTIR and online GC study of Fischer-Tropsch synthesis involving co-fed ethylene. *J. Catal.* **2018**, *362*, 10–17. [CrossRef]
94. Hutchings, G.J.; Copperthwaite, R.G.; van der Riet, M. Low methane selectivity using Co/MnO catalysts for the Fischer-Tropsch reaction: Effect of increasing pressure and Co-feeding ethene. *Top. Catal.* **1995**, *2*, 163–172. [CrossRef]

95. Zhang, Y.; Masuku, C.M.; Biegler, L.T. An MPCC Reactive Distillation Optimization Model for Multi-Objective Fischer–Tropsch Synthesis. In *Computer Aided Chemical Engineering*; Elsevier: Amsterdam, The Netherlands, 2019; Volume 46, pp. 451–456.
96. Ntelane, T.S.; Masuku, C.M.; Scurrell, M.S. TPSR study: Effect of microwave radiation on the physicochemical properties of Fe/ZSM-5 in the methanol to hydrocarbons (MTH) process. *J. Porous Mater.* **2019**. [CrossRef]
97. Yang, J.; Ledesma Rodriguez, C.; Holmen, A.; Chen, D. Ethylene co-feeding effect on Fischer–Tropsch synthesis to olefin over Co based catalysts. In Proceedings of the American Chemical Society National Meeting & Expo, San Diego, CA, USA, 25–29 August 2019.

© 2019 by the authors. Licensee MDPI, Basel, Switzerland. This article is an open access article distributed under the terms and conditions of the Creative Commons Attribution (CC BY) license (http://creativecommons.org/licenses/by/4.0/).

Article

Fischer–Tropsch: Product Selectivity–The Fingerprint of Synthetic Fuels

Wilson D. Shafer [1,*], Muthu Kumaran Gnanamani [2], Uschi M. Graham [3], Jia Yang [4], Cornelius M. Masuku [5], Gary Jacobs [6] and Burtron H. Davis [2]

1. Asbury University, One Macklem Drive, Wilmore, KY 40390, USA
2. Center for Applied Energy Research, University of Kentucky, 2540 Research Park Dr., Lexington, KY 40511, USA; muthu.gnanamani@uky.edu (M.K.G.); burtron.davis@uky.edu (B.H.D.)
3. Faraday Energy, 1525 Bull Lea Rd. Suite # 5, Lexington, KY 40511, USA; graham@topasol.com
4. Department of Chemical Engineering, Norwegian University of Science and Technology, N-7491 Trondheim, Norway; jia.yang@ntnu.no
5. Department of Civil and Chemical Engineering, University of South Africa, Private Bag X6, Florida 1710, South Africa; masukcm@unisa.ac.za
6. Chemical Engineering Program—Department of Biomedical Engineering, and Department of Mechanical Engineering University of Texas at San Antonio, 1 UTSA Circle, San Antonio, TX 78249, USA; gary.jacobs@utsa.edu
* Correspondence: wilson.shafer@asbury.edu

Received: 16 February 2019; Accepted: 7 March 2019; Published: 14 March 2019

Abstract: The bulk of the products that were synthesized from Fischer–Tropsch synthesis (FTS) is a wide range (C_1–C_{70+}) of hydrocarbons, primarily straight-chained paraffins. Additional hydrocarbon products, which can also be a majority, are linear olefins, specifically: 1-olefin, *trans*-2-olefin, and *cis*-2-olefin. Minor hydrocarbon products can include isomerized hydrocarbons, predominantly methyl-branched paraffin, cyclic hydrocarbons mainly derived from high-temperature FTS and internal olefins. Combined, these products provide 80–95% of the total products (excluding CO_2) generated from syngas. A vast number of different oxygenated species, such as aldehydes, ketones, acids, and alcohols, are also embedded in this product range. These materials can be used to probe the FTS mechanism or to produce alternative chemicals. The purpose of this article is to compare the product selectivity over several FTS catalysts. Discussions center on typical product selectivity of commonly used catalysts, as well as some uncommon formulations that display selectivity anomalies. Reaction tests were conducted while using an isothermal continuously stirred tank reactor. Carbon mole percentages of CO that are converted to specific materials for Co, Fe, and Ru catalysts vary, but they depend on support type (especially with cobalt and ruthenium) and promoters (especially with iron). All three active metals produced linear alcohols as the major oxygenated product. In addition, only iron produced significant selectivities to acids, aldehydes, and ketones. Iron catalysts consistently produced the most isomerized products of the catalysts that were tested. Not only does product selectivity provide a fingerprint of the catalyst formulation, but it also points to a viable proposed mechanistic route.

Keywords: Fischer–Tropsch synthesis (FTS); oxygenates; iron; cobalt; ruthenium; Anderson-Schulz-Flory (ASF) distribution

1. Introduction

1.1. The Heart of the X-to-Liquids Process

Fischer–Tropsch synthesis (FTS) is a pseudo-polymerization synthesis that is a key step in Gas-to-Liquids (GTL) [1–3], Coal-to-Liquids (CTL) [4–7], and Biomass-to-Liquids (BTL) [8–10]

processes, as displayed in Figure 1. Briefly, there are three main steps that are undertaken to produce materials. The first step is syngas production from natural resources, primarily materials that are of local natural abundance. To generate the raw syngas, these raw materials are first processed by gasification (in the case of coal or biomass) or steam reforming/partial oxidation (in the case of natural gas). The actual chemistry depends on which raw materials are to be locally utilized. The overall chemical reactions are provided below:

$$CTL/BTL = C + H_2O \rightleftharpoons CO + H_2 \text{ at a } H_2/CO \text{ ratio of } 0.7$$

$$GTL = CH_4 + H_2O \rightleftharpoons CO + 3 H_2$$

Given the complexities of natural resources, these equations do not include all of the extraneous components, such as heteroatoms (i.e., sulfur, nitrogen), metals, and CO_2. Thus, to obtain the needed pure syngas, a cleanup procedure is paired with the gasification process [11,12]. Furthermore, depending on the natural resource used, the off-gas from gasification needs to be treated by a versatile system, because a multitude of components must be eliminated, including: various alkali metals, transition metals, sulfur compounds (i.e., COS, H_2S), CO_2, nitrogen compounds (i.e., HCN, NH_3), and hydrohalic acids (i.e., HCl, HBr). Sulfur elimination is especially important, as it irreversibly binds to the FTS catalyst, poisoning active sites, and altering selectivity.

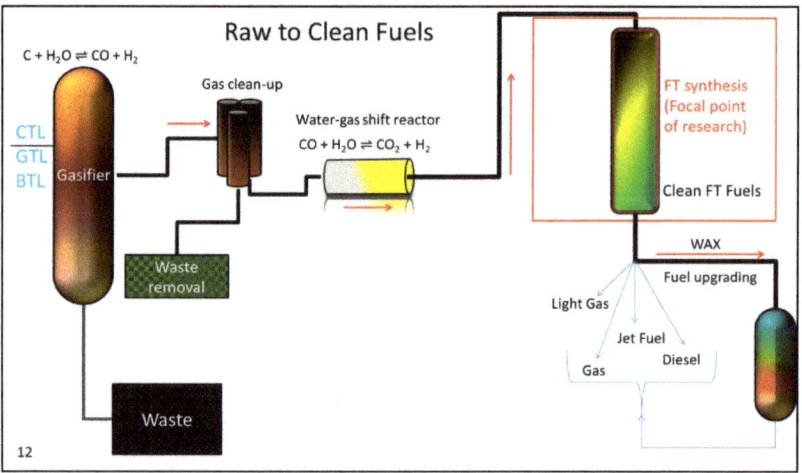

Figure 1. A simplified diagram of the Coal-to-Liquids (CTL), Gas-to-Liquids (GTL), and Biomass-to-Liquids (BTL) processes.

In the case of low H_2/CO syngas (i.e., from coal and/or biomass), water-gas shift (WGS):

$$WGS \quad H_2O + CO \rightleftharpoons H_2 + CO_2 \quad \Delta H = -41 \text{ kJ/mol}$$

may be used to increase the H_2/CO ratio [13]. Iron carbide catalyst is often used in this case, as it is active for both WGS and FTS, which are run simultaneously with a H_2/CO ratio that can be as low as ~0.7–1.0 for coal-derived syngas [14–16]. The last step is the catalytic upgrading of these hydrocarbons by the hydrocracking of waxes to target the desired hydrocarbon range of interest (e.g., gasoline, jet fuels, diesel, lubricants, etc.) [17].

1.2. Fischer–Tropsch Synthesis

At moderate pressures and temperatures, FTS is employed to convert the synthesis gas to a range of hydrocarbons, which, once upgraded, can be utilized as high quality, almost sulfur-free, fuels (namely diesel and jet fuels), lubricants, waxes, and chemicals [18,19]. The main representative reactions are given by:

$$\text{n-alkane production: } (2n + 1)H_2 \text{ (g)} + nCO \text{ (g)} \rightarrow C_nH_{2n+2} + nH_2O$$

$$\text{alkene production: } (2n)H_2 \text{ (g)} + nCO \text{ (g)} \rightarrow C_nH_{2n} + nH_2O$$

The product slate that is synthesized by FTS is remarkable, ranging from methane up to $C_{70}+$ [20,21]. Products vary based on reaction conditions (i.e., temperature, pressure, space velocity, and H_2/CO ratio), reactor type, and catalyst formulation. Though the identity of the C_1 monomer is still under scrutiny [22–25], FTS is generally assumed to follow a polymerization reaction, whereby:

1. reactants (H_2 and CO) chemisorb on active sites;
2. chain growth is initiated;
3. the carbon chain is then propagated;
4. the chain is terminated; and,
5. the final product desorbs from the catalyst surface (with the possibility of minor readsorption and reincorporation [26]).

1.3. Active FTS Metals

As of now, only four transitional metals are considered active for FTS: cobalt, iron, ruthenium, and nickel. Given that nickel has been plagued as a methanation catalyst, and simply due to the price of ruthenium, only cobalt and iron have typically been considered for industrial use to date [27,28]. Iron is the most abundant metal, and it is thus less expensive than cobalt. Iron is also preferred for industrial applications, where coal is the natural resource used to produce the syngas. The low level of H_2 in the syngas is due to the high abundance of naphthalene-based materials, molecules that contain more carbon than hydrogen. Iron is also different in that, as understood today, iron carbide, not the metal itself, is the active species. In addition, iron carbide alone is a very low-α (i.e., where α is chain growth probability) catalyst, and normally an alkali promoter, such as potassium, is added to increase the dissociation rate of CO, allowing for α to increase [29,30]. Copper is normally added as a promoter as well, to enhance the reducibility of iron oxides during activation [31].

Cobalt, the other active metal that is utilized in industry for FTS, is normally paired with the GTL process. Cobalt, unlike iron, possesses very low intrinsic WGS activity, and it is thus more useful for higher H_2/CO ratio syngas, such as that derived from natural gas. Because cobalt is expensive, it is typically supported on materials, such as Al_2O_3, SiO_2, and TiO_2, in order to produce nanoparticles that improve the fraction of exposed surface cobalt atoms [32]. The control of interactions between the active cobalt and the support is a balancing act. Poor interactions lead to the agglomeration of cobalt particles, resulting in lower CO conversion on a per gram catalyst basis [33]. However, if the interaction between the support and the cobalt is too strong, then fine Co particles can lead to low reducibility, and thus reduction promoters (e.g., Pt, Re, and Ru) are often added to facilitate their reduction by hydrogen dissociation and spillover or via a chemical effect [34].

1.4. Supports and Promoters

Research has shown that iron is not as efficient at hydrogenation as cobalt or ruthenium and, as a result, iron produces higher selectivities of olefin and oxygenated products [35]. Formulations of iron catalysts often include silica as a structural stabilizer to help prevent attrition; however, iron catalysts also rely on promoters, such as alkali metals and/or copper, to assist in FTS. Alkali promoters for iron catalysts also promote the formation of the active phase, iron carbide, while copper has been found to facilitate the reduction of iron oxides prior to carbide formation. The typical Fe compounds that are intermediates in the reduction/carburization process in CO are $Fe_2O_3 \rightarrow Fe_3O_4 \rightarrow FeO \rightarrow Fe_xC_y$ [36]. For cobalt, the promoters, which are often metals, like Pt, Re, and Ru, enhance the degree of reduction of cobalt oxides, and most significantly that of Co^{2+} to cobalt metal, with the typical reduction process being $Co_3O_4 + H_2 \rightarrow 3CoO + H_2O$ and $3CoO + 3H_2 \rightarrow 3Co^0 + 3H_2O$ [37]. The product distribution is much more diverse for the iron catalyst than for the cobalt or ruthenium catalysts. Furthermore, iron is the only catalyst that produces more than just one type of oxygenate. These changes seem to vary greatly based on the amount of promoter and promoter type. In fact, the products of the iron catalyst can become so diverse, such that, even before the oils are analyzed, one can physically look at the samples to see the color differences that vary from deep amber to light yellow, or even clear (Figure 2). Color, from early hydrogenation work, served as a basis of understanding the fraction of isomerized material present in the product selectivity [38], which was, in turn, driven by the olefin products formed. Initial indications led to a ratio of the amount of 1-olefin to the internal olefins, where the amber, dark colored products have significant amounts of 1-olefin when compared to all other products. The oil samples were mixed with Pd/C catalyst and then placed under 10 pounds of pressure for 5 min. Once separated, the oil was clear, yet all of the oxygenates were still present. During this process, olefins become hydrogenated [38]. However, more work is needed to fully confirm the relationship between the 1-olefin and the rest of the materials to fully understand what is truly responsible for the color. Further deviations in the selectivity can also be observed on the same catalyst, by merely observing the samples with time on-stream (TOS). The aim of this work is to review the product selectivity of oxygenated materials over FTS catalysts and to examine the effects of promoters (e.g., alkali metals, active supports), as well as to develop insights (from an analysis of oxygenates) into the reaction kinetics and mechanism.

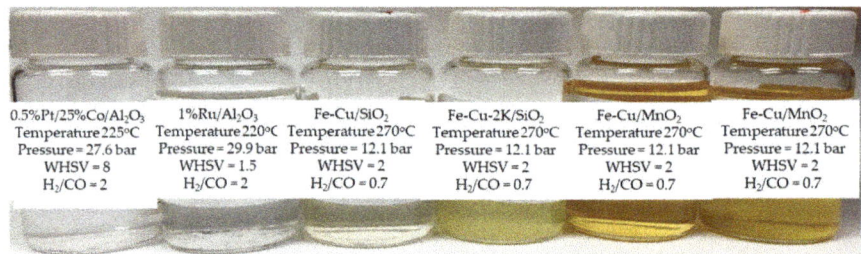

Figure 2. Deviations in the color of oils produced by Fischer–Tropsch synthesis (FTS).

Unlike iron, the promoters used for cobalt (e.g., Pt, Re, and Ru) do not alter the product selectivity to the same extent and they are primarily used to facilitate the reduction of cobalt oxides during catalyst activation. Because ruthenium and cobalt are expensive, they are typically supported (Al_2O_3, TiO_2, SiO_2), such that the number of active sites is determined, in part, by the interaction with the support, which governs the average size of cobalt nanoparticles and their percentage reduction while using a standard reduction procedure (e.g., 33%H_2:67% inert, 350 °C, 10 h). The purpose of adding the support for cobalt and ruthenium is to disperse the expensive metal to achieve higher active metal surface areas and to retard the sintering of the metal nanoparticles. The interaction between the metal and the support needs to be strong enough to stabilize small clusters, but not so strong that the cobalt surface is lost to the formation of cobalt support compounds; for example, cobalt/alumina catalysts that have low Co loading tend to form difficult to reduce cobalt aluminate species [39]. Furthermore, the support is often assumed to be an inert carrier with the FTS activity occurring on the active metal sites. However, recently, we made an attempt to make cobalt catalysts that mimic the behavior of Fe-based catalysts. Fe catalysts possess both a metal-like function in Fe carbide and a partially reducible oxide function in defect-laden Fe_3O_4. By adding an active partially reducible oxide—ceria—as the support, the oxygenate selectivity of cobalt FTS catalyst significantly increased [40–42]. This work summarizes the results for several reaction tests that were performed with cobalt and ruthenium serving as active components and various supports being utilized; the latter included relatively inert carriers, such as silica and alumina, as well as those possessing a higher density of reduced defect sites and that are considered to be more active, such as titania and ceria.

1.5. Known Mechanistic Routes

Bearing in mind that FTS products consist primarily of aliphatic molecules, FTS could be investigated as the synthesis of hydrocarbon molecules through a multi-step reaction. Therefore, several steps must occur in the synthesis route, which involves the addition of hydrogen in transforming CO bonding to C-C bonding [22]:

- associative/dissociative adsorption of CO (remains a point of contention);
- dissociative adsorption of hydrogen [43];
- transfer of 2 H* atoms to O* to form H_2O;
- allocation of 2 H* to C* to form -CH_2-;
- formation of a new C–C bond (unless methane is formed);
- desorption of H_2O; and,
- desorption of aliphatic product.

This list serves as a rough guide, but the specific order for all given steps remains elusive. CO must dissociate in order for aliphatic materials to be synthesized. This has led to various interpretations of the mechanistic steps of FTS.

Fischer, discerning that the synthesis route involved the formation of aliphatic carbon-hydrogen bonds and knowing the tendency of iron to form iron carbide, proposed a carbide mechanism, as shown in Figure 3. Although Fischer promoted this mechanism, it was not his first choice; he did not favor the carbide mechanism until results displayed hydrocarbons as the primary products for FTS [44]. Although this mechanism was set aside for a brief period, recent advances in surface science revealed that the catalysts displayed a high surface coverage of carbon, with little surface oxygen, and revived the idea that hydrocarbon products could be synthesized through the combination of methylene groups [45–49]. In the carbide mechanism, CO dissociatively adsorbs on the active FTS metal, covering the surface with carbon and oxygen atoms. The adatoms are sequentially hydrogenated, forming water and methylene monomer. Previous work that was done by Ojeda et al. discusses the quasi-equilibrium assumption for the dissociation and hydrogenation of CO within the first hydrogenation step; the second step is assumed to be rate-limiting [50]. Fischer first proposed a mechanistic route involving oxygenate intermediates for FTS, as very little aliphatic material was generated, and the products were mostly comprised of alcohols. Although the carbide mechanism gained ground after being proposed by Fischer, work that was conducted by Kummer and Emmett [46,47] once again gave rise to an oxygenate mechanism [49]. The detailed example within Figure 4 is currently known as the enol mechanism.

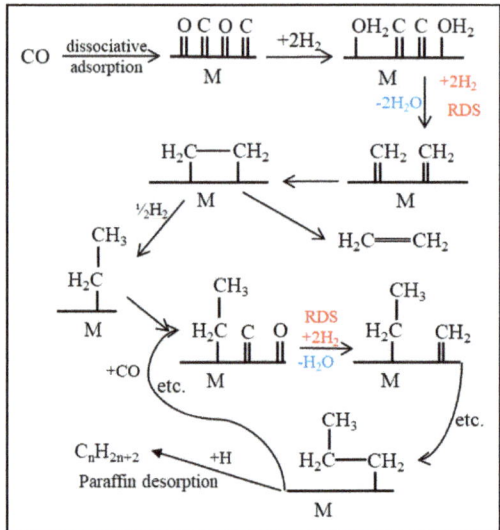

Figure 3. A proposed FTS route based on the carbide mechanism.

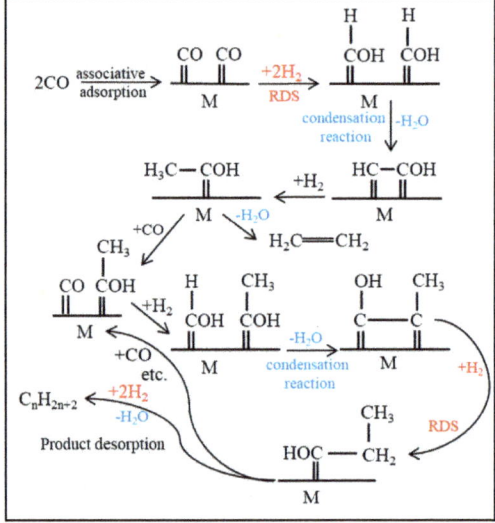

Figure 4. A proposed FTS route based on the enol mechanism.

The enol mechanism describes CO adsorbing without dissociation upon the FTS active metal surface. After chemisorption, CO reacts with adsorbed H atoms to create hydroxymethylene (M-CHOH). The enol structure grows by a sequence of condensation steps using the hydroxyl groups of adjacent hydroxymethylene species. This mechanism describes a route where the rate-controlling step is the hydrogenation of adsorbed CO.

Another commonly proposed mechanism is CO insertion [50], which is displayed in Figure 5, where the CO is molecularly adsorbed onto the active catalyst, and it undergoes bond scission only after being incorporated into the chain. Again, the proposed rate-limiting step is the hydrogenation of CO to the CH_2 methylene group. The assumed monomer for this mechanism is simply CO through its insertion into metal-carbon bonds. Interestingly, CO insertion is a common step in homogenously catalyzed reactions, such as hydroformylation [51]. The mechanism continued to be scrutinized, and even the primary versus secondary products are still subject to debate; several research groups are focused on shedding light on this topic [52–54].

1.6. Anderson, Schulz, and Flory

A polymerization model, as independently developed by Anderson et al. [55], Schulz [56], and Flory [57], which is known as the Anderson-Schulz-Flory (ASF) model, can describe the range of hydrocarbon products that were synthesized by FTS. The ideal FTS distribution is expressed by Equation (1), where the ASF model is the most commonly used to date:

$$\frac{M_n}{n} = (1 - \alpha)^2 \alpha^{n-1} \tag{1}$$

where M_n/n is the mole fraction of a hydrocarbon with carbon number n, and α is defined as the chain growth probability. Equation (2), which describes alpha (α), is defined by the molar rate at which the chain propagates (r_p) versus the rate at which it terminates (r_t).

$$\alpha = \frac{r_p}{(r_p + r_t)} \tag{2}$$

The ASF polymerization model assumes that the hydrocarbon chain lengths are independent, and thus, solely dependent on r_p and r_t. The distribution of FTS products can then be plotted in a linear manner with the natural log of the mole fraction as the dependent variable, y, and the carbon number as the independent variable, x, as shown in Figure 6 for iron, cobalt, and ruthenium catalysts.

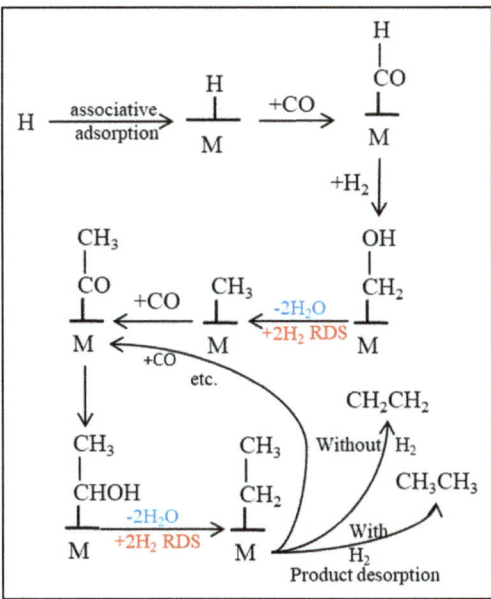

Figure 5. A proposed FTS route based on the CO insertion mechanism.

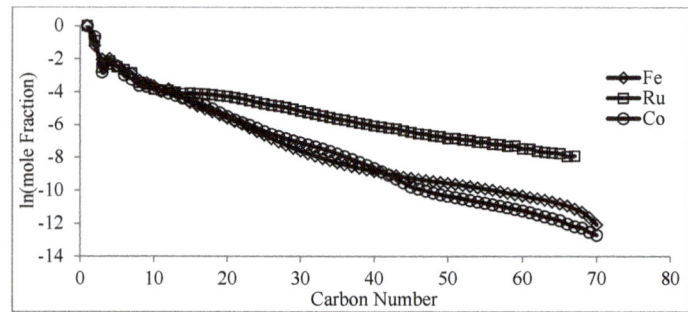

Figure 6. A general Anderson-Schulz-Flory (ASF) plot of the natural logarithm of the mole fraction for a cobalt, ruthenium, and iron FT product distribution. Fe-Cu/SiO$_2$, 270 °C, 12.1 bar, weight hourly space velocity (WHSV) = 2, H$_2$/CO = 0.7, 0.5%Pt/25%Co/Al$_2$O$_3$, Temperature 225 °C, Pressure = 27.6 bar, WHSV = 8, H$_2$/CO = 2; and 1%Ru/Al$_2$O$_3$, Temperature 241 °C, Pressure = 29.9 bar, WHSV = 1.5, H$_2$/CO = 2.

The ASF polymerization model is useful as a tool to describe a range of hydrocarbons when assessing a series of catalysts that portray certain characteristics. More importantly, trends can be displayed that attempt to understand the promoter, support, and reactor effects on FTS based on deviations from the ASF model. Figure 7 provides a general picture of an expected distribution for specific α values. A typical ASF plot totals all products (i.e., 1-olefin, n-paraffin, *cis/trans*-2-olefin, alcohol) to directly relate CO to a carbon number (i.e., C$_{10}$, C$_{11}$, etc.) and not to a specific product type. However, this review examines the types of products separately to compare the ASF model for more than one product classification. The products that are present in the oil and wax phases are so numerous that the retention times from standards could not be used to identify all of the products. Therefore, peaks for oxygenated compounds in the oil and wax are separated and identified using gas chromatography-mass spectrometry (GC-MS). Oxygenates in the gas and water samples are identified based on the retention times of standards. The ASF model is useful in describing the distributions of the different classifications of hydrocarbons and oxygenates. The apparent chain growth probability factor (i.e., α), as defined by the ASF model, depends on several factors, including vapor-liquid equilibrium (VLE), type of reactor, diffusional limitations, olefin re-adsorption, etc. [21,58–72]. Oxygenated compounds vaporize at higher temperatures (when compared at similar carbon chain length) and therefore will be more severely affected than hydrocarbons [73]. Oxygenates seem to display a negative deviation in ASF plots at approximately C$_{12}$, as opposed to hydrocarbons, and this is likely due to their higher boiling point in comparison with hydrocarbons of the same length.

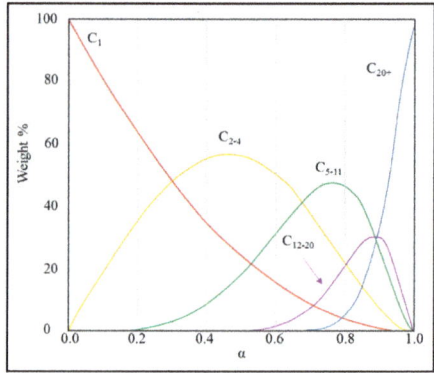

Figure 7. A distribution of products based upon α.

1.7. Product Distribution of Fischer–Tropsch Synthesis

The four primary classifications of FTS hydrocarbon products are: 1-olefins, paraffins, *trans*-2-olefins, and *cis*-2-olefins. To give an idea of the vast array of products, Figure 8 displays chromatograms of merely the oil phase (excluding gas, aqueous, and wax phases) for the distribution of products for iron (A) cobalt (B), and ruthenium (C) catalysts. General examples of a flame ionization detector (FID) scan (Figure 8) displays weight fractions of each major component, which can significantly vary, depending on the active metal. The effect of chain length on selectivity depends on the process conditions (e.g., temperature, pressure, H_2/CO ratio) and the type of reactor (e.g., slurry or fixed-bed). Chain length affects the olefin/paraffin (O/P) ratio [74–81] through differences in re-adsorption that, in turn, affect secondary reactions, such as hydrogenation, isomerization, and the re-initiation of chain growth [76,80,81]. Hydrocarbon chain length also influences the olefin/paraffin ratio by the disproportional accumulation of heavier products due to the decreasing vapor pressure with increasing chain length, particularly in slurry-based FTS systems [77,82]. As residence time increases, re-adsorption increases in a proportional manner; thus, an inverse relationship arises where, the longer the hydrocarbon chain, the lower the O/P ratio (as n increases, O/P decreases). Figure 8 visually describes this attribute (in this case using an iron catalyst), where the olefinic component more rapidly decreases with the carbon number in the FID chromatograph.

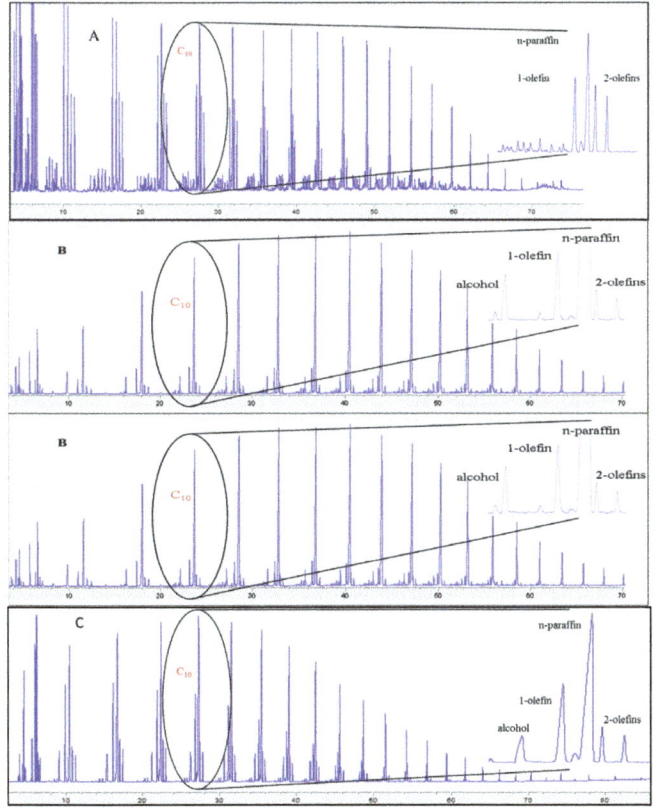

Figure 8. Typical flame ionization detector (FID) chromatographs for the products of the oil phase for Fe-Cu/SiO$_2$, 270 °C, 12.1 bar, WHSV = 2, H_2/CO = 0.7 (**A**), 0.5%Pt/25%Co/Al$_2$O$_3$, Temperature 225 °C, Pressure = 27.6 bar, WHSV = 8, H_2/CO = 2 (**B**); and, 1%Ru/Al$_2$O$_3$, Temperature 241 °C, Pressure = 19.2 bar, WHSV = 1.5, H_2/CO = 2 (**C**) FTS catalysts.

An advantage of the FID, specifically described later, is the ability to relatively assess peaks across chromatograms for different catalysts by weight %. This cannot be used to compare the total products produced, but the percentage of a specific product to the total. Here, this comparison works to quickly evaluate if a catalyst produces higher amounts of 1-olefin per carbon number. When calculating the actual mole of carbon %, the numbers agree with this relative amount per product, allowing the FID chromatogram to provide a quick "fingerprint" of product selectivity per catalyst.

The isomerized materials, although minor, can be observed from the gas range up to the longer chains in the oil fraction. Most of the branching that occurs in FTS is single methyl branched material, as shown in Figures 9 and 10. The diversity of these materials depend on the specific carbon number and, as the carbon number increases, the more diverse the methyl-branched paraffin become.

A minor fraction of the distribution, which consists of oxygenated products (Figures 11 and 12), is also important. For example, oxygenates can be used to investigate the mechanism of FTS [46,47,82] or serve as alternative chemicals. FTS catalysts have been altered in a specific manner with the intent of increasing the selectivity to oxygenates with the aim of producing alternative chemicals (e.g., chemical feedstocks) [40–42,83,84]. Two variables that affect product distribution [44] and that were controlled by changing catalyst formulation, include metal (e.g., alkali) promoters for iron, and support type (as well as the interfacial contact between support and metal) for supported cobalt and ruthenium catalysts.

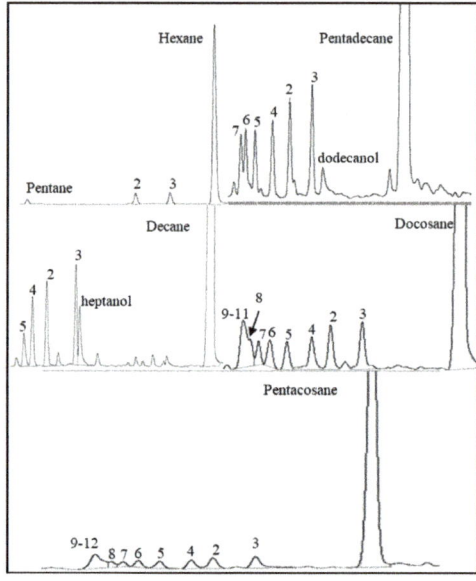

Figure 9. A chromatographic picture comparing different branching materials of specific carbon materials for a 0 K (i.e., no potassium promoter) active iron catalyst. This represents mixed oils from several iron runs where the oils were exposed to H_2 at 0.68 bar for 5 min using Pt/C catalyst converting all olefinic materials to alkanes. All runs at 230 °C, 12.1 bar, 0.7 H_2/CO with a WHSV = 2.

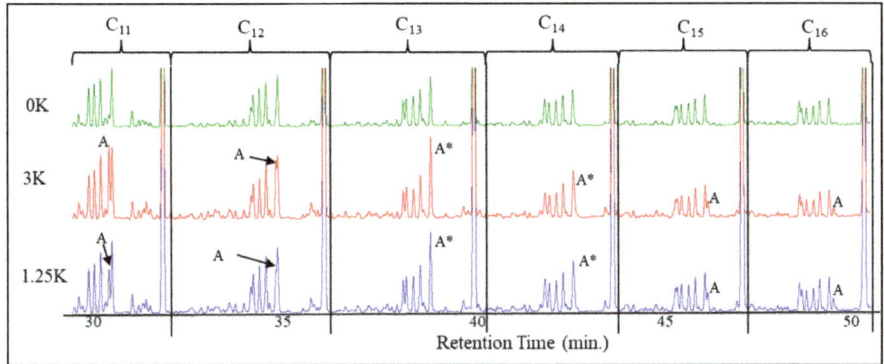

Figure 10. A chromatographic picture comparing different branched materials ranging from C_{11}–C_{16}. This is mixed oils from several iron runs where the oils were exposed to H_2 at 0.68 bar for 5 min. using Pt/C catalyst converting all olefinic materials to alkanes. A is the linear alcohol, A* is where the alcohol is buried behind the 3-methyl branched paraffin. Otherwise, the order of methyl paraffins—starting from the n-paraffin going right to left; 3-methyl, 2-methyl, 4-methyl, 5-methyl, etc. All runs at 230 °C, 12.1 bar, 0.7 H_2/CO, with a WHSV of 2.

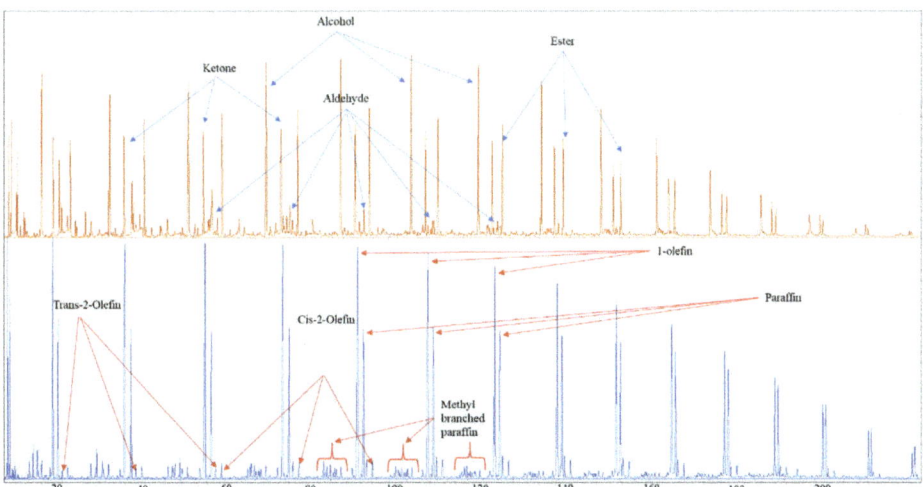

Figure 11. Two separate fractions of oil from a mixture of a multitude of iron FTS runs, separated by silica gel with N_2, and independently injected upon the MSD. The x-axis is retention time in minutes. The oxygenated chromatogram is only enlarged to display for identification purposes. These were not added to the overall analysis as the mole of carbon % were to small and difficult to quantify.

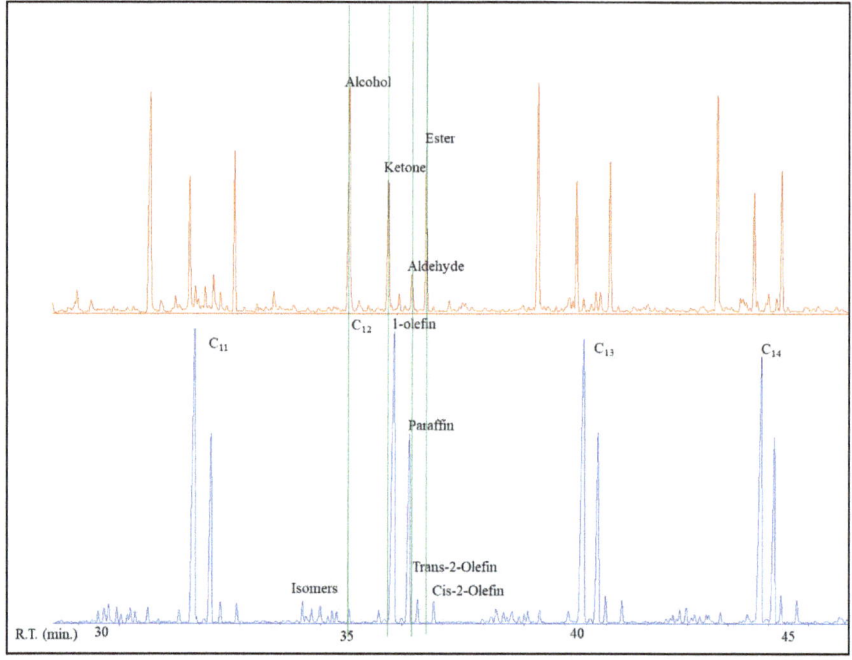

Figure 12. Two separate fractions of an iron FT oil separated by silica gel with N_2 and independently injected upon the Agilent MSD. The x-axis is retention time in minutes. The oxygenated chromatogram is only enlarged for the purpose of identification. The green lines display where the oxygenated material peaks would elute in relationship to the main hydrocarbon peaks—if these materials were not first separated.

The aim of this manuscript is to shed light on how product selectivity varies with FTS catalyst type for low-temperature FTS and, in turn, gain greater insight into the FTS mechanism.

2. Experimental

2.1. Catalyst Synthesis

2.1.1. Iron Catalysts

Fe–alkali/SiO_2 catalyst preparation: The preparation of a precipitated iron-silica catalyst with an atomic ratio 100Fe/4.6Si has been described elsewhere [85]. The alkali (Li, Na, K, Rb, and Cs) was then added by incipient wetness impregnation (IWI) with aqueous alkali carbonate to obtain an alkali/iron atomic ratio of 1.44/100 and copper, a known promoter, was not added in order to avoid complicating the analysis [85].

K/silica/Cu and K series iron catalysts: The preparation of the precipitated iron catalyst series was performed by dissolving iron (III) nitrate nonahydrate in deionized water, followed by the addition of tetraethylorthosilicate (TEOS) to achieve the desired Fe:Si ratio. This mixture was vigorously agitated until the TEOS had hydrolyzed. To make the final Fe:Si base, the slurry was then added to a continuously stirred tank reactor (CSTR), which served as a precipitation vessel, followed by a stream of ammonium hydroxide (14.8 M) that was added to maintain a pH of 9.0. Next, the iron base catalyst was vacuum filtered and then washed twice with deionized water. The final filter cake was dried for 24 h in an oven at 110 °C with flowing air. Lastly, the Fe:Si catalyst base powder was then impregnated by incipient wetness with specific amounts of aqueous potassium nitrate and copper nitrate solutions in order to produce the desired atomic compositions (0.5K:100Fe; 1K:100Fe, 1.25K:5.1Si:100Fe, 1.25K:2Cu:5.1Si:100Fe, 1.25K:5Cu:5.1Si:100Fe, 2K:100Fe, 5K:100Fe, 5K:5.1Si:100Fe). Finally, the Fe:Si:K:Cu catalysts were dried at 110 °C overnight, followed by impregnation, drying, and calcination in a muffle furnace at 350 °C for 4 h.

Fe-Cu/MnO_2 catalyst preparation: Manganese sulfate solution (1.7 M) was slowly added to the 0.4 M potassium permanganate solution, and after deposition, nitric acid (0.6 M) was added to this solution. This precipitate was aged for 10 h at room temperature, after which it was washed with deionized water and then dried at 120 °C for 15 h. Finally, the MnO_2 support was obtained after calcination at 200 °C for 4 h. The MnO_2 supported Fe-Cu catalysts were prepared by the IWI method using an aqueous solution of iron and copper nitrates. The iron loading of the catalyst was 20% and the Fe/Cu molar ratio was 100:15. After impregnation, the catalysts were dried at 120 °C for 15 h and then finally calcined in air at 350 °C for 4 h.

2.1.2. Cobalt Catalysts

Co/Silica—The 12.4 wt % Co/SiO_2 catalyst was prepared by the incipient wetness impregnation of a SiO_2 support (Davisil 634, pore volume of 0.75 cm^3/g, and surface area of 480 m^2/g) with a cobalt nitrate solution. Two impregnation steps were used to load the cobalt precursor. Between each step, the catalyst was dried under vacuum in a rotary evaporator at 80–100 °C. After the second impregnation/drying step, the catalyst was calcined in flowing air at 350 °C for 4 h.

Co/Alumina—Condea Vista Catalox high purity γ-alumina (150 m^2/g) was used as a support. Cobalt nitrate (Sigma Aldrich, St. Louis, MO, USA) served as the precursor to load 25% by weight cobalt onto the Al_2O_3 support. In this method, which follows a Sasol patent [86], the ratio of the volume of solution used to the weight of alumina was 1:1, such that approximately 2.5 times the pore volume of solution was used to prepare the loading solution. Two impregnation steps were used to load 12.5% of Co by weight for each step. Between each step, the catalyst was dried under vacuum in a rotary evaporator at 80 °C and the temperature was slowly increased to 100 °C. After the second impregnation/drying step, the catalyst was calcined at 350 °C for 4 h.

Co/Titania—Preparation of this catalyst is detailed in reference [87].

Co/Ceria—Preparation, and properties of these catalysts can be found in references [40–42].

Co/Carbon Spheres—Preparation and properties of these catalysts and reaction tests can be found in reference [59].

Unpromoted Co—A bulk cobalt catalyst was prepared while using cobalt nitrate solution as a cobalt precursor. In this method, a 1:1 stoichiometric amount of ammonium carbonate (aqueous solution) to cobalt, was added to an aqueous solution of $Co(NO_3)_2$*$6H_2O$ (Aldrich, 98%), in order to precipitate out Co as $CoCO_3$. The precipitate was filtered, and the resulting solids were washed with copious amounts of deionized water. The catalyst was then dried overnight at 120 °C and then calcined at 500 °C in air for 5 h.

2.1.3. Ruthenium Catalysts

Three ruthenium catalysts were prepared by IWI using an aqueous solution of ruthenium nitrosyl nitrate (Alfa Aesar, Haverhill, MA, USA). The samples were dried overnight at 120 °C and then calcined under air flow at 350 °C for 4 h. The catalysts included a 1.2 wt % Ru/TiO_2 catalyst with Degussa P-25 TiO_2 (72% anatase, surface area 45 m^2/g, calcined at 400 °C for 6 h), and a 3.0 wt. % Ru/SiO_2 catalyst that was prepared using PQ silica CS-2133 (surface area 352 m^2/g, which was calcined at 350 °C for 6 h) as the support. 1.0 wt% Ru/Al_2O_3 catalyst was prepared using Sasol-Catalox alumina (high purity γ-alumina, surface area 140 m^2/g, which was calcined at 350 °C for 6 h) as the support.

Ru/NaY—Specifics on the preparation of the NaY ruthenium catalyst are described in reference [87].

2.2. Catalyst Testing

All of the catalytic FTS reaction tests were performed in a 1-L CSTR; the mixing rate was 750 revolutions per minute (RPM) and the reaction temperature was monitored with an internal thermocouple. Each CSTR had three Brooks mass flow controllers, which were specifically calibrated and capable of delivering a continuous, very controlled flow of gas. The pressure was monitored with three gauges—pre-reactor, internal to the reactor, and post reactor—to ensure no plugging issues throughout the entire system. The effluent of the reactor was directed to two 500 mL pressure vessels, which served as traps, and they were set at 100 °C and 0 °C, respectively; these were used as a distillation setup to separate the vapor phase FTS products, oil and water. A third vessel set at 200 °C was used to periodically collect the heavier wax products that remained inside the reactor. A stainless steel (SS) filter that was located inside the reactor prevented catalyst loss during the collection of waxes. As wax products are liquid inside the reactor, collection was accomplished by setting a pressure difference between the reactor system and the offline pressure vessel. This created an internal vacuum that provided the driving force for wax product removal.

2.2.1. Iron Catalysts

Fe–Alkali/SiO_2: The catalyst (5 g) was mixed in the CSTR with 310 g of melted octacosane, which was used as the start-up solvent. The octacosane was treated to remove bromide impurity prior to mixing. The stirring speed was set at 750 rpm before the reactor was pressurized to 12.1 bar with carbon monoxide at a flow rate of 25 NL/h. The reactor temperature was then increased to 270 °C at a heating rate of 2 °C/min. The catalyst was pretreated at 270 °C for a total of 22 h. After this period, the reactor was then fed with syngas at a constant H_2/CO ratio of 0.67 by introducing and increasing the H_2 flow during a 2 h period. The pressure, temperature, and stirring speed were maintained at 12.1 bar, 270 °C, and 750 rpm, respectively.

Fe–Cu/MnO_2: 15 g of the iron catalyst were loaded into the CSTR with Durasyn 164 "C30" oil, a poly alpha-olefin startup oil. The catalyst was pretreated online in the same manner as described above for an iron catalyst (i.e., pretreatment for 22 h at 270 °C, and then running at 12.1 bar and 270 °C).

2.2.2. Cobalt Catalysts

Co/Silica: 18.3 g of the silica supported catalyst containing 20% Co were reduced in H_2 flow at 350 °C in a fixed bed reactor for 24 h. The treated catalyst was then pneumatically transferred under argon flow, without exposure to air, into the CSTR. To accomplish the transfer, the fixed-bed reactor was attached to the CSTR through a ball valve connection, the reduction vessel was then over-pressured with argon, and the catalyst powder was then forced through the ball valve. The reduction vessel was weighed before and after catalyst transfer to ensure that all of the catalyst was added to the CSTR. After transfer, the catalyst was reduced in flowing H_2 at 180 °C in the CSTR for another 24 h. The initial conditions of the reactor were 20.8 bar and 210 °C. The feed gas was initially set to 90.9 slph with a composition of 66.7% H_2, 33.3% CO (no inert gas was added), H_2:CO ratio of 2.0, weight hourly space velocity (WHSV) of 24.8, and a gas hourly space velocity (GHSV) of 5.0 SL/h/g-cat. An amount of 300 g of the startup Polywax 3000 (polyethylene fraction with average molecular weight of 3000) was used.

Co/Alumina: Another reactor was loaded with 9.0 g of an alumina supported catalyst. The calcined catalyst (ca. 20 g) was ex-situ reduced in a fixed bed reactor with a mixture of hydrogen and helium (1:2) at a flow rate of 70 SL/h at 350 °C. The reactor temperature was increased from room temperature to 100 °C at a rate of 2 °C/min and then held at 100 °C for 1 h; then, the temperature was increased to 350 °C at a rate of 1 °C/min and kept at 350 °C for 10 h. The catalyst was transferred to a CSTR to mix with 300 g of melted Polywax-3000. The catalyst was then reduced in situ in the CSTR in a flow of 30 SL/h H_2 at atmospheric pressure. The reactor temperature was increased to 220 °C at a rate of 1 °C/min and was maintained at this activation condition for 24 h. The initial conditions of the reactor were 18.9 bar and 200 °C. The feed gas was initially set to 45.0 slph with a composition of 66.7% H_2: 33.3% CO (i.e., no inert gas was added), a H_2:CO ratio of 2.0, WHSV of 20, and a GHSV of 5 SL/h/g-cat.

Co/Titania: The titania supported catalyst was activated and loaded in the same manner as the alumina supported catalyst—specific details are found in reference [20].

Co/Unsupported cobalt: The calcined unsupported catalyst (~14.0 g; weight was accurately known) was reduced ex-situ in a fixed bed reactor with a mixture of hydrogen and helium (1:2) at a flow rate of 70 SL/h at 350 °C for 10 h at atmospheric pressure. The catalyst was transferred under the protection of argon to a 1 L CSTR containing 300 g of melted Polywax 3000 (start-up solvent). The catalyst was then re-reduced in situ at 220 °C for 24 h in the CSTR in a flow of 30 SL/h hydrogen at atmospheric pressure. After the activation period, the reactor temperature was decreased to 170 °C and synthesis gas was introduced through a dip tube, while the reactor pressure was increased to 19.9 bar. The reactor temperature was then increased to 220 °C at a rate of 1 °C /min. The feed gas was initially set to 65 slph with a composition of 66.7% H_2, 33.3% CO, a H_2:CO ratio of 2.0, a WHSV of 4.8, and GHSV of 5 SL/g/h.

Co/Carbon: run specifics can be found in reference [59].

Co/Ceria: specific run parameters can be found in references [40–42].

2.2.3. Ruthenium Catalysts

The ruthenium catalysts were activated ex-situ in a one-inch plug flow reactor at 300 °C for 15 h under 3:1 He/H_2 mixture. The catalyst was then inserted into melted wax in the hood and allowed to cool. The catalyst/wax mixture was then added to the CSTR and again exposed to a reducing environment of pure H_2 for 24 h at 220°C before bringing it to the specified FTS conditions. Liquid samples were removed daily from three traps that were maintained at 200, 100, and 0 °C mounted after the CSTR. Water was separated, and the oil and wax samples were weighed separately and then combined (producing an oil phase). The FTS products were collected and then weighed and analyzed while using gas chromatography (GC).

The specifics on the run parameters of the NaY ruthenium catalyst are described elsewhere [87].

2.3. FTS Product Analysis

Instrumentation Methods

The FID instrument is used specifically for carbon based materials to quantify the products of FTS [88–91]. No specific sample preparation was necessary before the injection of the oil sample onto the GC column. One microliter of sample was injected onto an Agilent JW Scientific DB-5 GC Column (Agilent, Santa Clara, CA, USA) at 35 °C and 0.55 bar, with a 1.5 mL/min split flow. The quantification of the oil products was accomplished using a 6890 Agilent (Santa Clara, CA, USA) GC with flame ionization detector (FID). The typical range of hydrocarbons observed in this sampling ranged from C_4–C_{35}. Only very small amounts of the low end (C_4–C_6) and high end (C_{27}^+) products were observed).

A small aliquot of wax was taken from the sample vials and then placed into a small test tube that contained a small amount of o-xylene (Sigma Aldrich $\geq$ 99.0%) and warmed to around 80 °C. The samples were kept at the elevated temperature until all of the wax was dissolved. These liquids were then transferred to injection vials. The wax samples were also analyzed with a 6890 Agilent FID instrument. The column was a 30 m DB-1 (Agilent, Santa Clara, CA, USA) high temperature column); the injections occurred with a slow plunger at 50 °C and immediately increased to 390 °C and held at 390 °C for 30 min. The o-xylene was the first peak to elute and it was not included in the sample analysis. The hydrocarbon peaks that followed typically ranged from C_{11}–C_{70} (i.e., only very small amounts of the low end C_{11} and high end C_{50}^+ were observed).

The vapor (CO, CO_2, C_1–C_6, H_2, N_2) that exited the cold trap was directly sent to a Hewlett-Packard Quad Series Micro GC (also a Micro-GC 3000 from Agilent), a refinery gas analyzer

(RGA). This GC has four internal modules, each with their own injection port column and TCD that were run in parallel (A = mol sieve 10 m 5A, B = plot U (PPU) 8 m, C = alumina 10 m, D = OV-1 10 m) [92–95]. Two pumps in the GC allow for the same volume of sample to be injected into the four columns (being held at a constant temperature and pressure), quickly separating specific compounds in each column. Column A, which was the mol sieve, performed the separation of H_2, methane and CO and used Ar as the carrier gas. Column B, the plot U, separated the C_2 products and CO_2. Column C, the alumina plot, separated out C_3–C_5. Column D, the OV-1, separated C_6–C_8. Columns B-D used He as the carrier gas.

No sample preparation was necessary to inject the water samples. The water samples were injected onto the Hewlett Packard 5790 GC with a TCD and packed (1/8" × 6m Porapak) column at 2.0 bar and 100 °C, and also to an 8610C SRI GC while using the same column and method. Known standards were used for the peaks in the water phase, but the relative response factors were also used, as they provide information regarding the samples [96,97]. When any sample (e.g., water, gas, oil, or wax) needed identification, it was also run on the Agilent 5975N mass selective detector (MSD) that was directly connected to an Agilent 6890 GC. The method that was built in the GC-MSD uses the same method conditions and column as the 6890 Agilent GC-FID. In so doing, the peaks that were observed in the chromatographs could be matched between the FID and MSD, where the retention times (i.e., and patterns of specific compounds) were the same. This allows for specific product verification (by the MSD) and quantification (by FID) for as many of the FTS products as possible.

3. Results and Discussion

3.1. Product Selectivity Separation and Identification

The main FTS products elute in a specific order that is based on our analytical procedure (Figures 11 and 12), where the first peak is the 1-olefin, followed by the paraffin, *trans*-2-olefin, and lastly, the *cis*-2-olefin, which ranges from C_2 to greater than C_{70}. A variety of hydrocarbon products elute, in groups of the same carbon number, before the 4 major products. The majority of these are methyl branched paraffins and olefins that have the sp^2 carbon greater than C_2—i.e., more internally along the chain (e.g., *cis*-3-alkene, *trans*-3-alkene, *cis*-4-alkene, *trans*-4-alkene). These types of products tend to be more prevalent for iron than for ruthenium and cobalt. The products elute in the same order for each carbon number based on type (e.g., 1-olefin, paraffin, *trans*-2-olefin, *cis*-2-olefin). The carbon number distribution for the oxygenates also elute in a specific order (Figures 8, 11 and 12) (linear alcohol, ketone, aldehyde and ester).

For the entire spectrum of compounds, per carbon number, the first peak in the group is usually the branched material and the last compound to elute for each carbon number group is the *cis*-2-olefin. The major oxygenate group for all of the catalysts (with the exception of potassium doped iron catalyst with no silica) in the oil products is the alcohol, which elutes from the column before the 1-olefin. The carbon number of the alcohol that elutes has approximately three carbons less than the hydrocarbon peak (i.e., when the C_{12} hydrocarbons elute, the alcohol that elutes will be nonanol). Other oxygenates, such as esters, ketones, and aldehydes, are two carbon numbers less. Acids (when observed) are typically four carbons less.

Alcohol is the only oxygenate that is present in a carbon molar percentage that is sufficient for analysis of the cobalt and ruthenium catalysts. Depending on certain parameters (such as the metal promoter and support), the alcohol peak may be small, as with the lithium doped iron catalyst (Figure 13), the unsupported cobalt catalyst (Figure 14), and the alumina supported ruthenium catalyst (Figure 15); or, a dominant peak before the 1-olefin, as with a rubidium doped iron catalyst (Figure 16), a silica supported cobalt catalyst (Figure 17), and the titania supported ruthenium catalyst (Figure 18).

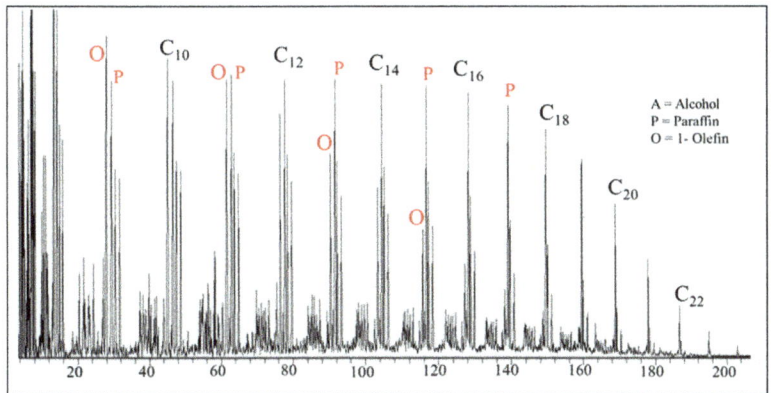

Figure 13. A chromatograph of the products produced with a lithium doped iron catalyst. The x-axis is retention time in minutes. Reaction Conditions: 270 °C, 12.1 bar, WHSV = 2, and $H_2/CO = 0.7$.

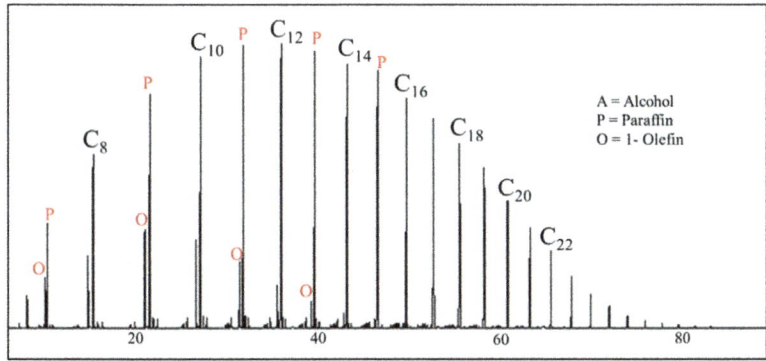

Figure 14. A chromatographic picture of an oil product from the unsupported cobalt catalyst. The x-axis is retention time in minutes. Reaction Conditions: 220 °C, 27.3 bar, WHSV = 3, and $H_2/CO = 2$.

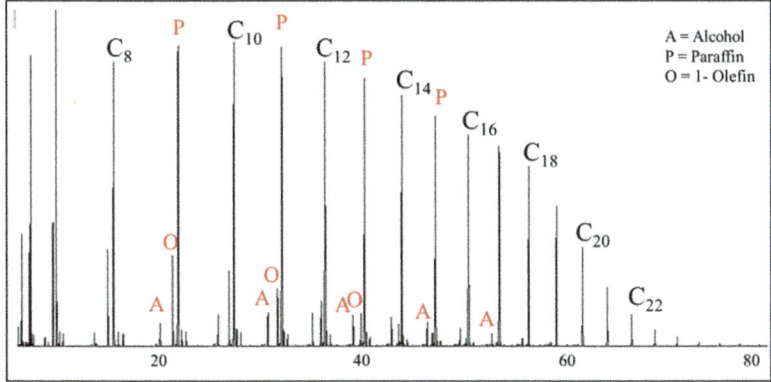

Figure 15. A chromatographic picture of the oil for the alumina supported ruthenium catalyst. The x-axis is retention time in minutes. Reaction Conditions: 241 °C, 19.2 bar, WHSV = 1.5, and $H_2/CO = 2$.

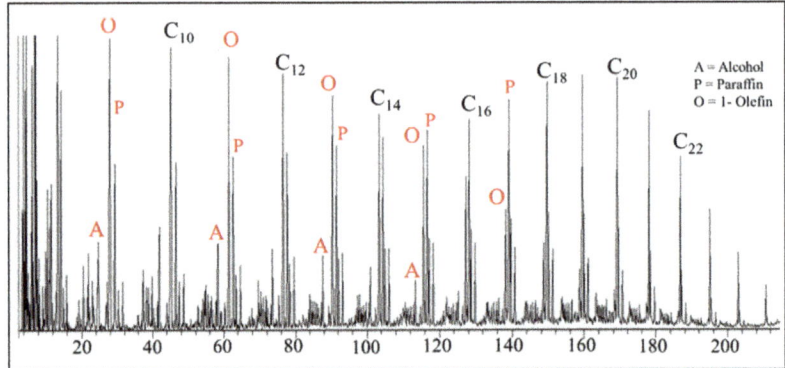

Figure 16. A chromatograph of the product produced with a rubidium doped iron catalyst. The x-axis is retention time in minutes. Reaction Conditions: 270 °C, 12.1 bar, WHSV = 2, and H_2/CO = 0.7.

A broad assortment of oxygenated compounds, such as alcohols, acids, ketones, esters, and aldehydes, were generated in the CSTR containing an iron catalyst. The range of oxygenates that was found in the aqueous phase was C_5 or lower, and any longer chained oxygenated compounds were present in the oil phase. The oxygenates that were in the oil product were ordered based on retention time (RT) (Figures 8 and 12):

- The ketone regularly falls near the 1-olefin and, as the carbon number increases, the peak moves toward the 1-olefin and it eventually merges with it. The major ketone observed is where the carbonyl is on the second carbon; every ketone that was observed in the GC-MS for Figures 5 and 9 has the second carbon as the carbonyl.
- Esters, by and large, appear between the *trans*-2-olefins and *cis*-2-olefins and, with increasing carbon number, move toward the *cis*-2-olefin. The major ester observed is also where the carbonyl is on the second carbon; again, every observed ester in the GC-MS for Figures 5 and 9 has the second carbon as the carbonyl and the oxygen between the second and third carbons.
- The aldehyde usually elutes between the paraffin and *trans*-2-olefin.
- Alcohol, which is the predominant oxygenate product in the oil phase, is the only oxygenate separated and verified for the higher carbon number fraction.

The main anomalies observed in this series are from these three catalysts:

- The Co/ceria catalyst (Figure 19): the linear alcohol peaks are more dominant than the paraffin and there is the presence of the two methyl alcohols.
- The unsupported K/Fe catalyst (Figure 20): the aldehyde is more dominant than the linear alcohol for some of the shorter hydrocarbons.
- The NaY supported ruthenium catalyst (Figure 21): the peak that is assigned to 3-olefin is intense enough to be observable for several carbon numbers.

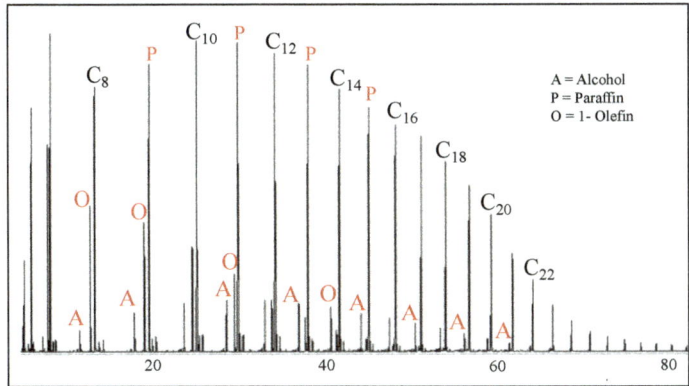

Figure 17. A chromatographic picture of a silica supported cobalt catalyst. The x-axis is retention time in minutes. Reaction Conditions: 220 °C, 27.6 bar, WHSV = 3, and $H_2/CO = 2$.

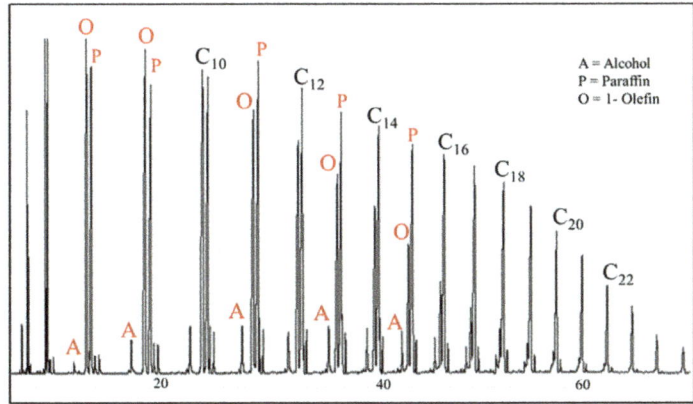

Figure 18. A chromatographic picture of a titania supported ruthenium catalyst. The x-axis is retention time in minutes. Reaction Conditions: 220 °C, 19.3 bar, WHSV = 3, and $H_2/CO = 2$.

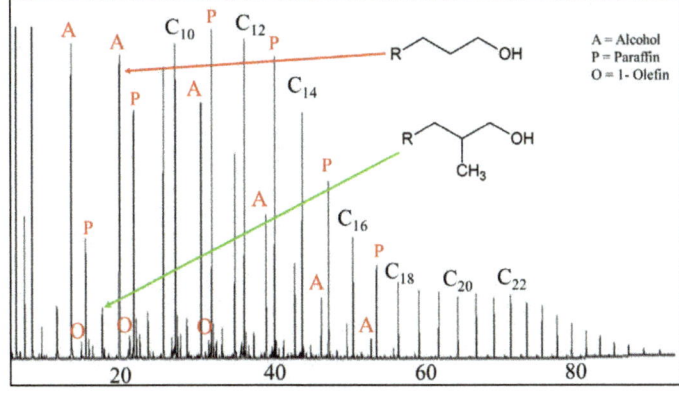

Figure 19. A chromatographic picture of a ceria support cobalt catalyst. The x-axis is retention time in minutes. Reaction Conditions: 230 °C, 27.6 bar, WHSV = 0.5, and $H_2/CO = 2$.

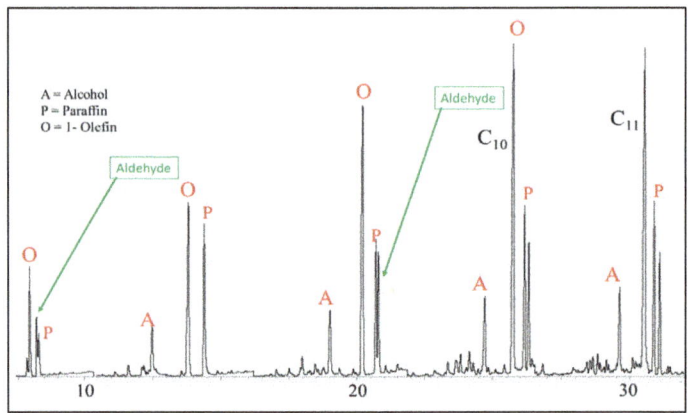

Figure 20. A chromatographic picture of potassium promoted iron catalyst. The x-axis is retention time in minutes. Reaction Conditions: 270 °C, 12.1 bar, WHSV = 2, and H_2/CO = 0.7.

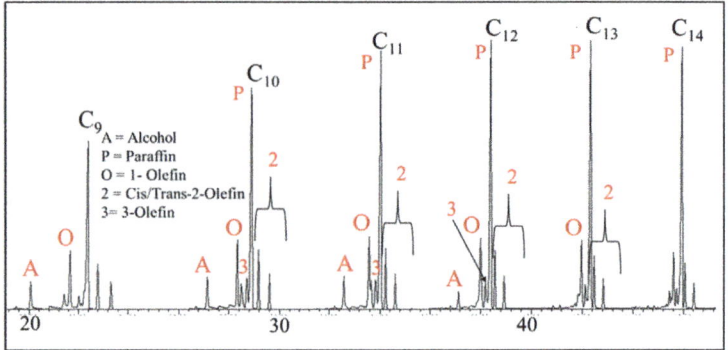

Figure 21. A chromatographic picture of a NaY supported ruthenium catalyst. The x-axis is retention time in minutes. Reaction Conditions: 220 °C, 19.2 bar, WHSV = 1.5, and H_2/CO = 0.7.

3.2. Mole Fraction

The mole fraction is calculated by means of the raw oil GC data, gas data, aqueous phase data, and the mass of the actual sample collected. The oil and water GC data provide a weight percentage, as the raw data are based on previous studies of hydrocarbon and oxygenate response factors [88–91]. The alcohol peaks in the oil phase were identified and then the hydrocarbon and alcohol peaks (including the branched hydrocarbons) were totaled based on the carbon number. From equation 3 and the weight fraction, the total moles for the oil with carbon number 1 to n could be found:

$$Total\ Mole = \sum_{i=1}^{n} \frac{(WF_n \times m_{oil})}{C_n} \qquad (3)$$

where C_n is the molecular weight of the paraffin with the chain being n carbons in length and m_{oil} is the mass of the oil that is removed from the reactor.

The aqueous phase was separated from the oil phase and each peak of the GC was identified. The raw data were obtained in weight percent and then combined according to carbon number. The calculation to find total moles for the aqueous phase follows the same procedure as for the oil phase.

The gas produced from the reactor can be divided by the molar volume (22.414 L/mol), providing the moles per h of gas produced, since the gas leaving the cold trap is at standard temperature and pressure. The total moles can then be calculated from the mol/h by determining the difference in the time-on-stream (TOS) of when the sample was taken as compared to the previous sample. The samples were collected as the run time increased for each of the catalysts. Once the total moles were found for the gas data, the oil data, and the aqueous data, the total distribution for that specific sample could be calculated. Thus, the moles of carbon of each product could be related to the moles of CO consumed.

3.3. Chain Growth Probability Factor (α)

3.3.1. Iron

The Na, K, Rb, and Cs run at low conversion produced enough oxygenate peaks to be visible in the oil phase. Examples of the ASF plot for cesium promoted iron (Figures 22 and 23) show the chain growth probability factor for the hydrocarbons and the oxygenated compounds. Some oxygenate peaks for specific iron runs (Li and unpromoted) could be identified by the MSD while using single ion monitoring (SIM) of specific fragments produced by the electron impact source (EIS). However, the oxygenated products were below the limits of detection (LOD) and could thus not be considered.

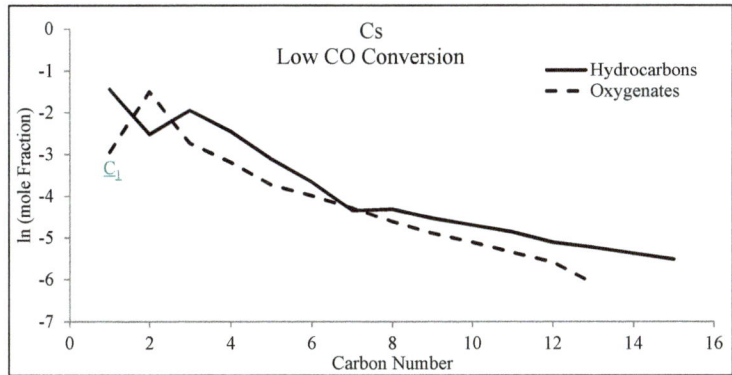

Figure 22. An ASF plot of hydrocarbons and the relative oxygenates using a cesium doped iron catalyst, ranging C_1 to C_{13} (oil, water, and wax combined) for a low CO conversion sample. Reaction Conditions: 270 °C, 12.1 bar, WHSV = 2, and H_2/CO = 0.7.

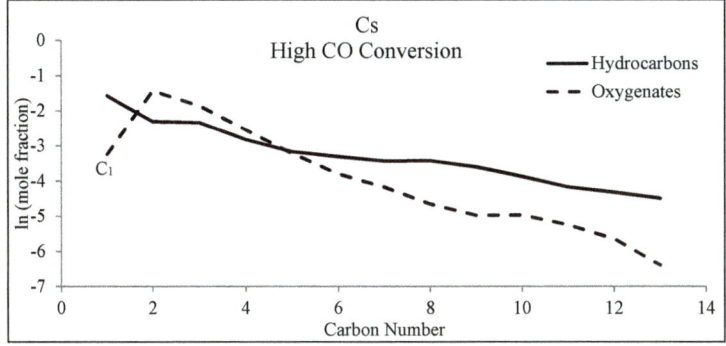

Figure 23. An ASF plot of hydrocarbons and the relative oxygenates using a cesium doped iron catalyst, ranging C_1 to C_{14} (oil, water, and wax combined) for a high CO conversion sample. Reaction Conditions: 270 °C, 12.1 bar, WHSV = 2, and H_2/CO = 0.7.

The chain growth probability factor of the hydrocarbons for alkali metal doped iron catalysts was lower ($\alpha = 0.75 \pm 0.07$) for the high CO conversion samples (Table 1) when compared to that of the low CO conversion samples ($\alpha = 0.86 \pm 0.04$). The oxygenated compounds additionally displayed a lower α value ($\alpha = 0.51 \pm 0.15$) for the high CO conversion samples as compared to the low CO conversion samples that have an average α of 0.56 ± 0.18. The variation between the average α plots for the oxygenates was smaller than that of the hydrocarbons for the different conversions. However, there was a larger deviation in α values for the oxygenates from the promoted catalysts when compared to that of the hydrocarbons.

α values of the oxygenates for the Na and K promoted Fe catalysts were the most affected by the change in CO conversion. The Li promoted Fe catalyst had the lowest α value, being significantly lower than that of other promoted catalysts, and it remained unchanged at different conversions. Rb and Cs promoted Fe catalysts consistently had the highest α values, which only changed slightly by changing CO conversion (i.e., average CO conversion across all the high CO conversion runs was $62.4\% \pm 2.56$ and average CO conversion across all low CO conversion runs was $23.5\% \pm 4.30$).

Table 1. Alpha (α) as compared across several FT synthesis catalysts.

	Support/Promoter	Hydrocarbons	Oxygenates
Cobalt 220 °C, 27.6 bar, WHSV = 2–5 $H_2/CO = 2$	Unsupported	0.82	0.68
	Alumina	0.88	0.75
	Silica	0.86	0.79
	Titania	0.79	0.79
	Carbon	0.76	0.60
	Ceria	0.80	0.64
Ruthenium 220 °C, 19.2 bar, WHSV = 1.5, $H_2/CO = 2$	Alumina	0.78	0.63
	Titania	0.79	0.52
	Silica	0.80	0.70
	NaY	0.75	0.70
Iron at High CO conversion supported with 5.1 silica. Reaction Conditions 270 °C, 12.1 bar, WHSV = 2, $H_2/CO = 0.7$	Unsupported	0.87	0.31
	LiHC	0.67	0.34
	NaHC	0.74	0.51
	KHC	0.78	0.60
	RbHC	0.74	0.64
	CsHC	0.72	0.66
Iron at Low CO conversion supported with 5.1 silica 270 °C, 12.1 bar, WHSV = 2, $H_2/CO = 0.7$	Unsupported	0.86	0.35
	LiLC	0.80	0.41
	NaLC	0.86	0.66
	KLC	0.82	0.74
	RbLC	0.92	0.69
	CsLC	0.89	0.73
Iron 270 °C, 12.1 bar, WHSV = 2, $H_2/CO = 0.7$	Manganese	0.78	0.56

Hydrocarbons Alpha	Average	Standard Deviation	Coefficient of Variance
Cobalt	0.82	0.04	5.49
Ruthenium	0.78	0.02	3.02
Iron High CO	0.75	0.07	8.93
Iron Low CO	0.86	0.04	5.22

Oxygenates Alpha	Average	Standard Deviation	Coefficient of Variance
Cobalt	0.71	0.08	11.34
Ruthenium	0.64	0.09	13.42
Iron High CO	0.51	0.15	30.05
Iron Low CO	0.60	0.17	29.02

The alphas here do not include the heavy wax materials. Thus, to make an accurate comparison, the hydrocarbons were compared across the range C_8–C_{15}. The oxygenates varied based on the mole of carbon % observed and are primarily in the range of ~C_4–C_{10}. A few catalysts displayed product selectivities where the oxygenated materials may have been less than the limits of detection. Low CO conversion for the iron is ~25% and high conversion is 60%.

3.3.2. Cobalt

Examples of ASF plots for cobalt catalysts are displayed in Figures 24 and 25. The α for the hydrocarbons that are produced from a cobalt catalyst is 0.82 ± 0.04 (Table 1). Two samples were taken as the run time increased, within each of the three runs. While there are variations in the CO conversions, they are within ($\pm4\%$). When considering the conversion rates among samples, the chain growth probability factor should remain somewhat the same from sample to sample. There are differences among the α values with support type, whereby the alumina supported catalyst (Table 1) had the highest α value of 0.88, whereas the carbon supported catalyst displayed a lower α of 0.76. All of the supported catalysts had higher α values for oxygenates in comparison with that of unsupported cobalt. The effect of the support for oxygenated materials has been corroborated [98–100].

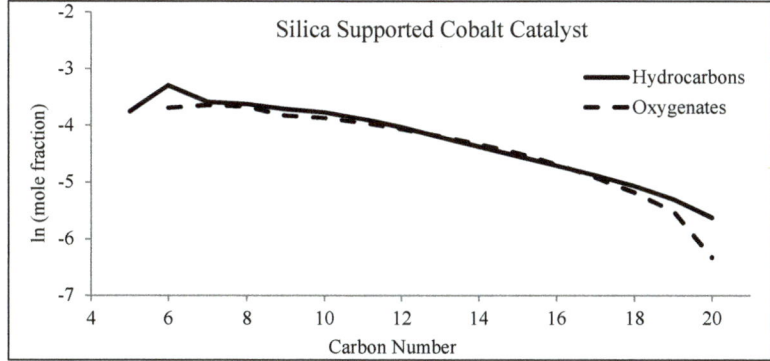

Figure 24. A carbon distribution of the cobalt silica supported catalyst. Reaction Conditions: 220 °C, 27.6 bar, WHSV = 3, and $H_2/CO = 2$.

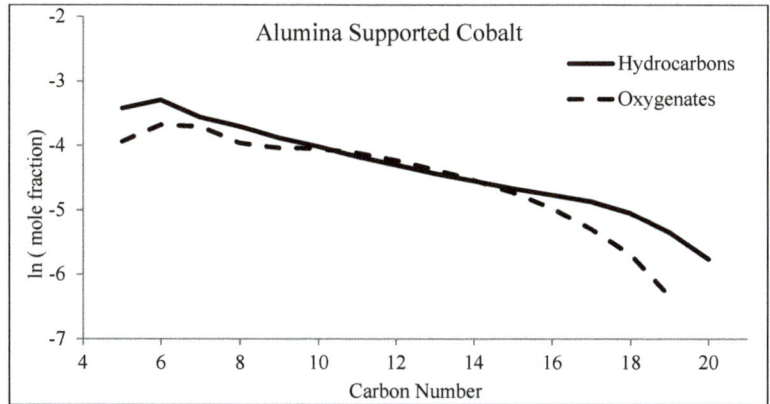

Figure 25. A carbon distribution of the cobalt alumina supported catalyst. Reaction Conditions: 220 °C, 27.6 bar, WHSV = 5, and $H_2/CO = 2$.

3.3.3. Ruthenium

Figures 26 and 27 display examples of the ASF plot for titania and alumina supported ruthenium catalysts. Based on the results in Table 1, the α values for the hydrocarbons and oxygenates were slightly higher for cobalt than ruthenium (Table 1). A reason for this is that the wax based materials were not added, causing larger deviation with the ruthenium catalyst than with cobalt. The overall α plots for the ruthenium catalysts were more uniform, independent of the support. Yet, in the case of oxygenates, α is quite different, indicating that the rate of growth for oxygenated materials may depend more so on the support-metal interfacial contact area (i.e., the perimeter of metal particles in contact with the support) than that of hydrocarbon growth.

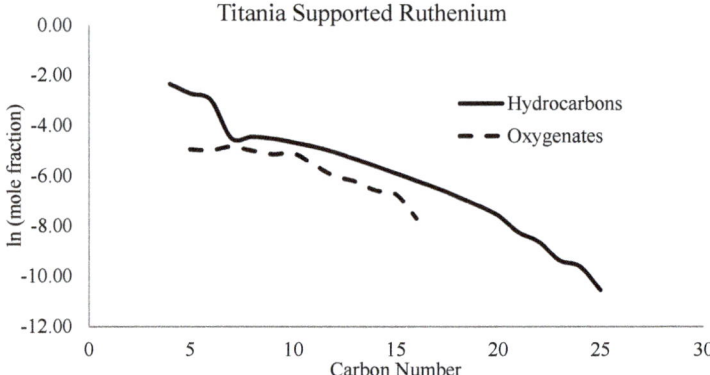

Figure 26. A carbon distribution of the titania supported ruthenium catalyst. Reaction Conditions: 220 °C, 19.2 bar, WHSV = 1.5, and $H_2/CO = 2$.

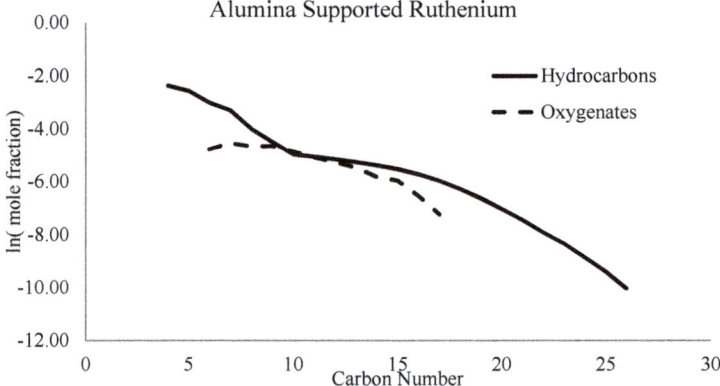

Figure 27. A carbon distribution of the alumina supported ruthenium catalyst. Reaction Conditions: 220 °C, 19.2 bar, WHSV = 3, and $H_2/CO = 2$.

3.4. Specific Product Selectivities

3.4.1. Iron

Ngantsoue-Hoc et al. [85] discussed the ratio of ethene to ethane, where Rb, Cs, and K promoted Fe catalysts produced the highest ratio. Extending this range to C_{25}, the trend continues to hold for the major components (1-olefin, paraffin, *cis*-2-olefin, and *trans*-2-olefin), where Rb, Cs, and K promoted Fe catalysts produced the highest selectivity of olefins (on a carbon molar percentage basis, Table 2). The undoped catalyst had the lowest selectivity to olefins, which was much lower than in the case of Li and Na promoted Fe catalysts. This pattern is consistent for both the high and low CO conversion samples, where Rb, Cs, and K promoted iron catalysts produced the highest olefin contents.

Table 2. Product Distribution in moles of C%.

	Support/Promoter	Methane	Oxygenate	Paraffin	1-Olefin	Trans-2-Olefin	Cis-2-Olefin	O/P	A/P
Cobalt 220 °C, 27.6 bar, WHSV = 2–5, H_2/CO = 2	Unsupported	42.20	1.44	42.79	10.56	2.33	0.68	0.16	0.02
	Alumina	14.98	2.51	66.10	13.12	1.00	2.30	0.20	0.03
	Silica	19.30	6.27	53.31	18.29	1.36	1.48	0.29	0.09
	Titania	19.40	2.93	41.03	30.26	3.57	2.81	0.61	0.05
	Carbon	18.00	10.07	58.38	7.89	3.26	2.40	0.18	0.13
	Ceria	28.80	46.59	11.92	8.81	2.04	1.84	0.31	1.14
Ruthenium 220 °C, 19.2 bar, WHSV = 1.5, H_2/CO = 2	Alumina	7.99	3.85	59.96	26.32	1.17	0.72	0.42	0.06
	Titania	8.36	3.40	54.90	24.64	4.90	3.80	0.53	0.05
	Silica	6.76	7.05	51.56	23.71	6.85	4.08	0.59	0.12
	NaY	7.53	2.24	53.70	25.13	7.11	4.30	0.60	0.04
Iron at High CO conversion supported with 5.1 silica. Reaction Conditions 270 °C, 12.1 bar, WHSV = 2, H_2/CO = 0.7	unsupported	46.26	0.25	37.22	12.90	2.33	1.04	0.19	0.00
	LiHC	12.46	2.50	33.71	40.19	6.92	4.22	1.11	0.05
	NaHC	12.87	3.81	34.69	44.27	2.51	1.83	1.02	0.08
	KHC	11.01	2.84	30.38	48.36	4.49	2.91	1.35	0.07
	RbHC	12.08	7.52	29.03	43.90	3.63	3.84	1.25	0.18
	CsHC	11.19	8.23	25.45	47.86	4.21	3.06	1.50	0.22
Iron at Low CO conversion supported with 5.1 silica 270 °C, 12.1 bar, WHSV = 2, H_2/CO = 0.7	Unsupported	17.66	0.57	54.66	18.81	4.52	3.78	0.37	0.01
	LiLC	13.41	5.01	29.99	45.11	3.73	2.75	1.19	0.12
	NaLC	10.43	3.16	31.24	45.19	5.51	4.48	1.32	0.08
	KLC	7.01	7.39	27.11	49.06	5.91	3.52	1.71	0.22
	RbLC	9.33	7.53	28.61	48.99	2.79	2.76	1.44	0.20
	CsLC	9.54	7.26	25.38	53.02	2.30	2.50	1.66	0.21
Iron 270 °C, 12.1 bar, WHSV = 2, H_2/CO = 0.7	Manganese	4.53	4.69	19.90	65.82	1.63	3.43	2.90	0.19

The data in Table 2 is normalized to compare only the oxygenates and four major peaks for FTS. CO_2 and the isomerized/unknown products are not included in this selectivity. Furthermore, only active iron catalysts display CO_2 above 1–2%, and isomerized products more than 5–10%. Lastly, the paraffin column does not include methane.

Overall, two main trends are observed in the FTS hydrocarbon products, as shown in Figure 28. The first trend "a" is observed with lithium, sodium, and low atomic ratios of K (less than 2 atomic ratio to iron) on Fe, and independent of the atomic ratios of copper and silica for the samples that were tested (Figure 29). The actual results are based on the weight fraction from the chromatography in the oil phase. Trend "a" is where the 1-olefin is lower or nearly the same as the relative paraffin material, and the *trans*-2-/*cis*-2-olefins are high, respectively. This trend holds for the lighter hydrocarbons (Figure 8: A, 13), but as the carbon length increases, the weight fraction of the paraffin increases. This could be due to secondary reactions, such as hydrogenation, reinsertion, etc. Trend "b" from Figure 28 is observed with Rb promoter, Cs promoter, high atomic ratios (larger than 2 atomic ratio of potassium to iron) of K promoter and supports like manganese. Trend "b" displays the 1-olefin as being much greater than the respective paraffin (as shown in Figures 16 and 30), and the *trans*-2-/*cis*-2-olefins are low or virtually absent. Once again, this mainly holds for the shorter chained hydrocarbons, but as the chain length increases, the paraffin starts to dominate. This is a general description of the chromatographic distribution of the products, as the selectivities may change depending on the catalyst composition (e.g., the atomic ratios of K). However, there appears to be an inverse relationship in the production of the olefinic material (i.e., as the selectivity of the 1-olefin increases, it decreases for the 2-olefins). This phenomenon can also be found on the single methyl branched material. As the 1-olefinic selectivity is increased, the branching decreases. These relationships can be fully observed when comparing all of the figures side by side (i.e., Figure 8: A, 13, 16, 28–30).

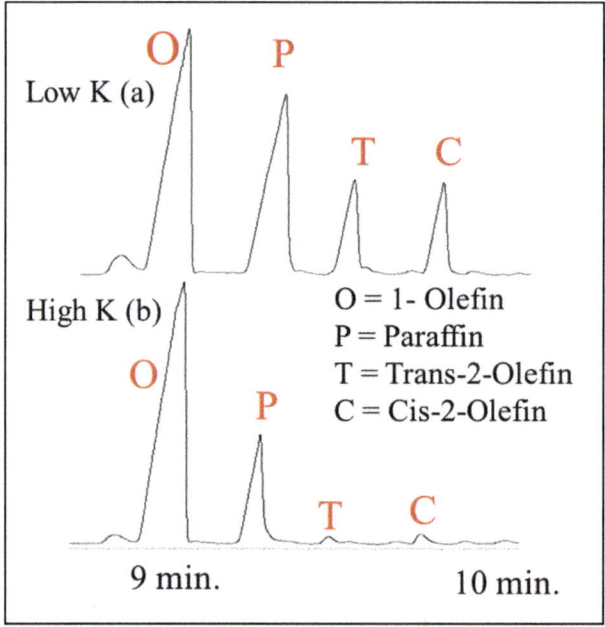

Figure 28. A zoomed in chromatographic picture to display the difference in the main products fraction by the addition of different atomic ratios of K. "Low" is less than 2, (i.e., 2K100Fe), whereas "High" is greater than 2. This figure was built from 9 iron runs (with combinations of Cu, Si and K) catalyst runs that deviated one component while leaving the other two constant (e.g. K was varied while Si and Cu remained constant). The O/P ratio and the 1-olefin/2-olefin ratios only changed by varying K in the recipe. Furthermore, only when K ≥ 2, (2 and 5 atomic ratios) and K < 2 (0.5, 1, and 1, 25 atomic ratios) did these ratios change. Reaction Conditions were 270 °C, 12.1 bar, WHSV = 2, and H_2/CO = 0.7.

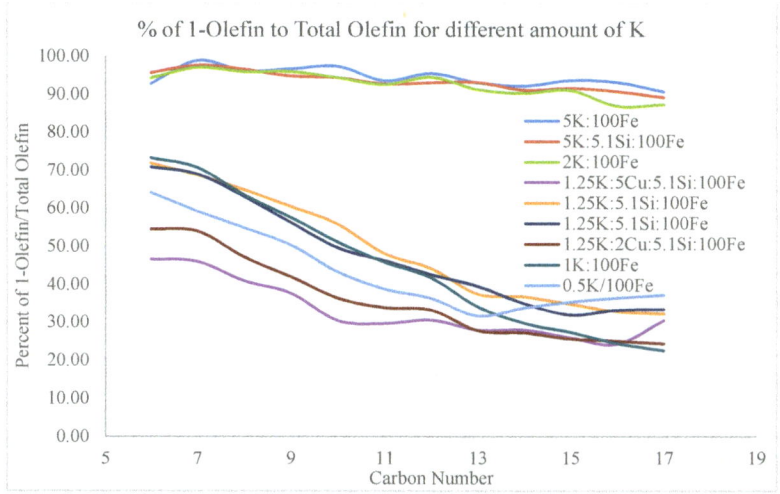

Figure 29. The observed deviations in the 1-olefin/total olefin ratio, by altering the iron catalyst recipe, are most likely due only to the different amounts of potassium. Copper and silica do not affect the olefin ratio observed in the product selectivity. Reaction Conditions: 270 °C, 12.1 bar, WHSV = 2, and $H_2/CO = 0.7$.

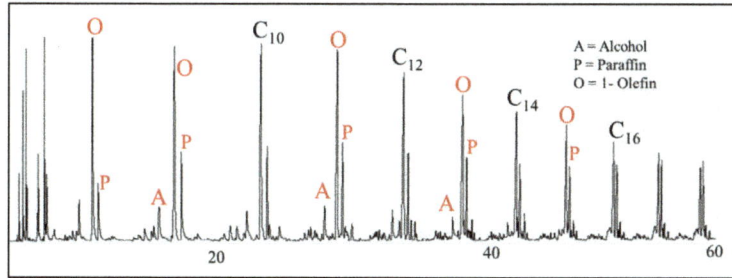

Figure 30. A chromatograph of the product produced with a manganese supported iron catalyst. The x-axis is retention time in minutes. Reaction Conditions: 270 °C, 12.1 bar, WHSV = 2, and $H_2/CO = 0.7$.

Iron, unlike cobalt (Figure 17) and ruthenium (Figure 18), also produces a higher selectivity of branched hydrocarbons (Figures 9–13, 16 and 31). The majority of branched hydrocarbons, as previously displayed in Figures 9 and 10, is the single methyl branched paraffin. The single branched materials always display the 2-methyl functional group and the branched materials contain mono-methyl groups that go all the way up to the middle of the chain (i.e., decane displays 2-methyl, 3-methyl, 4-methyl, 5-methyl; dodecane displays 2-methyl, 3-methyl, 4-methyl, 5-methyl, 6-methyl). The 2-methyl and 3-methyl single branched paraffins are normally the greatest and then decrease in the FID raw data—peak area (weight %)—as the methyl is located toward the center of the chain. The methyl branched isomerized materials, however, are not as evident for a typical cobalt catalyst. This is shown in Figure 14 for a 100% cobalt FTS run. These materials are detected more when CO binding is weak, due to insufficient back bonding. Route A, binding of the vinylic intermediate is weak, as displayed in Figures 9, 10, 13, 28, 29 and 31, display high amounts of the internal olefins, while the 1-olefin is lower. The selectivities of isomerized products and internal olefins both trend inversely to the 1-olefin.

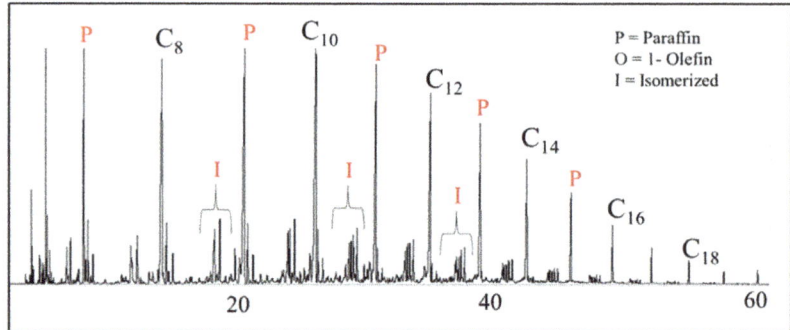

Figure 31. A chromatographic picture of the oil for an 100% iron catalyst. The *x*-axis is retention time in minutes. Reaction Conditions: 270 °C, 12.1 bar, WHSV = 2, and $H_2/CO = 0.7$.

Oxygenate production increases as the paraffin content decreases, where the Rb- and Cs-doped iron catalysts had the highest carbon mole % for oxygenate content. For example, the total carbon mole % of oxygenates for Rb- or Cs-promoted Fe is more than three times that of the K-promoted iron catalyst at high CO conversion. Furthermore, unlike the K promoted Fe catalyst, the total carbon mole % of the oxygenate content does not seem to depend on CO conversion (Table 2). The low conversion sample for the potassium doped iron catalyst produced a higher selectivity of oxygenates, which was more than three times the selectivity when compared to that of the high conversion sample. The high CO conversion sample for the Li doped iron catalyst displayed the lowest selectivity of oxygenates for the Group I series (i.e., more in line with cobalt alumina), whereas the sodium doped iron catalysts produced the lowest selectivity of the oxygenates at low CO conversion. The undoped catalyst produced a much lower selectivity of oxygenates (Table 2) than the promoted iron catalysts for the Group I alkali series (unsupported iron—i.e., 100%Fe—had the lowest selectivity of oxygenates of all the catalysts that were tested, even unsupported cobalt). K- and Li-doped iron catalysts exhibited the largest differences in carbon molar percentages for oxygenate selectivity between the high and low CO conversions. Cs and Na doped Fe catalysts had a smaller difference, whereas Rb and the undoped iron catalysts exhibited virtually no change in the selectivity of oxygenates that are produced at both high and low CO conversions.

The manganese supported iron catalyst displayed the highest fraction of olefins, mostly consisting of the 1-olefin (Figure 30), of all the catalysts tested. The oxygenated products were present, but manganese mainly promoted olefin selectivity.

3.4.2. Cobalt

Unlike typical iron catalysts, conventional cobalt catalysts consistently produce a very clean chromatogram where the straight chain paraffin is predominant. The product distribution that is displayed in the oil and wax products for cobalt seems to be more dependent upon the support, and less so on promoter type (Pt, Ru, Ag) and loading; the promoter does affect the distribution, but it mainly affects methane and the other hydrocarbon selectivities [101,102]. The addition of potassium to cobalt FT catalysts resulted in a promotion of chain growth [103], but the effect reached a maximum. The paraffin still dominates for the entire range of products (Figures 14, 17 and 32), with only slight fluctuations in the alcohol and olefin contents. From the results of catalyst testing, in hydrocarbons that were longer than C_8, the paraffin was more dominant than the olefin. In only one catalyst, cobalt promoted with ceria, was the paraffin dominated by another product—not the olefin, but rather the alcohol (Figure 19).

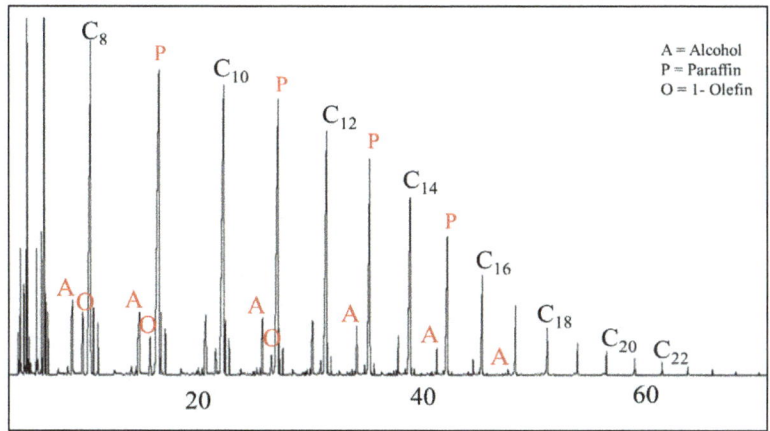

Figure 32. A chromatographic picture of a carbon supported cobalt catalyst. The x-axis is retention time in minutes. Reaction conditions: 230 °C, 27.6 bar, WHSV = 0.5, and $H_2/CO = 2$.

Recent work that was by the Davis group [40–42,104,105] focused on attempting to make cobalt catalysts behave more like iron catalysts (i.e., consisting of Fe carbides in close vicinity to defect-laden Fe oxides) by placing Co nanoclusters in close proximity to the reduced defect sites (e.g., O-vacancies and their associated Type II bridging OH groups). As depicted in Figure 33, one view is that (a) CO reacts with Type II bridging OH groups to form formate species and that these formate species can transfer molecular CO across the metal-support interface to terminate the FTS hydrocarbon chains. Or, (b) the mobile bridging OH groups may add directly. Hydrogenation of the added terminal -CO would lead to linear alcohols, while dehydration is proposed to result in linear olefins. Interestingly, as a function of time on-stream, Ce-containing Co catalysts produced significantly higher selectivities of olefins and oxygenates, with the oxygenates being almost exclusively linear alcohols. Surprisingly, as a function of time on-stream, the olefins selectivity decreased with a concomitant increase in the selectivity of linear alcohols, such that the sum of olefins + linear alcohols was virtually constant with time on-stream, suggesting that the two products were derived from the same adsorbed species. In examining Table 2 and summing up the total of olefins + linear alcohols, it is clear that ceria and titania supported cobalt catalysts exhibited the highest selectivity to these products (55.4% and 49.7%, respectively), lending support to the proposed mechanistic viewpoint that defects in the oxide promote total olefin + linear alcohol selectivity. On the other hand, unsupported Co and Co that was supported on relatively inert supports (Al_2O_3, carbon, and SiO_2) exhibited significantly lower selectivities to olefins + linear alcohols (15.6%, 18.0%, and 24.6%, respectively), as shown in Figure 33.

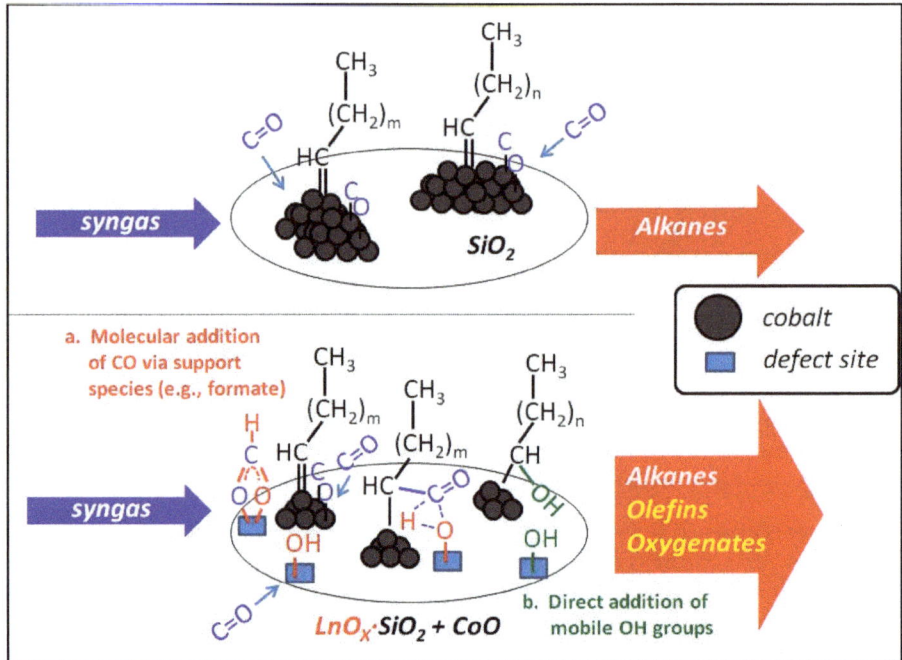

Figure 33. A mechanistic display of CO molecular addition through defect sites.

3.4.3. Ruthenium

Ruthenium catalysts more closely trended to the cobalt catalysts, where the bulk of the product observed was straight chained paraffins. Compared to cobalt catalysts, ruthenium catalysts produce a higher selectivity of olefins, which depend on the support. Furthermore, ruthenium on NaY produced sufficient 3-olefin to be observable over several hydrocarbon lengths (Figure 21). The titania supported ruthenium catalysts displayed a higher selectivity of olefins, which exceeded that of paraffins up to the diesel range. However, unlike iron, the four major products (by selectivity in molar carbon %: paraffin, 1-olefin, cis, and trans 2-olefin) were clearly observed, being similar to cobalt. Furthermore, as with cobalt, little to no methyl branched material was observed in the chromatograph, and no oxygenates were detected other than linear alcohols.

The selectivity of oxygenates for all of the ruthenium catalysts (Table 2) was slightly higher (1.3% higher for alumina, 0.8% for silica, and 0.5% for titania, respectively) than for the cobalt catalysts (alumina, titania, and unsupported). However, none of the ruthenium catalysts displayed oxygenate selectivity that was as high as the low CO conversion K-promoted Fe catalyst, the Rb-promoted Fe catalyst, the Cs-promoted Fe catalyst, the Co/carbon catalyst, and the 0.5%Pt/Co/ceria catalyst. However, the distribution was much like that produced by a typical cobalt catalyst, where the only oxygenates found were linear terminal alcohols.

3.5. A Mechanistic Description

3.5.1. The Electronic Nature of the Active Metal

The underlying concept behind the proposed mechanistic routes—to be discussed—is driven by the localized electronic environment of the active FTS metal (at low temperature), which determines the degree of back-donation of electron density. The nature of CO, a π-acceptor ligand, has the capability to interact with a metal in three ways, depending on its environment, through: (1) a weak back donation

-M-C≡O; (2) a semi-strong back donation -M=C=O; and/or, (3) a strong back-donation -M≡C-O [106]. The richer the electron density of the active metal, the more back donation to the $2\pi^*$ anti-bonding orbitals of CO (Figure 34) [106–108]. This localized electronic structure of the metal can be changed by the addition of a promoter (e.g., with iron catalyst, alkali is used). These additions to the active metal could potentially be a way of "tuning" the FTS catalyst for CO adsorption (with the potential to affect H_2 as well); too poor and CO adsorption is weak, causing issues with CO hydrogenation; too rich and CO adsorption is too strong where carbonaceous deposits begin to form. Typically, because of orbital overlap, even though metals, like copper, have more electron density to provide, the large d-orbitals from the metal do not pair well with the smaller carbon-carbon π orbitals and greatly limit this capability [109]. As a trend, the transition metals toward the left would yield better back bonding; yet, the rate at which oxidation occurs for those on the left apparently diminishes the capability for FTS [110]. Discussions here will only revolve around the differences among the active FTS catalysts. Further discussions of iron will be in the active carbide form, not the metallic form.

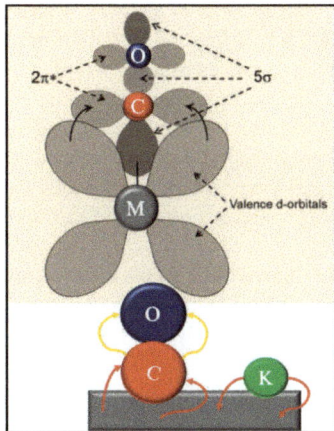

Figure 34. An atomic view of how the alkali could be affecting CO bonding.

Once the CO is adsorbed, vinylic species are described to form and serve as the active intermediates, where the FTS chains grow to eventually terminate. Many works from the literature suggest that the rate determining step involves hydrogen [45,49,100,110–113], but not the dissociation of H_2 [114–118]. As a result, the stabilization of chain growth intermediates may rely more on the tuning of the catalysts, specifically for the reactivity of CO. However, this does not discount observations regarding the competitive dissociation of H_2 in Ni and Co catalysts, where heavier oils and waxes that were produced from FTS interfere with hydrogen dissociation [119,120].

Since both the vinylic intermediate and CO exhibit π bonding, they will both be susceptible to electronic back-donation from the catalyst surface. Thus, if the CO adsorbs weakly, then the vinylic intermediates will not be stabilized, making their formation difficult, such that most of the carbon will form methane. The vinylic intermediates that do form will weakly adsorb, and the growth will not be as structured. This mechanism is proposed to be favored on iron carbide (Figure 31). If CO dissociates readily and H_2 is not significantly suppressed, again methane selectivity is observed to be high. Because CO readily dissociates, the formation of vinylic intermediates will be difficult, as the ones that do form could easily dissociate back to a single carbon atom. The active catalyst will have high surface coverage, which leads to coking, which is proposed to be the case with nickel catalyst. However, if CO adsorbs and remains stable enough (e.g., to form enol species or other relevant species), the vinylic intermediate that subsequently forms (e.g., from the condensation of two enol species) will also be stabilized, allowing for controlled chain growth. This is proposed for Ru and Co catalysts

(14, 15, 17, 18, 21, 33), where the main products (i.e., linear paraffins and linear 1-olefins) take up the bulk of the distribution, with few minor products being observed. Lastly, if CO dissociates readily, as observed with K promoted catalysts, and further observed with catalysts that are promoted by the heavier alkali metals (e.g., Rb, Cs), the main fraction of the products consists of linear paraffins and 1-olefins. However, mono-methyl branched paraffins and olefins, whose distributions appear to follow a specific trend (to be discussed), are contained in the remaining products. Since H_2 is limited, the vinylic intermediates will desorb without hydrogenating, resulting in alkenes and oxygenates (acids, aldehydes, esters, and alcohols), which limits paraffin production (Figures 11, 12 and 20). Cobalt catalysts that were modified by supporting the cobalt nanoparticles on defect-laden carriers were, as discussed previously, shown to produce linear olefins and linear alcohols; however, only iron FTS catalysts were found to produce a variety of oxygenated materials.

Deviations in the product selectivity are the highest for iron (i.e., iron selectivity appears to be more tunable) and much more so than for Co, Ru, and Ni. These deviations are apparent even in the physical attributes of the products (e.g., color, as shown in Figure 2). The deviation suggests diversity in the active phase (e.g., brought on by addition of K) [121]. A reason why different carbides are more active toward FTS (without K) could be due to the ability of iron to donate its electron density to CO. This could be a reason certain phases, such as cementite, are not as active as others, such as Hägg. However, there are several factors (e.g., morphology) that come into play [121].

The important intermediate considered in mechanistic schemes presented is adsorbed 2-carbon ethylene species, which was previously proposed in early work [122]. However, unlike previous work, chain growth likely does not proceed via ethylene, but rather by single carbon atoms. Evidence has shown that an ethylene like intermediate acts as a chain initiator, not as a chain propagator [122]. Furthermore, propylene was found to exhibit hydrogenation ability, but not a significant capability to initiate chain growth [122]. Because of the similarity in π bonding that the proposed vinylic intermediate and CO exhibit, when CO does not bind well to the catalyst surface, neither should the vinylic intermediate. As the binding of the vinylic intermediate is weak, the chain growth would not be as ordered (i.e., directional, forming linear chains); thus, methane and methyl-branched materials would be expected to form in higher selectivity. If CO dissociates too easily, then the adsorbed vinylic intermediate would follow suit, and thus methane would be expected to be a major product. If CO binds well, but H_2 is not as accessible, then the vinylic intermediate should be stabilized, but the termination by hydrogenation to paraffin would be limited, leading to higher selectivities of olefins and oxygenates. The fact that olefins have been found to reincorporate into growing chains on cobalt and ruthenium lends some support to the proposed vinylic intermediate [121].

The reason for the sp^2 carbon-carbon interaction is that olefins are π-acceptor ligands; their π orbitals interact with the d-orbitals of the metals, as shown in Figure 34. These alkene interactions are taken from the Dewar-Chatt-Duncanson model [123], which describes how alkene ligands interact with metals, and how these interactions between ligand and metal are dependent on the electronic density of the metal. The vinylic intermediate, as a π-acceptor, is affected in a similar manner to that of the adsorbed CO: as electronic donation from the metal increases, the carbons take on more sp^3 character, as shown in the pseudocyclopropane (defined as X_2) structure on the right. Here, the carbon atoms are more nucleophilic in nature, possessing greater electron density. As these carbons take on more of the X_2 configuration through increased back donation from the metal, growth (either by the addition of CO or CH_2) occurs solely at the C_x position. In contrast, with weak back donation from the metal, the alkene may interact in such a way that the carbons exhibit more sp^2 character, as shown by the simplified linear molecule on the left in Figures 35 and 36 (denoted as L), where the carbon atoms are more electrophilic in nature. This weak interaction would lead to less controlled growth (i.e., by addition of CO or CH_2) at either C position, resulting in internal olefins, branched paraffins, aldehydes, ketones, and acids.

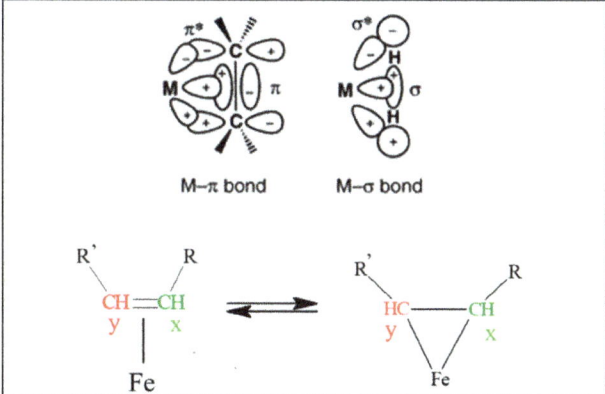

Figure 35. Alkenes are known as π donors. Above represents the Dewar-Chatt-Duncanson model for the orbital orientation between the π orbitals of the alkene and the d orbitals from the metal. The electronic density of the metal will determine the hybridization of the carbon, where the planar olefin adduct (left structure) displays more sp^2 hybridization for the carbon, and the metallocyclopropane exhibits more sp^3 character.

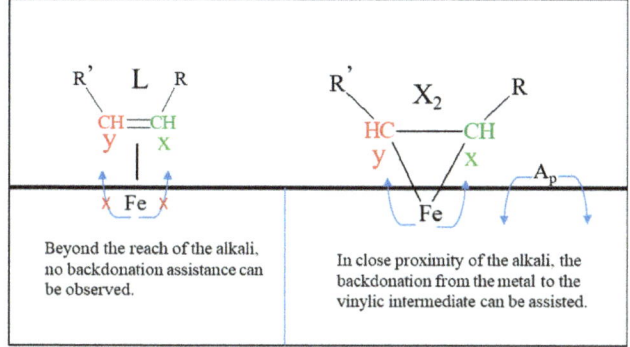

Figure 36. This figure describes types of possible bonding for alkenes in the electronic effects brought on by the addition of a promoter (A_p), such as the Group I alkali series. As the size of the promoter increases, so does the iron surface area that it affects; as the alkali basicity increases, so does its ability to assist in back donation of electron density.

3.5.2. A General Mechanistic Explanation

There is a large body of work available, including CO and H_2 adsorption studies [23,124–133] and isotopic kinetic modeling [20,22,24,46,47,52,53,98,110–120,122,132,133] investigations of FTS, which provide some insights into the mechanism. Furthermore, tremendous amounts of time and effort have been expended in investigating iron catalysts and the effect of promoters (i.e., mainly potassium) for iron FTS catalysts [134]. This review focuses on the product distribution of different types of FTS catalysts in an attempt to shed light on some aspects of the FTS mechanism and the fundamental differences in the nature of catalytic surfaces that affect the product distribution. Given the large volume of FTS literature, it is interesting that the question of how CO interacts with FTS surfaces is still under debate; this is not surprising, when considering the complexity and diversity of iron carbide surfaces that can be formed (Figure 37).

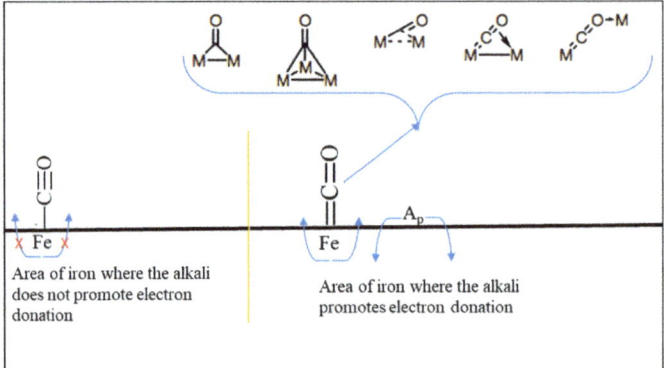

Figure 37. This figure describes types of possible bonding for CO depending on electronic effects that are caused by the addition of a promoter (A_p), such as the Group I alkali series. As observed with the vinylic intermediate, as the size of the promoter increases, so does the iron surface area that it affects. As the alkali basicity increases, so does its ability to assist in back donation of electron density.

Figure 38 depicts the first few steps of the various proposed mechanisms of CO adsorption and emphasizes the back-donation of electron density from d-orbitals of FTS metals, regardless of orientation or crystal structure. Variations in back-donation are observed in IR spectroscopy in the vibrational modes and frequencies of adsorbed CO, such that the following configurations are observed: Route A: M-C≡O; Route B: M=C=O; and, Route C: M≡C-O [135]. The growing chain has two carbons that are attached to the metal (until the chain is terminated), C_x and C_y, which have sp^2/sp^3 like character where the chain growth on C_x is preferred, as displayed in Figures 35 and 36, especially so in the case of the X_2 configuration. Each of the three routes is shown in Figure 38:

Route A: Based on CO insertion, and perhaps only occurring with the iron carbide catalyst, this mechanism may proceed when H_2 can readily adsorb onto the surface and dissociate, but when CO adsorbs molecularly, because scission of the CO bond without H*, is not kinetically favored [136]. The alkene takes on an L configuration, where chain growth at both C_x, and C_y is possible, as shown in Figures 35, 36 and 39. In that case, termination may produce internal olefins, branched paraffins (Figure 31), etc. This is not the primary pathway for any of the active metals used in FTS, such as cobalt or ruthenium; rather, it is only proposed to operate on the surface of unpromoted iron carbide (Figures 39–42).

Route B: CO scission is not as difficult as in Route A, but CO still adsorbs in an associative manner, followed by being dissociated with the aid of H* on the active metal surface. An example of a possible mechanism stemming from this is the enol mechanism. The X_2 configuration of the vinylic intermediate is dominant, providing well-ordered growth, decreasing the overall product diversity (Figure 43). This route is expected for cobalt and ruthenium catalysts.

Route C: Back donation from the metal is very strong and CO dissociates (Figure 44). Neither the L or X conformation of the vinylic intermediate exists as back donation is too strong, precluding its formation. This route is primarily expected for nickel catalysts.

This is a conceptual view, and complications arise given the heterogeneity of the active metal catalyst, and the proximity of the additives (promoters, active supports) to the active metal. This is all in the understanding that deviations in the expected product distribution may occur through reactor hold up, as previous research describes, such that secondary reactions may occur due to the re-adsorption of olefins, such as: reinsertion, hydrogenation, and isomerization [45,137–149].

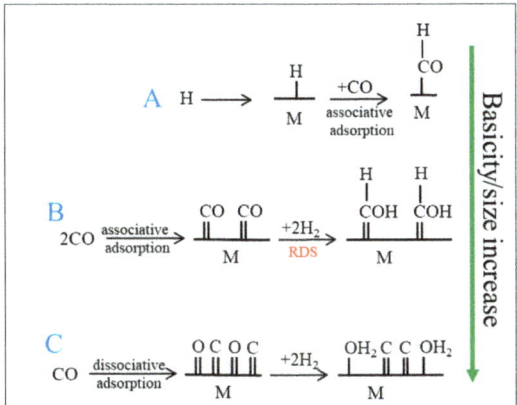

Figure 38. A mechanistic starting point used to describe the effects induced by changes in size and basicity of the alkali promoter on the iron catalyst.

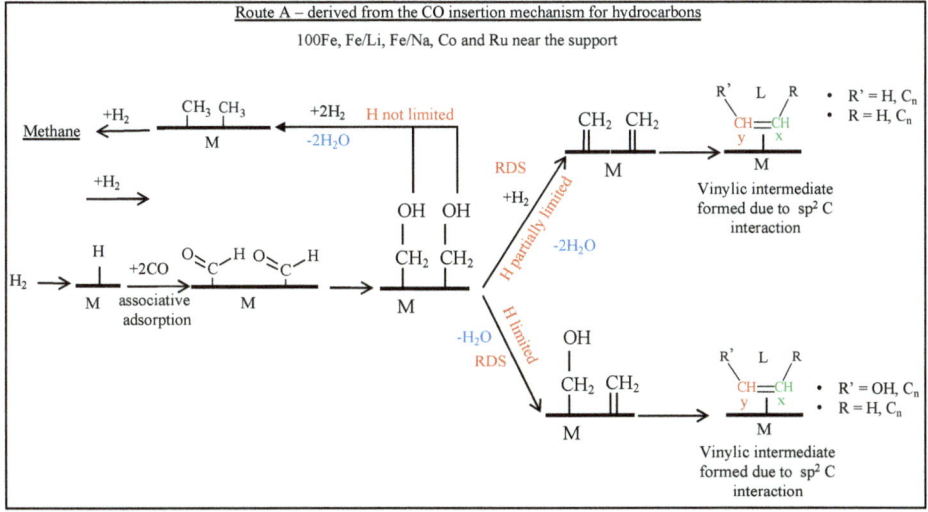

Figure 39. Proposed route A for the formation of several products observed in FTS, based on the CO insertion mechanism. Mechanistic routes for methane will be preferred when the back-donation of electron density from the metal is poor. This weak back donation does not easily allow for the vinylic intermediate to be formed or stabilized to the extent that is needed for chain growth. Furthermore, if the back-donation of electron density from the metal is too great (as with an activated nickel catalyst), then CO will dissociatively adsorb, and the vinylic intermediate, if formed, will be easily dissociated back to C_1, to again prefer methane formation. Thus, the careful tuning of electron back donation from the metal must be maintained to achieve desired chain growth.

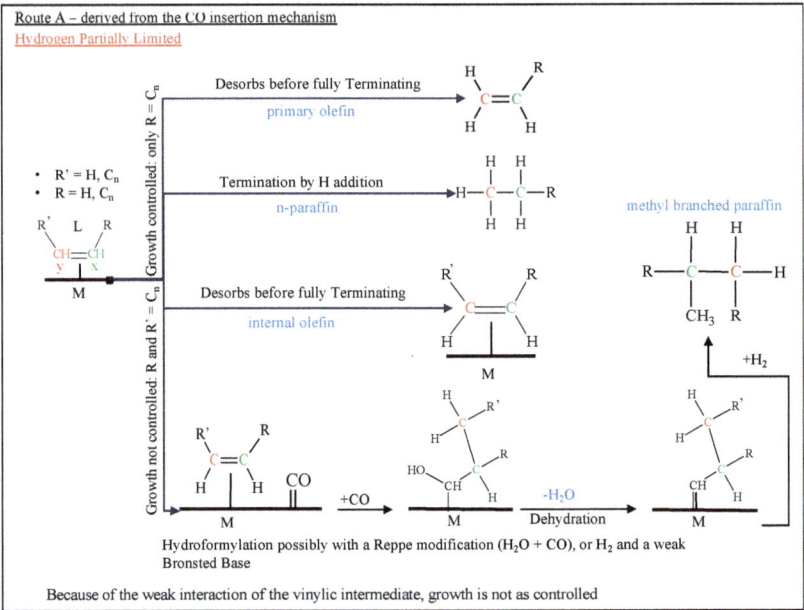

Figure 40. Taken from the previous route, where the vinylic intermediate takes on the L configuration, several products can be made because there is less controlled chain growth.

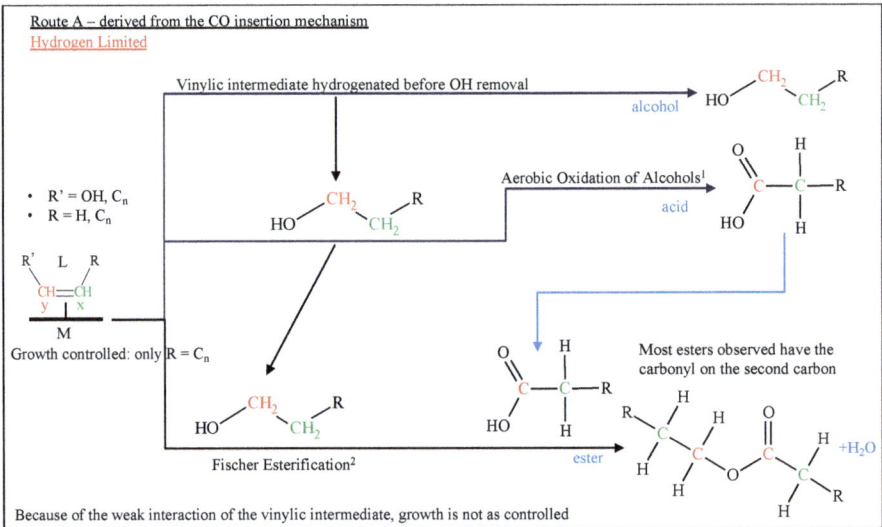

Figure 41. Proposed route A for the formation of several products observed in FTS, based on the CO insertion mechanism (Figure 38). This is the case where growth is somewhat controlled, but hydrogen is limited. However, given that growth occurs through CO insertion—alternative oxygenated materials could be made.

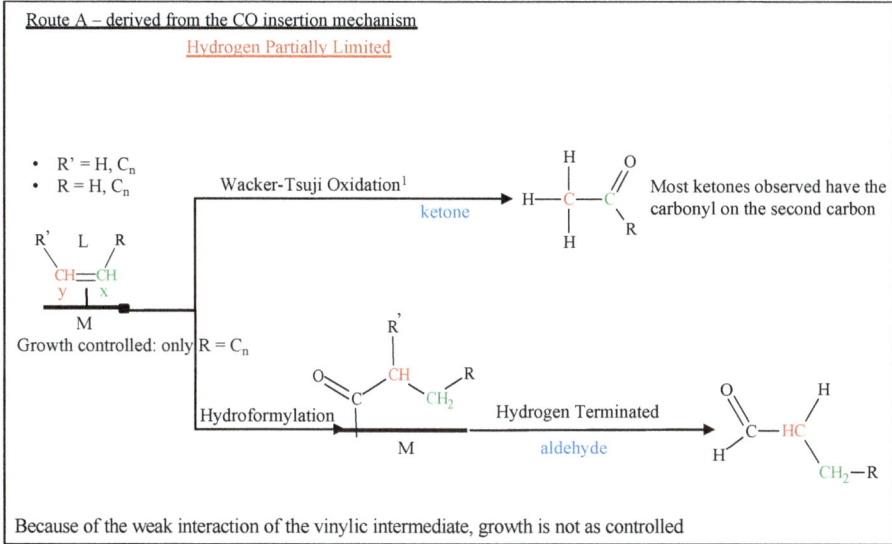

Figure 42. Proposed route A for the formation of several products observed in FTS, based on the CO insertion mechanism (Figure 38).

While the product distribution for primary versus secondary products is still under debate [20,22, 24,46,47,52,53,98,110–120,122,132,133], the findings from analytical work to date on various catalysts suggest the following points:

1. n-paraffins, 1-olefins, and linear alcohols are primary products. These products are observed through route B
2. n-paraffins, 1-olefins and linear alcohols come from the same active X_2-configuration, as depicted in Figures 35 and 36. Furthermore, 1-olefins and 2-olefins are, for the most part, not produced from each other (i.e., 2-olefins are not derived from 1-olefins and vice versa). Deviations are also possible with longer chained material through secondary reactions, especially under conditions where there is considerable reactor holdup. When X_2 is the dominant configuration, all other products, including branched paraffins, 2-olefins, etc., likely come from secondary reactions.

In active iron carbide FTS catalysts, the back-donation of electron density directly influences the bonding configuration of the vinylic intermediate. Without the addition of alkali promoter, poor back-donation results, Route A is dominant, and CO dissociation is kinetically unfavorable at these conditions, as displayed in Figure 30 where the major products are light n-paraffins, methane, and isomerized hydrocarbons. Once a promoter is added, there is a localized effect on the metal (Figure 37), and both Routes B and C will occur, depending on the promoter's capability to affect the electronic structure of iron carbide. The proximity of the promoter to the metal is likely a key factor in determining the extent to which each route occurs. However, given the capability of CO to bond in several different configurations, (e.g., linear, bent, and several distinct types of bridge bonded configurations), its bonding is much more complicated than that of the vinylic intermediate (Figure 36). As the promoter size/basicity or atomic ratio is increased, its capability to influence the metal is increased.

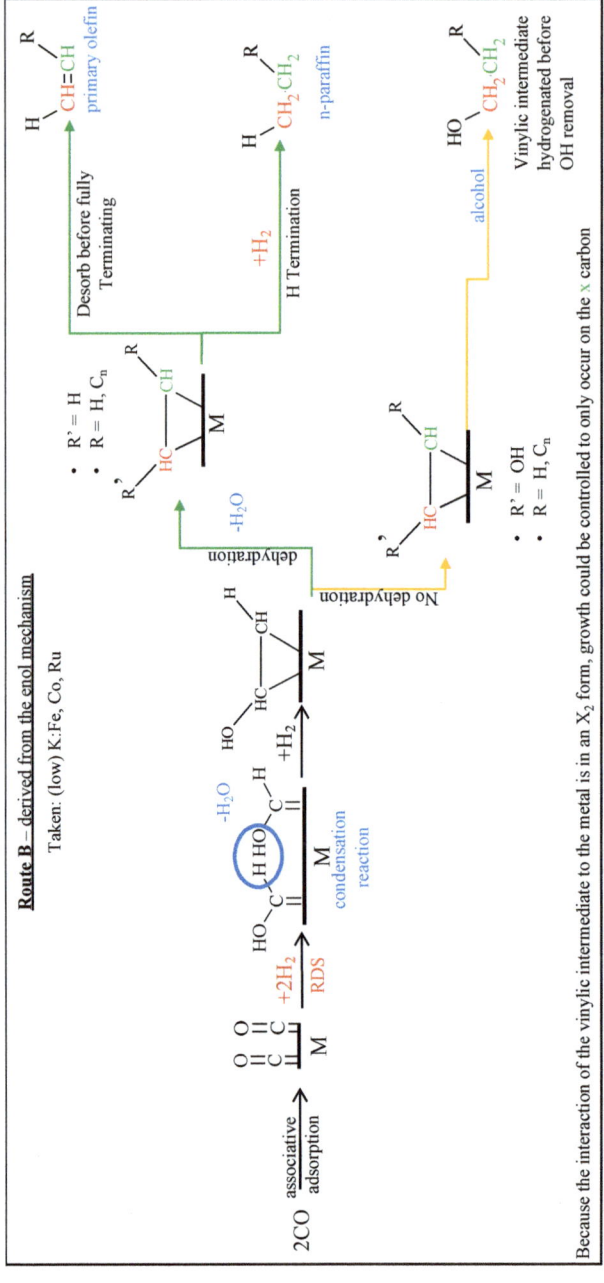

Figure 43. Proposed route B for the formation of several products observed in FTS, based on the enol mechanism.

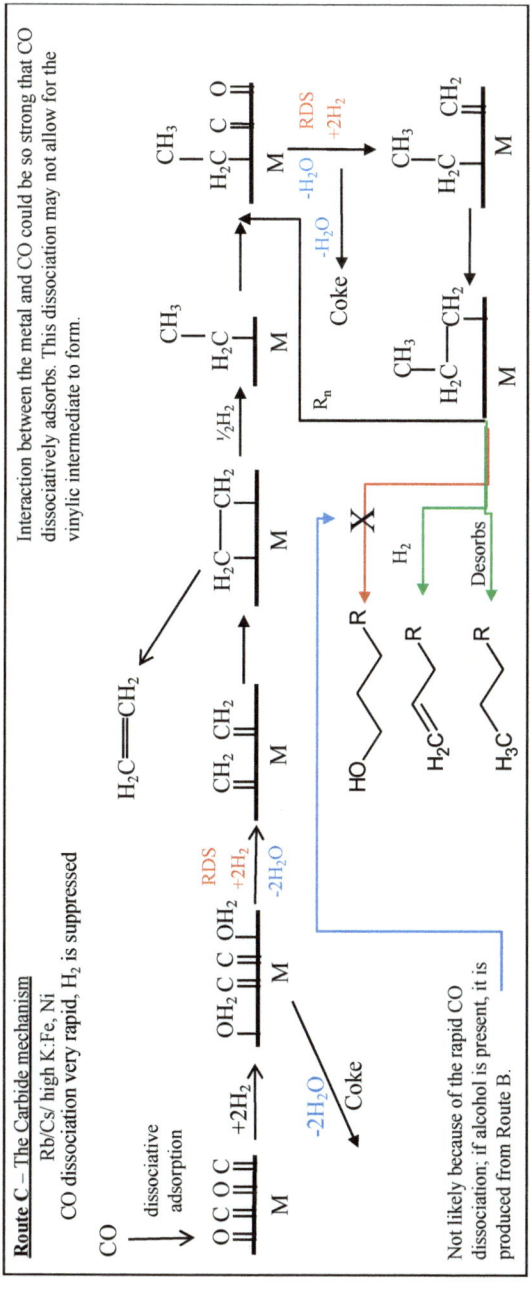

Figure 44. Proposed route C for the formation of several products that are observed in FTS, based on the carbide mechanism. Here, methane and coke are formed, where the back-donation of electron density is too strong; thus, the strong back-donation of electron density precludes the formation of the vinylic chain growth intermediate. With metals like nickel, where hydrogen is not as suppressed, methane would be the main product. However, as hydrogen becomes suppressed due to the high dissociation rate of CO (because of the increased interaction), the surface becomes covered with carbonaceous deposits.

Simply put, there needs to be a balance on the active surface of the CO, H, and chain growth intermediates. This balance is maintained by how the active metal reacts with CO through its back donation of electron density. Evidence of this balance could be further strengthened by kinetic isotope effect (KIE) investigations for FTS, where a feed having an equimolar mixture of H_2/D_2 resulted in the H/D ratio for conversion being typically less than 1 (and closer to 0.85—i.e., deuterium converted to a greater extent than hydrogen [20,22,24,46,47,52,53,98,110–120,122,132,133,150]). If the rate determining step is through termination by the addition of H, where the H adsorbed on the metal is taken up by carbon (as theoretically predicted in Figure 45), then the expected value of the KIE should be less than unity. To bolster this viewpoint, the calculations were completed by Shafer et al. [118] using the average infrared (IR) stretching frequencies of the M-H (in the range of 1700–2250 cm^{-1}) and C-H/C=H/C≡H. More isotopic work needs to be completed to better understand the KIE for FTS [20,22,24,46,47,52,53,98,110–120,122,132,133].

This concept, as simple as it is, could also explain why, in an FTS environment, the cobalt and ruthenium product distributions are so different from that of iron carbide.

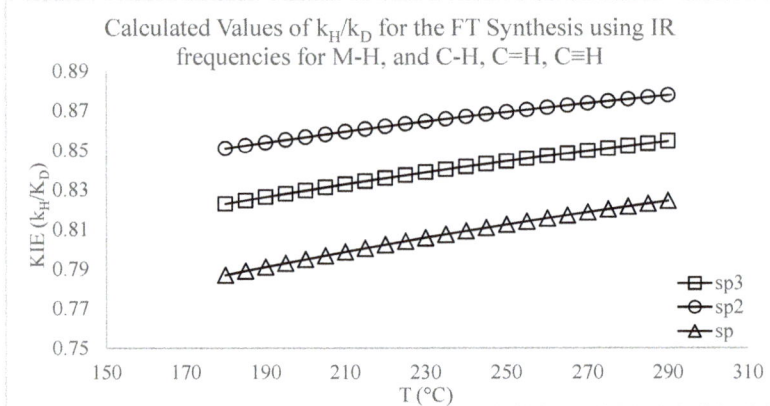

Figure 45. The calculated KIE value for FTS where H addition is the rate determining step, provided that the KIE for FTS was not significantly affected by hydrogen coverage. These calculations follow the ones that were performed by Shafer et al. [114–118]. Ratios are derived from the activation energy using IR stretching frequencies as the energy differences. The k_H/k_D values are taken from the average cm^{-1} for the M-H range (1700–2250 cm^{-1}). The standard deviations are ±0.06.

4. Product Distribution

4.1. Iron Catalysts

4.1.1. Hydrocarbons

The most telling feature in the addition of a base (i.e., an alkali like potassium) is the correlation between the increase in the 1-olefin selectivity and the decrease in methane selectivity, as shown in Table 2. 1-olefin selectivity tended to increase with increasing promoter basicity/size with iron catalysts. However, the internal 2-olefins show independence from the influence of the promoter (Table 2). The 2-olefins, being independent from the promoter, provide a good indication of the localized effects of the alkali on the iron, as shown in Figure 37. That is, some iron carbide surface has good contact with the alkali, such that vinylic intermediates exhibit the X_2 configuration (major products are linear alcohols, 1-olefins, and n-paraffins) and some patches on the surface lack promoter, resulting in vinylic intermediates that have more L character (major products are n-paraffins, monomethyl-paraffins).

Thermodynamic modeling of the catalyst independent of the FTS product distribution suggests that olefins and paraffins are the primary products [54,137–139,151]. Moreover, the influence of the promoter on the WGS rate, although it seems to increase with the alkali series, must also be considered, as it affects the partial pressures of H_2, CO, H_2O, and CO_2 in the reactor and could alter their surface coverages [152–156]. Pendyala et al. demonstrated the importance of the alkali on the carburization rate [157]; they found that adding water resulted in a positive effect in CO conversion at 270 °C. This showed that even under more oxidizing conditions, adding water did not oxidize the catalyst, because the carburization rate was sufficiently high. Yet, at 230 °C, which slowed the carburization rate of the catalyst, the addition of water oxidized the iron carbides, and FTS activity was lost. Recent TPR-EXAFS/XANES investigations by Ribeiro et al. [35] and Li et al. [158] indicate that the addition of alkali promotes the carburization rate of iron catalysts.

The findings are in line with the idea that adding potassium increases the back-donation of electron density for better adsorption and dissociation of CO; moreover, improving the back donation of electron density to the adsorbed vinylic intermediate results in more control, narrowing the product distribution by lessening the ability for internal olefins and branched hydrocarbons to form. Figure 39 describes chain growth via this proposed $C_x = C_y$ vinylic species. There has been good evidence that supports vinylic intermediates as chain propagators [118,159,160], which may also explain the capability of olefins to readsorb and potentially reincorporate into the growing chain.

In the current context, the authors are not considering secondary reactions, but rather the effect that the addition of promoter has on changing the electronic density of the active metal. This, in turn, may alter the prevailing pathway for the formation of primary products (Figures 38–44). This idea could indicate two different routes for the production of alkenes (Figures 38, 39 and 42), or possibly two different active sites that solely depend on the electronic structure of the iron. Shi et al. noted an independence of the formation of the 2-alkenes from the 1-alkenes when carrying out H/D exchange [161,162]. This is an indication that the formation of internal alkenes (i.e., again, excluding secondary reactions) is likely on a different pathway than the one that produces 1-alkenes. Shultz proposed, as shown in Figure 46, how secondary olefins may be formed on a separate pathway from that of the primary olefin. The same independence is observed in Tables 2 and 3 for the Group I alkali promoter series—the 1-olefins increase with the series, whereas no real trend is observed for the internal olefins. Instead of conceptualizing the differences in 1-olefin and internal olefin selectivities regarding different pathways per se, we emphasize that the differences stem from differences in energetics due to the location on the catalyst where the chain initiates. If, as displayed in Figure 37, the localized basicity that is due to the alkali is higher, then the 1-olefin selectivity would tend to increase. Moreover, if the basicity is higher in highly localized positions on the catalyst surface, then the internal olefin selectivity would be less sensitive to the alkali, as there would be significant surface available that is not in sufficient proximity to the alkali (i.e., where the internal olefin would form). This could explain the inverse relationship between the 1-olefin and internal olefins that has been observed from multiple runs across various the iron catalysts considered (Figures 28 and 29).

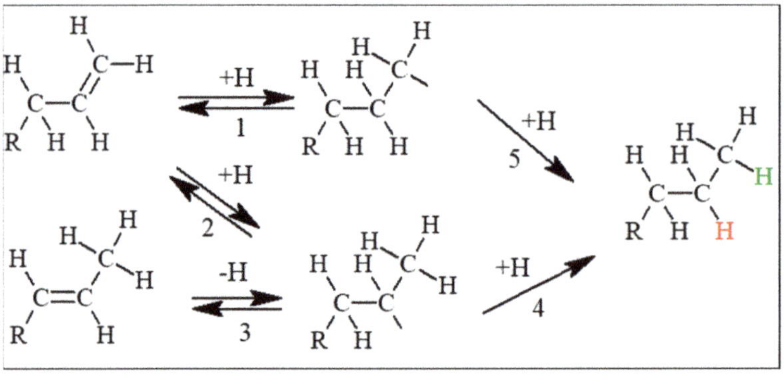

Figure 46. A redrawn mechanism taken from Shultz et al. [51] describing a possible route for the formation of the 2-olefins from the 1-olefin though reversible hydrogen addition.

Results herein (Figures 9–11, 31, 47 and 48, Tables 4 and 5) suggest that the predominant mechanism for branching depends on where on the surface of the catalyst the chain initiates. Several catalyst tests in a CSTR have probed the products formed from a variety of FTS catalyst series with iron (i.e., varying potassium atomic ratio, using different alkali metals, using an unpromoted catalyst, etc.) as summarized in Figures 8A, 9, 10, 11, 12 and 13, Figures 16, 20 and 28, Figures 29–31, as well as Tables 2–4). In each case, the selectivities of 2 and 3-methyl branched alkanes were the highest, followed by a decrease in the selectivity as the methyl was positioned closer to the central carbon (i.e., 4 methyl > 5 methyl > 6 methyl etc.). There is a pattern that was observed in the isomerized products, independent of potassium (Figures 10 and 48), suggesting that these are the products of the FTS process and not secondary reactions. These patterns are obtained from the different methyl branched products that were observed for each carbon number, not by plotting the selectivity of isomerized products having a specific methyl position across a series of products [163].

Table 3. Standard Deviations of the Products Based Upon the Change in CO Conversion.

Catalyst	Alpha		Product Distribution							
	Hydrocarbons	Oxygenates	Oxygenate	Paraffin	1-Olefin	Trans-2-Olefin	Cis-2-Olefin	O/P	A/P	Methane
No Alkali	0.01	0.03	0.23	7.90	4.18	1.54	1.94	0.13	0.00	20.22
Li	0.09	0.05	1.78	1.96	3.47	2.26	1.04	0.05	0.04	0.67
Na	0.09	0.11	0.46	4.17	0.65	2.12	1.87	0.21	0.00	1.73
K	0.03	0.09	3.22	5.14	0.49	1.00	0.43	0.26	0.10	2.83
Rb	0.13	0.04	0.01	2.25	3.60	0.59	0.76	0.13	0.01	1.95
Cs	0.12	0.05	0.69	1.22	3.65	1.34	0.39	0.11	0.01	1.17

The mole of carbon % of change in the product totals, and ASF plot alpha values by changing CO conversion; all numbers are displayed as standard deviations for values (provided in Tables 1 and 2) at high CO conversions (~63%) and low CO conversion (~23%). CO conversion values are displayed in Figure 32. The orange colored numbers denote a loss in value by the decrease in conversion (i.e., the alpha value for the hydrocarbons for the lithium promoted iron catalysts displays an orange 0.9 standard deviation; the value for the alpha at low conversion was lower than the alpha value at high CO conversion). Reactor Conditions: 270 °C, 12.1 bar, WHSV = 2, and H_2/CO = 0.7.

Table 4. The Percentage Difference in Methyl-branched to N-paraffins for a Range of K Promoted Iron Catalysts. Reactor Conditions: 230 °C, 12.1 bar, WHSV = 2, and H_2/CO = 0.7.

Totals	% Total Mole of Carbon					
	0%K	2%K	4%K	6%K	10%K	
Ph	79.80	76.86	78.13	76.60	62.79	
2Me	3.27	3.38	3.18	3.91	4.26	
3Me	3.53	3.28	3.03	3.99	4.37	
4Me	2.05	2.14	1.92	2.30	2.89	
5Me	1.34	1.20	1.72	1.48	2.10	
6Me	0.77	0.69	1.04	0.78	1.02	
7Me	0.34	0.30	0.54	0.32		
Unseparated	1.05	1.06	1.69	1.23	1.98	
Total	12.35	10.98	13.12	14.01	16.61	

Table 5. A Comparison of the Isomerized to Normal Mole Fraction over a Series of Alkali Promoted Iron catalysts at low CO conversions (15–20%). Reactor Conditions: 270 °C, 12.1 bar, WHSV = 2, and $H_2/CO = 0.7$.

	Iso/Normal Mole Fraction
No Alkali	0.21
Li	0.17
Na	0.16
Rb	0.16
K	0.16
Cs	0.13

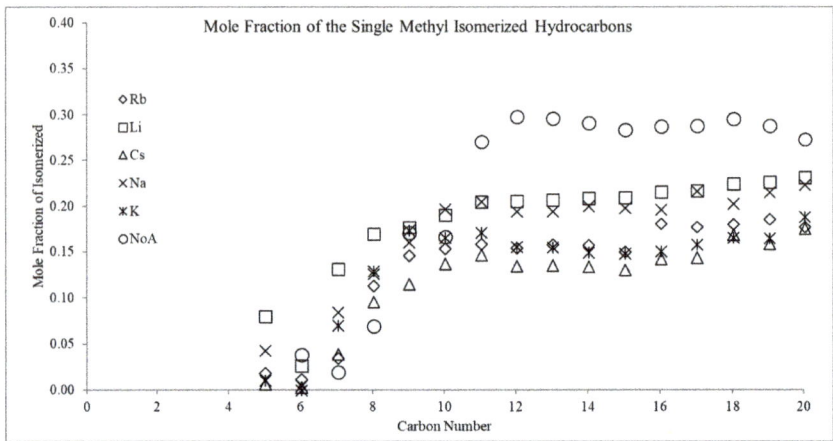

Figure 47. The molar carbon fraction of the total isomerized hydrocarbons per carbon number for the Group I series iron catalysts. As the basicity of the metal increases, allowing for more back donation, the overall branched material decreases. Reactor Conditions: 230 °C, 12.1 bar, WHSV = 2, and $H_2/CO = 0.7$.

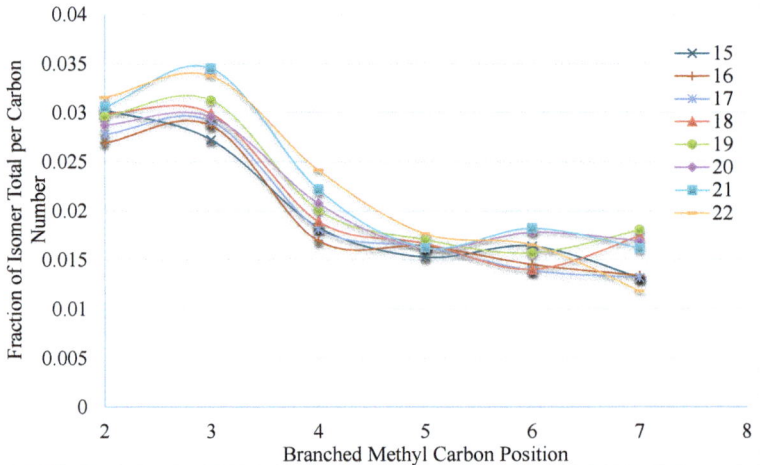

Figure 48. A graph of the single methyl branched products, arranged by carbon number, and not by the methyl group. This is a representation using a 2 K iron catalyst, where the oil was exposed to H_2 at 0.68 bar for 5 min. All of the runs at 230 °C, 12.1 bar, 0.7 H_2/CO with a WHSV of 2.

4.1.2. Oxygenates

Iron (i.e., iron carbide, with or without alkali promoter) seems to be the only active FTS catalyst that produces oxygenates in addition to linear alcohols in noteworthy amounts. Acids, ketones, esters (only in trace molar carbon % in the oil phase), and aldehydes were all detected in the oil and aqueous phases of the products that are produced by iron catalysts (Figures 20 and 30).

The largest change for the oxygenate distribution in the iron catalyst series is the difference between the unpromoted and promoted catalysts (Figures 49 and 50). There is a clear deviation that is brought on just by alkali addition, where there is a notable drop in the percentage of the aldehyde and ketone with respect to the total oxygenates. Interestingly, the drop in percentage by these two functional groups is offset by a percentage increase in the alcohols with respect to the total oxygenates. The acid functional group slightly increased in percentage with respect to the total oxygenates, by the addition of the smaller alkali metals (Li, Na). However, as basicity/size of the promoter increases, the acid, like the ketone and aldehyde, decreases with respect to the total oxygenates. Essentially, the ketone and aldehyde could be viewed as materials that only desorb as partially dissociated CO, because of an inability for CO to fully dissociate. This depends on the ability of H_2 to reach the growing chain, where termination proceeds through CO insertion.

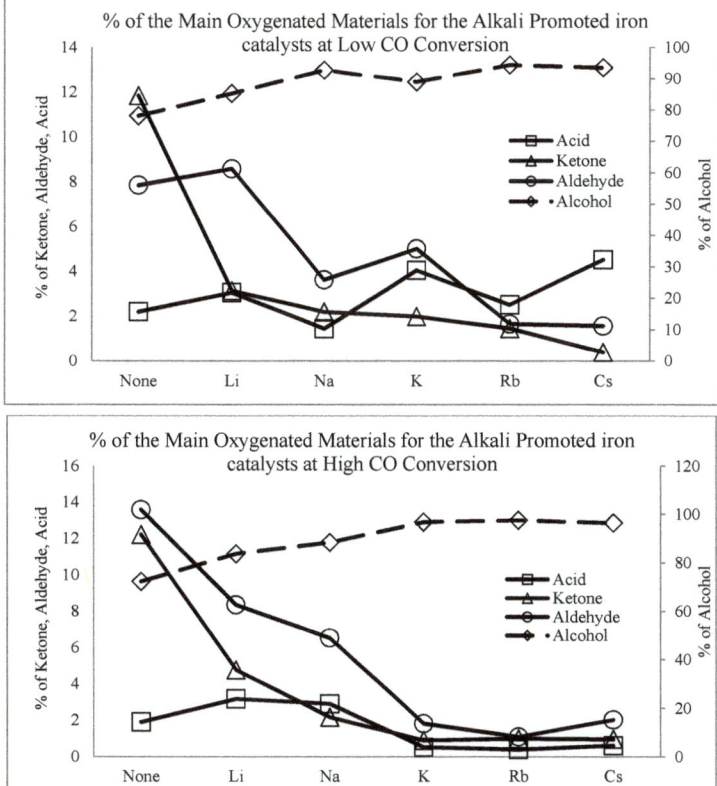

Figure 49. The % distribution of different oxygenated materials for each of the iron catalysts at low and high CO conversion. The percentage basis is taken as a % mole of carbon per functional group/%mole of C total oxygenates. Esters were observed, but they accounted for much less than one percent, with respect to the total oxygenates, and thus not added to this figure. Reactor conditions: 270 °C, 12.1 bar, WHSV = 2, and $H_2/CO = 0.7$.

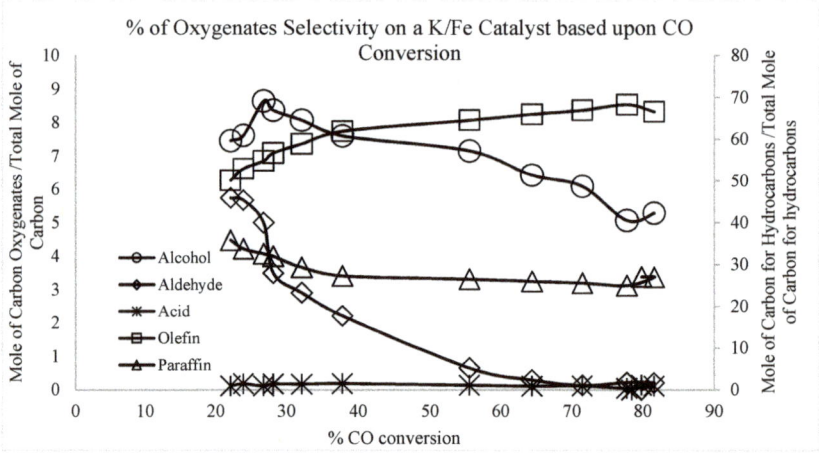

Figure 50. A product distribution of a K/Fe catalyst that shows the dependence of specific oxygenates with respect to the total oxygenates, on CO conversion. Reactor Conditions: 270 °C, 12.1 bar, WHSV = 2, $H_2/CO = 0.7$.

K not only can promote oxygenated materials, like the alcohol, but if enough is present, then aldehydes can also be produced (Figure 20). This difference can be observed when comparing the K promoted iron catalyst (Figure 20) with unpromoted iron (Figure 31). Thus, multiple routes are occurring as K donates electron density to promote CO dissociation, limiting hydrogen. If enough potassium is added, then hydrogen is so limited that the probability for termination by CO increases, resulting in the formation of these aldehydes. Regardless, provided that identical conditions exist for the Group I series, as Figure 48 and Table 2 show, there appears to be a direct relationship between the promoter series and the % of oxygenates observed. The correlation can only be attributed to the promoted iron series catalysts, as shown in Figure 28.

4.1.3. ASF Factors

Typically, both the hydrocarbon and oxygenate values (Figure 51) trended inversely (although less so for K promoted and unpromoted iron catalysts) with CO conversion. Residence time influences the secondary reactions, which is more significant for promoted catalysts. The K and Na doped Fe catalysts, as displayed in Tables 2 and 3, exhibited the greatest changes in oxygenate selectivity by changing the CO conversion. The Rb and Cs promoted Fe catalysts produced the largest carbon mole % and highest α values for oxygenated material. Unlike the use of K promoter (Figure 50, Table 2), where the selectivity of oxygenates is dependent upon CO conversion, the selectivity of oxygenates in Rb and Cs promoted Fe catalysts is independent of the CO conversion (Figure 49).

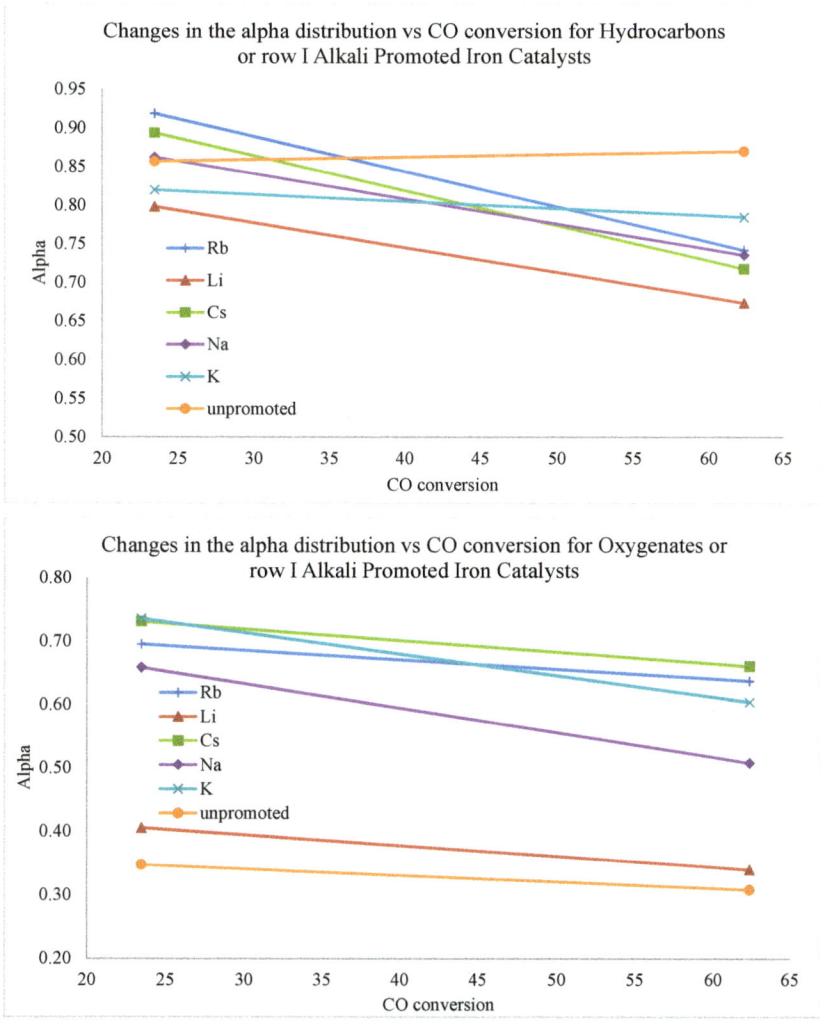

Figure 51. Changes in the alpha vs. CO conversion for Group I promoted iron catalysts. Reactor Conditions: 270 °C, 12.1 bar, WHSV = 2, and $H_2/CO = 0.7$.

4.1.4. Manganese Supported Iron Catalysts

The work here agrees with the literature, where supports, such as manganese (Figure 30) for iron, promote higher fractions of the 1-olefin, suppressing the 2-olefins [164–166]. However, the promotion of the 1-olefin is much more significant with the Fe/Mn catalyst that is shown in Table 2. Additionally, though not shown, the FeMn series of catalysts run under the same conditions as the alkali promoted iron series, displayed similar rates of WGS activity as the potassium promoted iron catalyst.

4.2. Cobalt and Ruthenium

Cobalt and ruthenium are very similar in their product selectivities (Figure 52). Unlike the iron series, the hydrocarbon product distribution of cobalt and ruthenium is much more specific to the n-paraffin and 1-olefin. Platinum is typically used with cobalt primarily to facilitate the reduction of cobalt. However, with the addition of reduction promoters, such as platinum, methane production is slightly increased [151,167,168]. This is to be expected, because platinum in a reduced state is a d^{10} metal, like nickel, and it would exhibit similar electron back-donation capability; thus, CO that was adsorbed on platinum would exhibit similar structure, as observed on nickel, M≡C–O. The addition of the alkali on cobalt, as observed with iron, increases the adsorption of CO, again most likely due to the increased ability for the d-orbitals of cobalt to interact with the carbon-carbon π orbitals [169].

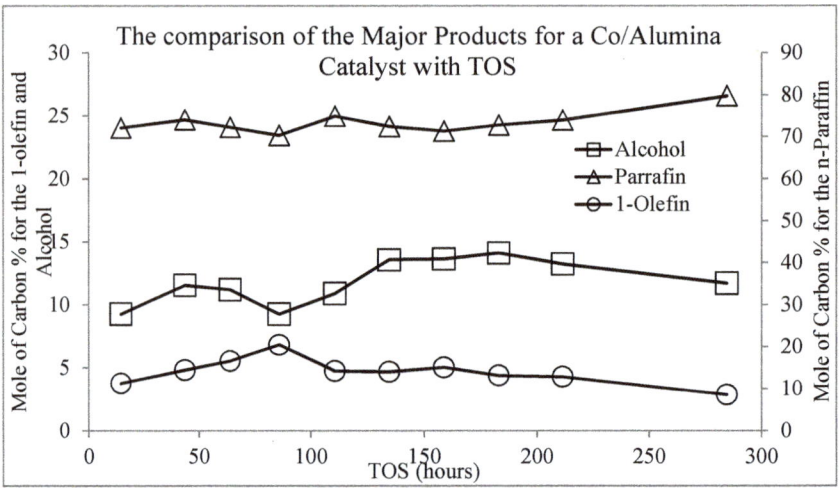

Figure 52. This figure tracks the changes of the major products with time on stream. Reactor conditions: 220 °C, 19.3 bar, WHSV = 5, H_2/CO = 2.

The first observation is the low selectivity of alcohols between the supported and unsupported cobalt catalysts. The low selectivity does not necessarily mean that the supports are directly involved in the synthesis of oxygenates, but that they influence their production. Ceria, like manganese and titania, is a partially reducible oxide, and unlike the silica and alumina, it has the potential for hydrogen spillover [170,171]. However, the spillover of hydrogen upon supports for cobalt, such as ceria, may create oxygen vacancy sites (or associated bridging OH groups) (Figure 53) through the removal of an oxygen atom in the support (or conversion of O to -OH groups), which changes the ceria from 4+ to 3+ (the underlined ceria atoms). This, in turn, has the potential to affect the product distribution in the FT process.

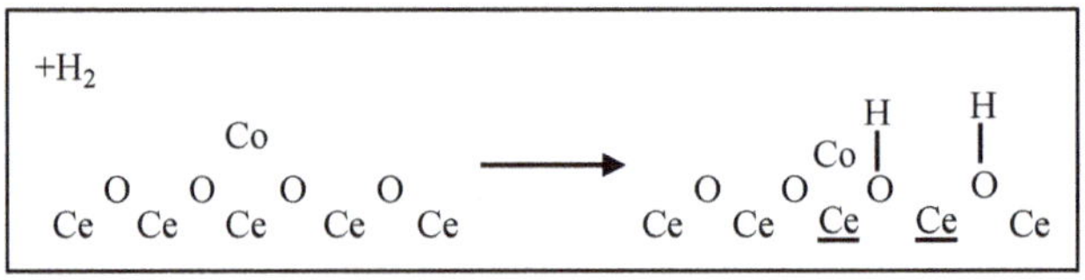

```
                                      H   H
  +H₂                                  |   |
         Co                         Co |   |
    O  O  O  O                   O  O  O   O
    Ce Ce Ce Ce  Ce      →       Ce Ce Ce  Ce  Ce
```

Figure 53. Bridging OH groups created in Ce due to addition of H_2 during the reduction process; ceria is a well-known partially reducible oxide.

Gnanamani et al. [40] conducted a H_2/D_2 switching study to examine the isotope effect on oxygenated products that are produced from a Co/Ce catalyst. Results (Table 6) from the switch showed an overall decrease in the oxygenated materials; however, the oxygenated products returned to their original composition upon switching back to hydrogen. These results suggest that the metal-support interface plays a significant role in the formation of oxygenated materials. In this case, CO could react at defect-associated bridging OH groups on ceria to form formate, which acts as a molecularly adsorbed CO species that is transferred across the interface to terminate chains; or, termination may proceed by the direct addition of mobile OH groups from ceria. In order to produce oxygenates, one view is that termination proceeds by the addition of molecularly adsorbed CO [171]. In either case, ceria could influence the termination pathway across the metal-support interface, with the growing chain being on metallic cobalt. Masuku et al. [172] completed similar H_2/D_2 switching experiments with an alumina supported cobalt catalyst to show no specific deviation in the products (Table 6).

Table 6. The difference in product distribution on a Cobalt/Alumina catalyst when H_2/CO is switched to D_2/CO. Reactor Conditions: 220 °C, 19.3 bar, WHSV = 2, and H_2/CO = 2.

Oil and Wax	D	H
Alcohol	3.96	3.25
1-Olefin	11.30	10.21
Paraffin	82.34	84.40
Cis-Olefin	0.93	0.95
Trans-Olefin	1.46	1.19
Total Olefin	13.69	12.35

Though the product distribution is similar, cobalt that is supported on carbon (Figure 32) could react differently from metal oxide supported cobalt catalysts [38]. Carbon has surface oxygenate functional groups and it is not necessarily inert. This gives rise to the potential for these supports to interfere in the chain growth process via functionalized groups, and further work is needed to shed light on this aspect [59].

5. Conclusions

The intention of this paper is to provide a brief overview of the product distribution across several FTS catalysts. Specific product selectivities (fingerprint) in the oil phase are highly informative, such that, by solely observing the oil phase, catalyst properties can largely be uncovered. The fingerprint of iron (iron carbide) is more diverse than cobalt and ruthenium and the surface of iron carbide catalyst has been much more tunable. To maximize the efficiency of an FTS catalyst, a sustained balance between CO, H_2, and chain growth intermediates must be maintained. If electron back-donation from the catalyst surface to the vinylic intermediate and CO is too weak, then the L configuration is dominant forming isomerized, non-alcohol oxygenates and high methane. If too strong, the vinylic intermediate (as does CO) dissociates too rapidly, such that methane production is high, and carbon deposits on the surface leading to coking. Both unbalanced routes occur at the cost of the overall process. As the fingerprint reveals clues to the catalyst, this work could serve as a starting point for future experiments that are aimed at tuning FTS catalysts.

Author Contributions: Analytical workup and manuscript primary writer, W.D.S.; Catalyst preparation, M.K.G., J.Y., C.M.M.; characterization, G.J., U.G.; Manuscript writing and primary editor—G.J.; Reactor testing, M.K.G., J.Y., C.M.M.; project manager; B.H.D.

Funding: This research received no external funding.

Acknowledgments: Shafer would like thank Asbury University, and Chelsea Parsons for assisting in the editing. Masuku would like to acknowledge partial financial support from the National Research Foundation of South Africa (Grant Number: 113652). Jacobs would like to acknowledge UTSA, the State of Texas, and the STARs program for financial support. The opinions, findings, and conclusions or recommendations expressed in this publication are those of the authors. The funding bodies accept no liability in this regard. Gnanamani would like to thank CAER at the University of Kentucky, and the Commonwealth of Kentucky.

Conflicts of Interest: The authors have no conflict of interest.

Glossary

ASF	Anderson-Schulz-Flory Plot
CSTR	Continuously Stirred Tank Reactor
EI	Electron Impact
FID	Flame Ionization Detector
FTS	Fischer–Tropsch Synthesis
GC	Gas Chromatograph
GHSV	Gas Hourly Space Velocity
HC	High CO Conversion
H_2/CO-	Hydrogen to Carbon Monoxide Ratio
IWI	Incipient Wetness Impregnation
LC	Low CO Conversion
LOD	Limit of Detection
MSD	Mass Selective Detector
O/P	Total Olefin to Paraffin Ratio
RT	Retention Time
SIM	Single Ion Monitoring
TCD	Thermal-conductivity Detector
WF	Weight Fraction
WHSV	Weight Hourly Space Velocity
VLE	Vapor Liquid Equilibrium

References

1. Wilhelm, D.J.; Simbeck, D.R.; Karp, A.D.; Dickenson, R.L. Syngas production for gas-to-liquids applications: Technologies, issues and outlook. *Fuel Process. Technol.* **2001**, *71*, 139–148. [CrossRef]
2. Wood, D.A.; Nwaoha, C.; Towler, B.F. Gas-to-liquids (GTL): A review of an industry offering several routes for monetizing natural gas. *J. Nat. Gas Sci. Eng.* **2012**, *9*, 196–208. [CrossRef]
3. Jessop, P.G.; Subramaniam, B. Gas-Expanded Liquids. *Chem. Rev.* **2007**, *107*, 2666–2694. [CrossRef]
4. Davis, B.H. Clean fuels from coal: The path to 1972. *Prepr. Symp. Am. Chem. Soc. Div. Fuel Chem.* **2003**, *48*, 141–143.
5. Zhang, Y.; Davis, B.H. Indirect coal liquefaction—Where do we stand? *Prepr. Am. Chem. Soc. Div. Pet. Chem.* **1999**, *44*, 20–24.
6. Whitehurst, D.D.; Mitchell, T.O.; Farcasiu, M. *Coal Liquefaction: The Chemistry and Technology of Thermal Processes*; Academic Press, Inc.: New York, NY, USA, 1980; 390p.
7. Liu, Z.; Shi, S.; Li, Y. Coal liquefaction technologies—Development in China and challenges in chemical reaction engineering. *Chem. Eng. Sci.* **2010**, *65*, 12–17. [CrossRef]
8. Horne, P.A.; Williams, P.T. Influence of temperature on the products from the flash pyrolysis of biomass. *Fuel* **1996**, *75*, 1051–1059. [CrossRef]
9. Swanson, R.M.; Platonc, A.; Satrio, J.A.; Brown, R.C. Techno-economic analysis of biomass-to-liquids production based on gasification. *Fuel* **2010**, *89*, S11–S19. [CrossRef]
10. Medrano, J.A.; Oliva, M.; Ruiz, J.; García, L.; Arauzo, J. Hydrogen from aqueous fraction of biomass pyrolysis liquids by catalytic steam reforming in fluidized bed. *Energy* **2011**, *36*, 2215–2224. [CrossRef]
11. Higman, C.; van der Burgt, M. *Gasification*; Gulf Professional Publishing: Houston, TX, USA, 2008; pp. 1–456.
12. Nikrityuk, P.A.; Meye, B. *Gasification Processes: Modeling and Simulation*; Wiley-VCH: New York, NY, USA, 2014; pp. 1–360.
13. Jacobs, G.; Crawford, A.C.; Davis, B.H. Water-gas shift: Steady state isotope switching study of the water-gas shift reaction over Pt/ceria using in-situ DRIFTS. *Catal. Lett.* **2005**, *100*, 147–152. [CrossRef]
14. Jacobs, G.; Crawford, A.; Williams, L.; Patterson, P.M.; Davis, B.H. Low temperature water-gas-shift: Comparison of thoria and ceria catalysts. *Appl. Catal. A Gen.* **2004**, *267*, 27–33. [CrossRef]
15. Martinelli, M.; Jacobs, G.; Graham, U.M.; Shafer, W.D.; Cronauer, D.C.; Kropf, J.A.; Marshall, C.L.; Khalid, S.; Visconti, C.G.; Lietti, L.; et al. Water-gas-shift: Characterization and testing of nanoscale YSZ supported Pt catalysts. *Appl. Catal. A Gen.* **2015**, *497*, 184–197. [CrossRef]

16. Fu, Q. Howard Saltsburg, Maria Flytzani-Stephanopoulos, Active Nonmetallic Au and Pt Species on Ceria-Based Water-Gas Shift Catalysts. *Science* **2003**, *301*, 935–938. [CrossRef] [PubMed]
17. Scherzer, J.; Gruia, A.J. *Hydrocracking Science and Technology*; Marcell Dekker Inc.: New York, NY, USA, 1996.
18. Fischer, F.; Tropsch, H. Synthesis of Petroleum at Atmospheric Pressure from Gasification Products of Coal. *Brennstoff-Chemie* **1926**, *7*, 97–104.
19. Fischer, F.; Tropsch, H. Development of the Benzene Synthesis from Carbon Monoxide and Hydrogen at Atmospheric Pressure. *Brennstoff-Chemie* **1930**, *11*, 489–500.
20. Yang, J.; Shafer, W.D.; Pendyala, V.R.R.; Jacobs, G.; Chen, D.; Holmen, A.; Davis, B.H. Fisher-Tropsch Synthesis: Using Deuterium as a Tool to Investigate Primary Product Distribution. *Catal. Lett.* **2014**, *144*, 524–530. [CrossRef]
21. Shi, B.; Davis, B.H. Fischer–Tropsch synthesis: Accounting for chain-length related phenomena. *Appl. Catal. A* **2004**, *277*, 61–69. [CrossRef]
22. Davis, B.H. Fischer–Tropsch synthesis: Current mechanism and futuristic needs. *Fuel Proc. Technol.* **2001**, *71*, 157–166. [CrossRef]
23. Tuxen, A.; Carenco, S.; Chintapalli, M.; Chuang, C.; Escudero, C.; Pach, E.; Jiang, P.; Borondics, F.; Beberwyck, B.; Alivisatos, A.P.; et al. Size-Dependent Dissociation of Carbon Monoxide on Cobalt Nanoparticles. *J. Am. Chem. Soc.* **2013**, *135*, 2273–2278. [CrossRef]
24. Qi, Y.; Yang, J.; Chen, D.; Holmen, A. Recent Progresses in Understanding of Co-Based Fischer–Tropsch Catalysis by Means of Transient Kinetic Studies and Theoretical Analysis. *Catal. Lett.* **2015**, *145*, 145–161. [CrossRef]
25. Jacobs, G.; Davis, B.H. Applications of isotopic tracers in Fischer–Tropsch synthesis. *Catal. Sci. Technol.* **2014**, *4*, 3927–3944. [CrossRef]
26. Shi, B.; Jacobs, G.; Sparks, D.E.; Davis, B.H. Fischer-Tropsch Synthesis: ^{14}C Labeled 1-alkene conversion using supercritical conditions with Co/Al_2O_3. *Fuel* **2005**, *84*, 1093–1098. [CrossRef]
27. Dry, M.E. The Fischer–Tropsch process: 1950–2000. *Catal. Today* **2002**, *71*, 227–241. [CrossRef]
28. Enger, B.C.; Holmen, A. Nickel and Fischer-Tropsch Synthesis. *Catal. Rev.* **2012**, *54*, 437–488. [CrossRef]
29. Milburn, D.R.; Chary, V.R.K.; Davis, B.H. Promoted iron Fischer-Tropsch catalysts: Characterization by nitrogen sorption. *Appl. Catal. A Gen.* **1996**, *144*, 121–132. [CrossRef]
30. Luo, M.; Hamdeh, H.; Davis, B.H. Potassium promoted iron Fischer-Tropsch Synthesis catalyst activation study with Mossbauer spectroscopy. *Prepr. Am. Chem. Soc. Div. Pet. Chem.* **2007**, *52*, 73–76.
31. Pendyala, V.R.R.; Jacobs, G.; Gnanamani, M.K.; Hu, Y.; MacLennan, A.; Davis, B.H. Selectivity control of Cu promoted iron-based Fischer-Tropsch catalyst by tuning the oxidation state of Cu to mimic K. *Appl. Catal. A Gen.* **2015**, *495*, 45–53. [CrossRef]
32. Rachid, O.; Singleton, A.H.; Goodwin, J.G., Jr. Comparison of patented Co F–T catalysts using fixed-bed and slurry bubble column reactors. *Appl. Catal. A Gen.* **1999**, *186*, 129–144.
33. Girardona, J.-S.; Gribovala, A.Co.; Gengembre, L.; Chernavskii, P.A.; Khodakov, A.Y. Optimization of the pretreatment procedure in the design of cobalt silica supported Fischer-Tropsch catalysts. *Catal. Today* **2005**, *106*, 161–165. [CrossRef]
34. Jacobs, G.; Das, T.K.; Zhang, Y.; Li, J.; Racoillet, G.; Davis, B.H. Fischer-Tropsch synthesis: Support, loading, and promoter effects on the reducibility of cobalt catalysts. *Appl. Catal. A Gen.* **2002**, *233*, 263–281. [CrossRef]
35. Ribeiro, M.C.; Jacobs, G.; Davis, B.H.; Cronauer, D.C.; Kropf, A.J.; Marshall, C.L. Fischer-Tropsch Synthesis: An In-Situ TPR-EXAFS/XANES Investigation of the Influence of Group I Alkali Promoters on the Local Atomic and Electronic Structure of Carburized Iron/Silica Catalysts. *J. Phys. Chem. C* **2010**, *114*, 7895–7903. [CrossRef]
36. Jacobs, G.; Ji, Y.; Davis, B.H.; Cronauer, D.; Kropf, A.J.; Marshall, C.L. Fischer-Tropsch synthesis: Temperature programmed EXAFS/XANES investigation of the influence of support type, cobalt loading, and noble metal promoter addition to the reduction behavior of cobalt oxide particles. *Appl. Catal. A* **2007**, *333*, 177–191. [CrossRef]
37. Wang, W.J.; Chen, Y.W. Influence of metal loading on the reducibility and hydrogenation activity of cobalt/alumina catalysts. *Appl. Catal.* **1991**, *77*, 223–233. [CrossRef]
38. Luo, M.; Shafer, W.D.; Davis, B.H. Fischer–Tropsch Synthesis: Branched Paraffin Distribution for Potassium Promoted Iron Catalysts. *Catal. Lett.* **2014**, *144*, 1031–1041. [CrossRef]

39. Davis, B.H. *Technology Development for Iron and Cobalt Fischer–Tropsch Catalysts*; Quarterly Report #1, Contract #DE-FC26-98FT40308; Department of Energy (DOE): Commonwealth of Kentucky, USA, 2002.
40. Gnanamani, M.K.; Jacobs, G.; Graham, U.M.; Ribeiro, M.C.; Noronha, F.B.; Shafer, W.D.; Davis, B.H. Influence of carbide formation on oxygenates selectivity during Fischer-Tropsch synthesis over Ce-containing Co catalysts. *Catal. Today* **2016**, *261*, 40–47. [CrossRef]
41. Ribeiro, M.C.; Gnanamani, M.K.; Rabelo-Neto, R.C.; Azevedo, I.R.; Pendyala, V.R.R.; Jacobs, G.; Davis, B.H.; Noronha, F.B. Fischer-Tropsch Synthesis: Studies on the effect of support reducibility on the selectivity to n-alcohols in Co/CeO$_2$.MO$_y$ (M = Si, Mn, Cr) catalysts. *Top. Catal.* **2014**, *57*, 550–560. [CrossRef]
42. Gnanamani, M.K.; Jacobs, G.; Shafer, W.D.; Ribeiro, M.C.; Pendyala, V.R.R.; Ma, W.; Davis, B.H. Fischer-Tropsch synthesis: Deuterium isotopic study for the formation of oxygenates over CeO$_2$-supported Pt-Co catalysts. *Catal. Commun.* **2012**, *25*, 12–17. [CrossRef]
43. Gaube, J.; Klein, H.F. Studies on the reaction mechanism of the Fischer–Tropsch synthesis on iron and cobalt. *J. Mol. Catal. A* **2008**, *283*, 60–68. [CrossRef]
44. Van Santen, R.A.; Markvoort, A.J.; Ghouri, M.M.; Hilbers, P.A.J.; Hensen, E.J.M. Monomer Formation Model versus Chain Growth Model of the Fischer–Tropsch Reaction. *J. Phys. Chem. C* **2013**, *117*, 4488–4504. [CrossRef]
45. Ojeda, M.; Li, A.; Nabar, R.; Nilekar, A.U.; Mavrikakis, M.; Iglesia, E. Kinetically Relevant Steps and H_2/D_2 Isotope Effects in Fischer−Tropsch Synthesis on Fe and Co Catalysts. *J. Phys. Chem. C* **2010**, *114*, 19761–19770. [CrossRef]
46. Kummer, J.T.; Podgurski, H.H.; Spencer, W.B.; Emmett, P.H. Mechanism studies of the Fischer-Tropsch synthesis. The addition of radioactive alcohol. *J. Am. Chem. Soc.* **1951**, *73*, 564–569. [CrossRef]
47. Kummer, J.T.; Dewitt, W.; Emmett, P.H. Some Mechanism Studies on the Fischer-Tropsch Synthesis Using C^{14}. *J. Am. Chem. Soc.* **1948**, *70*, 3632–3643. [CrossRef]
48. Storch, H.H.; Golumbic, N.; Anderson, R.B. *The Fischer–Tropsch and Related Syntheses*; Wiley: New York, NY, USA, 1951.
49. Pichler, H.; Schulz, H. Neuere Erkenntnisse auf dem Gebiet der Synthese von Kohlenwasserstoffen aus CO und H_2. *Chem. Ing. Tech.* **1970**, *42*, 1162–1174. [CrossRef]
50. Luo, M.; Bao, S.; Keogh, R.A.; Sarkar, A.; Jacobs, G.; Davis, B.H. Fischer-Tropsch Synthesis: A Comparison of Iron and Cobalt Catalysts. In Proceedings of the AIChE 2006 National Meeting, SanFrancisco, CA, USA, 12–17 November 2006.
51. Schulz, H. Principles of Olefin Selectivity in Fischer-Tropsch Synthesis on Iron and Cobalt Catalysts. In *Fischer-Tropsch Synthesis, Catalysts, and Catalysis: Advances and Applications*; Davis, B.H., Occelli, M.L., Eds.; CRC Press, Taylor & Francis Group: Boca Raton, FL, USA, 2015.
52. Jennifer, N. Deuterium Enrichment and the Kinetic Isotope Effect during Ruthenium Catalyzed Fischer-Tropsch Synthesis. Online Theses and Dissertations, Eastern Kentucky University, Lexington, Kentucky, 2015; p. 408.
53. Chakrabarti, D.; Gnanamani, M.K.; Shafer, W.D.; Ribeiro, M.C.; Sparks, D.E.; Prasad, V.; de Klerk, A.; Davis, B.H. Fischer–Tropsch Mechanism: $^{13}C^{18}O$ Tracer Studies on a Ceria–Silica Supported Cobalt Catalyst and a Doubly Promoted Iron Catalyst. *Ind. Eng. Chem. Res.* **2015**, *54*, 6438–6453. [CrossRef]
54. Schulz, H. Principles of Fischer–Tropsch synthesis—Constraints on essential reactions ruling FT-selectivity. *Catal. Today* **2013**, *214*, 140–151. [CrossRef]
55. Anderson, R.B.; Friedel, R.A.; Storch, H.H. Fischer-Tropsch Reaction Mechanism Involving Stepwise Growth of Carbon Chain. *J. Chem. Phys.* **1951**, *19*, 313. [CrossRef]
56. Schulz, G.V. Über die Beziehung zwischen Reaktionsgeschwindigkeit und Zusammensetzung des Reaktionsproduktes bei Makropolymerisationsvorgängen. *Z. Phys. Chem.* **1935**, *30*, 379–398. [CrossRef]
57. Flory, P.J. Molecular Size Distribution in Linear Condensation Polymers. *J. Am. Chem. Soc.* **1936**, *58*, 1877–1885. [CrossRef]
58. Schliebs, B.; Gaube, J. The Influence of the Promoter K_2CO_3 in Iron Catalysts on the Carbon Number Distribution of Fischer-Tropsch Products. *Ber. Der Bunsenges. Für Phys. Chem.* **1985**, *89*, 68–73. [CrossRef]
59. Graham, U.M.; Jacobs, G.; Gnanamani, M.K.; Lipka, S.M.; Shafer, W.D.; Swartz, C.R.; Jermwongratanachai, T.; Chen, R.; Rogers, F.; Davis, B.H. Fischer Tropsch synthesis: High oxygenate-selectivity of cobalt catalysts supported on hydrothermal carbons. *ACS Catal.* **2014**, *4*, 1662–1672. [CrossRef]

60. Dictor, R.A.; Bell, A.T. Fischer-Tropsch synthesis over reduced and unreduced iron oxide catalysts. *J. Catal.* **1986**, *97*, 121–136. [CrossRef]
61. Patzlaff, J.; Liu, Y.; Graffmann, C.; Gaube, J. Studies on product distributions of iron and cobalt catalyzed Fischer–Tropsch synthesis. *Appl. Catal. A* **1999**, *186*, 109. [CrossRef]
62. Madon, R.J.; Taylor, W.F. Fischer-Tropsch synthesis on a precipitated iron catalyst. *J. Catal.* **1981**, *69*, 32–43. [CrossRef]
63. Dictor, R.A.; Bell, A.T. Effects of potassium promotion on the activity and selectivity of iron Fischer-Tropsch catalysts. *Ind. Eng. Chem. Proc. Des. Dev.* **1983**, *22*, 97–103.
64. Satterfield, C.N.; Huff, G.A.; Longwell, J.P. Product distribution from iron catalysts in Fischer-Tropsch slurry reactors. *Ind. Eng. Chem. Process. Des. Dev.* **1982**, *21*, 465–470. [CrossRef]
65. Henrici-Olive, G.; Olive, S. The Fischer-Tropsch Synthesis: Molecular Weight Distribution of Primary Products and Reaction Mechanism. *Angew. Chem. Int. Ed. Eng.* **1976**, *15*, 136–141. [CrossRef]
66. Huff, G.A., Jr.; Satterfield, C.N. Evidence for two chain growth probabilities on iron catalysts in the Fischer-Tropsch synthesis. *J. Catal.* **1984**, *85*, 370–379. [CrossRef]
67. Donnelly, T.J.; Yates, I.C.; Satterfield, C.N. Analysis and prediction of product distributions of the Fischer-Tropsch synthesis. *Energy Fuels* **1988**, *2*, 734–739. [CrossRef]
68. Herzog, K.; Gaube, J. Kinetic studies for elucidation of the promoter effect of alkali in Fischer-Tropsch iron catalysts. *J. Catal.* **1989**, *115*, 337–346. [CrossRef]
69. Kuipers, E.W.; Vinkenburg, I.H.; Oosterbeek, H. Chain Length Dependence of α-Olefin Readsorption in Fischer-Tropsch Synthesis. *J. Catal.* **1995**, *152*, 137–146. [CrossRef]
70. Bukur, B.D.; Mukesh, D.; Patel, S.A. Promoter effects on precipitated iron catalysts for Fischer-Tropsch synthesis. *Ind. Eng. Chem. Res.* **1990**, *29*, 194–204. [CrossRef]
71. Masuku, C.M.; Hildebrandt, D.; Glasser, D. The role of vapour–liquid equilibrium in Fischer–Tropsch product distribution. *Chem. Eng. Sci.* **2011**, *66*, 6254–6263. [CrossRef]
72. Satterfield, C.N.; Huff, G.A., Jr. Carbon number distribution of Fischer-Tropsch products formed on an iron catalyst in a slurry reactor. *J. Catal.* **1982**, *73*, 187–197. [CrossRef]
73. Von Elbe, G. Internal Equilibria and Partial Vapor Pressures of Mixtures of Primary Normal Alcohols with Normal Paraffin Hydrocarbons. *J. Chem. Phys.* **1934**, *2*, 73–81. [CrossRef]
74. Iglesia, E.; Reyes, S.C.; Madon, R.J. Transport-Enhanced a-Olefin Readsorption Pathways in Ru-Catalyzed Hydrocarbon Synthesis. *J. Catal.* **1991**, *129*, 238–256. [CrossRef]
75. Davis, B.H. Fischer-Tropsch synthesis: relationship between iron catalyst composition and process variables. *Catalysis Today* **2003**, *84*, 83–98. [CrossRef]
76. Zhang, C.H.; Yang, Y.; Teng, B.T.; Li, T.Z.; Zheng, H.Y.; Xiang, H.W.; Li, Y.W. Study of an iron-manganese Fischer–Tropsch synthesis catalyst promoted with copper. *J. Catal.* **2006**, *237*, 405–415. [CrossRef]
77. Shi, B.; Davis, B.H. Fischer-Tropsch synthesis: The paraffin to olefin ratio as a function of carbon number. *Catal. Today* **2005**, *106*, 129–131. [CrossRef]
78. Shi, B.; Wu, L.; Liao, Y.; Jin, C.; Montavon, A. Explanations of the formation of branched hydrocarbons during Fischer-Tropsch synthesis by alkylidene mechanism. *Top. Catal.* **2014**, *57*, 451–459. [CrossRef]
79. Shi, B.; Davis, B.H. Fischer-Tropsch Synthesis: Evidence for Chain Initiation by Ethene and Ethanol for an Iron Catalyst. *Top. Catal.* **2003**, *26*, 157–161. [CrossRef]
80. Gnanamani, M.K.; Keogh, R.A.; Shafer, W.D.; Shi, B.; Davis, B.H. Fischer–Tropsch synthesis: Deuterium labeled ethanol tracer studies on iron catalysts. *Appl. Catal. A* **2010**, *385*, 46–51. [CrossRef]
81. Gnanamani, M.K.; Keogh, R.A.; Shafer, W.D.; Davis, B.H. Deutero-1-pentene tracer studies for iron and cobalt Fischer–Tropsch synthesis. *Appl. Catal. A Gen.* **2011**, *393*, 130–137. [CrossRef]
82. Masuku, C.M.; Hildebrandt, D.; Glasser, D.; Davis, B.H. Steady-State Attainment Period for Fischer–Tropsch Products. *Top. Catal.* **2014**, *57*, 582–587. [CrossRef]
83. Ao, M.; Pham, G.H.; Sunarso, J.; Tade, M.O.; Liu, S. Active Centers of Catalysts for Higher Alcohol Synthesis from Syngas: A Review. *ACS Catal.* **2018**, *8*, 7025–7050. [CrossRef]
84. Wang, Y.; Davis, B.H. Fischer–Tropsch synthesis. Conversion of alcohols over iron oxide and iron carbide catalysts. *Appl. Catal. A Gen.* **1999**, *180*, 277–285. [CrossRef]
85. Ngantsoue-Hoc, W.; Zhang, Y.; O'Brien, R.J.; Lou, M.; Davis, B.H. Fischer–Tropsch synthesis: Activity and selectivity for Group I alkali promoted iron-based catalysts. *Appl. Catal. A Gen.* **2002**, *236*, 77. [CrossRef]
86. Espinoza, R.L.; Visagie, J.L.; van Berge, P.J.; Bolder, F.H. Catalysts. U.S. Patent 5,733,839, 31 March 1998.

87. Yang, J.; Shafer, W.D.; Pendyala, V.R.R.; Jacobs, G.; Ma, W.; Chen, D.; Holmen, A.; Davis, B.H. Fischer-Tropsch synthesis: Deuterium kinetic isotopic effect for a 2.5%Ru/NaY catalysts. *Top. Catal.* **2014**, *57*, 508–517. [CrossRef]
88. Deitz, W.A. Response Factors for Gas Chromatographic Analysis. *J. Chromatogr. Sci.* **1967**, *5*, 68–71. [CrossRef]
89. Leveque, R.E. Determination of C3-C8 hydrocarbons in naphthas and reformates utilizing capillary column gas chromatography. *Anal. Chem.* **1967**, *39*, 1811–1818. [CrossRef]
90. Ackman, R.G. The Flame Ionization Detector: Further Comments on Molecular Breakdown and Fundamental Group Responses. *J. Gas Chromatogr.* **1968**, *6*, 497–5001. [CrossRef]
91. Ackman, R.G. Fundamental Groups in the Response of Flame Ionization Detectors to Oxygenated Aliphatic Hydrocarbons. *J. Gas Chromatogr.* **1964**, *2*, 173. [CrossRef]
92. Pendyala, V.R.R.; Jacobs, G.; Graham, U.M.; Shafer, W.D.; Martinelli, M.; Kong, L.; Davis, B.H. Fischer–Tropsch synthesis: Influence of acid treatment and preparation method on carbon nanotube supported ruthenium catalysts. *Ind. Eng. Chem. Res.* **2017**, *56*, 6408–6418. [CrossRef]
93. Holm, T. Aspects of the mechanism of the flame ionization detector. *J. Chromatogr. A* **1999**, *842*, 221–227. [CrossRef]
94. Down, R.D.; Lehr, J.H. (Eds.) *Environmental Instrumentation and Analysis Handbook*; John Wiley & Sons: Hoboken, NJ, USA, 2005.
95. Noltingk, B.E. *Measurement of Temperature and Chemical Composition: Jones' Instrument Technology*, 4th ed.; Butterworth-Heinemann: Oxford, UK, 2013; p. 192.
96. Barry, E.F.; Rosie, D.M. Response prediction of the thermal conductivity detector with light carrier gases. *J. Chromatogr. A* **1971**, *59*, 269–279. [CrossRef]
97. Heftmann, E. Chromatography: A Laboratory Handbook of Chromatographic and Electrophoretic Methods, 3rd edition. *J. Chromatogr. Sci.* **1978**, *16*, 15A.
98. Pendyala, V.R.R.; Chakrabarti, D.; de Klerk, A.; Keogh, R.A.; Sparks, D.E.; Davis, B.H. 14C-Labeled Alcohol Tracer Study: Comparison of Reactivity of Alcohols over Cobalt and Ruthenium Fischer–Tropsch Catalysts. *Top. Catal.* **2015**, *58*, 343–349.
99. Masuku, C.M.; Hildebrandt, D.; Glasser, D. Olefin pseudo-equilibrium in the Fischer–Tropsch reaction. *Chem. Eng. J.* **2012**, *181*, 667–676. [CrossRef]
100. Kellner, C.S.; Bell, A.T. Evidence for H_2/D_2 Isotope Effects on Fischer-Tropsch Synthesis over Supported Ruthenium Catalysts. *J. Catal.* **1981**, *67*, 175–185. [CrossRef]
101. Jacobs, G.; Ribeiro, M.C.; Ma, W.; Ji, Y.; Khalid, S.; Sumodjo, P.T.A.; Davis, B.H. Group 11 (Cu, Ag, Au) promotion of 15%Co/Al_2O_3 Fischer-Tropsch synthesis catalysts. *Appl. Catal. A Gen.* **2009**, *361*, 137–151. [CrossRef]
102. Ma, W.; Jacobs, G.; Keogh, R.A.; Bukur, D.B.; Davis, B.H. Fischer-Tropsch synthesis: Effect of Pd, Pt, Re, and Ru noble metal promoters on the activity and selectivity of a 25%Co/Al_2O_3 catalyst. *Appl. Catal. A Gen.* **2012**, *437–438*, 1–9. [CrossRef]
103. Eliseev, O.L.; Tsapkina, M.V.; Dement'eva, O.S. Promotion of cobalt catalysts for the Fischer-Tropsch synthesis with alkali metals. *Kinet. Catal.* **2013**, *54*, 207–212. [CrossRef]
104. Gnanamani, M.K.; Ribeiro, M.C.; Ma, W.; Shafer, W.D.; Jacobs, G.; Graham, U.M.; Davis, B.H. Fischer-Tropsch synthesis: Metal-support interfacial contact governs oxygenates selectivity over CeO2 supported Pt-Co catalysts. *Appl. Catal. A Gen.* **2011**, *393*, 17–23. [CrossRef]
105. Ribeiro, M.C.; Gnanamani, M.K.; Garcia, R.; Jacobs, G.; Rabelo-Neto, R.C.; Noronha, F.B.; Gomes, I.G.; Davis, B.H. Tailoring the product selectivity of Co/SiO_2 Fischer-Tropsch synthesis catalysts by lanthanide doping. *Catal. Today* **2018**. [CrossRef]
106. Blyholder, G.; Allen, M.C. Infrared spectra and molecular orbital model for carbon monoxide adsorbed on metals. *J. Am. Chem. Soc.* **1969**, *91*, 3158–3162. [CrossRef]
107. Blyholder, G. Molecular Orbital View of Chemisorbed Carbon Monoxide. *J. Phys. Chem.* **1964**, *68*, 2772–2777. [CrossRef]
108. Belosludov, R.V.; Sakahara, S.; Yajima, K.; Takami, S.; Kubo, M.; Miyamoto, A. Combinatorial computational chemistry approach as a promising method for design of Fischer–Tropsch catalysts based on Fe and Co. *Appl. Surf. Sci.* **2002**, *189*, 245–252. [CrossRef]

109. Hocking, R.K.; Hambley, T.W. Database Analysis of Transition Metal Carbonyl Bond Lengths: Insight into the Periodicity of δ Back-Bonding, ó Donation, and the Factors Affecting the Electronic Structure of the TM-CtO Moiety. *Organometallics* **2007**, *26*, 2815–2823. [CrossRef]
110. Betta, R.A.D.; Shelef, M. Heterogeneous Methanation: Absence of H_2-D_2, Kinetic Isotope Effect on Ni, Ru, and Pt. *J. Catal.* **1977**, *49*, 383–385. [CrossRef]
111. Shelef, M.; Betta, R.A.D. Reply to comments on heterogeneous methanation: Absence of H_2/D_2 kinetic isotope effect on Ni, Ru, and Pt. *J. Catal.* **1979**, *60*, 169–170. [CrossRef]
112. Van Nisselrooij, P.F.M.T.; Luttikholt, J.A.M.; van Meerten, R.Z.C.; de Croon, M.H.J.M.; Coenen, J.W.E. Hydrogen/deuterium kinetic isotope effect in the methanation of carbon monoxide on a nickel-silica catalyst. *Appl. Catal.* **1983**, *6*, 271–281. [CrossRef]
113. McKee, E.W. Interaction of hydrogen and carbon monoxide on platinum group metals. *J. Catal.* **1967**, *8*, 240–249. [CrossRef]
114. Shafer, W.D.; Jacobs, G.; Davis, B.H. Fischer–Tropsch Synthesis: Investigation of the Partitioning of Dissociated H_2 and D_2 on Activated Cobalt Catalyst. *ACS Catal.* **2012**, *2*, 1452–1456. [CrossRef]
115. Shafer, W.D.; Jacobs, G.; Selegue, J.P.; Davis, B.H. An Investigation of the Partitioning of Dissociated H_2 and D_2 on Activated Nickel Catalysts. *Catal. Lett.* **2013**, *143*, 1368–1373. [CrossRef]
116. Shafer, W.D.; Pendyala, V.R.R.; Gnanamani, M.K.; Jacobs, G.; Selegue, J.P.; Hopps, S.D.; Thomas, G.A.; Davis, B.H. Isotopic Apportioning of Hydrogen/Deuterium on the Surface of an Activated Iron Carbide Catalyst. *Catal. Lett.* **2015**, *145*, 1683–1690. [CrossRef]
117. Shafer, W.D.; Pendyala, V.R.R.; Jacobs, G.; Selegue, J.; Davis, B.H. *Fischer-Tropsch Synthesis, Catalysts, and Catalysis: Advances and Applications*; Davis, B.H., Occelli, M.L., Eds.; CRC Press, Taylor & Francis Group: Boca Raton, FL, USA, 2015.
118. Shafer, W.D. Investigation into the Competitive Partitioning Of Dissociated H2 and D2 on Activated Fischer-Tropsch Catalysts. Ph.D. Theses, University of Kentucky, Lexington, Kentucky, 2015.
119. Pendyala, V.R.R.; Shafer, W.D.; Jacobs, G.; Davis, B.H. Fischer-Tropsch synthesis: Effect of solvent on the H_2-D_2 isotopic exchange rate over an activated cobalt catalyst. *Can. J. Chem. Eng.* **2015**, *94*, 678–684. [CrossRef]
120. Pendyala, V.V.R.R.; Shafer, W.D.; Jacobs, G.; Davis, B.H. Fischer-Tropsch synthesis: Effect of solvent on the H_2-D_2 isotopic exchange rate over an activated nickel catalyst. *Catal. Today* **2016**, *270*, 2–8. [CrossRef]
121. Gnanamani, M.K.; Hamdeh, H.H.; Jacobs, G.; Shafer, W.D.; Sparks, D.E.; Davis, B.H. Fischer-Tropsch Synthesis: Activity and Selectivity of χ-Fe_5C_2 and θ-Fe_3C Carbides. In *Fischer-Tropsch Synthesis, Catalysts, and Catalysis: Advances and Applications*; Davis, B.H., Occelli, M.L., Eds.; CRC Press, Taylor & Francis Group: Boca Raton, FL, USA, 2015.
122. Raye, A.; Davis, B.H. Fischer-Tropsch Synthesis. Mechanism Studies Using Isotopes. *Catalysis* **1996**, *12*, 52–131.
123. Salvi, N.; Belpassi, L.; Tarantelli, F. On the Dewar–Chatt–Duncanson Model for Catalytic Gold(I) Complexes. *Chem. A Eur. J.* **2010**, *16*, 7231–7240. [CrossRef]
124. Weststrate, C.J.; van Helden, P.; van de Loosdrecht, J.; Niemantsverdriet, J.W. Elementary steps in Fischer-Tropsch synthesis: CO bond scission, CO oxidation and surface carbiding on Co(0001). *Surf. Sci.* **2016**, *648*, 60–66. [CrossRef]
125. Yang, J.; Frøseth, V.; Chen, D.; Holmen, A. Particle size effect for cobalt Fischer–Tropsch catalysts based on in situ CO chemisorption. *Surf. Sci.* **2016**, *648*, 67–73. [CrossRef]
126. Ozbek, M.O.; Niemantsverdriet, J.H. Niemantsverdriet Elementary reactions of CO and H2 on C-terminated χ-Fe5C2(0 0 1) surfaces. *J. Catal.* **2014**, *317*, 158–166. [CrossRef]
127. Nabaho, D.; Niemantsverdriet, J.H.; Claeys, M.; van Steen, E. Hydrogen spillover in the Fischer–Tropsch synthesis: An analysis of platinum as a promoter for cobalt–alumina catalysts. *Catal. Today* **2016**, *261*, 17–27. [CrossRef]
128. Bartholomew, C.H. Hydrogen adsorption on supported cobalt, iron, and nickel. *Catal. Lett.* **1990**, *7*, 27–51. [CrossRef]
129. Borg, Ø.; Frøseth, V.; Stors1ter, S.; Rytter, E.; Holmen, A. Fischer-Tropsch synthesis recent studies on the relation between the properties of supported cobalt catalysts and the activity and selectivity. *Surf. Sci. Catal.* **2007**, *167*, 117.

130. Zhang, R.; Hao, X.; Yong, Y.; Li, Y. Investigation of acetylene addition to Fischer–Tropsch Synthesis. *Catal. Commun.* **2011**, *12*, 1146–1148. [CrossRef]
131. Krishnamoorthy, S.; Li, A.; Iglesia, E. Pathways for CO_2 Formation and Conversion during Fischer–Tropsch Synthesis on Iron-Based Catalysts. *Catal. Lett.* **2002**, *80*, 77–86. [CrossRef]
132. Shi, B.; Davis, B.H. ^{13}C-tracer study of the Fischer–Tropsch synthesis: Another interpretation. *Catal. Today* **2000**, *58*, 255–261. [CrossRef]
133. Mann, B.E.; Turner, M.L.; Quyoum, R.; Marsih, N.; Maitlis, P.M. Demonstration by 13C NMR Spectroscopy of Regiospecific Carbon−Carbon Coupling during Fischer−Tropsch Probe Reactions. *J. Am. Chem. Soc.* **1999**, *121*, 6497–6498. [CrossRef]
134. Davis, B.H. Fischer–Tropsch Synthesis: Reaction mechanisms for iron catalysts. *Catal. Today* **2009**, *141*, 25–33. [CrossRef]
135. Bistoni, G.; Rampino, S.; Scafuri, N.; Ciancaleoni, G.; Zuccaccia, D.; Belpassi, L.; Tarantelli, F. How π back-donation quantitatively controls the CO stretching response in classical and non-classical metal carbonyl complexes. *Chem. Sci.* **2016**, *7*, 1174–1184. [CrossRef]
136. Ojeda, M.; Nabar, R.; Nilekar, A.U.; Ishikawa, A.; Mavrikakis, M.; Iglesia, E. CO activation pathways and the mechanism of Fischer–Tropsch synthesis. *J. Catal.* **2010**, *272*, 287–297. [CrossRef]
137. Schulz, H. Short history and present trends of Fischer–Tropsch synthesis. *Appl Catal A Gen.* **1999**, *186*, 3–12. [CrossRef]
138. Kuipers, E.W.; Scheper, C.; Wilson, J.H.; Vinkenburg, I.H.; Oosterbeek, H. Non-ASF product distributions due to secondary reactions during Fischer–Tropsch synthesis. *J. Catal.* **1996**, *158*, 288–300. [CrossRef]
139. Van der Laan, G.P.; Beenackers, A.A.C.M. Kinetics and Selectivity of the Fischer–Tropsch Synthesis: A Literature Review. *Catal. Rev.* **1999**, *41*, 255–318. [CrossRef]
140. Madon, R.J.; Reyes, S.C.; Iglesia, E. Primary and secondary reaction pathways in ruthenium-catalyzed hydrocarbon synthesis. *J. Phys. Chem.* **1991**, *95*, 7795–7804. [CrossRef]
141. Ji, Y.-Y.; Xiang, H.-W.; Yang, J.-L.; Xu, Y.-Y.; Li, Y.-W.; Zhong, B. Effect of reaction conditions on the product distribution during Fischer–Tropsch synthesis over an industrial Fe-Mn catalyst. *Appl. Catal. A* **2001**, *214*, 77–86. [CrossRef]
142. Iglesia, E.; Soled, S.L.; Fiato, R.A. Fischer-Tropsch synthesis on cobalt and ruthenium. Metal dispersion and support effects on reaction rate and selectivity. *J. Catal.* **1992**, *137*, 212–224. [CrossRef]
143. Novak, S.; Madon, R.J.; Suhl, H. Secondary effects in the Fischer-Tropsch synthesis. *J. Catal.* **1982**, *77*, 141–151. [CrossRef]
144. Schulz, H.; Claeys, M. Reactions of α-olefins of different chain length added during Fischer–Tropsch synthesis on a cobalt catalyst in a slurry reactor. *Appl. Catal. A* **1999**, *186*, 71–90. [CrossRef]
145. Kapteijn, F.; de Deugd, R.M.; Moulijn, J.A. Fischer–Tropsch synthesis using monolithic catalysts. *Catal. Today* **2005**, *105*, 350–356. [CrossRef]
146. Soled, S.; Iglesia, E.; Fiato, R.A. Activity and selectivity control in iron catalyzed Fischer-Tropsch synthesis. *Catal. Lett.* **1990**, *7*, 271–280. [CrossRef]
147. Bukur, D.B.; Nowicki, L.; Manne, R.K.; Lang, X.S. Activation Studies with a Precipitated Iron Catalyst for Fischer-Tropsch Synthesis: II. Reaction Studies. *J. Catal.* **1995**, *155*, 366–375. [CrossRef]
148. Donnelly, T.J.; Satterfield, C.N. Product distributions of the Fischer-Tropsch synthesis on precipitated iron catalysts. *Appl. Catal.* **1989**, *52*, 93–114. [CrossRef]
149. Tau, L.M.; Dabbagh, H.A.; Davis, B.H. Fischer-Tropsch synthesis: carbon-14 tracer study of alkene incorporation. *Energy Fuels* **1990**, *4*, 94–99. [CrossRef]
150. Ma, W.; Shafer, W.D.; Martinelli, M.; Sparks, D.E.; Davis, B.H. Fischer-Tropsch synthesis: Using deuterium tracer coupled with kinetic approach to study the kinetic isotopic effects of iron, cobalt and ruthenium catalysts. *Catal. Today* **2019**, in press. [CrossRef]
151. Schulz, H. Comparing Fischer-Tropsch Synthesis on Iron- and Cobalt Catalysts: The dynamics of structure and function. *Stud. Surf. Sci. Catal.* **2007**, *163*, 177–199.
152. Jacobs, G.; Ricote, S.; Patterson, P.M.; Graham, U.M.; Dozier, A.; Khalid, S.; Rhodus, E.; Davis, B.H.; Brookhaven National Laboratory. Low temperature water-gas shift: Examining the efficiency of Au as a promoter for ceria-based catalysts prepared by CVD of an Au precursor. *Appl. Catal. A Gen.* **2005**, *292*, 229–243. [CrossRef]

153. Ricote, S.; Jacobs, G.; Milling, M.; Ji, Y.; Patterson, P.M.; Davis, B.H. Low temperature water-gas shift: Characterization and testing of binary mixed oxides of ceria and zirconia promoted with Pt. *Appl. Catal. A Gen.* **2006**, *303*, 35–47. [CrossRef]
154. Jacobs, G.; Ricote, S.; Davis, B.H. Low temperature water-gas shift: Type and loading of metal impacts decomposition and hydrogen exchange rates of pseudo-stabilized formate over metal/ceria catalysts. *Appl. Catal. A Gen.* **2006**, *302*, 14–21. [CrossRef]
155. Jacobs, G.; Graham, U.M.; Chenu, E.; Patterson, P.M.; Dozier, A.; Davis, B.H. Low temperature water-gas shift: Impact of Pt promoter loading on the partial reduction of ceria, and consequences for catalyst design. *J. Catal.* **2005**, *229*, 499–512. [CrossRef]
156. Pendyala, V.R.R.; Jacobs, G.; Mohandas, J.C.; Luo, M.; Ma, W.; Gnanamani, M.K.; Davis, B.H. Fischer–Tropsch synthesis: Attempt to tune FTS and WGS by alkali promoting of iron catalysts. *Appl. Catal. A Gen.* **2010**, *389*, 131–139. [CrossRef]
157. Pendyala, V.R.R.; Jacobs, G.; Mohandas, J.C.; Luo, M.; Hamdeh, H.H.; Ji, Y.; Ribeiro, M.C.; Davis, B.H. Fischer–Tropsch Synthesis: Effect of Water Over Iron-Based Catalysts. *Catal. Lett.* **2010**, *140*, 98–105. [CrossRef]
158. Lia, J.; Cheng, X.; Zhang, C.; Chang, Q.; Wang, J.; Wang, X.; Lv, Z.; Dong, W.; Yang, Y.; Lia, Y. State Effect of alkalis on iron-based Fischer-Tropsch synthesis catalysts: Alkali-FeOx interaction, reduction, and catalytic performance. *Appl. Catal. A Gen.* **2016**, *528*, 131–141. [CrossRef]
159. Lee, S.W.; Coulombe, S.; Glavincevski, B. Investigation of methods for determining aromatics in middle-distillate fuels. *Energy Fuels* **1990**, *4*, 20–33. [CrossRef]
160. Tau, L.; Dabbagh, H.A.; Chawla, B.; Davis, B.H. Fischer-Tropsch synthesis with an iron catalyst: Incorporation of ethene into higher carbon number alkanes. *Catal. Lett.* **1990**, *7*, 141–150. [CrossRef]
161. O'Brien, B.S.R.; Davis, S.B.B.H. Mechanism of the Isomerization of 1-Alkene during Iron-Catalyzed Fischer–Tropsch Synthesis. *J. Catal.* **2001**, *199*, 202–208.
162. Shi, B.; Liao, Y.; Naumovitz, J.L. Formation of 2-alkenes as secondary products during Fischer–Tropsch synthesis. *Appl. Catal. A Gen* **2015**, *490*, 201–206. [CrossRef]
163. Shi, B.; Chunfen, J. Inverse kinetic isotope effects and deuterium enrichment as a function of carbon number during formation of C–C bonds in cobalt catalyzed Fischer–Tropsch synthesis. *J. Appl. Catal A Gen.* **2011**, *393*, 178–183. [CrossRef]
164. Van Dijk, W.L.; Niemantsverdriet, J.W.; Kraan, A.M.v.; van der Baan, H.S. Effects of Manganese Oxide and Sulphate on the Olefin Selectivity of Iron Catalysts in the Fischer Tropsch Reaction. *Appl. Catal.* **1982**, *2*, 273–288. [CrossRef]
165. Kreitman, K.M.; Baerns, M.; Butt, J.B. Manganese-oxide-supported iron Fischer-Tropsch synthesis catalysts: Physical and catalytic characterization. *J. Catal.* **1987**, *105*, 319–334. [CrossRef]
166. Ribeiro, M.C.; Jacobs, G.; Pendyala, R.; Davis, B.H.; Cronauer, D.C.; Kropf, A.J.; Marshall, C.L. Fischer-Tropsch Synthesis: Influence of Mn on the Carburization Rates and Activities of Fe-Based Catalysts by TPR-EXAFS/XANES and Catalyst Testing. *J. Phys. Chem. C* **2011**, *115*, 4783–4792. [CrossRef]
167. Diehl, F.; Khodakov, A.Y. Promotion of Cobalt Fischer-Tropsch Catalysts with Noble Metals: A Review. *Oil Gas Sci. Technol.* **2009**, *64*, 11–24. [CrossRef]
168. Hilmen, A.M.; Schanke, D.; Holmen, A. TPR study of the mechanism of rhenium promotion of alumina-supported cobalt Fischer-Tropsch catalysts. *Catal. Lett.* **1996**, *38*, 143–147. [CrossRef]
169. Vada, S.; Hoff, A.; Ådnanes, E.; Schanke, D.; Holmen, A. Fischer-Tropsch synthesis on supported cobalt catalysts promoted by platinum and rhenium. *Top. Catal.* **1995**, *2*, 155–162. [CrossRef]
170. Prins, R. Hydrogen Spillover. Facts and Fiction. *Chem. Rev.* **2012**, *112*, 2714–2738. [CrossRef]
171. Karim, W.; Spreafico, C.; Kleibert, A.; Gobrecht, J.; VandeVondele, J.; Ekinci, Y.; van Bokhoven, J.A. Catalyst support effects on hydrogen spillover. *Nature* **2017**, *541*, 68–71. [CrossRef]
172. Masuku, C.M.; Shafer, W.D.; Ma, W.; Gnanamani, M.K.; Jacobs, G.; Hildebrandt, D.; Glasser, D.; Davis, B.H. Variation of residence time with chain length for products in a slurry-phase Fischer–Tropsch reactor. *J. Catal.* **2012**, *287*, 93–101. [CrossRef]

© 2019 by the authors. Licensee MDPI, Basel, Switzerland. This article is an open access article distributed under the terms and conditions of the Creative Commons Attribution (CC BY) license (http://creativecommons.org/licenses/by/4.0/).

Article

The Preparation and Characterization of Co–Ni Nanoparticles and the Testing of a Heterogenized Co–Ni/Alumina Catalyst for CO Hydrogenation

Julián López-Tinoco [1,2], Rubén Mendoza-Cruz [3,*], Lourdes Bazán-Díaz [3], Sai Charan Karuturi [4], Michela Martinelli [5], Donald C. Cronauer [6], A. Jeremy Kropf [6], Christopher L. Marshall [6] and Gary Jacobs [4,7,*]

1. Department of Physics and Astronomy, University of Texas at San Antonio, One UTSA Circle, San Antonio, TX 78249, USA; jltzet2@gmail.com
2. Facultad de Ingeniería Química, Universidad Michoacana de San Nicolás de Hidalgo, Morelia, Michoacán 58030, Mexico
3. Instituto de Investigaciones en Materiales, Universidad Nacional Autónoma de México, Ciudad de México, 04510, Mexico; bazanlulu@materiales.unam.mx
4. Department of Mechanical Engineering, University of Texas at San Antonio, One UTSA Circle, San Antonio, TX 78249, USA; saicharan.karuturi@my.utsa.edu
5. Center for Applied Energy Research, University of Kentucky, 2540 Research Park Drive, Lexington, KY 40511, USA; michela.martinelli@uky.edu
6. Argonne National Laboratory, Argonne, IL 60439, USA; dccronauer@anl.gov (D.C.C.); kropf@anl.gov (A.J.K.); marshall@anl.gov (C.L.M.)
7. Department of Biomedical Engineering and Chemical Engineering, University of Texas at San Antonio, One UTSA Circle, San Antonio, TX 78249, USA
* Correspondence: rmendoza@materiales.unam.mx (R.M.-C.); gary.jacobs@utsa.edu (G.J.); Tel.: +1-210-458-7080 (G.J.)

Received: 7 November 2019; Accepted: 19 December 2019; Published: 21 December 2019

Abstract: Samples of well-controlled nanoparticles consisting of alloys of cobalt and nickel of different atomic ratios were synthesized using wet chemical methods with oleylamine as the solvent and the reducing agent. These materials were characterized by a variety of techniques, including high-angle annular dark-field scanning transmission electron microscopy (HAADF-STEM), X-ray energy dispersive spectroscopy (EDS), and X-ray diffraction (XRD). Small amounts of heterogenized catalysts were prepared using alumina as the support. However, the potential for use of Co–Ni catalysts in CO hydrogenation was explored using a larger amount of Co–Ni/alumina catalyst prepared from standard aqueous impregnation methods and tested in a continuously stirred tank reactor (CSTR) for Fischer–Tropsch synthesis (FTS). Results are compared to a reference catalyst containing only cobalt. The heterogenized catalysts were characterized using synchrotron methods, including temperature programmed reduction with extended X-ray absorption fine structure spectroscopy and X-ray absorption near edge spectroscopy (TPR-EXAFS/XANES). The characterization results support intimate contact between Co and Ni, strongly suggesting alloy formation. In FTS testing, drawbacks of Ni addition included decreased CO conversion on a per gram catalyst basis, although Ni did not significantly impact the turnover number of cobalt, and produced slightly higher light gas selectivity. Benefits of Ni addition included an inverted induction period relative to undoped Co/Al$_2$O$_3$, where CO conversion increased with time on-stream in the initial period, and the stabilization of cobalt nanoparticles at a lower weight % of Co.

Keywords: cobalt–nickel nanoparticles; cobalt–nickel alloys; cobalt; nickel; alumina; HAADF-STEM; TPR-EXAFS/XANES; Fischer–Tropsch synthesis; CO hydrogenation; CSTR

1. Introduction

Cobalt catalysts are important for the Fischer–Tropsch synthesis (FTS) reaction, which is at the heart of the gas-to-liquids (GTL) process [1]. Not only do they offer good activity and stability, as well as high selectivity toward heavier hydrocarbons, but they also possess low water–gas shift (WGS) activity relative to iron carbide catalysts, making them suitable for converting the higher H_2:CO ratio syngas associated with natural gas derived syngas [2].

Cobalt is an expensive metal, with a price that has ranged from $22 to $100 per kg within the past five years [3]. Therefore, in order to maximize its surface area, cobalt is typically supported on a metal oxide carrier such as γ-Al_2O_3 or TiO_2. Cobalt catalysts are typically prepared by aqueous impregnation methods, such as incipient wetness impregnation (IWI) or a preferred slurry impregnation method (SIM) [1] using a precursor such as cobalt nitrate, and then the catalyst is dried and calcined (e.g., 350 °C, 4 h). There are three major problems with heterogenized cobalt catalysts that are of interest to commercial developers.

The first problem is that of reducibility. Following calcination, cobalt is present as Co_3O_4 nanoclusters and, unlike unsupported cobalt oxide which reduces to the metal at 300–350 °C, the two-step reduction profile of Co_3O_4 on alumina, which involves CoO as an intermediate, is broadened. Co_3O_4 reduces to CoO in a relatively facile manner in the range of 300–350 °C, but to reduce CoO to Co^0, temperatures as high as 800 °C are needed [4,5]. In order to facilitate the reduction of cobalt oxides, promoters (e.g., Pt [4–8], Ru [4,7–10], Re [4,7,8,11–14], Ag [15,16]) are utilized, and researchers have suggested that the catalytic action is either by a direct chemical effect (e.g., alloying), or by reduction of the promoter at low temperature, followed by a H_2-dissociation and spillover mechanism that facilitates the reduction of cobalt oxides. One problem with these promoters is that they are very expensive. The price range of platinum, the most common promoter used, has been $25,000–$50,000 per kg within the past five years [3]. Fortunately, only small percentage loadings (<0.5% Pt) are needed to facilitate cobalt oxide reduction. Nevertheless, it is of interest to find a suitable base metal replacement. As nickel is in the same group as platinum and has a much lower price ($8–$20 per kg over the past five years [3]), investigations on Ni as a reduction promoter for cobalt oxides are important.

The second issue is selectivity. Cobalt is a d^7 metal and has an electronic configuration suitable for growing long-chained hydrocarbons. The metal has the right balance in that it is able to dissociate CO, as well as stabilize the vinylic intermediate in the appropriate sp^3 configuration, such that chain growth is controlled to one side, producing primarily straight chained paraffins, 1-olefins, and 1-alcohols [17]. Nickel, however, is a d^8 metal, and the electronic back-donation capability of Ni is excessive for the FTS application, such that the vinylic intermediate is insufficiently stabilized, resulting primarily in undesired light gas production [17]. Thus, there are questions related to how much nickel can be added to cobalt nanoparticles such that the resulting alloy still possesses adequate hydrocarbon selectivity for FT, and whether a catalyst possessing a nickel core and a cobalt shell might be developed such that the reaction occurs primarily on a cobalt shell.

Finally, there is the problem of stability with heterogenized cobalt catalysts [9,18–21]. While it is important to disperse cobalt on a carrier such as alumina to maximize the surface availability of Co^0, which provides the active sites, there is a thermodynamic limit regarding how small the cobalt nanoparticles can be in FTS and still maintain adequate stability [18]. In FTS, H_2O is a major product of the reaction, and studies have concluded that cobalt nanoparticles that are <2–4 nm will oxidize under commercially relevant reaction conditions; this renders the cobalt nanoparticles inactive. As such, catalysts are often made with cobalt nanoparticles that target the 6–10 nm range [19]. A consequence of this design is that the majority of the expensive cobalt is locked within the cobalt nanoparticle rather than being exposed to the surface where the catalysis occurs. Thus, it would be of interest to replace a fraction of the more expensive cobalt with a less expensive metal-like nickel.

The incorporation of a second metal to form alloyed metal nanoparticles has helped to increase activity on a per gram of catalyst basis, as well as improving stability, thus decreasing the cost of the process [22]. By controlling the synthesis conditions such as solvent, reducing agent, temperature and metal composition, it is also possible to control the size, shape and structure of the nanoparticles, tailoring in this way their final properties toward different applications [23–25].

In this work, we begin to explore the preparation of well-controlled nanoparticles consisting of alloys of cobalt and nickel having different atomic ratios. These are synthesized using wet chemical methods with oleylamine as both the solvent and the reducing agent. These materials were characterized by a variety of techniques, including HAADF-STEM, EDS, and XRD. Small amounts of heterogenized catalysts were also prepared using alumina as the support. However, because it was not possible at this time to prepare large enough batches for catalyst testing using a CSTR, the potential for use of Co–Ni catalysts in CO hydrogenation was explored using a larger amount of Co–Ni/alumina catalyst prepared from standard aqueous impregnation methods. This catalyst was tested in a CSTR for FTS and compared to a reference catalyst containing only cobalt. The heterogenized catalysts were characterized by, among other techniques, synchrotron methods including TPR-EXAFS and TPR-XANES.

2. Results

2.1. Cobalt–Nickel Nanoparticles

Co–Ni alloy nanoparticles were prepared by a colloidal wet-chemical method using oleylamine (OAm) as the solvent and reducing agent. Three samples were prepared with different initial atomic ratios, Co98.4Ni1.6 (Sample #1), Co92.7Ni7.3 (Sample #2), and Co89.7Ni10.3 (Sample #3). Figure 1a shows a low magnification HAADF image of the synthesized Co98.4Ni1.6 nanoparticles. The sample consisted of 20.5 ± 3.2 nm polycrystalline dendritic particles, formed by the apparent attachment of small crystallites. The high-magnification image (Figure 1b) confirmed that the particles were well crystallized.

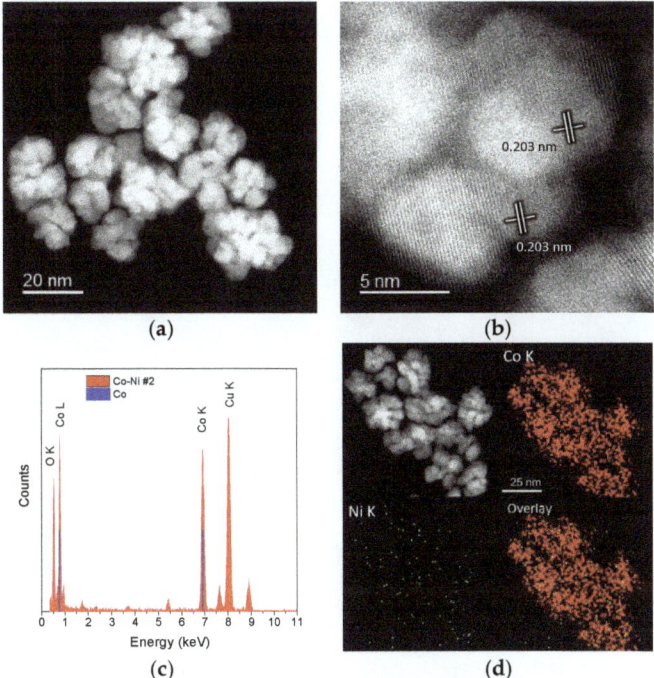

Figure 1. Sample #1. (**a**) Low-magnification and (**b**) high-magnification HAADF-STEM images of the Co98.4Ni1.6 nanoparticles. (**c**) EDS spectrum and (**d**) EDS maps taken from a group of nanoparticles. The Co K and Co L lines are plotted for reference.

The measured interplanar distance of the well-resolved lattice fringes was 0.203 nm, corresponding to the (002) planes of the hexagonal-close packed (HCP) phase or the (111) planes of a face-centered cubic (FCC) structure of cobalt [26]. Cobalt and nickel FCC have a very close atomic weight and similar lattice constants, differing by less than 1%, making it difficult to differentiate them within the measuring error. Therefore, EDS analysis was performed to estimate the Ni content that is present within the nanoparticle structure. Figure 1c corresponds to the EDS spectrum acquired from mapping a group of nanoparticles laying on the carbon support. It was observed that nickel was at an ultra-low concentration (Ni-K@7.477 keV), and its content was below detection limits. This is expected from the very low amount of Ni used during synthesis of the nanoparticles. Oxygen was also detected, such that the formation of an oxide layer on the nanoparticle surface, formed when exposing to air during purification, cannot be discounted. From the estimated Ni content, it was inferred that the synthesized nanoparticles crystallized in a hexagonal close packed structure, which forms at a relatively low temperature and low Ni content [27]. With increasing nickel content, a significant change in terms of shape and composition was achieved. Figure 2 corresponds to HAADF-STEM images of the nanoparticles when using a Co:Ni atomic ratio of 92.7:7.3. As observed from the low-magnification image in Figure 2a, the nanoparticles exhibit a very different morphology, presenting a core-shell feature with a mean diameter of 12.3 ± 1.5 nm core and 2.3 ± 0.2 nm shell. The low contrast of the outer layer surrounding the particles suggests a lower atomic density, attributed to the presence of an oxidized layer. The high-magnification images in Figure 2b reveal the multiply-twinned structure of the nanoparticles. Different regions of well-resolved lattice fringes are visible. The lattice distance measured at the nanoparticle core was 0.216 nm while at the nanoparticle surface the measured lattice distance was 0.150 nm, matching with the (100) planes of an HCP Co–Ni alloy (0.217 nm) and the (220) planes of the CoO FCC structure (0.151 nm) [27,28].

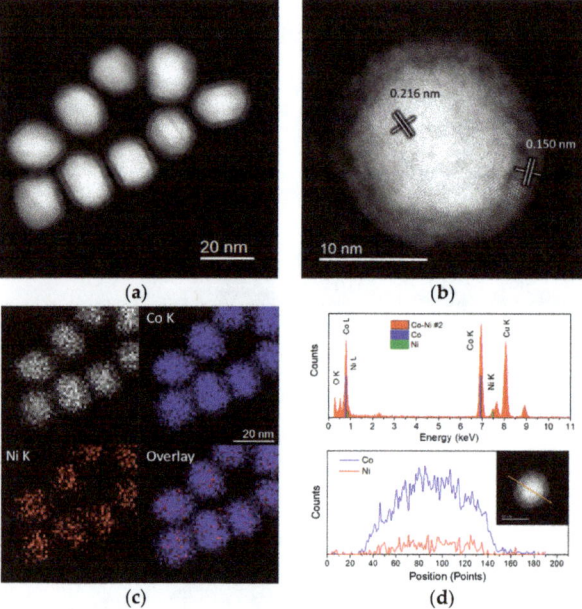

Figure 2. Sample #2. (**a**) Low-magnification and (**b**) high-magnification HAADF-STEM images of the synthesized Co–Ni 92.7:7.3 nanoparticles. (**c**) EDS maps and (**d**) EDS spectrum (up) taken from a group of nanoparticles. The Co K and Co L lines, as well as the Ni K and Ni L lines are plotted for reference. d) EDS line scan (down) crossing at the middle of the particle as shown in the inset, confirming the presence of both Co and Ni within the particle.

EDS analysis confirmed the presence of both Co and Ni elements within the nanoparticles, as shown from the EDS spectrum in Figure 2c. The estimated atomic composition was 92.7%Co–7.3%Ni, giving support to the presence of the HCP Co–Ni alloy. Additionally, an EDS line scan crossing at the center of an individual nanoparticle confirmed that Co and Ni are well distributed along the particle, as displayed in Figure 2d.

In the bulk, the Co–Ni alloy crystallizes in a complete solid solution having an FCC structure (α phase) for compositions >30% atomic nickel, while for compositions below ~30% atomic Ni and a relatively low temperature, a martensitic transformation from the FCC to HCP phase (ε phase) takes place. Since the nanoparticles were synthesized at a relatively low temperature (230 °C) and with a low Ni content, it was possible to obtain an HCP Co–Ni alloy. However, the boundaries of the bulk transition are not well established [27], and the temperature range of transitions lowers at the nanoscale, since it is strongly affected by particle size and shape [29,30].

In the absence of 1,2-hexadecanediol, larger nanoparticles were obtained by using only OAm as the reducing agent, as shown in the SEM image provided in Figure 3a (Sample #3). The nanoparticles presented cubic and octahedral shapes having a highly faceted surface. The average size was 85 ± 13 nm. Figure 3b,c show an HAADF-STEM image of a single particle at atomic resolution. From the measured lattice distances of 0.242 and 0.211 nm, which formed an angle of ~55°, the particle was indexed as a $Co_{1-x}Ni_xO$ FCC structure viewed along the [110] axis zone. The CoO-type phase crystallizes in a rock salt structure having the Fm3m space group, with a difference in the lattice parameters of ~2.3% between a CoO and NiO structure (a = 4.2667 Å for CoO, and a = 4.1684 Å for NiO). Hence, the measured distances between both structures suggest the formation of a $Co_{1-x}Ni_xO$ solid solution. The EDS analysis from a large group of nanoparticles (e.g., the cluster shown in Figure 3d) confirmed the presence of Co, Ni, and O, as presented in the spectrum and mapping in Figure 3c,d. The estimated average composition from the acquired EDS spectra was 89.7%Co–10.3%Ni, indicating an atomic percentage ratio of 8.7 between the cations.

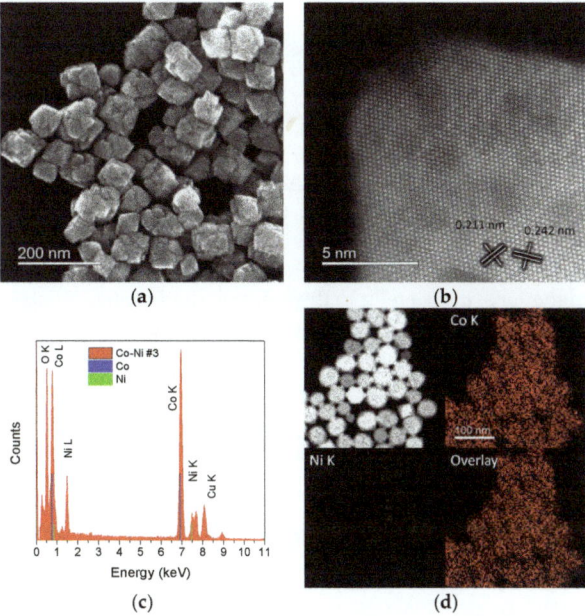

Figure 3. Sample #3. (**a**) SEM image of the synthesized Co89.7Ni10.3 sample. (**b**) High-magnification HAADF-STEM image of a single CoNi–O nanoparticle. (**c**) EDS spectrum and (**d**) EDS maps taken from a group of nanoparticles. The Co and Ni K and L lines are shown for reference in the spectrum.

To better determine the structure of the synthesized particles, the freshly prepared powders were characterized by XRD. Figure 4 shows the diffraction patterns of the Co–Ni alloy nanoparticles (Sample #1, #2, and #3). In Figure 4a,b, corresponding to the colloidal preparation method of Samples #1 and #2, the characteristic peaks of metallic nickel FCC (ICSD 518120) and metallic cobalt HCP (ICSD 15288) were taken as references. As shown, both patterns present similar features with peaks located at 44.5, 51.8, and 76.4° assigned to the (111), (200), and (220) planes of the FCC structure, with additional peaks at 41.7, 44.3, 62.6, and 76.1° assigned to the HCP phase of cobalt. The observed peaks presented a slight shift with respect to the reference; this is associated with the alloyed Co–Ni, indicating the coexistence of both FCC and HCP phases of cobalt within the samples [31,32]. Meanwhile, the XRD pattern of the colloidally prepared Sample #3 confirmed the oxidized nature of the nanoparticles, Figure 4c. The pattern is indexed as a CoO-type structure, confirming the $Co_{1-x}Ni_xO$ FCC solid solution phase.

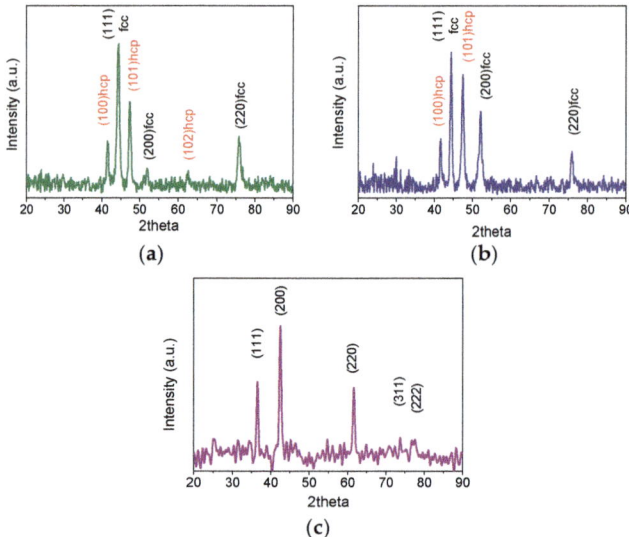

Figure 4. XRD patterns of (**a**) Co98.4Ni1.6 (Sample #1) and (**b**) Co92.7Ni7.3 (Sample #2) nanoparticles. The patterns showed the presence of both HCP (red labels) and FCC (black labels) cobalt structure. (**c**) XRD pattern of Co89.7Ni10.3 (Sample #3). The pattern matched with a $Co_{1-x}Ni_xO$ FCC solid solution.

2.2. Heterogenized Catalysts

For sample designations, please refer to the Materials and Methods Section. Figure 5 displays hydrogen TPR-XANES spectra at the Co K-edge of the different samples. Figure 5a,b reveal that the nanoparticles prepared by the non-conventional method (Samples #1 and 2) start as a mixture of Co^{2+} and Co^0, and the spectra resemble a mixture of CoO and Co^0, while that of Figure 5c corresponding to Sample #3 begins entirely in the Co^{2+} oxidation state, resembling CoO. As shown by the linear combination (LC) fittings of TPR-XANES spectra with reference compounds in Figure 6a,b, both Samples #1 and #2 started with high Co^0 content (75% and 80%, respectively), and achieved nearly complete reduction by 350 °C. Sample #3, as shown in Figure 6c, achieved nearly complete reduction from CoO to Co^0 by 400 °C. The point of 50% reduction of CoO to Co^0 occurred at ~330 °C.

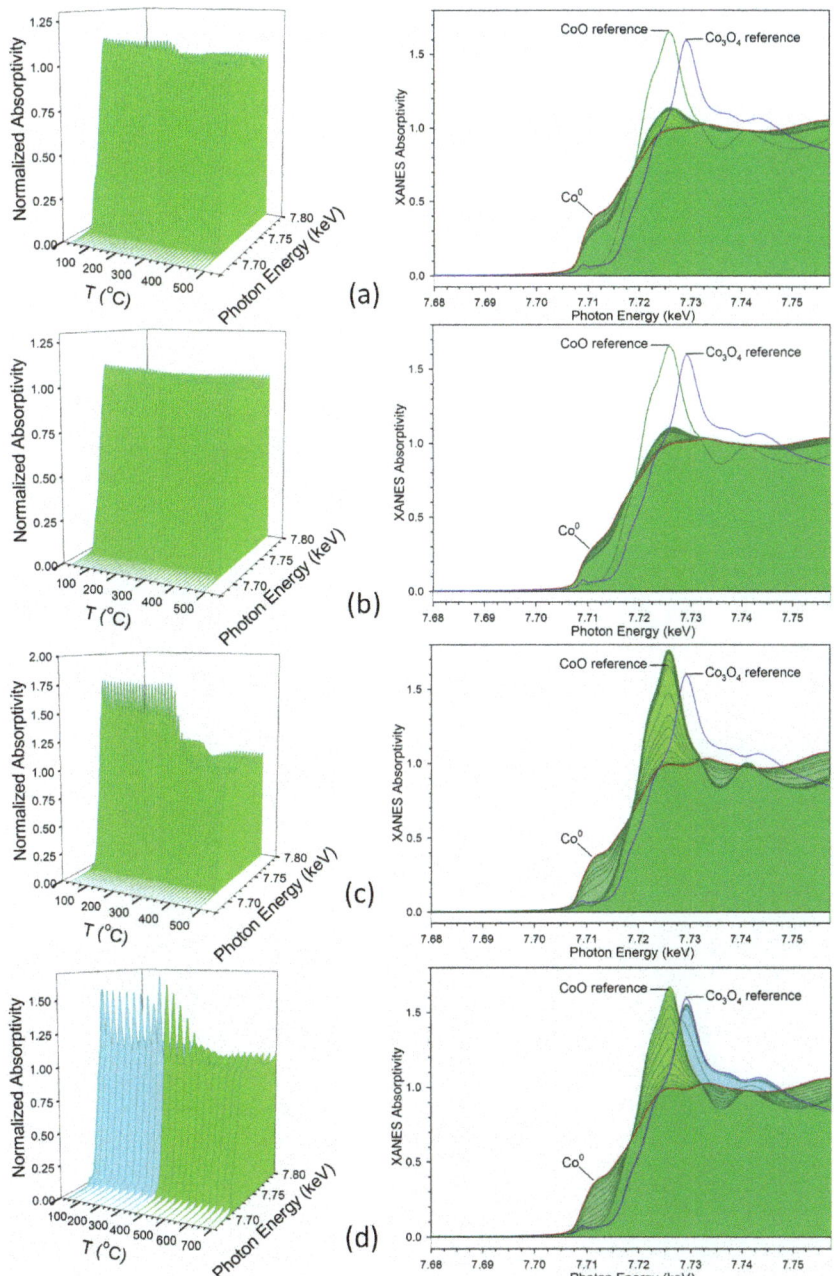

Figure 5. H_2-TPR-XANES spectra at the Co K-edge of Samples (**a**) #1, (**b**) #2, (**c**) #3, and (**d**) #4 (note difference in scale). (Cyan) is reduction of Co_3O_4 to CoO, and (Green) CoO to Co^0.

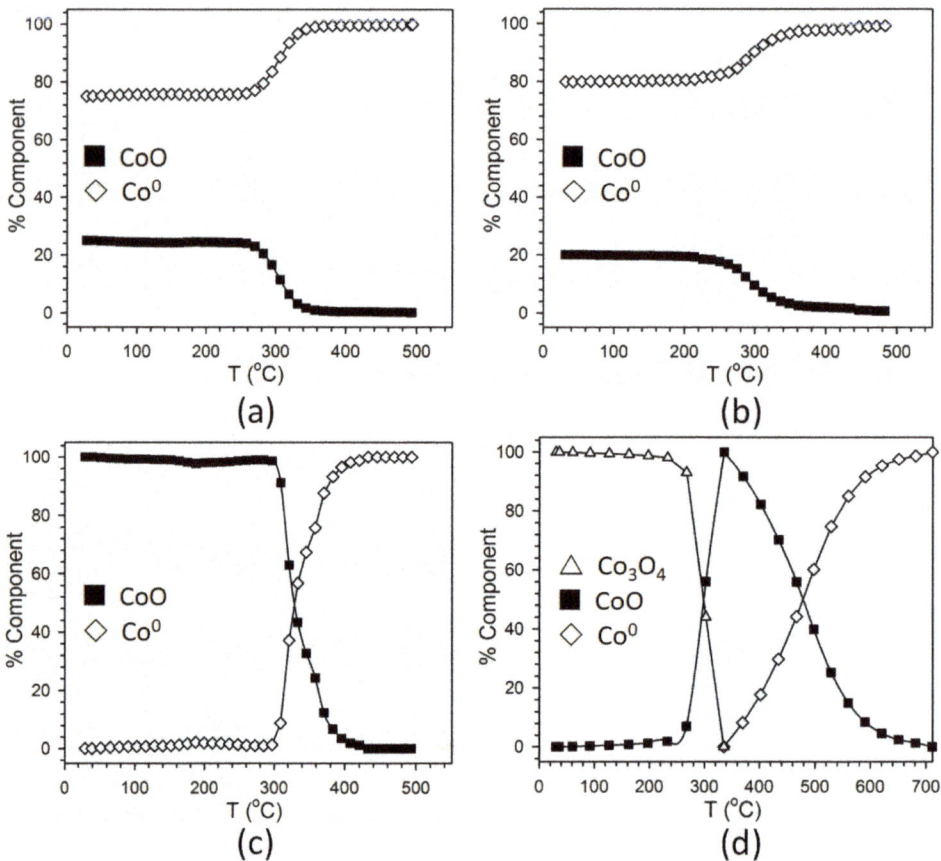

Figure 6. Linear combination fittings of H_2 TPR-XANES spectra of Samples (**a**) #1, (**b**) #2, (**c**) #3, and (**d**) #4. The reference compounds for Samples #1–3 were the first spectrum of Sample #3 (representing 100% Co^{2+}) and the final spectrum of each sample after remaining at 500 °C in H_2 (representing ~100% Co^0). Reference compounds for Sample #4 were the initial spectrum (representing a mixture of Co^{2+} and Co^{3+} similar to Co_3O_4), the point of maximum CoO content, and the final spectrum following TPR and holding (representing Co^0). Note difference in scale for (**d**).

Figure 5d, on the other hand, representing the catalyst prepared by the traditional slurry impregnation method (Sample #4), followed the conventional two-step reduction of cobalt oxides to metallic compounds, where: $Co_3O_4 + H_2 = 3CoO + H_2O$ and $3CoO + 3H_2 = 3Co^0 + 3H_2O$. Figure 6d shows linear combination fittings of the TPR-XANES spectra using reference compounds. The catalyst nears complete reduction to Co^0 by 580 °C. The point at which 50% of the Co_3O_4 converts to CoO occurs at ~290 °C, while the point at which 50% of the CoO converts to Co^0 occurs at ~480 °C. Note that Ni is clearly facilitating reduction, as the point of 50% of CoO conversion for unpromoted 25%Co/alumina was found from our prior work to be ~550 °C with nearly complete reduction at temperatures higher than 700 °C [5].

XANES snapshots along the TPR trajectory of key states of cobalt are provided in Figure 7. These snapshots highlight key differences among the different Co–Ni catalysts. For example, Co_3O_4 is only detected in Sample #4 (i.e., the catalyst prepared by the conventional method). While Samples #1 and #2 have very little CoO content initially, Sample #3 starts out as virtually all CoO, with no Co^0 or

Co$_3$O$_4$ being detected, in good agreement with the HAADF-STEM structural characterization. This suggests that Co–Ni nanoparticles can be tuned to have different oxidation state characteristics.

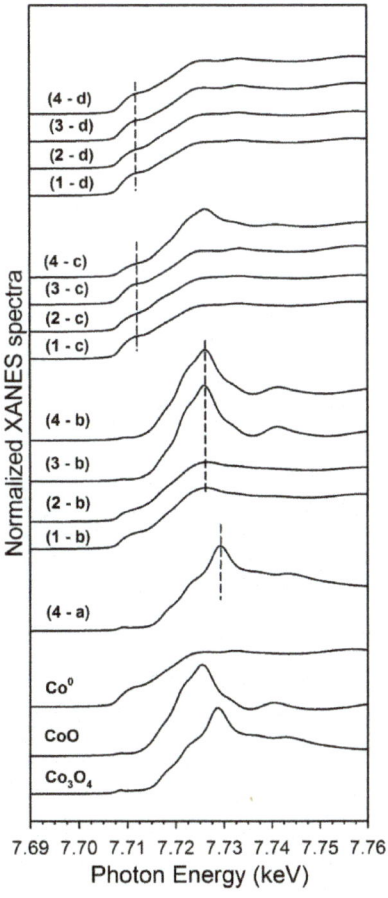

Figure 7. XANES snapshots at the Co K-edge of reference compounds, and (**a**) the point of maximum Co$_3$O$_4$ content in the H$_2$ TPR-XANES run, (**b**) the point of maximum CoO content in the H$_2$ TPR-XANES run, (**c**) the point at 500 °C, and (**d**) the point of maximum Co^0 content in the H$_2$ TPR-XANES run for each catalyst sample (following ramping and holding at T$_{max}$), including (1) Sample #1, (2), Sample #2, (3) Sample #3, and (4) Sample #4. Note that only Sample #4 displayed Co$_3$O$_4$ content (i.e., in the initial period).

Figure 8 displays hydrogen TPR-XANES spectra at the Ni K-edge of the different samples. Figure 8a,b reveal that the nanoparticles prepared by the nonconventional method (Samples #1 and 2) start as a mixture of primarily Ni^0 with very little Ni^{2+}, and the spectra resemble a mixture of Ni^0 and NiO, while that of Figure 8c,d corresponding to Sample #3 and Sample #4 begin entirely in the Ni^{2+} oxidation state, resembling NiO. Due to the low amount of Ni^{2+} in Samples #1 and #2, as well as the noise level of Sample #1 (due to low Ni content), LC fittings were not performed on these samples. However, LC fittings of TPR-XANES spectra with reference compounds were performed for both Samples #3 and #4. As shown in Figure 9, Sample #3 achieved nearly complete reduction of nickel by 350 °C, while Sample #4 achieved nearly complete reduction of nickel by 600 °C. Interestingly, the XANES spectra of Sample #4 display a shift in the NiO spectrum to lower energy when Co_3O_4 reduces to CoO; this suggests that Ni^{2+} is in intimate contact with cobalt oxides. The point at which the Ni^{2+} is at the halfway position between NiO associated with Co_3O_4 and NiO associated with CoO occurs at 290 °C, which matches the temperature at which 50% reduction of Co_3O_4 to CoO occurs. The point of 50% reduction of NiO to Ni^0 occurred at ~330 °C for Sample #3 and at ~480 °C for Sample #4. These temperatures match those of 50% CoO reduction to Co^0, suggesting once again that the Co and Ni are in intimate contact in both the oxide and metallic forms (e.g., alloys).

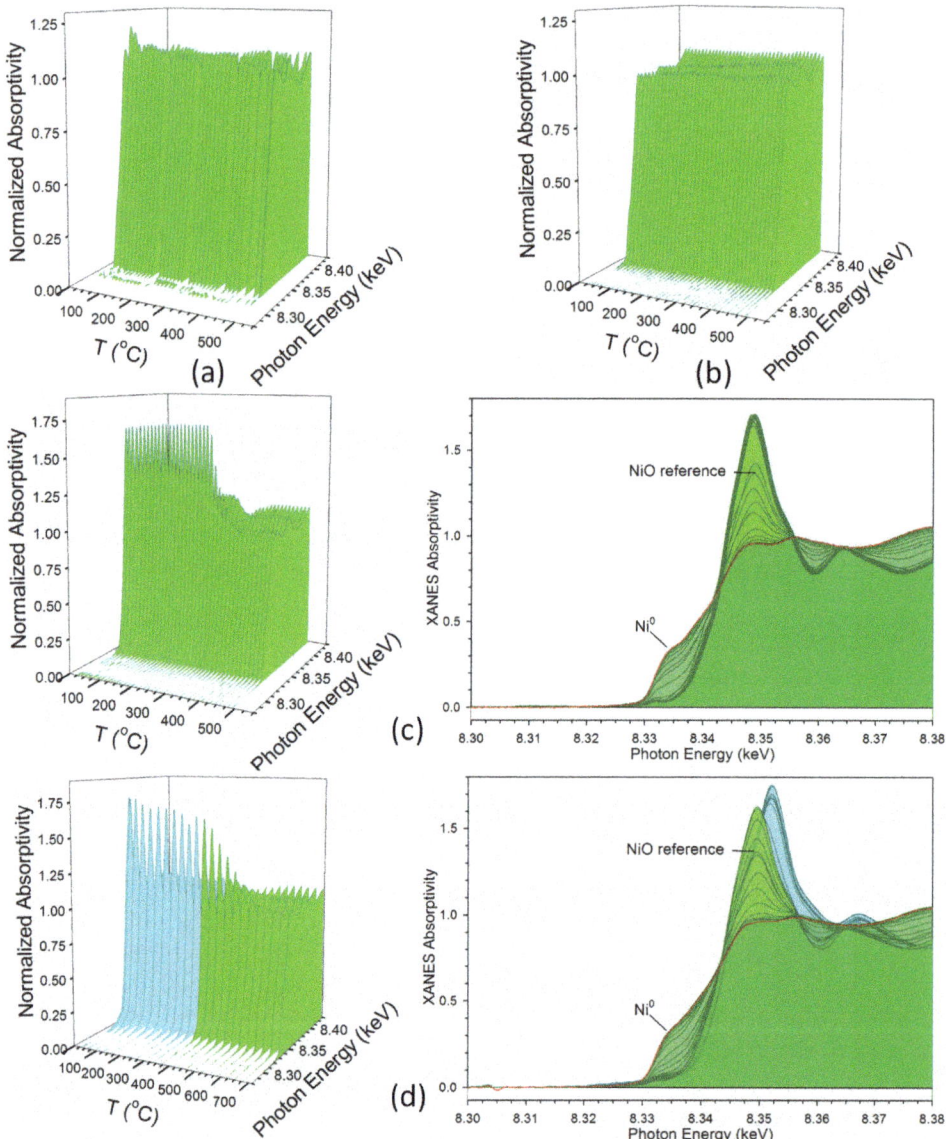

Figure 8. H_2-TPR-XANES spectra at the Ni K-edge of samples (**a**) #1, (**b**) #2, (**c**) #3, and (**d**) #4. (Cyan) is Ni^{2+} (e.g., NiO) associated with cobalt oxides during reduction of Co_3O_4 to CoO. (Green) is reduction of Ni^{2+} to Ni^0 when NiO reduction is associated with CoO reduction to Co^0.

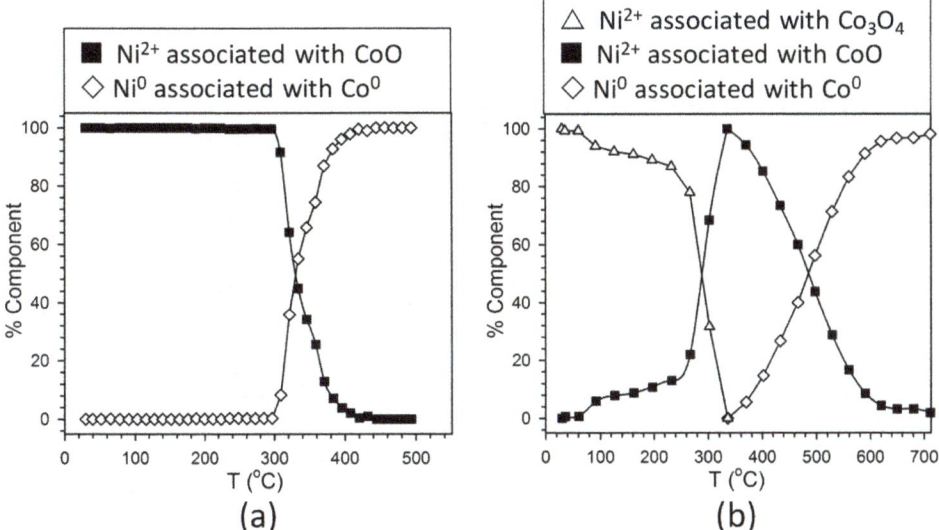

Figure 9. Linear combination fittings of H_2-TPR-XANES spectra of Samples (**a**) #3 and (**b**) #4. Reference compounds for Samples #3 were the first spectrum (Ni^{2+} associated with CoO) and the final spectrum after remaining at 500 °C in H_2 (representing ~100% Ni^0). Reference compounds for Samples #4 were the first spectrum (Ni^{2+} associated with Co_3O_4), the spectrum for Ni^{2+} associated with maximum CoO content, and the final spectrum after H_2 TPR (representing ~100% Ni^0).

XANES snapshots along the TPR trajectory of key states of nickel are provided in Figure 10. These spectra highlight key differences among the different Co–Ni catalysts. While Samples #1 and #2 have very little NiO content initially, Sample #3 and #4 start out as virtually all NiO, with no Ni^0 being detected. Once again, this is in good agreement with the structural characterization, and it suggests that Co–Ni nanoparticles can be tuned to have different oxidation state characteristics. Before (solid line) and after (dashed line) spectra denote the shift to lower energy for the Ni^{2+} spectra for Sample #4.

TPR-EXAFS data at the Co K-edge are provided in Figure 11. For Samples #1 and #2, which are represented by Figure 11a,b, there is already a strong peak for Co–Co coordination in Co^0 from the initial temperature. There is, however, a small amount of oxide initially present in Sample #1, which upon reduction causes slight shifts in the Fourier transform peak positions. This is best observed in the development of the third Co–Co coordination shell (see arrow). Figure 11c for Sample #3 starts with peaks of Co–O (low distance) and Co–Co (higher distance) consistent with CoO and then the peak for Co–Co for the metal starts to develop by ~325 °C. Figure 11d for Sample #4 shows similar contributions from our earlier work with Co/alumina catalysts [5]. The initial cyan spectra show a strong peak at low distance for Co–O and a broad asymmetric peak at higher distance consistent with multiple Co–Co distances in Co_3O_4, which is a spinel structure. However, a second oxide, CoO clearly develops as shown by the initial green spectrum. The Co–O peak is lower in intensity and the Co-Co peak has sharpened (consistent with a rock salt structure). Then, the Co–Co peak for the metal develops only at high temperature (~500 °C). The results are consistent with those of TPR-XANES. This is at a lower temperature than that observed in our previous work [5], where the peak develops at ~600 °C.

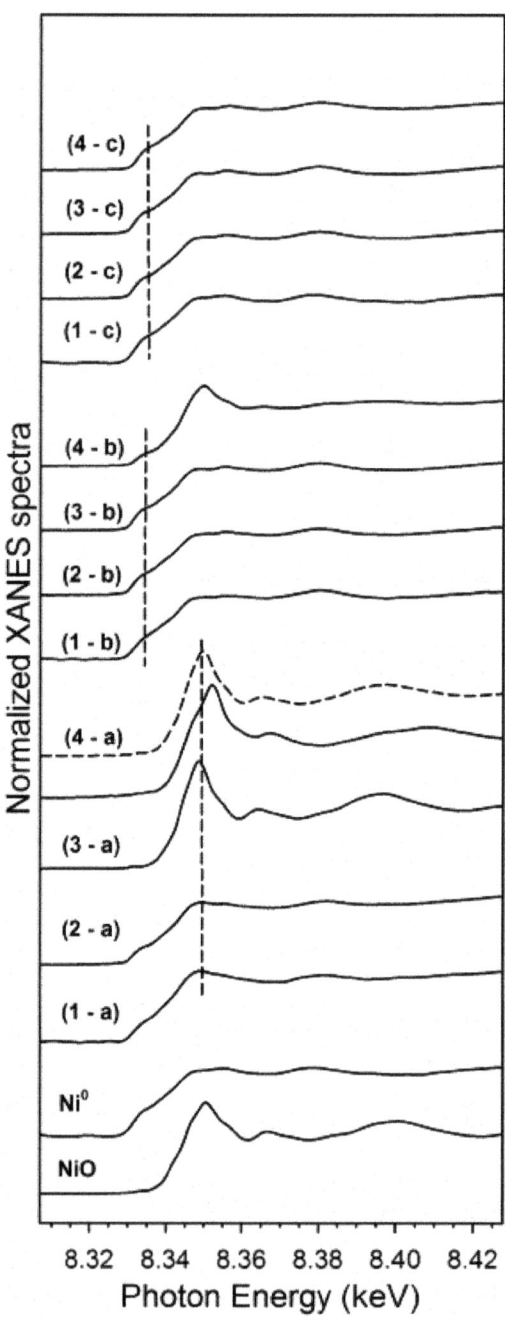

Figure 10. XANES snapshots at the Ni K-edge of reference compounds, and (**a**) the point of maximum NiO content in the H_2 TPR-XANES run, and (**b**) the point at 500 °C, and (**c**) the point of maximum Ni^0 content in the H_2 TPR-XANES run for each catalyst sample (following ramping and holding at T_{max}), including (1) Sample #1, (2), Sample #2, (3) Sample #3, and (4) Sample #4. Note that Samples #1 and #2 displayed very little NiO even in the initial spectra (compare 1a with 1c and compare 2a with 2c; they are very similar). Spectrum 4a: (Solid line) initial spectrum-NiO associated with Co_3O_4; (dashed line) spectrum at 335 °C-NiO associated with CoO.

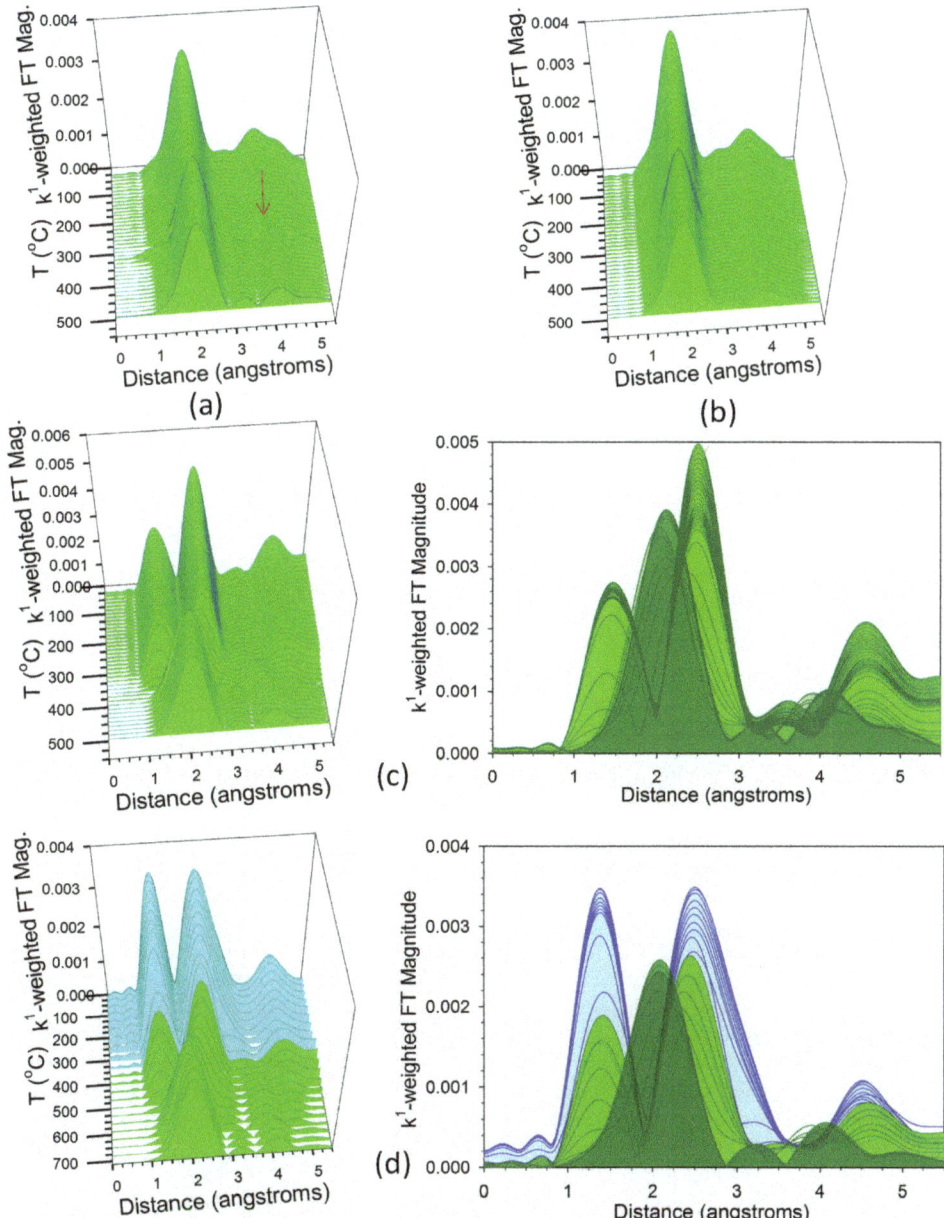

Figure 11. H$_2$-TPR-EXAFS spectra near the Co K-edge of samples (**a**) #1, (**b**) #2, (**c**) #3, and (**d**) #4. (Cyan) is reduction of Co$_3$O$_4$ to CoO, and (Green) is reduction of CoO to Co^0.

EXAFS snapshots along the TPR trajectory of key states of cobalt are provided in Figure 12. These snapshots highlight key differences among the different Co–Ni catalysts. For example, Co$_3$O$_4$ is only detected in Sample #4 (i.e., the catalyst prepared by the conventional method). While Samples #1 and #2 have very little CoO content initially (as shown by low peaks for Co–O and Co–Co coordination in CoO), Sample #3 starts out as virtually all CoO (strong peaks for Co–O and Co–Co coordination in CoO are detected at the outset), with no Co⁰ or Co₃O₄ being detected (Figure 8, left). The CoO in Sample #4 forms from the reduction of Co₃O₄ (Figure 8, top spectrum). The final spectra after TPR-EXAFS (Figure 8, right) shows the characteristic Co–Co coordination of Co⁰.

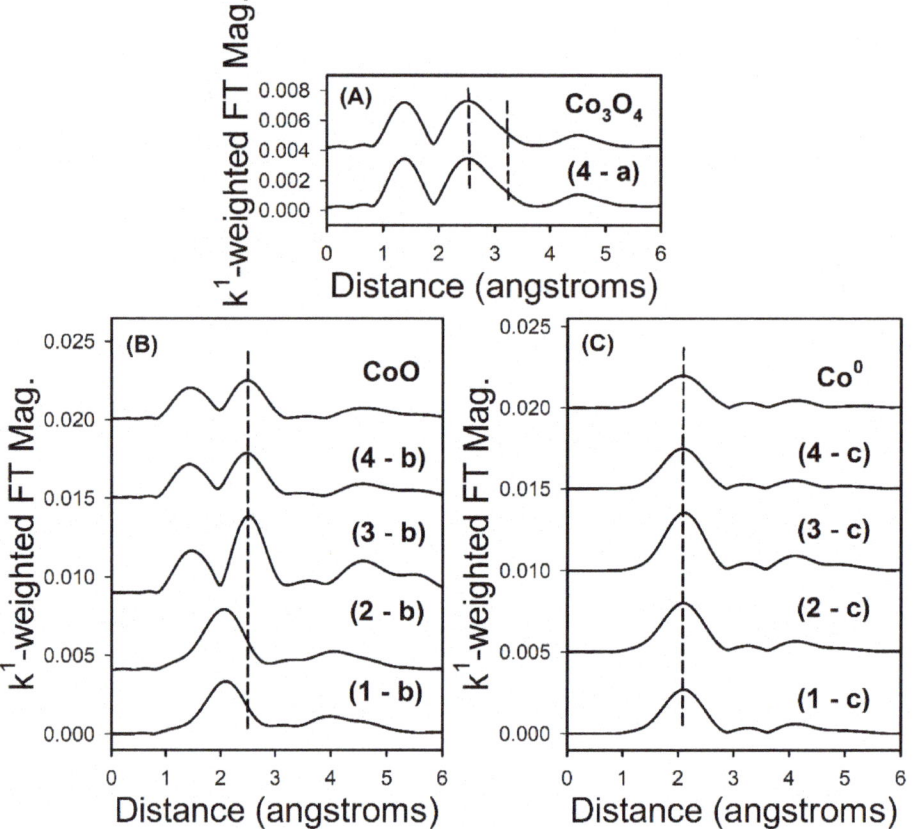

Figure 12. EXAFS snapshots near the Co K-edge of reference compounds, and (**A**) the point of maximum Co₃O₄ content, a, in the H₂ TPR-XANES run, (**B**) the point of maximum CoO content, b, in the H₂ TPR-XANES run, and (**C**) the point of maximum Co⁰ content, c, in the H₂ TPR-XANES run after the hold at T_{max}, for each catalyst sample, including (1) Sample #1, (2), Sample #2, (3) Sample #3, and (4) Sample #4. Note that only Sample #4 displayed Co₃O₄ content (i.e., in the initial period). Dashed lines denote Co–Co coordination in the oxides and metallic state, while the lower distance peak in (top) and (left) spectra at ~1.2 angstroms represents Co–O coordination.

TPR-EXAFS data at the Ni K-edge are provided in Figure 13. For Samples #1 and #2, the Ni was significantly reduced at the outset and these spectra are not provided (see Figure 10 left) moreover, there was significant noise in Sample #1 during much of the TPR. Figure 13a,b of Samples #3 and #4, respectively, starts with peaks of Ni–O (low distance) and Ni–Ni (higher distance) consistent with NiO and then the peak for Ni–Ni for the metal starts to develop by ~325 °C for Sample #3 and ~500 °C for Sample #4, respectively. These are similar temperatures to those of the development of the Co-Co metallic coordination peak, once again suggesting intimate contact between Co and Ni in the catalyst samples. For Sample #4, the cyan spectra denote where Ni^{2+} is associated with Co₃O₄ reduction to CoO and the light green spectra denote where Ni^{2+} is associated with CoO reduction to Co⁰. Dark green spectra indicate significant Co–Co metal coordination.

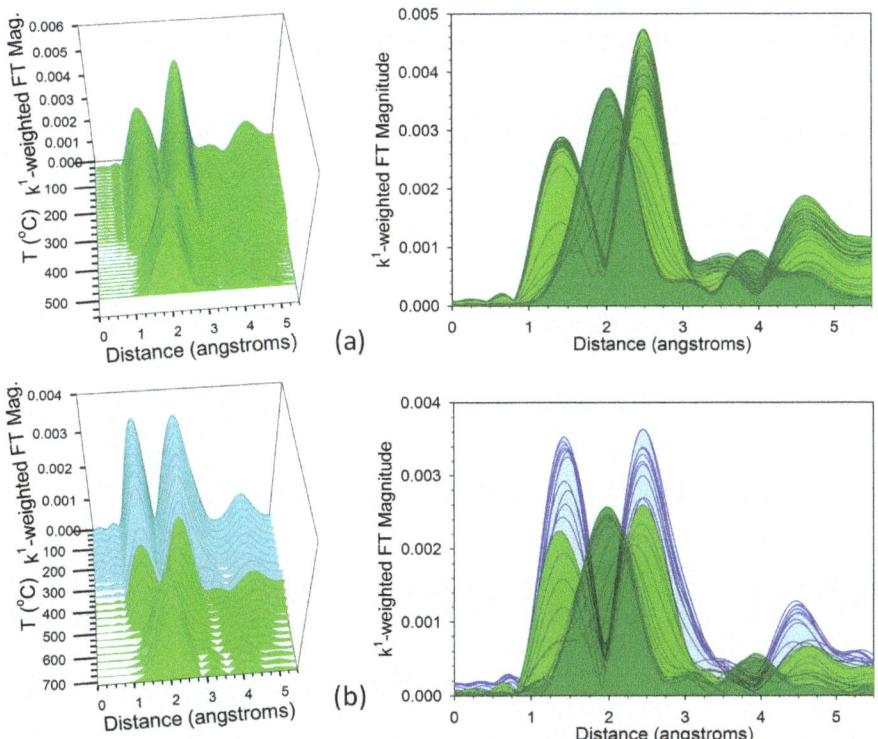

Figure 13. H$_2$-TPR-EXAFS spectra near the Ni K-edge of samples (**a**) #3, and (**b**) #4. (Cyan) is Ni^{2+} (e.g., NiO) associated with cobalt oxides during reduction of Co$_3$O$_4$ to CoO. (Green) is reduction of Ni^{2+} to Ni^0 when NiO associated with CoO reduction to Co^0.

EXAFS snapshots along the TPR trajectory of key states of nickel are provided in Figure 14. These snapshots highlight key differences among the different Co–Ni catalysts. While Samples #1 and #2 have very little NiO content initially (as shown by low peaks for Ni–O and Ni–Ni coordination in NiO in left hand spectra of Figure 14), Sample #3 and Sample #4 start out as virtually all NiO (strong peaks for Ni–O and Ni–Ni coordination in NiO are detected at the outset), with no Ni^0 being detected. For Sample #4, Ni^{2+} associated with Co$_3$O$_4$ is shown as a solid line, while Ni^{2+} associated with CoO is displayed as a dashed line. The final spectra (Figure 10, right) after TPR-EXAFS shows the characteristic Ni–Ni coordination of Ni^0.

Figures S1 and S2, as well as Table S1, provide EXAFS fittings for the Co K-edge and Ni K-edge data. To constrain the fitting model for EXAFS spectra following TPR-EXAFS and cooling (i.e., cobalt and nickel in metallic state), a simple model was developed where metal coordination to Ni (whether the core atom was Co or Ni) was given as a fraction, X, of metal–cobalt coordination. Using this approach, excellent fittings with low r-factors were obtained. The results are consistent with Co–Ni alloy formation, but do not prove Co–Ni alloy formation. Since Sample #4 showed large uncertainties with the model, a second constrained model (fitting #2) was performed that demonstrates that an excellent fitting can be obtained using X = 0.10.

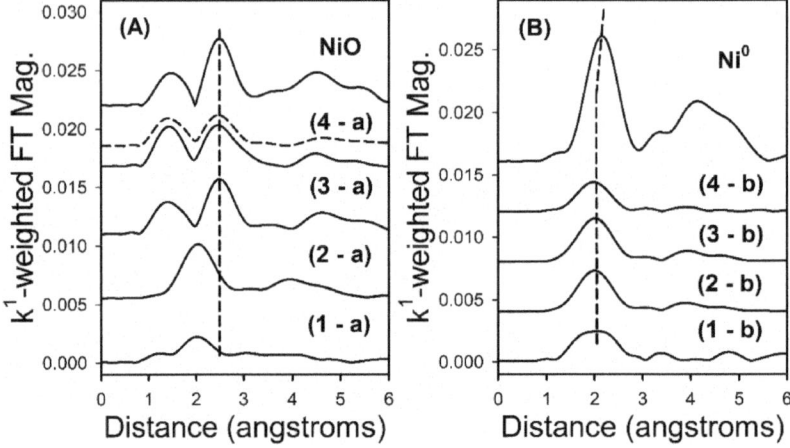

Figure 14. EXAFS snapshots near the Ni K-edge of reference compounds, (**A**) the point of maximum NiO content, a, in the H_2 TPR-XANES run, and (**B**) the point of maximum Ni^0 content, b, in the H_2 TPR-XANES run, for each catalyst sample, including (1) Sample #1, (2), Sample #2, (3) Sample #3, and (4) Sample #4. Vertical dashed lines denote Ni–Ni coordination in the oxides and metallic state, while the lower distance peak in the left-hand spectra at ~1.2 angstroms represents Ni–O coordination. Spectrum 4a: (Solid line) initial spectrum-NiO associated with Co_3O_4; (dashed line) spectrum at 335 °C-NiO associated with CoO.

2.3. Activity Data

Sample #4 and Sample #5 have similar surface areas of ~96 m^2/g suggesting that the nickel substitution does not affect catalyst morphology. The N_2 adsorption–desorption results are reported in Table 1.

Table 1. Brunauer-Emmett-Teller (BET) and Barrett-Joyner-Halenda (BJH) data of the samples.

Sample ID	A_s (BET) [m^2/g]	V_p [cm^3/g]	D_p [nm]
Sample #4	95.5	0.243	9.3
Sample #5	96.5	0.226	9.3

Results of hydrogen chemisorption with pulse O_2 titration are shown in Table 2. In agreement with the TPR-XANES results, the Ni-doped catalyst had a higher % reduction by 8–11%, depending on the method used. The Ni-doped catalyst had a slightly larger metal cluster size (approximately 2 to 3 nm larger diameter, depending on the method used).

XRD patterns for the support (Al_2O_3), the oxide, and reduced samples are shown in Figure 15. XRD profiles of the oxide samples are very similar showing characteristic reflection peaks of Co_3O_4 (i.e., $2\Theta = 36.8°$). No nickel diffraction peaks are distinguishable for Sample #4 suggesting the formation of Ni–Co solid solution. In contrast, the patterns for the reduced samples are quite different. In the case of the Ni-promoted catalyst (Sample #4), the cobalt seems to not be completely reduced as CoO (main diffraction peak at $2\Theta = 42.8°$) is still present in the catalyst, while the reflection peaks attributed to FCC metallic cobalt (main diffraction peak at $2\Theta = 44.6°$) can be observed for Co/Al_2O_3 (Sample #5). A quantitative crystallite size estimation for the reduced samples is not possible because of the poor signal/noise in the XRD patterns.

Table 2. H$_2$ chemisorption and pulse O$_2$ titration.

µmol H$_2$ Desorbed/g$_{cat}$	Uncorr.% Disp.	Uncorr.Diam. (nm)	O$_2$ Uptake (µmol/g$_{cat}$)	* % Red.	** % Red.	* Corr. % Disp.	** Corr. % Disp.	* Corr. Diam. (nm)	** Corr. Diam. (nm)
Sample #4—25%metal (90%Co-10%Ni)/Al$_2$O$_3$									
92.5	4.4	24	1495	54.6	40.5	8.0	10.8	12.9	9.6
Sample #5—25%Co/Al$_2$O$_3$									
91.3	4.3	24	1324	46.8	29.1	9.2	14.8	11.2	7.0

* method #1 assuming Co^0 oxidizes to Co$_3$O$_4$ and Ni^0 oxidizes to NiO. ** method #2 assuming all Co$_3$O$_4$ reduced to CoO and some CoO and NiO reduced to Co^0 and Ni^0. During oxidation, then, the Co^0 and Ni^0 oxidize to CoO and NiO, and all CoO oxidizes to Co$_3$O$_4$.

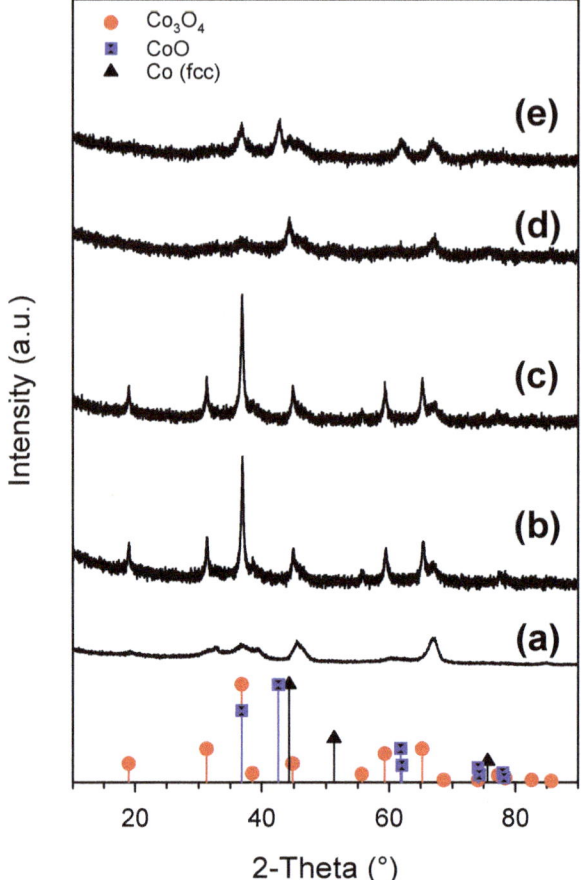

Figure 15. XRD patterns for (**a**) Al$_2$O$_3$, (**b**) oxide Sample #5, (**c**) oxide Sample #4, (**d**) reduced Sample #5 and (**e**) reduced Sample #4.

TEM and STEM of reduced Sample #4 are shown in Figure 16. The particle sizes are in the range of 18–23 nm, whereas EDS analysis confirms the presence of both Ni and Co. The uniform distribution of these two metals confirm the formation of Ni–Co solid solution as already observed from XRD. The measured Co/Ni weight ratio is ~10, very closed to the theoretical value. STEM micrographs for Sample #5 are reported in Figure 17, and the particle sizes fall within the range of 15–20 nm, similar to the nickel promoted catalyst. EDS shows that some areas are richer in cobalt (until 55 wt%), while in other areas the cobalt loading is lower than 10 wt%; however, the average loading for an extended area is close to 28 wt%.

CO conversion and product selectivity (CH$_4$, CO$_2$, C$_2$–C$_4$, and C$_{5+}$) versus Time on Stream (T.o.S.) are reported in Figures 18 and 19, respectively. CO conversion for Sample #4 starts at 21% and it progressively increases with T.o.S. until a steady-state value of 35.7%. On the other hand, the undoped catalyst, Sample #5, has a higher initial CO conversion of 42.5%, which slightly decreases until a value of 39.5%. The presence of nickel inverted the induction period. The performance of Sample #4 is stable after the induction period, even though the cobalt loading is lower compared to that of Sample #5.

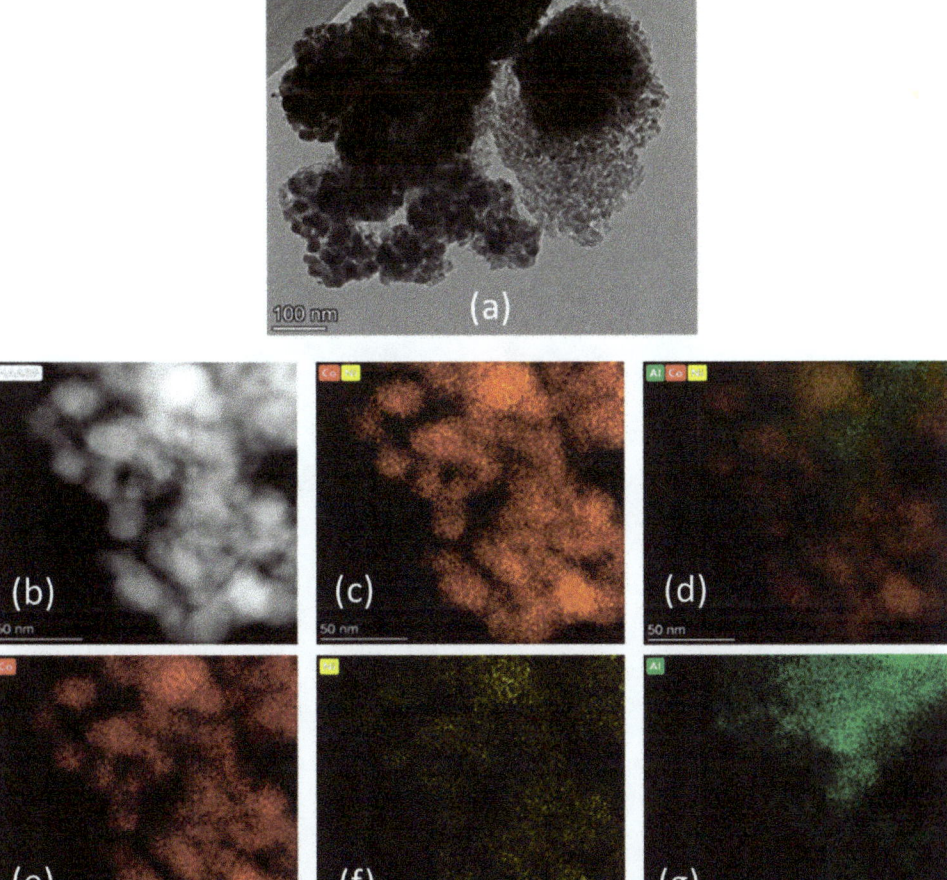

Figure 16. (**a**) TEM and (**b**) high-magnification HAADF-STEM image of Sample #4 with EDS maps. Elemental mapping legend: (**c**) cobalt and nickel, (**d**) aluminum, cobalt, and nickel, (**e**) cobalt, (**f**) nickel, and (**g**) aluminum. (Red) Cobalt, (Yellow) Nickel, and (Green) Aluminum.

The selectivities of Sample #4 evolve with the T.o.S. In particular, the methane selectivity decreases from 11% to 9.2%, C_2–C_4 selectivity drops from 18% to 8.5%, whereas the C_{5+} selectivity increases from 69% to 79% in the first 200 h (Figure 19). However, the methane selectivity is higher for Sample #4 than for Sample #5.

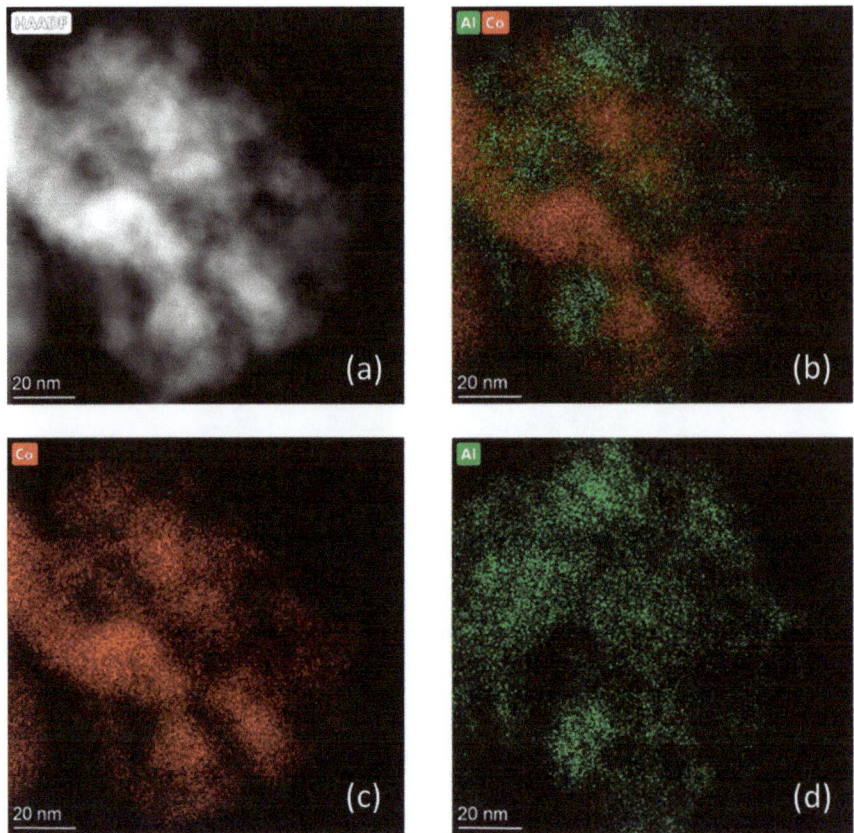

Figure 17. (**a**) High-magnification HAADF-STEM image of Sample #5 and EDS maps, including (**b**) aluminum and cobalt, (**c**) cobalt, and (**d**) aluminum. Elemental mapping legend: (Red) Cobalt and (Green) Aluminum.

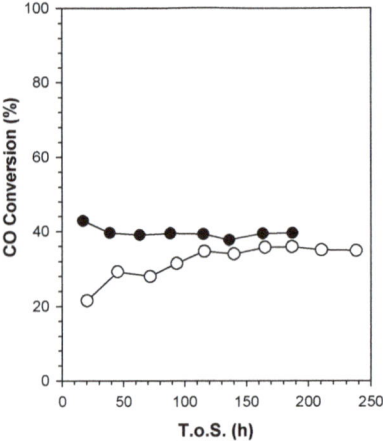

Figure 18. Evolution with T.o.S. of CO conversion for Sample #4 (open circle) and Sample #5 (filled circle), process conditions: T = 220 °C, P = 20.6 bar, H_2/CO = 2 mol/mol, SV = 3.4 slph per g_{cat}.

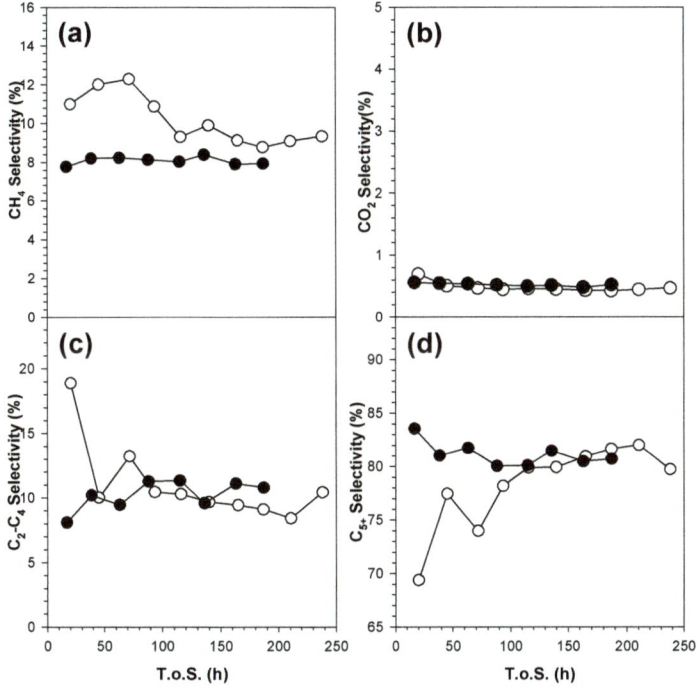

Figure 19. Evolution with T.o.S. of (**a**) CH_4, (**b**) CO_2, (**c**) C_2-C_4, and (**d**) C_{5+} selectivity for Sample #4 (open circle) and Sample #5 (filled circle), process conditions: T = 220 °C, P = 20.6 bar, H_2/CO = 2 mol/mol, SV = 3.4 slph per g_{cat}.

3. Discussion

From the results, it was observed that when a very low content of Ni during nanoparticle synthesis (Co–Ni 98.4:1.6) polycrystalline dendritic particles were formed with negligible Ni content. The XRD patterns showed the presence of a mixture of HCP and FCC metallic phases in the sample. However, by increasing the Ni content during the particle growth smaller Co–Ni alloy nanoparticles were obtained. The presence of Ni not only changes the atomic arrangement, but also enabled shape tuning of the particles. Furthermore, the aerobic synthesis protocol by using only oleylamine led to the formation of predominantly CoNi–O nanoparticles. The presence of 1,2-hexadecanediol seemed to have an effect. The addition of 1,2-hexadecanediol promotes the reduction of metal complexes at lower temperature compared to the synthesis protocol without using it [33], facilitating the higher percentage of Co and Ni reduction during their synthesis in Samples #1 and #2 to form metallic Co–Ni nanoparticles. Furthermore, not only did it restrict the oxidation of Co and Ni, but also the particle size in comparison with the protocol used without it [34]. This restricted oxidation was confirmed by the structural characterization through HAADF-STEM and XRD, and by the TPR-XANES profiles of Co and Ni, where the heterogenized unconventional catalysts exhibited high initial Co^0 and Ni^0 content (Samples #1 and 2) or high CoO/NiO content (Sample #3) in comparison with the catalyst prepared by traditional aqueous impregnation of nitrates; the conventional catalyst started in a higher oxidation state, with Co_3O_4 and Ni^{2+} associated with Co_3O_4. The TPR-XANES results at the Ni edge show that this Ni^{2+} undergoes a shift to lower energy when Co_3O_4 reduces to CoO in the case of the conventional catalyst prepared by aqueous impregnation. This result strongly suggests that Co and Ni are in intimate contact, supporting the alloy formation upon reduction; further evidence for alloy formation is that the NiO and CoO reduce over a similar range, such that the point of 50% reduction is identical. Although EXAFS results show that an alloy model provides an excellent fitting of the experimental results, the sizes of Co and Ni are so similar that a definitive conclusion on alloying should not be drawn from EXAFS results. The TPR-EXAFS/XANES and hydrogen chemisorption/O_2 titration results demonstrate that Ni has a positive effect on improving the percentage reduction (by 8–11%) of cobalt oxides relative to the undoped catalyst. For the conventional catalysts, chemisorption results indicate that adding Ni tended to slightly increase (by 2–3 nm) the metal cluster size relative to the undoped catalyst.

Cobalt-based catalysts are usually characterized by a decrease in the activity in the first few days. The possible reasons for this deactivation are: the re-oxidation of a small Co cluster to inactive CoO_X, sintering, cobalt-support mixed compound formation or some carbon deposition [35,36]. Thermodynamic calculations clearly show that spherical cobalt particle crystallites with a diameter lower than 4.4 nm may oxidize in steam–hydrogen environment [18]. This initial deactivation phenomenon was not observed for Sample #4. Thus, nickel has a stabilizing effect on the cobalt–nickel cluster; a similar stabilization was reported by Rytter et al. [37] and Nikparsa et al. [38]. Interestingly, the performance of Sample #4 is stable even after the induction period, despite having lower cobalt loading as compared to Sample #5 (i.e., the 25% Co/alumina reference catalyst).

Addition of Ni resulted in higher initial methane selectivity (~3–4% absolute); however, the difference in CH_4 selectivities between the two catalysts diminished with time to ~1%. The higher methane selectivity result is not surprising, because nickel has a higher hydrogenation capability. Shafer et al. [17] explain that the back-donation of nickel is so strong that the vinylic intermediate is not suitably stabilized, precluding the chain growth. Higher methane selectivity was also reported in previous studies where different Ni/Co ratios were investigated [39,40], although recent results suggest that lower amounts (e.g., <25% [41]) of Ni favor a selectivity closer to that of pure cobalt.

Finally, we have compared the activity/$ as the two metals have different market prices and the partial substitution of cobalt with a cheaper metal would be a benefit in terms of total catalyst cost. The price of nickel is reported to be 0.47 times the price of cobalt. At steady state the activity/$ for Sample #4 is 37.5, while it is 39.6 for Sample #5. However, additional studies to find the optimal loading of nickel are needed.

4. Materials and Methods

4.1. Sample Denomination

Colloidal Method:

Sample #1 20% metal (98.4% at. Co/1.6% at. Ni)/Al_2O_3
Sample #2 20% metal (92.7% at. Co/7.3% at. Ni)/Al_2O_3
Sample #3 20% metal (89.7% at. Co/10.3% at. Ni)/Al_2O_3

Impregnation Method:

Sample #4 25% metal (90% at. Co/10% at. Ni)/Al_2O_3 calcined 350 °C
Sample #5 25% Co/Al_2O_3 calcined 350 °C

4.2. Reagents

For the synthesis of Ni–Co nanoparticles, nickel (II) chloride hexahydrate (99.99%, MilliporeSigma, St. Louis, MO, USA), cobalt (III) acetylacetonate (99.99%, Aldrich), cobalt (II) acetate tetrahydrate (98%, Aldrich), oleylamine (70%, Aldrich), 1-dodecanethiol (≥98%, MilliporeSigma, St. Louis, MO, USA) and 1,2-hexadecanediol (90%, Aldrich) was used. All reagents were purchased from Sigma-Aldrich and were used without further purification.

4.3. Preparation of Ni and Co Precursors

Oleylamine was used in the preparation of the metallic precursors. For Ni, a mixture of 100 μL of nickel (II) chloride hexahydrate (dissolved in ethanol) and 6 mL of oleylamine was magnetically stirred for 30 min at 120 °C. The temperature of the mixture was increased to 230 °C for a period of 2.5 h with continuous agitation. The Co precursor was prepared by magnetic stirring of a 0.1 M mixture of cobalt (III) acetylacetonate dissolved in oleylamine for 30 min at 120 °C, and further increasing of temperature to 150 °C for 20 min. Both solutions changed their initial color to dark brown.

4.4. Preparation of Ni–Co Nanoparticles

(a) Co–Ni Sample #1 and #2: The Co–Ni nanoparticles were prepared as follows. 20 mL of oleylamine were previously heated to 120 °C and then the Co and Ni precursors were added to obtain two samples with a Co:Ni desired molar ratio. Then, 30 µL of a 1,2-hexadecanediol (0.3 M) is added and stirring was continued vigorously for 5 min at 170 °C. The temperature was slowly increased to 230 °C and the solution was kept 60 min more at this temperature. The resulting colloid was then cooled slowly to room temperature. Two samples were prepared with an atomic ratio of 98.4/1.6 (Sample #1) and 92.7/7.3 (Sample #2) of Co:Ni respectively. Finally, the nanoparticles were washed by centrifugation several times with a mixture of chloroform and ethanol at 4000 rpm and dried at 120 °C for 4 h.

(b) Co–Ni Sample #3: For this synthesis method of Co–Ni nanoparticles a Co/Ni atomic ratio of 89.7/10.3 was used. Briefly, 1 mmol of Co^{II} (acet) was completely dissolved in 4 mL of oleylamine by using magnetic stirring, and 150 µL of a $NiCl_2$ solution (0.1 M in ethanol) was injected. The mixture was stirred and heated at 100 °C for 30 min. Then, the temperature was raised to 230 °C for 1 h. Finally, the solution was let to cool down to room temperature. The samples were further purified by centrifuging several times, precipitating and redispersing the particles with a mixture of chloroform and ethanol. The final precipitated particles were collected and dried at 120 °C for 4 h.

(c) Co–Ni Sample #4: Sample #4 was prepared by a conventional slurry impregnation method and contained 25% metal by weight, with an atomic ratio of 90/10 of Co/Ni. Co/Alumina-Catalox 150 γ-alumina (150 m^2/g) was used as a support. Cobalt nitrate (Alfa Aesar) served as the precursor to load the cobalt onto the Al_2O_3 support. In this method, which follows a Sasol patent [1], the ratio of the volume of solution used to the weight of alumina was 1:1, such that approximately 2.5 times the pore volume of solution was used to prepare the loading solution. Two impregnation steps were used to load 12.5% of metal by weight for each step. Between each step, the catalyst was dried under vacuum in a rotary evaporator at 60 °C, and the temperature was slowly increased to 100 °C. After the second impregnation/drying step, the catalyst was calcined in air at 350 °C for 4 h.

(d) (d) Co Sample #5: Sample #5 was prepared in the same way as Sample #4, except no Ni was included. The weight was 25% Co.

4.5. Characterization

Characterization of Samples #1–3 was carried out in the core facilities at the UTSA. Scanning transmission electron microscopy (STEM) images of Ni–Co nanoparticles were obtained in an Ultra high-resolution cold-FEG scanning electron microscope (Hitachi 5500, Tokyo, Japan) equipped with bright field (BF) and annular dark field (ADF) detectors operating at 30 keV, and in an aberration-corrected electron microscope (JEOL ARM200F, Tokyo, Japan), operated at 200 kV in high-angle annular dark-field (HAADF) mode, spatial resolution of 0.78 Å. The instrument is equipped with a Silicon Drift Detector for energy dispersive X-ray spectrometry (EDS). Samples were prepared by dropping the nanoparticles suspension onto a carbon-coated copper grid and dried in air. X-ray diffraction (XRD) analysis was carried out using a Rigaku Ultima IV diffractometer (RIGAKU, Tokyo, Japan) equipped with a dual position graphite diffracted beam monochromator for Cu, using a Cu Kα radiation at 43 kV and 30 mA. Prior to XRD and STEM characterization, Sample #4 and Sample #5 were reduced in hydrogen (American Welding & Gas, Lexington, KY, USA) for 18 h at 350 °C, cooled down at 20 °C and then passivated with 1% O_2 in nitrogen (American Welding & Gas, Lexington,

KY, USA). XRD analysis was carried out with a Philips X'Pert diffractometer with monochromatic Cu Kα radiation (λ = 1.54 Å). The conditions employed included a 2Θ range of 10–90°, a scan rate of 0.01° per step, and a scan time of 4 s per step. STEM analysis was carried out with FEI Talos F200X (Thermo Scientific, Waltham, MA, USA) equipped with BF, DF2, DF4, and HAADF detectors. The imaging was performed using field emission gun and accelerating voltage of 200 kV and collected by high speed Ceta 16M camera. Data processing and analysis were carried out with Velox software (Thermo Scientific, Waltham, MA, USA). The samples were dispersed in ethanol (Alfa Aesar, Haverhill, MA, USA.), sonicated and then dropped into a carbon-coated copper grid and dried in air.

In situ H_2-TPR XAFS studies were performed at the Materials Research Collaborative Access Team (MR-CAT) beamline at the Advanced Photon Source, Argonne National Laboratory. A cryogenically cooled Si (1 1 1) monochromator selected the incident energy and a rhodium-coated mirror rejected higher order harmonics of the fundamental beam energy. The experiment setup was similar to that outlined by Jacoby [42]. A stainless-steel multi-sample holder (3.0 mm i.d. channels) was used to monitor the in situ reduction of six samples during a single TPR run. Approximately 6 mg of each sample was loaded as a self-supporting wafer in each channel. The catalyst to diluent weight was approximately 1:1. The diluent was alumina. The holder was placed in the center of a quartz tube, equipped with gas and thermocouple ports and Kapton windows. The amount of sample used was optimized for the Co K-edge, considering the absorption by Al of the support; however, the Ni K edge was also analyzed. The quartz tube was placed in a clamshell furnace mounted on the positioning table. Each sample cell was positioned relative to the beam by finely adjusting the position of the table to an accuracy of 20 μm (for repeated scans). Once the sample positions were fine-tuned, the reactor was purged with He for more than 5 min at 100 mL/min and then the reactant gas (H_2/He, 3.5%) was flowed through the samples (100 mL/min) and a temperature ramp of ~1.0 °C/min was initiated for the furnace to 500 °C and the sample was held at this temperature for 4 h. One exception to this was the catalyst prepared by standard impregnation methods, where the maximum temperature was 700 °C. The Co and Ni K-edge spectra were recorded in transmission mode and a Co metallic foil spectrum was measured simultaneously with each sample spectrum for energy calibration. X-ray absorption spectra for each sample were collected from 7500 to 9000 eV.

Data reduction of the EXAFS/XANES spectra was carried out using the WinXAS program [43]. The details of the XANES and EXAFS analyses for Co K-edge data are provided in the appendices of our previous article, and will not be repeated here, for the sake of brevity [5]. Ni K-edge data were processed in an analogous manner (i.e., same Δk and ΔR in fittings) with the exception of Sample #1, where the lower data quality due to the lower loading necessitated a shortening of the k-range to 3–9 Å^{-1}. For qualitative comparisons of XANES and EXAFS data, the references used for Co_3O_4, CoO, and Co^0 were the initial spectrum, the point of maximum CoO content, and the final spectrum of the TPR trajectory of 25% Co/Al_2O_3 from [5]. For NiO and Ni^0, the references were NiO (Alfa Aesar, Puratronic, 99.998%, Tewksbury, MA, USA) and a Ni^0 foil.

For XANES linear combination fittings, the reference compounds for Co K-edge data for Samples #1–3 were the first spectrum of Sample #3 (representing 100% Co^{2+}) and the final spectrum of each sample after remaining at 500 °C in H_2 (representing ~100% Co^0). Reference compounds for Sample #4 were the initial spectrum (representing a mixture of Co^{2+} and Co^{3+} similar to Co_3O_4), the point of maximum CoO content, and the final spectrum after H_2 TPR (representing Co^0). At the Ni K-edge, Samples #1 and #2 had a high amount of Ni^0, with only a small amount of Ni^{2+} initially; also, the data of Sample #1 were of lower quality due to the low Ni loading; LC fittings were not performed on these two samples. Reference compounds for Samples #3 were the first spectrum (Ni^{2+} associated with CoO) and the final spectrum after remaining at 500 °C in H_2 (representing ~100% Ni^0). Reference compounds for Samples #4 were the first spectrum (Ni^{2+} associated with Co_3O_4), the spectrum for Ni^{2+} associated with maximum CoO content, and the final spectrum after H_2 TPR (representing ~100% Ni^0). EXAFS data reduction and fitting were carried out using the catalysts in their final state following TPR and cooling in flowing H_2 using the WinXAS [43], Atoms [44], FEFF [45], and FEFFIT [45] programs. The k-range used for the fittings was 3–10 Å^{-1}. Fitting was confined to the first metallic coordination shell by applying a Hanning window in the Fourier transform magnitude spectra, and carrying out the back-transform to isolate that shell.

BET surface area and porosity characteristics were measured by Micromeritics 3-Flex system (Norcross, GA, USA) using N_2 physisorption (UHP N_2, Airgas, Lexington, KY, USA). Before testing, the temperature was slowly increased to 160 °C; then, a vacuum was pulled for at least 12 h until the sample pressure was approximately 50 mTorr. The BJH method was employed to determine the average pore diameter and pore volume. Additionally, pore size distribution was obtained as a function of pore diameter via the correlation dV/d (log D).

Hydrogen chemisorption/pulse reoxidation was performed using an Altamira (Altamira Instruments, Pittsburgh, PA, USA) AMI-300 unit. The catalyst was ramped at 2 °C/min and reduced at 350 °C for 10 h in 10 cm^3/min of UHP H_2 (Airgas, San Antonio, TX, USA) blended with 20 cm^3/min of UHP argon (Airgas, San Antonio, TX, USA). The catalyst was cooled to 100 °C, and 30 cm^3/min of UHP argon was flowed to prevent the adsorption of weakly bound hydrogen. Next, the catalyst was heated at 10 °C/min in flowing argon to 350 °C to desorb the chemisorbed hydrogen. The hydrogen temperature programmed desorption peak was integrated and compared to calibration pulses in order to determine the moles of hydrogen evolved. Pulses of UHP O_2 (Airgas, San Antonio, TX, USA) were then sent to reoxidize the catalyst until saturation was achieved. Two methods were used to estimate percentage of reduction. In the conventional approach, cobalt metal was assumed to oxidize to Co_3O_4 and nickel metal was assumed to oxidize to NiO. However, in a second approach, we assumed that during reduction, all Co_3O converted at least to CoO, and that a fraction of CoO and NiO converted to Co^0 and Ni^0. Thus, during reoxidation, the Co^0 and Ni^0 is first oxidized to CoO and NiO. Then, all CoO oxidizes to Co_3O_4. The two approaches set maximum and minimum limits for the metal cluster size when the uncorrected dispersion is corrected by taking into account the percentage of reduction by the metal, as follows: % Dispersion (Uncorrected) = (# metal atoms on the surface)/(# metal atoms in the sample)% Dispersion (Corrected) = (# metal atoms on the surface)/((# metal atoms in the sample)(% reduction)).

4.6. Reaction Testing

Activity tests were carried out in a lab scale rig equipped with a 1 L continuously stirred tank reactor (CSTR) (PPI, Warminster, PA, USA). Additional detail on the lab scale ring are reported in our previous work [15]. In a typical run, 9.6 g of calcined catalyst (63–125 μm) was loaded into a fixed bed reactor for ex situ reduction at 350 °C for 20 h, feeding 30 Nl/h H_2/He mixture (1:3 *v/v*, American Welding & Gas, Lexington, KY, USA) at atmospheric pressure. The reduced catalyst was transferred to a 1 L CSTR containing 310 g of melted Polywax 3000 (Baker Petrolite, Houston, TX, USA) by pneumatic transfer under the protection of nitrogen. The transferred catalyst was reduced in situ at 230 °C and at atmospheric pressure overnight feeding 30 Nl/h pure H_2 (American Welding & Gas, Lexington, KY, USA). In this work, the process conditions were: T = 220 °C, P = 300 psi, H_2/CO = 2 mol/mol and a stirring speed of 750 rpm. The products and the unconverted reactants leaving the reactor were sent to a warm trap, maintained at 100 °C, and then to a cold trap held at 0 °C. The uncondensed vapor stream was reduced to atmospheric, the flowrate was measured by a wet test meter, while the composition was analyzed by online 3000A micro-GC (Agilent, Santa Clara, CA, USA) equipped with four different columns (Molecular Sieve, Plot U, Alumina and OV-1) and TCD. The reaction products were collected in three traps maintained at different temperatures; a hot trap (200 °C), a warm trap (100 °C), and a cold trap (0 °C). The products were separated into different fractions (wax, oil, and aqueous) for quantification. The oil (C_4–C_{20}) were analysed with 7890 GC (Agilent, Santa Clara, CA, USA) equipped with DB-5 (60 m × 0.32 mm × 0.25 μm, Agilent J&W) column and FID, while waxes (C_{21}–C_{60}) were analyzed with an HP 6890 GC equipped with ZB-1HT column (30 m × 0.25 mm × 0.10 μm, Zebron) and FID.

5. Conclusions

A wet chemical approach using oleylamine as the solvent and reducing agent was applied to prepare Co–Ni nanoparticles. Nanoparticles were characterized by HAADF-STEM, EDS, and XRD. By increasing the Ni content of the alloy from Co98.4Ni1.6 to Co92.7Ni7.3, the size and shape of the nanoparticles change from 20.5 nm dendritic particles to a core-shell morphology with 12.2 nm HCP Co–Ni alloy within the core and a 2.3 nm shell showing evidence of oxidation. Small quantities of heterogeneous catalysts were prepared by supporting the nanoparticles on alumina. Interestingly, TPR-EXAFS/XANES measurements showed that the oxidation states of Co and Ni could be tuned from being completely 2+ to being up to 80% metallic, and the reduction of Ni and Co acted in tandem. This result is of interest, because much work in FTS research has been dedicated to facilitating cobalt reduction due to the presence of strong metal oxide-support interactions.

Due to the small quantity, it was not possible to test these catalysts in a CSTR reactor used to benchmark commercial slurry operations. We therefore tested a catalyst prepared by a conventional impregnation method with subsequent calcination. This catalyst offered a different TPR-XANES pattern in that it was the only catalyst to begin with Co_3O_4 with NiO being associated with this oxide. The Co_3O_4 reduced to CoO, and the NiO XANES spectra displayed a distinct shift to lower energy when reduction of Co_3O_4 to CoO occurred. Following this step, NiO and CoO reduced in tandem. This indicates that Co and Ni are in intimate contact, suggesting alloy formation. This explanation was also strongly favored by results of XRD, HAADF-STEM, and EDS, which indicate that a solid solution likely formed.

Finally, the catalyst prepared by impregnation was tested for Fischer–Tropsch synthesis using a CSTR. Compared to a catalyst prepared using cobalt, the CoNi alloy catalyst exhibited lower CO conversion on a per gram catalyst basis. However, Ni addition did not impact the turnover number of cobalt to a significant degree. Moreover, a slightly higher methane selectivity was observed initially (~3–4% absolute) with the CoNi catalyst; however, with time, the difference between the two catalysts became smaller, reaching ~1%. Despite these drawbacks, adding Ni had a pronounced impact on stability, resulting in an inverted induction period. Instead of a steep decline followed by a leveling off as observed with many conventional cobalt catalysts, the CO conversion increased with time on-stream for the CoNi catalyst. To stabilize typical cobalt catalysts, companies have relied on using high Co loadings. The results here indicate that the metal nanoparticles can be stabilized by Ni addition at a lower Co percentage loading.

Supplementary Materials: The following are available online at http://www.mdpi.com/2073-4344/10/1/18/s1. Figure S1: EXAFS fittings for Co K-edge data; Figure S2: EXAFS fittings for Ni K-edge data; Table S1: Results of EXAFS fittings for data acquired near the Co and Ni K-edges for catalysts following TPR-EXAFS after cooling.

Author Contributions: Conceptualization, catalyst preparation, catalyst characterization, formal analysis, writing, G.J. Catalyst preparation, catalyst characterization, formal analysis, S.C.K. Reaction testing, characterization, formal analysis, conceptualization, writing, M.M. Project administration, resources, C.L.M. Catalyst preparation, supervision, resources, D.C.C. Catalyst characterization, data curation, resources, supervision, A.J.K. Catalyst preparation, catalyst characterization, writing, J.L.-T. Catalyst preparation, catalyst characterization, writing, R.M.-C. Catalyst preparation, catalyst characterization, writing, L.B.-D. All authors have read and agreed to the published version of the manuscript.

Funding: This research received no external funding.

Acknowledgments: JLT gratefully acknowledges the scholarship from CONACyT for the postdoctoral stay, CVU No. 219621. Argonne's research was supported in part by the US Department of Energy (DOE), Office of Fossil Energy, National Energy Technology Laboratory (NETL). Advanced Photon Source was supported by the US Department of Energy, Office of Science, Office of Basic Energy Sciences, under Contract No. DE-AC02-06CH11357. MRCAT operations are supported by the Department of Energy and the MRCAT member institutions.

Conflicts of Interest: The authors declare no conflict of interest.

References

1. Espinoza, R.L.; Visagie, J.L.; van Berge, P.J.; Bolder, F.H. Fischer-Tropsch catalysts containing iron and cobalt. U.S. Patent 5,733,839, 31 March 1998.
2. van Berge, P.J.; Barradas, S.; van de Loosdrecht, J.; Visagie, J.L. Advances in the cobalt catalyzed Fischer-Tropsch synthesis. *Erdoel Erdgas Kohle* **2001**, *117*, 138–142.
3. Glacier Resource Management Group. InvestmentMine. Available online: http://www.infomine.com/investment/ (accessed on 15 November 2019).
4. Jacobs, G.; Das, T.K.; Zhang, Y.; Li, J.; Racoillet, G.; Davis, B.H. Fischer-Tropsch synthesis: Support, loading and promoter effects on the reducibility of cobalt catalysts. *Appl. Catal. A Gen.* **2002**, *233*, 263–281. [CrossRef]
5. Jacobs, G.; Ji, Y.; Davis, B.H.; Cronauer, D.C.; Kropf, A.J.; Marshall, C.L. Fischer-Tropsch synthesis: Temperature programmed EXAFS/XANES investigation of the influence of support type, cobalt loading, and noble metal promoter addition to the reduction behavior of cobalt oxide particles. *Appl. Catal. A Gen.* **2007**, *333*, 177–191. [CrossRef]
6. Jacobs, G.; Chaney, J.A.; Patterson, P.M.; Das, T.K.; Maillot, J.C.; Davis, B.H. Fischer-Tropsch synthesis: Study of the promotion of Pt on the reduction property of Co/Al_2O_3 catalysts by in situ EXAFS of Co K and Pt L_{III} edges and XPS. *J. Synchrotron Radiat.* **2004**, *11*, 414–422. [CrossRef] [PubMed]
7. Cook, K.M.; Perez, H.D.; Bartholomew, C.H.; Hecker, W.C. Effect of promoter deposition order on platinum-, ruthenium-, or rhenium-promoted cobalt Fischer-Tropsch catalysts. *Appl. Catal. A Gen.* **2014**, *482*, 275–286. [CrossRef]
8. Cook, K.M.; Hecker, W.C. Reducibility of alumina-supported cobalt Fischer-Tropsch catalysts: Effects of noble metal type, distribution, retention, chemical state, bonding, and influence on cobalt crystallite size. *Appl. Catal. A Gen.* **2012**, *449*, 69–80. [CrossRef]
9. Ma, W.; Jacobs, G.; Ji, Y.; Bhatelia, T.; Bukur, D.B.; Khalid, S.; Davis, B.H. Fischer-Tropsch synthesis: Influence of CO conversion on selectivities, H_2/CO usage ratios, and catalyst stability for a Ru promoted Co/Al_2O_3 catalyst using a slurry phase reactor. *Top. Catal.* **2011**, *54*, 757–767. [CrossRef]
10. Iglesia, E.; Soled, S.L.; Fiato, R.A.; Via, G.H. Bimetallic synergy in cobalt-ruthenium Fischer-Tropsch synthesis catalysts. *J. Catal.* **1993**, *143*, 345–368. [CrossRef]
11. Hilmen, A.M.; Schanke, D.; Holmen, A. TPR study of the mechanism of rhenium promotion of alumina-supported cobalt Fischer-Tropsch catalysts. *Catal. Lett.* **1996**, *38*, 143–147. [CrossRef]
12. Vada, S.; Hoff, A.; Ådnanes, E.; Schanke, D.; Holmen, A. Fischer-Tropsch synthesis on supported cobalt catalysts promoted by platinum and rhenium. *Top. Catal.* **1995**, *2*, 155–162. [CrossRef]
13. Ronning, M.; Nicholson, D.G.; Holmen, A. In situ EXAFS study of the bimetallic interaction in a rhenium-promoted alumina-supported cobalt Fischer-Tropsch catalyst. *Catal. Lett.* **2001**, *72*, 141–146. [CrossRef]
14. Jacobs, G.; Chaney, J.A.; Patterson, P.M.; Das, T.K.; Davis, B.H. Fischer-Tropsch synthesis: Study of the promotion of Re on the reduction property of Co/Al_2O_3 catalysts by in situ EXAFS/XANES of Co K and Re L_{III} edges and XPS. *Appl. Catal. A* **2004**, *264*, 203–212. [CrossRef]
15. Jacobs, G.; Ribeiro, M.C.; Ma, W.; Ji, Y.; Khalid, S.; Sumodjo, P.T.A.; Davis, B.H. Group 11 (Cu, Ag, Au) promotion of $15\%Co/Al_2O_3$ Fischer-Tropsch catalysts. *Appl. Catal. A Gen.* **2009**, *361*, 137–151. [CrossRef]
16. Jermwongratanachai, T.; Jacobs, G.; Ma, W.; Shafer, W.D.; Gnanamani, M.K.; Gao, P.; Kitiyanan, B.; Davis, B.H.; Klettlinger, J.L.S.; Yen, C.H.; et al. Fischer-Tropsch synthesis: Comparisons between Pt and Ag promoted Co/Al_2O_3 catalysts for reducibility, local atomic structure, catalytic activity, and oxidation-reduction (OR) cycles. *Appl. Catal. A* **2013**, *464–465*, 165–180. [CrossRef]
17. Shafer, W.D.; Gnanamani, M.K.; Graham, U.M.; Yang, J.; Masuku, C.M.; Jacobs, G.; Davis, B.H. Fischer-Tropsch: Product Selectivity—The Fingerprint of Synthetic Fuels. *Catalysts* **2019**, *9*, 259. [CrossRef]
18. van Steen, E.; Claeys, M.; Dry, M.E.; van de Loosdrecht, J.; Viljoen, E.L.; Visagie, J.L. Stability of nanocrystals: Thermodynamic analysis of oxidation and re-reduction of cobalt in water/hydrogen mixtures. *J. Phys. Chem. B* **2005**, *109*, 3575–3577. [CrossRef]
19. Jacobs, G.; Das, T.K.; Patterson, P.M.; Luo, M.; Conner, W.A.; Davis, B.H. Fischer-Tropsch synthesis: Effect of water on Co/Al_2O_3 catalysts and XAFS characterization of reoxidation phenomena. *Appl. Catal. A* **2004**, *270*, 65–76. [CrossRef]

20. Logdberg, S.; Boutonnet, M.; Walmsley, J.C.; Jaras, S.; Holmen, A.; Blekkan, E.A. Effect of water on the space-time yield of different supported cobalt catalysts during Fischer-Tropsch synthesis. *Appl. Catal. A* **2011**, *393*, 109–121. [CrossRef]
21. Hughes, N.A.; Gloriot, V.; Smiley, D.D.; Jacobs, G.; Pendyala, V.R.R.; Graham, U.M.; Ma, W.; Gnanamani, M.K.; Shafer, W.D.; Maclennan, A.; et al. Fischer-Tropsch synthesis: comparisons of Al2O3 and TiO2 supported Co catalysts prepared by aqueous impregnation and CVD methods. In *Fischer-Tropsch Synthesis, Catalysts and Catalysis: Advances and Applications*; Davis, B.H.; Occelli, M.L., Eds.; CRC Press, Taylor & Francis Group: Boca Raton, FL, USA, 2016; Ch. 6; pp. 85–106.
22. Singh, A.K.; Xu, Q. Synergistic catalysis over bimetallic alloy nanoparticles. *ChemCatChem* **2013**, *5*, 652–676. [CrossRef]
23. Nam, K.M.; Shim, J.H.; Han, D.W.; Kwon, H.S.; Kang, Y.M.; Li, Y.; Song, H.; Seo, W.S.; Park, J.T. Syntheses and characterization of wurtzite CoO, rocksalt CoO, and spinel Co3O4 nanocrystals: Their interconversion and tuning of phase and morphology. *Chem. Mater.* **2010**, *22*, 4446–4454. [CrossRef]
24. Morelos-Santos, O.; de la Torre, A.R.; Schacht-Hernández, P.; Portales-Martínez, B.; Soto-Escalante, I.; Mendoza-Martínez, A.M.; Mendoza-Cruz, R.; Velázquez-Salazar, J.J.; José-Yacamán, M. NiFe2O4 Nanocatalyst for Heavy Crude Oil Upgrading in Low Hydrogen/Feedstock Ratio. *Catal. Today* **2019**. [CrossRef]
25. Hu, L.; Wu, L.; Liao, M.; Hu, X.; Fang, X. Electrical transport properties of large, individual $NiCo_2O_4$ nanoplates. *Adv. Funct. Mater.* **2012**, *22*, 998–1004. [CrossRef]
26. Taylor, A.; Floyd, R.W. Precision measurements of lattice parameters of non-cubic crystals. *Acta Crystallogr.* **1950**, *3*, 285–289. [CrossRef]
27. Nishizawa, T.; Ishida, K. The Co-Ni (Cobalt-Nickel) System. *Bull. Alloy Phase Diagr.* **1983**, *4*, 390–395. [CrossRef]
28. Downs, R.T.; Bartelmehs, K.; Gibbs, G.; Boisen, M. Interactive Software for Calculating and Displaying X-Ray or Neutron Powder Diffractometer Patterns of Crystalline Materials. *Am. Mineral.* **1993**, *78*, 1104–1107.
29. Mendoza-Cruz, R.; Bazán-Diaz, L.; Velázquez-Salazar, J.J.; Samaniego-Benitez, J.E.; Ascencio-Aguirre, F.M.; Herrera-Becerra, R.; José-Yacamán, M.; Guisbiers, G. Order-disorder phase transitions in Au-Cu nanocubes: From nano-thermodynamics to synthesis. *Nanoscale* **2017**, *9*, 9267–9274. [CrossRef]
30. Guisbiers, G.; Mendoza-Pérez, R.; Bazán-Díaz, L.; Mendoza-Cruz, R.; Velázquez-Salazar, J.J.; José-Yacamán, M. Size and shape effects on the phase diagrams of nickel-based bimetallic nanoalloys. *J. Phys. Chem. C* **2017**, *121*, 6930–6939. [CrossRef]
31. Gallego, G.S.; Batiot-Dupeyrat, C.; Barrault, J.; Florez, E.; Mondragon, F. Dry Reforming of Methane over LaNi1−yByO3±δ (B = Mg, Co) Perovskites Used as Catalyst Precursor. *Appl. Catal. A Gen.* **2008**, *334*, 251–258. [CrossRef]
32. Li, H.; Liao, J.; Du, Y.; You, T.; Liao, W.; Wen, L. Magnetic-Field-Induced Deposition to Fabricate Multifunctional Nanostructured Co, Ni, And Coni Alloy Films as Catalysts, Ferromagnetic and Superhydrophobic Materials. *Chem. Commun.* **2013**, *49*, 1768–1770. [CrossRef]
33. Sun, S.; Zeng, H.; Robinson, D.B.; Raoux, S.; Rice, P.M.; Wang, S.X.; Li, G. Monodisperse mfe2o4 (m = fe, co, mn) nanoparticles. *J. Am. Chem. Soc.* **2004**, *126*, 273–279. [CrossRef]
34. Mourdikoudis, S.; Liz-Marzan, L.M. Oleylamine in nanoparticle synthesis. *Chem. Mater.* **2013**, *25*, 1465–1476. [CrossRef]
35. Tsakoumis, N.E.; Ronning, M.; Borg, O.; Rytter, E.; Holmen, A. Deactivation of cobalt based Fischer-Tropsch catalyst: A review. *Catal. Today* **2010**, *154*, 162–182. [CrossRef]
36. Jahangiri, H.; Bennet, J.; Mahjoubi, P.; Wilson, K.; Gu, S. A review of advanced catalyst development for Fischer-Tropsch synthesis of hydrocarbon from biomass derived syn-gas. *Catal. Sci. Technol.* **2014**, *4*, 2210–2229. [CrossRef]
37. Rytter, E.; Skagseth, T.H.; Eri, S.; Sjastad, A.O. Cobalt Fischer-Tropsch catalysts using nickel promoter as a rhenium substitute to suppress deactivation. *Ind. Eng. Chem. Res.* **2010**, *49*, 4140–4148. [CrossRef]
38. Nikparsa, P.; Mirzaei, A.A.; Rauch, R. Modification of Co/Al_2O_3 Fischer-Tropsch nanocatalysts by adding Ni: A kinetic approach. *Int. J. Chem. Kinet.* **2016**, *48*, 131–143. [CrossRef]
39. Shimura, K.; Miyazawa, T.; Hanaoka, T.; Hirata, S. Fischer-Tropsch synthesis over alumina supported bimetallic Co-Ni catalyst: Effect of impregnation sequence and solution. *J. Mol. Catal.* **2015**, *407*, 15–24. [CrossRef]

40. Yu, H.; Zhao, A.; Zhang, H.; Ying, W.; Fang, D. Bimetallic catalyst of Co and Ni for Fischer-Tropsch synthesis supported on alumina. *Energy Sources Part A Recovery Util. Environ. Eff.* **2015**, *37*, 47–54. [CrossRef]
41. van Helden, P.; Prinsloo, F.; van den Berg, J.A.; Xaba, B.; Erasmus, W.; Claeys, M.; van de Loosdrecht, J. Cobalt-nickel bimetallic Fischer-Tropsch catalysts: A combined theoretical and experimental approach. *Catal. Today* **2020**, *342*, 88–98. [CrossRef]
42. Jacoby, M. X-ray absorption spectroscopy. *Chem. Eng. News* **2001**, *79*, 33–38. [CrossRef]
43. Ressler, T. WinXAS: A Program for X-ray Absorption Spectroscopy Data Analysis under MS-Windows. *J. Synchrotron Radiat.* **1998**, *5*, 118–122. [CrossRef]
44. Ravel, B. ATOMS: Crystallography for the X-ray absorption spectroscopist. *J. Synchrotron Radiat.* **2001**, *8*, 314–316. [CrossRef] [PubMed]
45. Newville, M.; Ravel, B.; Haskel, D.; Rehr, J.J.; Stern, E.A.; Yacoby, Y. Analysis of multiple-scattering XAFS data using theoretical standards. *Phys. B Condens. Matter* **1995**, *208–209*, 154–156. [CrossRef]

© 2019 by the authors. Licensee MDPI, Basel, Switzerland. This article is an open access article distributed under the terms and conditions of the Creative Commons Attribution (CC BY) license (http://creativecommons.org/licenses/by/4.0/).

Article

Investigation of $C_1 + C_1$ Coupling Reactions in Cobalt-Catalyzed Fischer-Tropsch Synthesis by a Combined DFT and Kinetic Isotope Study

Yanying Qi, Jia Yang *, Anders Holmen and De Chen *

Department of Chemical Engineering, Norwegian University of Science and Technology, Sem Sælands vei 4, N-7491 Trondheim, Norway; yanying.qi@ntnu.no (Y.Q.); anders.holmen@ntnu.no (A.H.)
* Correspondence: jia.yang@ntnu.no (J.Y.); de.chen@ntnu.no (D.C.)

Received: 2 June 2019; Accepted: 17 June 2019; Published: 19 June 2019

Abstract: Understanding the chain growth mechanism is of vital importance for the development of catalysts with enhanced selectivity towards long-chain products in cobalt-catalyzed Fischer-Tropsch synthesis. Herein, we discriminate various $C_1 + C_1$ coupling reactions by theoretical calculations and kinetic isotope experiments. $CH_{x(x=0-3)}$, CO, HCO, COH, and HCOH are considered as the chain growth monomer respectively, and 24 possible coupling reactions are first investigated by theoretical calculations. Eight possible $C_1 + C_1$ coupling reactions are suggested to be energetically favorable because of the relative low reaction barriers. Moreover, five pathways are excluded where the C_1 monomers show low thermodynamic stability. Effective chain propagation rates are calculated by deconvoluting from reaction rates of products, and an inverse kinetic isotope effect of the $C_1 + C_1$ coupling reaction is observed. The theoretical kinetic isotope effect of $CO + CH_2$ is inverse, which is consistent with the experimental observation. Thus, the $CO + CH_2$ pathway, owing to the relatively lower barrier, the high thermodynamic stability, and the inverse kinetic isotope effect, is suggested to be a favorable pathway.

Keywords: Fischer-Tropsch synthesis; chain growth; CO insertion; kinetic isotope effect; DFT

1. Introduction

Fischer-Tropsch synthesis (FTS) is an alternative process to produce clean liquid fuels by converting syngas [1–4]. Cobalt-based catalysts, owing to the low water-gas shift activity and the remarkable stability, are favorable catalysts, for the synthesis of long-chain hydrocarbons from the synthesis gas with an H_2/CO ratio of about 2 [5–7]. Understanding the reaction mechanisms of chain propagation and termination of cobalt-catalyzed FTS is crucial to achieve a high selectivity of long-chain hydrocarbons. Considerable efforts have contributed to the investigation of the reaction mechanisms, and three prevailing mechanisms are proposed [8–15], that is, carbide mechanism, hydroxycarbene mechanism, and CO insertion mechanism. However, there is still no consensus about the chain growth mechanism because of the complexity of FTS reaction system.

The carbide mechanism was originally proposed by Fischer and Tropsch [16], suggesting that CO adsorbs dissociatively on the catalyst surface and the surface carbon is subsequently hydrogenated to form a methylene (CH_2) group, which is considered as a chain propagation monomer. The experimental results from Brady and Pettit [17,18] demonstrated that only the carbide mechanism could produce the product composition, and other mechanisms were thus excluded. However, an assumption was automatically made in their analysis, where the reactions of CH_2 with O and OH were ignored. Hu and coworkers suggested that $CH_3 + C$ and $CH_2 + CH_2$ are two major chain growth pathways at step sites based on the calculated reaction rates for all $C_1 + C_1$ coupling pathways in the carbide mechanism [19]. Furthermore, coupling reactions of $RCH_2 + C$ and $RCH + CH_2$ are proposed to be two major pathways for all the hydrocarbons with different carbon numbers [15]. However, the CO/HCO insertion mechanism is not considered in their calculations. The carbide mechanism is also demonstrated to be a favorable pathway on the Co(10−10) surface [20] and Co(10−11) [21] based on density functional theory(DFT) calculations. Recently, steady state isotopic transient kinetic analysis(SSITKA) measurements proved that chain growth rate is much higher compared with the rate of monomer formation [22], which means the rate of monomer formation or the site coverage of monomer need to be taken into account in the discrimination of chain growth mechanisms. Notably, the carbide mechanism cannot account for all the products' formation, especially oxygenates.

As an alternative to the carbide mechanism, the CO insertion mechanism is reported to be a preferred mechanism, considering that its reaction barrier is low on Co(0001) and the theoretical turnover frequency(TOF) fits well with the experimental value [12,23]. Meanwhile, transient kinetic studies between the H_2/inert and H_2/CO forward and backward switch provide an indication of CO involved in the C_{2+} formation [24]. The authors also found that the chain growth probability is proportional to the partial pressure of CO rather than the surface C coverage, which further supports their argument that CO is the chain growth monomer. However, further DFT and experimental studies are needed to distinguish the nature of the monomer, such as CO and CH_xO_y, in the chain growth pathway. The CHO insertion pathway is demonstrated to be more energetically favorable on a clean Co(0001) surface [11]. It is reported that the chain growth proceeds via the CO insertion mechanism on Co(0001) and the stepped surfaces, but via the carbide mechanism on a more open Co(10−11) surface at lower CO coverage [25]. It is noted that the chain growth mechanism is sensitive to both the surface structure and the CO coverage. Thus, the different surface models and CO site coverages employed in the various theoretical calculations result in different conclusions. A considerable debate still remains and, therefore, a systematic comparison of all the possible reaction pathways under certain CO coverage is highly desired.

It is difficult to discriminate reaction mechanisms of complex reaction networks such as FTS by solely employing theoretical or experimental methods. In the current study, a combined approach of kinetic isotope experiments and DFT calculations was employed to study the chain growth mechanism in cobalt-catalyzed Fischer-Tropsch synthesis. The combined method has proved very useful in the mechanism investigation of CO activation and methane formation [26,27]. We firstly calculated the reaction barriers of all the possible coupling reactions with CH_x, CO, HCO, COH, and HCOH as the chain growth monomers by DFT calculations. The thermodynamic stability of C_1 monomers was analyzed to further discriminate various reactions. Moreover, we performed kinetic isotope experiments using H_2 and D_2. The reaction rates of the products were deconvoled to get the effective chain propagation rate, and thus the kinetic isotope effect (KIE) of the $C_1 + C_1$ coupling reaction was obtained. Kinetic isotope effects of the energetically favorable pathways were calculated and compared with the experimental result to further identify the favorable reaction pathway.

2. Results and Discussion

2.1. Adsorption of C_2 Species

The adsorption characteristics of C_2 surface species involved in the reaction were first evaluated. The most stable adsorption configurations are summarized in Figure 1, and the corresponding adsorption energies and structural parameters are reported in Table 1. The favorable adsorption site of CO was the top site, which is consistent with the experimental observation [28,29]. It is noted that one CO spectator is clearly seen in Figure 1, while the other CO spectator is covered by the inserted side view. The detailed positions of the CO spectators are shown in Figure S1, Supplementary Materials. For C-$CH_{x(x=0-3)}$ species, the carbon atom adsorbs on the hcp hollow site and $CH_{x(0-2)}$ bonds to the nearby fcc/top site, while CH_3 points away from the surface. CH adsorbs on the fcc hollow site in CH-$CH_{x(x=1-3)}$ species. The distance between two carbon atoms increases with an increasing x value in C (CH, CH_2, CO, HCO, COH, or HCOH)-$CH_{x(x=1-3)}$. For most oxygenates, CH_x prefers to adsorb on the hollow site. The C-C distances are similar for the same x value in the different surface species $CH_{x(1-3)}$-CO(HCO, COH, HCOH). The adsorption energies of most surface species are slightly smaller compared with the literature values, which could be the result of different functional and CO site coverage employed. The adsorption energies of CH_2CO and CH_2HCO calculated by employing Perdew-Burke-Ernzerhof functional are −0.85 eV and −2.12 eV, respectively, which are similar to the literature values of −0.98 eV and −2.18 eV, respectively [23].

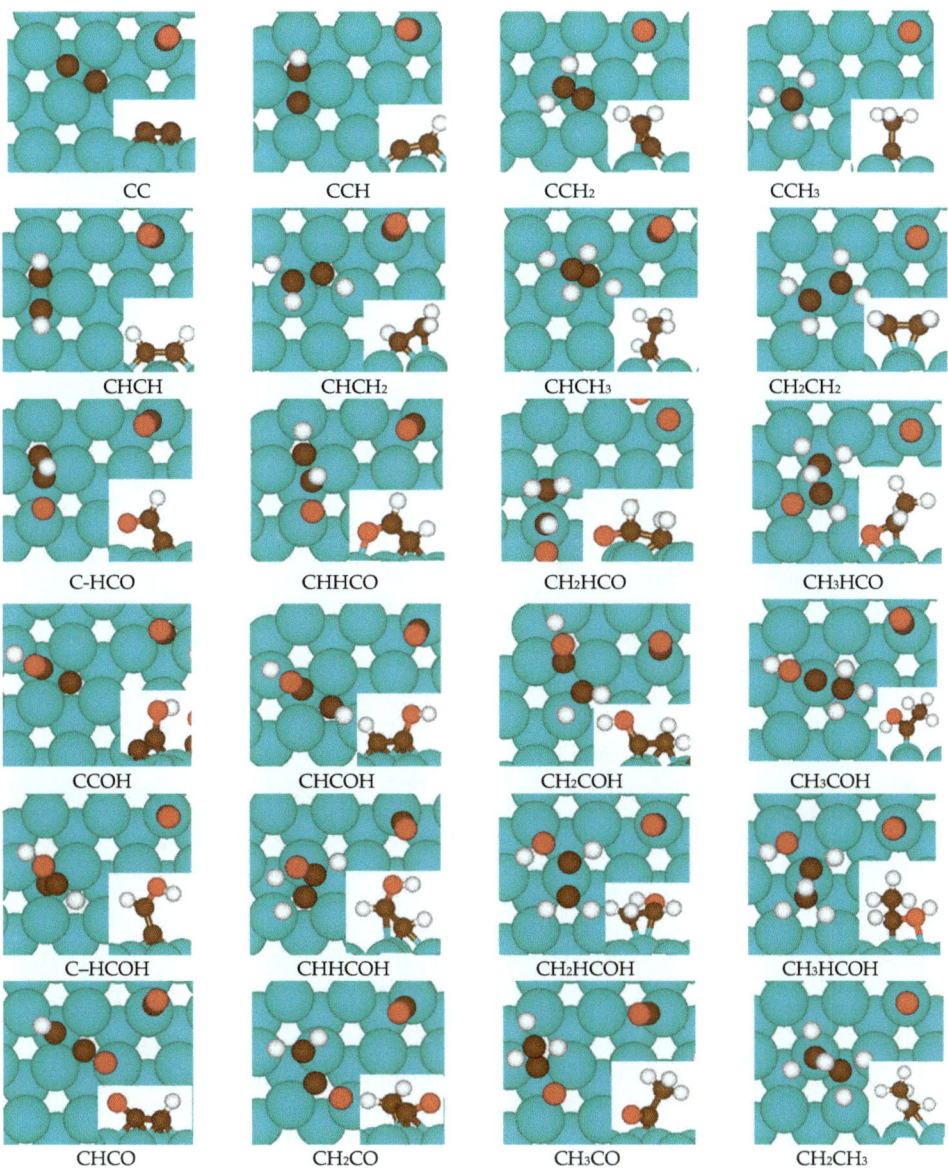

Figure 1. Top views and side views (inserted) of C_2 species adsorbed on the surface.

Table 1. Adsorption energies and structure parameters of C_2 species.

Surface Species	E_{ads} (eV)	d_{C-C} (Å)	d_{C-O} (Å)	d_{C-H} (Å)
CC	−6.49	1.33	–	–
CCH	−4.83	1.35	–	1.09
CCH_2	−3.68	1.38		1.09
CCH_3	−5.13	1.50		1.10
CHCO	−2.82	1.39	1.25	1.09
CH_2CO	−0.77	1.43	1.28	1.09
CH_3CO	−1.86	1.51	1.31	1.10
CHCH	−2.12	1.38		1.09
$CHCH_2$	−2.46	1.41		1.09
$CHCH_3$	−3.35	1.52		1.15, 1.09
CH_2CH_2	−0.77	1.42		1.09
CH_2CH_3	−1.44	1.55		1.10
CH_3CH_3	−0.35	1.53		1.09
C-HCO	−4.67	1.42	1.28	1.10
CHHCO	−3.96	1.41	1.29	1.10
CH_2HCO	−2.01	1.43	1.19	1.04
CH_3HCO	−0.59	1.50	1.35	1.09
C-COH	−4.79	1.39	1.34	–
CHCOH	−2.01	1.41	1.37	1.09
CH_2COH	−2.34	1.45	1.35	1.09
CH_3COH	−2.36	1.50	1.34	1.09
C-HCOH	−4.43	1.38	1.34	1.09
CHHCOH	−2.77	1.39	1.35	1.11
CH_2HCOH	−0.65	1.40	1.38	1.09
CH_3HCOH	−1.43	1.52	1.46	1.10

2.2. DFT Calculations of C_1 + C_1 Coupling Reactions

The most favorable pathway of CO activation is the hydrogenation-assisted CO dissociation on Co(0001), according to the previous results [26,27,30,31]. Herein, the intermediates such as CH_x, CO, HCO, COH, and HCOH were considered as the chain growth monomer, respectively. The transition states were located for the coupling reactions involving CH_x, CO, HCO, COH, and HCOH.

The configurations of the transition states are plotted in Figure 2 and the reaction barriers are summarized in Table 2, where the vertical row is arranged by the sequence of the intermediates in the reaction from syngas towards methane and the horizontal row is the possible intermediates involved in the reactions. For the transition states of coupling reactions between C and $CH_{x(0-3)}$, the C atom always bonds to the hcp hollow site and the $CH_{x(0-3)}$ bonds to the nearby bridge or top site, which is consistent with the results reported by Cheng et al. [19]. With respect to the coupling reactions CH + CH, CH + CH_2, CH + CH_3, CH_2 + CH_2, and CH_2 + CH_3, CH and CH_2 adsorb on the hollow sites. For CO, HCO, COH, and HCOH insertion reactions, C, CH, and CH_2 adsorb on the hollow sites, while CH_3 bonds to the off-top site. The barrier of the C + C coupling reaction (i.e., 0.97 eV) is the highest and that of C + CH_2 is the lowest among all the coupling reactions between C and $CH_{x(0-3)}$. The barriers of C + CH, C + CH_3, CH + CH_3, and CH_2 + CH_3 are around 0.70 eV. It is noted that the barrier of CH_2 + CH_2 (i.e., 0.12 eV) is the lowest, while that of CH_3 + CH_3 (i.e., 1.84 eV) is the highest among all the $CH_{x(0-3)}$ + $CH_{x(0-3)}$ coupling reactions. For the CO insertion reactions, the barrier of the CH_3 + CO coupling reaction (i.e., 1.19 eV) is higher than the others, which is attributed to the strong and directional CH_3 surface bonds, as reported previously [12]. The CH_2 + CO coupling shows the lowest barrier (i.e., 0.56) among all the CO insertion pathways. With regards to HCO insertion into CH_x, the barriers are lower than the corresponding CO insertion reactions, and thus the HCO insertion is more favourable over the CO insertion by solely comparing the reaction barriers. The smaller gap between the highest occupied molecular orbital and the lowest unoccupied molecular orbital of HCO compared with that of CO greatly facilitates the charge transfer and hybridization between HCO and catalysts, as reported by Li and co-workers [11], which results in the higher activity of HCO insertion. The relative barriers of HCO insertion reactions follow a similar pattern to the CO insertion pathways, and the barrier of CH_3 + HCO is the largest. CH_2 + HCO owns a significantly lower barrier, which is 0.05 eV. For COH insertion into CH_x, the barriers are higher than that for HCO insertion because of the relatively strong adsorption of COH. The pathways of HCOH insertion into CH_x also exhibit low barriers, especially for HCOH + CH_2.

Figure 2. Top views and side views (inserted) of transition states for the $C_1 + C_1$ coupling reactions. The number represents the activation energy in eV.

Table 2. Activation energies of the coupling reactions.

	+CO	+HCO	+HCOH	+CH	+CH$_2$	+CH$_3$	+C	+COH
HCO	–	–	–	0.34	0.05	0.80	–	–
HCOH	–	–	–	0.29	0.16	0.77	0.32	–
CH	0.90	0.34	0.29	0.57	0.49	0.75	0.71	1.40
CH$_2$	0.56	0.05	0.16	0.49	0.12	0.62	0.57	0.48
CH$_3$	1.19	0.80	0.77	0.75	0.62	1.84	0.74	1.10

The barriers of coupling reactions are summarized in a special style in Table 3, where both vertical and horizontal rows are in an order of low to high adsorption heat of the intermediates. It clearly shows that the central area of the table gives the lower activation energies, which possibly suggests that there is a volcano curve between the activation barriers and the adsorption energies. For the steps between HCO and HCOH, the transition states are difficult to locate. Besides, the site coverage of HCOH is low because the hydrogenation of HCO is rate determining [26,27]. Considering the low site coverage of HCOH, the activity of the steps involving HCOH will be low, and thus these steps are not favorable.

Table 3. Comparison of the activation energies of main coupling reactions.

	CH$_3$	HCO	HCOH	CH$_2$	CH
CH$_3$	1.84	0.80	0.77	0.62	0.75
HCO	0.80	–	–	0.05	0.34
HCOH	0.77	–	–	0.16	0.29
CH$_2$	0.62	0.05	0.16	0.12	0.49
CH	0.75	0.34	0.29	0.49	0.57

By analyzing the activation barriers, the reactions with CH$_3$ display relative high barriers compared with other CH$_x$ species, and thus we exclude the possibility with CH$_3$ participating in the chain growth mechanism. We estimate that an increase of the barrier by 0.1 eV could roughly result in a decrease of reaction rate by an order of magnitude at 500 K, when keeping the site coverages of surface species constant. Thus, we could also exclude the reactions with barriers larger than 0.6 eV. As we know, the reaction rate of a surface reaction is a combined result from the reaction barrier and the site coverage of reactants. We exclude the possibility of HCOH as a chain initiator or monomer considering the low site coverage of HCOH. Therefore, the remaining possibilities are as follows: CO + CH$_2$, HCO + CH, HCO + CH$_2$, CH + CH, CH + CH$_2$, CH$_2$ + CH$_2$, C + CH$_2$, and CH$_2$ + COH.

2.3. Thermodynamic Analysis

The thermodynamic stability of the possible chain growth monomers is another important factor to identify the favourable reaction mechanism. Figure 3 illustrates the relative free energy of each C$_1$ monomer at 483 K with the pressure at 1.85 bar and 20 bar, respectively. The calculation details are summarized in S1, Supplementary Materials. We can get a clear understanding of which species are likely to be present on the cobalt surface by analysing the relative free energies. HCOH is the most unstable one, which is expected. CO is the most stable one, which is consistent with the SSITKA result, observing the CO coverage of 0.5 [26]. The relative free energy decreases with the increase of the pressure, which means the adsorption of CO is stronger and the site coverage of CO is larger at a higher pressure. The combined relative free energy of two reactants for the eight energetic favourable C$_1$ + C$_1$ reactions are summarized in Table 4. The CO + CH$_2$ coupling reaction shows extremely high thermodynamics stability, owing to the low free energy of CO. The relative free energy reflects the thermodynamic stability of the reactants and the high value indicates thermodynamic instability, and thus will be unlikely present under the reaction condition. Thus, we exclude the coupling reactions involving C$_1$ species with relatively low thermodynamic stability and select three

C–C coupling reactions with low relative free energy (i.e., CO + CH$_2$, HCO + CH, CH + CH) to be further discriminated by the analysis of kinetic isotope effect.

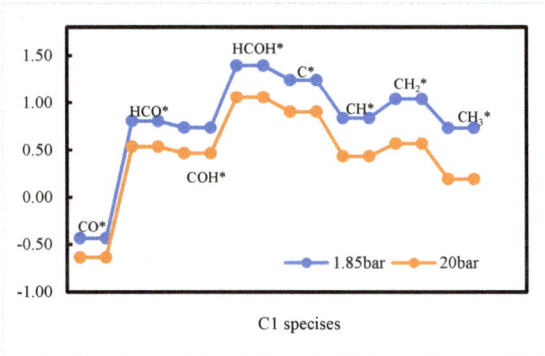

Figure 3. Free energies of each C$_1$ monomer, referenced to CO(g), H$_2$O(g), and H$_2$(g) for Fischer-Tropsch synthesis at the pressure of 1.85 bar and 20 bar with H$_2$/CO = 2 and 483 K.

Table 4. The combined relative free energies of two reactants.

Reactants	CO + CH$_2$	HCO + CH	HCO + CH$_2$	CH + CH	CH + CH$_2$	CH$_2$ + CH$_2$	C + CH$_2$	CH$_2$ + COH
G(eV)	0.60	1.64	1.84	1.67	1.87	2.07	2.28	1.77

2.4. Analysis of Kinetic Isotope Effect

The kinetic isotope effect (KIE) is defined as the ratio of the rates with CO/H$_2$ and CO/D$_2$ as reactants (r_H/r_D), which could provide important kinetic information, assisting in the discrimination of reaction mechanisms. The KIE was measured for the product formation on a 20% Co/CNT catalyst, as shown in Table 5. The product with a certain carbon number (C$_n$) includes both olefin and paraffin products.

Table 5. Kinetic isotope effects (KIE) for the product formation (483 K, 1.85 bar, H$_2$/CO = D$_2$/CO = 10).

Reaction Rate [μmol/(g$_{cat}$s)]	$r_{t,H}$	$r_{t,D}$	KIE = $r_{t,H}/r_{t,D}$
r_{CH4}	2.349	2.383	0.99
r_{C2}	0.158	0.180	0.88
r_{C3}	0.128	0.192	0.66
r_{C4}	0.076	0.152	0.50
r_{C5+}	0.046	0.121	0.38

The magnitude of inverse KIE increases with an increasing carbon number, as shown in Table 5. To study the chain growth step by monomer addition, an effective chain propagation rate is introduced and its KIE is investigated. Scheme 1 describes a simplified reaction network in this study.

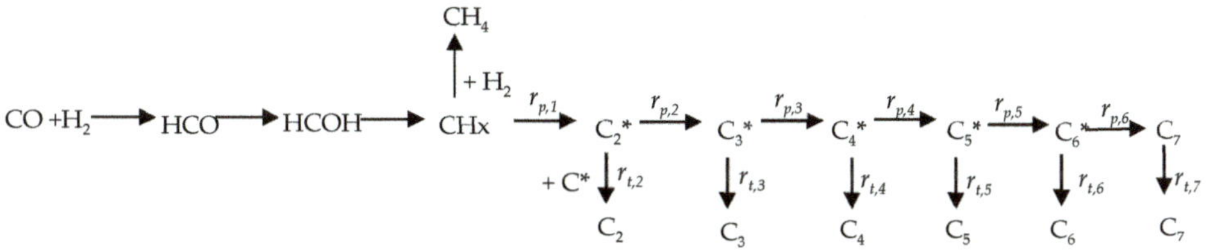

Scheme 1. Reaction network of cobalt-catalyzed Fischer-Tropsch synthesis.

The effective chain propagation rate ($r_{p,n}$) is calculated in such a way that the effect of the chain propagation reaction after particular carbon number (C_7) is excluded owing to methanation conditions, as shown in Equations (1) and (2).

$$r_{p,6} = r_{t,7} \tag{1}$$

$$r_{p,n} = r_{t,n+1} + \frac{n+1}{n+2}r_{p,n+1}, n = 1\text{–}5, \tag{2}$$

where $r_{t,n}$ and $r_{p,n}$ are the carbon based termination and propagation rate, respectively, of the hydrocarbon with carbon number n. According to the equations, we can get the $r_{p,n}$ with CO/H_2 and CO/D_2 as reactants, and the results are summarized in Table 6. The KIE values for the propagation reactions decrease with the increasing carbon number. A similar phenomenon was observed for CO_2 hydrogenation on both Co and Fe catalysts [32].

Table 6. Kinetic isotope effects for the propagation rate with different carbon numbers (483 K, 1.85 bar, $H_2/CO = D_2/CO = 10$).

Reaction Rate [mol/(g_{cat}s)]	r_H	r_D	KIE
$r_{p,4}$	0.05	0.12	0.38
$r_{p,3}$	0.11	0.25	0.45
$r_{p,2}$	0.21	0.38	0.56
$r_{p,1}$	0.30	0.43	0.69

An inverse KIE (i.e., 0.69) of the $C_1 + C_1$ coupling reaction is observed experimentally. It is generally assumed that the KIE is expressed as the difference in rate constants in the calculations, which involves the contribution from the zero-point energy (ZPE) differences and the entropy differences (ΔS) between H and D isotopic isomers. The details of KIE calculation are reported in S2, Supplementary Materials [27,31]. The results of different coupling reactions are summarized in Table 7. The KIEs of three possible coupling reactions are 0.84, 0.98, and 0.95 for $CH_2 + CO$, $CH + HCO$, and $CH + CH$ at 500 K, respectively. $CO + CH_2$ shows inverse KIE, which is mainly attributed to the contribution of zero-point energy. The differences of KIE between the theoretical calculation and the experimental observation could be attributed to the accuracy of second-derivation methods to estimate the vibrational frequency [31]. Small inaccuracy in theoretical calculations may result in deviation of the kinetic isotope effect, as KIE is highly sensitive to vibrational frequencies. Besides, the catalyst surface is more complicated than the model employed. The KIEs of CH + HCO and CH + CH are around 1, which do not show inverse isotope effects and could be excluded. Therefore, $CO + CH_2$ could be the possible reaction pathway owing to its reverse kinetic isotope effect.

Table 7. The kinetic isotope effect at 500 K for three possible coupling routes.

Elementary Step	ZPE Contribution	Entropic Contribution	KIE
$CO + CH_2 \rightarrow CH_2CO$	0.78	1.08	0.84
$HCO + CH \rightarrow CHHCO$	0.96	1.02	0.98
$CH + CH \rightarrow CHCH$	0.91	1.04	0.95

3. Method

3.1. Theoretical Method and Model

All DFT calculations were implemented with the Vienna Ab initio Simulation Package (VASP) [33] using Bayesian error estimation functional with van der Waals correlation (BEEF-vdW) [34] and projector augmented wave (PAW) pseudopotentials [35]. Spin polarization was involved in all calculations. Large-sized Co nanoparticles were used in our experiments. On the basis of a geometrical, cuboctahedral model from Van Hardeveld and Hartog, terrace surfaces are dominant for large particles [36]. Besides, the (111) face-centered-cubic (fcc)-cobalt is detected in the X-ray powder diffraction pattern and the high-resolution transmission electron microscopy image of reduced cobalt-based catalysts [37]. The Co(0001) surface of hexagonal close-packed (hcp) is close to the (111) surface of fcc metals. Therefore, Co(0001) was selected as the model. A 3 × 3 supercell of five layers and 12 Å of vacuum spacing between the successive metal slabs was used. Two CO molecules were pre-adsorbed on the surface as spectators in this work. An electronic plane wave cut-off of 500 eV was used, with the Brillouin zone being sampled by a mesh of 5 × 5 × 1. For the Fermi-surface smearing, the first-order Methfessel–Paxton scheme was applied with a smearing width of 0.2 eV. The nudged elastic band (NEB) method [38] was employed to locate the initial transition states, which were subsequently optimized by the dimer method [39]. Vibrational frequencies were calculated to verify the transition states, with one negative mode corresponding to the desired reaction coordinate. The optimizations were converged until all forces on the atoms were lower than 0.01 eV/Å.

$$E_{ads} = E_{A+slab} - E_A - E_{ads}$$

$$E_a = E_{TS} - E_{IS}$$

Adsorption energies of surface species were defined as the difference between the minimum total energy of the surface species on the slab and total energy of surface plus gas phase species. Activation energies of elementary steps were calculated as the energy difference between the transition states and the separately adsorbed reactants.

Kinetic isotope effects have contributions from zero-point energy (ZPE) corrections and entropies (S) (see details in S2, Supplementary Materials). For adsorbed species, the frustrated translational and rotational modes are treated as special cases of vibrational modes. Accordingly, the entropy is evaluated by the following:

$$S = S_{vib} = R \sum_i^{3N} \left(\frac{x_i}{e^{x_i} - 1} - \ln(1 - e^{-x_i}) \right),$$

where $x_i = \frac{hc}{k_B T} \frac{1}{\lambda_i}$, c is speed of light, k_B is Boltzmann constant, and $1/\lambda_i$ is the wavenumber corresponding to each vibrational frequency. Zero-point energies were calculated according to the following equation:

$$E_{ZPE} = \sum_{i=1}^{3N-6(5)} \frac{N_A h v_i}{2},$$

where N_A is Avogadro's number, h is Plank's constant, v_i is the frequency of the normal mode, and N is the number of atoms. All of the frequencies, except those for two CO* spectators and the imaginary one for the transition state, were accounted for by the ZPE and entropy calculations. Vibrational frequencies were calculated by numerical differentiation of the forces using a second-order finite difference approach with a step size of 0.015 Å. The frequencies were converged with respect to the step size. The frequencies are summarized in Table S1, Supplementary Materials. It is reported that the frustrated rotation and translation are with low frequency modes, and those modes contribute largely to the entropy; thus, they need to be computed accurately [40]. Besides, small vibrational frequencies combined with high temperatures could result in larger errors from the harmonic oscillator approximation, as reported by Bai et.al [41]. Most of the frequencies calculated here are larger than 200 cm^{-1} and for these cases, the harmonic oscillator approximation is reasonably accurate.

3.2. Kinetic Isotope Experiment

A 20% Co/CNT (carbon nanotube) catalyst was tested at methanation conditions (483 K, 1.85 bar, H$_2$/CO = 10). The details of catalyst preparation and characterization have been published in a previous paper [26]. The experiment was performed using a fixed-bed quartz reactor (4 mm i.d.).

The catalyst (25 mg; 53–90 μm) was diluted with 50 mg of inert silicon carbide (75–150 μm) to improve the isothermal conditions along the catalyst bed. The catalyst was reduced in situ in 10 NmL/min H_2 with a ramping rate of 1 K/min to 623 K. After 16 h reduction, the catalyst was cooled down to 443 K. The temperature was then increased at a ramping rate of 1 K/min to 463 K. At 463 K, syngas H_2/CO/Ar (15/1.5/33.5 NmL/min) was introduced and the pressure was adjusted to 1.85 bar. The system was further heated to 483 K at a rate of 1 K/min. The concentrations of H_2, CO, Ar, and C_1–C_7 hydrocarbons were analyzed with an online gas chromatograph (HP5890) equipped with a thermal conductivity detector and a flame ionization detector.

The kinetic isotope effect was determined by switching the feed from H_2/CO to D_2/CO at 483 K, 1.85 bar, and H_2/CO = 10. The kinetic isotope effect measurement was performed after pseudo steady state was reached, about 5 h time on stream.

4. Conclusions

The possible reaction pathways of C_1 + C_1 coupling were firstly explored by calculating the adsorption energies and the activation barriers on the Co(0001) surface with two CO as spectators. The activation energies of the coupling reactions varied from 0.05 eV to 1.84 eV. An increase of the barrier by 0.1 eV could lead to a decrease of reaction rate by an order of magnitude at 500 K, when keeping the site coverages of surface species constant. Thus, we excluded the reactions with barriers larger than 0.6 eV and the remaining possibilities were as follows: CO + CH_2, HCO + CH, HCO + CH_2, CH + CH, CH + CH_2, CH_2 + CH_2, C + CH_2, and CH_2 + COH. It is noted the reaction rate results from both the activation barrier and the site converge of the reactants. The thermodynamic stability of the reactants was analyzed at two different reaction pressures. CO is the most stable specie at both pressures. The chain growth monomers are more stable at a high pressure. Three reaction pathways (i.e., CH_2 + CO, HCO + CH, and CH + CH), with their relatively low barriers and high thermodynamic stability, were further discriminated by analyzing the kinetic isotope effect. Kinetic isotope experiments demonstrated that the magnitude of inverse KIE for the products increases with increasing carbon number. The inverse kinetic isotope effect was observed for the C_1 + C_1 coupling reaction. The theoretical KIE of CH_2 + CO is inverse, which agrees with the experimental result. Therefore, the CH_2 + CO coupling reaction is suggested to be a favorable chain propagation pathway by a combined theoretical calculation and kinetic experiment approach.

Supplementary Materials: The following are available online at http://www.mdpi.com/2073-4344/9/6/551/s1, S1: The calculation method of free energy, Figure S1: The stable configurations of two spectators on the surface, Table S1: Frequencies for the calculation of kinetic isotope effects, S2: The formulas for KIE calculation.

Author Contributions: Conceptualization, Y.Q., J.Y., A.H., and D.C.; methodology, Y.Q., J.Y., A.H., and D.C.; validation, Y.Q. and J.Y.; formal analysis, Y.Q. and J.Y.; investigation, Y.Q. and J.Y.; writing—original draft preparation, Y.Q.; writing—review and editing, Y.Q., J.Y., A.H., and D.C.; supervision, J.Y. and D.C.; funding acquisition, D.C.

Funding: The financial support from Norwegian Research Council through ISP program 209337 and the industrial Catalysis Science and Innovation (iCSI) center is gratefully acknowledged.

Acknowledgments: The computational time provided by the Notur project NN4685K is highly acknowledged.

Conflicts of Interest: The authors declare no conflict of interest.

References

1. Ledesma, C.; Yang, J.; Chen, D.; Holmen, A. Recent Approaches in Mechanistic and Kinetic Studies of Catalytic Reactions Using SSITKA Technique. *ACS Catal.* **2014**, *4*, 4527–4547. [CrossRef]
2. Van Santen, R.A.; Markvoort, A.J.; Filot, I.A.W.; Ghouri, M.M.; Hensen, E.J.M. Mechanism and microkinetics of the Fischer-Tropsch reaction. *Phys. Chem. Chem. Phys.* **2013**, *15*, 17038–17063. [CrossRef] [PubMed]
3. Markvoort, A.J.; van Santen, R.A.; Hilbers, P.A.J.; Hensen, E.J.M. Kinetics of the Fischer-Tropsch Reaction. *Angew. Chem. Int. Ed.* **2012**, *51*, 9015–9019. [CrossRef] [PubMed]
4. Dry, M.E. The Fischer-Tropsch process: 1950–2000. *Catal. Today* **2002**, *71*, 227–241. [CrossRef]

5. Yang, J.; Tveten, E.Z.; Chen, D.; Holmen, A. Understanding the Effect of Cobalt Particle Size on Fischer-Tropsch Synthesis: Surface Species and Mechanistic Studies by SSITKA and Kinetic Isotope Effect. *Langmuir* **2010**, *26*, 16558–16567. [CrossRef]
6. Den Breejen, J.P.; Radstake, P.B.; Bezemer, G.L.; Bitter, J.H.; Froseth, V.; Holmen, A.; de Jong, K.P. On the Origin of the Cobalt Particle Size Effects in Fischer-Tropsch Catalysis. *J. Am. Chem. Soc.* **2009**, *131*, 7197–7203. [CrossRef] [PubMed]
7. Bezemer, G.L.; Bitter, J.H.; Kuipers, H.; Oosterbeek, H.; Holewijn, J.E.; Xu, X.D.; Kapteijn, F.; van Dillen, A.J.; de Jong, K.P. Cobalt particle size effects in the Fischer-Tropsch reaction studied with carbon nanofiber supported catalysts. *J. Am. Chem. Soc.* **2006**, *128*, 3956–3964. [CrossRef] [PubMed]
8. Dry, M.E. Practical and theoretical aspects of the catalytic Fischer-Tropsch process. *Appl. Catal. A Gen.* **1996**, *138*, 319–344. [CrossRef]
9. Roferdepoorter, C.K. A comprehensive mechanism for the Fischer-Tropsch synthesis. *Chem. Rev.* **1981**, *81*, 447–474. [CrossRef]
10. Ledesma, C.; Yang, J.; Blekkan, E.A.; Holmen, A.; Chen, D. Carbon Number Dependence of Reaction Mechanism and Kinetics in CO Hydrogenation on a Co-Based Catalyst. *ACS Catal.* **2016**, *10*, 6674–6686. [CrossRef]
11. Zhao, Y.-H.; Sun, K.; Ma, X.; Liu, J.; Sun, D.; Su, H.-Y.; Li, W.-X. Carbon Chain Growth by Formyl Insertion on Rhodium and Cobalt Catalysts in Syngas Conversion. *Angew. Chem. Int. Ed.* **2011**, *50*, 5335–5338. [CrossRef] [PubMed]
12. Zhuo, M.K.; Tan, K.F.; Borgna, A.; Saeys, M. Density Functional Theory Study of the CO Insertion Mechanism for Fischer-Tropsch Synthesis over Co Catalysts. *J. Phys. Chem. C* **2009**, *113*, 8357–8365. [CrossRef]
13. Inderwildi, O.R.; Jenkins, S.J.; King, D.A. Fischer-tropsch mechanism revisited: Alternative pathways for the production of higher hydrocarbons from synthesis gas. *J. Phys. Chem. C* **2008**, *112*, 1305–1307. [CrossRef]
14. Cheng, J.; Hu, P.; Ellis, P.; French, S.; Kelly, G.; Lok, C.M. Chain growth mechanism in Fischer-Tropsch synthesis: A DFT study of C-C coupling over Ru, Fe, Rh, and Re surfaces. *J. Phys. Chem. C* **2008**, *112*, 6082–6086. [CrossRef]
15. Cheng, J.; Hu, P.; Ellis, P.; French, S.; Kelly, G.; Lok, C.M. A DFT study of the chain growth probability in Fischer-Tropsch synthesis. *J. Catal.* **2008**, *257*, 221–228. [CrossRef]
16. Fischer, F.; Tropsch, H. The Synthesis of Petroleum at Atmospheric Pressures from Gasification Products of Coal. *Brennstoff Chem.* **1926**, *7*, 97–116.
17. Brady, R.C.; Pettit, R. On the mechanism of the Fischer-Tropsch reaction-the chain propagation step. *J. Am. Chem. Soc.* **1981**, *103*, 1287–1289. [CrossRef]
18. Brady, R.C.; Pettit, R. Reactions of diazomethane on transition metal surfaces and their relationship to the mechanism of the Fischer-Tropsch reaction. *J. Am. Chem. Soc.* **1980**, *102*, 6181–6182. [CrossRef]
19. Cheng, J.; Gong, X.Q.; Hu, P.; Lok, C.M.; Ellis, P.; French, S. A quantitative determination of reaction mechanisms from density functional theory calculations: Fischer-Tropsch synthesis on flat and stepped cobalt surfaces. *J. Catal.* **2008**, *254*, 285–295. [CrossRef]
20. Zhang, R.; Kang, L.; Liu, H.; He, L.; Wang, B. Insight into the CC chain growth in Fischer-Tropsch synthesis on HCP Co(10-10) surface: The effect of crystal facets on the preferred mechanism. *Comput. Mater. Sci.* **2018**, *145*, 263–279. [CrossRef]
21. Liu, H.; Zhang, R.; Ling, L.; Wang, Q.; Wang, B.; Li, D. Insight into the preferred formation mechanism of long-chain hydrocarbons in Fischer-Tropsch synthesis on Hcp Co(10-11) surfaces from DFT and microkinetic modeling. *Catal. Sci. Technol.* **2017**, *7*, 3758–3776. [CrossRef]
22. Chen, W.; Filot, I.A.W.; Pestman, R.; Hensen, E.J.M. Mechanism of Cobalt-Catalyzed CO Hydrogenation: 2. Fischer-Tropsch Synthesis. *ACS Catal.* **2017**, *7*, 8061–8071. [CrossRef] [PubMed]
23. Zhuo, M.K.; Borgna, A.; Saeys, M. Effect of the CO coverage on the Fischer-Tropsch synthesis mechanism on cobalt catalysts. *J. Catal.* **2013**, *297*, 217–226. [CrossRef]
24. Schweicher, J.; Bundhoo, A.; Kruse, N. Hydrocarbon Chain Lengthening in Catalytic CO Hydrogenation: Evidence for a CO-Insertion Mechanism. *J. Am. Chem. Soc.* **2012**, *134*, 16135–16138. [CrossRef] [PubMed]
25. Su, H.-Y.; Zhao, Y.; Liu, J.-X.; Sun, K.; Li, W.-X. First-principles study of structure sensitivity of chain growth and selectivity in Fischer-Tropsch synthesis using HCP cobalt catalysts. *Catal. Sci. Technol.* **2017**, *7*, 2967–2977. [CrossRef]

26. Yang, J.; Qi, Y.; Zhu, J.; Zhu, Y.-A.; Chen, D.; Holmen, A. Reaction mechanism of CO activation and methane formation on Co Fischer–Tropsch catalyst: A combined DFT, transient, and steady-state kinetic modeling. *J. Catal.* **2013**, *308*, 37–49. [CrossRef]
27. Qi, Y.; Yang, J.; Duan, X.; Zhu, Y.-A.; Chen, D.; Holmen, A. Discrimination of the mechanism of CH_4 formation in Fischer-Tropsch synthesis on Co catalysts: A combined approach of DFT, kinetic isotope effects and kinetic analysis. *Catal. Sci. Technol.* **2014**, *4*, 3534–3543. [CrossRef]
28. Lahtinen, J.; Vaari, J.; Kauraala, K.; Soares, E.A.; Van Hove, M.A. LEED investigations on Co(0001): The (3 × 3)R30°-CO overlayer. *Surf. Sci.* **2000**, *448*, 269–278. [CrossRef]
29. Lahtinen, J.; Vaari, J.; Kauraala, K. Adsorption and structure dependent desorption of CO on Co(0001). *Surf. Sci.* **1998**, *418*, 502–510. [CrossRef]
30. Ojeda, M.; Nabar, R.; Nilekar, A.U.; Ishikawa, A.; Mavrikakis, M.; Iglesia, E. CO activation pathways and the mechanism of Fischer-Tropsch synthesis. *J. Catal.* **2010**, *272*, 287–297. [CrossRef]
31. Ojeda, M.; Li, A.W.; Nabar, R.; Nilekar, A.U.; Mavrikakis, M.; Iglesia, E. Kinetically Relevant Steps and H-2/D-2 Isotope Effects in Fischer-Tropsch Synthesis on Fe and Co Catalysts. *J. Phys. Chem. C* **2010**, *114*, 19761–19770. [CrossRef]
32. Gnanamani, M.K.; Jacobs, G.; Shafer, W.D.; Sparks, D.; Davis, B.H. Fischer–Tropsch Synthesis: Deuterium Kinetic Isotope Study for Hydrogenation of Carbon Oxides Over Cobalt and Iron Catalysts. *Catal. Lett.* **2011**, *141*, 1420. [CrossRef]
33. Kresse, G.; Furthmuller, J. Efficiency of ab-initio total energy calculations for metals and semiconductors using a plane-wave basis set. *Comput. Mater. Sci.* **1996**, *6*, 15–50. [CrossRef]
34. Wellendorff, J.; Lundgaard, K.T.; Mogelhoj, A.; Petzold, V.; Landis, D.D.; Norskov, J.K.; Bligaard, T.; Jacobsen, K.W. Density functionals for surface science: Exchange-correlation model development with Bayesian error estimation. *Phys. Rev. B* **2012**, *85*, 235149. [CrossRef]
35. Kresse, G.; Joubert, D. From ultrasoft pseudopotentials to the projector augmented-wave method. *Phys. Rev. B* **1999**, *59*, 1758–1775. [CrossRef]
36. Van Hardeveld, R.; Hartog, F. The statistics of surface atoms and surface sites on metal crystals. *Surf. Sci.* **1969**, *15*, 189–230. [CrossRef]
37. Tsakoumis, N.; Dehghan-Niri, R.; Rønning, M.; Walmsley, J.; Borg, Ø.; Rytter, E.; Holmen, A. X-ray absorption, X-ray diffraction and electron microscopy study of spent cobalt based catalyst in semi-commercial scale Fischer-Tropsch synthesis. *Appl. Catal. A Gen.* **2014**, *479*, 59–69. [CrossRef]
38. Jonsson, H.; Mills, G.; Jacobsen, K.W. Nudged elastic band method for finding minimum energy paths of transitions. In *Classical and Quantum Dynamics in Condensed Phase Simulations*; World Scientific: Singapore, 2011; pp. 385–404.
39. Henkelman, G.; Jonsson, H. A dimer method for finding saddle points on high dimensional potential surfaces using only first derivatives. *J. Chem. Phys.* **1999**, *111*, 7010–7022. [CrossRef]
40. Gunasooriya, G.T.K.K.; van Bavel, A.P.; Kuipers, H.P.C.E.; Saeys, M. CO adsorption on cobalt: Prediction of stable surface phases. *Surf. Sci.* **2015**, *642*, 6–10. [CrossRef]
41. Bai, Y.; Chen, B.W.J.; Peng, G.; Mavrikakis, M. Density functional theory study of thermodynamic and kinetic isotope effects of H_2/D_2 dissociative adsorption on transition metals. *Catal. Sci. Technol.* **2018**, *8*, 3321–3335. [CrossRef]

© 2019 by the authors. Licensee MDPI, Basel, Switzerland. This article is an open access article distributed under the terms and conditions of the Creative Commons Attribution (CC BY) license (http://creativecommons.org/licenses/by/4.0/).

Article

Cobalt-Based Fischer–Tropsch Synthesis: A Kinetic Evaluation of Metal–Support Interactions Using an Inverse Model System

Anna P. Petersen, Michael Claeys, Patricia J. Kooyman and Eric van Steen *

DST-NRF Centre of excellence in catalysis: c*change, Catalysis Institute, Department of Chemical Engineering, University of Cape Town, Private Bag X3, Rondebosch 7701, South Africa;
anna.petersen@alumni.uct.ac.za (A.P.P.); michael.claeys@uct.ac.za (M.C.); patricia.kooyman@uct.ac.za (P.J.K.)
* Correspondence: eric.vansteen@uct.ac.za

Received: 30 August 2019; Accepted: 21 September 2019; Published: 24 September 2019

Abstract: Metal–support interactions in the cobalt–alumina system are evaluated using an inverse model system generated by impregnating Co_3O_4 with a solution of aluminum sec-butoxide in n-hexane. This results in the formation of nano-sized alumina islands on the surface of cobalt oxide. The activated model systems were kinetically evaluated for their activity and selectivity in the Fischer–Tropsch synthesis under industrially relevant conditions (220 °C, 20 bar). The kinetic measurements were complemented by H_2-chemisorption, CO-TPR, and pyridine TPD. It is shown that the introduction of aluminum in the model system results in the formation of strong acid sites and enhanced CO dissociation, as evidenced in the CO-TPR. The incorporation of aluminum in the model systems led to a strong increase in the activity factor per surface atom of cobalt in the rate expression proposed by Botes et al. (2009). However, the addition of aluminum also resulted in a strong increase in the kinetic inhibition factor. This is accompanied by a strong decrease in the methane selectivity, and an increase in the desired C_{5+} selectivity. The observed activity and selectivity changes are attributed to the increase in the coverage of the surface with carbon with increasing aluminum content, due to the facilitation of CO dissociation in the presence of Lewis acid sites associated with the alumina islands on the catalytically active material.

Keywords: Fischer–Tropsch; cobalt; alumina; strong metal support interactions

1. Introduction

In the Fischer–Tropsch process, a transition metal-containing catalyst is used to produce hydrocarbons from the very basic starting materials hydrogen and carbon monoxide, which can be derived from various carbon-containing resources such as coal, natural gas, biomass, and even waste [1–3]. Metallic cobalt is often regarded as the catalyst of choice, due to its high conversion rate, high selectivity toward linear hydrocarbons, and limited tendency to convert carbon monoxide into carbon dioxide [2,4]. The high cost of metallic cobalt necessitates a sustained, high surface area of the catalytically active metal, keeping in mind the size dependency of the turnover frequency [5,6]. Thus, nano-sized cobalt crystallites of the optimum size [7] need to be distributed on a support to reduce pressure drop. In general, high surface area materials are used as supports, which provide enough space to disperse nano-sized cobalt crystallites and ensure minimal contact between the crystallites. This will reduce the likelihood of sintering via coalescence, and thus increase the thermal stability of the catalyst [8,9].

Support materials are typically considered to be inert, although it is well known that they affect the catalytic active phase in various ways [10–14]. For instance, the reducibility of the precursor(s) of the catalytic active phase [10–12] and the particle morphology [13] may change upon changing the support. Furthermore, the support material may provide novel active sites at the perimeter of the catalytically active phase [14].

Model systems can give specific insights in the role of the support in catalysis. Inverse models [15–19], in which the support (or its precursor) is deposited on the catalytically active material (or its precursor), may be used to distinguish between interface effects on the one hand and changes in morphology and degree of reduction on the other hand. We have shown that the deposition of small amounts of the precursor of a support compound in the form of an alkoxide on Co_3O_4 (as the precursor to metallic cobalt) results in the formation of small islands of support-like structures on the surface of the precursor of the catalytically active material [18,19]. The controlled presence of these nano-sized islands (with a typical size of less than 1 nm) on the catalytically active phase may mimic the metal–support interface and may thus give insights in the role of the metal–support interactions on the activity of the catalytic active phase. Here, we explore the kinetic effect of the modification of Co_3O_4 (as a precursor for metallic cobalt) with aluminum sec-butoxide (as a precursor for alumina) on the activity, stability, and selectivity in the Fischer–Tropsch synthesis, as studied in a slurry reactor.

2. Results

Cobalt oxide (Co_3O_4) with an average crystallite size of 14 nm was prepared by the calcination of a cobalt carbonate precursor at 300 °C [18], which was impregnated with aluminum sec-butoxide in dry n-hexane to achieve a weight loading of 0.1, 0.5 or 2.5 wt.% Al [18]. The formation of a separate, alumina phase could not be observed using X-ray diffraction (XRD), although analysis of transmission electron microscopy (TEM) images showed the presence of small alumina islands (<1 nm) on the surface of Co_3O_4 [18].

Metallic cobalt is the catalytic active phase in the Fischer–Tropsch synthesis [4], and the precursor needs to be reduced prior to the FT synthesis. Temperature programmed reduction profiles of the calcined material showed a significant broadening of the reduction peak upon modification with aluminum, which was accompanied by a shift of the reduction peaks to higher reduction temperatures [18]. This was ascribed to a blockage of defect sites by alumina, thereby hindering hydrogen activation and the nucleation of reduced cobalt phases. The appearance of reduction processes taking place at temperatures in excess of 400 °C were ascribed to the reduction of cobalt ions in proximity to alumina in the catalyst. These materials were reduced prior to the Fischer–Tropsch synthesis isothermally at 350 °C for 10 h. The degree of reduction (DoR) decreased gradually from 96% to 86% upon increasing alumina loading from 0 to 2.5 wt.% (see Table 1). This further implies that the modification of alumina impedes the reduction of Co_3O_4 as observed in the profiles obtained from temperature-programmed reduction (TPR) [18].

Table 1. Physicochemical characterization of used materials.

Sample	Al Loading wt.%	D_{Co3O4} nm	DoR [1,2] %	H_2 Uptake [1] cm^3 (STP)/g	Dispersion [1,3] %	D_{Co} [1,4] nm
Al0	0	14	96	0.32	0.22	427
Al0.1	0.1	14	96	0.88	0.62	155
Al0.5	0.5	14	90	2.07	1.47	65
Al2.5	2.5	14	86	3.94	2.85	33

[1] Reduction: 350 °C in pure H_2 for 10 hrs; [2] Degree of reduction; [3] Dispersion of cobalt in the reduced catalyst taking into account the degree of reduction $D[\%] = \frac{N_{Co,surface}}{N_{Co} \cdot DOR} \times 100\%$); [4] Cobalt particle size estimated from $d_{Co}[nm] = \frac{96}{D[\%]}$.

The catalysts studied in this work were essentially bulk cobalt catalysts, and these materials were expected to sinter extensively during catalyst activation, despite the low reduction temperature (the chosen reduction temperature (350 °C) is above the Hüttig temperature of metallic cobalt (253 °C), implying the significant mobility of cobalt atoms at defect sites [20], which may lead to sintering if cobalt particles are in contact with each other). The metal dispersion after reduction as determined using H_2 chemisorption varied between 0.23% (particle size approximately 412 nm) for Al0 and 3.30% (particle size approximately 29 nm) for Al2.5, showing severe sintering for specifically the sample containing no aluminum (Al0). Interestingly, the introduction of even a small amount of aluminum to Co_3O_4 can drastically reduce the extent of catalyst sintering, whilst still achieving a high degree of reduction. The alumina islands on the surface of cobalt oxide [18] may act here as a spacer preventing direct contact between cobalt particles and thereby reducing the extent of sintering. This also indicates that contact between alumina and the cobalt phase remains intact throughout the reduction of Co_3O_4 to metallic cobalt.

The catalyst activity of the materials was evaluated in a slurry reactor at a pressure of 20 bar and at 220 °C. The initial space velocity was so that the conversion as a function of time could be followed. The CO conversion for all catalysts decreases as a function of time on stream (TOS) during the start-up of the reaction (see Figure 1) due to loss in the activity of the catalyst. Catalyst deactivation has been reported to take place in two distinct stages [21]. The first stage of deactivation is typically ascribed to catalyst sintering. This is often followed by a second stage, in which slower deactivation processes take place, such as catalyst poisoning, catalyst re-oxidation, or build-up of polymeric carbon deposits. The conversion of CO drops rather quickly as a function of time with the samples containing none or a very small amount of aluminum. A steady conversion level was achieved with these samples after less than 50 h on-line. The samples with more aluminum require longer times to stabilize.

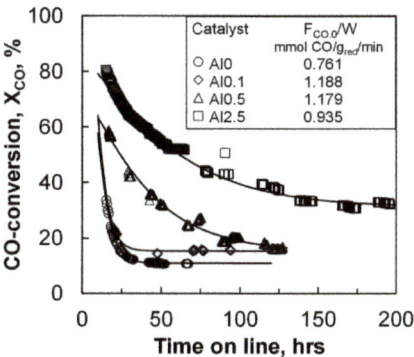

Figure 1. Initial CO conversion in the Fischer–Tropsch synthesis in a slurry reactor at 220 °C, 20 bar and a feed containing H_2:CO:Ar = 2:1:1 as a function of time on line (solid lines represent fit to the model – see text).

The initial activity was evaluated after 16 h on-line (see Table 2), which corresponds to four times the residence time. The initial activity increased with increasing aluminum loading from 0.24 to 0.70 $mmol_{CO}/g_{red.\ cat}$/min, corresponding to a cobalt time yield (CTY) of 2.3×10^{-4} s^{-1} and 6.9×10^{-4} s^{-1} for the samples Al0 and Al2.5, respectively. However, the initial turnover frequency (TOF) was found to decrease with increasing alumina loading from 0.21 s^{-1} to 0.04 s^{-1} for catalyst Al0 and Al2.5, respectively. The obtained turnover frequencies for the sample without aluminum (Al0) resembles the turnover frequency obtained over model catalyst supported on alumina with minimal metal support interactions [6], whereas the obtained turnover frequency for the sample Al2.5 resembles the turnover frequency obtained with a typical impregnated cobalt on alumina catalyst [22]. The decrease in the turnover frequency with increasing aluminum content despite the increase in the rate of reaction per unit mass is a consequence of the strong effect of the increasing aluminum content on the cobalt dispersion (see Table 1). The activity after ca. 95 h on stream as a function of the aluminum content shows the same trend.

The CO-conversion as a function of time on-line was modeled using the rate expression proposed by Botes et al. [23] (see Equation (1)). The loss in activity with time on-line due to deactivation, a, was modeled based on a first-order generalized power-law expression (GPLE) [21,24], taking into account a finite, residual activity (see Equation (2)):

$$\frac{F_{CO,inlet}}{W} \cdot X_{CO} = -r_{CO} = \frac{A_0 \cdot p_{H_2}^{0.75} \cdot p_{CO}^{0.5}}{\left(1 + B \cdot p_{CO}^{0.5}\right)^2} \cdot a(t) \qquad (1)$$

$$\text{with} \quad -\frac{da}{dt} = k_d \cdot (a - a_\infty) \qquad (2)$$

where $F_{CO,inlet}$ is the molar inlet flow of CO, W is the mass of reduced cobalt, X_{CO} is the conversion of CO, A_0 is the rate constant per unit mass of reduced catalyst for the fresh catalyst, B is the kinetic inhibition factor, a is the relative activity, and a_∞ is the final residual activity relative to the initial activity. The partial pressures of hydrogen and carbon monoxide within the reactor are functions of the conversion of CO as well. The conversion of CO as a function of time on-line was fitted to this set of equations, and the obtained kinetic parameters are shown in Table 2.

Table 2. Initial activity and deactivation in the Fischer–Tropsch synthesis at 220 °C and 20 bar conducted in a slurry reactor with a feed H_2:CO:Ar = 2:1:1 (kinetic evaluation of the data according to the rate expression proposed by Botes et al. [23] (see Equation (1)) including a first-order deactivation model according to generalized power-law expression (GPLE) [21,23] (see Equation (2)); uncertainty intervals determined using the bootstrap method [25] with 400 simulations). CTY: cobalt time yield, TOF: turnover frequency.

	Sample	Al0	Al0.1	Al0.5	Al2.5
	Al-loading, wt.%	0.0	0.1	0.5	2.5
TOS = 16 h	$-r_{CO}$, mmol/g/min	0.22	0.40	0.72	0.74
	CTY, 10^{-4} s^{-1}	2.2	3.9	7.1	7.2
	TOF, s^{-1}	0.098	0.063	0.048	0.026
TOS = 95 h	$-r_{CO}$, mmol/g/min	0.08	0.19	0.24	0.40
	CTY, 10^{-4} s^{-1}	0.8	1.9	2.4	3.9
	TOF, s^{-1}	0.035	0.030	0.016	0.014
	A_0, mmol/g Co/min/bar$^{1.25}$	22.1 ± 0.2	12.0 ± 3.0	$(1.9 ± 0.0) \times 10^3$	$(8.06 ± 0.02) \times 10^4$
	$A_{0'}$, mol/mol Co$_{surface}$/s/bar$^{1.25}$	9.9 ± 0.1	1.9 ± 0.5	$(1.3 ± 0.0) \times 10^2$	$(2.8 ± 0.02) \times 10^3$
	B, bar$^{-0.5}$	6.1 ± 0.1	3.0 ± 1.1	68 ± 1	441 ± 2
	k_d, hr^{-1}	0.18 ± 0.01	0.20 ± 0.07	0.02 ± 0.00	0.02 ± 0.00
	a_∞	0.06 ± 0.00	0.08 ± 0.07	0.17 ± 0.01	0.27 ± 0.00
	R^2	0.9984	0.9791	0.9829	0.9902

The mass-specific, kinetic constant A_0 increases quite strongly with increasing aluminum content in the catalyst. Correcting the kinetic constant for the change in metal dispersion A_0' (see Table 2) shows an increase in the initial kinetic constant with increasing aluminum loading, implying that alumina on the surface facilitates the activity of the catalyst beyond increasing the dispersion (sample Al0.1 seems to be the exception here, possibly due to the difficulty in adequately determining the initial activity). The kinetic inhibition factor, B, also increases with increasing aluminum loading, implying that the inhibition of the rate of the Fischer–Tropsch synthesis by carbonaceous species on the surface becomes stronger upon increasing the aluminum loading. These observations are consistent with our DFT (density functional theory) calculations [26], which indicate that alumina near the metallic cobalt surface (e.g., as a ligand) would facilitate CO dissociation by providing an alternative pathway on the dense Co(111) surface.

The rate constant for deactivation is ca. 10 times smaller for the samples containing 0.5 and 2.5 wt.% aluminum than the rate constants for deactivation for the samples containing none or a very small amount of aluminum, indicating that the presence of small alumina islands on cobalt [18] retards the fast initial deactivation process. Furthermore, the final activity, a_∞, increases with increasing aluminum content. Aluminum modification was found to prevent sintering during reduction, as evident from the increased metal dispersion of the activated catalyst determined by H_2 chemisorption (see Table 1). The modification of the cobalt surface with alumina may thus also retard and prevent further sintering during the Fischer–Tropsch synthesis.

The CO hydrogenation over cobalt-based catalysts results mainly in the formation of organic product compounds. Some CO_2 is formed, but typically with a selectivity of less than 1–2% C. The origin of CO_2 in cobalt-based Fischer–Tropsch synthesis has not been established beyond doubt and may be attributed to the water–gas shift reaction (i.e., with the co-generation of hydrogen), further reduction of the catalyst using CO as the reductant, carbon deposition via the Boudouard reaction, or as an alternative pathway to remove surface oxygen.

Figure 2 shows the selectivity for the formation of methane and for the formation of the desired liquid product, C_{5+} as a function of the aluminum content in the catalyst after ca. 250 h on-line. The conversion level was kept constant at ca. 14.5 ± 2% by varying the space velocity. The methane selectivity and C_{5+} selectivity show as expected, an opposite trend. The seesaw behavior for these product groups is expected, due to the normalization of the product selectivity to 100% and the relatively small amount of C_2–C_4 formed in cobalt-based Fischer–Tropsch synthesis [27]. The methane selectivity compared at the same time on-line shows a clear decrease with increasing aluminum content. The observed change in the methane selectivity cannot be attributed to a particle size effect [5,6,28] seeing the large metallic cobalt particles, which are obtained after reduction (see Table 1). Hence, the change in product selectivity must be attributed to the presence of alumina islands on the cobalt surface. The methane selectivity is known to decrease with an increase in the 'surface' inventory of active carbon [29,30]. An increase in the coverage with surface carbon will decrease the availability of surface hydrogen [31,32]; this may (in a limited region [33]) result in a decrease in the methane selectivity. Alumina in close proximity to the catalytically active, metallic cobalt surface may facilitate CO dissociation [26], thus resulting in an increase in the coverage with surface carbon and reduced hydrogen availability on the surface. This could thus explain the observed decrease in the methane selectivity with increasing aluminum content. The seesaw relationship between the C_{5+} selectivity and the methane selectivity implies an increase in the C_{5+} selectivity with increasing carbon coverage on the surface.

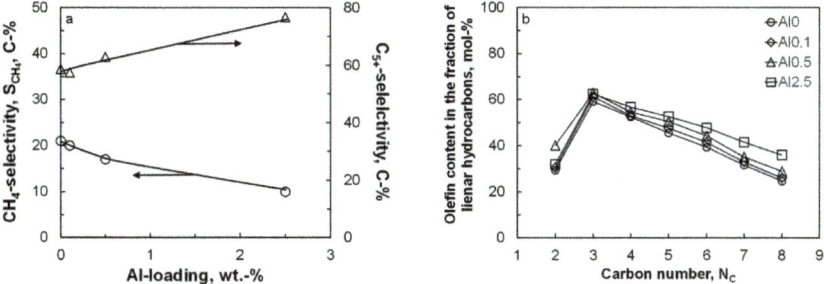

Figure 2. Selectivity for methane and C_{5+} as a function of the aluminum content (**a**) and the olefin content in the fraction of linear hydrocarbons as a function of carbon number (**b**) in the Fischer–Tropsch synthesis (220 °C, 20 bar feed; H_2:CO:Ar = 2:1:1; X_{CO} = 14.5 ± 2%).

Olefins are the major, primary product of the Fischer–Tropsch synthesis, even when cobalt is used as the catalytically active material [27]. The primarily formed olefins can re-adsorb and undergo secondary reactions, such as hydrogenation, double-bond isomerization, and reincorporation in the chain growth scheme [31]. Figure 2 shows the expected behavior for the olefin content in the fraction of linear hydrocarbons as a function of carbon number: a rather low olefin content in C_2 due to the high reactivity of ethene, and a decrease in the olefin content in the fraction of linear hydrocarbons with increasing carbon numbers from C_3 onwards due to the increased concentration of these compounds in the liquid phase [34]. The olefin content in the fraction of linear hydrocarbons increases with increasing aluminum content (except for the olefin content in the fraction of C_2 hydrocarbons). This indicates a reduced extent of secondary hydrogenation of the primarily formed long-chain olefins in the Fischer–Tropsch synthesis possibly due to an increased coverage of the surface with carbonaceous species and thus lower hydrogen availability on the surface.

The changes in the initial kinetic parameters, A_0 and B (see Table 2), and the change in the selectivity due to the modification of the catalytically active material with aluminum are thought to be related to the facilitation of the CO dissociation and the resulting increase in the coverage of the surface with carbon. The ease of CO dissociation may be monitored by CO treatment of these materials, or by reducing Co_3O_4 in CO rather than H_2 [35,36]. CO activation may initially result in the formation of cobalt carbide [36–38]. The reduction of Co_3O_4 using carbon monoxide has been reported to proceed sequentially [36]:

$$Co_3O_4 + CO \rightarrow 3\, CoO + CO_2 \tag{3}$$

$$3\, CoO + 6\, CO \rightarrow 3/2\, Co_2C + 3\, CO_2 \tag{4}$$

Cobalt carbide is thermodynamically not stable at a low chemical potential of carbon (e.g., in the absence of CO), and will thus decompose, yielding metallic cobalt. In hydrogen, cobalt carbide decomposes rapidly at temperatures as low as 150–200 °C [36,38]. The decomposition of cobalt carbide is retarded in inert atmosphere, but it will decompose into metallic cobalt and graphitic carbon at temperatures between 300–350 °C [38]:

$$Co_2C \rightarrow 2\, Co + C \tag{5}$$

Figure 3 shows the profiles obtained from the temperature-programmed reduction in the presence of CO (CO-TPR) for the alumina-modified catalysts. The profiles for the samples with an aluminum content of 0.5 wt.% or less are characterized by a minor peak at ca. 240 °C, a sharp peak at 295 °C, a small feature at 320 °C, and a broad feature above 500 °C. The CO-TPR was followed using in situ XRD (see Figure 3, right). It can be seen that the reduction of Co_3O_4 in the presence of carbon monoxide starts at ca. 210 °C with the formation of CoO, showing that the first feature in the CO-TPR can be ascribed to CO consumption for the reduction of Co_3O_4 to CoO. This is subsequently transformed into Co_2C (at ca. 295 °C). This will be accompanied by a large amount of CO consumed, thus appearing as a large peak in the CO-TPR at these temperatures. It can thus be concluded that the transformation of Co_3O_4 into Co_2C proceeds sequentially, $Co_3O_4 \rightarrow CoO \rightarrow Co_2C$, as discussed before [36]. At temperatures around 350 °C, the transformation of Co_2C to metallic cobalt can be observed in the in situ XRD experiment. This does not appear to be associated with any additional consumption of carbon monoxide. It is thus concluded that at these temperatures, in the presence of carbon monoxide, cobalt carbide transforms into metallic cobalt (with the co-generation of carbon and the simultaneous disappearance of cobalt carbide). The decomposition of Co_2C in the presence of carbon monoxide into metallic cobalt and carbon is observed with this sample at ca. 365 °C, i.e., at slightly higher temperatures than in the presence of argon [38]. It has been argued that cobalt carbide can be observed if the decomposition of the meta-stable Co_2C phase is kinetically hindered [38]. The presence of CO in these experiments would have resulted in an increase in the chemical potential of carbon and thus an increase in the decomposition temperature.

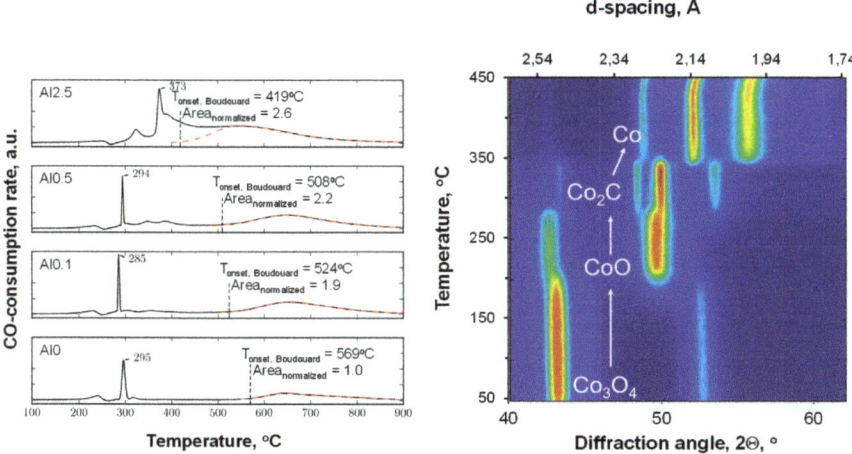

Figure 3. CO-TPR profiles (left) and the corresponding in situ XRD reduction experiments of Al0 using carbon monoxide as the reducing agent (right).

Exposure of metallic cobalt to CO-containing gases at elevated temperatures will result in carbon formation [39] via the Boudouard reaction:

$$2\,CO \rightarrow C + CO_2 \tag{6}$$

This will be seen at temperatures higher than 400 °C, when metallic cobalt is available.

The CO-TPR profile of sample containing 0.1 wt.% aluminum (Al0.1) shows essentially the same features as the sample containing no aluminum (Al0). However, the onset of the high-temperature peak, which is associated with carbon deposition due to the Boudouard reaction, has shifted to a lower temperature. At the same time, the area under this peak is almost twice as large as the area under the high temperature peak in the CO-TPR of Al0. The on-set for the high temperature peak in the CO-TPR for the sample containing 0.5 wt.% aluminum (Al0.5) was shifted to an even lower temperature. The amount of CO consumed for the Boudouard reaction is larger for the sample Al0.5 than for the sample Al0.1. The observed shift in the on-set temperature for the Boudouard reaction implies that carbon deposition becomes more facile upon increasing the aluminum content in the sample. This could be ascribed to a change in the degree of reduction within the catalyst, but it was shown (see Table 1) that the reducibility of samples in hydrogen reduces with increasing aluminum content. Hence, it is proposed [26] that the interface between metallic cobalt and alumina islands on the surface [18] may be the origin of the higher propensity for CO dissociation. Carbon deposition from CO is thermodynamically limited at high temperatures, and thus the variation in the amount of carbon deposition upon changing the aluminum content in the sample, as visualized by the differences in the area associated with the high-temperature feature in the CO-TPR, may be a consequence of the observed shift in the on-set of the Boudouard reaction to lower temperatures with increasing aluminum content.

The CO-TPR profile of the sample containing 0.5 wt.% aluminum contains additional features with maxima at ca. 345 °C and 380 °C, which are not observed in the CO-TPR profiles of the samples containing less aluminum. This may be considered CO consumption due to the retarded reduction of cobalt, e.g., of cobalt covered by the alumina islands. The reduction of this cobalt will yield directly metallic cobalt at 380 °C, as Co_2C does not seem to be stable at this temperature, as evidenced by the in situ XRD study (see Figure 3).

The CO-TPR profile of the sample with the highest weight loading, 2.5 wt.% aluminum (Al2.5), differed strongly from the CO-TPR profiles of the samples containing less aluminum. The sharp peak that appeared at ca. 290 °C for the samples with lower aluminum contents was shifted toward 373 °C. This may be attributed to the presence of alumina islands covering the cobalt oxide surface and hindering the overall reduction process. The high-temperature feature, which was separated by almost 300 °C for the other samples, overlaps with the reduction peaks around 400 °C.

Carbon deposition by the Boudouard reaction could be visualized using TEM imaging of the sample Al0.5 after CO-TPR (see Figure 4). The image shows metal particle sizes within the expected range if only minimal sintering has taken place. It appears that a reduction in CO inhibits the sintering process observed for H_2 reduction (see Table 1). The metal particles are encapsulated by material appearing such as multi-wall carbon nanotubes. The d-spacing of this material was determined to be 3.4 Å, which is consistent with graphitic carbon [40,41]. The internal diameter of the carbon nanotubes appears to be in the range of 9–11 nm. The internal diameter is thought to be linked to the metal particle size [42] (the reduction of Co_3O_4 particles with a size of 14 nm is expected to yield in the absence of sintering metal particles with a size of 10–11 nm).

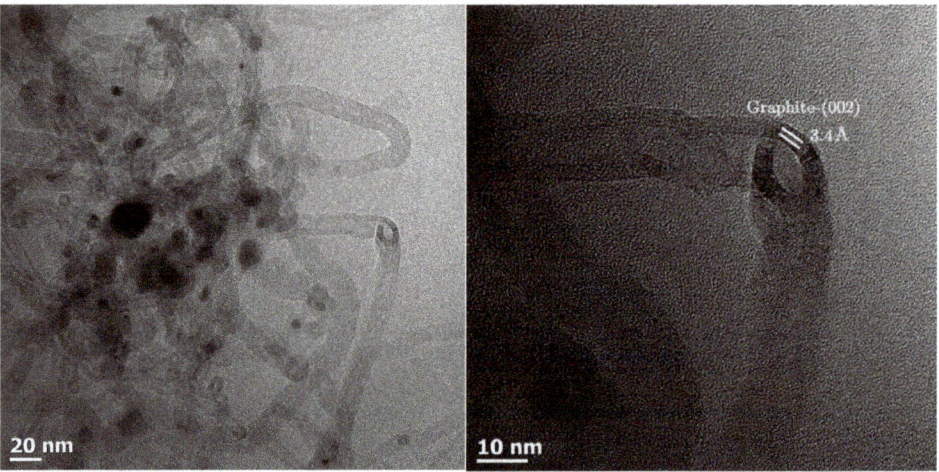

Figure 4. TEM image of the sample containing 0.5 wt.% aluminum (Al0.5) after CO-TPR showing metal particles encapsulated in multi-wall carbon nanotubes.

The enhanced activity for CO dissociation as indicated by the shift in the on-set of the Boudouard reaction to lower temperatures upon increasing the aluminum content in the samples and as argued to be the origin of the change in the kinetic coefficients and the change in the selectivity has been linked to the Lewis acid property of the promoter [25,42,43]. Carbon monoxide may act as a weak Lewis base, which can donate electrons from the 4σ molecular orbital to a Lewis acid [15]. The 4σ molecular orbital is a bonding orbital, and electron withdrawal results in weakening of the C–O bond.

Alumina is known to exhibit Lewis acidity, although the presence of Brønsted acid sites cannot be ruled out [44]. Hence, alumina modification may have introduced acidity in these samples. The acidity of the activated catalysts was investigated by pyridine uptake and by the temperature-programmed desorption of pyridine (py-TPD; see Figure 5). The pyridine uptake of the activated sample containing no aluminum (Al0) during pulse chemisorption was 30 μmol/g. The pyridine uptake on the sample containing aluminum increases roughly linearly with increasing alumina loading over the activated catalyst, and the pyridine uptake relative to the aluminum loading was determined to be 0.2 mol pyridine/mol Al.

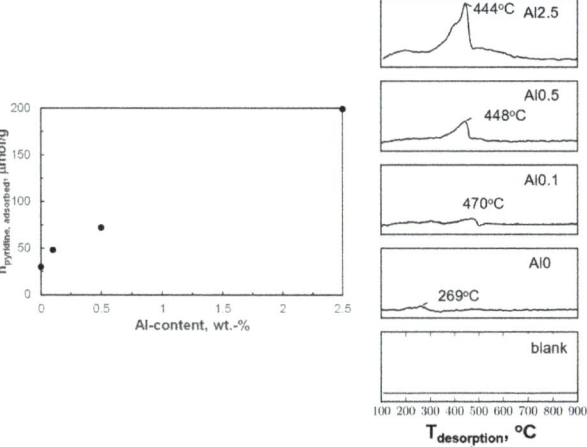

Figure 5. Uptake (left) and temperature programmed desorption (right) of pyridine from the reduced samples (reduction in pure H_2, 300 ml$_n$/g/min; $T_{reduction}$ = 350 °C; $t_{reduction}$ = 0.5 h).

Pyridine desorbed at a relatively low temperature of 269 °C from the sample containing no aluminum (Al0), indicating rather weakly adsorbed pyridine. A sharp peak between 350–500 °C was observed during pyridine TPD for the alumina-modified samples. Sharp peaks above 350 °C have been reported to correspond to the decomposition of pyridine to N_2 and CO_2 [45,46]. The catalyst had also darkened after the experiment, which may suggest carbon deposition [45] indicating that pyridine decomposition may have occurred during the pyridine TPD. Hence, the position of the peak is expected to correlate with the rate of pyridine decomposition rather than the rate of pyridine desorption (and thus the peak area with the thermal conductivity of the decomposition products). The decomposition of pyridine at temperatures below 500 °C, in an inert atmosphere, is only catalyzed by strong acid sites [46,47]. The pyridine TPD experiments indicate that the acid sites on the activated modified samples were strong enough to catalyze the decomposition of pyridine at temperatures below 500 °C.

It can be concluded that pyridine adsorption is associated with the presence of strong acid sites upon modification of the samples with alumina, and that the amount of pyridine adsorbed increases roughly linearly with the aluminum content. Thus, it seems plausible that these strong acid sites are associated with the presence of alumina islands on the surface of the metallic cobalt [18]. The relative constant amount of pyridine adsorbed per aluminum may indicate that the strong acid site is associated with a cluster of ca. 5 Al atoms. Alternatively, these acid sites are located at the interface between the alumina islands and metallic cobalt, and an increase in the aluminum content is associated with an increase in the number of alumina islands, but not necessarily their size.

3. Discussion

Figure 6 shows schematically the formation of a reverse model system to investigate metal–support interactions by the impregnation of Co_3O_4 with a well-defined crystallite size with a solution of aluminum sec-butoxide in n-hexane [18] (see Figure 6). Upon drying and calcination, sub-nanometer-sized alumina islands on cobalt oxide are formed [18]. Lewis acid sites are being introduced in the catalyst, as evidenced by pyridine uptake and pyridine TPD, due to the presence of aluminum possibly at the interface between metallic cobalt and the alumina islands, which will affect the performance of cobalt in the Fischer–Tropsch synthesis

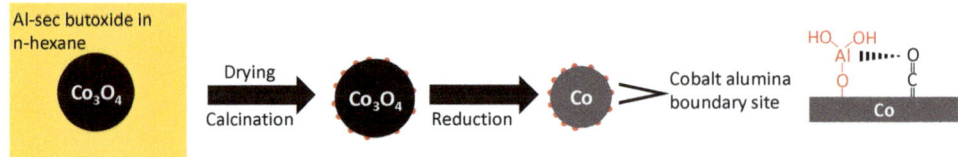

Figure 6. Schematic depiction of the formation of sub-nanometer-sized alumina islands upon impregnating a cobalt oxide surface with Al-sec butoxide in n-hexane, and their interaction with CO adsorbed on the reduced metal surface.

It may be argued that metal support interactions are exaggerated in this model system with small alumina islands on top of a larger, nano-sized cobalt oxide crystallite. However, these sub-nanometer-sized islands may be present in real catalysts as well. During impregnation, the support will dissolve slightly upon contact with a metal–salt-containing aqueous solution. The dissolved species emanating from the support will reprecipitate during the drying and calcination step. A random reprecipitation process will result in some support material being deposited on/near the metal precursor.

The reduction of cobalt in these model systems is impeded by the presence of alumina islands on top of cobalt oxide (although cobalt aluminate could not be detected using XANES, X-ray absorption near edge structure [18]). This was caused by a change in the pre-exponential factor, implying the inhibition of hydrogen activation on these model systems or the inhibition of the nucleation process.

The reduction of these model systems in hydrogen results in sintering. These alumina islands may act as spacers, and seem to reduce the extent of sintering via coalescence, as evidenced by the increase in the H_2 uptake upon increasing the aluminum content. Support materials are thought to aid the thermal stability of catalysts by dispersing the catalytically active phase over a large surface area, since the limited particle mobility on support materials under typical reaction conditions will reduce sintering due to coalescence. Here, we highlight an additional route to stabilize nano-sized particles against sintering by decorating the surface of the catalytically active phase with support-like structures. This would impede the coalescence of particles even in catalysts with a high metal loading.

The presence of alumina islands in our model system may affect the resulting catalytic activity in the Fischer–Tropsch synthesis in two ways, viz. by providing an alternative pathway for CO dissociation and by changing the strength of CO adsorption [26]. The presence of an alternative pathway for CO dissociation will be experimentally observed by an overall, more facile CO dissociation in the presence of alumina. This was evidenced by an increase in the kinetic activity factor, A_0', in the rate expression proposed by Botes et al. [23] and by the lowering of the on-set temperature for the Boudouard reaction in the CO-TPR. The increase in the strength of adsorption due to the presence of alumina was evidenced by an up to 70-fold increase in the kinetic inhibition, B, in the rate expression proposed by Botes et al. [23] upon increasing the aluminum content. This was further substantiated by the change in the product selectivity. The product selectivity in the Fischer–Tropsch synthesis is strongly controlled by the hydrogen availability on the surface [31]. An increase in the strength of CO adsorption will reduce the hydrogen availability and thus (within a limited region [33]), reduce methane selectivity, increase C_{5+} selectivity, and reduce the extent of secondary reactions.

4. Materials and Methods

Cobalt oxide, Co_3O_4, with an average crystallite size of 14 nm, was prepared by the calcination of a cobalt carbonate precursor at 300 °C. The detailed synthesis procedure has been elsewhere [18]. The resulting Co_3O_4 crystallites were impregnated with aluminum sec-butoxide in dry n-hexane to achieve a weight loading of aluminum of 0.1, 0.5, or 2.5 wt.%, respectively. The obtained materials were calcined at 300 °C for 4 h. This results in the formation of highly dispersed, island-like alumina structures on cobalt oxide [18]. The catalysts are labeled according to their loading with aluminum as Al0, Al0.1, Al0.5, and Al2.5.

Hydrogen chemisorption was performed using an ASAP 2020 (Micromeritics, Norcross, GA, USA). The catalyst (100 mg) was dried for 12 h at 120 °C and reduced in hydrogen by increasing the temperature at a heating rate of 1 °C/min to 350 °C and holding for 10 h. After reduction, the chamber was evacuated for 30 min both at 330 °C and 140 °C. The hydrogen uptake was measured at a temperature of 120 °C.

The degree of reduction (DoR) was determined in the temperature-programmed reduction (TPR) cell (Autochem 2950; Micromeritics, Norcross, GA, USA) by activation the catalyst (50 mg) in a flow of 10 ml_n/min pure H_2 at 350 °C by (heating rate: 1 °C/min) for 10 h. Afterwards, the temperature was decreased to 60 °C in flowing hydrogen, and the catalyst bed was purged for 60 min in a flow of 50 ml_n/min of 5 vol% H_2 in Ar. The catalyst bed temperature was increased to 900 °C at 10 °C/min whilst monitoring the hydrogen consumption with a thermal conductivity detector (TCD). The degree of reduction was determined from the ratio of the hydrogen consumption in the TPR after reduction at 350 °C for 10 h (the hydrogen consumption was calibrated using the reduction of Ag_2O) with the amount of Co_3O_4 loaded in the TPR cell.

Temperature programmed reduction studies using CO (CO-TPR) were performed in an Autochem 2950 (Micromeritics, Norcross, GA, USA). The temperature of the catalyst bed (50 mg) was increased to 120 °C at a heating rate of 10 °C/min in a flow of 50 ml_n/min Ar, and the catalyst was dried for 1 h. The catalyst temperature was decreased to 60 °C, and the gas was switched to CO (10 ml_n/min). Subsequently, the temperature was increased to 900 °C (heating rate 10 °C/min), and the TCD signal was recorded.

X-ray diffraction (XRD) was performed using a Bruker AXS D8 advance diffractometer (Karlsruhe, Germany) with Co-K_α radiation (1.78897 Å) operating at 40 kV and 30 mA between 20–120° 2θ with a scanning speed of 0.0562° 2θ/s. The phase transitions occurring during the temperature-programmed reduction experiment were explored using in situ reduction with CO (4 ml_n/min) as the reducing agent. The catalyst was loaded into a borosilicate capillary. The capillary was fitted into an in situ cell, which was developed in house [48,49]. The capillary was heated with infrared heaters. The temperature was increased at a heating rate of 1 °C/min from 50 to 450 °C, and X-ray diffractograms were taken every 5 min.

Transmission electron microscopy (TEM) was performed using a FEI Tecnai T20 (Thermo Fisher Scientific, Waltham, MA, USA) equipped with a LaB_6 filament operated at 200 kV. Catalysts were dispersed in methanol, ultrasonicated for 2 min, and transferred to a carbon coated copper grid.

Pyridine temperature-programmed desorption (Py-TPD) was performed using an Autochem 2920 (Micromeritics, Norcross, GA, USA) equipped with a vapor. The catalyst (100 mg) was activated in a stream of 30 ml_n/min of pure H_2 by increasing the temperature to 350 °C at a heating rate of 10 °C/min and reducing the catalyst for 30 min. The temperature was decreased to 120 °C at a heating rate of 10 °C/min under a flow of 10 ml_n/min He and kept there for 1 h. The sample temperature was decreased to 110 °C, where pyridine pulse adsorption was performed. Helium saturated with pyridine at 60 °C was pulsed 15 times over the activated catalyst in 5-min intervals. Each dose contained 27 μmol pyridine. The pyridine uptake was monitored using a TCD. Weakly adsorbed pyridine was removed by flowing 50 ml_n/min of pure helium over the catalyst for 1 h. Py-TPD was carried out by increasing the temperature to 900 °C at a heating rate of 10 °C/min in a stream of 50 ml_n/min He, whilst monitoring the tail gas using a TCD detector.

The Fischer–Tropsch synthesis was performed in a 1-L slurry reactor. The catalyst (3–5 g) was activated ex situ in a flow of hydrogen with a space velocity of 12 $l_n/h/g_{cat}$. The temperature was increased to 350 °C at a constant heating rate of 1 °C/min, and subsequently, the catalyst was reduced for 10 h. The activated catalyst was cooled to room temperature, and then transferred to 30 g of molten wax under an argon atmosphere. The wax was cooled to room temperature, and the mass of the reduced catalyst ($m_{cat,red}$) was determined by differential weighing of the wax tablet before and after addition of the catalyst. The reactor was filled with 250 g of molten wax, and the catalyst wax tablet was loaded into the reactor. The reactor was pressurized with argon to 20 bar, and the temperature was increased to 220 °C at a stirring speed of 350 rpm. When the desired temperature and pressure were reached, diluted syngas (H_2:CO:Ar = 2:1:1) was introduced. The permanent gases were analyzed on-line with a Varian CP-4900 micro GC equipped with a TCD detector to determine the CO conversion and methane selectivity using argon as an internal standard. Gaseous samples were collected in pre-evacuated ampoules [50] at regular intervals, and the samples were analyzed off-line by GC-FID.

5. Conclusions

Metal–support interactions can be investigated using a reverse model system created by impregnating Co_3O_4 with aluminum sec-butoxide. This results in the formation of alumina islands on the surface of cobalt oxide. The presence of these islands strongly affects the performance of these materials in the Fischer–Tropsch CO hydrogenation. The overall activity per unit mass of catalyst increases with increasing aluminum content, but this is due to the increase in the metal surface area as a consequence of reduced sintering in the presence of alumina islands. A kinetic analysis shows that the kinetics activity factor per the surface cobalt atoms and the kinetic inhibition factor increases with increasing aluminum content. The increase in the kinetic activity factor is ascribed to the enhanced ability to dissociate CO in the presence of the Lewis acid alumina on the surface of the catalytically active metal. The increase in the inhibition factor is ascribed to the increase in the strength of CO adsorption in the presence of alumina. The latter does affect the selectivity of the Fischer–Tropsch synthesis favorably with a decrease in the methane selectivity and an increase in the C_{5+} selectivity with increasing aluminum content.

Author Contributions: Conceptualization of research by E.v.S.; investigation—A.P.P.; formal analysis, E.v.S., A.P.P.; TEM analysis, P.J.K.; writing—original draft preparation, E.v.S., A.P.P.; writing—review and editing, E.v.S, P.J.K; visualization, E.v.S, A.P.P. and P.J.K.; supervision, E.v.S., M.C.

Funding: Financial support from the DST-NRF Centre of Excellence in Catalysis c*change is gratefully acknowledged. This work is based on the research supported in part by the National Research Foundation of South Africa (Grant numbers: 114606, EVS and 94878 PJK).

Acknowledgments: In this section, you can acknowledge any support given which is not covered by the author contribution or funding sections. This may include administrative and technical support, or donations in kind (e.g., materials used for experiments).

Conflicts of Interest: The authors declare no conflict of interest. The funders had no role in the design of the study; in the collection, analyses, or interpretation of data; in the writing of the manuscript, or in the decision to publish the results.

References

1. Dry, M.E. Present and future applications of the Fischer-Tropsch process. *Appl. Catal. A Gen.* **2004**, *276*, 1–3. [CrossRef]
2. Van Steen, E.; Claeys, M. Fischer-Tropsch catalysts for the Biomass-to-Liquid (BTL)-process. *Chem. Eng. Technol.* **2008**, *31*, 655–666. [CrossRef]
3. Tucker, C.L.; van Steen, E. Activity and selectivity of a cobalt-based Fischer-Tropsch catalysts operating at high conversion for once-through biomass-to-liquid operations. *Catal. Today* **2018**. [CrossRef]
4. Iglesia, E. Design, synthesis, and use of cobalt-based Fischer-Tropsch synthesis catalysts. *Appl. Catal. A Gen.* **1997**, *161*, 59–78. [CrossRef]
5. Bezemer, G.L.; Bitter, J.H.; Kuipers, H.P.C.E.; Oosterbeek, H.; Holewijn, J.E.; Xu, X.; Kapteijn, F.; van Dillen, A.J.; de Jong, K.P. Cobalt particle size effects in the Fischer–Tropsch reaction studied with carbon nanofiber supported catalysts. *J. Am. Chem. Soc.* **2006**, *128*, 3956–3964. [CrossRef] [PubMed]
6. Fischer, N.; van Steen, E.; Claeys, M. Structure sensitivity of the Fischer-Tropsch activity and selectivity on alumina supported catalysts. *J. Catal.* **2013**, *299*, 67–80. [CrossRef]

7. Munnik, P.; de Jongh, P.E.; de Jong, K.P. Control and impact of the nanoscale distribution of supported cobalt particles using in the Fischer-Tropsch synthesis. *J. Am. Chem. Soc.* **2014**, *136*, 7333–7340. [CrossRef]
8. Andrew, S.P.S. Theory and practice of the formulation of heterogeneous catalysts. *Chem. Eng. Sci.* **1988**, *36*, 1431–1445. [CrossRef]
9. Wanke, S.E.; Flynn, P.C. The sintering of supported metal catalysts. *Catal. Rev.* **1976**, *12*, 93–135. [CrossRef]
10. Jacobs, G.; Da, T.K.; Zhang, Y.; Li, J.; Racoillet, G.; Davis, B.H. Fischer-Tropsch synthesis: Support, loading and promoter effects on the reducibility of cobalt catalysts. *Appl. Catal. A Gen.* **2002**, *233*, 263–281. [CrossRef]
11. Rane, S.; Borg, Ø.; Yang, J.; Rytter, E.; Holmen, A. Effect of alumina phases on hydrocarbon selectivity in Fischer-Tropsch synthesis. *Appl. Catal. A Gen.* **2010**, *338*, 160–167. [CrossRef]
12. Saib, A.M.; Claeys, M.; van Steen, E. Silica supported cobalt Fischer-Tropsch synthesis: Influence of support pore diameter. *Catal. Today* **2002**, *71*, 395–402. [CrossRef]
13. Vaarkamp, M.; Miller, J.T.; Modica, F.S.; Koningsberger, D.C. On the relation between particle morphology, structure of the metal-support interface and catalytic properties of Pt/γ-Al$_2$O$_3$. *J. Catal.* **1996**, *163*, 294–305. [CrossRef]
14. Briggs, N.M.; Barreti, L.; Wegener, E.C.; Herrera, L.V.; Gomez, L.A.; Miller, J.T.; Crossley, S.P. Identification of active sites on supported metal catalysts with carbon nanotube hydrogen highways. *Nature Commun.* **2018**, *9*, 3827. [CrossRef] [PubMed]
15. Hayek, K.; Fuchs, M.; Klötzer, B.; Reichl, W.; Rupprechter, G. Studies of metal–support interactions with "real" and "inverted" model systems: Reactions of CO and small hydrocarbons with hydrogen on noble metals in contact with oxides. *Top. Catal.* **2000**, *13*, 55–66. [CrossRef]
16. Mogorosi, R.P.; Fischer, N.; Claeys, M.; van Steen, E. Strong metal support interaction by molecular design: Fe-silicate interactions in Fischer-Tropsch catalysts. *J. Catal.* **2012**, *289*, 140–150. [CrossRef]
17. Mogorosi, R.P.; Claeys, M.; van Steen, E. Enhanced activity via surface modification of Fe-based Fischer-Tropsch catalyst precursor with titanium butoxide. *Top. Catal.* **2014**, *57*, 572–581. [CrossRef]
18. Petersen, A.P.; Forbes, R.P.; Govender, S.; Kooyman, P.J.; van Steen, E. Effect of alumina modification on the reducibility of Co$_3$O$_4$ crystallites studied on inverse model catalysts. *Catal. Lett.* **2018**, *148*, 1215–1227. [CrossRef]
19. Macheli, L.; Roy, A.; Carleschi, E.; Doyle, B.P.; van Steen, E. Surface modification of Co$_3$O$_4$ nanocubes with TEOS for an improved performance in the Fischer-Tropsch synthesis. *Catal. Today* **2018**. [CrossRef]
20. Moulijn, J.A.; van Diepen, A.E.; Kapteijn, F. Catalyst deactivation: Is it predictable? What to do? *Appl. Catal. A Gen.* **2001**, *212*, 3–16. [CrossRef]
21. Argyle, M.D.; Frost, T.S.; Bartholomew, C.H. Cobalt Fischer–Tropsch catalyst deactivation modeled using generalized power law expressions. *Top. Catal.* **2014**, *57*, 415–429. [CrossRef]
22. Nabaho, D.; Niemantsverdriet, J.W.; Claeys, M.; van Steen, E. Hydrogen spillover in the Fischer-Tropsch synthesis: An analysis of platinum as a promoter for cobalt-alumina catalysts. *Catal. Today* **2016**, *261*, 17–27. [CrossRef]
23. Botes, F.G.; van Dyk, B.; McGregor, C. The development of a macro kinetic model for a commercial Co/Pt/Al$_2$O$_3$ Fischer-Tropsch catalyst. *Ind. Eng. Chem. Res.* **2009**, *48*, 10439–10447. [CrossRef]
24. Fuentes, G.A. Catalyst deactivation and steady-state activity: A generalized power-law equation model. *Appl. Catal.* **1985**, *15*, 33–40. [CrossRef]
25. Hu, W.; Xie, J.; Chau, H.W.; Si, B.C. Evaluation of parameter uncertainties in nonlinear regression using Microsoft Excel spreadsheet. *Environ. Syst. Res.* **2015**, *4*, 4. [CrossRef]
26. Van Heerden, T.; van Steen, E. Metal-support interaction on cobalt-based FT catalysts - a DFT study of model inverse catalysts. *Faraday Discuss.* **2017**, *197*, 87–99. [CrossRef] [PubMed]
27. Van Steen, E.; Claeys, M.; Möller, K.P.; Nabaho, D. Comparing a cobalt-based catalyst with iron-based catalysts for the Fischer-Tropsch XTL-process operating at high conversion. *Appl. Catal A Gen.* **2018**, *549*, 51–59. [CrossRef]
28. Borg, Ø.; Dietzel, P.D.C.; Spjelkavik, A.I.; Tveten, E.Z.; Walmsley, J.C.; Diplas, S.; Holmen, A.; Rytter, E. Fischer-Tropsch synthesis: Cobalt particle size and support effects on intrinsic activity and product distribution. *J. Catal.* **2008**, *259*, 161–164. [CrossRef]
29. Bertole, C.J.; Mims, C.A.; Kiss, G. The effect of water on the cobalt-catalyzed Fischer–Tropsch synthesis. *J. Catal.* **2002**, *210*, 84–96. [CrossRef]

30. Bertole, C.J.; Kiss, G.; Mims, C.A. The effect of surface-active carbon on hydrocarbon selectivity in the cobalt-catalyzed Fischer–Tropsch synthesis. *J. Catal.* **2004**, *223*, 309–318. [CrossRef]
31. Schulz, H.; van Steen, E.; Claeys, M. Selectivity and mechanism of Fischer-Tropsch synthesis with iron and cobalt catalysts. *Stud. Surf. Sci. Catal.* **1994**, *81*, 455–460.
32. Weststrate, C.J.; Niemantsverdriet, J.W. Understanding FTS selectivity: The crucial role of surface hydrogen. *Faraday Discuss.* **2017**, *197*, 101–116. [CrossRef] [PubMed]
33. Van Santen, R.A.; Ciobîcă, I.M.; van Steen, E.; Ghouri, M.M. Mechanistic issues in Fischer-Tropsch catalysis. *Adv. Catal.* **2011**, *54*, 127–187.
34. Schulz, H.; Claeys, M. Reactions of α-olefins of different chain length added during Fischer–Tropsch synthesis on a cobalt catalyst in a slurry reactor. *Appl. Catal. A Gen.* **1999**, *186*, 71–90. [CrossRef]
35. Yang, J.; Jacobs, G.; Jermwongratanachai, T.; Pendyala, V.R.R.; Ma, W.; Chen, D.; Holmen, A.; Davis, B.H. Fischer–Tropsch synthesis: Impact of H_2 or CO activation on methane selectivity. *Catal. Lett.* **2014**, *144*, 123–132. [CrossRef]
36. Paterson, J.; Peacock, M.; Ferguson, E.; Purven, R.; Ojeda, M. In situ diffraction of Fischer-Tropsch catalysts: Cobalt reduction and carbide formation. *ChemCatChem* **2017**, *9*, 3463–3469. [CrossRef]
37. Nakamura, J.; Toyoshima, I.; Tanaka, K. Formation of carbidic and graphite carbon from CO on polycrystalline cobalt. *Surf. Sci.* **1988**, *201*, 185–194. [CrossRef]
38. Claeys, M.; Dry, M.E.; van Steen, E.; du Plessis, E.; van Berge, P.J.; Saib, A.M.; Moodley, D.J. In-situ magnetometer study on the formation and stability of cobalt carbide in Fischer-Tropsch synthesis. *J. Catal.* **2014**, *318*, 193–202. [CrossRef]
39. Bremmer, M.; Zacharaki, E.; Sjåstad, A.O.; Navarro, V.; Frenken, J.W.M.; Kooyman, P.J. In situ TEM observation of the Boudouard reaction: Multi-layered graphene formation from CO on cobalt nanoparticles at atmospheric pressure. *Faraday Discuss.* **2017**, *197*, 337–351. [CrossRef]
40. Kharissova, O.V.; Kharisov, B.I. Variations of interlayer spacing in carbon nanotubes. *RSC Adv.* **2014**, *4*, 30807–30815. [CrossRef]
41. Takenaka, S.; Ishida, M.; Serizawa, M.; Tanabe, E.; Otsuka, K. Formation of carbon nanofibers and carbon nanotubes through methane decomposition over supported cobalt catalysts. *J. Phys. Chem. B* **2004**, *108*, 11464–11472. [CrossRef]
42. Boffa, A.B.; Lin, C.; Bell, A.T.; Somorjai, G.A. Lewis acidity as an explanation for oxide promotion of metals: Implications of its importance and limits for catalytic reactions. *Catal. Lett.* **1994**, *27*, 243–249. [CrossRef]
43. Johnson, G.R.; Bell, A.T. Effects of Lewis acidity of metal oxide promoters on the activity and selectivity of Co-based Fischer–Tropsch synthesis catalysts. *J. Catal.* **2016**, *338*, 250–264. [CrossRef]
44. Rascon, F.; Wischert, R.; Coperet, C. Molecular nature of support effects in single-site heterogeneous catalysts: Silica vs. alumina. *Chem. Sci.* **2011**, *2*, 1449–1456. [CrossRef]
45. Stevens, R.W., Jr.; Chuang, S.S.C.; Davis, B.H. Temperature-programmed desorption/decomposition with simultaneous DRIFTS analysis: Adsorbed pyridine on sulfated ZrO_2 and Pt-promoted sulfated ZrO_2. *Thermochim. Acta* **2003**, *407*, 61–71. [CrossRef]
46. Pribylova, L.; Dvorak, B. Gas chromatography/mass spectrometry analysis of components of pyridine temperature-programmed desorption spectra from surface of copper-supported catalysts. *J. Chromatogr. A* **2009**, *1216*, 4046–4050. [CrossRef]
47. Barzetti, T.; Selli, E.; Moscotti, D.; Forni, L. Pyridine and ammonia as probes for FTIR analysis of solid acid catalysts. *J. Chem. Soc. Faraday Trans.* **1996**, *92*, 1401–1407. [CrossRef]
48. Claeys, M.; Fischer, N. PCT Patent WO 2013/005180 A1. Sample presentation device for radiation-based analytical equipment. 10 January 2013.
49. Fischer, N.; Clapham, B.; Feltes, T.; van Steen, E.; Claeys, M. Size-dependent phase transformation of catalytically active nanoparticles captured in situ. *Angew. Chem. Int. Ed.* **2014**, *53*, 1342–1345. [CrossRef]
50. Schulz, H.; Böhringer, W.; Kohl, C.P.; Rahman, N.M.; Will, A. *Entwicklung und Anwendung der Kapillar-GC Gesamtproben-Technik für Gas/Dampf Vielstoffgemische*; DGMK Forschungsbericht: Hamburg, Germany, 1984; Volume 320.

 © 2019 by the authors. Licensee MDPI, Basel, Switzerland. This article is an open access article distributed under the terms and conditions of the Creative Commons Attribution (CC BY) license (http://creativecommons.org/licenses/by/4.0/).

Article

The Effect of Potassium on Cobalt-Based Fischer–Tropsch Catalysts with Different Cobalt Particle Sizes

Ljubiša Gavrilović, Jonas Save and Edd A. Blekkan *

Department of Chemical Engineering, Norwegian University of Science and Technology, Sem Sælands vei 4, 7491 Trondheim, Norway; ljubisa.gavrilovic@ntnu.no (L.G.); jonassave@gmail.com (J.S.)
* Correspondence: edd.a.blekkan@ntnu.no; Tel.: +47-7359-4157

Received: 15 February 2019; Accepted: 8 April 2019; Published: 10 April 2019

Abstract: The effect of K on 20%Co/0.5%Re/γ-Al_2O_3 Fischer–Tropsch catalysts with two different cobalt particle sizes (small, in the range 6–7 nm and medium size, in the range 12–13 nm) was investigated. The catalyst with the smaller cobalt particle size had a lower catalytic activity and C_{5+} selectivity while selectivities towards CH_4 and CO_2 were slightly higher than over the catalyst with larger particles. These effects are ascribed to lower hydrogen concentration on the surface as well as the lower reducibility of smaller cobalt particles. Upon potassium addition all samples showed decreased catalytic activity, reported as Site Time Yield (STY), increased C_{5+} and CO_2 selectivities, and a decrease in CH_4 selectivity. There was no difference in the effect of potassium between the sample with small cobalt particles compared to the sample with medium size particles). In both cases the specific activity (STY) fell and the C_{5+} selectivity increased in a similar fashion.

Keywords: Fischer–Tropsch; cobalt; catalyst deactivation; potassium

1. Introduction

Biomass to liquid (BTL) via gasification and integrated Fischer–Tropsch synthesis (FTS) is an attractive process for the production of green diesel and jet fuel [1]. The first step involves biomass gasification to produce syngas (CO + H_2) [2]. It was reported [3] that impurities in biomass feedstocks and partial gasification can lead to contaminants (such as: tars, alkali, HCN, H_2S, etc.) in the produced syngas, which might cause catalyst deactivation and other problems downstream. Therefore, a series of cleaning steps must be applied to fulfil the requirement for pure gas before entering the FTS reactor [4]. Alkali salts (mainly potassium) are the dominant salts in fly ash composition after biomass gasification [5]. Due to poor plant design or imperfections in the cleaning section, these alkali salts might be present in the gas when entering the FTS section [6]. It was previously reported that alkali has a strong negative effect on FTS activity on Co-based catalyst [7–12]. Poisoned samples were characterized carefully, but none of the standard techniques applied showed significant differences between the reference (unpoisoned) and poisoned samples. Hence, the catalyst loss in activity and changes in selectivity that was observed were explained by the hypothesis that potassium is able to move to the cobalt sites responsible for FTS when reactions or the pre-treatment conditions are reached [11–13]. The mobility of potassium species under reaction conditions has been demonstrated in other systems [14], where the transport mechanism was suggested to involve OH-groups on the surface. Once the potassium reaches the cobalt particles there are very low barriers against transport on the cobalt surface, as calculated using DFT, and the adsorption of K is favourable on all sites, including sites such as the B_5 and B_6 sites often considered active sites for the FTS [15].

The effect of Co particle size on cobalt-based FT catalysts have been studied extensively [16–19]. A decreased turnover frequency (TOF) and increased methane selectivity has been found for FTS experiments (performed at 1 and 35 bar ($H_2/CO = 2$)) for catalysts with Co particles smaller than around 6 nm [18]. With decreasing Co particle size the fraction of sites that are less coordinated (like steps, kinks, edges, and corners [20]) increases.

To better understand the nature of the effects reported on the cobalt-based FTS catalysts upon alkali addition, here we report an investigation of the effect of potassium on alumina-supported cobalt catalysts containing small or medium sized cobalt particles. A difference in the behavior could give information on the nature of K poisoning, and on the role of various sites on the cobalt surface.

2. Results

Two alumina-supported catalysts, with different cobalt particle sizes, were prepared using the incipient wetness impregnation technique. During the impregnation process, the pores of the support are filled with the active metal solution by means of capillary forces. Thus, the particle size of the active metal is determined during the drying procedure [21]. To be able to alter the particle size of the active metal, one needs to use support with the wider pores, like α-Al_2O_3, since the wider pores will consequently give a bigger particle size. By mixing the solvent (water) with ethylene glycol (EG) during the impregnation step, a smaller cobalt particle size was achieved. In this way, the cobalt particle size was successfully altered without changing the composition of the active material or the physical properties of the support material. A summary of the characterization results is presented in Table 1. Both the X-ray diffraction (XRD) and H_2-chemisorption experiments showed decreased particle size for the catalyst prepared using EG (Cat2). This catalyst (Cat2) showed higher dispersion (15.6 %) and consequently reduced particle size compared to the Cat1 (D = 7.7 %) (prepared without EG) measured by H_2-chemisorption. XRD measurements confirmed the reduction in Co_3O_4 particle size from ~15 nm to ~5 nm upon ethylene glycol addition. The XRD profiles of all the catalyst with different potassium loadings are illustrated in Figures 1 and 2. All the samples showed existence of the cubic Co_3O_4 and Al_2O_3 phases, while potassium is not visible due to the low concentration. No difference can be detected in the size or shape of the peaks regardless of potassium concentration, Figure 1. Upon EG addition, all the Co_3O_4 peaks are broadened and less pronounced, while the Al_2O_3 peaks become sharpened and more pronounced, Figure 2. This is in agreement with previously published work where cobalt particle size was also altered using EG [16,17]. It was explained by EG acting mainly as a surfactant, thus increasing the wetting ability of the cobalt salt solutions. Borg et al. [17] showed, using transmission electron microscopy (TEM) images, that Co_3O_4 crystallites appeared in aggregates with dimensions above 100 nm with pure water as a solvent, while the aggregates were absent when EG was used.

Table 1. Characterization results.

Catalyst	K Impurity (ppm)	D^a (%)	$d(Co^0)^b$ (nm)	BET Surface Area^c (m^2/g)	Average Pore Diameter (nm)	Pore Volume (cm^3/g)	$d(Co_3O_4)^d$ (nm)	$d(Co^0)^d$ (nm)
Cat1	6	7.7	12.5	145	13.7	0.60	14.8	11.1
Cat1	551	7.7	12.5	137	12.2	0.46	15	11.3
Cat1	902	7.3	13.3	137	12.4	0.47	17.2	12.9
Cat2	15	15.6	6.2	143	12.2	0.52	4.9	3.7
Cat2	471	14.3	6.7	143	12.4	0.53	4.6	3.5
Cat2	886	15.1	6.4	152	10.6	0.48	4.8	3.6

^{a} Standard deviation 3.3%; ^{b} Found by H_2-chemisorption; ^{c} Standard deviation 4.7%; ^{d} Found by X-ray diffraction (XRD) experiments.

Upon alkali addition, no significant differences are observed between Cat1 and Cat2. The measured potassium loadings are close to the nominal values. The low level measured for the samples without added potassium is due to the support containing traces of K. The particle size estimates did not change upon the addition of potassium. All the catalysts showed almost the same surface area (~140 m^2/g), pore size (~0.5 nm), pore volume (~12 cm^3/g), regardless of the EG or K loading, the deviations being within experimental error. These catalyst morphological findings are similar to our previously published work with alkali poisoning [11].

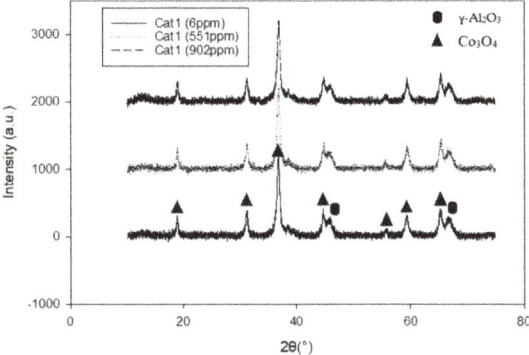

Figure 1. XRD patterns for 20%Co/0.5%Re/γ-Al$_2$O$_3$ with different K loadings.

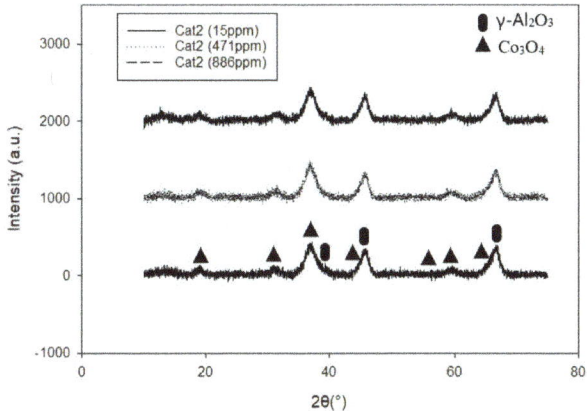

Figure 2. XRD patterns for 20%Co/0.5%Re/γ-Al$_2$O$_3$ with different K loadings.

Temperature-programmed reduction (TPR) profiles of the catalysts Cat1 and Cat2 with different K loadings are presented in Figure 3. All the samples showed a typical reduction profile often observed for alumina supported Co catalyst [22]. The small first peak (around 220 °C) visible for some of the catalysts represents the reduction and removal of residual nitrates from the cobalt precursor [23]. Two main peaks at ~330 °C and ~430 °C are referred to the transition from Co$_3$O$_4$ to CoO and CoO to metallic Co, respectively [22]. The peak temperature of the first reduction peak was slightly decreased to ~290 °C, while the second reduction peak was shifted to slightly higher temperatures (~490 °C) for the catalyst prepared with EG. This is related to the particle size effect, indicating a more difficult reduction to metallic cobalt, which was earlier reported by Jacobs et al. [24]. A lower degree of reduction has also been reported for cobalt-based FT catalysts of similar particle sizes [17,25]. A slight shoulder can be seen on the second reduction peak for the catalyst prepared with EG. Borg et al. [26] ascribed this phenomenon to a larger spread in the particle size distribution, indicating a large variation in the degree of interaction between particles and the support. There is no difference upon K addition, all the samples follow the same trend as the unpoisoned catalysts, which is a similar observation as in previously published work focusing on medium-sized cobalt particles only [9,12].

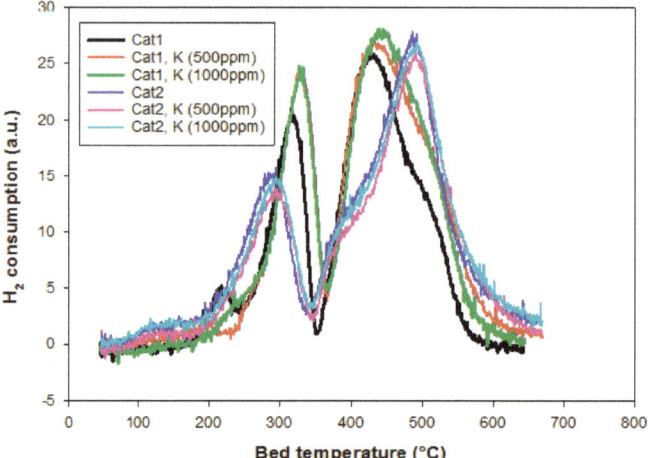

Figure 3. Temperature-programmed reduction (TPR) profiles of the catalysts Cat1 and Cat2 with different K loadings.

The Fischer–Tropsch activity and selectivity results are presented in Table 2. Catalyst activity measured as STY was found to be 0.054 to 0.027 s^{-1} for the Cat1 and Cat2 catalysts with Co particle sizes of ~13 and ~6 nm respectively. The smaller cobalt particles also give significantly lower C_{5+} selectivity and higher CH_4 and CO_2. It is previously reported that catalyst activity on supported cobalt catalysts is considered independent of cobalt dispersion [27] and support identity [28], but this only holds for larger cobalt particles. Bezemer et al. [18] reported a strong influence of the cobalt particle size (in the range 3 to 8 nm) on both the cobalt site-time yield and the C_{5+} selectivity using carbon nanofibers as a support. The cobalt particle size effect was explained as a combination of CO-induced surface reconstruction and non-classical structure sensitivity. In the present work, γ-alumina is used as the catalyst support, and we observe the same particle size effect on the catalyst activity and selectivity, in agreement with Borg et al. [17]. Wang et al. [29] showed that for very small Co particles (1.4–2.5 nm), the Co surface was readily oxidized by water vapor, while for the larger Co particles (3.5–10.5 nm), such oxidation was not evident. This was confirmed by Azzam et al. [30] where the poor catalyst activity and higher CH_4 and CO_2 selectivities for the very small particles (1 nm) was ascribed to catalyst oxidation by water. However, in the present work the particle size is in the range 6–13 nm indicating that oxidation is not the reason for the dramatic effect in the specific activity. Yang et al. [16] also reported increased turnover frequency (TOF) with increasing cobalt particle size. They explained the effect by the increased CO coverage which is directly proportional with increasing Co particle size. Den Brejeen et al. [31] reported, using SSITKA experiments, that small cobalt particles have higher coverage of 'irreversible' adsorbed CO, which can block the surface and lead to lower activity, while the coverage of H is increased which leads to more methane production.

The activity decreases severely upon increased K loading for both catalysts. The loss in specific activity with the potassium loading is similar for the two samples, losing approximately 30% of the activity with the highest K loading. The changes in catalytic activity are similar to those observed previously [9–13].

Also in terms of selectivity is there little difference in the response to the potassium loading between the two samples. Both catalysts showed increased C_{5+} and CO_2, and decreased CH_4 selectivity upon K addition. Hence, the effect of potassium is the same independent of the cobalt particle size in the range investigated here.

The main conclusion from this work is that K effects the cobalt based catalyst in the same way for two distinctly different cobalt particle sizes. Although the catalyst with smaller particle size (Cat2) showed lower activity and a poorer selectivity, the K effect appears to be independent of Co particle size. We have previously shown that step sites are preferred sites for CO and K adsorption [15], but the difference between facets is small and the adsorption of K is energetically favored on all sites on the cobalt surface. The smaller particles investigated here are likely to have a slightly larger fraction of step and edge sites, but the difference is perhaps not large enough to be significant.

Table 2. Catalytic activity reported as site time yield (STY) and selectivities to C_{5+}, CH_4 and CO_2 with different catalyst and potassium loadings (Calculation procedures see the Supplementary Materials).

Catalyst	K (ppm)	STY (s^{-1}) [a]	C_{5+} (%)	CH_4 (%)	CO_2 (%)
Cat1	6	0.054	85.4	7.3	0.16
Cat1	551	0.041	86.6	6.8	0.23
Cat1	902	0.039	86.3	6.7	0.40
Cat2	15	0.027	79.3	10	0.23
Cat2	471	0.022	83.1	9.5	0.39
Cat2	886	0.019	81.0	9.2	0.55

[a] Standard deviation ± 7%.

3. Materials and Methods

The 20%Co/0.5%Re/γ-Al_2O_3 catalyst was prepared using a one-step incipient wetness impregnation of an aqueous solution of $Co(NO_3)_2·6H_2O$ and $HReO_4$ [10]. The catalyst support was from Sasol (Puralox γ-alumina). The catalyst was then dried in a stationary oven at 120 °C for 1 h. The calcination procedure in flowing air in a fixed bed quartz reactor at 300 °C for 16 h, using a ramp rate of 2 °C/min was applied after drying. Finally, the oxidized catalyst precursors were sieved to a particle range of 53–90 µs. The prepared catalyst should be free from impurities, but by analysis a small amount of K was found in the support (6 and 15 ppm) [26]. In order to achieve different particle sizes, a method using mixtures of distilled water and ethylene glycol (EG) in the impregnation solution was used [17]. One sample was prepared using only water as the solvent (Cat1), the second sample was prepared using a mixture of 80 wt% water and 20 wt% EG (Cat2), the latter sample is expected to have a smaller cobalt particle size. These two samples (in their calcined state) were post-impregnated with 500 and 1000 ppm of potassium in the form of KNO_3 dissolved in deionized water. Inductively coupled plasma–mass spectrometry (ICP–MS) was used to analyze the amount of potassium present in the catalysts. The catalysts samples were dried and calcined using the same conditions both after the initial preparation and after post-impregnation with potassium.

In order to determine surface area, pore volume and average pore diameter of the prepared catalysts, a Tristar II 3020 was used to perform a volumetric adsorption of N_2. Prior to the measurement at liquid nitrogen temperature, the catalyst samples (~70 mg, 53-90 µm) were outgassed in vacuum, first at room temperature for 1h and then at 200 °C overnight. For calculation of the surface area the Brunauer–Emmet–Teller (BET) [32] isotherm was used while to determine a pore volumes and average pore diameters of the samples the Barret–Joyner–Halenda (BJH) [33] method was applied.

H_2-chemisorption was carried out using a Micromeritics ASAP2010 unit. The catalyst sample (0.2 g) was placed between quartz wool wads and loaded in the chemisorption reactor. Before the measurements, the sample was reduced in flowing hydrogen at 350 °C for 16h with a ramping rate of 60 °C/h After reduction, the analysed samples were cooled to 30 °C under vacuum. Chemisorption data was obtained at 30 °C between 0.020 and 0.667 bar H_2 pressure. It was presumed that for each surface cobalt atom there was one H chemisorption site and that neither Re, K nor the support contributed to chemisorption. The particle sizes (in nm) were estimated using the following equation [34], where D is cobalt dispersion in %:

$$d(Co^0) = \frac{96.2}{D}$$

TPR experiments were carried out in the Altamira AMI-300RHP. The catalyst sample (100 mg) was loaded between wads of quartz wool and placed in a quartz u-tube reactor. Prior the measurement, the catalyst was treated in inert gas at 200 °C. The catalyst was then reduced in hydrogen flow (7% H_2/Ar 50 ml/min) to a temperature of 700 °C with a ramp rate of 10 °C/min. Finally, the samples were cooled down to ambient temperature.

A D8 DaVinci-1 X-ray Diffractometer with CuKα radiation was used for all XRD experiments. The analysis was run for 60 min for each catalyst sample, examining a range of 2θ from 10 to 75° at a step size of 0.013° while using the X-ray source at 40 kV and 40 mA. The average cobalt oxide crystallite thickness was calculated by Scherrer's equation [35] using the (311) Co_3O_4 peak located at 2θ = 36.8°, and applying a K-factor of 0.89. Subsequently, the metallic cobalt particle size was estimated using the relative volume contraction between Co_3O_4 and metallic cobalt [36]:

$$d(Co^0) = 0.75 d(Co_3O_4)$$

Fischer–Tropsch experiments wer perfomed in 10 mm ID steel tube fixed bed reactor at industrially relevant conditions (210 °C, 20 bar and a H_2/CO ratio of 2.1). The catalyst samples (1 g) were mixed with inert SiC (20 g) and loaded between quartz wool wads to fix the location in the reactor. To improve the heat distribution, aluminium blocks were placed around the reactor. The reactor was then placed in an electrically heated furnace. Before hydrogen reduction, a leak test with He was performed. The *in situ* reduction was carried out at 350 °C for 10 h with a ramp rate of 1 °C/min from ambient temperature. After reduction, the samples were cooled to 170 °C. Prior to the syngas introduction (250 Nml/min), He was used to build up a pressure to 20 bar. The temperature program was set to increase the temperature first from 170 to 190 °C and then to the final temperature of 210 °C with ramping rates of 20 °C/min and 5 °C/min, respectively. Liquid Fischer–Tropsch products were separated in a hot trap at ~87 °C and a cold trap at ambient temperature. Wax was collected in the hot trap while other liquids (water and light hydrocarbons) were collected in the cold pot at. The C_1-C_4 gases were analyzed using a HP 6890 gas chromatograph. N_2 (internal standard), H_2, CO, CH_4 and CO_2 were analyzed on a TCD following separation on a Carbosieve column. Hydrocarbon products were separated with a GS-Alumina PLOT column and detected on a flame ionization detector (FID). CH_4 was used to combine TCD and FID analysis in the calculations. The syngas contained 3 % N_2 which is used as an internal standard for quantification of the products to close the mass balance. Activity data were reported based on measurements at constant feed rate (250 Nml/min) after 24 h time on stream. Then the syngas flow was changed to obtain ~50 % CO conversion. Selectivity data are reported at 50 ± 5% CO conversion based on the analysis of C_1–C_4 hydrocarbons in gas phase after ~48 h time-on-stream. Since the focus is on the amount of higher hydrocarbons, the selectivity is reported in the usual way as C_{5+} and CH_4 selectivity.

4. Conclusions

The effect of potassium on Co-based FT catalysts was examined on two different particle sizes of cobalt, small and medium. The catalyst activity decreased with decreased particle size, while the morphological characteristics are unchanged. The C_{5+} selectivity decreased, while CH_4 and CO_2 increased for the catalyst with smaller particle size. Upon potassium addition there was no change in dispersion, surface area, pore size, pore volume, regardless of the particle size or potassium concentration. However, all the catalysts showed a negative effect of K on catalyst activity. C_{5+} selectivity was increased, while methane and CO_2 decreased with increasing K concentration. There was no difference in the potassium effect regarding the particle size, indicating that the catalytic performance was affected in the same fashion.

Supplementary Materials: The following are available online at http://www.mdpi.com/2073-4344/9/4/351/s1, Calculation procedures, 1.1. Site time yield; 1.2. Selectivities.

Author Contributions: Conceptualization, E.A.B. and L.G.; methodology, J.S. and L.G.; investigation, J.S.; data curation, J.S.; original draft preparation, L.G.; writing—review and editing E.A.B.; supervision, E.A.B.; project administration, E.A.B.; funding acquisition, E.A.B.

Funding: We thank the Research Council of Norway for funding (contracts no: 228741, 257622, 280846).

Conflicts of Interest: The authors declare no conflict of interest.

References

1. Rauch, R.; Kiennemann, A.; Sauciuc, A. Fischer-Tropsch Synthesis to Biofuels (BtL Process). In *The Role of Catalysis for the Sustainable Production of Bio-Fuels and Bio-Chemicals*; Elsevier Inc.: Amsterdam, The Netherlands, 2013; pp. 397–443.
2. van Steen, E.; Claeys, M. Fischer-Tropsch catalysts for the biomass-to-liquid process. *Chem. Eng. Technol.* **2008**, *31*, 655–666. [CrossRef]
3. Woolcock, P.J.; Brown, R.C. A review of cleaning technologies for biomass-derived syngas. *Biomass Bioenergy* **2013**, *52*, 54–84. [CrossRef]
4. Boerrigter, H.; Calis, H.P.; Slor, D.J.; Bodenstaff, H. Gas Cleaning for Integrated Biomass Gasification (Bg) and Fischer-Tropsch (Ft) Systems; Experimental Demonstration of Two Bg-Ft Systems. In Proceedings of the 2nd World Conference and Technology Exhibition on Biomass for Energy, Industry and Climate Protection, Rome, Italy, 10–14 May 2004.
5. Norheim, A.; Lindberg, D.; Hustad, J.E.; Backman, R. Equilibrium calculations of the composition of trace compounds from biomass gasification in the solid oxide fuel cell operating temperature interval. *Energy Fuels* **2009**, *23*, 920–925. [CrossRef]
6. Boerrigter, H.; Calis, H.-P.; Uil, H.D. Green Diesel from Biomass via Fischer-Tropsch synthesis: New Insights in Gas Cleaning and Process Design. In Proceedings of the Pyrolysis and Gasification of Biomass and Waste, Strasbourg, France, 30 September–1 October 2002; pp. 1–13.
7. Tristantini, D.; Lögdberg, S.; Gevert, B.; Borg, Ø.; Holmen, A. The effect of synthesis gas composition on the Fischer-Tropsch synthesis over Co/γ-Al$_2$O$_3$ and Co-Re/γ-Al$_2$O$_3$ catalysts. *Fuel Process. Technol.* **2007**, *88*, 643–649. [CrossRef]
8. Borg, Ø.; Hammer, N.; Enger, B.C.; Myrstad, R.; Lindvåg, O.A.; Eri, S.; Skagseth, T.H.; Rytter, E. Effect of biomass-derived synthesis gas impurity elements on cobalt Fischer-Tropsch catalyst performance including in situ sulphur and nitrogen addition. *J. Catal.* **2011**, *279*, 163–173. [CrossRef]
9. Balonek, C.M.; Lillebø, A.H.; Rane, S.; Rytter, E.; Schmidt, L.D.; Holmen, A. Effect of alkali metal impurities on Co-Re catalysts for Fischer-Tropsch synthesis from biomass-derived syngas. *Catal. Lett.* **2010**, *138*, 8–13. [CrossRef]
10. Lillebø, A.H.; Patanou, E.; Yang, J.; Blekkan, E.A.; Holmen, A. The effect of alkali and alkaline earth elements on cobalt based Fischer-Tropsch catalysts. *Catal. Today* **2013**, *215*, 60–66. [CrossRef]
11. Gavrilović, L.; Brandin, J.; Holmen, A.; Venvik, H.J.; Myrstad, R.; Blekkan, E.A. Deactivation of Co-Based Fischer–Tropsch Catalyst by Aerosol Deposition of Potassium Salts. *Ind. Eng. Chem. Res.* **2018**, *57*, 1935–1942. [CrossRef]
12. Gavrilović, L.; Brandin, J.; Holmen, A.; Venvik, H.J.; Myrstad, R.; Blekkan, E.A. Fischer-Tropsch synthesis—Investigation of the deactivation of a Co catalyst by exposure to aerosol particles of potassium salt. *Appl. Catal. B Environ.* **2018**, *230*, 203–209. [CrossRef]
13. Patanou, E.; Lillebø, A.H.; Yang, J.; Chen, D.; Holmen, A.; Blekkan, E.A. Microcalorimetric studies on Co-Re/γ-Al$_2$O$_3$ catalysts with na impurities for fischer-tropsch synthesis. *Ind. Eng. Chem. Res.* **2014**, *53*, 1787–1793. [CrossRef]
14. Olsen, B.K.; Kügler, F.; Castellino, F.; Jensen, A.D. Poisoning of vanadia based SCR catalysts by potassium: Influence of catalyst composition and potassium mobility. *Catal. Sci. Technol.* **2016**, *6*, 2249–2260. [CrossRef]
15. Chen, Q.; Svenum, I.-H.; Qi, Y.; Gavrilovic, L.; Chen, D.; Holmen, A.; Blekkan, E.A. Potassium adsorption behavior on hcp cobalt as model systems for the Fischer–Tropsch synthesis: A density functional theory study. *Phys. Chem. Chem. Phys.* **2017**, *19*, 12246–12254. [CrossRef]
16. Yang, J.; Frøseth, V.; Chen, D.; Holmen, A.; Frøseth, V.; Chen, D.; Holmen, A. Particle size effect for cobalt Fischer-Tropsch catalysts based on in situ CO chemisorption. *Surf. Sci.* **2016**, *648*, 67–73. [CrossRef]
17. Borg, Ø.; Dietzel, P.D.C.; Spjelkavik, A.I.; Tveten, E.Z.; Walmsley, J.C.; Diplas, S.; Eri, S.; Holmen, A.; Rytter, E. Fischer-Tropsch synthesis: Cobalt particle size and support effects on intrinsic activity and product distribution. *J. Catal.* **2008**, *259*, 161–164. [CrossRef]

18. Bezemer, G.L.; Bitter, J.H.; Kuipers, H.P.C.E.; Oosterbeek, H.; Holewijn, J.E.; Xu, X.; Kapteijn, F.; Van Diilen, A.J.; De Jong, K.P. Cobalt particle size effects in the Fischer-Tropsch reaction studied with carbon nanofiber supported catalysts. *J. Am. Chem. Soc.* **2006**, *128*, 3956–3964. [CrossRef]
19. Pendyala, V.R.R.; Jacobs, G.; Ma, W.; Klettlinger, J.L.S.; Yen, C.H.; Davis, B.H. Fischer-Tropsch synthesis: Effect of catalyst particle (sieve) size range on activity, selectivity, and aging of a Pt promoted Co/Al$_2$O$_3$ catalyst. *Chem. Eng. J.* **2014**, *249*, 279–284. [CrossRef]
20. Nakhaei Pour, A.; Housaindokht, M. Fischer-Tropsch synthesis over CNT supported cobalt catalysts: Role of metal nanoparticle size on catalyst activity and products selectivity. *Catal. Lett.* **2013**, *143*, 1328–1338. [CrossRef]
21. Arslan, I.; Walmsley, J.C.; Rytter, E.; Bergene, E.; Midgley, P.A. Toward three-dimensional nanoengineering of heterogeneous catalysts. *J. Am. Chem. Soc.* **2008**, *130*, 5716–5719. [CrossRef]
22. Rønning, M.; Tsakoumis, N.E.; Voronov, A.; Johnsen, R.E.; Norby, P.; Van Beek, W.; Borg, Ø.; Rytter, E.; Holmen, A. Combined XRD and XANES studies of a Re-promoted Co/γ-Al$_2$O$_3$ catalyst at Fischer-Tropsch synthesis conditions. *Catal. Today* **2010**, *155*, 289–295. [CrossRef]
23. Bao, A.; Li, J.; Zhang, Y. Effect of barium on reducibility and activity for cobalt-based Fischer-Tropsch synthesis catalysts. *J. Nat. Gas Chem.* **2010**, *19*, 622–627. [CrossRef]
24. Jacobs, G.; Das, T.K.; Zhang, Y.; Li, J.; Racoillet, G.; Davis, B.H. Fischer-Tropsch synthesis: Support, loading, and promoter effects on the reducibility of cobalt catalysts. *Appl. Catal. A. Gen.* **2002**, *233*, 263–281. [CrossRef]
25. Khodakov, A.Y.; Griboval-Constant, A.; Bechara, R.; Zholobenko, V.L. Pore size effects in Fischer Tropsch synthesis over cobalt-supported mesoporous silicas. *J. Catal.* **2002**, *206*, 230–241. [CrossRef]
26. Borg, Ø.; Eri, S.; Blekkan, E.A.; Storsæter, S.; Wigum, H.; Rytter, E.; Holmen, A. Fischer-Tropsch synthesis over γ-alumina-supported cobalt catalysts: Effect of support variables. *J. Catal.* **2007**, *248*, 89–100. [CrossRef]
27. Iglesia, E.; Soled, S.L.; Fiato, R.A. Fischer-Tropsch synthesis on cobalt and ruthenium. Metal dispersion and support effects on reaction rate and selectivity. *J. Catal.* **1992**, *137*, 212–224. [CrossRef]
28. Storsæter, S.; Borg, Ø.; Blekkan, E.A.; Holmen, A. Study of the effect of water on Fischer-Tropsch synthesis over supported cobalt catalysts. *J. Catal.* **2005**, *231*, 405–419. [CrossRef]
29. Wang, Z.-J.; Skiles, S.; Yang, F.; Yan, Z.; Goodman, D.W. Particle size effects in Fischer–Tropsch synthesis by cobalt. *Catal. Today* **2012**, *181*, 75–81. [CrossRef]
30. Azzam, K.; Jacobs, G.; Ma, W.; Davis, B.H. Effect of cobalt particle size on the catalyst intrinsic activity for fischer-tropsch synthesis. *Catal. Lett.* **2014**, *144*, 389–394. [CrossRef]
31. Den Breejen, J.P.; Radstake, P.B.; Bezemer, G.L.; Bitter, J.H.; Frøseth, V.; Holmen, A.; De Jong, K.P. On the Origin of the Cobalt Particle Size Effects in Fischer-Tropsch. *J. Am. Chem. Soc.* **2009**, *131*, 7197–7203. [CrossRef]
32. Brunauer, S.; Emmett, P.H.; Teller, E. Adsorption of Gases in Multimolecular Layers. *J. Am. Chem. Soc.* **1938**, *60*, 309–319. [CrossRef]
33. Barrett, E.P.; Joyner, L.G.; Halenda, P.P. The Determination of Pore Volume and Area Distributions in Porous Substances. I. Computations from Nitrogen Isotherms. *J. Am. Chem. Soc.* **1951**, *73*, 373–380. [CrossRef]
34. Jones, R.D.; Bartholomew, C.H. Improved flow technique for measurement of hydrogen chemisorption on metal catalysts. *Appl. Catal.* **1988**, *39*, 77–88. [CrossRef]
35. Patterson, A.L. The Scherrer formula for X-ray particle size determination. *Phys. Rev.* **1939**, *56*, 978–982. [CrossRef]
36. Schanke, D.; Vada, S.; Blekkan, E.A.; Hilmen, A.M.; Hoff, A.; Holmen, A. Study of Pt-Promoted Cobalt CO Hydrogenation Catalysts. *J. Catal.* **1995**, *156*, 85–95. [CrossRef]

© 2019 by the authors. Licensee MDPI, Basel, Switzerland. This article is an open access article distributed under the terms and conditions of the Creative Commons Attribution (CC BY) license (http://creativecommons.org/licenses/by/4.0/).

Comparative Studies of Fischer-Tropsch Synthesis on Iron Catalysts Supported on Al$_2$O$_3$-Cr$_2$O$_3$ (2:1), Multi-Walled Carbon Nanotubes or BEA Zeolite Systems

Pawel Mierczynski [1,*], Bartosz Dawid [1], Karolina Chalupka [1], Waldemar Maniukiewicz [1], Izabela Witoska [1], Krasimir Vasilev [2] and Malgorzta I. Szynkowska [1]

[1] Institute of General and Ecological Chemistry, Lodz University of Technology, Zeromskiego 116, Lodz 90–924, Poland
[2] School of Engineering, University of South Australia, Mawson Lakes 5095, Australia
* Correspondence: pawel.mierczynski@p.lodz.pl; Tel.: +48-42-631-31-25

Received: 30 May 2019; Accepted: 8 July 2019; Published: 15 July 2019

Abstract: The main goal of the presented paper is to study the influence of a range of support materials, i.e., multi-walled carbon nanotubes (MWCNTs), Al$_2$O$_3$-Cr$_2$O$_3$ (2:1), zeolite β-H and zeolite β-Na on the physicochemical and catalytic properties in Fischer-Tropsch (F-T) synthesis. All tested Fe catalysts were synthesized using the impregnation method. Their physicochemical properties were extensively investigated using various characterization techniques such as the Temperature-Programmed Reduction of hydrogen (TPR-H$_2$), X-ray diffraction, Temperature-Programmed Desorption of ammonia (TPD-NH$_3$), Temperature-Programmed Desorption of carbon dioxide (TPD-CO$_2$), Fourier transform infrared spectrometry (FTIR), Brunauer Emmett Teller method (BET) and Thermogravimetric Differential Analysis coupled with Mass Spectrometer (TG-DTA-MS). Activity tests were performed in F-T synthesis using a high-pressure fixed bed reactor and a gas mixture of H$_2$ and CO (50% CO and 50% H$_2$). The correlation between the physicochemical properties and reactivity in F-T synthesis was determined. The highest activity was from a 40%Fe/Al$_2$O$_3$-Cr$_2$O$_3$ (2:1) system which exhibited 89.9% of CO conversion and 66.6% selectivity toward liquid products. This catalyst also exhibited the lowest acidity, but the highest quantity of iron carbides on its surface. In addition, in the case of iron catalysts supported on MWCNTs or a binary oxide system, the smallest amount of carbon deposit formed on the surface of the catalyst during the F-T process was confirmed.

Keywords: hydrogenation of CO; iron catalysts; syngas; monometallic iron catalysts

1. Introduction

Fischer-Tropsch synthesis is a promising catalytic route for the environmentally friendly production of fuels from biomass, coal and natural gas [1–5]. However, on an industrial scale, a very efficient and stable catalyst is desirable. The catalyst activity and selectivity are influenced by the nature and structure of the carrier material, metal dispersion, the nature of metal, metal loading, and to a large extent, it depends on the methodology of preparation [6]. Metallic catalysts in Fischer-Tropsch synthesis are often supported on oxides, zeolites or binary oxides systems and, more recently, on carbon-based systems such as activated carbon and carbon nanotubes (CNT) [7–9]. In particular, there is a growing interest in carbon nanotubes and their application as a F-T catalyst support [10,11]. Carbon nanotubes are exciting due to their unique original chemical, physical, optical and electron properties. Their special structure offers a high surface area and the presence of various defects which offer the possibility to introduce functional groups. The introduced functional groups can generate specific physicochemical properties of the material and improve the interaction between the metal particles incorporated into their surfaces, thereby forming new catalytic systems. Zhang et al. [12] studied iron catalysts supported on the MWCNTs system in the Fischer-Tropsch process and they reported a high selectivity towards C_2–C_7 hydrocarbons but low CO conversions values. In addition, the usage of carbon nanofibers as a catalyst support in the F-T process was also investigated [13] and reported to have high stability. It is well known that acidic zeolite or other acidic supports present good performance in the F-T reaction and high selectivity to isoparaffin and unsaturated compounds generated in the final product [4,14,15]. In our previous works, iron catalysts supported on BEA zeolites and binary oxide (Al_2O_3-Cr_2O_3 (2:1)) were investigated in Fischer-Tropsch synthesis. It was proven that the dealumination process of zeolite BEA and the increase of the content of iron in the catalytic material improves the catalyst activity in the case of the process carried out on the Fe catalyst supported on the zeolite BEA [4]. In addition, the influence of the binary oxide carrier composition of iron catalysts on their reactivity properties was studied and the results also showed that the increase of Fe loading and the content of Al_2O_3 in the catalytic system lead to an increase of CO conversion in the hydrogenation process [2]. The obtained results in the F-T process also showed that the activation condition has a large impact on the catalytic properties of iron catalysts in the F-T reaction. It was proven that the Fe catalyst which was previously activated in a mixture of 50% CO and 50% H_2 exhibited the highest activity and selectivity towards liquid product formation and this effect was explained by the highest quantity of iron carbide phases present on the catalyst surface after the activation process [16]. The novelty of this work is related to the comparison studies of high active iron catalysts supported on various supports such as MWCNTs, zeolite BEA and binary oxide in Fischer-Tropsch synthesis. Additionally, the correlation between their reactivity and physicochemical properties in the investigated process was found. In addition, in the presented paper, a highly active and stable Fe/MWCNTs catalytic system was prepared and tested in the hydrogenation process of CO.

The product of the Fischer-Tropsch (F-T) process mostly contains linear paraffin, whose distribution is varied from CH_4 to long-chain hydrocarbons and obeys the Anderson–Schulz–Flory (ASF) law [17]. It is particularly desirable to obtain a final hydrocarbon product having 5 to 11 carbon atoms in a molecule containing mostly isoparaffin and olefin as gasoline range liquid fuels with a high-octane value are an important fuel in the petrochemical industry. In addition, the liquid fraction containing hydrocarbons with 10 to 21 carbon atoms in a molecule is also attractive because they are the main components in diesel. That is why the main idea of the Fischer-Tropsch reaction is to generate fuel fraction (gasoline or diesel) on a highly active, stable and selective heterogeneous catalyst, which is beneficial from the point of view of generating clean fuel. In order to achieve a liquid product containing gasoline or diesel fractions, it is strongly recommended to use them in the process of bi-functional catalysts which should have acid and metallic centres on the surface. A high yield of the fuel fraction production along with a satisfactory selectivity of the desired products is also required [18–20]. Both of these features can be achieved using bifunctional catalysts. One of the main reasons for the deactivation of Fischer-Tropsch catalysts is the deposition of carbon [5,21,22]. The formations on the catalyst surface carbon deposit blocks the active sites of the catalyst, leading to its deactivation [23]. Menon reported that the Fischer-Tropsch synthesis is a carbon insensitive reaction due to the fact that there is sufficient hydrogen on the catalyst surface for the hydrogenation of the carbon present on the catalyst surface in order to keep the surface a free surface of the catalyst [24]. Therefore, in order to regenerate the deactivated catalyst, it should be previously oxidized and then reduced to remove the carbidic and surface carbon [25].

Based on the above-presented suggestions, the main aim of the work was to carry out comparative investigations of iron systems for the Fischer-Tropsch process and reveal the link between their physicochemical and catalytic properties. It is well known that the catalytic activity of heterogeneous catalysts in the chemical process depends on the nature, dispersion and reactivity of the active centres in relation to the substrates; on the steric effects and diffusion of the reagents; and on the created products in the catalytic material pores [5]. That is why, in our studies, we used various support materials whose structure can affect the transport of reagents and products from and to the catalyst bed with different specific surface areas and various reducing and acidic properties. In order to avoid steric effects, the carrier material should have a unified pore system which plays a crucial role during the process and directly influences the catalyst selectivity in the F-T process.

2. Results and Discussion

2.1. Influence of the Catalytic Material Composition on the Catalytic Reactivity of Monometallic Iron Catalysts in the Hydrogenation of CO

The catalytic measurements performed in the hydrogenation process showed that the distribution of the hydrocarbons in the final product depends strongly on the support used for the catalytic system. The catalytic activity results expressed as CO conversion and product selectivity in Fischer-Tropsch synthesis are given in Table 1. The obtained catalytic activity results confirmed that the most active catalyst was the 40%Fe/Al_2O_3-Cr_2O_3 (2:1) system which exhibited approximately 90% of CO conversion and high selectivity towards the liquid product (66.6%). In addition, the smallest amount of CO_2 was created in the final product obtained during the F-T reaction using this catalyst, which means that the WGS process runs in a limited way. Furthermore, in the case of the 40%Fe/Zeolite β-H and 40%Fe/MWCNT catalysts, we observed CO conversion values above 71% and selectivities towards liquid product above 67%. The only difference which was observed in the final product was the amount of carbon dioxide and methane formed during the process.

Table 1. The results of the activity measurements expressed as CO conversion and selectivity towards all products of the Fischer-Tropsch (F-T) process performed at 280 °C under elevated pressure (3 MPa) from a mixture of H_2 and CO (molar ratio 1:1).

Catalysts	Conv. (%)	S_{CO2}	S_{CH4} in HCs	S_{C2-C4} in HCs	S_{C5-C9} in HCs	$S_{C10-C21}$ in HCs	Liquid	Paraffin/Olefins	Linear/Branched
40%Fe/MWCNT	71.1	7.6	14.3	5.9	13.0	66.8	67.5	-	4.3
40%Fe/Zeolite β-H	75.7	4.5	18.4	5.6	44.5	31.5	69.6	-	18.2
40%Fe/Zeolite β-Na	65.2	11.0	32.6	9.1	26.0	32.3	52.2	-	2.0
40%Fe/Al_2O_3-Cr_2O_3 (2:1)	89.9	2.4	16.3	5.4	17.4	60.9	66.6	17.2	4.3

In the case of the iron catalyst supported on MWCNTs, about 7.6% of the CO_2 and about 14.3% of the CH_4 were observed in the final product. In the case of the iron catalyst supported on Zeolite β-Na, the lowest value of CO conversion and the selectivity towards liquid product was observed at 65.2% and 52.2%, respectively. It is also worth noticing that in the case of this system, the highest selectivity to CH_4 (32.6%) and towards CO_2 (11%) was detected. The latter suggests that the reaction of methanation and the conversion of carbon monoxide with water vapour ran more efficiently. In addition, the stability tests for the iron catalysts were also carried out in the hydrogenation of the CO process. Figure 1 presents the carbon monoxide conversion values as a function of the reaction time for 40%Fe/Al_2O_3-Cr_2O_3 (2:1), 40%Fe/MWCNT and the iron catalysts supported on both types of zeolite BEA. The results clearly showed that all the tested catalysts exhibited stable values during 96 h of the reaction. Only in the case of the 40%Fe/Al_2O_3-Cr_2O_3 (2:1) and 40%Fe/Zeolite β-H catalysts, was a slight decrease in CO conversion values observed during the first 48 h of the process [16].

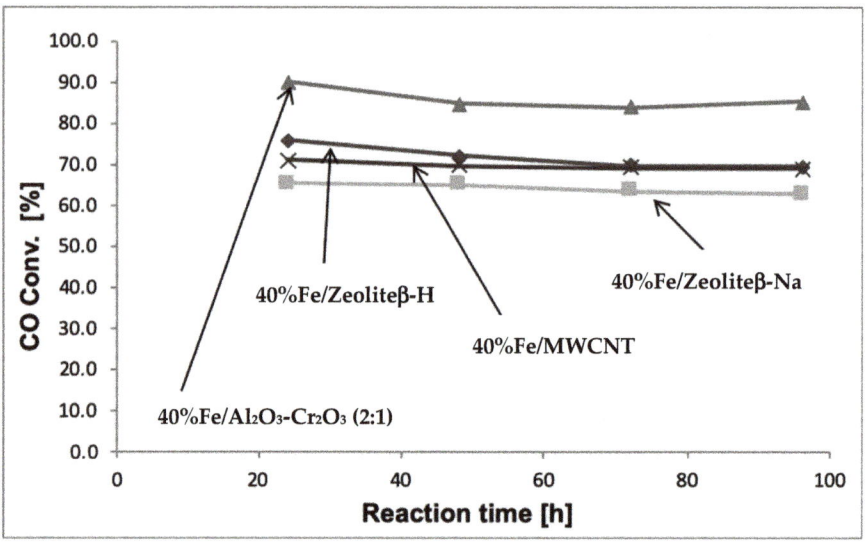

Figure 1. The carbon monoxide conversion values of iron catalysts supported on various supports obtained in the Fischer-Tropsch (F-T) process performed in a reaction mixture at 280 °C (50% H2:50% CO) at 30 atmospheres using a 2 g sample in each test.

Liquid products formed during the hydrogenation of the CO process were also analysed using GC-MS and FTIR techniques (see Figure 2). The obtained results showed that the saturated, unsaturated and branched hydrocarbons arise within the investigated process on the 40%Fe/Al$_2$O$_3$-Cr$_2$O$_3$ (2:1) catalyst. Table 1 presents the selectivity of the hydrocarbons created during the Fischer-Tropsch reaction. It is worth emphasizing that in the case of the 40%Fe/Al$_2$O$_3$-Cr$_2$O$_3$ (2:1) catalyst, the formation of unsaturated hydrocarbons was detected which were not observed when the process was carried out on the remaining catalysts. In addition, the highest quantity of the linear hydrocarbons formed in the liquid product was observed for the 40%Fe/Zeolite β-H catalyst as evidenced by the highest ratio between the linear and branched hydrocarbons (18.2). The GC-MS analysis of the liquid product obtained during the hydrogenation of CO on the 40%Fe/Zeolite β-H catalyst material gave evidence that it had a high selectivity to linear hydrocarbons. Whereas, the chromatographic analysis performed for the 40%Fe/MWCNT and 40%Fe/Al$_2$O$_3$-Cr$_2$O$_3$ (2:1) catalysts showed the ratio between linear and branched hydrocarbons to be equal to 4.3. In the case of the 40%Fe/Zeolite β-Na catalyst, the formation of lower quantities of linear hydrocarbons was observed. Figure 2 shows the FTIR analysis of the products of the hydrogenation of the CO process performed on the iron catalysts supported on various types of supports. FTIR measurements recorded for the studied Fe catalysts confirmed the presence on their surface of the following functional groups assigned to –OH stretching, –C–H stretching in alkanes, –CH$_2$, –CH, –CH$_3$ and (CH$_2$)n > 4. The presence of the bands attributed to the previously mentioned functional groups confirmed the occurrence of linear and branched hydrocarbons in the final liquid product and the GC-MS results obtained for these catalysts. In addition, the occurrence of the specific bands located on the FTIR spectrum of 40%Fe/Al$_2$O$_3$-Cr$_2$O$_3$ (2:1) assigned to C=C in alkenes, C–H in alkenes, and R–CH=CH–R were confirmed. The presence of the above listed functional groups on the spectra confirmed the chromatographic analysis and the formation of unsaturated compounds formed via the F-T process [2].

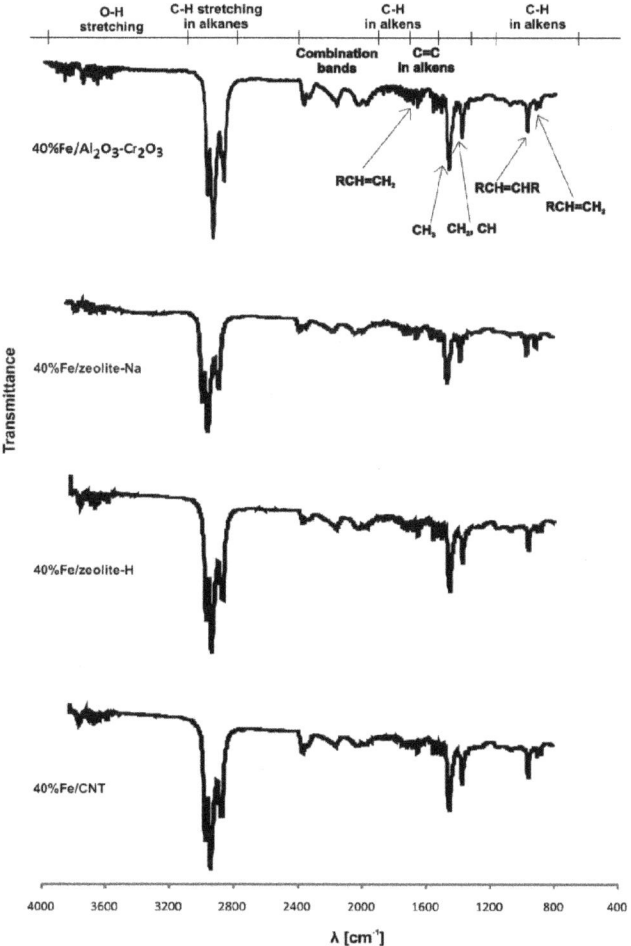

Figure 2. Fourier transform infrared spectrometry (FTIR) spectra's of the product formed in the hydrogenation of the CO process at 280 °C under elevated pressure (3 MPa) from a mixture of H_2 and CO (50% CO and 50% H_2) using Fe catalysts supported on various carriers (binary oxides, zeolite-Na, zeolite β-H and MWCNTs) calcined in an air atmosphere at 500 °C for 4 h and activated in a reaction mixture.

Table 1 also shows the distributions of the hydrocarbon fractions produced within the F-T process into gasoline and diesel expressed as S_{C5-C9} and $S_{C10-C21}$. The results presented in Table 1 clearly show that in the case of the 40%Fe/MWCNT and 40%Fe/Al_2O_3-Cr_2O_3 catalysts, a diesel fraction is formed mainly. The higher quantity of the C_5–C_9 fraction in the final product was formed for the 40%Fe/Zeolite β-H and 40%Fe/Zeolite β-Na catalysts equal to 44.5% and 26.0%, respectively. It is also worth mentioning that the F-T process carried out on the 40%Fe/Zeolite β-H catalyst leads mainly to the production of a C_5–C_9 hydrocarbons fraction which means that it can be used for petrol generation. On the other hand, the reaction which was performed on iron catalysts supported on the Na form of the zeolite BEA leads mainly to the production of the same quantity of hydrocarbons, which are components of gasoline and diesel.

One of the main problems in the hydrogenation of CO is the carbon deposit formed during the process which blocks access to the active centres of the reaction. In order to study the carbon deposit formed in the F-T process, the thermal analysis of the iron catalysts (40%Fe/Zeolite β-H, 40%Fe/Zeolite β-H, 40%Fe/MWCNT and 40%Fe/Al$_2$O$_3$-Cr$_2$O$_3$) after calcination and F-T reaction was done. The obtained TG-DTA-MS measurements are given in Figure 3 and Figure S1–S3, respectively. Figure 3 shows the mass spectrum of the gaseous products formed during the decomposition of all investigated catalysts. The amount and the form of the carbon deposit on the iron catalyst surface (40%Fe/Zeolite β-H and 40%Fe/Zeolite β-Na) are presented in Figure S1 and in Table 2. As it is seen in Figure S1, the carbon deposition is higher for the protonic form of zeolite (36.50%) than for the sodium form of zeolite (31.20%). The oxidation process of carbon occurs in two steps for both iron zeolite catalysts. Both forms of carbon deposits are easily oxidised. The oxidation proceeds up to 600 °C (Figure S1). Two visible peaks of CO$_2$ (MS profile, Figure 3) with a maximum at 370 °C are attributed to the removal of the surface carbides forms (FeC$_2$ and FeC$_3$) and/or filamentous carbon. This type of carbon is probably α carbon, whereas the peak with a maximum at 510 °C may be attributed to the removal of the oligomerized carbon species, which is signed as β carbon (C$_\beta$) or bulks carbides, in line with earlier reports [26,27]. In the case of the iron catalysts supported on the carbon nanotubes material (MWCNTs), we carried out TG-DTA-MS measurements for calcined and used catalysts in order to calculate the quantity of the carbon deposited on the catalyst surface (Figure S2). The TG-DTA measurements showed that iron catalyst Fe/MWCNT used in F-T had a total carbon of 67.22% while for the calcined catalyst, the total carbon which was oxidized to CO$_2$ was 47.11%. These results confirm that the smallest amount of carbon deposit formed after the F-T reaction (20.11) was in the case of the Fe/MWCNT system. These TG-DTA-MS results indicate that this catalytic material should exhibit the highest stability in the investigated process of the hydrogenation of CO. For iron supported on binary oxides catalysts (Figure S3), two peaks of CO$_2$ formation are also observed and this type of carbon is easily oxidised too. Moreover, the amount of carbon deposition is much lower than those of previously investigated catalysts and is equal to 23.45% (Table 2 and Figure S3). In all studied cases, the type of carbon deposition with a maximum at 310 °C is assigned to the removal of surface carbides forms and/or filamentous carbon (C$_\alpha$), while the second peak with a maximum located at 520 °C can be assigned to the oxidation of iron carbides (C$_\beta$) [26]. It is worth noting that the main type of carbon deposition is β carbon.

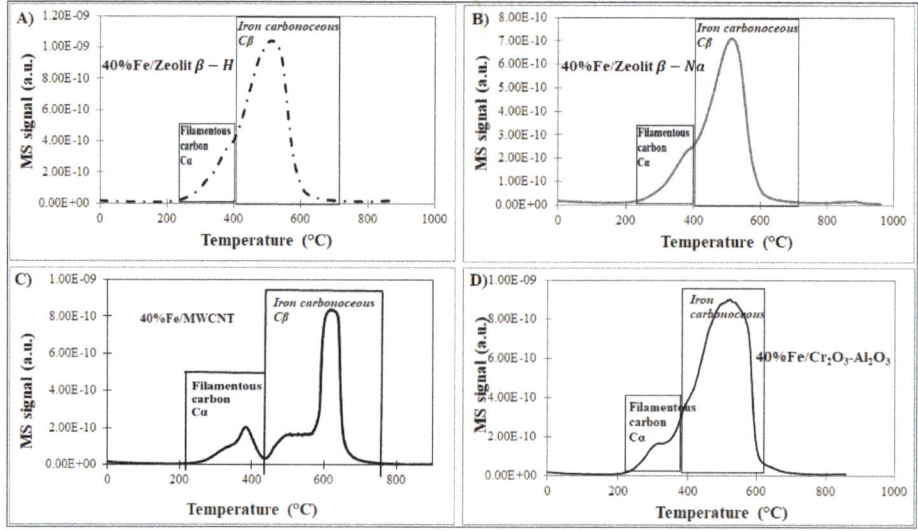

Figure 3. The mass spectrum of the gaseous decomposition products in (**A**) Fe/Zeolite β–H; (**B**) Fe/Zeolite β–Na; (**C**) Fe/MWCNT; (**D**) Fe/Al$_2$O$_3$-Cr$_2$O$_3$.

Table 2. The amount of carbon deposition on iron catalysts.

Catalyst	Amount of Carbon Deposition (%)
40%Fe/H-zeolite	36.50
40%Fe/Na-zeolite	31.20
40%Fe/MWCNT	47.11
40%Fe/MWCNT (after reaction)	67.22 (20.11 *) see Figure S2
40%Fe/Cr_2O_3-Al_2O_3	23.45

* the value in the bracket is a difference between the total carbon in the case of the 40%Fe/MWCNTs catalyst after the reaction and in the case of calcined catalyst. All measurements were done for the catalysts after 20 h of conducting the process.

2.2. Specific Surface Area (SSA) and Pore Size of the Investigated Catalytic Systems

The specific surface area (SSA) and pore sizes of the prepared systems were investigated by the BET method. The SSA measurements of Cr_2O_3-Al_2O_3 and iron catalysts are given in Table 3. The results of the SSA results showed that 40%Fe/Al_2O_3-Cr_2O_3 (2:1) catalyst exhibited the lowest values of specific surface area and pore sizes equal to 118 $m^2 \cdot g^{-1}$ and 2.6 nm, respectively. The SSA measurements obtained for the rest of the catalytic systems showed that the highest specific surface area exhibited 40%Fe/Zeolite β-Na catalyst. Iron catalysts supported on Zeolite β-H had an SSA of 256 $m^2 \cdot g^{-1}$. The pore sizes of the iron catalysts supported on the zeolite systems were practically the same, while, in the case of the 40%Fe/MWCNT system, the SSA measurements showed that the specific surface area of this system was 217 $m^2 \cdot g^{-1}$ and the pore size of this material was equal to 9.3 nm.

Table 3. The results of the specific surface area and average pore radius of iron catalysts.

Catalyst	Specific Surface Area ($m^2 \cdot g^{-1}$)	Average Pore Radius (nm)
40%Fe/MWCNT	217	9.3
40%Fe/Zeolite β-H	256	4.5
40%Fe/Zeolite β-Na	278	4.2
40%Fe/Al2O3-Cr2O3 (2:1)	118	2.6

2.3. Reduction Behaviour of Catalytic Material

The interaction between the components of the catalytic system used in Fischer-Tropsch synthesis was determined using the TPR-H_2 technique. The reduction profiles recorded for supports (MWCNTs, Al_2O_3-Cr_2O_3 (2:1)) and iron catalysts are shown in Figure 4. The TPR-H_2 profile of the MWCNTs did not show any reduction stages. In the case of the binary oxide system, only one reduction peak located in the temperature range 300–580 °C was observed, which was assigned to the reduction of Cr(VI) species formed from the previously oxidized phase of α–Cr_2O_3 [28–32].

The TPR-H_2 curves recorded for the 40%Fe/MWCNT system showed four reduction steps with the maxima of the peaks at 250, 300, 500 and 640 °C, respectively. The first reduction peak located on the TPR profile (240–260 °C) is assigned to the reduction of the surface species of Fe_2O_3 to Fe_3O_4. The second effect with the maximum situated at about 300 °C is connected with the partial reduction of Fe_3O_4 (Fe(II, III)) to the wustite phase (FeO–Fe(II)) [33–35] together with the reduction of Fe_2O_3 strongly interacting with MWCNTs. The third TPR-H_2 effect situated at a higher temperature of about 500 °C is connected with the reduction of the magnetite and wustite phases to metallic iron. The last reduction stage visible on the TPR curve above 550 °C is related to the methanation process of the MWCNTs. The TPR-H_2 profile of 40%Fe/Al_2O_3-Cr_2O_3 (2:1) presents three reduction steps with maxima at about 250, 350 and 550 °C. The first reduction stage is connected with the reduction of Fe_2O_3 to Fe_3O_4. The second reduction stage is attributed to the reduction of the Cr(VI) species and Fe_3O_4 to wustite [2]. The last reduction step presents the reduction of the wustite phase to metallic iron. The TPR-H_2 profiles of zeolite-based catalysts present two reduction stages independently on the support form. The observed reduction stages are connected with the reduction of Fe_2O_3 to Fe_3O_4. The second peak was attributed to the reduction of magnetite to Fe^0 [33–35].

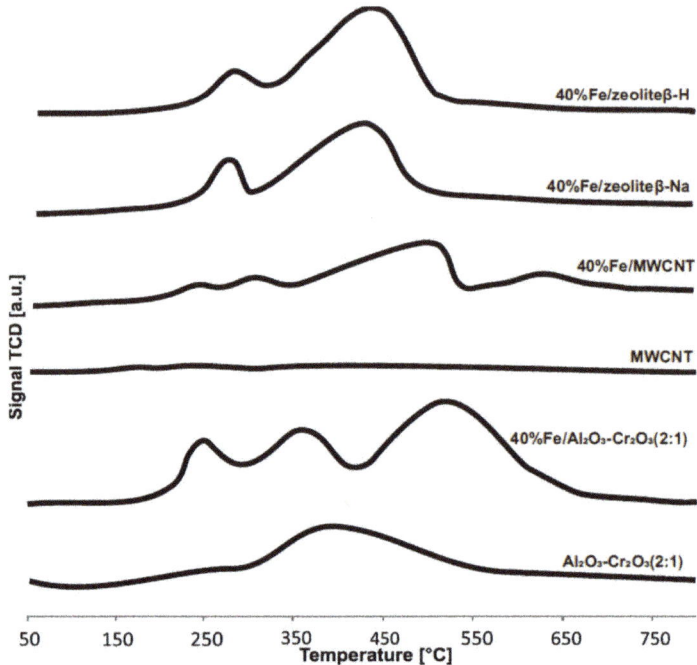

Figure 4. The temperature programmed reduction of hydrogen (TPR-H_2) curves of supports and iron catalysts supported on two types of zeolite β, MWCNTs and Al_2O_3-Cr_2O_3 (2:1).

To better understand the reducibility of the Fe/MWCNT catalyst, XRD measurements in the temperature range of 50–50 °C using a mixture of 5% H_2-95% Ar were performed. The results of the phase composition studies are given in Figure 5. The results clearly showed that starting from 50 °C, only the hematite phase on the diffraction curve recorded for this system was observed. Raising the temperature to 300 °C results in the appearance of reflexes on the XRD curve assigned to the magnetite and wustite phases. This result confirms the first reduction step of the catalytic material. A further increase of the temperature from 300 to 450 °C leads to the increase of the intensity of the diffraction peaks of wustite and the decrease of the intensity of the reflexes of magnetite. This means that at this temperature, the reduction of α-Fe_2O_3 strongly interacts with MWCNTs. Above 550 °C, on the XRD curves recorded for the investigated catalyst, only diffraction peaks assigned to the metallic iron phase was visible. This confirms the total reduction of iron oxides to Fe^0. In addition, this result also confirms the mechanism of the reduction of this system which was postulated based on the TPR reduction measurements that the last reduction stage was connected with the methanation process [36,37] of the MWCNTs. This reduction stage observed on the catalyst surface is confirmed by the lack of other crystallographic phases on the XRD curves recorded above 550 °C.

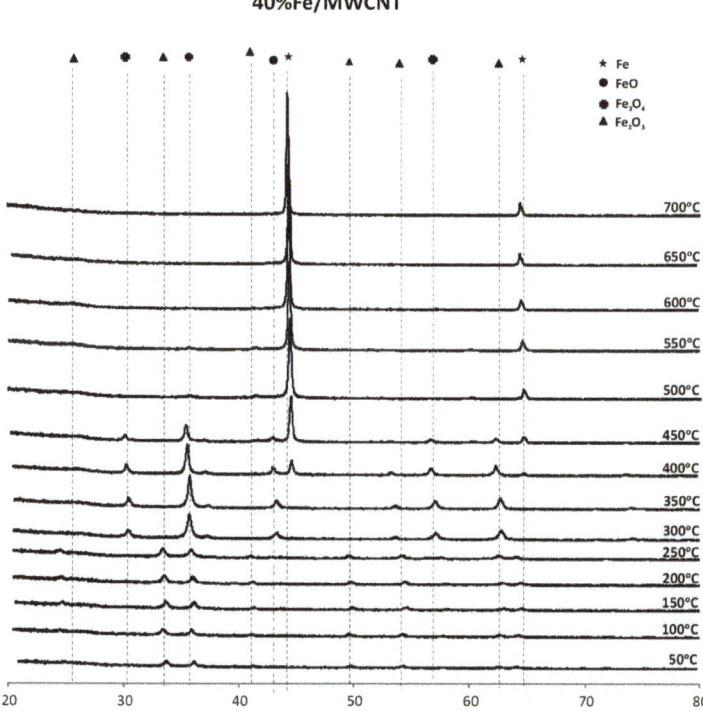

Figure 5. The X-ray diffraction patterns of the Fe/multi-walled carbon nanotube (MWCNT) catalyst collected during the reduction process carried out at a temperature range of 50–700 °C using a reducing mixture of 5% H_2-95% Ar.

2.4. Phase Composition Research of Iron Supported Catalysts

The analysis of the X-ray diffraction studies of the iron catalysts after calcination and after the reaction were studied and the results are presented in Figures 6 and 7. The X-ray analysis of the calcined samples is presented in Figure 6. The diffraction curves recorded for the calcined 40% Fe/MWCNT, 40%Fe/Zeolite β-Na and 40%Fe/Zeolite β-H catalysts confirmed the presence of the hematite and magnetite phases. Moreover, the XRD diffraction curve of the calcined 40% Fe/Al2O3-Cr2O3 (2:1) catalyst showed diffraction peaks assigned to magnetite, $FeCr_2O_4$, $FeAl_2O_4$ phases, respectively. The X-ray diffractograms recorded for the spent Fe catalysts supported on MWCNTs, Al_2O_3-Cr_2O_3 (2:1), zeolite β-H -Na and zeolite β-H are given in Figure 7. The obtained XRD results showed the occurrence of the η-Fe_2C, o-Fe_3C, h-Fe_3C, χ-Fe_5C_2 and magnetite phases on the diffraction patterns recorded for iron catalysts. XRD analysis performed for the spent catalysts did not confirm the presence of the hematite phase on the surface of the catalysts. In addition, depending on the catalyst composition, the $FeCr_2O_3$ and $FeAl_2O_4$ phases were also detected on the diffraction curves recorded for the investigated spent catalysts. It is also worth noting that the previously mentioned phases of carbides were not observed on the XRD curves recorded for the calcined catalysts. These results confirm that iron carbides are the active species of the F-T process and are created during the activation process performed in a mixture of CO and H_2 (50%CO and 50% H_2). Furthermore, the quantity of the carbides determines the catalytic activity in the investigated process. The activity tests performed in the F-T process clearly indicate that iron carbides such as η-Fe_2C, h-Fe_3C, χ-Fe_5C_2 which were created on the surface of the catalyst, are the active species in the hydrogenation of the CO reaction [38,39].

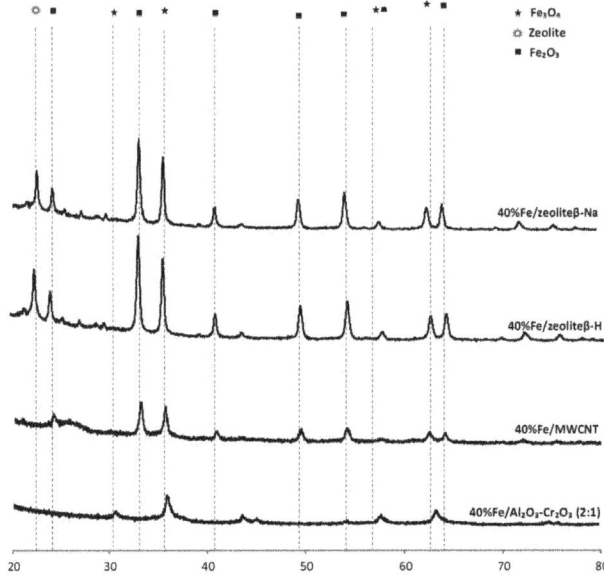

Figure 6. The X-ray diffraction patterns recorded for calcined iron catalysts supported on MWCNTs and binary oxide support.

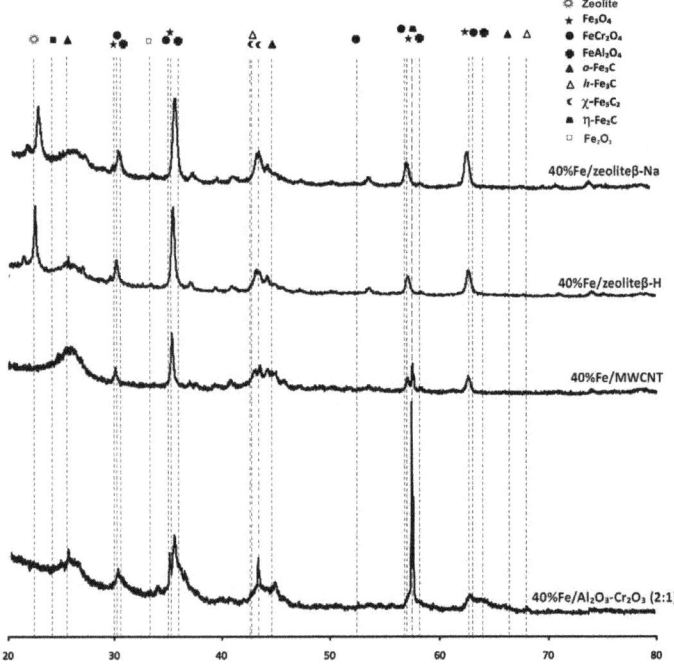

Figure 7. The X-ray diffraction patterns recorded for the spent iron catalysts supported on binary oxide, MWCNTs, zeolite β-H -Na and zeolite β-H.

2.5. The Influence of the Catalytic Support on the Acidity Properties of the Fe Based Catalysts

The acidity and the distribution of acidic centres of the investigated catalysts were determined by the TPD-NH$_3$ method. The distribution of acid centres on the surface of the reduced catalyst has been estimated based on the area under the desorption peaks in the temperature range 100–600 °C and the results are presented in Table 4 and Figure 8. The results of acidity measurements confirmed the existence of three types of acid centres present on the surface of the tested catalysts. The TPD-NH$_3$ results obtained for the investigated catalysts show that 40%Fe/MWCNT had the highest acidity and 40%Fe/Zeolite β-H had the lowest total acidity. In addition, the catalyst based on MWCNTs was characterized by a small number of weak centres and the highest quantity of medium and strong centres compared to the rest of the studied catalytic systems. Furthermore, the catalysts which were characterized by the lowest total acidity of the surface showed the highest CO conversion and high selectivity towards liquid products [4,40].

Table 4. The amount of NH$_3$ adsorbed on the surface of the reduced 40%Fe/MWCNT (multi-walled carbon nanotube), 40%Fe/Zeolite β-H, 40%Fe/Zeolite β-Na and 40%Fe/Al$_2$O$_3$-Cr$_2$O$_3$ (2:1) catalysts (reduction for 1 h in a mixture of 5%H$_2$-95%Ar at 500 °C) calculated from the temperature programmed desorption (TPD-NH$_3$) measurements.

Catalysts/Supports	Weak Centres (mmol g^{-1})	Medium Centres (mmol·g^{-1})	Strong Centres (mmol·g^{-1})	The Total Amount of Desorbed NH$_3$ (mmol·g^{-1})
40%Fe/Al$_2$O$_3$-Cr$_2$O$_3$ (2:1)	0.07	0.09	0.09	0.25
40%Fe/Zeolite β-H	0.22	0.11	0.07	0.40
40%Fe/Zeolite β-Na	0.34	0.23	0.08	0.65
40%Fe/MWCNT	0.03	0.73	0.22	0.98

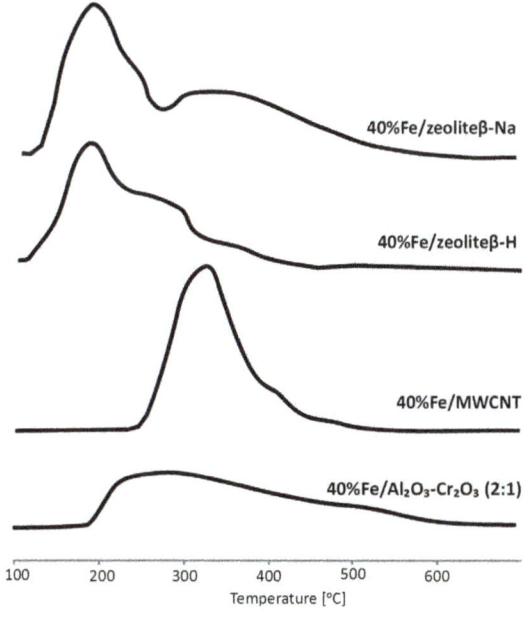

Figure 8. The temperature programmed desorption (TPD-NH$_3$) profiles collected for the reduced catalysts supported on binary oxide, MWCNTs, zeolite β-H -Na and zeolite β-H.

2.6. The Influence of the Carrier Type on the Catalyst Basicity

TPD-CO$_2$ measurements were done for all iron catalysts in the temperature range 100–900 °C and the results are presented in Table 5 and Figure 9, respectively. The obtained measurements performed for the iron systems showed that the catalysts which exhibited the highest total acidity had the lowest basic properties. This tendency was observed for the rest of the investigated catalysts. One exception was the iron catalyst supported on multi-walled carbon nanotubes. This system exhibited the highest acidity as well as the highest basicity compared to other investigated catalysts. The obtained results may be related to the decomposition of the MWCNTs which took place during the temperature treatment of the studied catalyst [41,42].

Table 5. The amount of carbon dioxide adsorbed on the reduction at 500 °C in a mixture of 5%H_2-95%Ar and iron catalysts calculated from the surfaces under the TPD-CO_2 profiles.

Catalysts	Total Basicity (mmol·g^{-1})
40%Fe/$Al_2O_3Cr_2O_3$ (2:1)	1.16
40%Fe/Zeolite-H	1.10
40%Fe/Zeolite-Na	0.91
40%Fe/MWCNT	1.32

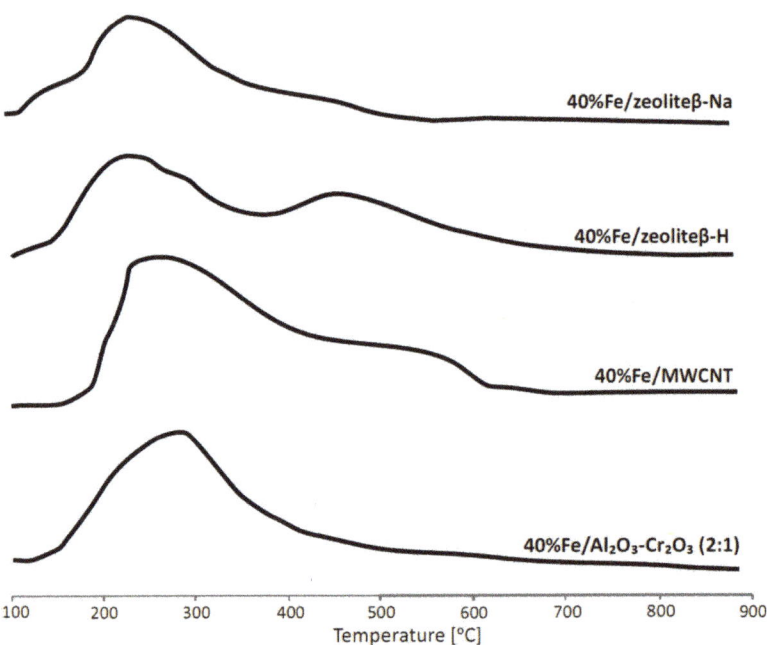

Figure 9. The TPD-CO_2 profiles recorded for reduction at 500 °C in a mixture of 5%H_2-95%Ar and iron catalysts supported on various carriers.

2.7. SEM-EDS Measurements of Iron Catalysts Supported on Two Types of Zeolites

The morphology of the prepared iron-based catalysts supported on Al_2O_3-Cr_2O_3 (2:1), zeolite β-H, and zeolite β-Na was investigated in order to elucidate the reasons for the differences in the activity of monometallic Fe catalysts in the F-T process. Both monometallic catalysts supported on the two types of zeolite were investigated. The investigated catalysts were calcined in an air atmosphere for 4 h and the results are presented in Figure 10. The performed SEM-EDS analysis of the surfaces of the catalyst confirmed the presence of Al, Si, O and Fe elements on their surface. The SEM-EDS analysis of the iron catalyst supported on zeolite β–Na confirmed the lack of sodium on the catalyst surface. This result can be explained by the detection limits of SEM-EDS technique. The only observed differences in the tested surfaces were the contents of the individual elements present on the catalyst surface. In the case of the catalyst deposited on the sodium form of BEA zeolite, larger amounts of Fe were observed, which was confirmed by the XRD method. The 40%Fe/Al$_2$O$_3$-Cr$_2$O$_3$ (2:1) catalyst was also studied by this technique [2].

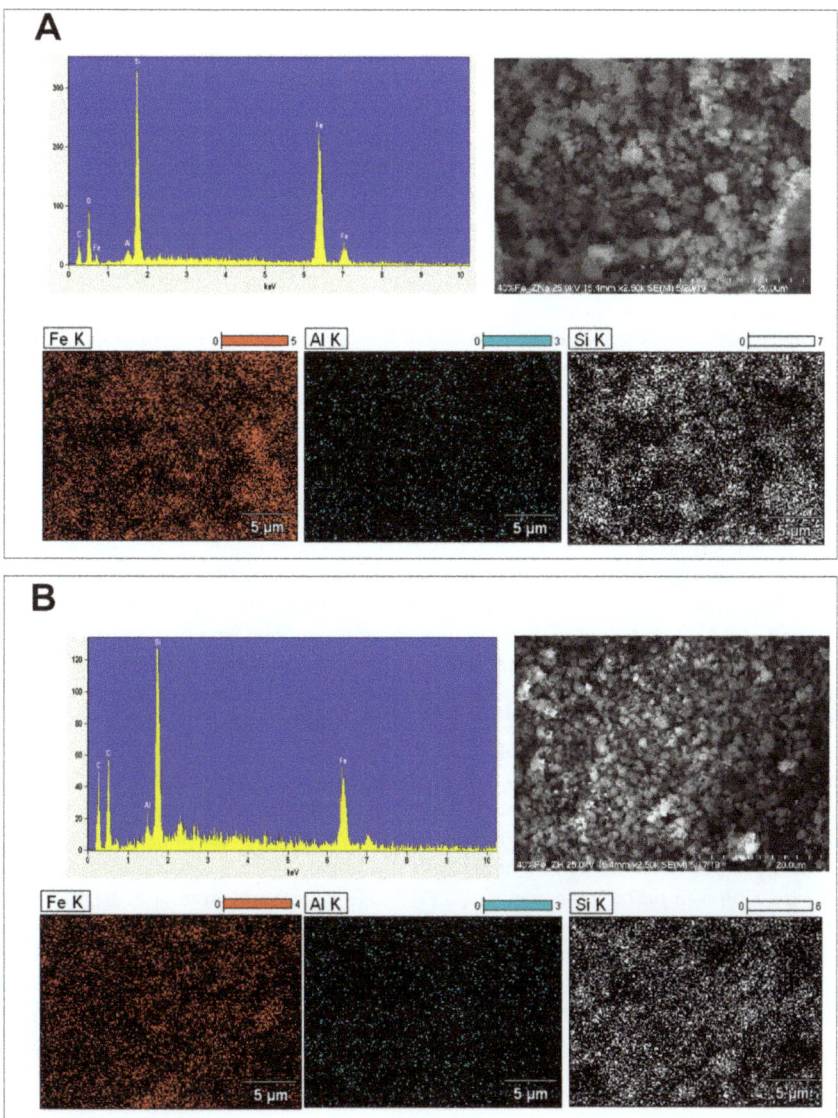

Figure 10. Scanning electron microscope images and EDX spectra collected for the investigated (**A**) 40%Fe/zeolite-β-Na and (**B**) 40%Fe/zeolite-β-H catalysts.

The SEM images collected for the 40%Fe/Al$_2$O$_3$-Cr$_2$O$_3$ (2:1) supported catalyst (results not shown in this paper) confirmed the composition of the prepared catalytic systems. Analysis of the elemental composition of the iron catalyst confirmed the highest quantity of the iron species present on the catalyst surface compared to the rest of the catalysts investigated in this work [2]. This phenomenon explained the activity row observed for the iron catalysts. It is also worth emphasizing that a similar tendency was observed when comparing the discussed catalyst with other iron catalytic systems supported on binary oxide systems which are characterized by various molar ratios between Al and Cr [2].

3. Experimental

3.1. Supports and Catalysts Preparation

Binary oxide support was prepared according to our previous work [2]. In the case of the zeolite supports, the Zeolite β-Na form was purchased from HUTONG GLOBAL CO while the Zeolite β-H form was prepared using the ion exchange of the Zeolite β-Na which was transformed into a hydrogen form by the ion exchange method using ammonium nitrate (V). Zeolite β-Na was treated with a 1 M ammonium nitrate (V) solution at 80 °C for 2 h. Then the residue was filtered and washed with distilled water. Afterwards, the obtained material was dried at 105 °C for 4 h and then the obtained carrier material was calcined 4 h at 400 °C in an air atmosphere. MWCNTs were purchased from Sigma Aldrich (CAS:308068-56-6). Monometallic iron catalyst supported on MWCNTs, binary oxide Al_2O_3-Cr_2O_3 (2:1), Zeolite β-Na and Zeolite β-H were prepared by the conventional impregnation method using $Fe(NO_3)_3 \times 9H_2O$ as a precursor of the active phase. The monometallic Fe catalyst supported on Al_2O_3-Cr_2O_3 (2:1) was then dried in an air atmosphere at 120 °C and then calcined in the same atmosphere at 400 °C for 4 h. The Fe/MWCNT system was calcined in an argon stream for 4 h at 300 °C in order to avoid the oxidation process of the MWCNTs. Monometallic Fe catalysts supported on two kinds of zeolite were prepared analogously to the Fe/Al_2O_3-Cr_2O_3 (2:1) catalyst.

3.2. Characterization of the Catalytic Material

The specific surface area (SSA) and the pore size measurements of catalytic materials were investigated by the BET method based on the low temperature (77 K) nitrogen adsorption using ASAP 2020 Micrometrics (Surface Area and Porosity Analyzer, Micromeritics Instrument Corporation, Norcross, GA, USA). The pore size distributions were determined based on the BJH method. The distributions of the acids centres and total catalytic acidity of catalytic materials were studied using the TPD-NH_3 technique. Each catalytic system (about 0.2 g of the sample) was previously reduced in a mixture of 5%H_2-95%Ar at 500 °C for 1 h and then purified in the helium stream at 600 °C for 30 min^{-1}. Then the sample was cooled down to room temperature in a helium stream. In the next step, the physically adsorbed NH_3 was removed from the surface of the sample in the same atmosphere up to 100 °C. Then the chemisorbed NH_3 desorbed from the catalyst surface was monitored using a thermal-conductivity detector (TCD) during the heating of the sample from 100 to 600 °C. Temperature-programmed desorption of carbon dioxide measurements was performed in a quartz reactor using CO_2 as a probe molecule. The TPD-CO_2 experiments were done in the temperature range of 100–900. The TPD measurements were carried out after removing the CO_2 physisorbed on the catalyst surface. The reduction behaviour of the iron-containing system was investigated by TPR - H_2 technique. In all cases, the TPR-H_2 measurements were carried out using automatic AMI-1 equipment from the ambient to 900 °C. The heating rate in all measurements was 10 and 1 °C min^{-1} for the supports and Fe catalysts, respectively. The mass of the sample was about 0.1 g. During the TPR-H_2 measurements, the mixture of 5%H_2-95%Ar was used. To monitor the concentration of hydrogen, a thermal conductivity detector was applied. Phase composition studies of the calcined and reduced Fe catalysts were performed using a PANalyticalX'Pert Pro MPD diffractometer in the Bragg-Brentano reflecting geometry. Cu Kα radiation was used during all measurements. A PANalyticalX'Celerator detector was used in each experiment in the 2 Theta angles 5–90. Phase composition studies of the reduced catalytic system (5%H_2-95%Ar for 1 h at 500 °C) were performed using "ex situ" XRD measurements (Malvern Panalytical Ltd., Malvern, United Kingdom). In addition, XRD measurements were performed in the reaction chamber in the temperature range of 50–750 °C in a mixture of 5% H_2-95% Ar with a heating rate 10 °C per minute for a selected 40% Fe/MWCNT sample in order to understand its reduction mechanism. A Scanning Electron Microscope (SEM) (S-4700 HITACHI, Tokyo,

Japan) with an energy-dispersive detector (EDX) (ThermoNoran, Madison, WI, USA) was used in order to investigate the surface morphology of the Fe catalysts supported on various carrier materials. The X-ray spectra were used to determine the qualitative analysis of the surface of the catalytic materials. The distributions of each element on the catalytic surface were studied. The carbon deposition formed on the catalyst surface was determined by the TG-DTA-MS technique. Thermogravimetric analysis was carried out for catalysts after 20 h of operation in the reaction using a SETSYS 16/18 thermal analyzer from Setaram (Caluire, France) and a quadrupole mass spectrometer Balzers (Germany). The measurements were carried out in the temperature range of 20–1000 °C using a linear temperature rate of 10 °C/min. In all cases, the sample weight was about 10 mg in the dynamic conditions of a gas stream—Air (Air Products). The flow rate of the gas mixture was 40 cm^3/min in each experiment. The samples were outgassed before each measurement in order to be purified.

3.3. Catalytic Activity Measurements in the F-T Process

Hydrogenation of CO was performed using a high-pressure fixed bed reactor. A gas mixture of hydrogen and carbon monoxide with a molar ratio of 1 to 1 was used in the studies. The total flow of the reaction mixture in each catalytic test was 90 mL min^{-1}. The catalyst mass during the catalytic test was 2 g. The Fischer-Tropsch process was performed at 280 °C and at 30 atmospheres. Before the catalytic activity measurement, the catalyst was activated in a reaction mixture (50%H_2-50%CO for 1 h at 500 °C. The reactivity measurements were carried out after 20 h of the reaction. CO conversion and product distribution were determined using the GC-MS (Hewlett Packard 5890 SERIES II GCMS, United States of America) and GC techniques (equipped with a TCD or FID detector (Inco GC-505M Chromatograph, Poland)). All necessary information about the catalytic tests were also presented in our previous work [2, 16]. In addition, in order to confirm our activity measurements analysis of the liquid hydrocarbons formed via the hydrogenation of CO, an IRTracer-100 FTIR (Shimadzu, Kyoto, Japan) spectrometer was used. The FTIR spectra's of the analysed products were obtained using an MCT detector. A resolution of 4.0 cm^{-1} and 128 scans were applied in each measurement. The "Specac" ATR accessory was used in each analysis.

4. Conclusions

F-T process performed on iron catalysts supported on various carriers prepared by the impregnation method was investigated. The results of the reactivity tests showed that the type of hydrocarbons formed in the hydrogenation of CO reaction depended strongly on the catalyst composition. The catalytic activity measurements indicated that the most active system in the studied process was the 40%Fe/Al_2O_3-Cr_2O_3 (2:1) catalyst. The hydrogenation of CO process performed at 30 atmospheres at a temperature of 280 °C confirmed that this system had high selectivity to linear hydrocarbons. In the case of this catalyst, the small quantity of unsaturated hydrocarbons formation was also detected. In addition, the 40%Fe/zeolite β-H catalyst was characterized by very high selectivity to linear hydrocarbons (94.8%) and to gasoline fraction. Iron catalyst supported on the Na-form of zeolite BEA was characterized by lowest selectivity to linear hydrocarbons (66.6%), but a higher yield for diesel fraction. The reaction performed on 40%Fe/MWCNT system mainly led to the diesel fraction production. The Fischer-Tropsch process conducted on the iron catalysts supported on MWCNTs and Al_2O_3-Cr_2O_3 (2:1) systems had the lowest amount of carbon deposit created on these catalytic material surfaces. These results confirm the possibility of using these systems for the CO hydrogenation process. It was also proven that the catalysts which showed the highest total acidity exhibited the highest CO conversion in the hydrogenation process and also high selectivity towards liquid product.

Supplementary Materials: The following are available online at http://www.mdpi.com/2073-4344/9/7/605/s1, Figure S1. Analysis of catalytic systems carried out in an air atmosphere in the temperature range 25-900°C for A) 40%Fe/Zeolite β-H and B) 40%Fe/Zeolite β-Na, Figure S2. Analysis of catalytic systems carried out in an air atmosphere in the temperature range 25-900°C for A) 40%Fe/MWCNT, B) 40%Fe/MWCNT (after reaction) and C) the mass spectrum of the gaseous decomposition products, Figure S3. Analysis of catalytic systems carried out in an air atmosphere in the temperature range 25-900°C for A) 40%Fe/ Al_2O_3-Cr_2O_3 (after reaction).

Author Contributions: Data curation, P.M.; Investigation, B.D., W.M., K.C.; P.M., W.M. performed XRD measurements, P.M. performed specific surface area measurements, K.C. performed TG-DTA-MS measurements, B.D. performed activity tests, TPR and TPD experiments; Methodology, Research consultation, M.I.S., K.V., I.W., Analysis of the SEM-EDS measurements, M.I.S.; Supervision, P.M., K.V. and M.I.S.; Visualization, P.M. and B.D.; Writing—Original draft, P.M.; Writing—Review & editing, P.M.

Funding: This work was partially funded from NCBiR-Grant no. BIOSTRATEG2/297310/13/NCBiR/2016.

Conflicts of Interest: The authors declare no conflict of interest.

References

1. Tijmensen, M.J.A.; Faaij, A.P.C.; Hamelinck, C.N.; van Hardeveld, M.R.M. Exploration of the possibilities for production of Fischer Tropsch liquids and power via biomass gasification. *Biomass Bioenergy* **2002**, *23*, 129–152. [CrossRef]
2. Mierczynski, P.; Dawid, B.; Maniukiewicz, W.; Mosinska, M.; Zakrzewski, M.; Ciesielski, R.; Kedziora, A.; Dubkov, S.; Gromov, D.; Rogowski, J.; et al. Fischer–Tropsch synthesis over various Fe/Al_2O_3–Cr_2O_3 catalysts. *React. Kinet. Mech. Catal.* **2018**, *124*, 545–561. [CrossRef]
3. Bukur, D.B.; Nowicki, L.; Manne, R.K.; Lang, X.S. Activation Studies with a Precipitated Iron Catalyst for Fischer-Tropsch Synthesis: II. Reaction Studies. *J. Catal.* **1995**, *155*, 366–375. [CrossRef]
4. Chalupka, K.A.; Maniukiewicz, W.; Mierczynski, P.; Maniecki, T.; Rynkowski, J.; Dzwigaj, S. The catalytic activity of Fe-containing SiBEA zeolites in Fischer–Tropsch synthesis. *Catal. Today* **2015**, *257*, 117–121. [CrossRef]
5. Peña, D.; Cognigni, A.; Neumayer, T.; van Beek, W.; Jones, D.S.; Quijada, M.; Rønning, M. Identification of carbon species on iron-based catalysts during Fischer-Tropsch synthesis. *Appl. Catal. A Gen.* **2018**, *554*, 10–23. [CrossRef]
6. Zhang, J.; Chen, J.; Ren, J.; Li, Y.; Sun, Y. Support effect of Co/Al_2O_3 catalysts for Fischer–Tropsch synthesis. *Fuel* **2003**, *82*, 581–586. [CrossRef]
7. Bezemer, G.L.; Bitter, J.H.; Kuipers, H.P.C.E.; Oosterbeek, H.; Holewijn, J.E.; Xu, X.; Kapteijn, F.; van Dillen, A.J.; de Jong, K.P. Cobalt Particle Size Effects in the Fischer–Tropsch Reaction Studied with Carbon Nanofiber Supported Catalysts. *J. Am. Chem. Soc.* **2006**, *128*, 3956–3964. [CrossRef]
8. Guczi, L.; Stefler, G.; Geszti, O.; Koppány, Z.; Kónya, Z.; Molnár, É.; Urbán, M.; Kiricsi, I. CO hydrogenation over cobalt and iron catalysts supported over multiwall carbon nanotubes: Effect of preparation. *J. Catal.* **2006**, *244*, 24–32. [CrossRef]
9. Bahome, M.C.; Jewell, L.L.; Hildebrandt, D.; Glasser, D.; Coville, N.J. Fischer–Tropsch synthesis over iron catalysts supported on carbon nanotubes. *Appl. Catal. A Gen.* **2005**, *287*, 60–67. [CrossRef]
10. Serp, P.; Corrias, M.; Kalck, P. Carbon nanotubes and nanofibers in catalysis. *Appl. Catal. A Gen.* **2003**, *253*, 337–358. [CrossRef]
11. Rodríguez-reinoso, F. The role of carbon materials in heterogeneous catalysis. *Carbon* **1998**, *36*, 159–175. [CrossRef]
12. Zhang, Z.; Zhang, J.; Wang, X.; Si, R.; Xu, J.; Han, Y.F. Promotional effects of multiwalled carbon nanotubes on iron catalysts for Fischer-Tropsch to olefins. *J. Catal.* **2018**, *365*, 71–85. [CrossRef]
13. Van Steen, E.; Prinsloo, F.F. Comparison of preparation methods for carbon nanotubes supported iron Fischer–Tropsch catalysts. *Catal. Today* **2002**, *71*, 327–334. [CrossRef]
14. Tao, Y.; Hattori, Y.; Matumoto, A.; Kanoh, H.; Kaneko, K. Comparative Study on Pore Structures of Mesoporous ZSM-5 from Resorcinol-Formaldehyde Aerogel and Carbon Aerogel Templating. *J. Phys. Chem. B* **2005**, *109*, 194–199. [CrossRef]
15. Chang, C.D.; Lang, W.H.; Silvestri, A.J. Synthesis gas conversion to aromatic hydrocarbons. *J. Catal.* **1979**, *56*, 268–273. [CrossRef]
16. Mierczynski, P.; Dawid, B.; Chalupka, K.; Maniukiewicz, W.; Witonska, I.; Szynkowska, M.I. Role of the activation process on catalytic properties of iron supported catalyst in Fischer-Tropsch synthesis. *J. Energy Inst.* **2019**. [CrossRef]
17. Cho, H.S.; Ryoo, R. Synthesis of ordered mesoporous MFI zeolite using CMK carbon templates. *Microporous Mesoporous Mater.* **2012**, *151*, 107–112. [CrossRef]

18. Caesar, P.D.; Brennan, J.A.; Garwood, W.E.; Ciric, J. Advances in Fischer-Tropsch chemistry. *J. Catal.* **1979**, *56*, 274–278. [CrossRef]
19. Nguyen-Ngoc, H.; Moller, K.; Ralek, M. Liquid Phase Synthesis of Aromates and Isomers on Polyfunctional Zeolitic Catalyst Mixtures. In *Studies in Surface Science and Catalysis*; Jacobs, P.A., Jaeger, N.I., Jírů, P., Kazansky, V.B., Schulz-Ekloff, G., Eds.; Elsevier: Amsterdam, The Netherlands, 1984; Volume 18, pp. 291–297.
20. Chum, H.L.; Overend, R.P. Biomass and renewable fuels. *Fuel Process. Technol.* **2001**, *71*, 187–195. [CrossRef]
21. Moodley, D.J.; van de Loosdrecht, J.; Saib, A.M.; Overett, M.J.; Datye, A.K.; Niemantsverdriet, J.W. Carbon deposition as a deactivation mechanism of cobalt-based Fischer–Tropsch synthesis catalysts under realistic conditions. *Appl. Catal. A Gen.* **2009**, *354*, 102–110. [CrossRef]
22. Maniecki, T.; Stadnichenko, A.; Maniukiewicz, W.; Bawolak, K.; Mierczynski, P.; Boronin, A.; Jozwiak, W. An active phase transformation on surface of Ni-Au/Al_2O_3 catalyst during partial oxidation of methane to synthesis gas. *Kinet. Catal.* **2010**, *51*, 573–578. [CrossRef]
23. Niemantsverdriet, J.W.; Van der Kraan, A.M.; Van Dijk, W.L.; Van der Baan, H.S. Behavior of metallic iron catalysts during Fischer-Tropsch synthesis studied with Moessbauer spectroscopy, x-ray diffraction, carbon content determination, and reaction kinetic measurements. *J. Phys. Chem.* **1980**, *84*, 3363–3370. [CrossRef]
24. Menon, P.G. Coke on catalysts-harmful, harmless, invisible and beneficial types. *J. Mol. Catal.* **1990**, *59*, 207–220. [CrossRef]
25. Sai Prasad, P.S.; Bae, J.W.; Jun, K.-W.; Lee, K.W. Fischer–Tropsch Synthesis by Carbon Dioxide Hydrogenation on Fe-Based Catalysts. *Catal. Surv. Asia* **2008**, *12*, 170–183. [CrossRef]
26. De Bokx, P.K.; Kock, A.J.H.M.; Boellaard, E.; Klop, W.; Geus, J.W. The formation of filamentous carbon on iron and nickel catalysts: I. Thermodynamics. *J. Catal.* **1985**, *96*, 454–467. [CrossRef]
27. Zhang, C.; Zhao, G.; Liu, K.; Yang, Y.; Xiang, H.; Li, Y. Adsorption and reaction of CO and hydrogen on iron-based Fischer–Tropsch synthesis catalysts. *J. Mol. Catal. A Chem.* **2010**, *328*, 35–43. [CrossRef]
28. Maniecki, T.; Mierczynski, P.; Maniukiewicz, W.; Bawolak, K.; Gebauer, D.; Jozwiak, W. Bimetallic Au-Cu, Ag-Cu/$CrAl_3O_6$ Catalysts for Methanol Synthesis. *Catal. Lett.* **2009**, *130*, 481–488. [CrossRef]
29. Maniecki, T.P.; Bawolak, K.; Gebauer, D.; Mierczynski, P.; Jozwiak, W.K. Catalytic activity and physicochemical properties of Ni-Au/Al_3CrO_6 system for partial oxidation of methane to synthesis gas. *Kinet. Catal.* **2009**, *50*, 138–144. [CrossRef]
30. Maniecki, T.; Mierczynski, P.; Bawolak, K.; Lesniewska, E.; Rogowski, J.; Jozwiak, W. Characterization of Cu-(Ag, Au)/$CrAl_3O_6$ Methanol Synthesis Catalysts by TOF-SIMS and SEM-EDS Techniques. *Pol. J. Chem.* **2009**, *83*, 1643–1651.
31. Mierczynski, P.; Maniecki, T.; Maniukiewicz, W.; Jozwiak, W. Cu/Cr_2O_3 center dot 3Al_2O_3 and Au-Cu/Cr_2O_3 center dot 3Al_2O_3 catalysts for methanol synthesis and water gas shift reactions. *React. Kinet. Mech. Catal.* **2011**, *104*, 139–148. [CrossRef]
32. Maniecki, T.P.; Bawolak, K.; Mierczyński, P.; Jozwiak, W.K. Development of Stable and Highly Active Bimetallic Ni–Au Catalysts Supported on Binary Oxides $CrAl_3O_6$ for POM Reaction. *Catal. Lett.* **2008**, *128*, 401. [CrossRef]
33. Jozwiak, W.K.; Kaczmarek, E.; Maniecki, T.P.; Ignaczak, W.; Maniukiewicz, W. Reduction behavior of iron oxides in hydrogen and carbon monoxide atmospheres. *Appl. Catal. A Gen.* **2007**, *326*, 17–27. [CrossRef]
34. Maniecki, T.; Mierczynski, P.; Maniukiewicz, W.; Gebauer, D.; Jozwiak, W. The effect of spinel type support $FeAlO_3$, $ZnAl_2O_4$, $CrAl_3O_6$ on physicochemical properties of Cu, Ag, Au, Ru supported catalysts for methanol synthesis. *Kinet. Catal.* **2009**, *50*, 228–234. [CrossRef]
35. Jozwiak, W.; Maniecki, T.; Mierczynski, P.; Bawolak, K.; Maniukiewicz, W. Reduction Study of Iron-Alumina Binary Oxide $Fe_{2-x}Al_xO_3$. *Pol. J. Chem.* **2009**, *83*, 2153–2162.
36. Mierczynski, P.; Vasilev, K.; Mierczynska, A.; Maniukiewicz, W.; Szynkowska, M.I.; Maniecki, T.P. Bimetallic Au-Cu, Au-Ni catalysts supported on MWCNTs for oxy-steam reforming of methanol. *Appl. Catal. B Environ.* **2016**, *185*, 281–294. [CrossRef]
37. Mierczynski, P.; Vasilev, K.; Mierczynska, A.; Maniukiewicz, W.; Ciesielski, R.; Rogowski, J.; Szynkowska, I.M.; Trifonov, A.Y.; Dubkov, S.V.; Gromov, D.G.; et al. The effect of gold on modern bimetallic Au-Cu/MWCNT catalysts for the oxy-steam reforming of methanol. *Catal. Sci. Technol.* **2016**. [CrossRef]
38. Lu, Y.; Yan, Q.; Han, J.; Cao, B.; Street, J.; Yu, F. Fischer–Tropsch synthesis of olefin-rich liquid hydrocarbons from biomass-derived syngas over carbon-encapsulated iron carbide/iron nanoparticles catalyst. *Fuel* **2017**, *193*, 369–384. [CrossRef]

39. Wezendonk, T.A.; Sun, X.; Dugulan, A.I.; van Hoof, A.J.F.; Hensen, E.J.M.; Kapteijn, F.; Gascon, J. Controlled formation of iron carbides and their performance in Fischer-Tropsch synthesis. *J. Catal.* **2018**, *362*, 106–117. [CrossRef]
40. Chalupka, K.A.; Casale, S.; Zurawicz, E.; Rynkowski, J.; Dzwigaj, S. The remarkable effect of the preparation procedure on the catalytic activity of CoBEA zeolites in the Fischer–Tropsch synthesis. *Microporous Mesoporous Mater.* **2015**, *211*, 9–18. [CrossRef]
41. Mierczynski, P.; Ciesielski, R.; Kedziora, A.; Nowosielska, M.; Kubicki, J.; Maniukiewicz, W.; Czylkowska, A.; Maniecki, T. Monometallic copper catalysts supported on multi-walled carbon nanotubes for the oxy-steam reforming of methanol. *React. Kinet. Mech. Catal.* **2015**, 1–17. [CrossRef]
42. Mierczynski, P.; Mierczynska, A.; Maniukiewicz, W.; Maniecki, T.P.; Vasilev, K. MWCNTs as a catalyst in oxy-steam reforming of methanol. *RSC Adv.* **2016**, *6*, 81408–81413. [CrossRef]

© 2019 by the authors. Licensee MDPI, Basel, Switzerland. This article is an open access article distributed under the terms and conditions of the Creative Commons Attribution (CC BY) license (http://creativecommons.org/licenses/by/4.0/).

Article

Fischer–Tropsch Synthesis: Computational Sensitivity Modeling for Series of Cobalt Catalysts

Harrison Williams [1], Muthu K. Gnanamani [2], Gary Jacobs [3], Wilson D. Shafer [1] and David Coulliette [1,*]

[1] Asbury University, One Macklem Drive, Wilmore, KY 40390, USA; harrison.williams@asbury.edu (H.W.); wilson.shafer@asbury.edu (W.D.S.)
[2] Center for Applied Energy Research, University of Kentucky, 2540 Research Park Dr., Lexington, KY 40511, USA; muthu.gnanamani@uky.edu
[3] Department of Biomedical Engineering and Chemical Engineering, University of Texas at San Antonio/Department of Mechanical Engineering, 1 UTSA Circle, San Antonio, TX 78249, USA; gary.jacobs@utsa.edu
* Correspondence: david.coulliette@asbury.edu; Tel.: +859-858-3511

Received: 22 July 2019; Accepted: 5 October 2019; Published: 15 October 2019

Abstract: Nearly a century ago, Fischer and Tropsch discovered a means of synthesizing organic compounds ranging from C_1 to C_{70} by reacting carbon monoxide and hydrogen on a catalyst. Fischer–Tropsch synthesis (FTS) is now known as a pseudo-polymerization process taking a mixture of CO as H_2 (also known as syngas) to produce a vast array of hydrocarbons, along with various small amounts of oxygenated materials. Despite the decades spent studying this process, it is still considered a black-box reaction with a mechanism that is still under debate. This investigation sought to improve our understanding by taking data from a series of experimental Fischer–Tropsch synthesis runs to build a computational model. The experimental runs were completed in an isothermal continuous stirred-tank reactor, allowing for comparison across a series of completed catalyst tests. Similar catalytic recipes were chosen so that conditional comparisons of pressure, temperature, SV, and CO/H_2 could be made. Further, results from the output of the reactor that included the deviations in product selectivity, especially that of methane and CO_2, were considered. Cobalt was chosen for these exams for its industrial relevance and respectfully clean process as it does not intrinsically undergo the water–gas shift (WGS). The primary focus of this manuscript was to compare runs using cobalt-based catalysts that varied in two oxide catalyst supports. The results were obtained by creating two differential equations, one for H_2 and one for CO, in terms of products or groups of products. These were analyzed using sensitivity analysis (SA) to determine the products or groups that impact the model the most. The results revealed a significant difference in sensitivity between the two catalyst–support combinations. When the model equations for H_2 and CO were split, the results indicated that the CO equation was significantly more sensitive to CO_2 production than the H_2 equation.

Keywords: Fischer–Tropsch synthesis; cobalt; modeling; kinetics

1. Introduction

Fischer–Tropsch synthesis (FTS) is a process used to produce a vast range of organic compounds. Discovered nearly a century ago, FTS reacts H_2 and CO over a metal catalyst to produce C_1-C_{70}-chain compounds, including gasoline and alternate materials [1,2]. Thus, it is often used to generate fuels, monomers for common polymers, rubber, etc. Despite the long history of research and industrial use of FTS, it is still unknown how this process works on a microscopic level [3–22]. There are at least three proposed mechanisms, namely, the carbide originally proposed by Fischer and Tropsch, CO insertion, and formate mechanisms. Yet, none of these can fully describe the entire FTS process, mainly because of the diversity of materials generated by different active metals [1].

To further comprehend FTS, many have attempted to use mathematical modeling as a useful means of describing the process [23–28]. Density functional theory (DFT) calculations, paired with experimental procedures, have provided further insight into the FTS mechanism [18–21]. These models generally fall into one of two categories—probabilistic models and kinetic models. Probabilistic models attempt to describe the propagation of the carbon chain to generate the products observed [23–25]. Kinetic models focus more specifically on the rate of the reaction, determining a mechanism to describe the reaction as a whole [26–28]. While these models have significant application for understanding FTS, they do not exist without limitation. Kinetic models depend on the mechanism of the reaction, which, as stated earlier, is not fully understood due to the diverse array of products. Furthermore, they become quite mathematically complex. Probabilistic models focus on the likelihood of a specific carbon chain propagating or terminating. This type of modeling is limited by the monomer used for propagation, which is not fully understood.

This research sought to create a model that describes the rate of change of hydrogen and carbon monoxide as the reaction progresses. Using a series of industrially relevant cobalt-catalyst runs with varying supports, a system of differential equations was created to model the unreacted syngas in terms of the products generated. Each chemical product output term (such as moles of methane, water, carbon dioxide, etc.) was scaled by an impact parameter, a value in units of hr-1. The model was then tested using sensitivity analysis (SA) to determine which impact parameter and, consequently, which associated chemical product term altered the solution of the model the most. From the results of this analysis, conclusions were drawn about the chemical implications, which may then be used to further examine catalytic runs.

2. Computational Modeling

The computational model was established using chemical engineering process principles from the stirred-tank slurry reactor and basic chemical kinetics. The rate of H_2 and CO coming out of the reactor (unreacted syngas) was modeled using ordinary first-order differential equations (ODEs) [29]. More specifically, the rate of change of H_2 in the reaction can be described by the hydrogen flow into the reactor minus everything coming out of the reactor (H_2, water, methane, hydrocarbon products). All the products (excluding unreacted hydrogen) were scaled by an "impact parameter". This model was applied to CO as well to produce two ODEs as follows:

$$\frac{d(M_{H_2}(t))}{dt} = Q_{H_2} - q_{H_2}M_{H_2}(t) - Z_1 M_{H_2O}(t) - Z_2 M_{CH_4}(t) - Z_3 M_{liq}(t) - Z_9 M_{C_2C_4}(t)\left[-Z_{10}M_{CO_2}(t)\right]* \quad (1)$$

$$\frac{d(M_{CO}(t))}{dt} = Q_{CO} - q_{CO}M_{CO}(t) - Z_4 M_{H_2O}(t) - Z_5 M_{CO_2}(t) - Z_6 M_{CH_4}(t) - Z_7 M_{liq}(t) - Z_8 M_{C_2C_4}(t) \quad (2)$$

where Q_{H2} and Q_{CO} are input rates of H_2 and CO in moles per hour, respectively; q_{H2} and q_{CO} are the output rates for H_2 and CO in units of hr-1, respectively; $MH_2(t)$, $MCO(t)$, $MH_2O(t)$, $MCO_2(t)$, $MCH_4(t)$, $MC_2C_4(t)$, and $Mliq(t)$ are the time-dependent functions for the moles of H_2, CO, CO_2, H_2O, CH_4, C_2–C_4 products, and C_5^+ (or liquid) products, respectively; and Z1–10 are the impact parameters in units of hr-1. It is important to note that $MH_2(t)$ and $MCO(t)$ are the unknown functions for this system. All other mole functions were interpolated from the experimental data (see Supplementary Materials for data), but these two are the functions for which the system is solved. The $MH_2(t)$ and $MCO(t)$ solutions of the ODE system were compared to the experimentally observed values of these functions (see Supplementary Materials) and the resulting sum of squared error was the "response" for the sensitivity analysis. The * symbol at the end of the hydrogen equation is to note that the bracketed term was added later, thus explaining the variance in sequence of impact parameters. Originally, the carbon dioxide term was not included in the hydrogen equation but was later added to explore the possibility of a water–gas shift (WGS).

Due to the innovative nature of the differential equations modeling the reactor, the solver used to evaluate the solutions to this model was the Livermore Solver for Ordinary Differential Equations (LSODA). LSODA performs a test for numerical stiffness as the time steps evolve. A stiff differential equation is one which exhibits solutions with different time scales. For example, stiffness may occur in chemical modeling when simulating reactions with very different reaction rate constants. To solve stiff equations numerically, one must use an implicit solver that requires time-consuming nonlinear equation solutions (e.g., Newton's method for nonlinear systems) for each time step. The LSODA solver primarily uses a non-stiff solver for speed but dynamically checks the solution to determine if a stiff method is required and applies an implicit solver for those time steps. Since the values of the impact parameters were unknown, this approach provided the most flexibility in the numerical solutions. Initial testing of the solution showed good qualitative agreement with the data, indicating that the LSODA solver worked properly with this model. Once the solution was established, sensitivity analysis (SA) was performed. SA is a branch of computational methods that provide insight into how much the total change in the output of a model can be divided up and allocated to the "input" variables/parameters of the model [30]. In other words, SA measures the effect on the output of the model that results from changing the input factors. While there are two types of SA, local and global, local was not used in this work since it involves only measures the change in output at a specified point. Global SA measures the output across a specified range for the input parameters, which was more practical in analyzing this model [30].

While there are many methods for global SA, the Sobol method and Morris method [31–33] were of primary interest for application to the model. These are commonly used for such systems and they are both well validated in multiple applications [32,34,35]. The Sobol method is a variance-based global SA method that generates sensitivity indices from ratios of variance terms [33]. The limitation of Sobol is that calculating the variances is very complex mathematically, and the method requires a large number of samples (often more than 10,000) to counteract any error from negative indices. Furthermore, the primary function of the Sobol method is to analyze the interactions by looking at the second- and higher-order indices, but since the primary focus of this work was the impact of a factor on the model output, the Morris method was used as the SA tool for this preliminary study.

The Morris method is known as a one-at-a-time (OAT) global SA method [31]. This method measures the impact of each input factor on the model by changing only one input factor at a time and observing the change in the model output. Specifically, Morris generates a set of elementary effects (EEs) for each input factor across the given range [31]. An EE is generated by starting with the model evaluated at a specific set of input factors. Then, the model is evaluated again with only one factor increased or decreased by a specified amount, D. The unchanged model output is subtracted from the altered model output, the absolute value is taken for the difference, and this absolute value is divided by D, as follows:

$$EE_i = \frac{|Y(X_1, X_2, \ldots, X_i + D) - Y(X_1, X_2, \ldots, X_i)|}{D} \quad (3)$$

In the equation above, Y represents the model, $X_{1\ldots i}$ is the set of parameters, and i = {1, 2, ..., n}, n being the number of parameters in the model. A set of these EEs using multiple values for D is generated for each input parameter across the given range. Since each set, known as the distribution of EEs, can be quite large, they are sampled. The original Morris sampling technique was a type of random sampling that did not use a high number of samples (10–50), but Campolongo et al. discovered a better sampling technique [31]. In this new technique, 500–1000 original samples are taken from the distribution of EEs. These are then analyzed to determine the optimal 10–50 samples (those that most closely describe the set as a whole). The optimal samples, or optimal trajectories, create a new set that is much smaller and easier to analyze statistically. The average, μ^*, and the standard deviation, σ, are taken for each optimal sample set. μ^* represents the impact of the specific parameter on the model, and σ represents the interaction between a parameter and the other input factors [31]. The greater the magnitude for these values, the greater the impact (μ^*) and interaction (σ) for the corresponding input parameter. Since Morris is a relatively simple method that focuses more on the impact and less on the interactions between parameters, it was chosen as the global SA method for this work. It is important to note that, when performing SA on this model, the impact parameters were the input parameters. While each impact parameter has a corresponding output product (e.g., k_1 corresponds to the hydrogen term for H_2O output), there is not a necessary relationship between the degree of impact and the product itself. However, this work hypothesizes that there is at least a small direct correlation between the magnitude of impact for an impact parameter and the amount of corresponding product produced.

3. Results and Discussion

To ensure the effectiveness of this modeling, the mass closure for these runs had to be accounted for. Since this work was to understand, specifically, how the reactants affected each product through the sensitivity analysis, our entire process would collapse and be non-functional if mass closure was not accounted for. To do this, the mass flow controllers were calibrated beforehand, the flow in and out was measured regularly, and all liquids pulled from the system were weighed and accounted for. The difference between output and input displayed almost 96% closure for mass balance for all our work.

Using the data collected from the experimental runs, the mole functions were interpolated, the differential equations were solved, and the solution was then tested using Morris method SA. Initially, the DEs were loosely coupled, meaning that the solutions for $MH_2(t)$ and $MCO(t)$ could be somewhat dependent on one another and the input parameters (Z1–Z9) could potentially interact. Each set of experimental data was analyzed nine times with Morris, on varying ranges from 0.1–50 to 0.1–500. The number of samples generated varied between 500 and 1000, and the optimal trajectories varied between 0 and 50, as described by Campolongo et al. [31]. The μ^* and σ values were collected from the Morris output and the average for all nine analyses was taken. The μ^* and σ averages for each data set were compared to the other data sets with the same catalyst–support combination and then compared to the values for the other category of data to discern any variance between the two types of supports.

Next, the two DEs were uncoupled. The CO_2 term for the hydrogen equation, along with k_{10}, was added to explore the possibility of WGS occurring in the 0.5 wt% Pt, 25 wt% Co/Al_2O_3 catalyst. Both equations were analyzed using Morris, following the same procedure as described above. The μ^* and σ values were recorded for both the H_2 and the CO equations. These were then compared, primarily the μ^* values for k_5 and k_{10}, the CO_2 impact parameters for the CO and H_2, respectively. In theory, should μ^* for k_{10} consistently be greater than that of $k5$, it could be indicative of hydrogen having a more substantial role in producing CO_2. This could only happen through WGS, and thus if Z10 > Z5, it could indicate the possibility of WGS. Otherwise, if Z5 > Z10, this would most likely perpetuate the prevailing theory, primarily for cobalt catalyst FTS, that the Boudouard reaction is producing the CO_2 [36,37].

Listed below are the data collected from the outlined modeling and sensitivity analysis procedure. As a key, Table 1 is the list of the five product categories and the corresponding impact parameters for the two ODEs in the model. For example, Z_4 is the impact parameter for water in the carbon monoxide equation.

Recall that the μ^* value for each parameter indicates how much that parameter influences the model. In other words, changing the impact parameter with the highest μ^* value will alter the output of the model more than any of the other parameters. The μ^* value does not indicate any level of interaction among the parameters, only how significantly that parameter affects the model. To examine the interactions between the parameters effectively, Sobol method SA should be used. Tables 2 and 3 show the data collected from Morris method SA performed on the coupled DEs.

Table 1. Key for understanding the impact parameters and their corresponding equation (columns 2 and 3) and product (column 1).

Product	CO	H_2
H_2O	Z_4	Z_1
CH_4	Z_6	Z_2
CO_2	Z_5	Z_{10}*
C_2–C_4	Z_8	Z_9
C_{5+}	Z_7	Z_3

The * is to note that Z_{10} was added during the second set of sensitivity analysis runs with the split equations. It does not appear in the first set of data.

Table 2. Averages of the μ^* values across the nine sensitivity analysis (SA) runs of each data set in Category 1, the data from experiments using 20 wt% Co/SiO_2 catalyst.

Parameters	Data Set 1A	Data Set 1B	Data Set 1C	Average
Z_1	1633.96	1866.03	2117.57	1872.52
Z_2	137.167	125.673	135.359	132.733
Z_3	21,848.6	27,452.0	32,795.5	27,365.4
Z_4	3216.69	3743.59	4242.44	3734.24
Z_5	25.4884	21.0851	44.2910	30.2882
Z_6	276.976	253.349	274.167	268.164
Z_7	45,393.1	56,067.5	67,022.8	56,161.1
Z_8	77.2750	82.0301	95.3840	84.8964
Z_9	38.4781	40.2210	47.2408	41.9800

The last column is an average of μ^* across all three runs for each impact parameter.

Table 3. Averages of the μ^* values across the nine SA runs of each data set in Category 2, the data from experiments using 0.5 wt% Pt, 25 wt% Co/Al_2O_3 catalyst.

Parameters	Data Set 2A	Data Set 2B	Data Set 2C	Average
Z_1	2628.02	2147.25	2356.76	2377.34
Z_2	196.136	166.964	179.690	180.930
Z_3	39,224.1	30,725.4	34,196.2	34,715.2
Z_4	5269.89	4190.98	4644.67	4701.85
Z_5	94.4548	64.4021	65.7956	74.8841
Z_6	401.960	338.096	364.823	368.293
Z_7	80,809.9	62,518.7	70,123.6	71,150.8
Z_8	129.969	315.635	129.089	191.564
Z_9	62.2522	57.5090	63.4687	61.0766

The last column is an average of μ^* across all three runs for each impact parameter.

From these data, the order of impact for the nine impact parameters can be observed for both supports. For 20 wt% Co/SiO_2, the order of impact is $Z_7 > Z_3 > Z_4 > Z_1 > Z_6 > Z_2 > Z_8 > Z_9 > Z_5$. In contrast, the order for 0.5 wt% Pt, 25 wt% Co/Al_2O_3 is $Z_7 > Z_3 > Z_4 > Z_1 > Z_6 > Z_2 > Z_8 > Z_5 > Z_9$. Although a subtle difference, the order of k_9 and k_5 is inverted, indicating that k_5 had a greater impact on the model with the cobalt–alumina catalyst than the cobalt–silica. These orders of impact held constant across all three sets of data for each type of support, as can be seen in Tables 2 and 3.

Table 4 lists the averages for each category side-by-side in order to compare the two supports more closely. From the hypothesis, this variation in the order of the parameter impact indicates that CO_2, the corresponding product for Z_5, is more impactful on the alumina support than the silica.

Table 4. Comparison of the average µ* values between the two supports.

Parameters	Data Category 1	Data Category 2
Z_1	1872.52	2377.34
Z_2	132.733	180.930
Z_3	27,365.4	34,715.2
Z_4	3734.24	4701.85
Z_5	30.2882	74.8841
Z_6	268.164	368.293
Z_7	56,161.1	71,150.8
Z_8	84.8964	191.564
Z_9	41.9800	61.0766

Category 1 is 20 wt% Co/SiO_2, Category 2 is 0.5 wt% Pt, 25 wt% Co/Al_2O_3.

Furthermore, based on the proposed correlation between the products and corresponding parameters, alumina produces a greater amount of carbon dioxide than does silica. To verify this hypothesis, the original data from the data sets were examined, and the theory was confirmed. As Table 5 shows, the cobalt–alumina catalysts did in fact produce consistently more carbon dioxide than the cobalt–silica.

Table 5. Average moles of CO_2 produced per sample period for each data set.

Data Set	Category 1	Category 2
A	0.0317108	0.0791931
B	0.0315982	0.0824212
C	0.0642815	0.0675372
Average	**0.0425302**	**0.0763838**

Recall that Category 1 contains data sets with 20 wt% Co/SiO_2 and Category 2 contains data sets with 0.5 wt% Pt, 25 wt% Co/Al_2O_3. The bold values are the averages for each category.

Due to this CO_2 production difference, the two DEs were separated and examined individually using the Morris method. Only the Category 2 data were explored since the CO_2 parameter was more impactful for the cobalt–alumina support. The additional carbon dioxide must come from an additional source. According to the hypothesis, that source could be either water–gas shift (WGS) or the Boudouard reaction. In order to account for any CO_2 produced as a result of hydrogen interaction, thus allowing for the possibility of WGS, the parameter Z_{10} was added to the hydrogen equation along with the CO_2 mole function. Tables 6 and 7 show the data collected from Morris method SA for the separated equations.

Table 6. Average across the nine SA runs for the µ* values of each data set with the 0.5 wt% Pt, 25 wt% Co/Al_2O_3 catalyst.

H_2 Parameters	Data Set 2A	Data Set 2B	Data Set 2C	Average
Z_1	2415.85	1932.34	2146.77	2164.99
Z_2	181.013	154.389	165.666	167.023
Z_3	35,685.7	28,044.22	31,052.3	31,594.1
Z_9	57.3919	53.2704	58.4463	56.3696
Z_{10}	**41.2276**	**29.2660**	**29.3326**	**33.2754**

Hydrogen equation only. The last column is an average of µ* across all three runs for each parameter. The row in bold is the parameter of interest (i.e., the parameter for CO_2).

The parameters Z_5 and Z_{10} were of primary interest. As is evident from the data, the impact of k_5 was invariably greater in magnitude for the CO equation than the impact of Z_{10} for the H_2 equation across all three data sets in Category 2. Since $Z_5 > Z_{10}$, the likelihood of carbon dioxide being produced from WGS is very minimal. The results indicate that the extra CO_2 is a product of the Boudouard reaction.

Table 7. Average across the nine SA runs for the μ* values of each data set with the 0.5 wt% Pt, 25 wt% Co/Al_2O_3 catalyst.

CO Parameters	Data Set 2A	Data Set 2B	Data Set 2C	Average
Z_4	4867.07	3805.30	4251.97	4308.11
Z_5	**86.3071**	**59.4121**	**60.4160**	**68.7117**
Z_6	371.787	310.962	339.218	340.656
Z_7	74,015.7	57,129.3	64,064.8	65,069.9
Z_8	118.735	107.272	119.497	115.168

Carbon monoxide equation only. The last column is an average of μ* across all three runs for each parameter. The row in bold is the parameter of interest (i.e., the parameter for CO_2).

4. Materials and Methods

4.1. Catalyst Preparation

The cobalt and platinum starting materials were cobalt and platinum nitrate purchased from Alfa Aesar (Haverhill, MA, USA). A 25% cobalt alumina catalyst was prepared through a slurry impregnation procedure using Catalox 150 (high purity γ-alumina, ≈150 m^2/g, Sasol, Hamburg, Germany) as a support. Next, the calcination of the catalysts was performed under air at 623 K for 4 h using a tube furnace (Lindberg/MPH, Riverside Road Riverside, MI 49084). The alumina was calcined for 10 h before impregnation and then cooled under an inert gas to room temperature. $Co(NO_3)_2 \cdot 6H_2O$ (99.9% purity) was used as the precursor for Co. In this method, which follows a Sasol patent [38], the ratio of the volume of solution used to the weight of alumina was 1:1, such that approximately 2.5 times the pore volume of solution was used to prepare the loading solution. A two-step impregnation method was used allowing a loading of two portions of 12.5% Co by weight. After the impregnation, the catalyst was then dried under vacuum by means of a rotary evaporator (Buchi R-100, Flawil, Switzerland) at 353 K, then slowly increased to 368 K. After the second impregnation/drying step, the catalyst was calcined under air flow at 623 K for 4 h.

The 20% Co/SiO_2 catalyst was also prepared using the aqueous slurry phase impregnation method, with cobalt nitrate as the cobalt precursor. The support was PQ-SiO$_2$ CS-2133, surface area about 352 m^2/g. The catalyst was calcined in flowing air or flowing ~5% nitric oxide [39] in nitrogen at a rate of 1 L/min for 4 h at 623 K.

4.2. Characterization

4.2.1. BET Surface Area and Porosity Measurements

Surface area, pore volume, and average pore radius of the catalyst calcined at 623 K was evaluated through Brunauer-Emmett-Teller (BET) by using a Micromeritics Tri-Star 3000 gas adsorption analyzer system (Norcross, GA, USA) (Table 8). About 0.35 g of the catalyst was first weighed out, then added to a 3/8 in. o.d. sample tube. The adsorption gas was N_2; sample analysis was performed at the boiling temperature of liquid nitrogen. Prior to the measurement, the chamber temperature was gradually increased to 433 K, then evacuated for several hours to approximately 6.7 Pa.

Table 8. Surface area and pore size distribution of silica-supported cobalt catalysts.

Support/Catalysts	Co, wt%	Reduction Temperature (K)	S_{BET}, m^2/g	Total Pore Volume, cm^3/g	Average Pore Radius (nm)
20% Co/SiO_2	20	623	248	0.856	5.6
0.5%Pt-25%Co/Al_2O_3	25	623	98.5	0.218	4.4

4.2.2. Hydrogen Chemisorption

Hydrogen chemisorption was conducted using temperature-programmed desorption (TPD), Table 8, with a Zeton-Altamira AMI-200 instrument (Pittsburgh, PA, USA). The Co/Al$_2$O$_3$ ($\approx$0.22 g) was activated in a flow of 10 cm^3/min of H$_2$ mixed with 20 cm^3/min of argon at 553 K for 10 h and then cooled under flowing H$_2$ to 373 K. The sample was held at 373 K under flowing argon to remove and/or prevent adsorption of weakly bound species prior to increasing the temperature slowly to 623 K, the temperature at which oxidation of the catalyst was carried out.

The TPD spectrum was integrated and the number of moles of desorbed hydrogen determined by comparing its area to the areas of calibrated hydrogen pulses. The loop volume was first determined by establishing a calibration curve with syringe injections of hydrogen into a helium flow. Dispersion calculations were based on the assumption of a 1:1 H:CO stoichiometric ratio and a spherical cobalt cluster morphology. After TPD of hydrogen, the sample was reoxidized at 623 K using pulses of oxygen. The percentage of reduction was calculated by assuming that the metal reoxidized to Co$_3$O$_4$. Results of chemisorption are reported in Table 9.

Table 9. H$_2$ chemisorption (temperature-programmed desorption (TPD)) and pulse reoxidation of metallic phases of cobalt-supported catalysts (cat.).

Support/Catalysts	µmol H$_2$ Desorbed Per g Cat.	Uncorrected% Dispersion	Uncorrected Diameter (nm)	µmol O$_2$ Consumed Per g Cat.	% Reduction	Corrected% Dispersion	Corrected Diameter (nm)
20% Co/SiO$_2$	41.2	2.43	42.5	1086	48	5.06	20.4
0.5%Pt-25%Co/Al$_2$O$_3$	132.9	6.2	16.5	1733	61.3	10.3	10

4.2.3. Catalyst Activity Testing

Reaction experiments were conducted using a 1 L stirred-tank slurry reactor (STSR; Fort Worth, TX, USA) equipped with a magnetically driven stirrer with turbine impeller, a gas inlet line, and a vapor outlet line with a stainless steel (SS) fritted filter (2 mm) placed external to the reactor. A tube fitted with a SS fritted filter (0.5 mm opening) extending below the liquid level of the reactor was used to withdraw reactor wax (i.e., wax that is solid at room temperature), thereby maintaining a relatively constant liquid level in the reactor. Separate Brooks Instrument mass flow controllers (Hatfield, PA, USA) were used to control the flow rates of hydrogen and carbon monoxide. Carbon monoxide, prior to use, was passed through a vessel containing lead oxide on alumina to remove traces of iron carbonyls. The gases were premixed in an equalization vessel and fed to the STSR below the stirrer, which was operated at 750 rpm. The reactor temperature was maintained constant ($\approx$274 K) using a temperature controller.

A 12 g amount of the oxide cobalt catalyst (sieved 63–106 µm) was loaded into a fixed-bed reactor for 12 h of ex situ reduction at 623 K and atmospheric pressure using a gas mixture of H$_2$/He (60 L/h) with a molar ratio of 1:3. The reduced catalyst was transferred to an already capped 1 L STSR containing 310 g of melted Polywax 3000 (Baker Hughes, Houston, TX) under the protection of inert nitrogen gas. The reactor used for the reduction was weighed three separate times, before and after reduction, and after catalyst transfer, in order to obtain an accurate amount of reduced catalyst that was added. The transferred catalyst was again reduced in situ at 503 K at atmospheric pressure using pure hydrogen (20 L/h) for another 24 h before starting the FT reaction. Each run was held at 493 K and 18.7 atm (275 psig). The CO:H$_2$ ratio was 1:2 for each experiment. The weight hourly space velocity (WHSV) varied from 3.0 to 6.0 slph/g catalyst, depending on the purpose of the experiment.

Gas, water, oil, light wax, and heavy wax samples were collected daily and analyzed. Heavy wax samples were collected in a 473 K hot trap connected to the filter-containing dip tube. The vapor phase in the region above the reactor slurry passed continuously to the warm (373 K) and then the cold (273 K) traps located external to the reactor. The light wax and water mixture were collected daily from the warm trap and an oil plus water sample from the cold trap. Tail gas from the cold trap was analyzed with an online HP Quad Series Micro gas chromatograph (GC) (Palo Alto, CA, USA), providing molar compositions of C_1–C_5 olefins and paraffins, as well as H_2, CO and CO_2. The analysis of the aqueous phase used an SRI 8610C GC (20720 Earl St. Torrance, CA 90503, USA) with a thermal conductivity detector (TCD).

Products were analyzed with an Agilent 6890 GC (Santa Clara, CA, USA) with a flame ionization detector (FID) and a 60 m DB-5 column. Hydrogen and carbon monoxide conversions were calculated based on GC analysis of the gas products, the gas feed rate, and the gas flow that was measured at the outlet of the reactor.

5. Conclusions

In conclusion, the mathematical model created in this work accurately described the FTS process in terms of rate of change of syngas. When the model was analyzed using Morris method SA, a significant difference was observed between cobalt–silica and cobalt–platinum–alumina catalysts, namely a difference in the effect of the impact parameters k_5 and k_9. Based on the initial hypothesis, this difference indicated that the cobalt–platinum–alumina generated more carbon dioxide than did the cobalt–silica. This was verified by the raw data as well. These conclusions are highly specific to the two catalyst–support types in question, and thus more general conclusions must come from further research. Thus, the equations were separated, and the cobalt–platinum–alumina catalyst data were reexamined, now with an additional CO_2 term on the hydrogen equation to test for the possibility of WGS. The observed results showed that the carbon monoxide contribution to CO_2 was consistently higher than the hydrogen contribution. Since WGS is the only pathway in which hydrogen produces CO_2 during FTS, WGS was excluded as a source of CO_2 production. Hence, the Boudouard reaction was determined to be the more probable source of additional carbon dioxide. Though the Boudouard reaction yields a high amount of carbon on the surface, to the best of our knowledge, there is no direct evidence that this reaction leads directly and solely to inert graphitic carbon (coke). Moreover, one of the main mechanistic routes proposed for FT is the carbide mechanism, where CO completely dissociates before reacting with hydrogen. In order to understand this rate of formation for C, evidence of more than one type of carbon on the catalyst surface, an active phase and an unreactive phase, has been observed on multiple catalysts such as cobalt [40] and Fe [41].

Further indications using sensitivity analysis show that both the H_2 and CO equations displaying Z_3 and Z_7 obtained the highest value, indicating the liquids are most sensitive to any deviations brought on by H_2 and CO. Though expected, these details could further indicate the importance for a specific balance in the FTS system between the two reactants, suggesting the difficulty in tuning a specific catalyst for C_5^+ products as a whole. Additionally, the C_5^+ products are more sensitive in the CO reaction than the H_2, which implies that the balance for the FTS system lies more heavily in the CO than for H_2, primarily describing the importance of CO, possibly due to the metal's ability to back donation [42] in the FTS process.

While this research is preliminary, the model and analysis produced some excellent descriptive results for FTS. Although this work only examined cobalt catalyst data, future work could include the incorporation of multiple catalysts and supports. Impact parameters could potentially be optimized in future work to find the ideal values for a given catalyst. The model can be adapted to include terms for olefins and paraffins, and the liquid (C_{5+}) term could be broken down into individual terms for each number of carbons per chain (e.g., C_5, C_6, C_7, etc.). The equations could be changed to predict specific product amounts based on initial conditions such as catalyst, support, temperature, pressure, flowrate, etc. Ultimately, this type of FTS modeling not only describes the process in various and versatile ways but can also be adapted and altered to potentially produce a predictive model.

Supplementary Materials: The following are available online at http://www.mdpi.com/2073-4344/9/10/857/s1.

Author Contributions: H.W. gathered information and wrote most of the paper. W.D.S. provided most of the analytical results and oversaw part of the process, G.J. provided the catalyst characterization and finalized edits,

M.K.G. provided experimental oversight, data collection/analysis and editing. D.C. designed and coded the algorithm and oversaw the entire project.

Funding: This research received no external funding.

Conflicts of Interest: The authors declare no conflict of interest.

References

1. Fischer, F.; Tropsch, H. Synthesis of Petroleum at Atmospheric Pressure from Gasification Products of Coal. *Brennstoff-Chemie* **1926**, *7*, 97–104.
2. Fischer, F.; Tropsch, H. Development of the Benzene Synthesis from Carbon Monoxide and Hydrogen at Atmospheric Pressure. *Brennstoff-Chemie* **1930**, *11*, 489–500.
3. Schulz, H. Short history and present trends of Fischer–Tropsch synthesis. *Appl. Catal.* **1999**, *186*, 3–12. [CrossRef]
4. Davis, B.H. Fischer–Tropsch Synthesis: Reaction mechanisms for iron catalysts. *Catal. Today* **2009**, *141*, 25–33. [CrossRef]
5. Navarro, V.; van Spronsen, M.A.; Frenken, J.W.M. In situ observation of self-assembled hydrocarbon Fischer–Tropsch products on a cobalt catalyst. *Nat. Chem.* **2016**, *8*, 929–934. [CrossRef]
6. Chakrabarti, D.; Gnanamani, M.K.; Shafer, W.D.; Ribeiro, M.C.; Sparks, D.E.; Prasad, V.; de Klerk, A.; Davis, B.H. Fischer–Tropsch Mechanism: $^{13}C^{18}O$ Tracer Studies on a Ceria–Silica Supported Cobalt Catalyst and a Doubly Promoted Iron Catalyst. *Ind. Eng. Chem. Res.* **2015**, *54*, 6438–6453. [CrossRef]
7. Tuxen, A.; Carenco, S.; Chintapalli, M.; Chuang, C.; Escudero, C.; Pach, E.; Jiang, P.; Borondics, F.; Beberwyck, B.; Alivisatos, A.P.; et al. Size-Dependent Dissociation of Carbon Monoxide on Cobalt Nanoparticles. *J. Am. Chem. Soc.* **2013**, *135*, 2273–2278. [CrossRef]
8. Qi, Y.; Yang, J.; Chen, D.; Holmen, A. Recent Progresses in Understanding of Co-Based Fischer–Tropsch Catalysis by Means of Transient Kinetic Studies and Theoretical Analysis. *Catal. Lett.* **2015**, *145*, 145–161. [CrossRef]
9. Van Santen, R.A.; Markvoort, A.J.; Filot, I.A.W.; Ghouriab, M.M.; Hensen, E.J.M. Mechanism and Microkinetics of the Fischer–Tropsch reaction. *Phys. Chem. Chem. Phys.* **2013**, *15*, 17038–17063. [CrossRef]
10. Banerjee, A.; van Bavel, A.V.; Kuipers, H.P.C.E.; Saeys, M. CO Activation on Realistic Cobalt Surfaces: Kinetic Role of Hydrogen. *ACS Catal.* **2017**, *7*, 5289–5293. [CrossRef]
11. Todic, B.; Bhatelia, T.; Froment, G.F.; Ma, W.; Jacobs, G.; Davis, B.H.; Bukur, D.B. Kinetic Model of Fischer-Tropsch Synthesis in a Slurry Reactor on Co-Re/Al_2O_3 Catalyst. *Ind. Eng. Chem. Res.* **2013**, *52*, 669–679. [CrossRef]
12. Keyvanloo, K.; Fisher, M.J.; Hecker, W.C.; Lancee, R.J.; Jacobs, G.; Bartholomew, C.H. Kinetics of deactivation by carbon of a cobalt Fischer–Tropsch catalyst: Effects of CO and H_2 partial pressures. *J. Catal.* **2015**, *327*, 33–47. [CrossRef]
13. Ma, W.; Jacobs, G.; Sparks, D.E.; Spicer, R.L.; Davis, B.H.; Klettlinger, J.L.S.; Yen, C.H. Fischer-Tropsch synthesis: Kinetics and water effect study over 25%Co/Al_2O_3 catalysts. *Catal. Today* **2014**, *228*, 158–166. [CrossRef]
14. Botes, F.G.; van Dyk, B.; McGregor, C. The Development of a Macro Kinetic Model for a Commercial Co/Pt/Al_2O_3 Fischer-Tropsch Catalyst. *Ind. Eng. Chem. Res.* **2009**, *48*, 10439–10447. [CrossRef]
15. Brady, R.C.; Pettit, R. Mechanism of the Fischer-Tropsch reaction. The chain propagation step. *J. Am. Chem. Soc.* **1981**, *103*, 1287–1289. [CrossRef]
16. Yang, J.; Shafer, W.D.; Pendyala, V.R.R.; Jacobs, G.; Chen, D.; Holmen, A.; Davis, B.H. Fischer–Tropsch Synthesis: Using Deuterium as a Tool to Investigate Primary Product Distribution. *Catal. Lett.* **2014**, *144*, 524–530. [CrossRef]
17. Shafer, W.D.; Jacobs, G.; Davis, B.H. Fischer–Tropsch Synthesis: Investigation of the Partitioning of Dissociated H_2 and D_2 on Activated Cobalt Catalyst. *ACS Catal.* **2012**, *2*, 1452–1456. [CrossRef]
18. Yang, J.; Qia, Y.; Zhu, J.; Zhu, Y.-A.; Chen, D.; Holmen, A. Reaction mechanism of CO activation and methane formation on Co Fischer–Tropsch catalyst: A combined DFT, transient, and steady-state kinetic modeling. *J. Catal.* **2013**, *308*, 37–49. [CrossRef]

19. Ojeda, M.; Nabar, R.; Nilekar, A.U.; Ishikawa, A.; Mavrikakis, M.; Iglesia, E. CO activation pathways and the mechanism of Fischer–Tropsch synthesis. *J. Catal.* **2010**, *272*, 287–297. [CrossRef]
20. Ojeda, M.; Li, A.; Nabar, R.; Nilekar, A.U.; Mavrikakis, M.; Iglesia, E. Kinetically Relevant Steps and H_2/D_2 Isotope Effects in Fischer−Tropsch Synthesis on Fe and Co Catalysts. *J. Phys. Chem. C* **2010**, *114*, 19761–19770. [CrossRef]
21. Gracia, J.M.; Prinsloo, F.F.; Niemantsverdriet, J.W. Mars-van Krevelen-like Mechanism of CO Hydrogenation on an Iron Carbide Surface. *Catal. Lett.* **2009**, *133*, 257–261. [CrossRef]
22. Davis, B.H. Fischer-Tropsch synthesis: Current mechanism and futuristic needs. *Fuel Process. Technol.* **2001**, *71*, 157–166. [CrossRef]
23. Filip, L.; Zámostný, P.; Rauch, R. Mathematical model of Fischer-Tropsch synthesis using variable alpha-parameter to predict product distribution. *Fuel* **2019**, *243*, 603–609. [CrossRef]
24. Ahón, V.R.; Costa, E.F.; Monteagudo, J.E.P.; Fontes, C.E.; Biscaia, E.C.; Lage, P.L.C. A comprehensive mathematical model for the Fischer–Tropsch synthesis in well-mixed slurry reactors. *Chem. Eng. Sci.* **2005**, *60*, 677–694. [CrossRef]
25. Yu, F.; Lin, T.; Wang, X.; Li, S.; Lu, Y.; Wang, H.; Zhong, L.; Sun, Y. Highly selective production of olefins from syngas with modified ASF distribution model. *Appl. Catal. A Gen.* **2018**, *563*, 146–153. [CrossRef]
26. Nikbakht, N.; Mirzaei, A.A.; Atashi, H. Kinetic modeling of the Fischer-Tropsch reaction over a zeolite supported Fe-Co-Ce catalyst prepared using impregnation procedure. *Fuel* **2018**, *229*, 209–216. [CrossRef]
27. Ostadi, M.; Rytter, E.; Hillestad, M. Evaluation of kinetic models for Fischer–Tropsch cobalt catalysts in a plug flow reactor. *Chem. Eng. Res. Des. Trans. Inst Chem. Eng. Part A* **2016**, *114*, 236–246. [CrossRef]
28. Nabipoor Hassankiadeh, M.; Haghtalab, A. Product Distribution of Fischer-Tropsch Synthesis in a Slurry Bubble Column Reactor Based on Langmuir-Freundlich Isotherm. *Chem. Eng. Commun.* **2013**, *200*, 1170–1186. [CrossRef]
29. Seborg, D.E.; Mellichamp, D.A.; Thomas, F. *Process Dynamics and Control*; John Wiley & Sons, Inc.: Hoboken, NJ, USA, 2011.
30. Papraćanin, E. Local and Global Sensitivity Analysis of Modeling Parameters for Composting Process. *Technol. Acta* **2019**, *11*, 9–16.
31. Campolongo, F.; Cariboni, J.; Saltelli, A. An effective screening design for large models. *Environ. Model. Softw.* **2007**, *22*, 1509–1518. [CrossRef]
32. Zhang, X.; Trame, M.; Lesko, L.; Schmidt, S. Sobol Sensitivity Analysis: A Tool to Guide the Development and Evaluation of Systems Pharmacology Models. *CPT Pharmacomet. Syst. Pharmacol.* **2015**, *4*, 69–79. [CrossRef]
33. Sobol, I. Global sensitivity indices for nonlinear mathematical models and their Monte Carlo estimates. *Math. Comput. Simul.* **2001**, *55*, 271–280. [CrossRef]
34. Cariboni, J.; Gatelli, D.; Liska, R.; Saltelli, A. The role of sensitivity analysis in ecological modeling. *Ecol. Model.* **2007**, *203*, 167–182. [CrossRef]
35. Nossent, J.; Elsen, P.; Bauwens, W. Sobol' sensitivity analysis of a complex environmental model. *Environ. Model. Softw.* **2011**, *26*, 1515–1525. [CrossRef]
36. Hunt, J.; Ferrari, A.; Lita, A.; Crosswhite, M.; Ashley, B.; Stiegman, A.E. Microwave-Specific Enhancement of the Carbon-Carbon Dioxide (Boudouard) Reaction. *J. Phys. Chem.* **2013**, *117*, 26871–26880. [CrossRef]
37. Bremmer, G.M.; Zacharaki, E.; Sjåstad, A.O.; Navarro, V.; Frenken, J.W.M.; Kooyman, P.J. In situ TEM observation of the Boudouard Reaction: Multi-layered graphene formation from CO on cobalt nanoparticles at atmospheric pressure. *Faraday Discuss.* **2017**, *197*, 337–351. [CrossRef]
38. Espinoza, R.L.; Visagie, J.L.; van Berge, P.J.; Bolder, F.H. Catalysts. US Patent 5,733,839, 31 March 1998.
39. Jacobs, G.; Ma, W.; Davis, B.H.; Cronauer, D.C.; Kropf, J.; Marshall, C.L. Fischer–Tropsch Synthesis: TPR-XAFS Analysis of Co/Silica and Co/Alumina Catalysts Comparing a Novel NO Calcination Method with Conventional Air Calcination. *Catal. Lett.* **2010**, *140*, 106–115. [CrossRef]
40. Chen, W.; Kimpel, T.F.; Song, Y.; Chiang, K.-F.; Zijlstra, B.; Pestman, R.; Wang, P.; Hensen, E.J.M. Influence of Carbon Deposits on the Cobalt-Catalyzed Fischer–Tropsch Reaction: Evidence of a Two-Site Reaction Model. *ACS Catal.* **2018**, *8*, 1580–1590. [CrossRef]

41. Zhang, C.; Zhao, G.; Liu, K.; Yang, Y.; Xiang, H.; Li, Y. Adsorption and reaction of CO and hydrogen on iron-based Fischer–Tropsch synthesis catalysts. *J. Mol. Catal. A* **2010**, *328*, 35–43. [CrossRef]
42. Shafer, W.D.; Gnanamani, M.K.; Graham, U.M.; Yang, J.; Masuku, C.M.; Jacobs, G.; Davis, B.H. Fischer–Tropsch: Product Selectivity–The Fingerprint of Synthetic Fuels. *Catalysts* **2019**, *9*, 259. [CrossRef]

© 2019 by the authors. Licensee MDPI, Basel, Switzerland. This article is an open access article distributed under the terms and conditions of the Creative Commons Attribution (CC BY) license (http://creativecommons.org/licenses/by/4.0/).

Article

Fischer-Tropsch Synthesis: Cd, In and Sn Effects on a 15%Co/Al₂O₃ Catalyst

Wenping Ma [1,*], Gary Jacobs [1,†], Wilson D. Shafer [1,‡], Yaying Ji [1], Jennifer L. S. Klettlinger [2], Syed Khalid [3], Shelley D. Hopps [1] and Burtron H. Davis [1,§]

[1] Center for Applied Energy Research, University of Kentucky, 2540 Research Park Drive, Lexington, KY 40511, USA; gary.jacobs@utsa.edu (G.J.); wilson.shafer@asbury.edu (W.D.S.); Yaying.Ji@uky.edu (Y.J.); Shelley.hopps@uky.edu (S.D.H.)
[2] NASA Glenn Research Center, 21000 Brookpark Rd., Cleveland, OH 44135, USA; j.klettlinger@nasa.gov
[3] NSLS, Brookhaven National Lab., Brookhaven Ave., Upton, NY 11973, USA; khalid@bnl.gov
* Correspondence: wenping.ma@uky.edu
† Current address: Department of Chemical Engineering and Biomedical Engineering/Department of Mechanical Engineering, University of Texas at San Antonio, 1 UTSA Circle, San Antonio, TX 78249, USA.
‡ Current address: Asbury University, One Macklem Drive, Wilmore, KY 40390, USA.
§ Decreased.

Received: 30 August 2019; Accepted: 11 October 2019; Published: 16 October 2019

Abstract: The effects of 1% of Cd, In and Sn additives on the physicochemical properties and Fischer-Tropsch synthesis (FTS) performance of a 15% Co/Al₂O₃ catalyst were investigated. The fresh and spent catalysts were characterized by BET, temperature programmed reduction (TPR), H_2-chemisorption, NH_3 temperature programmed desorption (TPD), X-ray absorption near edge spectroscopy (XANES), and X ray diffraction (XRD). The catalysts were tested in a 1 L continuously stirred tank reactor (CSTR) at 220 °C, 2.2 MPa, H_2/CO = 2.1 and 20–55% CO conversion. Addition of 1% of Cd or In enhanced the reduction degree of 15%Co/Al₂O₃ by ~20%, while addition of 1% Sn slightly hindered it. All three additives adversely impacted Co dispersion by 22–32% by increasing apparent Co cluster size based on the H_2-chemisorption measurements. However, the decreased Co active site density resulting from the additives did not result in a corresponding activity loss; instead, the additives decreased the activity of the Co catalysts to a much greater extent than expected, i.e., 82–93%. The additional detrimental effect on catalyst activity likely indicates that the Cd, In and Sn additives migrated to and covered active sites during reaction and/or provided an electronic effect. XANES results showed that oxides of the additives were present during the reaction, but that a fraction of metal was also likely present based on the TPR and reaction testing results. This is in contrast to typical promoters that become metallic at or below ~350 °C, such as noble metal promoters (e.g., Pt, Ru) and Group 11 promoters (e.g., Ag, Au) on Co catalysts in earlier studies. In the current work, all three additives remarkably increased CH_4 and CO_2 selectivities and decreased C_{5+} selectivity, with the Sn and In additives having a greater effect. Interestingly, the Cd, In, or Sn additives were found to influence hydrogenation and isomerization activities. At a similar conversion level (i.e., In the range of 40–50%), the additives significantly increased 2-C_4 olefin content from 3.8 to 10.6% and n-C_4 paraffin from 50 to 61% accompanied by decreases in 1-C_4 olefin content from 48 to 30%. The Sn contributed the greatest impact on the secondary reactions of 1-olefins, followed by the In and Cd. NH_3-TPD results suggest enhanced acid sites on cobalt catalysts resulting from the additives, which likely explains the change in selectivities for the different catalysts.

Keywords: Fischer-Tropsch synthesis; Co; Al₂O₃; Pt; Cd; In; Sn; hydrocarbon selectivity; synergic effect; GTL; additives; reducibility; XANES

1. Introduction

Supported cobalt catalysts have received renewed attention in converting natural gas to liquid fuels (GTL) due to their high activity, high selectivity toward heavier hydrocarbons and excellent stability in long term operation. A large number of studies in past decades have focused on developing various supported Co catalysts aimed at high productivity of heavier hydrocarbons and good stability. Al_2O_3, SiO_2, and TiO_2 supports have been commonly used since the energy crisis of the 1970s. Because the reducibility and dispersion of cobalt, which remarkably affect Fischer-Tropsch synthesis (FTS) performance of cobalt catalysts, are closely related to the interaction between the support and cobalt, appropriate reduction additives such as noble metals (e.g., Pt, Ru, and Re) or Group 11 metals (e.g., Au, Ag), structural modifiers (e.g., Zr, Ce, K, Mn), or a high Co loading (e.g., >20%) are used to overcome the strong interaction issue, consequently leading to increased productivity [1–25]. Additives may also provide additional benefits by producing hydrocarbons with a desired product spectrum, thus significantly decreasing the catalyst cost. Generally, cobalt-based catalysts are reduced at 350 °C under hydrogen for several hours to activate cobalt metal sites. Under this standard reduction scheme, Co/Al_2O_3 catalysts were found to have a limited reducibility but a relatively higher Co dispersion due to the strong interaction between the support and cobalt oxides [3–8], whereas Co/SiO_2 and Co/TiO_2 catalysts usually exhibited higher reducibility but with a lower Co dispersion because of the weaker interactions between the support and cobalt oxides [3,4]. Therefore, additive effect studies have been an important topic ever since the discovery of FTS in the 1920s.

The impact of noble metal additives (e.g., Pt, Ru, Re and Pd) having an atomically equivalent loading as that of 0.5% by weight Pt, has been researched carefully in order to understand the catalyst structure-performance relationships [4,6,7,16,17]. It was found that all noble metal additives significantly improved the Co reduction degree in $25\%Co/Al_2O_3$ from 55 to 68–72% and increased Co dispersion from 5.5 to 9–10% [7], which was likely through a H_2 dissociation and spillover mechanism [3,14]. The increased Co site density due to the reduction promoting capability of the noble metal additives led to a near doubling of the FTS activity in comparison with the unpromoted cobalt catalyst. However, Pd was an exception; Pd addition resulted in nearly unchanged FTS activity, and it accelerated the deactivation rate of the cobalt catalyst.

The noble metal additives were also found to alter hydrocarbon selectivity in different ways. At atomically equivalent loadings to 0.5% by weight Pt and about 50% conversion, the Ru and Re additives improved C_{5+} selectivity and suppressed methane formation, while the Pd additive prohibitively worsened selectivities by increasing the formation of methane and light hydrocarbons at the expense of losing heavier hydrocarbons [7]; Pt also tended to worsen the selectivities, but this effect was only slight. Further studies by XANES/EXAFS indicated different structures of the noble metal additives. Both Pd-Pd and Pd-Co coordination were found in the spent Pd-Co catalysts, but for the other promoted catalysts (Pt, Re, and Ru-which performed significantly better than the Pd promoted one), only coordination from the additive to Co was detected by EXAFS [6,16]. Thus, the small Pd particles or Pd patches were deemed to be responsible in part for losses in catalyst performance.

The effect of Group 11 metals (e.g., Cu, Ag, and Au with a loading range of 0.5–5.0%) on $15\%Co/Al_2O_3$ catalysts have been also investigated with the aim of (1) finding a substitute for Pt and (2) possibly lowering light gas selectivity, thus potentially reducing the costs of the cobalt catalyst and process [5]. It was found that all levels of Group 11 metals improved the reducibility of cobalt oxides (50 to 70–90%), leading to improved catalyst activity and stability especially for Ag and low levels of Au. At a similar CO conversion level of ca. 50%, the addition of less than 1.5% or 2.8% Ag not only improved the CO rate, but also improved the selectivity towards heavier hydrocarbons relative to the unpromoted cobalt catalyst (81 to 83%). However, Cu (>0.5%) and higher loadings of Au significantly poisoned the cobalt catalyst and remarkably promoted CH_4 formation (9 to 21%) in comparison to the unpromoted cobalt catalyst. This could be due to Cu metal being present on the Co surface and blocking Co sites and affecting the relative hydrogenation rates. Similar to the Pd additive effect addressed previously, EXAFS results revealed only Co-Co and Me-Me (Me = Cu, Ag and Au) structures in the cobalt catalysts after the standard H_2 reduction. This study suggested that Ag is a promising potential additive as a substitute for the Pt that is used for commercial cobalt catalysts due to the better FTS performance of the Ag-Co catalyst and lower price compared with Au.

The effect of up to 5% Zr structural modifier on 25%Co/Al_2O_3 catalysts has been studied using wide and narrow pore Al_2O_3 supports (Puralox HP14/150 and Catalox 150) [8]. The cobalt catalyst prepared with the wide pore support was found to perform much better than the cobalt prepared with the narrow pore Al_2O_3. The addition of Zr made further improvement of performance of both cobalt catalysts by facilitating Co reduction or reducing cobalt cluster size, but the Zr additive slightly increased CH_4 selectivity of the cobalt catalysts supported on a wide pore alumina.

In addition to the important additives reviewed above, other additives such as V, Mg and Ce [15,18,19], K [15], Cr, Ti, Mn, and Mo [9,20,21], Nb [22] and P [23] have been also studied. These additives served as structural or electronic additives to modify cobalt catalyst behavior. Most of them at low loadings were reported to enhance cobalt reduction and dispersion, and promoted activity and selectivity toward heavier hydrocarbons for cobalt catalysts.

According to this large number of additive effect studies, some of them such as Ag and Zr showed great benefits to the performance of the cobalt catalyst and are promising steps toward a potential replacement for Pt. Efforts continue to find potential substitutes for Pt additive and to better understand structure-performance relationships. In this work, we explore the effects of the Group 12–14 elements as potential additives and based on our success with Ag, have selected the Row 5 elements Cd, In, and Sn. The effect of these additives on physiochemical properties and FTS performance of 15%Co/Al_2O_3 catalysts were carefully studied using various characterization and testing techniques.

2. Results and Discussion

2.1. BET and Porosity Measurements

BET and porosity results of unpromoted and 1%Cd, 1%In, and 1%Sn catalysts are summarized in Table 1. BET of Catalox SBA 150 γ-Al_2O_3 is 149 m^2/g [6–8,16,17,25]. After loading 15% Co, BET surface area decreased to 116 m^2/g. A weight % loading of 15% is equivalent to ca. 20% by weight of Co_3O_4. If the support is the only contributor to the surface area, then the area of the 15%Co/Al_2O_3 catalysts should be 0.8 × 150 m^2/g = 120 m^2/g, which is close to the experimental values listed in Table 1. Thus, no significant decrease in surface area was observed due to pore blocking by Co oxide particles. The addition of 1%Cd, 1%In, and 1%Sn further decreased the BET of SBA 150 supported catalysts to 110, 111 and 115 m^2/g, respectively. The pore volume (0.31 cm^3/g) and the pore size (radius = 4.7 nm) remained nearly unchanged for the unpromoted and the three promoted 15% Co/Al_2O_3 catalysts, which were lower than those of the SBA-150 support (0.5 cm^3/g and 5.4 nm from [8]). The results further suggest that the addition of Co and the Cd, In and Sn additives did not significantly block the pores of the alumina support.

Table 1. Results of BET surface area and H_2 chemisorption pulse re-oxidation.

Support/Catalyst	BET SA	Pore Volume (BJH Adsorp)	Average Pore Radius (BJH Adsorp)	H_2 Desorbed	Co Reduction Degree	Apparent Corrected Co Disperion	Apparent Corrected Average Co Diamter
	(m^2/g)	(cm^3/g)	(nm)	(μmol/g$_{cat}$)	(%)	(%)	(nm)
15% Co/Al_2O_3-150	116.2	0.3096	4.72	57.3	48.5	9.28	11.1
1%Cd 15% Co/Al_2O_3	110.4	0.3024	4.75	52.6	58.8	7.19	14.3
1%In 15% Co/Al_2O_3	111.6	0.3053	4.73	45.6	56.6	6.34	16.3
1%Sn 15% Co/Al_2O_3	115.0	0.3051	4.68	36.6	46.2	6.23	13.8

2.2. TPR and Hydrogen Chemisorption/Pulse Reoxidation

TPR profiles of the unpromoted and 1%Cd, 1%In, and 1%Sn cobalt catalysts are presented in Figure 1. Two peaks occurred at the temperature ranges of 220–380 °C and 400–700 °C, which represent the standard two-step reduction of cobalt: $Co_3O_4 + H_2 \rightarrow 3CoO + H_2O$ and $3CoO + 3H_2 \rightarrow Co^0 + 3H_2$, where the second step consumes three times as much hydrogen as the first step. The broad high temperature peak for the reduction of CoO to Co^0 indicated strong interactions between CoO and the alumina support. It is interesting that the addition of 1%Cd, 1%In and 1% Sn did not significantly change the temperature for the first reduction step of Co_3O_4 to CoO, and the peak temperature for the unpromoted and three promoted cobalt catalysts remained at 320–330 °C (Figure 1). However, the addition of 1% Cd shifted the second reduction peak to lower temperature by 80 °C relative to the unpromoted 15%Co/Al_2O_3 catalyst (540 vs. 460 °C), and thus, as the atomic number of the additive is increased, the broad reduction peak occurs at higher temperatures, e.g., 535 °C for In and 570 °C for Sn, which indicated that only Cd and, to a lesser extent, In, facilitated reduction of CoO; on the other hand, Sn addition slightly hindered the reduction of CoO.

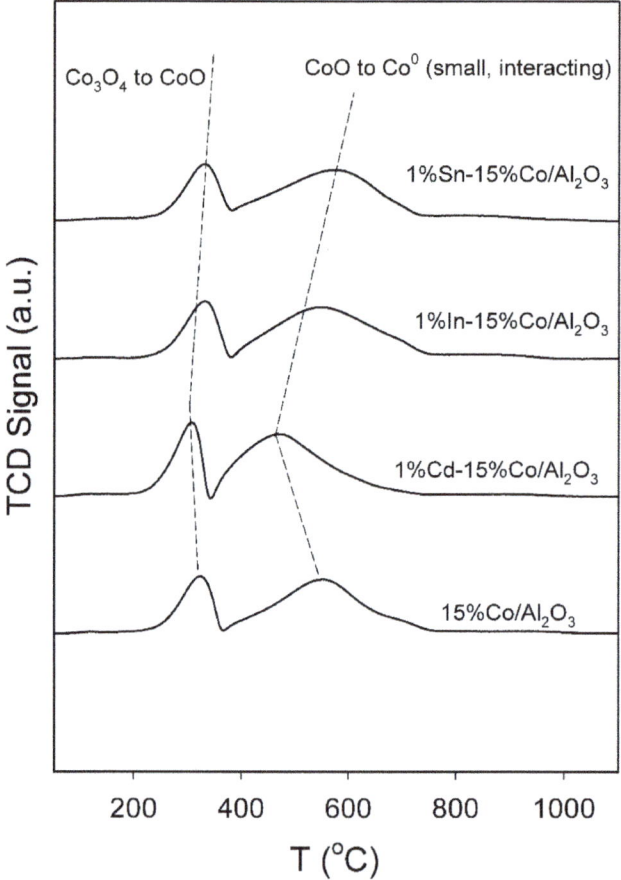

Figure 1. TPR profiles of, moving upward, 15%Co/Al_2O_3, 1%Cd-15%Co/Al_2O_3, 1%In-15%Co/Al_2O_3, and 1%Sn-15%Co/Al_2O_3. Catalysts prepared by SPI using 150 m²/g γ-Al_2O_3.

H_2 chemisorption and oxygen titration results are also shown in Table 1. The cobalt reduction degree increased from 48.5% to 57–59% for the 1%Cd and 1% In promoted cobalt catalysts, but it slightly decreased to 46% for the 1%Sn promoted catalyst, as compared to that of the unpromoted catalyst. This could be due to the formation of cobalt-additive coordination, as this was shown for the noble metals by EXAFS and suggested for the Group 11 metal Cu based on chemisorption results [4–6,16,17].

The reduction results as determined by H_2 chemisorption pulse re-oxidation are in agreement with the TPR results above. In terms of the TPR and H_2 chemisorption/pulse oxidation results, it is likely that the Cd, In and Sn additives are highly dispersed on the catalyst surface, with a fraction strongly interacting with the support and resulting in little reduction (consistent with the XANES results in the next section which showed the presence of oxides), while a fraction was likely metallic and/or coordinated with Co to increase cobalt reduction. Interestingly, the hydrogen chemisorption/pulse reoxidation results show that all Cd, In, and Sn additives lowered cobalt site densities compared with the unpromoted catalyst, with H_2 desorption amounts decreased from 57 to 37–53%. Assuming the desorption of hydrogen comes only from the surface of Co^0, the "apparent" average cluster size is increased for the Cd, In and Sn promoted catalysts (11 to 14–16 nm). However, more likely, the decreased cobalt site density by Cd, In and Sn could arise by (1) the additives being located on the cobalt surface and covering some cobalt sites, or (2) blocking of pores. However, the BET results discussed above tend to rule out the latter explanation. That is, it is likely that the Co particles remained the same size but that Cd, In and Sn addition blocked surface sites.

2.3. NH_3 TPD Study of Cd, In and Sn Promoted Catalysts

The NH_3-TPD profiles of the unpromoted and the Cd, In and Sn promoted 15%/Al_2O_3 catalysts are shown in Figure 2. Two types of acidic sites (weak and mild to strong) are clearly identified with the corresponding NH_3 desorption peaks present at ca. 90 °C and 200–260 °C, respectively. For the unpromoted cobalt catalyst, the first NH_3 desorption peak is more intense than the second one, indicating the unpromoted cobalt catalyst surface was mainly covered by the weak acid sites (likely L sites). In the case of adding 1%Cd, the NH_3 desorption amounts in both peaks increased greatly, but the first peak area exhibited a greater change, implying that addition of Cd resulted in a greater fraction of weak acid sites. However, for the In-promoted cobalt catalyst, the first NH_3 desorption peak shows only a slight increase in the intensity, but the second one became more intense; moreover, the second peak slightly shifted to lower temperatures (i.e., 240 vs. 260 °C). Thus, addition of 1%In primarily promoted the strong acid sites. Interestingly, the Sn promoted cobalt catalyst displayed a similar NH_3-TPD profile to that of the In promoted cobalt catalyst, suggesting a similar acid site density (e.g., B-sites) on the Sn and In promoted cobalt catalysts. However, the second peak moved to even lower temperatures relative to the Cd promoted and unpromoted cobalt catalysts. It can be observed that the higher the atomic number, the lower the peak temperature (i.e., 220 °C for the In-Co and 200 °C for the Sn-Co catalysts). The total amount of NH_3 desorbed follows the order: Cd (119.3 µmol/g) > In (110.5 µmol/g) > Sn (104 µmol/g) > unpromoted (73.2 µmol/g).

2.4. XRD Study of Unpromoted and Cd, In and Sn Promoted Cobalt Catalysts

The XRD spectra of the four reduced cobalt catalysts are depicted in Figure 3. The unpromoted and the Cd, In and Sn promoted cobalt catalysts show similar XRD patterns in the 2θ range of 30°–70°. Five intense reflections are observed at 2θ of 36.9° and 61.9°, 42.9°, and 45.9°, 67°, representing the characteristic peaks of cobalt oxide, Co, and alumina, respectively. Interestingly, regardless of whether the unpromoted catalyst was used, or whether additives were used, the intensities for each peak for the four different cobalt catalysts are quite similar, suggesting that average cobalt cluster sizes are similar. Based on the Scherrer equation, cobalt cluster size for the unpromoted and Cd, In and Sn calculated at the 2θ of 42.9° are 12.7 nm, 11.4 nm, 12.5 nm and 11.1 nm, respectively, which suggests that addition of Cd, In and Sn on the cobalt catalyst only slightly decreased the cobalt cluster size, if at all. This result is not consistent with the H_2-chemisorption results, which suggest an apparent increased Co cluster size with the addition of Cd, In and Sn. The discrepancy strongly suggests that the additives cover some surfaces of cobalt nanoparticles or block pores of the support. However, the latter reason can be excluded based on the BET results as discussed in Section 2.1.

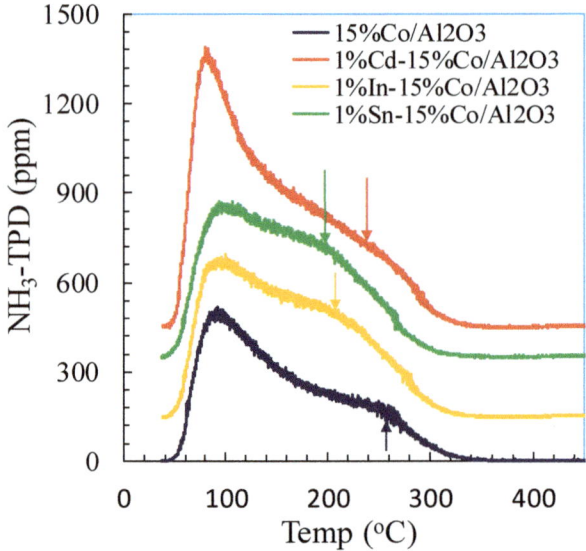

Figure 2. NH$_3$–TPD profile of unpromoted and Cd, In and Sn promoted 15%Co/Al$_2$O$_3$ catalysts.

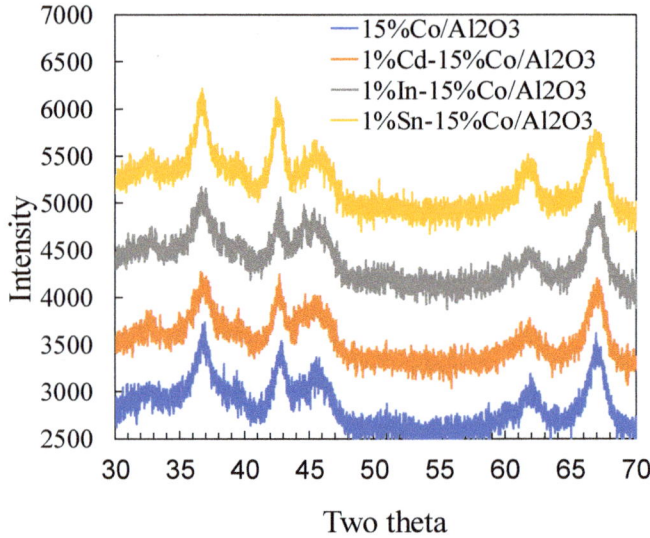

Figure 3. XRD patterns unpromoted and Cd, In and Sn promoted 15%Co/Al$_2$O$_3$ catalysts.

2.5. XANES Study of Cd, In and Sn Promoted Catalysts

To explore the electronic structure of the Cd, In, and Sn additives by XANES, 1%Cd, 1%In and 1%Sn supported on high cobalt loading (25%) catalysts were used because the additives had a greater effect on the more highly loaded Co catalysts. After the catalysts were reduced ex-situ, the cobalt catalysts were transferred to a 1L-CSTR under inert gas where 3000 Polywax was previously charged and melted. In this way the in-situ state of the Cd, In and Sn additives in the catalyst following H$_2$ activation were able to be reviewed by XANES. Figure 4 shows the XANES spectra of 1%Cd-25%Co/Al$_2$O$_3$ (left), 1%In-25%Co/Al$_2$O$_3$ (middle) and 1%Sn-25%Co/Al$_2$O$_3$ (right), respectively, along with the spectra of reference Cd, In and Sn metal foils. The results show that metal oxides were observed for the three additives after activation. This result is different from the results of the noble metal additives Pt, Pd, and Ru [6,16,17] and Group 11 metals (Cu, Ag, and Pt) studied previously [4–6,17], In which only the metallic state rather than the oxidized state was found for the additives by XANES/EXAFS.

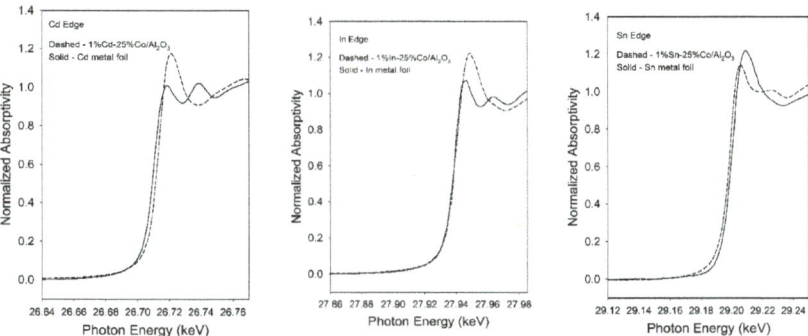

Figure 4. XANES spectra of 1%Cd-25%Co/Al$_2$O$_3$ (**left**), 1%In-25%Co/Al$_2$O$_3$ (**middle**) and 1%Sn-25%Co/Al$_2$O$_3$ (**right**) compared to the reference Cd, In and Sn metal foils.

2.6. Effect of Cd, In and Sn Additives on Fischer-Tropsch Synthesis

The effects of 1%Cd, 1%In and 1%Sn additives and time on CO rate, CH$_4$, C$_{5+}$ and CO$_2$ selectivities are shown in Figure 5a–d and Table 2. The promoted catalysts were running at 20–30% CO conversion in the first ca. 50 h; afterwards, CO conversion was adjusted to the 40–50% level in order to better compare catalyst selectivities. The CO rate for the unpromoted catalyst was 0.022 mol/g-cat/h. The addition of 1%Cd, 1%In or 1%Sn additive drastically decreased the catalyst activity to 0.0015, 0.004 and 0.003 mol/g-cat/h, respectively, corresponding to very high rate loss percentages of 93%, 82% and 86% for the Cd, In and Sn additives, respectively. After adjusting CO conversion to 40–50% level after 50 h, the unpromoted cobalt catalyst and 1%Cd promoted catalyst displayed better stability, and the CO rate over the next time period of 100–150 h remained at ca. 0.017 mol/gcat/h for the unpromoted catalyst and ca. 0.0015 mol/gcat/h for the Cd promoted catalyst. However, the In and Sn promoted catalysts were slowly deactivating with time, with CO rate changes from 0.004 to 0.0028 mol/gcat/h and from 0.003 to 0.0025 mol/gcat/h for the In and Sn promoted catalysts, respectively. The significant decreases in cobalt catalyst activity caused by the additives are not consistent with the changes in the H$_2$ chemisorption results as discussed in Table 1. In re-examining the H$_2$ desorption results, the H:Co ratio is assumed to be 1:1, where "Co" refers to surface Co^0 atoms. The addition of 1%Cd, 1%In and 1%Sn only led to cobalt site density decreases of 8.2%, 20.4% and 36.1%, respectively. Thus, it was expected to have similar activity losses percentages for the promoted catalysts, since the change in FT activity has been generally consistent with changes in H$_2$ chemisorption capacities of cobalt catalysts resulting from different loadings of Pt, Re, Ru, Ag, and Au, and even Zr additives [5–8,16,17]. The activity of the Cd, In and Sn promoted catalysts being 3 to 10 times lower than the expected results strongly suggests that other reasons may account for the lowering in catalyst activity. During an investigation of Group 11 additives (i.e., the coinage metals–Cu, Ag, and Au), while all the additives facilitated cobalt oxide reduction, only the Ag and Au additives increased the catalyst activity on a per g of catalyst basis. All the catalysts had higher metal site densities by H$_2$-TPD relative to the unpromoted 15%Co/Al$_2$O$_3$ catalyst, but Cu decreased catalyst activity; however, the difference in the adverse effects of the Cu additive were much less pronounced than that of the Cd, In and Sn additives in this study. Thus, H$_2$-TPD only reports metal site density, and in the case of Cu, Cu^0 was likely on the surface of the cobalt particles so that, while it on the one hand promoted reduction of cobalt oxides, it decreased the cobalt surface site density by blocking sites on the surface. However, BET results do not suggest pore blocking by cobalt and the additives.

Table 2. Activity and selectivity of unpromoted and Cd, In, Sn and Pt promoted 15%Co/Al$_2$O$_3$ catalysts [a].

TOS (hrs)	CO Conversion (%)	GHSV (NL/gcat/h)	H$_2$/CO	Selectivity (C atom %)			CO$_2$ Selectivity (%)
				CH$_4$	C$_2$-C$_4$	C$_{5+}$	
15%Co/Al$_2$O$_3$-150							
11–57	48–51	2.78	2.10	7.44	7.02	85.54	0.47
59–193	50–52	2.44	2.10	8.00	8.63	83.37	0.52
1%Cd-15%Co/Al$_2$O$_3$-150							
25–72	18–20	0.50	2.10	10.45	6.88	82.67	2.17
97–146	38–40	0.26	2.10	11.50	7.39	81.11	1.59
1%In-15%Co/Al$_2$O$_3$-150							
49–97	47–52	0.50	2.10	13.77	10.95	75.28	3.28
146–170	40–44	0.50	2.10	14.25	10.59	75.16	2.07
1%Sn-15%Co/Al$_2$O$_3$-150							
25.3	27.07	1.00	2.10	13.05	9.31	77.64	2.93
33–52	47–51	0.51	2.27	13.46	9.98	76.56	4.21
74–98	37	0.51	2.17	14.56	11.34	74.09	3.83
106–122	47	0.47	2.11	14.06	10.92	75.01	3.59

[a] Reaction conditions: 220 °C, 2.2 MPa, and H$_2$/CO = 2.1.

Another possibility, however, is that the catalysts have the active site density as measured by chemisorption after activation, but that the site density is decreased due to reoxidation once FTS is started. A previous study of Co catalysts in an in-situ EXAFS/XANES flow cell by Huffman et al. [24], reported that Co catalysts promoted with K were much more susceptible to reoxidation (i.e., even at low conversion) compared to catalysts having no K, which only oxidized under high H$_2$O partial pressure at high conversion. However, carefully examining the FTS activity data as shown in Figure 5a, the Cd, In and Sn promoted cobalt catalysts did not deactivate in the first 50 h; instead, the activities slowly increased with time for all cases. Therefore, rapid reoxidation of cobalt particles facilitated by Cd, In and Sn additives is only a possible explanation if it occurred prior to measurement of the first point.

According to the above discussion, blocking of pores by the additives can be excluded; thus, it was not a cause for the unexpected low activity of the promoted cobalt catalysts; the fast oxidation of cobalt particles is uncertain. The formation of M-Co (M = Cd, In and Sn) coordination or alloying can also be ruled out, because if this were the case, the H$_2$ chemisorption/pulse re-oxidation results would show similar large differences between the unpromoted and the promoted cobalt catalysts-i.e., ~90%, but this was not observed. Furthermore, the XRD experiment did not show Co-M alloys peaks. Based on the XRD results, addition of Cd, In and Sn on the cobalt catalyst should have led to slightly increased cobalt site density by decreasing cobalt cluster size; thus, the most likely reason that explains the unexpected low activity for the Cd, In and Sn promoted catalysts is that the additives covered cobalt sites and poisoned the surfaces of the cobalt catalysts. Larger Sn atoms might contribute more significantly to cobalt site poisoning, resulting in the lowest catalyst activity. However, an electronic effect of the additives such as Cd, In and Sn on catalyst performance cannot be excluded. Additional study is needed to clarify the assumption.

From Table 2 and Figure 5b,c, CH$_4$ selectivity for the unpromoted 15%Co/Al$_2$O$_3$ catalyst is 7.4% at 50% CO conversion, but it increased dramatically to about 10.5%, 13.8% and 13.5% for the 1%Cd, and 1%In or 1%Sn promoted catalysts, respectively. This caused corresponding drops in C$_{5+}$ selectivity to ca. 82.7 %, 75.3, and 76.6%. Thus, the In and Sn additives have a greater impact on increasing CH$_4$ and suppressing heavier hydrocarbon formation relative to Cd. The greater increase in CH$_4$ and light hydrocarbon selectivities for the In and Sn promoted cobalt catalysts than that of Cd is likely associated with a greater density of strong acid sites on the two cobalt catalysts as determined by NH_3-TPD (Figure 2). This conclusion is further evidenced by the In and Sn promoted cobalt catalysts having similar amounts of strong acid sites (located after 200 °C) and displaying essentially the same hydrocarbon selectivities at about 50% CO conversion, i.e., CH_4 selectivity 13.5–13.8%, C_2–C_4 selectivity 10–11% and C_{5+} selectivity 74–75%. It is likely that the strong acid sites on the surface of cobalt promoted H_2 adsorption and promoted methane and light hydrocarbon formation.

The addition of 1%Cd, 1%In or 1%Sn additives also led to significantly increased CO_2 selectivity (0.5 to 2–5%). The significant changes in catalyst selectivity also indicated that species other than metallic Co are present in the catalyst, since metallic cobalt does not possess intrinsic water-gas shift (WGS) activity. Note that higher WGS activity leads to higher methane selectivity, since WGS promotes the formation of hydrogen. The XANES results showed that the Cd, In and Sn additives were present in oxidized form after reduction. It is well known that a synergy between a partially reducible oxide and a metal results in WGS activity, and this might explain the higher CO_2 selectivity during FTS.

The effects of 1%Cd, 1%In and 1%Sn additives and time on the contents of propylene, propane, 1-butene, 2-butene, total butene and butane are shown in Figure 6a–f, respectively. Table 3 also summarized mean values of these parameters at different time ranges. At 40–50% CO conversion level, C_3 olefin content for the unpromoted 15%Co/Al_2O_3 catalyst was 60.7%. Doping 1%Cd, 1%In and 1%Sn to the catalyst decreased C_3 olefin content to 58.6%, 53.9%, and 46%, respectively, but C_3 paraffin content increased to 41.4%, 46.1% and 48.9% from 39.3% (Table 3, Figure 6a,b). Moving to C_4 hydrocarbons, precisely the same trend is observed at 40–50% CO conversion. The Cd, In and Sn additives resulted in decreases in 1-C_4 olefin selectivity to 45.4%, 40.1% and 25.6%, respectively, from 48.1% as compared to the unpromoted catalyst. This also led to a measurable increase in 2-C_4 olefin content to 5.2%, 5.5%, 10.7% from 3.5% and C_4 paraffin contents to 49.4%, 54.4%, 63.9% from 48.4%, respectively (Table 3 and Figure 6c–f). The results are interesting, as they clearly indicate that the Cd, In and Sn additives enhanced the hydrogenation and isomerization of 1-olefin reactions on the cobalt catalysts. The extent of these secondary reactions increased with increases in the atomic number of the additive. The Sn apparently is the most effective one among all three additives to greatly increase the hydrogenation and isomerization reactions. As discussed in terms of the TPR and XANES results, a fraction of the additives might be in a metallic state, while the remaining additive may be highly dispersed and strongly interacting with the support or cobalt surface, which are in oxide form and/or Co-M (M=Cd, In and Sn) coordinated states. Thus, it is postulated that addition of Cd, In and Sn to the cobalt catalyst likely creates new acid sites on catalyst surface, leading to a higher activity of secondary reactions of olefins relative to the unpromoted cobalt catalyst. This hypothesis is consistent with the NH_3-TPD results as discussed in Section 3.3, which indicated a greater abundance of mild acid sites on the Sn-Co catalyst occurring at low temperature (200 °C) relative to the Cd and In promoted cobalt catalysts. Much higher 2-C_4 olefin content for the Sn-Co catalyst suggests that mild acid sites are more active for the secondary reaction of olefins.

Table 3. Olefins and paraffins contents of unpromoted and Cd, In, Sn and Pt promoted 15%Co/Al_2O_3 catalysts [a].

TOS (hrs)	CO conv. (%)	Olefin Content (%)				Total C_4 Olefin Content (%)	Paraffin Content (%)		
		1-C_2	2-C_3	1-C_4	2-C_4		C_2	C_3	C_4
15%Co/Al_2O_3-150									
11–57	48–51	7.43	60.67	48.09	3.47	51.56	92.57	39.33	48.44
59–193	50–52	7.38	60.24	46.96	3.83	50.79	92.62	39.76	49.21
1%Cd-15%Co/Al_2O_3-150									
25–72	18–20	27.89	62.72	57.25	0.00	57.25	72.11	37.28	42.75
97–146	38–40	13.18	58.59	45.43	5.19	50.62	86.82	41.41	49.38

Table 3. *Cont.*

TOS	CO	Olefin Content (%)				Total C$_4$ Olefin	Paraffin Content (%)		
(hrs)	conv. (%)	1-C$_2$	2-C$_3$	1-C$_4$	2-C$_4$	Content (%)	C$_2$	C$_3$	C$_4$
1%In-15%Co/Al$_2$O$_3$-150									
49–97	47–52	5.52	53.87	40.06	5.51	45.57	94.48	46.13	54.43
146–170	40–44	6.79	56.68	44.42	2.78	47.20	93.21	43.32	52.80
1%Sn-15%Co/Al$_2$O$_3$-150									
25.3	27.07	10.61	51.23	30.90	10.69	41.59	89.39	48.77	58.41
33–52	47–51	9.30	46.07	25.57	10.58	36.15	90.70	53.93	63.85
74–98	37	8.37	47.78	28.78	9.03	37.81	91.63	52.22	62.19
106–122	47	7.30	47.59	30.10	8.53	38.63	92.70	52.41	61.37

Reaction conditions: 220 °C, 2.2 MPa, and H$_2$/CO = 2.1. C$_4$ olefin selectivity, % = 100 × rates of all C$_4$ olefins/rates of all C$_4$ hydrocarbons; 1-C$_4$ olefin selectivity, % = 100 × rate of 1-C$_4$ olefin/rates of all C$_4$ hydrocarbons; 2-C$_4$ olefin selectivity, % = 100 × rate of 2-C$_4$ olefin/rates of all C$_4$ hydrocarbons.

In our previous studies, Pd and Pt were found to increase CH$_4$ (8–12%) and suppress heavier hydrocarbon formation (83–76%); and, Pd displayed much higher hydrogenation and isomerization activities (1-C$_4$ olefin: 47–25%, 2-C$_4$ olefin: 7–14.8%) [7,16]. Thus, the impact of Sn on the formation of olefins and paraffins resembles that of Pd.

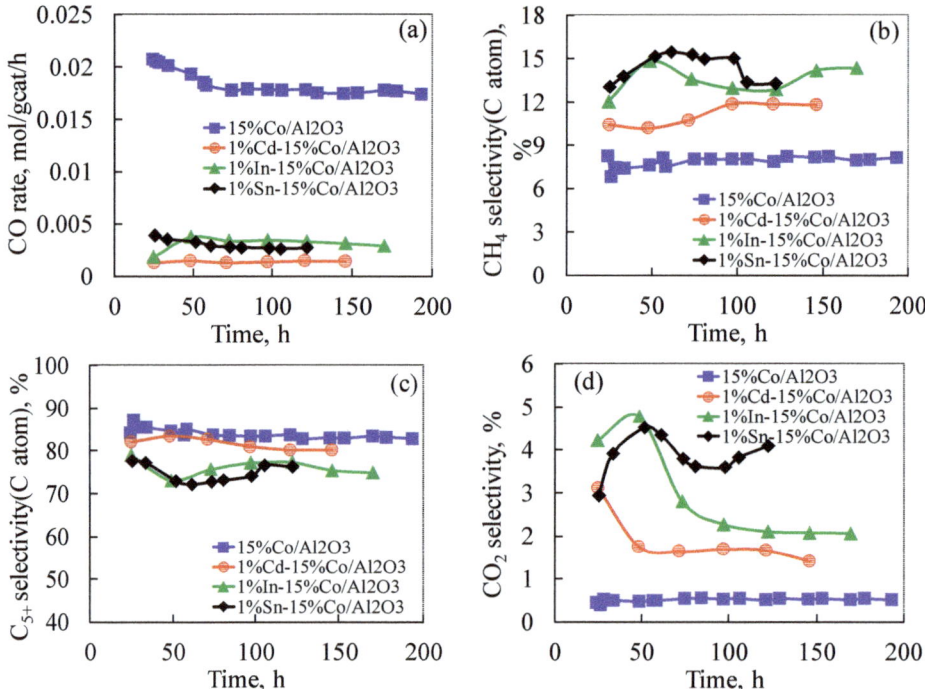

Figure 5. Change in (**a**) CO rate, (**b**) CH$_4$ selectivity, (**c**) C$_{5+}$ selectivity and (**d**) CO$_2$ olefin selectivity over unpromoted and Cd, In and Sn promoted 15%Co/Al$_2$O$_3$ catalysts. Reaction conditions: 220 °C, 2.2 MPa, H$_2$/CO = 2.1, X$_{CO}$ = 40–50%.

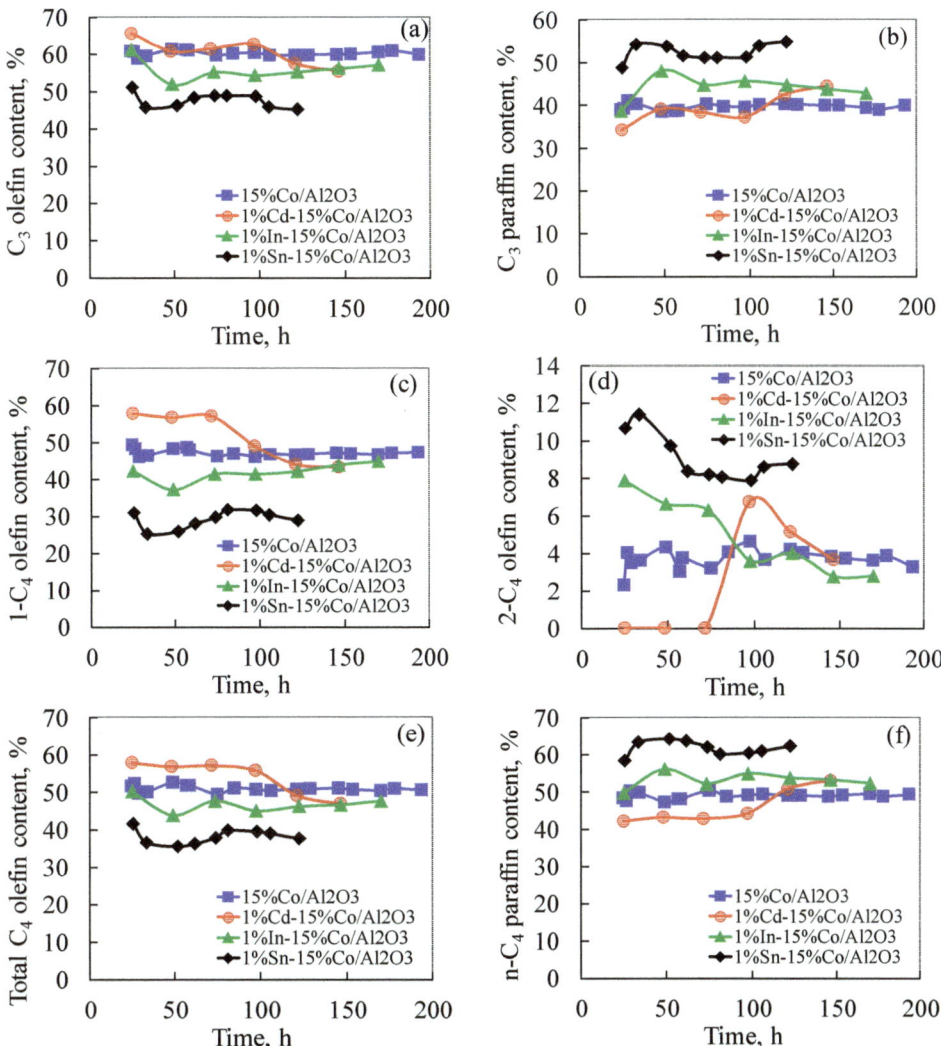

Figure 6. Change of (**a**) C_3 olefin selectivity, (**b**) C_3 paraffin selectivity, (**c**) 1-C_4 olefin selectivity, (**d**) 2-C_4 olefin selectivity, (**e**) total-C_4 olefin selectivity and (**f**) C_4 paraffin selectivity with time over npromoted and Cd, In and Sn promoted 15%Co/Al_2O_3 catalysts. Reaction conditions: 220 °C, 2.2 MPa, H_2/CO = 2.1, X_{CO} = 40–50%.

The Cd, In and Sn elements used as additives to modify FTS cobalt catalyst performance have been scarcely reported. However, some studies have employed them (e.g., as bimetallics such as Pt-In, Pt-Sn) for other associated reactions. Cho et al. [26] reported that Sn supported on a mesoporous zeolite (3Dom-I MFI) offered significant improvements for the isomerization of C_5 and C_6 sugars such as xylose and glucose. Srinivasan et al. [27] found that Sn at different loadings changed the activity and aromatics selectivity of Sn-Pt/Al_2O_3 for n-octane conversion. Passos et al. [28] studied In-Pt/Al_2O_3, and Sn-Pt/Al_2O_3 bimetallic catalysts for cyclohexane dehydrogenation, methylcyclopentane hydrogenolysis, and *n*-heptane conversion. It was found that after the catalyst was reduced by 1.5%H_2/Ar at 500 °C for 30 min., 50–80% In and 25–50% Sn was in a zero-valent state in the bimetallic system. During the methylcyclopentane hydrogenolysis and *n*-heptane conversion reactions at 500 °C, the In and Sn additives were reported to decrease the activity of the Pt/Al$_2$O$_3$ catalyst, which was explained by the In and Sn additives diluting Pt active sites on catalyst. Furthermore, addition of In or Sn led to decreases in the selectivity of hydrogenolysis, and an increase in the selectivity for dehydrogenation and aromatization products, but Sn was reported to greatly enhance the isomerization activity for the Pt/Al$_2$O$_3$ catalyst in the conversion of *n*-heptane. Coleto et al. [29] studied the transformation of 1-pentene, and 1-hexene over bimetallic Pt-Re/Al$_2$O$_3$, Pt-Sn/Al$_2$O$_3$ and Pt-Ge/Al$_2$O$_3$ catalysts. The Pt-Sn/Al$_2$O$_3$ catalyst was also reported to have high hydrogenation activity to n-pentane at low temperature (200 °C), while high isomerization activity was observed at the expense of hydrogenation at a high temperature of 500 °C, which is consistent with the study of Passos et al. [28]. The investigation showed by comparing with the result of the Al$_2$O$_3$ support that the double bond shift and skeletal isomerization of olefins are both acid-catalyzed reactions, while hydrogenation sites are metallic in nature. Thus, the hydrogenation sites and isomerization likely changed with temperature, with higher temperatures yielding more acid sites for the Pt-Sn catalyst. Mazzieri et al. [30] reported the same role of Sn in increasing isomerization activity for the reaction of naphtha reforming over a trimetallic Pt-Re-Sn catalyst supported on chlorided Al$_2$O$_3$. In this study, the improvements in secondary reactions of 1-olefins observed with the addition of Group 12, Group 13 and Group 14 elements (i.e., Cd, In and Sn) are consistent with the studies of Coleto et al. [26] and Passos et al. [28], which suggests that the additives not only boosted the hydrogenation rate of olefins, but they also produced new acid sites for isomerization, for example possibly as MO$_x$, M-Co (M = Cd, In and Sn).

3. Experimental

3.1. Catalyst Preparation

The unpromoted 15%Co/Al$_2$O$_3$ catalysts and 1% Cd, 1% In and 1%Sn promoted cobalt catalysts were prepared by the slurry phase impregnation (SPI) method, as previously described in [3–8,16,17,25]. Catalox SBA 150 -Al$_2$O$_3$ was used as the catalyst support. The salts used for the Cd and In were nitrates, while SnCl$_2$ served as the salt for Sn. The additives were incorporated into the catalyst by incipient wetness impregnation (IWI) following the addition of cobalt. The catalysts were calcined in air for 4 h at 350 °C.

3.2. BET Surface Area and Porosity Measurements

A Micromeritics 3-Flex system was used to measure BET surface area and porosity characteristics. Before testing, the temperature was slowly increased to 160 °C; then, a vacuum was pulled for at least 12 h until the sample pressure was approximately 50 mTorr. The BJH method was employed to determine the average pore diameter and pore volume. Additionally, pore size distribution was obtained as a function of pore diameter via the correlation dV/d(log D).

3.3. Temperature Programmed Reduction

Temperature programmed reduction (TPR) was carried out with a Zeton Altamira AMI-200 unit (Pittsburgh, PA, USA) with a flow rate of 30 cm^3/min of 10%H$_2$/Ar. The heating rate was 5 °C/min from 50 °C to 1100 °C, with a final 30 min hold.

3.4. Hydrogen Chemisorption/Pulse Reoxidation

Chemisorption with hydrogen was conducted in a Zeton Altamira AMI-200 unit using a thermal conductivity detector (TCD). The mass of the catalyst was ~0.220 g. Each catalyst was reduced at 350 °C for 10 h in 30 cm^3/min of 33%H$_2$ in He and the temperature was decreased to 100 °C. Argon was flowed to remove any physisorbed species and the temperature was increased back to 350 °C in argon to desorb the chemisorbed hydrogen. The temperature programmed desorption peaks were integrated and the # of moles of hydrogen evolved was calculated.

After TPD of hydrogen, pulses of pure O_2 in He were sent to oxidize the catalyst. The extent of reduction was determined on the assumption that Co^0 reoxidized to Co_3O_4. Uncorrected % dispersion assumes (erroneously) complete reduction while corrected % dispersion includes the extent of reduction:

$$\%D_{uc} = (\text{\# of } Co^0 \text{ atoms on surface} \times 100\%)/(\text{total \# Co atoms})$$

$$\%D_c = (\text{\# of } Co^0 \text{ atoms on surface} \times 100\%)/[(\text{total \# Co atoms})(\text{fraction reduced})]$$

3.5. NH_3 Temperature Programmed Desorption

A microreactor loaded with ca. 200 mg of powder catalyst was employed to analyze the acid sites on the surface of the catalyst by means of NH_3-TPD. A total flow rate of 50 sccm was used for NH_3 adsorption and desorption with effluent gases being analyzed with a quadrupole mass spectrometer (QMS 200, Pfeiffer Vacuum, Asslar, Germany). The catalysts were first reduced at 350 °C in a flow of 10% H_2 in He for 2 h, followed by the gas mixture being replaced with pure He for purging and cooling prior to NH_3 adsorption at 40 °C. When the catalyst was saturated with NH_3 by flowing 2000 ppm NH_3 in N_2 (ca. 30 min), the He flow was switched back again for purging at 40 °C in order to remove weakly adsorbed NH_3 (ca. 30 min). The temperature-programmed desorption was then carried out in He at 50 cc/min using a ramp rate of 10 °C/min up to 600 °C.

3.6. X-ray Absorption Near Edge Spectroscopy (XANES)

X-ray absorption near edge spectroscopy was carried out using transmission mode in the vicinity of the Cd, In, and Sn K-edges at the National Synchrotron Light Source (NSLS) at Brookhaven National Laboratory, Upton, New York, Beamline X18-b. The beamline utilized a Si (111) channel-cut monochromator. The catalysts were prepared and activated in the same way as if conducting a reaction test except that following treatment in hydrogen, the catalyst was cooled so that it became fixed in the solid startup wax. Samples were made into self-supporting disks. XANES spectra were analyzed using WinXAS software [31] by comparing the spectra qualitatively once normalized.

3.7. X-ray Diffraction (XRD)

X-ray diffraction (XRD) on powder samples was performed for freshly reduced cobalt catalysts at room temperature using a Rigaku Diffractometer (DMAX-B, Tokyo, Japan) operating with Cu Kα radiation (1.54 Å), In order to identify cobalt structure and cluster size. All cobalt catalyst samples were reduced at 350 °C by 25%H_2/He for 15 h followed by passivation using 1%O_2/N_2 gas mixture prior to conducting the XRD measurement.

3.8. Catalytic Activity Testing

Catalyst reaction tests were carried out using a continuously stirred tank reactor (CSTR, PPI, Fort Worth, TX, USA)) that makes use of a mag drive stirrer with turbine impeller, gas-inlet outlet lines with a stainless steel (SS) fritted filter (7 μm) placed outside the reactor. To withdraw wax, a stainless steel tube with a 2 micron fritted filter was placed below the liquid level of the reactor. Mass flow controllers controlled the H_2 and CO flow rates. Reactant gases were thoroughly mixed prior to the reactor. CO was scrubbed of iron carbonyls using lead oxide-alumina. Reactants entered the CSTR below the impeller, which had a stirring speed of 750 rpm. Temperature was well controlled by a temperature controller.

The amount of catalyst used was 12–18 g in the size range of 45–90 μm. The catalyst was first reduced ex-situ in a tubular reactor at 350 °C at 1 atm for 15 h using a gas mixture of H_2/He (60 NL/h) with a volume ratio of 1:3. Reduced catalyst was transferred by forcing the catalyst out with N_2 to the CSTR, which held 315 g of melted Polywax 3000. The reactor was weighed prior to and following catalyst transfer. The transferred catalyst was further exposed to pure hydrogen (30 NL/h) for another 10 h at 230 °C to ensure reduction, prior to commencing FTS.

FTS reaction conditions were 220 °C, 2.2 MPa, H_2/CO = 2.1. Space velocity was controlled to achieve a CO conversion of 50%. Reaction products were continuously removed from the reactor head space and sent to two collection vessels, a trap maintained at 100 °C and a trap held at 0 °C. Uncondensed vapor was decreased to atmospheric pressure. Gas flow was measured by a wet test meter and the gas was analyzed using online gas chromatography. Accumulated liquids in the CSTR were removed daily through a 2 micron sintered metal filter. CO conversion was determined on the basis of GC data using a micro-GC equipped with thermal conductivity detectors. Wax, oil and the water phase products were also collected and analyzed by three different gas chromatographs. To investigate the effect of Group 12–14 elements, Cd, In and Sn were selected from Row 5 based on our earlier success with Ag as an additive. The activity and product selectivities (e.g., CH_4, C_{5+}, CO_2, 1-olefin, 2-olefin and paraffin) of unpromoted 15%Co/Al_2O_3 catalysts and 1% of Cd-, In, and Sn-supported cobalt catalysts were studied at a reference CO conversion of about 40–50%.

4. Conclusions

TPR and hydrogen chemisorption/pulse reoxidation results showed that only a fraction of Cd, In and Sn was reduced. NH_3-TPD results indicated that addition of Cd, In and Sn promoted mild to strong acid sites, which might be responsible for the enhancement of hydrogenation of 1-olefin during the FTS. Cd and In were found to promote CoO reduction to Co^0, while Sn slightly hindered it, resulting in more unreduced cobalt than the other two additives. The XANES results showed oxidized states for the Cd, In and Sn additives after the catalysts were activated at 350 °C by H_2. The TPR, XANES and reaction results suggest that M-M, M_xO_y and M-Co coordination (M refers Cd, In and Sn) may be present in the catalysts. XRD results showed only a slight decrease in cobalt cluster size with addition of 1% of Cd, In or Sn.

The FTS reaction was carried out on all research catalysts at 220 °C, 2.2 MPa, H_2/CO = 2.1 and 25–50% CO conversion using a 1-L CSTR for about 200 h. Space velocity was adjusted if needed during testing. Addition of 1% the additives resulted in 3 to 10 fold activity losses relative to the unpromoted cobalt in comparison to the expected activity losses in terms of the decreases in cobalt sites as determined by H_2 chemisorption capacities. The significant catalyst activity losses were explained based on the additives covering and poisoning cobalt and possible electronic effects resulting from the interaction of the additives.

Addition of Cd, In or Sn greatly modified the selectivity of the cobalt catalyst. All the additives remarkably promoted the formation of methane, light hydrocarbons and CO_2, and suppressed heavier hydrocarbon formation. However, addition of Cd, In and Sn greatly improved secondary reactions of 1-olefins. The extent of the improvement increased with increasing atomic number (Cd < In < Sn). The selectivity changes are linked with acid sites on the cobalt catalysts, which were found to be promoted by the additives, while a fraction of reduced additives in the metallic phase might improve the rate of hydrogenation during FTS. It is concluded that mild to strong acid sites on the cobalt catalysts (i.e., In-Co and Sn-Co) enhanced H_2 adsorption to a greater extent and promoted methane and light hydrocarbon selectivities, while mild acid sites on the cobalt catalysts (i.e., Sn-Co), enhanced the isomerization reaction of 1-olefins to a greater extent relative to other types of acid sites on the cobalt catalyst.

Author Contributions: Conceptualization, W.M., G.J., B.H.D., writing—original draft preparation, W.M., G.J., writing—review and editing, J.L.S.K., B.H.D., investigation, W.M., G.J., W.D.S., Y.J., S.D.H., resources, J.L.S.K., B.H.D, S.K., supervision, G.J., J.L.S.K., B.H.D., project administration, J.L.S.K., B.H.D., funding acquisition, G.J., B.H.D., formal analysis, W.M., G.J., W.D.S., Y.J., S.D.H., S.K., data curation, W.D.S., S.K., visualization, G.J., B.H.D., validation, W.M., W.D.S., Y.J., S.D.H., reaction testing, W.M., catalyst preparation, G.J., catalyst characterization, G.J., Y.J., S.D.H., product analysis, W.D.S., synchrotron beamline operation, S.K.

Funding: This research was funded by NASA (grant number NNX11A175A) and the Commonwealth of Kentucky. The APC was funded by UK-CAER.

Acknowledgments: This paper is dedicated to the late Professor Burtron H. Davis.

Conflicts of Interest: The authors declare no conflict of interest.

References

1. Espinoza, R.L.; van Berge, P.J.; Bolder, F.H. Catalysts. U.S Patent 5,733,839, 31 March 1998.
2. van Berge, P.J.; Barradas, S.; van de Loosdrecht, J.; Visagie, J.L. Advances in the cobalt catalyzed Fischer-Tropsch synthesis. *Erdoel Erdgas Kohle* **2001**, *117*, 138–142.
3. Jacobs, G.; Das, T.K.; Zhang, Y.; Li, J.; Racoillet, G.; Davis, B.H. Fischer-Tropsch synthesis: Support, loading and promoter effects on the reducibility of cobalt catalysts. *Appl. Catal. A Gen.* **2002**, *233*, 263–281. [CrossRef]
4. Jacobs, G.; Ji, Y.; Davis, B.H.; Cronauer, D.C.; Kropf, A.J.; Marshall, C.L. Fischer-Tropsch synthesis: Temperature programmed EXAFS/XANES investigation of the influence of support type, cobalt loading, and noble metal promoter addition to the reduction behavior of cobalt oxide particles. *Appl. Catal. A Gen.* **2007**, *333*, 177–191. [CrossRef]
5. Jacobs, G.; Ribeiro, M.C.; Ma, W.; Ji, Y.; Khalid, S.; Sumodjo, P.T.A.; Davis, B.H. Group 11 (Cu, Ag, Au) promotion of 15%Co/Al$_2$O$_3$ Fischer-Tropsch catalysts. *Appl. Catal. A: Gen.* **2009**, *361*, 137–151. [CrossRef]
6. Jacobs, G.; Ma, W.; Davis, B.H. Influence of reduction promoters on stability of cobalt/γ-alumina Fischer-Tropsch synthesis catalysts. *Catalysts* **2014**, *4*, 49–76. [CrossRef]
7. Ma, W.; Jacobs, G.; Keogh, R.A.; Bukur, D.B.; Davis, B.H. Fischer-Tropsch synthesis: Effect of Pd, Pt, Re, and Ru noble metal promoters on the activity and selectivity of a 25%Co/Al$_2$O$_3$ catalyst. *Appl. Catal. A Gen.* **2012**, *437–438*, 1–9. [CrossRef]
8. Ma, W.; Jacobs, G.; Gao, P.; Jermwongratanachai, T.; Shafer, W.D.; Pendyala, V.R.R.; Chia, H.Y.; Klettlinger, J.L.S.; Davis, B.H. Fischer-Tropsch synthesis: Pore size and Zr promotional effects on the activity and selectivity of 25%Co/Al$_2$O$_3$ catalysts. *Appl. Catal. A Gen.* **2014**, *475*, 314–324. [CrossRef]
9. Chonco, Z.H.; Nabaho, D.; Claeys, M.; van Steen, E. The role of reduction promoters in Fischer-Tropsch catalysts for the production of liquid fuels. In Proceedings of the 23rd Meeting of the North American Catalysis Society, Louisville, KY, USA, 2–7 June 2013.
10. Hilmen, A.M.; Schanke, D.; Holmen, A. TPR study of the mechanism of rhenium promotion of alumina-supported cobalt Fischer-Tropsch catalysts. *Catal. Lett.* **1996**, *38*, 143–147. [CrossRef]
11. Cook, K.M.; Perez, H.D.; Bartholomew, C.H.; Hecker, W.C. Effect of promoter deposition order on platinum-, ruthenium-, or rhenium-promoted cobalt Fischer-Tropsch catalysts. *Appl. Catal. A Gen.* **2014**, *482*, 275–286. [CrossRef]
12. Vada, S.; Hoff, A.; Ådnanes, E.; Schanke, D.; Holmen, A. Fischer-Tropsch synthesis on supported cobalt catalysts promoted by platinum and rhenium. *Top. Catal.* **1995**, *2*, 155–162. [CrossRef]
13. Cook, K.M.; Hecker, W.C. Reducibility of alumina-supported cobalt Fischer-Tropsch catalysts: Effects of noble metal type, distribution, retention, chemical state, bonding, and influence on cobalt crystallite size. *Appl. Catal. A Gen.* **2012**, *449*, 69–80. [CrossRef]
14. Jacobs, G.; Chaney, J.A.; Patterson, P.M.; Das, T.K.; Maillot, J.C.; Davis, B.H. Fischer-Tropsch synthesis: Study of the promotion of Pt on the reduction property of Co/Al$_2$O$_3$ catalysts by in situ EXAFS of Co K and Pt LIII edges and XPS. *J. Synchrotron Radiat.* **2004**, *11*, 414–422. [CrossRef]
15. Ma, W.; Ding, Y.J.; Lin, L.W. Fischer-Tropsch Synthesis over Activated-Carbon-Supported Cobalt Catalysts: Effect of Co Loading and Promoters on Catalyst Performance. *Ind. Eng. Chem. Res.* **2004**, *43*, 2391–2398. [CrossRef]
16. Jacobs, G.; Ma, W.; Gao, P.; Todic, B.; Bhatelia, T.; Bukur, D.B.; Khalid, S.; Davis, B.H. Fischer-Tropsch Synthesis: Differences Observed in Local Atomic Structure and Selectivity with Pd Compared to Typical Promoters (Pt, Re, Ru) of Co/Al$_2$O$_3$ Catalysts. *Top. Catal.* **2012**, *55*, 811–817. [CrossRef]
17. Jermwongratanachai, T.; Jacobs, G.; Ma, W.; Shafer, W.D.; Gnanamani, M.K.; Gao, P.; Kitiyanan, B.; Davis, B.H.; Klettlinger, J.L.S.; Yen, C.H.; et al. Studies on the regeneration of sulfur-poisoned NOx storage and reduction catalysts, including a Ba composite oxide. *Appl. Catal. A Gen.* **2013**, *464–465*, 165–180. [CrossRef]
18. Guerrero-Ruiz, A.; Sepulveda-Escribano, A.; Rodriguez-Ramos, I. Carbon monoxide hydrogenation over carbon supported cobalt or ruthenium catalysts. Promoting effects of magnesium, vanadium and cerium oxides. *Appl. Catal. A Gen.* **1994**, *120*, 71–83. [CrossRef]
19. Zhang, Y.; Xiong, H.; Liew, K.Y.; Li, J. Effect of magnesia on alumina-supported cobalt Fischer-Tropsch synthesis catalysts. *J. Mol. Catal.* **2005**, *237*, 172–181. [CrossRef]

20. Kikuchi, E.; Sorita, R.; Takahashi, H.; Matsuda, T. Catalytic performances of cobalt-based ultrafine particles prepared by chemical reduction in slurry-phase Fischer-Tropsch synthesis. *Appl. Catal. A Gen.* **1999**, *186*, 121–128. [CrossRef]
21. Van der Riet, M.; Hutchings, G.J.; Copperthwaite, R.G. Selective formation of C_3 hydrocarbons from carbon monoxide and hydrogen using cobalt-manganese oxide catalysts. *J. Chem. Soc. Chem. Commun.* **1986**, *10*, 798–799. [CrossRef]
22. Mendes, F.M.T.; Perez, C.A.C.; Noronha, F.B.; Schmal, M. TPSR of CO hydrogenation on $Co/Nb_2O_5/Al_2O_3$ catalysts. *Catal. Today* **2005**, *101*, 45–50. [CrossRef]
23. Gnanamani, M.K.; Jacobs, G.; Graham, U.M.; Pendyala, V.R.R.; Martinelli, M.; MacLennan, A.; Hu, Y.; Davis, B.H. Effect of sequence of P and Co addition over silica for Fischer-Tropsch synthesis. *Appl. Catal. A Gen.* **2017**, *538*, 190–198. [CrossRef]
24. Huffman, G.P.; Shah, N.; Zhao, J.M.; Huggins, F.E.; Hoost, T.E.; Halvorsen, S.; Goodwin, J.G. In situ XAFS investigation of K-promoted Co catalysts. *J. Catal.* **1995**, *151*, 17–25. [CrossRef]
25. Ma, W.; Jacobs, G.; Keogh, R.A.; Yen, C.H.; Klettlinger, J.L.S.; Davis, B.H. Fischer-Tropsch synthesis: Effect of Pt promoter on activity, selectivities to hydrocarbons and oxygenates, and kinetic parameters over 15%Co/Al_2O_3. In *Synthetic Liquids Production and Refining*; Klerk, A.D., King, D.L., Eds.; American Chemical Society: Washington, DC, USA, 2011; pp. 127–153.
26. Cho, H.J.; Dornath, P.; Fan, W. Synthesis of Hierarchical Sn-MFI as Lewis Acid Catalysts for Isomerization of Cellulosic Sugars. *ACS Catal.* **2014**, *4*, 2029–2037. [CrossRef]
27. Srinivasan, R.; Rice, L.A.; Davis, B.H. Electron microdiffraction study of platinum-tin-alumina reforming catalysts. *J. Catal.* **1991**, *129*, 257–268. [CrossRef]
28. Passos, F.B.; Aranda, D.A.G.; Schmal, M. Characterization and catalytic activity of bimetallic Pt-In/Al_2O_3 and Pt-Sn/Al_2O_3 catalysts. *J. Catal.* **1998**, *178*, 478–488. [CrossRef]
29. Coleto, I.; Roldan, R.; Jimenez-Sanchidrian, C.; Gomez, J.P.; Romero-Salguero, F.J. Transformation of α-olefins over Pt-M (M = Re, Sn, Ge) supported chlorinated alumina catalysts. *Fuel* **2007**, *86*, 1000–1007. [CrossRef]
30. Mazzieri, V.A.; Grau, J.M.; Vera, C.R.; Yori, J.C.; Parera, J.M.; Pieck, C.L. Role of Sn in Pt-Re-Sn/Al_2O_3-Cl catalysts for naphtha reforming. *Catal. Today* **2005**, *107–108*, 643–650. [CrossRef]
31. Ressler, T. WinXAS: A program for X-ray absorption spectroscopy data analysis under MS-Windows. *J. Synchrotron Radiat.* **1998**, *5*, 118–122. [CrossRef]

© 2019 by the authors. Licensee MDPI, Basel, Switzerland. This article is an open access article distributed under the terms and conditions of the Creative Commons Attribution (CC BY) license (http://creativecommons.org/licenses/by/4.0/).

Article

Selective CO Hydrogenation Over Bimetallic Co-Fe Catalysts for the Production of Light Paraffin Hydrocarbons (C_2–C_4): Effect of Space Velocity, Reaction Pressure and Temperature

Seong Bin Jo [1,†], Tae Young Kim [2,†], Chul Ho Lee [2], Jin Hyeok Woo [2], Ho Jin Chae [1], Suk-Hwan Kang [3], Joon Woo Kim [4], Soo Chool Lee [1,*] and Jae Chang Kim [2,*]

[1] Research Institute of Advanced Energy Technology, Kyungpook National University, Daegu 41566, Korea; santebin@knu.ac.kr (S.B.J); hwman777@nate.com (H.J.C.)
[2] Department of Chemical Engineering, Kyungpook National University, Daegu 41566, Korea; tyoung0218@knu.ac.kr (T.Y.K.); cjfgh38@knu.ac.kr (C.H.L.); wjh8865@knu.ac.kr (J.H.W.)
[3] Institute for Advanced Engineering, Yongin 41718, Korea; shkang@iae.re.kr
[4] Research Institute of Industrial Science and Technology, Pohang 37673, Korea; realjoon@rist.re.kr
* Correspondence: soochool@knu.ac.kr (S.C.L.); kjchang@knu.ac.kr (J.C.K.);
 Tel.: +82-53-950-5622 (S.C.L. & J.C.K.)
† Seong Bin Jo and Tae young Kim contributed equally to this work.

Received: 6 August 2019; Accepted: 16 September 2019; Published: 19 September 2019

Abstract: Synthetic natural gas (SNG) using syngas from coal and biomass has attracted much attention as a potential substitute for fossil fuels because of environmental advantages. However, heating value of SNG is below the standard heating value for power generation (especially in South Korea and Japan). In this study, bimetallic Co-Fe catalyst was developed for the production of light paraffin hydrocarbons (C_2–C_4 as well as CH_4) for usage as mixing gases to improve the heating value of SNG. The catalytic performance was monitored by varying space velocity, reaction pressure and temperature. The CO conversion increases with decrease in space velocities, and with an increase in reaction pressure and temperature. CH_4 yield increases and C_{2+} yield decreases with increasing reaction temperature at all reaction pressure and space velocities. In addition, improved CH_4 yield at higher reaction pressure (20 bar) implies that higher reaction pressure is a favorable condition for secondary CO_2 methanation reaction. The bimetallic Co-Fe catalyst showed the best results with 99.7% CO conversion, 36.1% C_2–C_4 yield and 0.90 paraffin ratio at H_2/CO of 3.0, space velocity of 4000 mL/g/h, reaction pressure of 20 bar, and temperature of 350 °C.

Keywords: Synthetic natural gas (SNG); Cobalt; Iron; Fischer-Tropsch synthesis; C_2–C_4 hydrocarbons; paraffin ratio

1. Introduction

At present, the production of synthetic natural gas (SNG), mainly consisting of methane, has aroused extensive attention and been commercially produced from different starting materials, including coal and solid dry biomass (e.g., wood and straw) [1–5]. CH_4 via synthesis gas (syngas, CO + $3H_2$) is an effective and environmentally friendly method, because it emits the smallest amount of CO_2 per energy unit among all fossil fuels. However, the heating value of CH_4 is typically below the standard heating value for power generation (especially in South Korea and Japan) [6–12]. For power generation, liquefied petroleum gas (LPG, C_3–C_4 hydrocarbons) must be added to SNG to enhance its heating value; however, the price of LPG is strongly correlated with that of oil. In principle, synthetic light hydrocarbons (C_1–C_4 ranges) via Fischer–Tropsch (FT) reaction could be added to SNG as a substitute for LPG by using the same syngas source (H_2/CO ratio = 3.0) for the SNG process. Furthermore, the gas products must maintain a high paraffin ratio, because olefins exhibit a low heating value, as well as being more susceptible to hydration with CH_4 and liquefaction than paraffins of the same carbon chain length under pipeline conditions (-5 °C, 70 bar) [13]. Therefore, the FT product gas must have a high paraffin ratio in C_2–C_4 ranges, as well as a high light hydrocarbon yield (CH_4 and C_2–C_4) if it is to be used to replace LPG for power generation.

Inui et al. reported a "high calorific methanation" process using Co-Mn-Ru/Al_2O_3 catalyst for the production of high-calorie gas comparable to natural gas with added C_2–C_4 hydrocarbons [6]. The Co-Mn-Ru/Al_2O_3 catalyst afforded high CO conversion (98.8%) and C_2–C_4 selectivity (19.1%). Lee et al. elucidated the role of each component in the Co-based catalysts, and proposed the 10Co-6Mn-2.5Ru/Al_2O_3 and 20Co-16Mn/Al_2O_3 as optimum catalysts for high heating value of SNG [7]. They also developed Fe-Zn and Fe-Cu catalysts, and the Fe-based catalysts were evaluated after caburization and reduction pretreatment [8–10]. In an earlier report, bimetallic Co-Fe catalysts supported on γ-Al_2O_3 were developed for the production of light hydrocarbons (C_2–C_4 ranges) at high CO conversion [11]. It was found that the reducibility of the iron phase was enhanced in the presence of cobalt, leading to enhanced catalytic activity. Of all catalysts, 5Co-15Fe/γ-Al_2O_3 exhibited the highest C_2–C_4 paraffin selectivity at high CO conversion. The high CO conversion and similar hydrocarbon distribution of 5Co-15Fe/γ-Al_2O_3 compared to 20Fe/γ-Al_2O_3 is due to improved iron reducibility. Moreover, the effects of the H_2/CO gas ratio and the reaction temperature on the catalytic performance over 5Co-15Fe/γ-Al_2O_3 catalyst were investigated: the FT catalyst showed high paraffinic C_2–C_4 selectivity and CO conversion at H_2/CO = 3.0, reaction temperature of 300 °C and pressure of 10 bar; but this led to substantial byproduct formation, such as C_{5+} liquid and waxy hydrocarbons and CO_2. Despite the high C_2–C_4 yield, a considerable amount of byproducts (C_{5+} hydrocarbons and CO_2) need to be condensed or separated to be used as mixing gases in SNG for practical processing. To overcome this problem, hybrid catalysts (FT + cracking) in a double-layered bed reactor system were introduced to minimize C_{5+} and CO_2 [12]. The layer of cracking catalysts (SAPO-34 zeolite and Ni catalysts) was loaded underneath the FT catalyst (5Co-15Fe/γ-Al_2O_3) layer in the double-layered bed reactor system. Compared with the FT catalyst in a single-layered bed reactor, cracking catalysts (SAPO-34 and Ni catalysts) convert C_{5+} hydrocarbons into light hydrocarbons (CH_4 and C_2–C_4) in the double-layered bed reactor system. In addition, the Ni catalyst improved the CO conversion and reduced the CO_2 yield via methanation.

At present, few studies have made an effort to improve the heating value of SNG by producing paraffinic C_2–C_4 hydrocarbons and minimizing byproducts (C_{5+} and CO_2). Although catalytic performance, including CO conversion and hydrocarbon distribution, is strongly dependent on the operation conditions such as space velocity, reaction pressure and temperature in the practical process, these effects were not investigated in detail. Herein, catalytic performance over bimetallic Co-Fe catalyst (5Co-15Fe/γ-Al_2O_3) under different reaction conditions (SV, P and T) is evaluated to determine the optimum operating conditions for the production of high paraffinic C_2–C_4 yield, as well as reduction of byproduct (C_{5+} and CO_2). In addition, characterization of the catalysts was performed using inductively coupled plasma optical emission spectroscopy (ICP-OES), X-ray diffraction (XRD) techniques and Brunauer-Emmett-Teller (BET) analysis.

2. Results

Table 1 shows the metal content and textural properties, such as BET surface area, pore volume, and average pore size, of support material (γ-Al_2O_3) and FT catalyst (5Co-15Fe/γ-Al_2O_3). As listed in Table 1, the metal content in FT catalyst (5Co-15Fe/γ-Al_2O_3) is 5.2 wt.% Co and 14.1 wt.% Fe, which is almost consistent with the intended metal loading. The γ-Al_2O_3 shows a BET surface area of 156.9 m^2/g, pore volume of 0.23 cm^3/g and average pore size of 5.9 nm, and those values decreased after impregnation of γ-Al_2O_3 with cobalt and iron. In addition to γ-Al_2O_3 (JCPDS No. 10-0425), fresh 20Co/γ-Al_2O_3 and 20Fe/γ-Al_2O_3 catalysts showed Co_3O_4 phase (JCPDS No. 43-1003) and Fe_2O_3 phase (JCPDS No. 52-1449), respectively (Figure S1) [11]. However, the XRD peaks of Fe phase are very broad compared to those of Co_3O_4 phase, due to the high dispersion of iron phase on γ-alumina [11,12,14,15]. In the case of the bimetallic Co-Fe catalysts, the XRD peaks of Co_3O_4 decreased and those of Fe_2O_3 increased slightly with increasing iron-to-cobalt ratio. However, fresh 5Co-15Fe/γ-Al_2O_3 showed XRD patterns of CoO (JCPDS No. 48-1719) and Fe_2O_3 phases, indicating the incorporation of Co into Fe_2O_3 after calcination, as shown in Figure 1 [11,12,14,15]. On the other hand, the reduced 5Co-15Fe/γ-Al_2O_3 showed XRD peaks of Co metal (JCPDS No. 01-1259) and Fe metal (JCPDS No. 87-0722), respectively. The crystallite size of CoO phase is 4.7 nm, while that of Fe_2O_3 could not be calculated because of the too-broad peaks of Fe_2O_3, as mentioned above. In reduced states of the catalyst, crystallite sizes of the Fe metal are ~20 nm. Based on the H_2-TPR results, reduction temperatures of cobalt phase increased, and those of iron decreased with the increasing iron-to-cobalt ratio (Figure S2). Thus, it can be concluded that the incorporation of Co into the Fe_2O_3 phase results in a weaker interaction between iron and alumina, which enhance the reducibility of iron species and catalytic activity (Figure S3). Therefore, the 5Co-15Fe/γ-Al_2O_3 catalyst was chosen as the catalyst with optimum mass ratio for the following studies.

Table 1. Characterization of γ-Al_2O_3 support material and FT catalyst (5Co-15Fe/γ-Al_2O_3).

	Metal Content (wt.%) [a]		Textural Properties			Crystallite Size (nm) [c]			
						Fresh		Reduced	
	Co	Fe	BET Surface Area (m^2/g)	Pore Volume (cm^3/g)	Average Pore Size (nm) [b]	CoO	Fe_2O_3	Co^0	Fe^0
γ-Al_2O_3	-	-	156.9	0.23	5.9	-	-	-	-
5Co-15Fe/γ-Al_2O_3	5.2	14.1	40.6	0.09	4.6	4.7	-	-	20

[a] Metal contents were determined by ICP-OES. [b] Average pore size were measured following the Barrett-Joyner-Halenda (BJH) method. [c] Crystallite sizes of metal phase were calculated using the Scherer equation.

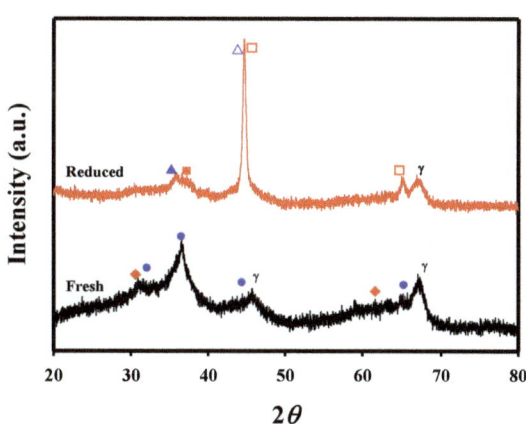

Figure 1. XRD patterns of the FT catalyst (5Co-15Fe/γ-Al_2O_3) in fresh and reduced states; (●) Co_3O_4, (▲) CoO, (△) Co metal, (♦) Fe_2O_3, (■) Fe_3O_4, and (□) Fe metal.

Figure 2 shows the CO conversion of FT catalyst (5Co-15Fe/γ-Al_2O_3) as a function of time on stream after reduction at 500 °C for 1 h under a 10% H_2/N_2 gas mixture. Activity tests were conducted under conditions of H_2/CO = 3.0 at different reaction parameters, such as space velocity, reaction pressure and temperature. Overall, CO conversion of the FT catalyst increased with the increase in reaction temperature. Under almost all operating conditions, the FT catalyst maintains its CO conversion and yield of hydrocarbons. However, the CO conversion of the FT catalyst decreased below

10 bar, with a space velocity of 8000 mL/g/h, at 300 and 350 °C (Figure 2c). CO conversion decreased from 64.9 to 55.9% at 300 °C, and from 97.9 to 81.4% at 350 °C as the reaction progressed. It is well known that the deactivation of the catalysts in CO hydrogenation is due to several factors, including sintering, re-oxidation of active materials, poisoning, coke formation on the surface of active materials, etc., but its causes and effects are not elucidated in this paper. The deactivation of the FT catalysts was compensated by the high reaction temperature (Figure 2c) and pressure (Figure 2f), leading to an increase in CO conversion [16].

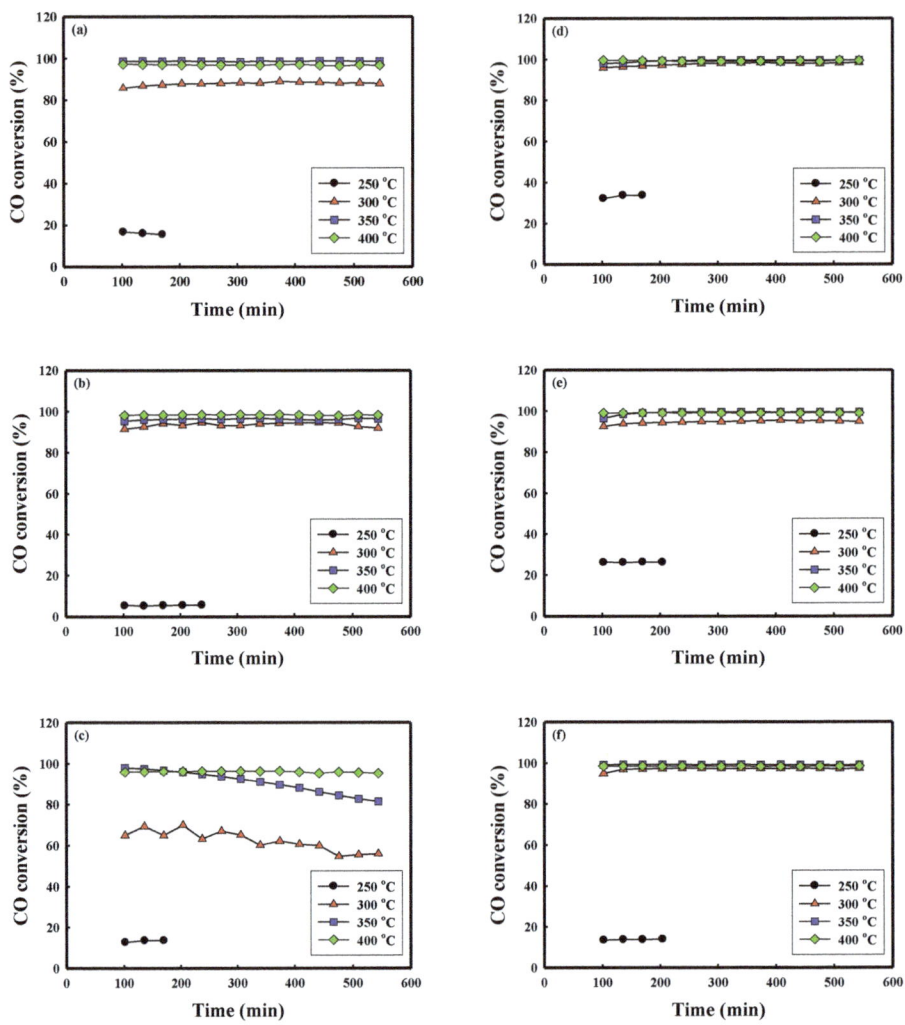

Figure 2. CO conversion of FT catalyst at different reaction temperature under 10 bar (left) and 20 bar (right) at space velocity of (**a,d**) 4000, (**b,e**) 6000, and (**c,f**) 8000 mL/g/h as a function of time on stream.

The results of the catalytic behavior, such as the initial CO conversion and initial hydrocarbon yield of the FT catalyst, are shown in Figures 3 and 4, and summarized in Table 2. Figure 3 shows the initial CO conversion of the FT catalyst as a function of reaction temperature. As shown in Figure 3, the initial CO conversion of the catalyst increased dramatically to 300 °C, and then increased slightly above

350 °C at all pressures and space velocities. Furthermore, it was also found that the CO conversion increased with the increase in reaction pressure, and with the decrease in space velocity. At high temperature (≥300 °C), CO conversion appears to be largely independent of space velocity and reaction pressure. It was reported that CO conversion initially increased dramatically, and then decreased with increasing reaction temperatures (>400 °C) [17,18]. In addition, CO conversion increased with increase in reaction pressure and decrease in space velocity [18,19]. However, space velocity and reaction pressure have little effect on CO conversion, because high reaction temperature (≥300 °C) has a significant influence on CO conversion.

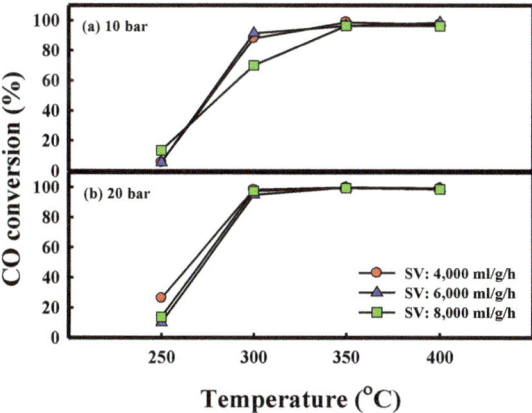

Figure 3. The initial CO conversion of the FT catalyst (5Co-15Fe/γ-Al$_2$O$_3$) under (**a**) 10 and (**b**) 20 bar at different space velocities as a function of reaction temperature.

Table 2. Summarization of catalytic performance over FT catalyst (5Co-15Fe/γ-Al$_2$O$_3$) at different space velocity, reaction pressure and temperature.

P (bar)	SV (ml/g/h)	T (°C)	Conversion		Yield (%)				(C$_2$–C$_4$)/(C$_1$–C$_4$)	P/(P+O)
			CO	H$_2$	CH$_4$	C$_2$–C$_4$	C$_{5+}$	CO$_2$		
10	4000	300	88.0 ± 1.2	43.1 ± 0.5	26.7 ± 0.9	26.9 ± 1.8	15.4 ± 0.3	19.1 ± 0.1	0.50	0.96
		350	98.7 ± 0.1	49.3 ± 0.2	31.7 ± 2.1	31.8 ± 0.6	13.9 ± 2.5	21.2 ± 0.2	0.50	0.89
		400	97.2 ± 0.2	51.7 ± 0.2	42.5 ± 1.5	23.5 ± 1.8	11.5 ± 0.5	19.7 ± 0.4	0.36	0.82
	6000	300	91.5 ± 1.0	38.2 ± 0.9	21.5 ± 0.7	25.8 ± 0.6	23.8 ± 1.6	20.4 ± 0.4	0.55	0.98
		350	96.4 ± 0.2	40.0 ± 0.8	32.3 ± 0.4	23.2 ± 0.3	14.3 ± 0.7	26.6 ± 0.2	0.42	0.91
		400	98.5 ± 0.0	47.6 ± 0.2	43.9 ± 0.5	19.1 ± 1.5	12.4 ± 2.1	23.0 ± 0.2	0.30	0.87
	8000	300	77.1 ± 15.1	35.8 ± 6.6	22.8 ± 1.1	22.2 ± 2.2	9.2 ± 3.2	15.7 ± 3.3	0.49	0.88
		350	96.3 ± 4.9	31.7 ± 2.0	29.9 ± 1.4	31.4 ± 1.1	15.5 ± 1.6	19.5 ± 0.8	0.51	0.81
		400	96.2 ± 0.2	55.3 ± 1.2	48.4 ± 0.3	15.7 ± 0.3	14.4 ± 0.9	17.7 ± 0.3	0.24	0.83
20	4000	300	98.2 ± 0.1	50.8 ± 0.1	29.8 ± 0.0	33.1 ± 0.0	16.3 ± 0.1	19.1 ± 0.1	0.53	0.93
		350	99.7 ± 0.1	57.0 ± 0.2	31.2 ± 0.4	36.1 ± 0.6	17.6 ± 0.4	14.9 ± 0.2	0.54	0.90
		400	99.2 ± 0.1	62.2 ± 0.6	48.0 ± 1.6	25.2 ± 1.6	12.2 ± 0.6	13.8 ± 0.1	0.34	0.91
	6000	300	90.5 ± 0.3	42.6 ± 0.2	26.4 ± 0.3	31.9 ± 0.2	15.6 ± 0.5	21.0 ± 0.0	0.55	0.93
		350	99.6 ± 0.1	53.0 ± 0.1	35.5 ± 0.2	33.8 ± 0.2	11.8 ± 0.4	18.7 ± 0.0	0.49	0.87
		400	99.1 ± 0.0	60.3 ± 0.1	48.7 ± 0.7	26.5 ± 0.6	8.5 ± 1.0	15.5 ± 0.1	0.35	0.87
	8000	300	97.2 ± 0.2	48.8 ± 0.2	30.2 ± 0.4	29.8 ± 0.8	17.7 ± 0.4	19.6 ± 0.0	0.50	0.89
		350	99.3 ± 0.1	53.7 ± 0.3	33.7 ± 0.8	32.6 ± 1.4	15.6 ± 1.8	17.5 ± 0.4	0.49	0.86
		400	98.5 ± 0.1	58.7 ± 0.3	52.6 ± 1.3	22.4 ± 0.9	7.8 ± 1.7	15.6 ± 0.1	0.30	0.86

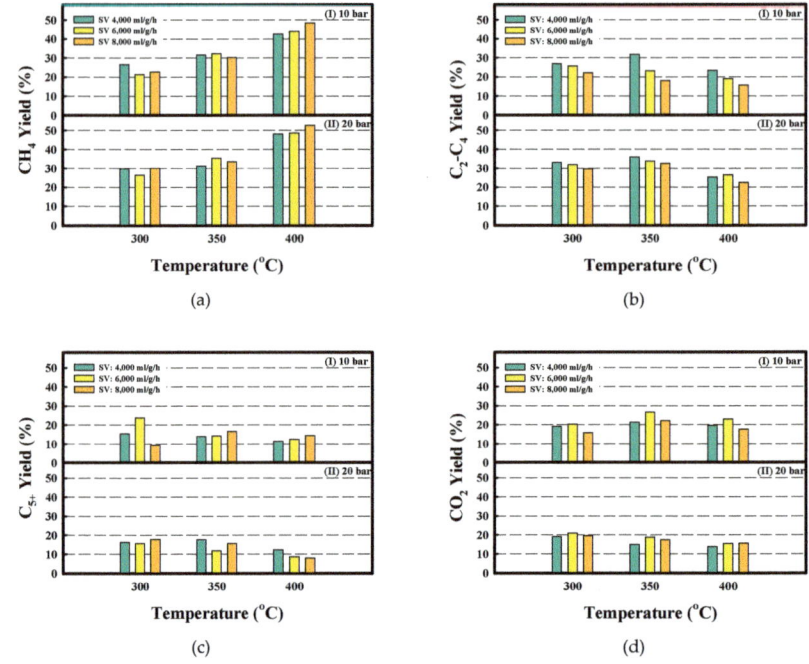

Figure 4. The initial yield of (**a**) CH_4, (**b**) C_2–C_4, (**c**) C_{5+} and (**d**) CO_2 over the FT catalyst (5Co-15Fe/ γ-Al_2O_3) at different space velocity as a function of reaction temperature at (I) 10 bar and (II) 20 bar.

Figure 4 shows the initial product yields for (a) CH_4, (b) C_2–C_4, (c) C_{5+} and (d) CO_2 under high CO conversion conditions (≥300 °C). The effects of reaction parameters such as space velocity, reaction pressure and temperature on product distribution are strongly dependent on the secondary reactions of primary products including hydrogenation, reinsertion, hydrogenolysis, and isomerization [19,20]. It is well known that CH_4 increases and C_{5+} hydrocarbons decreases with increasing space velocity under typical FTS conditions. In addition, an increase in reaction temperature shifts the hydrocarbon distribution towards light hydrocarbon products, whereas an increase in reaction pressure shifts the hydrocarbon distribution to heavier products in typical FTS reaction [19]. As shown in Figure 4a, CH_4 yield increases with the increase in reaction temperature under all reaction pressures and space velocities. In contrast to typical FTS conditions, improved CH_4 yield at high reaction pressure (20 bar) indicates that the higher reaction pressure is favorable for secondary CO_2 methanation reaction, as discussed below in Figure 4d [21–24]. In addition, space velocity did not affect methane yield at relatively low temperature (300 and 350 °C), whereas CH_4 yield increases with the increase in space velocity at 400 °C. As shown in Figure 4b, C_2–C_4 yield reached the highest values at 300 or 350 °C, and then decreased at higher reaction temperatures. Furthermore, reaction pressure enhanced C_2–C_4 yield, and space velocity is inversely proportional to C_2–C_4 yield. In the case of C_{5+} hydrocarbon yield, although it is difficult to confirm the effects of reaction parameters, C_{5+} yield decreased with the increase in reaction pressure and temperature (Figure 4c). As shown in Figure 4d, CO_2 yield exhibited almost the same values (ca. 20%) at reaction temperatures between 350 and 400 °C. This is due to the fact that the increase in CO_2 production with increasing CO conversion can mainly be attributed to increased water gas-shift reaction (WGS, $CO + H_2O \leftrightarrow CO_2 + H_2$) at high water partial pressures [11,25]. At 20 bar, however, it is notable that CO_2 yield decreases with the increase in reaction temperature, despite almost 100% CO conversion at all temperatures. These results show that higher reaction pressure and temperature are favorable conditions for secondary CO_2 methanation reaction.

Figure 5 shows the yield of light hydrocarbons in C_1–C_4 range and ratio of C_2–C_4 to C_1–C_4 yield under different reaction parameters, such as space velocity, reaction temperature and pressure. At 10 bar, sum of CH_4 and C_2–C_4 yield increased (45–53 to 63–66%), and ratio of C_2–C_4 to C_1–C_4 yield decreased (0.5–0.55 to 0.25–0.35) with the increase in reaction temperature between 300 to 400 °C, as shown in Figure 5a. The sum of CH_4 and C_2–C_4 yield decreased from 53 to 45% at 300 °C, but these values did not change a lot at higher reaction temperature (350 and 400 °C) with an increase in space velocity. In addition, the ratio of C_2–C_4 to C_1–C_4 yield decreased from 0.35 to 0.25 at 400 °C, but its value remained constant (ca. 0.5–0.55) at lower temperature (300 and 350 °C). At 20 bar, the sum of CH_4 and C_2–C_4 yield increased (58–63% to 73–75%), and the ratio of C_2–C_4 to C_1–C_4 yield decreased (0.49–0.55 to 0.30–0.35), with the increase in reaction temperature between 300 to 400 °C, as shown in Figure 5b. The sum of CH_4 and C_2–C_4 yield remained constant at all reaction temperatures between 300 and 400 °C. In addition, the ratio of C_2–C_4 to C_1–C_4 yield decreased from 0.35 to 0.30 at 400 °C, but its value remained constant (ca. 0.5–0.55) at lower temperature (300 and 350 °C). Overall, the yield of light hydrocarbons in C_1–C_4 range increased with increasing in reaction temperature between 300 to 400 °C, because CH_4 yield increases dramatically and C_2–C_4 yield decreases slightly, resulting in reduction of $(C_2$–$C_4)/(C_1$–$C_4)$ ratio. With increasing in space velocity, on the other hand, the sum of CH_4 and C_2–C_4 yield and $(C_2$–$C_4)/(C_1$–$C_4)$ ratio decreased slightly. Furthermore, high reaction pressure enhanced the light hydrocarbon (C_1–C_4) yield and the $(C_2$–$C_4)/(C_1$–$C_4)$ ratio. At both 10 and 20 bar, a space velocity of 4000 mL/g/h and 350 °C are considered to be appropriate conditions for high calorific methanation, since the 5Co-15Fe/γ-Al_2O_3 catalyst exhibit the highest light hydrocarbon (CH_4 and C_2–C_4) yield and $(C_2$–$C_4)/(C_1$–$C_4)$, respectively.

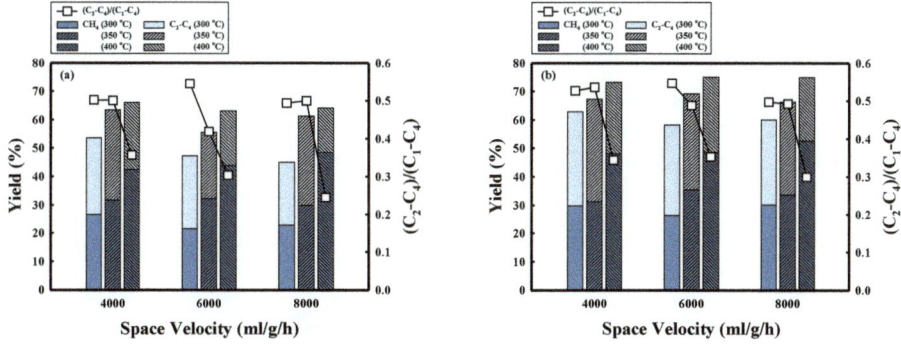

Figure 5. C_1–C_4 yield and $(C_2$–$C_4)/(C_1$–$C_4)$ ratio of the FT catalyst (5Co-15Fe/γ-Al_2O_3) as a function of space velocity at (**a**) 10 bar and (**b**) 20 bar

Figure 6 shows the paraffin ratio (P/(P+O)) with different reaction parameters, such as H_2/CO ratio, space velocity, reaction pressure and temperature, where P and O represent the yields of paraffins and olefins in the C_2–C_4 range, respectively. As shown in Figure 6a, the paraffin ratio increased with increasing H_2/CO ratio at a space velocity of 6000 mL/g/h, although this effect is not noticeable at H_2/CO ratios above 2.0, due to the adjustment of H_2/CO ratio by WGS conditions [11,25]. In addition, it was found that the paraffin ratio showed a positive correlation with CO conversion. The fact that the paraffin ratio increases with the CO conversion suggests that increasing the H_2/CO ratio and temperatures leads to a second hydrogenation of the olefins into paraffin and an increase of the CO conversion. As shown in Figure 6b, space velocity, reaction pressure and temperature have little effect on paraffin ratio at H_2/CO ratio of 3.0, but CO conversion is strongly dependent on reaction temperature. Furthermore, it is found that a higher reaction pressure (20 bar) improved the CO conversion above a reaction temperature of 300 °C, because higher reaction pressure enhanced hydrogen adsorption on support of the catalyst and improved second hydrogenation [19].

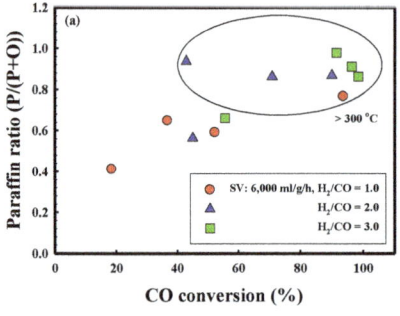 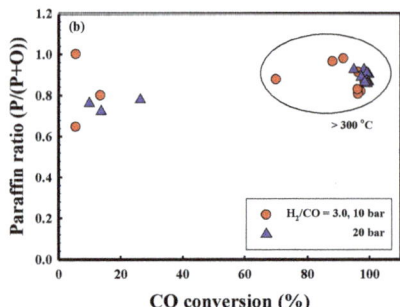

Figure 6. Paraffin ratio of the FT catalyst (5Co-15Fe/γ-Al$_2$O$_3$) as a function of CO conversion at different operating conditions: (**a**) effect of H$_2$/CO ratio at SV of 6000 ml/g/h, and (**b**) effect of reaction pressure at H$_2$/CO ratio of 3.0.

Table 3 shows the comparison of the 5Co-15Fe/γ-Al$_2$O$_3$ catalyst under optimum conditions with different catalysts published in other papers. As listed in Table 3, Co-Mn-Ru catalysts showed high CO conversion and CH$_4$ yield, but relatively low C$_{2+}$ hydrocarbon yield, compared to Fe-based catalysts [7,8]. Lee et al. [7] reported that Mn promoter in cobalt catalysts acted as a Lewis acid, which increased the carbon chain growth and C$_{2+}$ hydrocarbon yield, but suppressed CO conversion. Ru promoter, one of the noble metals, provided the catalyst with hydrogen spillover ability, enhancing the reducibility of the cobalt site and CO conversion, but decreasing the C$_{2+}$ hydrocarbon yield. In addition, Ru promoters were substituted to save cost for the catalysts, and 20Co-16Mn/γ-Al$_2$O$_3$ catalyst reduced at 700 °C for 1 h showed similar reactivity to 10Co-6Mn-2.5Ru/γ-Al$_2$O$_3$ catalyst reduced at 400 °C for 1 h. The Fe-based bulk catalysts promoted by Cu and Zn showed high CO conversion and C$_2$–C$_4$ yield, but low paraffin ratio and high byproduct yield (C$_{5+}$ and CO$_2$) [8,9]. On the other hand, the FT catalyst (5Co-15Fe/γ-Al$_2$O$_3$) affords high CO conversion (99.7%), high light paraffinic hydrocarbon yield (31.2% CH$_4$ and 36.1% C$_2$–C$_4$), and low byproduct formation (C$_{5+}$ and CO$_2$) under optimum conditions (SV: 4000 mL/g/h, T: 350 °C and P: 20 bar).

Table 3. Comparison of catalytic performance of FT catalyst (5Co-15Fe/γ-Al$_2$O$_3$) with different catalysts published in other papers.

Catalysts	H$_2$/CO	Reaction Condition	CO Conv. (%)	Yield				P/(P+O)	Ref
				CH$_4$	C$_2$–C$_4$	C$_{5+}$	CO$_2$		
5Co-15Fe/ γ-Al$_2$O$_3$	3.0	SV: 6000 ml/g/h, 300 °C, 10 bar	91.5	21.5	25.8	23.8	20.4	0.98	This study
	3.0	SV: 4000 ml/g/h, 350 °C, 20 bar	99.7	31.2	36.1	17.6	14.9	0.90	This study
10Co-6Mn-2Ru/ γ-Al$_2$O$_3$ [a]	3.0	SV: 6000 ml/g/h, 300 °C, 10 bar	99.7	60.6	24.5	4.8	10.9	0.96	[8]
10Co-6Mn-2.5Ru/ γ-Al$_2$O$_3$ [b]	3.0	SV: 6000 ml/g/h, 250 °C, 10 bar	100.0	53.0	23.0	8.6	n/a [g]	n/a [g]	[7]
20Co-16Mn/ γ-Al$_2$O$_3$ [c]	3.0	SV: 6000 ml/g/h, 250 °C, 10 bar	92.0	53.0	24.0	5.8	n/a [g]	n/a [g]	[7]
FC15 [d]	3.0	SV: 6000 ml/g/h, 300 °C, 10 bar	97.5	21.5	35.7	12.9	27.4	0.58	[9]
FZ5 [e]	3.0	SV: 6000 ml/g/h, 300 °C, 10 bar	89.9	19.1	35.3	24.1	24.1	0.76	[8]
FZ10 [f]	3.0	SV: 6000 ml/g/h, 300 °C, 10 bar	98.2	23.7	40.0	12.7	21.9	0.70	[8]

[a] 10 wt.% Co, 6 wt.% Mn, and 2 wt.% Ru, reduced at for 400 °C for 1h. [b] 10 wt.% Co, 6 wt.% Mn, and 2.5 wt.% Ru, reduced at for 400 °C for 1h. [c] 20 wt.% Co, 16 wt.% Mn, reduced at for 700 °C for 1h. [d] Fe/Cu atomic ratio = 15, reduced at 500 °C for 1 h. [e] Fe/Zn atomic ratio = 5, reduced at 500 °C for 1 h. [f] Fe/Zn atomic ratio = 10, caburized at 500 °C for 1 h. [g] n/a: Not applicable.

3. Materials and Methods

3.1. Catalyst Synthesis

The bimetallic Co-Fe catalyst was synthesized by wet impregnation of γ-alumina (Sigma–Aldrich, St. Louis, MO, USA) with $Co(NO_3)_2 \cdot 6H_2O$ (Sigma–Aldrich, St. Louis, MO, USA) and $Fe(NO_3)_3 \cdot 9H_2O$ (Sigma–Aldrich, St. Louis, MO, USA) according to the same method as our previous papers [11,12]. During the impregnation procedure of the FT catalyst (5Co-15Fe/γ-Al_2O_3), γ-Al_2O_3 was added to an anhydrous ethanol solution containing cobalt and iron nitrates. The weight percentages of the cobalt and iron metal based on the catalyst were 5% and 10%, respectively. After stirring for 24 h, the solvent was vaporized in a rotary evaporator at 40–60 °C. The samples were dried at 120 °C for 12 h, and subsequently calcined at 400 °C for 8 h.

3.2. Characterization

The metal contents in FT catalyst were measured using inductively coupled plasma optical emission spectroscopy (ICP-OES; Perkin-Elmer, Waltham, MA, USA). Nitrogen adsorption-desorption isotherms at -196 °C were measured using a Micrometrics ASAP 2020 instrument (Norcross, GA, USA) to acquire the textural properties of the materials. Average pore size were measured following the Barrett-Joyner-Halenda (BJH) method. The crystal structure of the FT catalyst was analyzed via X-ray diffraction (XRD; PANalytical, Amsterdam, Netherlands) using a Cu Kα radiation source at the Korea Basic Science Institute in Daegu. Crystallite sizes of metal phase were calculated using the Scherer equation.

3.3. Activity Tests

Prior to the reaction, the catalysts (0.5 g) were placed in a fixed-bed stainless steel reactor (1/2 inch I.D.) and reduced with a 10 vol% H_2/N_2 gas mixture at 500 °C for 1 h. Then, the gas stream (H_2, CO, N_2) was fed to the reactor at a different total gas flow (33, 50 and 60 mL/min); N_2 gas was used as an internal standard in the feed gas. The reactor was pressurized to 10 or 20 bar with the feed gas stream using a back-pressure regulator at constant pressure and heated to 200 °C. Then, the temperature was increased to 250, 300, 350, or 400 °C, and maintained during the FT reaction. All volumetric gas flows were measured at standard temperature and pressure (S.T.P). To prevent the condensation of water vapor and hydrocarbons, the inlet and outlet lines of the reactor were maintained at temperatures above 250 °C, and the liquid and wax products were collected in a cold trap (0 °C) before injection of the gas into the reactor and GC column. The outlet gases were analyzed using a gas chromatograph (Agilent 6890; Agilent, Santa Clara, CA, USA) equipped with both a thermal conductivity detector (TCD), and a flame ionization detector (FID). A packed column (Carboxen 1000; Bellefonte, PA, USA) was connected to the TCD to analyze the CO, H_2, N_2, and CO_2 gases, and a capillary column (GS Gas Pro; Agilent, Santa Clara, CA, USA) was connected to the FID to analyze the hydrocarbon gases.

CO conversion, selectivity and yield for each product were calculated using Equations (1)–(4).

$$\text{CO conversion (carbon mole \%)} = \left(1 - \frac{\text{CO in the product gas (mol/min)}}{\text{CO in the feed gas (mol/min)}}\right) \times 100 \quad (1)$$

$$\text{Selectivity for hydrocarbons with carbon number } n \text{ (carbon mole \%)}$$
$$= \frac{n \times C_n \text{ hydrocarbon in the product gas (mol/min)}}{\text{(total carbon−unreacted CO) in the product gas (mol/min)}} \times 100 \quad (2)$$

$$\text{Selectivity for carbon dioxide (carbon mole \%)}$$
$$= \frac{CO_2 \text{ in the product gas (mol/min)}}{\text{(total carbon−unreacted CO) in the product gas (mol/min)}} \times 100 \quad (3)$$

$$\text{Yield for hydrocarbons and carbon dioxide} = \frac{\text{CO conversion} \times \text{Selectivity}}{100} \quad (4)$$

4. Conclusions

At present, few studies have made an effort to produce mixing gases consisting of paraffinic C_2–C_4 hydrocarbons into SNG for power generation. In this study, bimetallic Co-Fe catalysts supported on γ-alumina were developed, and the effects of operating parameters such as space velocity, reaction pressure and temperature on catalytic performance were elucidated for the production of light paraffin hydrocarbon yield (C_2–C_4 range) with high paraffin ratio, as well as reduction of byproduct formation (C_{5+} and CO_2). It was found that CO conversion increases with a decrease in space velocity, and with an increase in reaction pressure and temperature. CH_4 yield increases and C_{2+} yield decreases with increasing reaction temperature at all reaction pressures and space velocities. In addition, improved CH_4 yield at higher reaction pressure (20 bar) implies that higher reaction pressure is a favorable condition for secondary CO_2 methanation reaction. While paraffin ratio shows a positive correlation with the CO conversion according to increasing H_2/CO ratio, reaction pressure and temperature have little effect on paraffin ratio at a H_2/CO ratio of 3.0. Based on these results, the optimum conditions were determined to be H_2/CO of 3.0, space velocity of 4000 mL/g/h, reaction pressure of 20 bar, and temperature of 300 °C, and the FT catalyst (5Co-15Fe/γ-Al_2O_3) affords a high light hydrocarbon yield (31.2 % CH_4, and 36.1 % C_2–C_4) with high paraffin ratio (0.90). Based on these results, the bimetallic Co-Fe catalysts can be used for production of high paraffinic light hydrocarbons.

Supplementary Materials: The following are available online at http://www.mdpi.com/2073-4344/9/9/779/s1, Figure S1: XRD patterns of (I) fresh and (II) reduced monometallic and bimetallic Co-Fe catalysts; (●) Co_3O_4, (▲) CoO, (△) Co metal, (◆) Fe_2O_3, (■) Fe_3O_4, and (□) Fe metal; Figure S2. H_2-TPR profiles of the monometallic and bimetallic catalysts supported on γ-alumina: (a) 20Co/γ-Al_2O_3, (b) 15Co-5Fe/γ-Al_2O_3, (c) 10Co-10Fe/γ-Al_2O_3, (d) 5Co-15Fe/γ-Al_2O_3, and (e) 20Fe/γ-Al_2O_3 (5 °C/min, pure hydrogen); Figure S3: CO conversion and hydrocarbon distribution of the monometallic and bimetallic catalysts supported on γ-alumina: 20Co/γ-Al_2O_3, 15Co-5Fe/γ-Al_2O_3, 10Co-10Fe/γ-Al_2O_3, 5Co-15Fe/γ-Al_2O_3, and 20Fe/γ-Al_2O_3 at H_2/CO ratio = 3.0, 300 °C, and 10 bar.

Author Contributions: Conceptualization, S.B.J., T.Y.K., S.-H.K. and J.W.K.; Data curation, S.B.J., T.Y.K., C.H.L. and J.H.W.; Formal analysis, T.Y.K., C.H.L., J.H.W. and H.J.C.; Investigation, S.B.J. and T.Y.K.; Project administration, S.C.L. and J.C.K.; Supervision, S.C.L. and J.C.K.; Writing—original draft, S.B.J.; Writing—review & editing, S.B.J. and T.Y.K.

Funding: This research was funded by the Korea Institute of Energy Technology Evaluation and Planning (KETEP) and the Ministry of Trade, Industry & Energy (MOTIE) of the Republic of Korea (No.20173010050110 and the Basic Science Research Program through the National Research Foundation of Korea (NRF) funded by the Ministry of Science, ICT & Future Planning (No.2017R1A2B4008275).

Acknowledgments: This work was supported by the Korea Institute of Energy Technology Evaluation and Planning (KETEP) and the Ministry of Trade, Industry & Energy (MOTIE) of the Republic of Korea. (No.20173010050110). This research was also supported by the Basic Science Research Program through the National Research Foundation of Korea (NRF) funded by the Ministry of Science, ICT & Future Planning (No.2017R1A2B4008275).

Conflicts of Interest: The authors declare no conflict of interest.

References

1. Zhao, B.; Chen, Z.; Chen, Y.; Ma, X. Syngas methanation over Ni/SiO_2 catalyst prepared by ammonia-assisted impregnation. *Int. J. Hydrog. Energy* **2017**, *42*, 27073–27083. [CrossRef]
2. Hwang, S.; Lee, J.; Hong, U.G.; Gil Seo, J.; Jung, J.C.; Koh, D.J.; Lim, H.; Byun, C.; Song, I.K. Methane production from carbon monoxide and hydrogen over nickel–alumina xerogel catalyst: Effect of nickel content. *J. Ind. Eng. Chem.* **2011**, *17*, 154–157. [CrossRef]
3. Gao, J.; Jia, C.; Li, J.; Zhang, M.; Gu, F.; Xu, G.; Zhong, Z.; Su, F. Ni/Al_2O_3 catalysts for CO methanation: Effect of Al_2O_3 supports calcined at different temperatures. *J. Energy Chem.* **2013**, *22*, 919–927. [CrossRef]
4. Kopyscinski, J.; Schildhauer, T.J.; Biollaz, S.M. Production of synthetic natural gas (SNG) from coal and dry biomass—A technology review from 1950 to 2009. *Fuel* **2010**, *89*, 1763–1783. [CrossRef]
5. Liu, Y.; Zhu, L.; Wang, X.; Yin, S.; Leng, F.; Zhang, F.; Lin, H.; Wang, S. Catalytic methanation of syngas over Ni-based catalysts with different supports. *Chin. J. Chem. Eng.* **2017**, *25*, 602–608. [CrossRef]

6. Inui, T.; Sakamoto, A.; Takeguchi, T.; Ishigaki, Y. Synthesis of highly calorific gaseous fuel from syngas on cobalt-manganese-ruthenium composite catalysts. *Ind. Eng. Chem. Res.* **1989**, *28*, 427–431. [CrossRef]
7. Lee, Y.H.; Kim, H.; Choi, H.S.; Lee, D.-W.; Lee, K.-Y. Co-Mn-Ru/Al_2O_3 catalyst for the production of high-calorific synthetic natural gas. *Korean J. Chem. Eng.* **2015**, *32*, 2220–2226. [CrossRef]
8. Lee, Y.H.; Lee, D.-W.; Kim, H.; Choi, H.S.; Lee, K.-Y. Fe–Zn catalysts for the production of high-calorie synthetic natural gas. *Fuel* **2015**, *159*, 259–268. [CrossRef]
9. Lee, Y.H.; Lee, D.-W.; Lee, K.-Y. Production of high-calorie synthetic natural gas using copper-impregnated iron catalysts. *J. Mol. Catal. A Chem.* **2016**, *425*, 190–198. [CrossRef]
10. Lee, Y.H.; Lee, K.-Y. Effect of surface composition of Fe catalyst on the activity for the production of high-calorie synthetic natural gas (SNG). *Korean J. Chem. Eng.* **2017**, *34*, 320–327. [CrossRef]
11. Jo, S.B.; Chae, H.J.; Kim, T.Y.; Lee, C.H.; Oh, J.U.; Kang, S.-H.; Kim, J.W.; Jeong, M.; Lee, S.C.; Kim, J.C. Selective CO hydrogenation over bimetallic Co-Fe catalysts for the production of light paraffin hydrocarbons (C_2-C_4): Effect of H_2/CO ratio and reaction temperature. *Catal. Commun.* **2018**, *117*, 74–78. [CrossRef]
12. Jo, S.B.; Kim, T.Y.; Lee, C.H.; Kang, S.-H.; Kim, J.W.; Jeong, M.; Lee, S.C.; Kim, J.C. Hybrid catalysts in a double-layered bed reactor for the production of C_2–C_4 paraffin hydrocarbons. *Catal. Commun.* **2019**, *127*, 29–33. [CrossRef]
13. Lee, J.; Kang, S. Formation behaviours of mixed gas hydrates including olefin compounds. *Chem. Eng. Trans.* **2013**, *32*, 1921–1926.
14. Griboval-Constant, A.; Butel, A.; Ordomsky, V.V.; Chernavskii, P.A.; Khodakov, A.; Khodakov, A. Cobalt and iron species in alumina supported bimetallic catalysts for Fischer–Tropsch reaction. *Appl. Catal. A Gen.* **2014**, *481*, 116–126. [CrossRef]
15. Lögdberg, S.; Tristantini, D.; Borg, Ø.; Ilver, L.; Gevert, B.; Järås, S.; Blekkan, E.A.; Holmén, A. Hydrocarbon production via Fischer–Tropsch synthesis from H_2-poor syngas over different Fe-Co/γ-Al_2O_3 bimetallic catalysts. *Appl. Catal. B Environ.* **2009**, *89*, 167–182. [CrossRef]
16. Rytter, E.; Holmen, A. Deactivation and Regeneration of Commercial Type Fischer-Tropsch Co-Catalysts—A Mini-Review. *Catalysts* **2015**, *5*, 478–499. [CrossRef]
17. Gao, J.; Gu, F.; Zhong, Z.; Liu, Q.; Su, F. Recent advances in methanation catalysts for the production of synthetic natural gas. *RSC Adv.* **2015**, *5*, 22759–22776. [CrossRef]
18. Meng, F.; Li, X.; Lv, X.; Li, Z. Co hydrogenation combined with water-gas-shift reaction for synthetic natural gas production: A thermodynamic and experimental study. *Int. J. Coal Sci. Technol.* **2018**, *5*, 439–451. [CrossRef]
19. Yang, J.; Ma, W.; Chen, D.; Holmén, A.; Davis, B.H. Fischer–Tropsch synthesis: A review of the effect of CO conversion on methane selectivity. *Appl. Catal. A Gen.* **2014**, *470*, 250–260. [CrossRef]
20. Novak, S. Secondary effects in the Fischer-Tropsch synthesis. *J. Catal.* **1982**, *77*, 141–151. [CrossRef]
21. Frontera, P.; Macario, A.; Ferraro, M.; Antonucci, P. Supported Catalysts for CO_2 Methanation: A Review. *Catalysts* **2017**, *7*, 59. [CrossRef]
22. Kirchner, J.; Anolleck, J.K.; Lösch, H.; Kureti, S. Methanation of CO_2 on iron based catalysts. *Appl. Catal. B Environ.* **2018**, *223*, 47–59. [CrossRef]
23. Le, T.A.; Kim, M.S.; Lee, S.H.; Kim, T.W.; Park, E.D. CO and CO_2 methanation over supported Ni catalysts. *Catal. Today* **2017**, *293*, 89–96. [CrossRef]
24. Stangeland, K.; Kalai, D.; Li, H.; Yu, Z. CO_2 Methanation: The Effect of Catalysts and Reaction Conditions. *Energy Procedia* **2017**, *105*, 2022–2027. [CrossRef]
25. Galvis, H.M.T.; De Jong, K.P. Catalysts for Production of Lower Olefins from Synthesis Gas: A Review. *ACS Catal.* **2013**, *3*, 2130–2149. [CrossRef]

© 2019 by the authors. Licensee MDPI, Basel, Switzerland. This article is an open access article distributed under the terms and conditions of the Creative Commons Attribution (CC BY) license (http://creativecommons.org/licenses/by/4.0/).

Article

Effect of Operating Temperature, Pressure and Potassium Loading on the Performance of Silica-Supported Cobalt Catalyst in CO_2 Hydrogenation to Hydrocarbon Fuel

Rama Achtar Iloy and Kalala Jalama *

Department of Chemical Engineering, Doornfontein Campus, University of Johannesburg, Doornfontein 2028, Johannesburg, South Africa; achtar2006@yahoo.fr
* Correspondence: kjalama@uj.ac.za

Received: 31 August 2019; Accepted: 10 September 2019; Published: 26 September 2019

Abstract: Potassium (1–5 wt.%)-promoted and unpromoted Co/SiO_2 catalysts were prepared by impregnation method and characterized by nitrogen physisorption, temperature-programmed reduction (TPR), CO_2 temperature-programmed desorption (TPD), X-ray diffraction (XRD) and X-ray photoelectron spectroscopy (XPS) techniques. They were evaluated for CO_2 hydrogenation in a fixed bed reactor from 180 to 300 °C within a pressure range of 1–20 bar. The yield for hydrocarbon products other than methane (C_{2+}) was found to increase with an increase in the operating temperature and went through a maximum of approximately 270 °C. It did not show any significant dependency on the operating pressure and decreased at potassium loadings beyond 1 wt.%. Potassium was found to enhance the catalyst ability to adsorb CO_2, but limited the reduction of cobalt species during the activation process. The improved CO_2 adsorption resulted in a decrease in surface H/C ratio, the latter of which enhanced the formation of C_{2+} hydrocarbons. The highest C_{2+} yield was obtained on the catalyst promoted with 1 wt.% of potassium and operated at an optimal temperature of 270 °C and a pressure of 1 bar.

Keywords: CO_2 hydrogenation; cobalt; potassium; pressure; temperature

1. Introduction

The promoting capabilities of alkali metals, namely potassium, have been investigated for a variety of catalysts and reactions, including steam reforming of bioethanol [1], water gas shift [2], N_2O decomposition [3], Fischer-Tropsch synthesis (FTS) [4–6] and CO_2 hydrogenation [7–11]. One of the earliest studies on the use of potassium as a promoter for the catalyst used in CO_2 hydrogenation to hydrocarbons is that of Russell and Miller [12]. They investigated several copper-activated cobalt catalysts at atmospheric pressure from 448 to 573 K with H_2/CO_2 ratio varied from 2 to 3. All the catalysts mainly produced methane and liquid hydrocarbons were observed only after potassium addition to the catalyst in the form of either potassium carbonate or phosphate. Potassium was believed to selectively poison methane forming centres, and therefore, promote methylene radicals polymerization by the repression of the competitive hydrogenation reaction. Similarly, Owen et al. [13] studied the effect of potassium, along with that of lithium and sodium, on the performance of Co/SiO_2 catalysts. The catalytic testing was carried out at 643 K, atmospheric pressure and using an H_2/CO_2 ratio of 3. They showed that with an alkali loading as low as 1 wt.%, the products distribution shifts towards longer chain hydrocarbons. Furthermore, C_2 and C_3 olefins, which did not form over the unpromoted catalyst, were detected in relatively significant amounts over the promoted catalysts. The authors attributed this behaviour to the ability of potassium to enhance the surface to molecule charge transfer, resulting in increased CO and reduced hydrogen binding strength. These findings were further corroborated by a more recent investigation by Shi et al. [8] on a CoCu/TiO$_2$ system containing 1.5–3.5 wt.% K. Using CO_2 temperature-programmed desorption, the authors were able to link an improved yield of liquid hydrocarbons (C_{5+}) to the increased CO_2 adsorption capacity of the catalyst, when loaded with potassium.

It appears that potassium has an enormous potential in the conversion of CO_2 to liquid hydrocarbons. To derive most of the benefit from this promoter, the study of its effect on the reaction must be integrated with that of the effect of operating conditions. Most studies have reported the effect of potassium on cobalt-based catalysts under pre-selected operating conditions that were not optimized. Hence, the present study aims at systematically evaluating the promoting effect of potassium on a Co/SiO$_2$ system used in CO_2 hydrogenation under optimized temperature and pressure conditions.

2. Results and Discussion

2.1. Surface Area and Porosity

The information on the surface area and porosity of the catalysts investigated is presented in Table 1. The data show that cobalt incorporation into the silica support results in a significant drop in the surface area from 186.6 to 133.1 m^2/g. This behaviour is generally explained by the growth of cobalt oxide particles within the pores of the support during catalyst calcination, leading to some level of pore obstruction. This agrees well with the pore volume data, which show a decrease from 1.5 to 1.0 cm^3/g. The introduction of potassium, in amounts above 3% in the catalyst, further amplifies this phenomenon.

Table 1. Surface area and porosity data.

Sample	BET Surface Area (m^2/g)	Pore Volume (cm^3/g)	Pore Diameter (nm)
SiO$_2$	186.6	1.5	31.2
15%Co/SiO$_2$	133.1	1.0	30.2
15%Co-1%K/SiO$_2$	137.8	1.2	33.4
15%Co-3%K/SiO$_2$	123.4	1.1	35.5
15%Co-5%K/SiO$_2$	70.0	0.8	45.1

2.2. X-ray Diffraction

Figure 1 shows the XRD patterns of unpromoted and promoted catalysts before and after reduction. All the unreduced catalysts showed diffraction peaks at 2θ values of approximately 18°, 30°, 36.6°, 39°, 44.5°, 55.3°, 60° and 65°, attributed to Co$_3$O$_4$ [14].

After catalysts reduction, the diffraction peaks for Co$_3$O$_4$, which were present in the unreduced catalysts disappeared (Figure 1b). The only visible peaks are those of the lower oxide of cobalt (CoO) at 42.4° and metallic cobalt at 44.5°.

The Scherrer equation was used to calculate the average crystallite sizes of cobalt species in the catalyst, using two theta values of 36.6°, 42.4° and 44.5° for Co$_3$O$_4$, CoO and Co respectively. The data are reported in Table 2. Although there is no observable trend in the data with respect to Co$_3$O$_4$ and Co, it appears that the average crystallite size for CoO decreases with increasing potassium loading in the catalyst. This suggests that potassium controls the size of CoO in the catalyst.

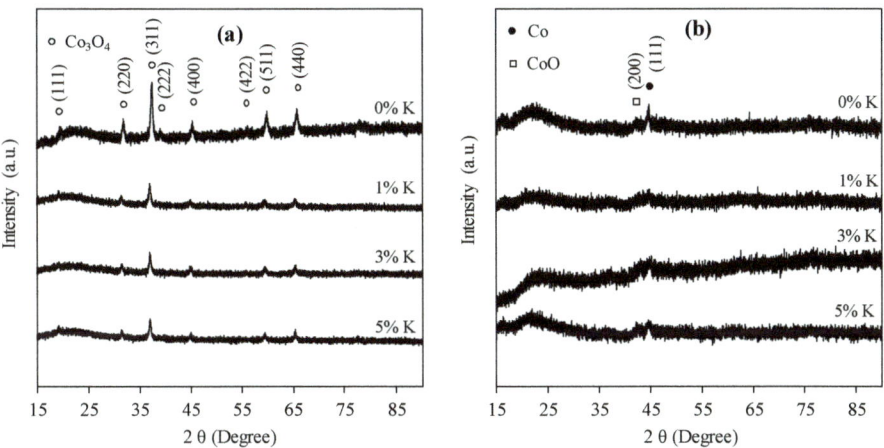

Figure 1. XRD patterns for unpromoted and potassium-promoted 15%Co/SiO$_2$ catalysts: (**a**) Before reduction, and (**b**) after reduction.

Table 2. The particle size of the calcined catalysts.

Catalyst	Freshly Calcined [a]	Reduced and Passivated [a]	
	Co$_3$O$_4$	CoO	Co
15%Co/SiO$_2$	18.16	9.36	9.44
15%Co-1%K/SiO$_2$	23.52	7.30	7.34
15%Co-3%K/SiO$_2$	24.35	6.70	6.72
15%Co-5%K/SiO$_2$	19.13	2.46	8.85

[a] Particle size in nm.

2.3. Temperature-Programmed Reduction (TPR)

The effect of potassium addition on the reducibility of silica-supported cobalt catalysts was investigated using TPR analysis. TPR profiles of various potassium-promoted catalysts, along with that of the unpromoted sample are presented in Figure 2. For the unpromoted catalyst, an early and slow reduction process was observed from ca. 170 °C. It became significant from ca. 290 °C, where a fast reduction peak was observed to start and went through a maximum at 365 °C. Subsequent overlapping reduction peaks, with respective maxima at ca. 395, 425 and 466 °C, were also observed. These peaks can be attributed to a two-step reduction of Co$_3$O$_4$ species in the catalyst to CoO and Co^0. The presence of more than two peaks observed for this reduction process could indicate that not all the cobalt species in the catalyst underwent reduction at the same time. For example, as N$_2$ adsorption data suggest some level of pore obstruction in the catalyst, it is possible that some cobalt species only got reduced after the reduction of some of those that blocked some pores. Adding potassium to the catalyst reduced the reducibility of cobalt species as per the following observations: (i) The reduction temperatures for the catalysts shifted to higher values. For example, the start of the reduction process moved from 170 °C for the unpromoted catalyst to 210 and 255 °C for catalysts containing 1% and 3–5% K respectively; (ii) the area under the TPR profile below 500 °C decreased, indicating lower degree of catalyst reduction as the amount of potassium increased in the catalyst and (iii) the formation of cobalt species in strong interaction with the support, as shown by a broad reduction peak, with a maximum at ca. 512 °C, observed in the catalyst containing 5% K. The negative effect of potassium on the reduction of cobalt catalyst was also reported by Jacobs et al. [6] who found that (0.5–5%) K shifted the reduction peak temperatures to higher values and lowered the extent of catalyst reduction. This suggests that potassium interacts with the cobalt species and possibly the silica support [15].

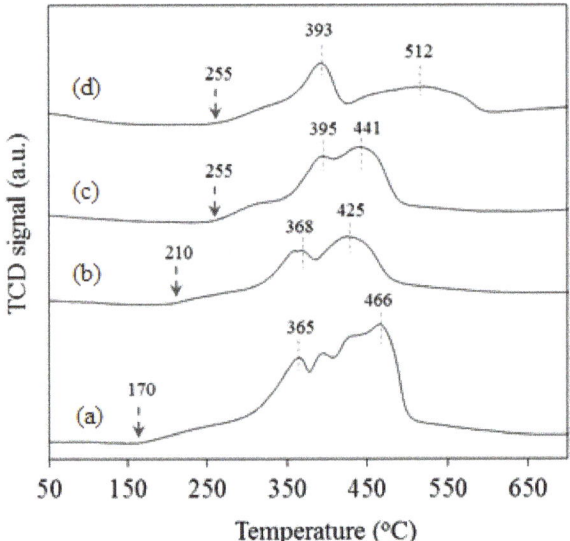

Figure 2. TPR profiles for (a) 15%Co/SiO$_2$, (b) 15%Co-1%K/SiO$_2$, (c) 15%Co-3%K/SiO$_2$, and (d) 15%Co-5%K/SiO$_2$.

2.4. CO$_2$ Temperature-Programmed Desorption (CO$_2$-TPD)

CO$_2$-TPD profiles for 15%Co/SiO$_2$ and 15%Co-1%K/SiO$_2$ are presented in Figure 3. Both catalysts showed two desorption peaks, with the first one centred at 65 °C with near-identical areas. This low-temperature peak can be attributed to the desorption of physically adsorbed CO$_2$. A second peak, observed for each catalyst, was attributed to the desorption of chemisorbed CO$_2$ and was used as an indication of the strength and amounts of basic sites in the catalyst. As expected, the data show that the addition potassium to the catalyst increases the strength and amounts of basic sites in the catalyst. This is indicated by the large and extended CO$_2$ desorption peak, which goes through its maximum at ca. 187 °C, compared to a corresponding small peak, with a maximum at ca. 134 °C, for the unpromoted catalyst. These data agree with earlier studies [8,16] that also reported an improvement in CO$_2$ adsorption in cobalt-based catalyst upon potassium addition.

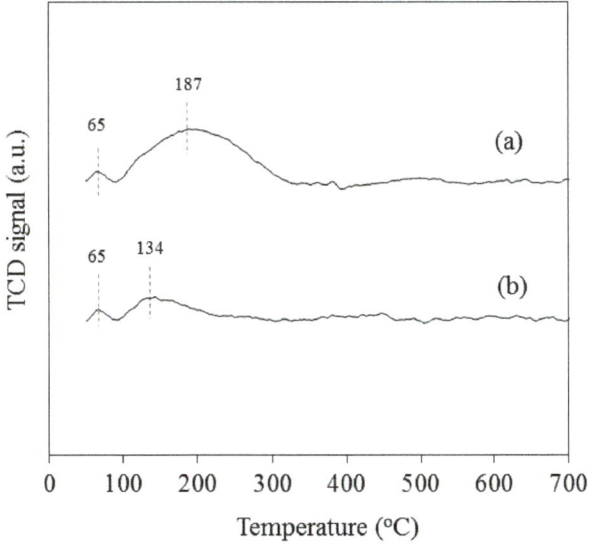

Figure 3. CO_2-TPD profiles of (**a**) 15%Co-1%K/SiO_2 and (**b**) 15%Co/SiO_2.

2.5. X-ray Photoelectron Spectroscopy (XPS)

XPS spectra, in the Co 2p region, for calcined and activated catalysts are shown in Figure 4.

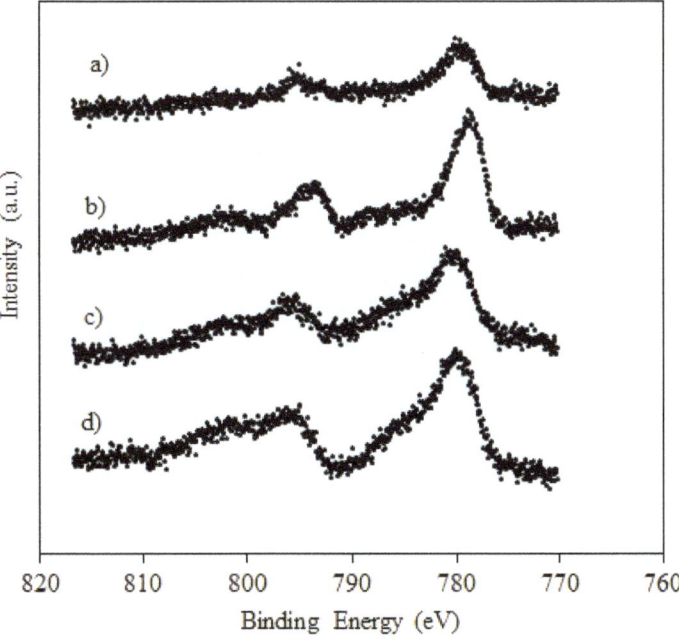

Figure 4. XPS data for calcined catalysts (**a**) 10%Co/SiO_2-calc., (**b**) 10%Co/1%K/SiO_2-calc.; and reduced catalysts (**c**) 10%Co/SiO_2-red., (**d**) 10%Co/1%K/SiO_2-red.

The Co $2p_{1/2}$ and $2p_{3/2}$ peaks for the calcined and unpromoted catalyst were respectively observed at ca. 795.2 and 779.6 eV and are characteristic of Co_3O_4 [14,17], in agreement with XRD data, discussed in Section 2.2. A shift to lower binding energies can be observed for Co $2p_{1/2}$ (to 793.5 eV) and $2p_{3/2}$ (to 778 eV) following catalyst promotion with potassium. This suggests an electronic donation by potassium as also observed by other studies, where potassium was added to Co/Al_2O_3 [18] and Pd/Co_3O_4 and Co_3O_4 [19] catalysts.

Spectra of reduced catalysts (Figure 4c,d) display features of CoO with broader Co $2p_{1/2}$ and $2p_{3/2}$ peaks and increased intensities of the shake-up satellite features [17,20]. They look similar for both the unpromoted and the potassium-promoted catalysts. These findings indicate that the electronic properties of cobalt in the catalyst were modified by potassium during the calcination process, not during catalyst reduction, causing a different reduction behaviour for the promoted catalyst.

2.6. Catalyst Testing

2.6.1. Effect of Temperature

In order to study the effect of temperature, CO_2 hydrogenation was carried out over a 15%Co-3%K/SiO_2 catalyst at atmospheric pressure from 180 to 300 °C. The temperature dependency of CO_2 conversion and its corresponding Arrhenius plot are reported in Figure 5. As expected, the CO_2 conversion continuously increased from 0.6 to 18.4% as the temperature was raised from 180 to 300 °C, in agreement with other earlier studies [21–23].

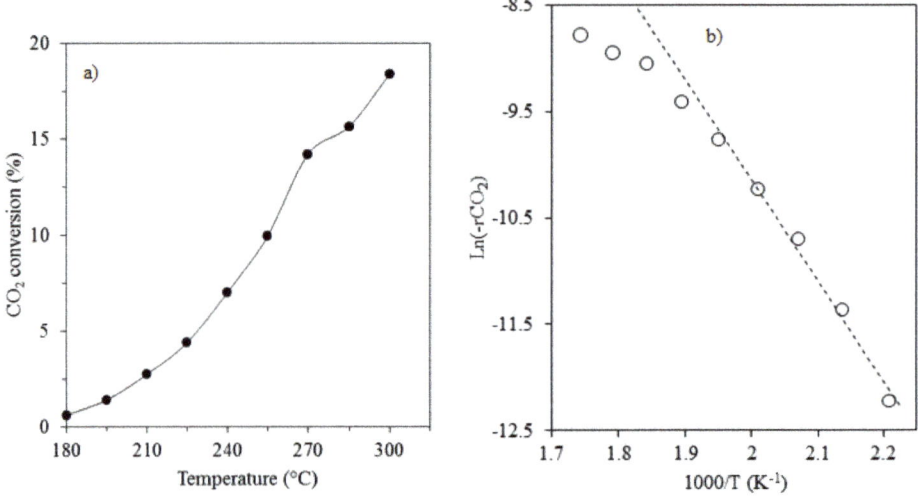

Figure 5. CO_2 conversion during hydrogenation vs. reaction temperature (1 bar, SV = 0.92 NL/g_{cat}/h, H_2/CO_2 = 3.1/1): (**a**) CO_2 conversion vs. temperature; (**b**) Arrhenius plot (Ea = 78 kJ/mol).

From the Arrhenius plot, activation energy of 78 kJ/mol was obtained in a temperature range of 180 to 240 °C. A marked curvature in the Arrhenius plot can be observed at temperatures above 240 °C, suggesting that the catalyst surface underwent some changes, possibly including deactivation by carbon [23]. For comparison, activation energies reported by earlier studies involving cobalt catalysts for CO_2 hydrogenation are summarised in Table 3.

The value of the activation energy obtained in this study is similar to most of those reported in earlier studies. Exceptions can be noticed for the data reported by Weatherbee and Bartholomew [23], who reported activation energies of 93 and 171 kJ/mol over 15%Co/SiO2 at 1 and 10 bar respectively.

The effect of the operating temperature on products selectivity and yields is summarized in Figure 6. The methane selectivity showed relatively little dependency on temperature from 180 to 225 °C, but continuously increased from 240 to 300 °C, while the selectivity to CO decreased almost linearly with increasing temperature (Figure 6a). Both C_2-C_4 and C_{5+} selectivities increased as the temperature was raised and went through a maximum at 240 °C before decreasing. Figure 6c shows that up to 270 °C, the yields for C_{2+}, CH_4 and CO all increased with the rise in temperature, with the yield of CO remaining the highest of the three. The yield of methane and C_{2+} hydrocarbons were similar up to 240 °C, above which the yield of methane quickly surpassed that of C_{2+} in an exponential manner.

Table 3. The activation energy for CO_2 hydrogenation over cobalt catalysts.

Catalyst	H_2/CO_2	P (atm.)	T (°C)	Ea (kJ/mol)	References
15%Co/3%K/SiO$_2$	3/1	1	180–240	78	This work
Pristine Co	4/1	1	207–237	77	[24]
100% Co	4/1	1	190–230	79	[25]
4.5%Co/S1 *	4/1	1	210–260	79	[25]
4.6%Co/S3 *	4/1	1	200–240	76	[25]
15%Co/SiO$_2$	4/1	1	183–203	93	[23]
15%Co/SiO$_2$	4/1	11	180–222	171	[23]
3%Co/SiO$_2$	4/1	1	227–277	79	[23]

* S1 and S3 are carbon supports obtained from saran copolymer. The difference between the two is in the burn-off percentage, i.e., 0 and 20% for S1 and S3 respectively [26].

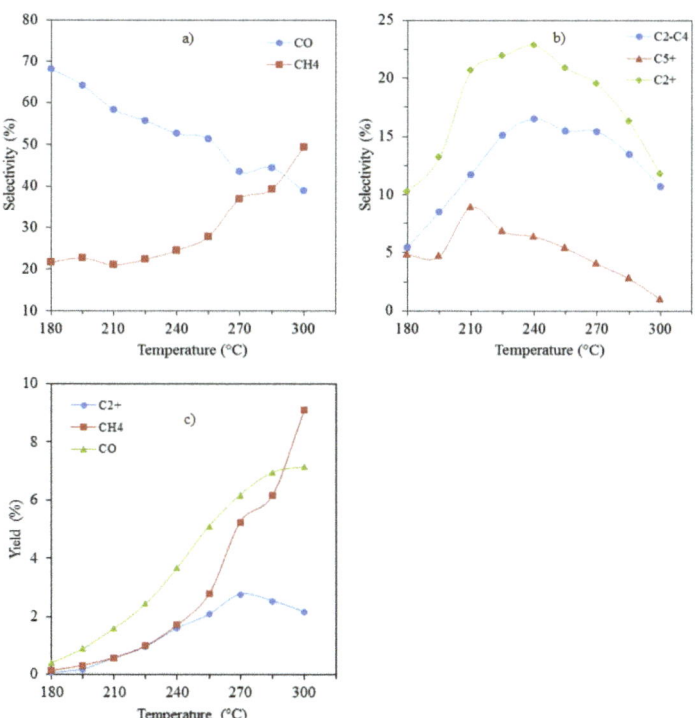

Figure 6. Effect of temperature on products selectivity: (**a**) CH_4 and CO; (**b**) C_{2+} hydrocarbons; and products yields: (**c**) C_{2+}, CH_4 and CO.

The rise in CO yield flattened off around 285 °C and was surpassed by the fast-rising yield of methane around 290 °C. The C_{2+} yield went through a maximum at 270 °C, indicating that, above this temperature, the reaction is turning into a preferential methanation process.

The mechanism of CO_2 hydrogenation to hydrocarbons is still subject of some controversies. However, since CO formed during CO_2 hydrogenation, it is most likely that hydrocarbons formed via a typical Fischer-Tropsch mechanism. Indeed, this is a plausible explanation, since some studies [27–29] have shown that, in the presence of CO, on cobalt-based catalysts, CO_2 behaves like an inert gas and only reacts when CO is depleted. Also, the rapid increase in methane yields with the temperature at values above 240 °C is typical to FT reaction [22,23]. An operating temperature of 270 °C was selected as optimal for the rest of the study.

2.6.2. Effect of Pressure

The effects of pressure on CO_2 conversion, and products selectivities and yields are reported in Figure 7. An increase in operating pressure, from 1 to 15 bar, resulted in an increase in CO_2 conversion. As can be seen from Figure 7a, the CO_2 conversion measured at 1 bar was ca. 12.5%; it increased to ca. 21%, 22%, and 27% when the pressure was increased to 5, 10 and 15 bar, respectively. Further increase in the operating pressure to 20 bar resulted in a slight decrease of CO_2 conversion to ca. 26%. The increase in CO_2 conversion with the operating pressure from 1 to 15 bar was expected because of an increase in reactants partial pressures. However, the decrease in CO_2 conversion observed when the operating pressure was increased from 15 to 20 bar was not expected; it is possible that some CO_2 or reaction intermediate species irreversibly adsorbed on the catalyst surface, blocking some active sites.

An increase in operating pressure from 1 to 5 bar significantly decreased the selectivities to CO and C_{2+} hydrocarbons from ca. 48% and 21% to 8% and 11%, respectively (Figure 7b). Further increase in pressure only resulted in slight decreases in CO and C_{2+} hydrocarbons selectivities. An opposite behaviour was observed for CH_4 selectivity, which increased from 30 to 81% when the operating pressure was increased from 1 to 5 bar. Further increase in operating pressure resulted in a relatively slight increase in CH_4 selectivity. Similar trends can be observed for CO and CH_4 yields as a function of the operating pressure (Figure 7c); however, the C_{2+} yield was not significantly affected by changes in operating pressure. It remained between 2.1% and 2.7% over the range of pressure used. Under these conditions, operating at 1 bar is optimal.

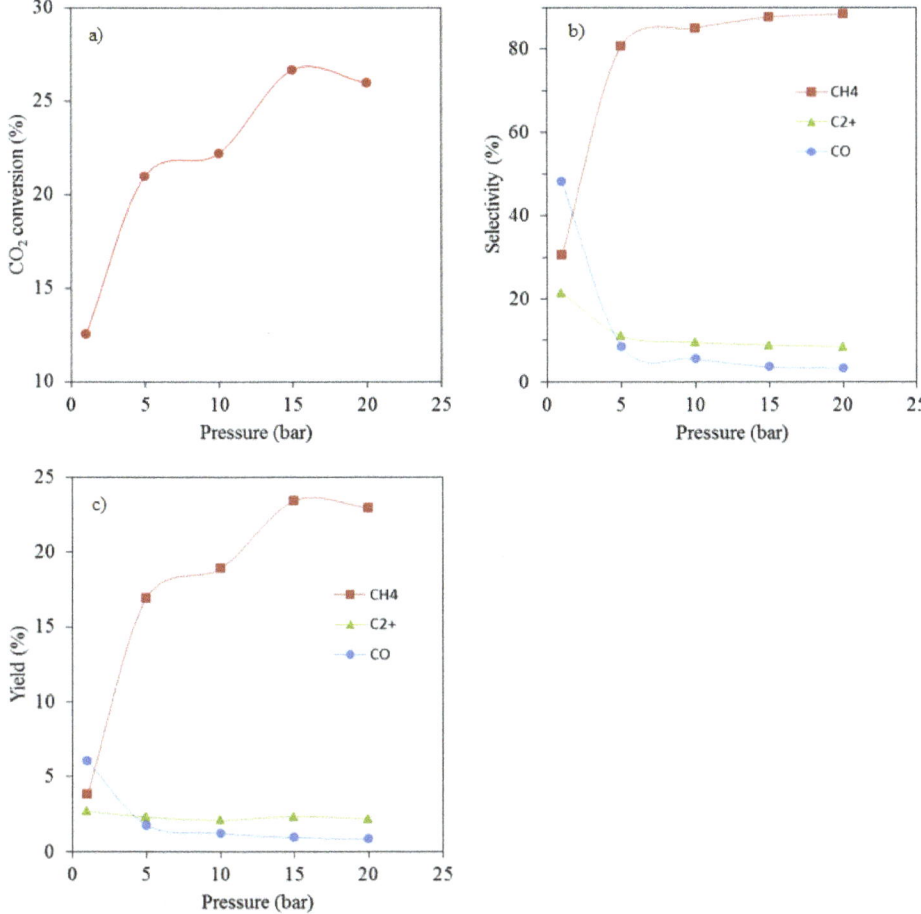

Figure 7. Effect of the operating pressure on (**a**) CO_2 conversion, (**b**) products selectivity and (**c**) products yields (catalyst—15%Co/3%K/SiO$_2$, 270 °C, SV = 0.92 NL/g$_{cat}$/h, H$_2$/CO$_2$ = 3.1/1).

2.6.3. Effect of Potassium Addition

Figure 8 shows the effect of potassium addition on CO_2 conversion, and products selectivity and yields. It is observed that the presence of potassium at a loading of as low as 1% results in a significant drop in CO_2 conversion from 39 to 16% when compared to the unpromoted catalyst (Figure 8a).

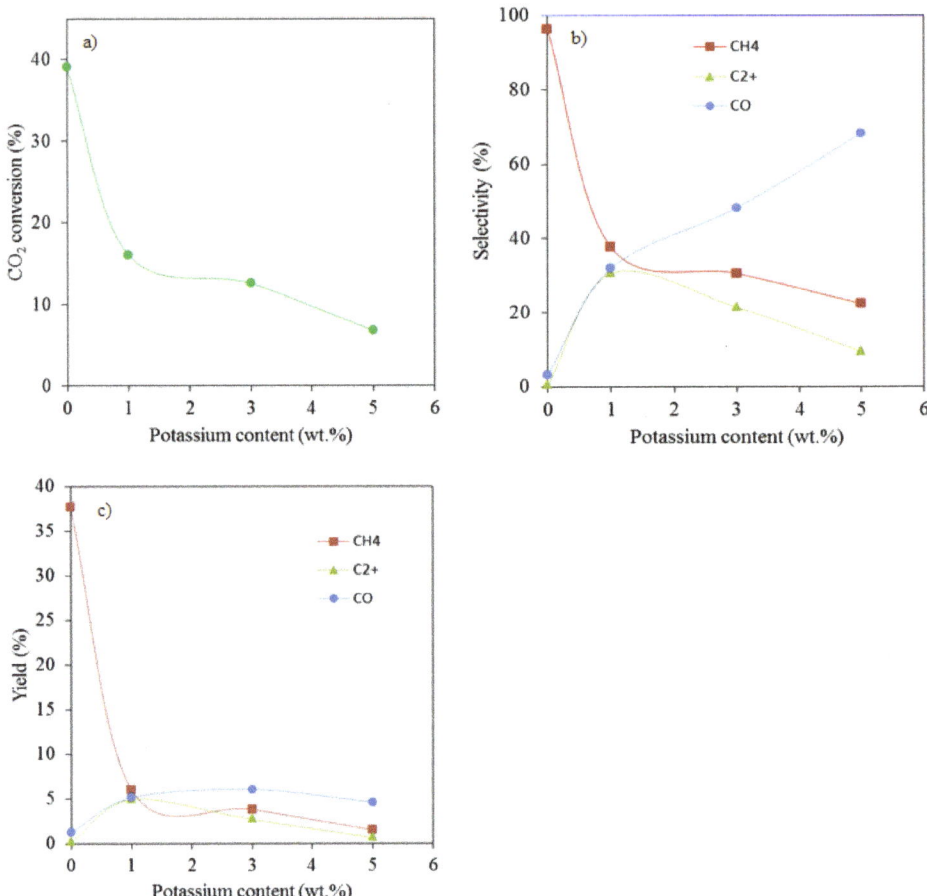

Figure 8. Effect of potassium loading on (**a**) CO_2 conversion, (**b**) product selectivity and (**c**) product yields.

Adding more potassium further exacerbates this behaviour, but with an attenuated effect. The following can explain these observations: (i) Coverage of active sites by potassium, although considered to happen at a low extent because of the low potassium loading employed; (ii) increase in CO_2 adsorption capacity: As discussed earlier in Section 2.4, CO_2-TPD results have shown that the CO_2 adsorption capacity for the catalyst improves upon potassium addition. As a consequence, the H/C molar ratio on the catalyst surface is decreased, leading to a drop in CO_2 conversion. This is in agreement with numerous experimental [30–33] and theoretical [34,35] investigations that reported a drop in CO_2 conversion with a decrease in H_2/CO_2 molar ratio. The decrease in CO_2 conversion with potassium addition was also reported by Owen et al. [13] and Shi et al. [8] on Co/SiO_2 and $CoCu/TiO_2$ catalysts, respectively. Using H_2 and CO_2 temperature-programmed desorption analyses, Shi et al. [8] were able to show that potassium promotion decreases the H_2 adsorption capacity of the catalyst, while that of CO_2 is enhanced; (iii) the oxidation state of cobalt in the catalyst: XRD results have shown the presence of CoO and metallic cobalt phases in all the reduced and passivated catalysts. TPR analyses, on the other hand, confirmed that these catalysts have different reducibility properties. Promotion with potassium limits the reducibility of the catalysts, resulting in limited amounts of metallic cobalt sites for CO_2 conversion. A similar relationship between catalyst reducibility and CO_2 activity was reported by Melaet et al. [36] who conducted CO_2 hydrogenation on Co/SiO_2 catalysts activated at 523 K and 723 K. They established, by means of XPS, that CoO and metallic cobalt formed upon activation at 523 K and 723 K respectively. The catalyst reduced at 723 K showed higher activity.

The selectivity towards methane decreased from 96 to 37.6% (Figure 8b) upon adding 1 % of potassium to the catalyst. Meanwhile, the selectivity of both CO and C_{2+} hydrocarbons increased. Additional amounts of potassium resulted in a further decrease in methane selectivity and increase in CO selectivity. The selectivity of C_{2+} hydrocarbons, on the other hand, decreased with further increase in potassium loading above 1%. The improvement of C_{2+} hydrocarbons selectivity at 1% potassium can be attributed to a decreased surface H/C ratio as discussed earlier. This implies that carbon-containing species from CO_2 dissociation can polymerize rather than being hydrogenated as is often the case in a hydrogen-rich environment.

The decreased C_{2+} hydrocarbons selectivity at 3% and 5% potassium loading is also explained by an increased CO_2 adsorption capacity of the catalyst, causing a decrease in the surface H/C ratio. Since the CO yield is shown to increase and undergo little variations with further increase in potassium loading, while both CH_4 and the C_{2+} yields decrease (Figure 8c), it is possible that there is not enough surface hydrogen to readily react with both CO_2 and CO on the catalyst surface.

The highest C_{2+} yield achieved in this study was ca. 5% and was measured on the 15%Co/1%K/SiO_2 catalyst at 1 bar and 270 °C. This condition is compared to results reported in other studies that used cobalt-based catalysts for CO_2 hydrogenation under various conditions, as summarized in Table 4. Of the catalysts that produced C_{5+} products at low pressure (<2 bar), our catalyst had the lowest methane selectivity under the optimized operating temperature and pressure. This is particularly important, since it offers opportunities to limit the formation of the undesirable methane without the need for excessive operating pressures that will make the process more energy-intensive.

3. Materials and Methods

3.1. Catalyst Synthesis

The catalyst synthesis consisted essentially of two steps, namely support preparation and metal loading. Fumed silica with an average particle size range of 0.2–0.3 µm, supplied by Sigma-Aldrich South Africa, served as the catalyst supporting material. Given its small particle size, it was pre-treated with deionized water and agglomerated by drying overnight at 120 °C in the air before crushing and sieving to obtain a powder with particles within the size range of 212–500 µm. The powder so obtained was subsequently calcined at 400 °C for 6 h in the air to lock its properties before loading the metals. The addition of cobalt and potassium was done through co-impregnation with solutions of cobalt and potassium nitrates—both purchased from Sigma-Aldrich. After impregnation, the catalysts were dried overnight at 120 °C and calcined at 400 °C in the air for 6 h. All the prepared catalysts contained 15 wt.% cobalt with varying potassium loading (0–5 wt.%). The amount of silica used in catalyst preparation was reduced to account for the addition of potassium. This allowed for the cobalt loading to be kept constant for all the catalysts.

Table 4. Summary of catalytic performance data for CO_2 hydrogenation over cobalt-based catalysts.

Entry	Catalyst	Prep. Method	H_2:CO_2	T [K]	P [bar]	SV	Conv. [%]	%Selectivity CH$_4$	CO	C_2+	C_5+	References
1	100Co/5Cu	Coprecip.	3:1	473		0.2 L/gCat/h [a]	49				1.12 [b]	[12]
2	100Co/5Cu/2K$_2$CO$_3$	Coprecip.	3:1	498		0.16	56				2.32	
3	100Co/5Cu/2K$_2$CO$_3$	Coprecip.	3:1	498		0.16	40				1.91	
4	100Co/5Cu/2K$_2$CO$_3$	Coprecip.	3:1	473		0.16	22				2.81	
5	100Co/5Cu/2K$_2$CO$_3$	Coprecip.	3:1	498		0.16	10				0.19	
6	100Co/5Cu/2K$_2$CO$_3$	Coprecip.	3:1	473		0.16	44				1.59	
7	100Co/5Cu/2K$_2$CO$_3$	Coprecip.	3:1	473		0.16	54				1.43	
8	100Co/5Cu/5CeO$_2$/2K$_2$CO$_3$	Coprecip.	3:1	498		0.16	34				2.11	
9	100Co/5Cu/1CeO$_2$/2K$_2$CO$_3$	Coprecip.	2:1	498		0.15	40				0.10	
10	100Co/5Cu/1CeO$_2$	Coprecip.	2:1	498		0.15	40				1.42	
11	100Co/5Cu/1CeO$_2$/3K$_3$PO$_4$	Coprecip.	3:1	498		0.16	21				0.29	
12	100Co/5Cu/1CeO$_2$/4.5K$_2$CO$_3$/100MgO	Coprecip.	2:1	513		0.3	8				0.51	
13	100Co/5Cu/1CeO$_2$/4.5K$_2$CO$_3$/100MgO	Coprecip.	2:1	498		0.15	19				2.71	
14	100Co/5Cu/1CeO$_2$/6K$_2$CO$_3$/100H.S.C.	Coprecip.	2:1	518		0.15	23				1.61	
15	100Co/5Cu/100CeO$_2$/7K$_2$CO$_3$	Coprecip.	2:1	523		0.15	23				0.24	
16	100Co/5Cu/1CeO$_2$/4.5K$_2$CO$_3$/100F.C.	Coprecip.	2:1	513		0.075	23				1.90	
17	100Co/5Cu/1CeO$_2$/3.8K$_2$CO$_3$/50H.S.C.	Coprecip.	2:1	498		0.12	22					
18	3%Co/SiO$_2$	Impregnation	4:1, 95%N$_2$	500	1.4	4340/h	9.6	71	25	4.6		[23]
19				500		8480	6.5	54	35	11		
20				525		8480	12.3	59	33	8.2		
21				525		16,400	9.4	42	49	8.9		
22				550		16,400	13.7	42	52	5.9		
23				550		24,600	12	32	52	17		
24	15%Co/SiO$_2$	Impregnation	4:1, no N$_2$	476	1	2050–3850	10.5	86.9	12.6	0.7	0	
25				478	11	450–9620	11.2	89	10.7	0.34	0	
26	100%Co	Reduction	4:1	493	1	500–3000 h^{-1}	1.9	98	2			[25]
27	4.5%Co/S1	Impregnation		493			1.8	40	60			
28	4.6%Co/S3	Impregnation		493			6.3	66	34			
29	100 Co/60 MnO/147 SiO$_2$/0.15Pt	Precip. and Impregnation	2:1	463	10	30 mL/min/g of Co	18	95				[37]
30	15%Co/Al$_2$O$_3$	Impregnation	2.45:1	493	20	4800 cm^3(STP)/h/gcat	33	>90				[38]
31	20%Co/SSP	Impregnation	20:2	493	1	18 L/gcat/h	27	89.5	10.5			[39]
32	20%Co/MCM-41						28	91.4	8.6			
33	20%Co/TiSSP						16	92.1	7.9			
34	Co/TiMCM-41						34	94.9	5.1			
35	0.5% Pt–25% Co/γ-Al$_2$O$_3$	Impregnation	3:1	493	19.9	5.0 L/g cat/h		93.3		6.66	5.16	[28]
36	5%Co/Al$_2$O$_3$ [c]	Impregnation	6:1	533	1	13.5 mL/min/(63 to 70 mg of cat)	0.21	35.7				[40]
37	10%Co/Al$_2$O$_3$ [c]						0.91	74.2				
38	15%Co/Al$_2$O$_3$ [c]						2.45	87.8				
39	20%Co/Al$_2$O$_3$ [c]						2.1	85.7				
40	Co/Al$_2$O$_3$	Solid state reaction of gibbsite and CoNT	10:1	543	1	150 mL/min/gcat	76	82.2	17.8			[41]

Table 4. *Cont.*

Entry	Catalyst	Prep. Method	H$_2$:CO$_2$	T [K]	P [bar]	SV	Conv. [%]	%Selectivity											References
41	Co/Al$_2$O$_3$	Solid state reaction of gibbsite and CoAc					48.7	76.7	23.3										
42	Co/Al$_2$O$_3$	Solid state reaction of gibbsite and CoAA					20.3	76.4	23.6										
43	Co/Al$_2$O$_3$	Solid state reaction of gibbsite and CoCL					6.1	100	0										
44	Co/Al$_2$O$_3$	Impregnation using CoNT					32.2	86.5	13.5										
45	20%Co/SiO$_2$		3:1	643	Atmospheric		67.4	95.3	4.2	0.6	0								[13]
46	20%Co/1%Pd/SiO$_2$						50.7	93.4	6.3	0.3	0								
47	10%Co/1%Pd/1%K/SiO$_2$						36.4	89.3	8.0	2.8	0								
48	20%Co/1%Pd/1%K/SiO$_2$						63.4	80.3	13.9	5.9	0								
49	10%Co/1%Pd/1%K/SiO$_2$						39.1	82.9	9.5	7.6	0.09								
50	20%Co/1%Pd/0.5%K/SiO$_2$						62.8	76.0	15.3	8.8	0								
51	20%Co/1%Pd/1.5%K/SiO$_2$						59.1	64.7	16.2	19.1	1.26								
52	20%Co/1%Pd/3%K/SiO$_2$						43.2	53.1	24.3	22.6	2.73								
53	20%Co/1%K/SiO$_2$						36.1	45.3	16.9	37.8	7.87								
54	20%Co/1%Pt/1%K/SiO$_2$						36.5	41.5	20.8	37.7	9.58								
55	20%Co/1%Ru/1%K/SiO$_2$						45.1	52.6	12.6	34.8	5.68								
56	20%Co/1%Pd/1%Li/SiO$_2$						39.5	56.1	19.2	24.6	1.94								
57	20%Co/1%Pd/1%Na/SiO$_2$						41.9	48.4	20.3	31.3	7.33								
58	20%Co/1%Li/SiO$_2$						39.3	58.4	21.4	20.2	0.47								
59	20%Co/1%Na/SiO$_2$						51.2	42.1	21.7	36.3	5.01								
60	20%Co/1%K/SiO$_2$						47.6	50.1	17.0	32.9	3.65								
61	20%Co/1%Mo/SiO$_2$						64.8	88.7	6.5	4.8	0								
62	20%Co/1%Cr/SiO$_2$						60.9	75.9	22.8	1.2	0								
63	20%Co/1%Mn/SiO$_2$						62	91.1	6.9	2.0	0								
64	20%Co/1%Na/1%Mn/SiO$_2$						42.7	58.2	19.7	22.2	0.80								
65	20%Co/1%Na/1%Mo/SiO$_2$						43.9	38.3	15.7	45.9	8.76								
66	CoCu/TiO$_2$	Deposition-precipitation	73:24	523	50	3000 mL/g/h	23.1	87.0	1.3	10.2	4.76								[8]
67	1.5 K-CoCu/TiO$_2$						21.2	59.3	4.7	36.5	13.21								
68	2.0 K-CoCu/TiO$_2$						13.8	37.1	19.7	44.6	17.39								
69	2.5 K-CoCu/TiO$_2$						13	22.4	35.1	43.3	23.08								
70	3.0 K-CoCu/TiO$_2$						12.8	21.9	35.9	41.5	19.53								
71	3.5 K-CoCu/TiO$_2$						11.9	18.9	45.9	35.1	16.81								
72	15%Co–1%K/SiO$_2$	Impregnation	3:1	543	1 bar	0.92 NL/gcat/h	16	37.6	31.9	30.5	7.8								This study

[a] Calculated from reported flow of CO$_2$ over 24 h, H$_2$/CO$_2$ ratio and the mass of catalyst. [b] Calculated from the reported milliliters of oil that formed during the reaction, assuming an average chain length of 7 (density of 0.684). [c] Data read from graphs.

3.2. Catalyst Characterization

The surface area and the porosity of the catalysts were measured by nitrogen physisorption at −196 °C using an Accelerated Surface Area and Porosimetry System (ASAP 2460) from Micromeritics. Each analysis was preceded by degassing the sample at 150 °C for 4 h. The multipoint Brunauer-Emmett-Teller (BET) method was used to determine the surface area of the materials analysed.

The reducibility of the catalysts was studied by means of temperature-programmed reduction (TPR). An in-house built instrument, equipped with a thermal conductivity detector (TCD), was used for this purpose. In a typical analysis, 100 mg of catalyst was loaded in a stainless-steel reactor and heated to 300 °C for one hour under 70 NmL/min of helium to remove traces of moisture and other ambient contaminants. This step was referred to as degassing. After allowing the reactor to cool to room temperature, helium was switched with a gas mixture containing 5% H_2 in argon at a flow of 65 NmL/min. In the final step, the temperature was raised from room temperature to 700 °C at a heating rate of 10 °C/min, while recording the signal of the TCD.

Temperature-programmed desorption (TPD) of CO_2 was carried out using the same instrument as described for TPR analysis. Different to TPR analysis, the catalysts used in this analysis were first reduced at 335 °C for 17 h, using the same reactor and conditions as for the reduction of catalyst samples used in the CO_2 hydrogenation testing as will be described in Section 2.3. The reduced catalysts were passivated using 5% O_2 in helium for 2 h at ambient temperature before their transfer from the CO_2 hydrogenation reactor to the TPD apparatus. Two hundred milligrams of catalyst sample was degassed in a similar manner as for TPR analysis. After degassing and cooling to room temperature, the temperature was raised to 335 °C at a heating rate of 10 °C/min and maintained at this value for 30 min under a flow of 5% H_2 in argon. This step was necessary for the removal of the cobalt oxide layer formed during catalyst passivation. Thereafter, the reactor was cooled and maintained at 50 °C for at least 10 min before replacing H_2 (5% in argon) with CO_2 (10% in helium). CO_2 adsorption was performed at 50 °C for 1 h before re-introducing helium, but this time to remove the physically adsorbed CO_2 molecules. TPD was then performed under helium flow, after stabilization of the TCD signal, from 50 to 700 °C at a heating rate of 5 °C/min.

X-ray diffraction (XRD) analysis was performed to identify the oxidation state of cobalt species in the unreduced and reduced catalyst samples. The instrument used for this purpose was a Rigaku Ultima IV equipped with a copper target. The voltage and current at which the diffractometer was operated were 40 kV and 30 mA respectively. Spectra were acquired in the range of 2θ from 10° to 90° with a step size of 0.01° at the scanning speed of 1°/min.

X-ray photoelectron spectroscopy (XPS) was used to determine the oxidation states of the elements present on the surface of the catalysts. This analysis was performed on a Specs Phoibos 150 spectrometer with a monochromatic X-ray source Al Kα at 1486.71 eV. A low-energy electron flood gun operated at 2.0–2.5 eV and 20 μA was used to stabilize the sample surface charge. The spectrometer was operated at constant pass energy of 40 eV. The shift in binding energy peaks position, due to the surface charging effect was corrected by setting the C 1s binding energy to 284.8 eV [14].

3.3. Catalyst Testing

Carbon dioxide hydrogenation was carried out in a system which consisted mainly of a stainless steel fixed-bed reactor (16 mm i.d. × 220 mm length) mounted in an electrical furnace, a mass flow controller (Aalborg), a back-pressure regulator and a product collection pot. The furnace temperature was controlled using a programmable temperature controller connected to a K-type thermocouple and the furnace heating element. Accurate reaction temperatures were measured by means of another K-type thermocouple in direct contact with the catalyst bed held in place by plugs of quartz wool. Any liquid product formed was collected in a cold pot mounted at the bottom of the reactor. There was no need for a hot trap since the products were mainly light hydrocarbons. The reactor outlet was connected to a three-way valve, which made it possible to either send the reaction products to vent or to an online gas chromatograph (GC) for analysis. The Dani Master GC used in this study was equipped with a flame ionization detector (FID) connected to a capillary column (Supel-Q^{TM} PLOT) that separated hydrocarbons and oxygenates, and a thermal conductivity detector (TCD) connected to a packed column (60/80 Carboxen 1000) for the separation of H_2, N_2, CO and CO_2.

Prior to testing, 500 mg of catalysts were reduced in flowing hydrogen (23 NmL/min) at 335 °C and atmospheric pressure for 17 h. Catalyst testing was done at temperatures ranging from 180 to 300 °C with an increment of 15 °C and at pressures within a range of 1–20 bar at a space velocity of 0.92 NL/g$_{cat}$/h. The feed gas was premixed and contained 21.8% CO_2, 68.6% H_2 and 9.6% N_2. After testing, all catalysts were passivated in 5% O_2 in helium (23 NmL/min) at room temperature for 2 h. The nitrogen present in the feed gas was used as an internal standard for mass balance calculations. The CO_2 conversion, the rate of CO_2 conversion, the rate of products formation, selectivity and yield were calculated according to Equations (1)–(5), where F and X indicate the total molar gas flow rate and mole fraction respectively. The subscripts "in" and "out" refer to the gas streams entering or leaving the reactor.

$$CO_2 \text{ conversion } (\%) = \frac{X_{CO_2,\,in} - \frac{X_{N_2,\,in}}{X_{N_2,\,out}} \times X_{CO_2,\,out}}{X_{CO_2,in}} \times 100, \qquad (1)$$

$$\text{Rate of } CO_2 \text{ conversion } = \frac{F_{in}\left[X_{CO_2,\,in} - \frac{X_{N_2,\,in}}{X_{N_2,\,out}} \times X_{CO_2,\,out}\right]}{\text{Catalyst mass}}, \qquad (2)$$

$$\text{Rate of formation of product } i = \frac{F_{out} \times X_{i,\,out}}{\text{Catalyst mass}}, \qquad (3)$$

$$\text{Selectivity of product } i\ (\%) = \frac{\text{moles of carbon in product i per unit time}}{\text{Rate of } CO_2 \text{ conversion } \times \text{ Catalyst mass}} \times 100, \qquad (4)$$

$$\text{Yield of product } i\ (\%) = \frac{\text{Selectivity of product } i \times CO_2 \text{ conversion}}{100}. \qquad (5)$$

After a change in operating conditions or in catalyst sample, the reactor was allowed to reach a steady state and maintained at the new conditions for at least two days. At least two data points were generated per day. To ensure reproducibility, each data point was an average of three independent measurements that were closer to each other within 5% error range.

4. Conclusions

The aim of this study was to investigate the effects of operating conditions (temperature, pressure) and potassium loading on the performance of silica-supported cobalt catalysts in CO_2 hydrogenation. The highest yield in C_{2+} hydrocarbons was measured at 1 bar and 270 °C. Potassium was found to negatively affect the reducibility of the catalyst, while enhancing its CO_2 adsorption capacity. The improved CO_2 adsorption capacity of the catalyst leads to a lower surface H/C ratio, which promotes chain growth reactions. The limited catalyst reducibility resulted in low catalyst activity and is explained by an electric donation of potassium to cobalt species during the calcination process of the catalyst. The optimal operating pressure and temperature determined in this study, combined with catalyst promotion with 1 wt.% of potassium, significantly lowered the undesirable methane selectivity when compared to other cobalt-based catalysts that also produced some C_{5+} hydrocarbons at low pressures (<2 bar). This constitutes a significant further step in the development of efficient catalysts for CO_2 hydrogenation to liquid fuels.

Author Contributions: Project conceptualization and methodology: R.A.I. and K.J.; Materials synthesis, experiments and data collection: R.A.I.; Data analysis and interpretation: R.A.I. and K.J.; Manuscript writing and editing: R.A.I. and K.J.; Project administration and supervision: K.J.

Funding: This project was funded by the National Research Foundation (Grant: UID 90757) and the University of Johannesburg Global Excellence Stature (GES) program.

Conflicts of Interest: The authors declare no conflicts of interest.

References

1. Espinal, R.; Taboada, E.; Molins, E.; Chimentao, R.J.; Medina, F.; Llorca, J. Cobalt hydrotalcites as catalysts for bioethanol steam reforming. The promoting effect of potassium on catalyst activity and long-term stability. *Appl. Catal. B Environ.* **2012**, *127*, 59–67. [CrossRef]
2. Xie, X.; Yin, H.; Dou, B.; Huo, J. Characterization of a potassium-promoted cobalt-molybdenum/alumina water-gas shift catalyst. *Appl. Catal.* **1991**, *77*, 187–198. [CrossRef]
3. Asano, K.; Ohnishi, C.; Iwamoto, S.; Shioya, Y.; Inoue, M. Potassium-doped Co_3O_4 catalyst for direct decomposition of N_2O. *Appl. Catal. B Environ.* **2008**, *78*, 242–249. [CrossRef]
4. Trépanier, M.; Tavasoli, A.; Dalai, A.K.; Abatzoglou, N. Co, Ru and K loadings effects on the activity and selectivity of carbon nanotubes supported cobalt catalyst in Fischer–Tropsch synthesis. *Appl. Catal. A Gen.* **2009**, *353*, 193–202. [CrossRef]
5. Tavasoli, A.H.M.A.D.; Khodadadi, A.; Mortazavi, Y.; Sadaghiani, K.; Ahangari, M.G. Lowering methane and raising distillates yields in Fischer–Tropsch synthesis by using promoted and unpromoted cobalt catalysts in a dual bed reactor. *Fuel Process. Technol.* **2006**, *87*, 641–647. [CrossRef]
6. Jacobs, G.; Das, T.K.; Zhang, Y.; Li, J.; Racoillet, G.; Davis, B.H. Fischer–Tropsch synthesis: Support, loading, and promoter effects on the reducibility of cobalt catalysts. *Appl. Catal. A Gen.* **2002**, *233*, 263–281. [CrossRef]
7. Calafat, A.; Vivas, F.; Brito, J.L. Effects of phase composition and of potassium promotion on cobalt molybdate catalysts for the synthesis of alcohols from CO_2 and H_2. *Appl. Catal. A Gen.* **1998**, *172*, 217–224. [CrossRef]
8. Shi, Z.; Yang, H.; Gao, P.; Li, X.; Zhong, L.; Wang, H.; Liu, H.; Wei, W.; Sun, Y. Direct conversion of CO_2 to long-chain hydrocarbon fuels over K–promoted $CoCu/TiO_2$ catalysts. *Catal. Today* **2018**, *311*, 65–73. [CrossRef]
9. Petala, A.; Panagiotopoulou, P. Methanation of CO_2 over alkali-promoted Ru/TiO_2 catalysts: I. Effect of alkali additives on catalytic activity and selectivity. *Appl. Catal. B Environ.* **2018**, *224*, 919–927. [CrossRef]
10. Panagiotopoulou, P. Methanation of CO_2 over alkali-promoted Ru/TiO_2 catalysts: II. Effect of alkali additives on the reaction pathway. *Appl. Catal. B Environ.* **2018**, *236*, 162–170. [CrossRef]
11. Numpilai, T.; Witoon, T.; Chanlek, N.; Limphirat, W.; Bonura, G.; Chareonpanich, M.; Limtrakul, J. Structure–activity relationships of $Fe-Co/K-Al_2O_3$ catalysts calcined at different temperatures for CO_2 hydrogenation to light olefins. *Appl. Catal. A Gen.* **2017**, *547*, 219–229. [CrossRef]
12. Russell, W.W.; Miller, G.H. Catalytic hydrogenation of carbon dioxide to higher hydrocarbons. *J. Am. Chem. Soc.* **1950**, *72*, 2446–2454. [CrossRef]
13. Owen, R.E.; O'Byrne, J.P.; Mattia, D.; Plucinski, P.; Pascu, S.I.; Jones, M.D. Cobalt catalysts for the conversion of CO_2 to light hydrocarbons at atmospheric pressure. *Chem. Commun.* **2013**, *49*, 11683–11685. [CrossRef] [PubMed]
14. Ernst, B.; Bensaddik, A.; Hilaire, L.; Chaumette, P.; Kiennemann, A. Study on a cobalt silica catalyst during reduction and Fischer-Tropsch reaction: In situ EXAFS compared to XPS and XRD. *Catal. Today* **1998**, *39*, 329–341. [CrossRef]
15. Huffman, G.P.; Shah, N.; Zhao, J.M.; Huggins, F.E.; Hoost, T.E.; Halvorsen, S.; Goodwin, J.G. In-situ XAFS investigation of K-promoted Co catalysts. *J. Catal.* **1995**, *151*, 17–25. [CrossRef]
16. De la Osa, A.; De Lucas, A.; Valverde, J.L.; Romero, A.; Monteagudo, I.; Coca, P.; Sánchez, P. Influence of alkali promoters on synthetic diesel production over Co catalyst. *Catal. Today* **2011**, *167*, 96–106. [CrossRef]
17. Biesinger, M.C.; Payne, B.P.; Grosvenor, A.P.; Lau, L.W.; Gerson, A.R.; Smart, R.S.C. Resolving surface chemical states in XPS analysis of first row transition metals, oxides and hydroxides: Cr, Mn, Fe, Co and Ni. *Appl. Surf. Sci.* **2011**, *257*, 2717–2730. [CrossRef]
18. Hu, X.; Dong, D.; Shao, X.; Zhang, L.; Lu, G. Steam reforming of acetic acid over cobalt catalysts: Effects of Zr, Mg and K addition. *Int. J. Hydrog. Energy* **2017**, *42*, 4793–4803. [CrossRef]
19. Kono, E.; Tamura, S.; Yamamuro, K.; Ogo, S.; Sekine, Y. Pd/K/Co-oxide catalyst for water gas shift. *Appl. Catal. A Gen.* **2015**, *489*, 247–254. [CrossRef]
20. Petitto, S.C.; Langell, M.A. Surface composition and structure of Co_3O_4 (110) and the effect of impurity segregation. *J. Vac. Sci. Technol. A* **2004**, *22*, 1690–1696. [CrossRef]
21. Melaet, G.; Lindeman, A.E.; Somorjai, G.A. Cobalt particle size effects in the Fischer–Tropsch synthesis and in the hydrogenation of CO_2 studied with nanoparticle model catalysts on silica. *Top. Catal.* **2014**, *57*, 500–507. [CrossRef]

22. Iablokov, V.; Beaumont, S.K.; Alayoglu, S.; Pushkarev, V.V.; Specht, C.; Gao, J.; Alivisatos, A.P.; Kruse, N.; Somorjai, G.A. Size-controlled model Co nanoparticle catalysts for CO_2 hydrogenation: Synthesis, characterization, and catalytic reactions. *Nano Lett.* **2012**, *12*, 3091–3096. [CrossRef] [PubMed]
23. Weatherbee, G.D.; Bartholomew, C.H. Hydrogenation of CO_2 on group VIII metals: IV. Specific activities and selectivities of silica-supported Co, Fe, and Ru. *J. Catal.* **1984**, *87*, 352–362. [CrossRef]
24. Mutschler, R.; Moioli, E.; Luo, W.; Gallandat, N.; Züttel, A. CO_2 hydrogenation reaction over pristine Fe, Co, Ni, Cu and Al_2O_3 supported Ru: Comparison and determination of the activation energies. *J. Catal.* **2018**, *366*, 139–149. [CrossRef]
25. Guerrero-Ruiz, A.; Rodriguez-Ramos, I. Hydrogenation of CO_2 on carbon-supported nickel and cobalt. *React. Kinet. Catal. Lett.* **1985**, *29*, 93–99. [CrossRef]
26. Fernández-Morales, I.; Guerrero-Ruiz, A.; López-Garzón, F.J.; Rodríguez-Ramos, I.; Moreno-Castilla, C. Hydrogenolysis of n-butane and hydrogenation of carbon monoxide on Ni and Co catalysts supported on saran carbons. *Appl. Catal.* **1985**, *14*, 159–172. [CrossRef]
27. Yao, Y.; Hildebrandt, D.; Glasser, D.; Liu, X. Fischer–Tropsch synthesis using H2/CO/CO2 syngas mixtures over a cobalt catalyst. *Ind. Eng. Chem. Res.* **2010**, *49*, 11061–11066. [CrossRef]
28. Gnanamani, M.K.; Shafer, W.D.; Sparks, D.E.; Davis, B.H. Fischer–Tropsch synthesis: Effect of CO_2 containing syngas over Pt promoted Co/γ-Al_2O_3 and K-promoted Fe catalysts. *Catal. Commun.* **2011**, *12*, 936–939. [CrossRef]
29. Habazaki, H.; Yamasaki, M.; Zhang, B.P.; Kawashima, A.; Kohno, S.; Takai, T.; Hashimoto, K. Co-methanation of carbon monoxide and carbon dioxide on supported nickel and cobalt catalysts prepared from amorphous alloys. *Appl. Catal. A Gen.* **1998**, *172*, 131–140. [CrossRef]
30. Dorner, R.W.; Hardy, D.R.; Williams, F.W.; Davis, B.H.; Willauer, H.D. Influence of Gas Feed Composition and Pressure on the Catalytic Conversion of CO_2 to Hydrocarbons Using a Traditional Cobalt-Based Fischer-Tropsch Catalyst. *Energy Fuels* **2009**, *23*, 4190–4195. [CrossRef]
31. Gnanamani, M.K.; Jacobs, G.; Shafer, W.D.; Sparks, D.; Davis, B.H. Fischer–Tropsch synthesis: Deuterium kinetic isotope study for hydrogenation of carbon oxides over cobalt and iron catalysts. *Catal. Lett.* **2011**, *141*, 1420–1428. [CrossRef]
32. Fröhlich, G.; Kestel, U.; Łojewska, J.; Łojewski, T.; Meyer, G.; Voß, M.; Borgmann, D.; Dziembaj, R.; Wedler, G. Activation and deactivation of cobalt catalysts in the hydrogenation of carbon dioxide. *Appl. Catal. A Gen.* **1996**, *134*, 1–19. [CrossRef]
33. Lahtinen, J.; Anraku, T.; Somorjai, G.A. C, CO and CO_2 hydrogenation on cobalt foil model catalysts: Evidence for the need of CoO reduction. *Catal. Lett.* **1994**, *25*, 241–255. [CrossRef]
34. Torrente-Murciano, L.; Mattia, D.; Jones, M.D.; Plucinski, P.K. Formation of hydrocarbons via CO_2 hydrogenation–A thermodynamic study. *J. CO_2 Util.* **2014**, *6*, 34–39. [CrossRef]
35. Gao, J.; Wang, Y.; Ping, Y.; Hu, D.; Xu, G.; Gu, F.; Su, F. A thermodynamic analysis of methanation reactions of carbon oxides for the production of synthetic natural gas. *RSC Adv.* **2012**, *2*, 2358–2368. [CrossRef]
36. Melaet, G.; Ralston, W.T.; Li, C.S.; Alayoglu, S.; An, K.; Musselwhite, N.; Kalkan, B.; Somorjai, G.A. Evidence of highly active cobalt oxide catalyst for the Fischer–Tropsch synthesis and CO_2 hydrogenation. *J. Am. Chem. Soc.* **2014**, *136*, 2260–2263. [CrossRef] [PubMed]
37. Riedel, T.; Claeys, M.; Schulz, H.; Schaub, G.; Nam, S.-S.; Jun, K.-W.; Choi, M.-J.; Kishan, G.; Lee, K.-W. Comparative study of Fischer–Tropsch synthesis with H2/CO and H2/CO_2 syngas using Fe- and Co-based catalysts. *Appl. Catal. A Gen.* **1999**, *186*, 201–213. [CrossRef]
38. Visconti, C.G.; Lietti, L.; Tronconi, E.; Forzatti, P.; Zennaro, R.; Finocchio, E. Fischer–Tropsch synthesis on a Co/Al_2O_3 catalyst with CO_2 containing syngas. *Appl. Catal. A Gen.* **2009**, *355*, 61–68. [CrossRef]
39. Janlamool, J.; Praserthdam, P.; Jongsomjit, B. Ti-Si composite oxide-supported cobalt catalysts for CO_2 hydrogenation. *J. Nat. Gas Chem.* **2011**, *20*, 558–564. [CrossRef]

40. Das, T.; Deo, G. Synthesis, characterization and in situ DRIFTS during the CO_2 hydrogenation reaction over supported cobalt catalysts. *J. Mol. Catal. A Chem.* **2011**, *350*, 75–82. [CrossRef]
41. Srisawad, N.; Chaitree, W.; Mekasuwandumrong, O.; Shotipruk, A.; Jongsomjit, B.; Panpranot, J. CO_2 hydrogenation over Co/Al_2O_3 catalysts prepared via a solid-state reaction of fine gibbsite and cobalt precursors. *React. Kinet. Mech. Catal.* **2012**, *107*, 179–188. [CrossRef]

© 2019 by the authors. Licensee MDPI, Basel, Switzerland. This article is an open access article distributed under the terms and conditions of the Creative Commons Attribution (CC BY) license (http://creativecommons.org/licenses/by/4.0/).

Article

Kinetics of Fischer–Tropsch Synthesis in a 3-D Printed Stainless Steel Microreactor Using Different Mesoporous Silica Supported Co-Ru Catalysts

Nafeezuddin Mohammad [1], Sujoy Bepari [2], Shyam Aravamudhan [1] and Debasish Kuila [1,2,*]

[1] Department of Nanoengineering, Joint School of Nanoscience and Nanoengineering, North Carolina A&T State University, Greensboro, NC 27411, USA; nmohammad@aggies.ncat.edu (N.M.); saravamu@ncat.edu (S.A.)
[2] Department of Chemistry, North Carolina A&T State University, Greensboro, NC 27411, USA; sbepari@ncat.edu
* Correspondence: dkuila@ncat.edu; Tel.: +1-336-285-2243

Received: 14 September 2019; Accepted: 15 October 2019; Published: 21 October 2019

Abstract: Fischer–Tropsch (FT) synthesis was carried out in a 3D printed stainless steel (SS) microchannel microreactor using bimetallic Co-Ru catalysts on three different mesoporous silica supports. CoRu-MCM-41, CoRu-SBA-15, and CoRu-KIT-6 were synthesized using a one-pot hydrothermal method and characterized by Brunner–Emmett–Teller (BET), temperature programmed reduction (TPR), SEM-EDX, TEM, and X-ray photoelectron spectroscopy (XPS) techniques. The mesoporous catalysts show the long-range ordered structure as supported by BET and low-angle XRD studies. The TPR profiles of metal oxides with H_2 varied significantly depending on the support. These catalysts were coated inside the microchannels using polyvinyl alcohol and kinetic performance was evaluated at three different temperatures, in the low-temperature FT regime (210–270 °C), at different Weight Hourly Space Velocity (WHSV) in the range of 3.15–25.2 kgcat.h/kmol using a syngas ratio of H_2/CO = 2. The mesoporous supports have a significant effect on the FT kinetics and stability of the catalyst. The kinetic models (FT-3, FT-6), based on the Langmuir–Hinshelwood mechanism, were found to be statistically and physically relevant for FT synthesis using CoRu-MCM-41 and CoRu-KIT-6. The kinetic model equation (FT-2), derived using Eley–Rideal mechanism, is found to be relevant for CoRu-SBA-15 in the SS microchannel microreactor. CoRu-KIT-6 was found to be 2.5 times more active than Co-Ru-MCM-41 and slightly more active than CoRu-SBA-15, based on activation energy calculations. CoRu-KIT-6 was ~3 and ~1.5 times more stable than CoRu-SBA-15 and CoRu-MCM-41, respectively, based on CO conversion in the deactivation studies.

Keywords: Fischer-Tropsch synthesis; mesoporous silica based catalysts; kinetic studies; 3-D printed microchannel microreactor

1. Introduction

Although Fischer–Tropsch (FT) synthesis was discovered by Franz Fischer and Hans Tropsch in the 1920s in Germany [1], it has gained immense attention in last few years due to depletion of non-renewable energy sources. FT synthesis is an environmental friendly route for alternative fuels and can produce liquid fuels from carbon sources by coal-to-liquid (CTL), natural gas-to-liquid (GTL) and biomass-to-liquid (BTL) [2] processes. Three types of reactors have been utilized commercially for FT synthesis: Fixed bed, fluidized bed, and slurry bubble column bed by leading GTL companies like Shell, Sasol, Exxon Mobil, and Energy Int. [3]. There is a minimum scale limit of this FT process to be economical; for example, the Pearl GTL, a collaboration between Shell and Qatar petroleum, producing 140,000 bpd (barrels per day) is considered as a profitable economic scale for the FT GTL process [4]. The limitation of the scale-up considerations to commercialize small scale plants to be more profitable has driven industries and researchers to pursue an interest in alternative technologies. Since FT synthesis is highly exothermic in nature, there is a need for much process intensification technologies. The microreactor platform, which contains microstructured units called microreactors, uses a large number of small, parallel channels with different channel designs. This technology provides an alternative platform for controlling highly exothermic reactions like FT synthesis with enhanced mass and heat transfer. It has gained much attention in process intensification of FT synthesis [5,6] as isothermal operating conditions are well maintained in a microreactor with good control over process parameters which favors quick screening of catalyst for different chemical reactions. The reaction zone for these microreactors are several parallel microchannels with small geometry. The specific surface area of the reaction zone is greatly enhanced by the design of microchannels resulting in an efficient FT synthesis. In addition to efficient heat and mass transfer with good heat dissipation, microreactors also have advantages such as high reaction throughput, easy scale-up, good portability, and lower cost over conventional reactors [6–10]. This has been demonstrated commercially and in R&D by Velocys and Micrometrics Corporations [11–14].

Iron, cobalt, and ruthenium catalysts have been extensively used for FT synthesis [15]. To increase the performance of catalysts, different supports have been used; some of the previous studies examined the role of Al_2O_3 [16–20], TiO_2 [21–28], SiO_2 [29–33], and CNTs [34–36] as supporting materials for the formation of higher alkanes. These supports tend to enhance the FT process by increasing the active number of catalytic sites and good metal dispersion with the high surface area. Therefore, the selection of support and study of its interaction with the incorporated metal ion plays an important role in catalysis. In our previous studies, sol-gel encapsulated catalysts were used in silicon microreactors for FT synthesis [37–39]. While Al_2O_3 and SiO_2 sol-gel supports show similar behavior in formation of higher alkanes such as ethane, propane, and butane for the reactions at 1 atm, TiO_2 has a profound effect on FT synthesis [40] and the stability of the catalysts are observed in reverse order from that observed with SiO_2 and Al_2O_3. However, in all these studies, sol-gel coated catalysts in silicon microreactors tend to have challenges such as low surface area, clogging of microchannels and difficulty in reducing the metal oxides to expose active sites. In addition, the Si-microreactors are fragile and they break easily and require a large infrastructure for fabrication. Further, it's more difficult to increase pressure for FT studies using Si-microreactors. Thus, we have turned our attention to 3D printed stainless steel (SS) microreactors which are easy to fabricate by direct metal laser sintering *layer-by-layer* additive manufacturing technique. Recently, 3D printed microreactors have been used as flow devices in many chemical reactions such as fast difluoromethylation [41], a customizable Lab-On-Chip device for optimization of carvone semicarbazon [42], a micro fuel cell [43,44], and wide range of organic and inorganic reactions [45–47]. Further, these metal printed microreactors have been used for high pressure and temperature chemical reactions providing a new fast developing reactor technology in process development to industrial scale [47], which makes them suitable for reactions like FT synthesis due to its good mechanical and thermal properties. Although the specific surface area of stainless steel microreactor is less when compared to silicon microreactors used in our previous studies [40], the use of stainless steel material increases heat transfer, its chemical and mechanical resistances play a major role in process intensification of chemical processes. In order to increase specific surface area of the reaction zone in microreactors, we synthesized catalysts with surface area greater than 1000 m^2/g using mesoporous MCM-41 support. The use of high surface area MCM-41 for FT catalysis stems from our previous studies, which can be prepared easily by one-pot hydrothermal procedure and are extremely stable, for steam reforming of methanol to produce hydrogen [48–51]. Bimetallic catalysts containing Co and one other metal—Fe, Ru, or Ni—were prepared to investigate the synergistic effect of bi-metallic species on the FT performance (manuscript submitted), The results show that CoRu-MCM-41 is more active than other bimetallic catalysts in producing longer-chain hydrocarbons at one atmosphere.

In this manuscript, we have focused on the kinetics of FT synthesis in a 3-D printed Stainless Steel(SS) microreactor using CoRu bimetallic catalysts supported by MCM-41, and two other mesoporous silica supports: SBA-15 and KIT-6. In order to translate new advancements in the laboratory as well as industry on both catalysis and microreactors for FT synthesis, chemical kinetics is a key issue in developing mathematical models for the reactors. However, to our knowledge, the kinetics of FT synthesis using mesoporous materials in a microreactor is relatively unknown in the literature. So, in order to understand more about the interaction between silica mesoporous materials and the metal, and especially kinetics, three different types of silica mesoporous materials (MCM-41, SBA-15, and KIT-6) containing cobalt and ruthenium metals were synthesized by one-pot hydrothermal method. To address the thermodynamic stability of the catalysts, CO-conversion using these three catalysts in the 3-D printed SS microchannel microreactor was also investigated.

2. Results and Discussion

2.1. Catalyst Characterization

2.1.1. Textural Evaluation of Catalysts

The textural properties of the catalysts were evaluated using nitrogen Brunner–Emmett–Teller (BET) physisorption analysis. Table 1 shows the BET surface area, pore volume and the average pore diameter of all three catalysts. The surface areas of the catalysts are different depending upon the type of silica support. While the surface area of CoRu-MCM-41 was 1025 m^2/g, that of CoRu-SBA-15 and CoRu-KIT-6 was around 691 m^2/g and 690 m^2/g, respectively. The general trend is consistent with that reported in the literature [48,52,53]. The pore diameter in the range of 3.2–5.3 nm was obtained from BJH desorption plot. The pore volume was in the range of 0.77–0.92 cm^3/g. Figure 1a shows nitrogen adsorption–desorption isotherms for all the catalysts with pore size distribution of mesoporous materials. These isotherms represent the category of Type IV isotherms which is typical for mesoporous materials as mentioned in IUPAC classification [54]. These isotherms are classified into three different types of regions. The initial part of isotherm is a linear increment of nitrogen uptake at lower relative pressures (P/P$_0$ = 0–0.2) called Type II isotherm. This is due to the adsorption of N$_2$ on monolayer and multilayer within the pore walls of the catalyst. For relative pressure in the range of P/P$_0$ = 0.2–0.4, there is an exponential increment in the isotherms which indicates the ordered mesoporous structure of the catalysts. Especially, the steepness of CoRu-MCM-41 is sharp when compared to the other two samples which indicate that MCM-41 support is more ordered in the nature of all the catalysts. Finally, the third region, in the relative pressure range of P/P$_0$ = 0.4–0.95, has a long plateau for all the catalysts and it corresponds to the multilayer adsorption on the outer surface of the catalyst. The hysteresis loop for the samples is associated with condensation of N$_2$ uptake in the interstitial voids of mesopores of the support [55]. Figure 1b shows pore size distribution obtained from BJH desorption plots. A sharp single peak for pore size with narrow distribution is observed for all the catalysts covering uniformly the pores with sizes in the range of 3.2 to 5.3 nm as shown in Table 1. The pore sizes of KIT-6 support appear to be larger and wider when compared to that of MCM-41 and SBA-15 supports. Furthermore, the pore distribution of MCM-41 is bi-modal, having major pores distributed in the range of 2 to 3 nm and minimal pore distribution between 3–4 nm.

Table 1. Brunner–Emmett–Teller (BET) surface area, pore size, pore volume and EDX metal loadings of synthesized catalysts.

Mesoporous Silica Supported Catalyst with Intended Metal Loadings	Surface Area [a] (m^2/g)	Pore Volume [b] (cm^3/g)	Pore Size [c] (nm)	Metal Loadings Obtained from SEM-EDX (wt %)
10% Co5%Ru-MCM-41	1025	0.77	3.2	9%Co3.9%Ru-MCM-41
10% Co5%Ru-SBA-15	691	0.73	4.2	8.4%Co4.5%Ru-KIT-6
10%Co5%Ru-KIT-6	690	0.92	5.3	11.1%Co5.6%Ru-SBA-15

[a] = Variation range ±2%, [b] = Variation range ±3%, [c] = Variation range ±5%.

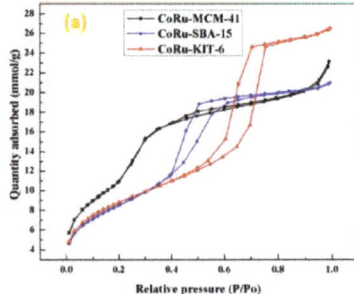 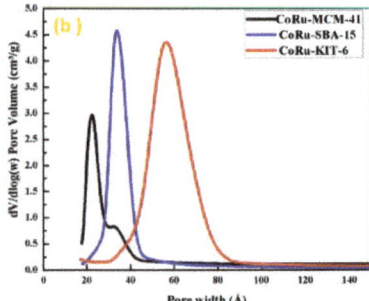

Figure 1. (a) N$_2$ adsorption–desorption isotherms of CoRu-S (S = MCM-41, SBA-15, KIT-6) and (b) pore size distribution of the catalyst.

2.1.2. SEM-EDX Analysis

The metal loadings in the catalysts (wt%) and the surface morphology were obtained by SEM-EDX analysis. Figure S1 shows the SEM-EDX images of a typical MCM-41 catalyst showing uniform metal distribution with porous morphology. The actual and intended metal loadings are quite similar (Table 1) and suggest that the one-pot hydrothermal synthesis is one of the best routes to prepare mesoporous materials with uniform metal distribution. This uniformity plays a key role in the activity of FT catalysts; it not only decreases sintering but also increases the thermal stability of the catalysts for long-term studies.

2.1.3. Transmission Electron Microscopic (TEM) Imaging

The size of the metal particles and the structure of the mesoporous support in all catalysts were obtained from TEM studies. The high magnification images in Figure 2 show the uniform ordered hexagonal pores present in the support. It is also worth noting that MCM-41 support has well defined hexagonal pores when compared to KIT-6 and SBA-15 and this is consistent with the BET surface area and the low angle XRD studies (discussed below). A uniform metal distribution with black dots, as shown in Figure S2, having almost circular in shape is clearly evident in the mesoporous silica matrix.

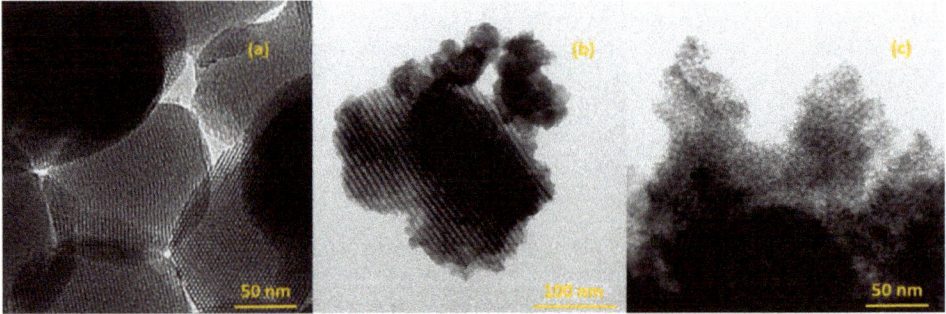

Figure 2. High magnification TEM images of CoRu-S Catalysts (S = (**a**) MCM-41, (**b**) SBA-15, (**c**) KIT-6.

2.1.4. Powder X-Ray Diffraction (XRD) Studies of Calcined Catalysts

In order to obtain information about the structural phases of the catalysts, XRD studies were carried out. Figure 3 shows the small angle XRD diffraction patterns for different mesoporous silica supported catalysts. The variations of peaks are probably due to the presence of metal nanoparticles present in the catalyst. For CoRu-MCM-41 catalyst, a sharp intense peak between 2-theta values 2–3°

and two broad peaks between 2-theta values 4–5.5° corresponds to (100), (110), and (200) reflections of hexagonal mesoporous structure. This confirms that these catalysts are highly ordered mesoporous in nature with no deformation of hexagonal framework even after the addition of metals and this is consistent with the observed TEM images. For CoRu-SBA-15 catalyst, the peak between 2-theta value 1–2° indicates the mesoporous structure with 2D hexagonal symmetry with p6mm space group and long range ordered mesoporous structure [56]. For CoRu-KIT-6 catalyst, the peak at 2-theta value 0.94° corresponds to (211) plane and two low intensity peaks between 1.5–2° ascribes to (420) and (332) diffraction planes. These planes confirmed the characteristic three-dimensional nature of mesoporous KIT-6 reported in the literature [57].

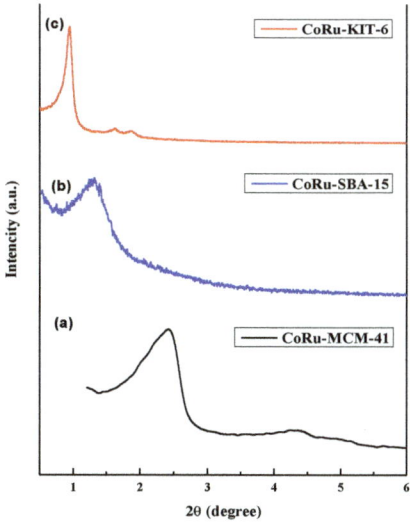

Figure 3. Low angle XRD of three bimetallic mesoporous catalysts: (**a**) CoRu-MCM41; (**b**) CoRu-SBA15; (**c**) CoRu-KIT6.

The wide-angle XRD (WAXRD) analysis was carried out to determine the crystallinity of metal oxides in different mesoporous supports. Figure 4 shows the WAXRD patterns of these samples. The observed 2θ angles are compared with the JCPDS (Joint Committee on Powder Diffraction Standards) database. For all catalysts, the peaks at 18.90° (111), 31.09° (220), 36.74° (311), 38.36° (222), 44.72° (400), 59.25° (511), and 65.26° (440) correspond to the cubic structure of Co_3O_4 (JCPDS-42-1467) [58,59]. The orthorhombic structure of RuO_2 (JCPDS-88-0323) is consistent with the observed peaks at 28.18° (110), 35.27° (101), and 54.56° (211) in all the catalysts.

2.1.5. X-Ray Photoelectron Spectroscopy (XPS)

To determine the oxidation states of Co and Ru in MCM-41, KIT-6, and SBA-15, XPS studies were performed. Figure S3 shows the XPS spectra of Si 2p and O 1s containing a single spectrum which is centered at 104 eV and 532 eV, respectively, and confirms the presence of silicates in the sample. Figure 5a shows the Co 2p spectra for all the samples; the Co $2p_{3/2,}$ and Co $2p_{1/2}$ peaks are clearly observed to indicate the presence of cobalt in two oxidation states in the silica matrix [60–62]. The peaks centered at 780.5 eV and 796.2 eV are associated with Co $2p_{3/2}$ and Co $2p_{1/2}$, respectively in the MCM-41 matrix. Whereas, in the case of KIT-6, the peaks for Co $2p_{3/2}$ and Co $2p_{1/2}$ are observed at 779.6 eV and 795.8 eV, respectively. For SBA-15, the similar peaks are noticed at 779.7 eV and 794.8 eV, respectively. It is clear from these data that the binding energy for cobalt in the MCM-41 matrix is distinctly higher when compared to that of cobalt in KIT-6 and SBA-15. This suggests that cobalt in two oxidation states in MCM-41 is in a different environment from that of SBA-15 and KIT-6. This is also consistent with temperature programmed reduction (TPR) profile showing much higher reduction temperatures for CoRu-MCM-41 catalyst as discussed below. Similar XPS spectra for Co 2p were observed and analyzed by Bhoware et al., [63]. Figure 5b shows the XPS spectra for the ruthenium metal in the catalyst. The presence of Ru in the sample is confirmed by the Ru 3d spectra which is centered almost at 284.8 eV and it is associated with the Ru $3d_{3/2}$ oxidation state [64]. However, in contrast to cobalt, there is no significant difference in the binding energy of the Ru metal ions in different mesoporous silica supports.

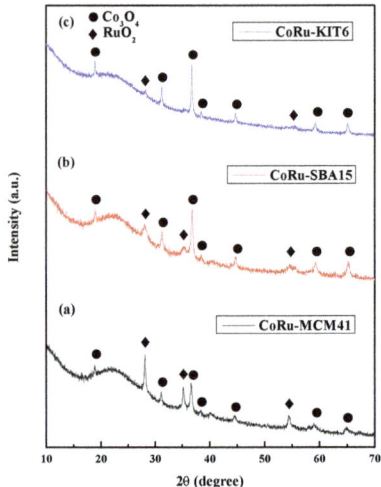

Figure 4. Wide angle XRD patterns of three bimetallic catalysts: (**a**) CoRu-MCM41; (**b**) CoRu-SBA15; (**c**) CoRu-KIT6.

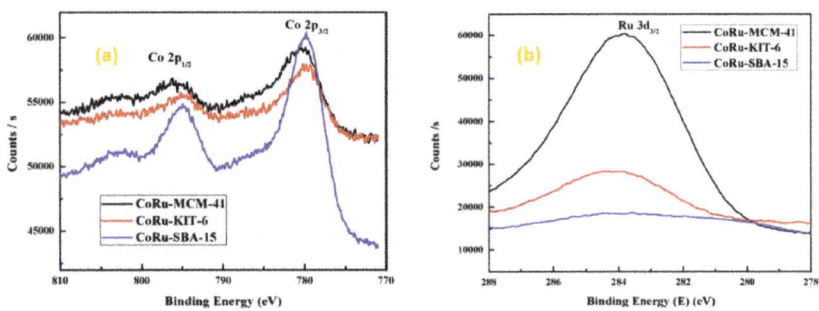

Figure 5. XPS spectra of metals incorporated MCM-41, KIT-6, and SBA-15: (**a**) Co 2p spectra and (**b**) Ru 3d spectra.

2.1.6. H_2 Temperature Programmed Reduction (H_2 TPR)

Temperature programmed reduction (TPR) is an ideal technique to analyze the reduction behavior of metal oxides in mesoporous silica. It helps to investigate the interaction between metal and the support by providing information on physiochemical properties of the material. All the calcined catalysts are treated with 10%H_2 to record TPR profiles for Co and Ru metal oxides shown in Figure 6. All the samples contain well defined peaks for ruthenium at low reduction temperatures and cobalt at much higher reduction temperatures. The TPR profiles of all the catalysts show that the ruthenium oxide is reduced with H_2 at a relatively lower temperature between 100 °C and 250 °C with main hydrogen consumption peaks for Ru^{3+} to Ru^0 which is also reported by Panpranot et al. [65]. However, the reduction behavior of Co_3O_4 ($Co_3O_4 \rightarrow CoO \rightarrow Co^0$) [66] in three samples to Co^0 is remarkably different depending on the type of support. For MCM-41, as reported by Lim et al., the small peak centered around 310 °C ascribes the reduction of cobalt to CoO, while the second main peak corresponds to the reduction of CoO to metal ions Co^{2+} into the silica network [67]. The last hydrogen uptake has a peak centered almost 780 °C which suggests that the cobalt and the MCM-41 support have strong interaction which is also confirmed by the binding energy obtained from XPS in Figure 5a [48,68]. This could be due to the formation of a spinel structure as cobalt silicates [69] and consistent with the XPS and XRD data. Unlike MCM-41, the reduction temperatures of Co inSBA-15 and KIT-6 were quite low around 365 °C and 375 °C, respectively, confirming weaker metal interactions with SBA-15 and KIT-6 supports [53]. However, no separate three peaks were observed for the reduction of cobalt, this may be due to the absence of silicates in SBA-15 and KIT-6 samples. The shift in the reduction peaks to the lower temperatures can also be due to the incorporation of Ru metal in the support [70]. Although the 5% weight of Ru is maintained in the catalyst sample, there might be a slight difference in the actual loadings of the Ru metal as shown in EDX Table 2. Qin et al., have studied the effect of the Ru metal on Co-SBA-15 catalyst at different loading and found a remarkable effect on the activity of catalyst during the FT synthesis [70]. Thus, the overall interactions of metal–metal and metal–support have a strong influence on the reducibility and reactivity of the catalysts for FT synthesis. Since the operating temperature zone of the FT synthesis is less than 350 °C, the activity of the catalyst is more dependent on the ease of reducibility of the metal oxides to pure metals (active sites) in the support.

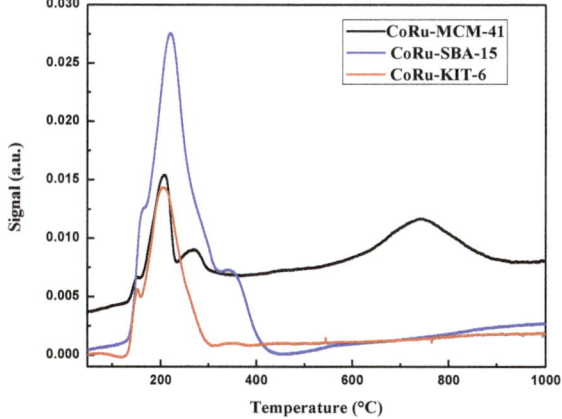

Figure 6. H_2-TPR(temperature programmed reduction) profiles of CoRu-S (S = MCM-41, KIT-6 and SBA-15) Catalysts.

The amount of hydrogen consumed by CoRu-MCM-41, CoRu-SBA-15, and CoRu-KIT-6 catalysts was calculated and quantified to be 0.108, 0.05, and 0.032 mmol H_2/gm, respectively in the temperature range of 25 to 1000 °C. The amount of hydrogen consumed by CoRu-MCM-41 was found almost two times more than that of CoRu-SBA-15 and 3 times more than by CoRu-KIT-6. Higher hydrogen uptake by CoRu-MCM-41 is most likely due to the reduction of Co-silicates ~750 °C.

Although MCM-41 exhibits higher hydrogen consumption than other catalysts, the amount of hydrogen adsorbed by CoRu-KIT-6 in the temperature range of 25–311 °C is higher when compared to that by CoRu-MCM-41 and CoRu-SBA-15 in the same range of temperature.

2.2. FT Reaction Mechanism

In order to have a better understanding of the effect of metal and support interaction on catalysts, kinetic studies were carried out in the SS microchannel microreactors. The main difficulty to describe the FT kinetics is the complexity of its mechanism and a larger number of possible chemical species involved. Kinetic models of FT synthesis using Co based catalysts are less abundant than Fe based catalysts in literature [71]. Most of the existing models are mainly based on power-law models where Langmuir–Hinselwood (LH) type equations have been used by different researchers [72–76]. Although the simple power-law expression is widely recognized in the field of catalysis, it was recognized to have limited application in FT synthesis due to the narrow range of reaction conditions [77,78]. However, LH type equations are widely used for prediction of rates over a wide range of reaction conditions. As an example, Yates and Satterfield [72] worked on Co-catalyst and fitted the rate data obtained at 220–240 °C. They found that the rate data were best fitted with simple LH expression. Rautavuoma and van dar Baan [79] reported the rate of reaction at 1 atm pressure and 250 °C. They observed that reaction proceeds through CO dissociation and formation of -CH_2- surface intermediate.

2.2.1. Reaction Mechanism

In the microchannels of the microreactor, the flow of reactants is basically laminar. The complexity of the microchannel microreactor increases due to the parabolic type velocity profile; so, an average velocity profile is approximated during the development of the model for the microreactor system [80]. The outlet concentrations of the limiting reactant (CO), which was related to the rate of reaction, were calculated by an in-line GCMS. The following differential equation was used for a reactor model defined as Equation (1):

$$\frac{W_{cat}}{F_{in,CO}} = \int_{X_{CO,in}}^{X_{CO,out}} \frac{dX_{CO}}{-r_{CO}} \tag{1}$$

W_{cat} = Wt. of the catalyst
$F_{in,CO}$ = Molar feed rate of CO
X_{CO} = Conversion of CO
$-r_{CO}$ = Disappearance rate of CO

Equation (2), below, is used to calculate the disappearance rate of CO

$$-r_{CO} = \frac{X_{CO} F_{in,CO}}{W_{cat}} \tag{2}$$

The following boundary conditions (BC) were used:

$W = 0; F_i = F_{i(inlet)}$
$W = W_{cat}; F_i = F_{i(exit)}$

The partial pressure of the compound was calculated using the following equations:

$$p_i = \frac{m_i}{\sum_{i=1}^{N_c} m_i} P_T \tag{3}$$

where p_i is the partial pressure of the component, P_T is the total pressure of the reactor at the inlet (1 atm) and N_c is the total number of components. m_i is the number of moles of component i.

2.2.2. Mechanism and Kinetics

In order to determine the most suitable kinetic model for a particular catalyst, all possible combinations of FT reactions were considered and rate equations were developed based on CO conversion. A number of Langmuir–Hinshelwood and Eley–Rideal models have been developed for kinetics of CO hydrogenation to hydrocarbons in different types of reactors over the past few years [80–82]. In this study, it was assumed that the FT reactions occur only at active sites and proposed six possible mechanisms for FT reactions to develop kinetic models in the microchannel microreactor as shown in Table 2. In order to derive an appropriate model that describes a suitable FT equation, we considered six cases with different elementary reaction steps for each case as shown in Table 2.

Table 2. Elementary reaction steps for Fischer–Tropsch synthesis.

Model	No	Elementary Reaction
FT-1	1	$CO + * \underset{k_{-1}}{\overset{k_1}{\rightleftharpoons}} CO* \quad K_1 = \frac{k_1}{k_{-1}}$
	2	$CO* + H_2 \overset{k}{\rightarrow} C + D + *$
FT-2	1	$H_2 + * \underset{k_{-1}}{\overset{k_1}{\rightleftharpoons}} H_2* \quad K_1 = \frac{k_1}{k_{-1}}$
	2	$H_2* + CO \overset{k}{\rightarrow} C + D + *$
FT-3	1	$CO + * \underset{k_{-1}}{\overset{k_1}{\rightleftharpoons}} CO* \quad K_1 = \frac{k_1}{k_{-1}}$
	2	$H_2 + * \underset{k_{-2}}{\overset{k_2}{\rightleftharpoons}} H_2* \quad K_2 = \frac{k_2}{k_{-2}}$
	3	$CO* + H_2* \overset{k}{\rightarrow} C + D + 2*$
FT-4	1	$CO + * \underset{k_{-1}}{\overset{k_1}{\rightleftharpoons}} CO* \quad K_1 = \frac{k_1}{k_{-1}}$
	2	$CO* + H_2 \underset{k_{-2}}{\overset{k_2}{\rightleftharpoons}} COH_2* \quad K_2 = \frac{k_2}{k_{-2}}$
	3	$COH_2* \overset{k}{\rightarrow} C + D + *$
FT-5	1	$H_2 + * \underset{k_{-1}}{\overset{k_1}{\rightleftharpoons}} H_2* \quad K_1 = \frac{k_1}{k_{-1}}$
	2	$H_2* + CO \underset{k_{-2}}{\overset{k_2}{\rightleftharpoons}} COH_2* \quad K_2 = \frac{k_2}{k_{-2}}$
	3	$COH_2* \overset{k}{\rightarrow} C + D + *$
FT-6	1	$CO + * \underset{k_{-1}}{\overset{k_1}{\rightleftharpoons}} CO* \quad K_1 = \frac{k_1}{k_{-1}}$
	2	$H_2 + * \underset{k_{-2}}{\overset{k_2}{\rightleftharpoons}} H_2* \quad K_2 = \frac{k_2}{k_{-2}}$
	3	$CO* + H_2* \underset{k_{-3}}{\overset{k_3}{\rightleftharpoons}} COH_2* + * \quad K_3 = \frac{k_3}{k_{-3}}$
	4	$COH_2* \overset{k}{\rightarrow} C + D + *$

In FT-1, CO is adsorbed on active site (*) of catalyst to form a CO* intermediate. Then, CO* reacts with H_2 to give the products C (hydrocarbons) and D (H_2O). Similarly, H_2 can be adsorbed on the catalyst site (*) to form H_2* intermediate and this intermediate subsequently reacted with CO to yield products in the FT-2 mechanism. There are two steps of adsorption in the FT-3 mechanism. In 1st and 2nd steps, CO and H_2 both are adsorbed on catalyst active site (*) to form two intermediates (CO* and H_2*). In the last step (surface reaction), these two intermediates react with each other to give products C and D. The FT-4 model consisted of three different steps. In the 1st step (adsorption), CO is adsorbed on catalyst site (*) to form CO*. In 2nd step (surface reaction), the intermediate (CO*) reacts with H_2 to form another intermediate (COH_2*). In the last step (desorption), the final intermediate gave products (C and D). Like FT-4, FT-5 also consists of three different steps—adsorption, surface reaction, and desorption. In the 1st step (adsorption), H_2 is adsorbed on catalyst site (*) to form the H_2* intermediate. This intermediate reacts with CO to form another intermediate (COH_2*) in the 2nd step (surface reaction). In 3rd step, the products (C and D) are formed from the intermediate (COH_2*). In contrast to other models, FT-6 consists of four different steps. The 1st and 2nd steps are like that of the FT-3 mechanism. The 3rd step is the surface reaction where two intermediates (CO* and H_2*) react with each other to give another intermediate COH_2* and released one active site (*). In last step (desorption), the intermediate (COH_2*) yields products (C and D).

Using the six models described above, six rate equations can be deduced for FT reactions by considering surface reaction and rate-limiting desorption as shown in Table 3. (See Appendix A for the rate equation derived using the FT-3 kinetic model).

Table 3. Proposed kinetic equations for Fischer–Tropsch synthesis.

Model	Rate Controlling Step (RCS)	Kinetic Equation
FT-1	(2)	$-r_{CO} = \dfrac{kK_1 p_{CO} p_{H_2}}{(1+K_1 p_{CO})}$
FT-2	(2)	$-r_{CO} = \dfrac{kK_1 p_{CO} p_{H_2}}{(1+K_1 p_{H_2})}$
FT-3	(3)	$-r_{CO} = \dfrac{kK_1 K_2 p_{CO} p_{H_2}}{(1+K_1 p_{CO} + K_2 p_{H_2})^2}$
FT-4	(3)	$-r_{CO} = \dfrac{kK_1 K_2 p_{CO} p_{H_2}}{(1+K_1 K_2 p_{CO} p_{H_2} + K_1 p_{CO})}$
FT-5	(3)	$-r_{CO} = \dfrac{kK_1 K_2 p_{CO} p_{H_2}}{(1+K_1 K_2 p_{CO} p_{H_2} + K_1 p_{H_2})}$
FT-6	(4)	$-r_{CO} = \dfrac{kK_1 K_2 K_3 p_{CO} p_{H_2}}{(1+K_1 K_2 K_3 p_{CO} p_{H_2} + K_1 p_{CO} + K_2 p_{H_2})}$

All the models presented in Table 3 were verified against experimental data to obtain the best suitable mechanism with the best fit. The models FT-1, FT-2, FT-4, and FT-5 are based on Eley–Rideal-type mechanism. In this case, one reactant gets adsorbed and another reactant reacts directly from the gas phase to form intermediates that yield products. Other models FT-3 and FT-6 are based on the Langmuir–Hinshelwood mechanism, which means all reactants are adsorbed on the catalyst surface before the products are formed. The kinetic parameter (k) and equilibrium constants (K_1, K_2, K_3) at each temperature were evaluated by non-linear regression analysis based on Levenberg–Marquart algorithm in POLYMATH software by minimizing the sum of squared residuals of reaction rates [83,84]. The objective function is defined as:

$$F = \sum_{i=1}^{N} \left(r_{cal_i} - r_{exp_i}\right)^2 \quad (4)$$

where, N is the number of total observations, r_{cal_i} and r_{exp_i} are calculated from the model equation and experimental rates at the different reaction temperatures.

The rate constants and the equilibrium constants can be related to The Arrhenius equation and van't Hoff laws as shown below:

$$k_i(T) = A \exp\left(-\frac{E_{ai}}{RT}\right) \quad (5)$$

$$K_i(T) = K \exp\left(-\frac{\Delta H_i}{RT}\right) \quad (6)$$

where k_i and K_i are reaction and equilibrium constants, respectively. E_{ai} and ΔH_i are the apparent activation energy and standard enthalpy change of i species.

2.3. Effect of Space Velocity on CO Conversion

The influence of the weight hourly space velocity (WHSV) on the CO conversion for three different catalysts at 1 atm and H_2/CO molar ratio 2 with error bar is shown in Figure 7a–c. The reactions were carried out at three different temperatures (210 °C, 240 °C, and 270 °C). CO conversion increases with the increase of space velocity and temperature. While CO conversion increases quickly with the increase of space velocity at the beginning, it remains almost constant at higher space velocity as the reaction reaches the equilibrium state. The variation of CO conversion was within 10% as observed during these reactions.

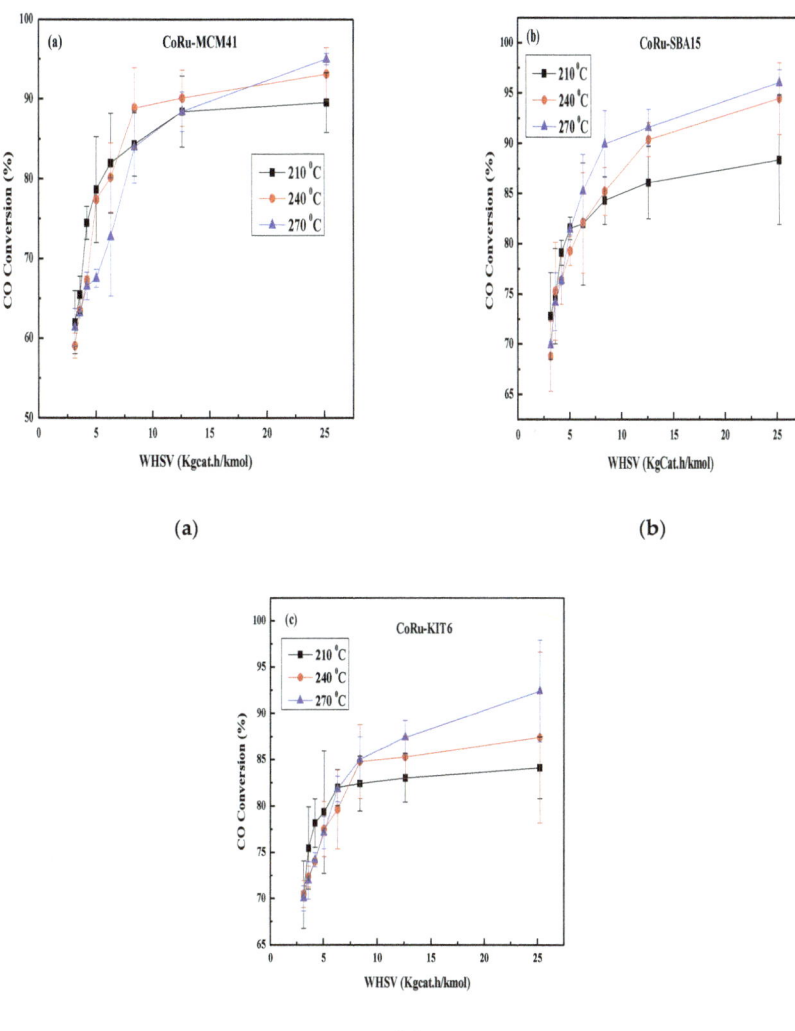

Figure 7. Effect of space velocity on CO conversion: (**a**) CoRu-MCM-41; (**b**) CoRu-SBA-15; (**c**) CoRu-KIT-6.

2.4. FT Kinetic Model

The kinetic models derived from Langmuir–Hinshelwood and Eley–Rideal mechanisms consider elementary reactions consuming CO and H_2 to produce hydrocarbons and water. Recently, the mechanistic aspects of FT synthesis were well investigated by computational catalysis studies using DFT-based quantum chemical models [85–87]. However, the use of Langmuir–Hinshelwood and Eley–Rideal models facilitates understanding of the FT mechanism more easily. In this work, all the kinetic models derived based on these mechanisms are investigated and fitted with experimental data to check the feasibility of the proposed mechanism. The objective function (F) in Equation (4) was utilized to measure the goodness of the model to select the best-fitted mechanism for FT synthesis. Table 4 shows the experimental data obtained for all catalysts at 210 °C, 240 °C, and 270 °C.

Table 4. Kinetic experimental data for all the catalysts.

Temperature	W/F (Kgcat.h/kmol)	CoRu-MCM-41		CoRu-SBA-15		CoRu-KIT-6	
		Reaction Rate (Kmol/Kg.h)	Conversion (%)	Reaction Rate (Kmol/Kg.h)	Conversion (%)	Reaction Rate (Kmol/Kg.h)	Conversion (%)
210 °C	25.2	0.035	89.53	0.035	88.35	0.033	84.12
	12.6	0.070	88.43	0.068	86.08	0.065	83.03
	8.39	0.100	84.29	0.1	84.29	0.098	82.41
	6.30	0.130	81.97	0.13	81.97	0.13	81.97
	5.04	0.156	78.64	0.162	81.52	0.158	79.34
	4.20	0.177	74.45	0.188	79.11	0.186	78.15
	3.60	0.181	65.46	0.208	74.74	0.21	75.45
	3.15	0.196	61.98	0.231	72.79	0.224	70.43
240 °C	25.2	0.036	93.09	0.037	94.43	0.034	87.39
	12.6	0.071	90.09	0.071	90.36	0.067	85.27
	8.39	0.105	88.84	0.01	85.18	0.101	84.79
	6.30	0.127	80.14	0.013	82.07	0.126	79.63
	5.04	0.153	77.34	0.157	79.24	0.154	77.48
	4.20	0.160	67.23	0.182	76.32	0.176	74.02
	3.60	0.176	63.42	0.209	75.25	0.201	72.35
	3.15	0.187	59.02	0.218	68.76	0.224	70.44
270 °C	25.2	0.037	94.95	0.038	95.98	0.036	92.40
	12.6	0.070	88.39	0.072	91.55	0.069	87.39
	8.39	0.100	83.96	0.107	89.90	0.101	85.01
	6.30	0.115	72.70	0.135	85.22	0.130	81.80
	5.04	0.133	67.48	0.162	81.42	0.153	77.08
	4.20	0.158	66.53	0.182	76.32	0.177	74.18
	3.60	0.175	63.20	0.206	74.19	0.2	71.93
	3.15	0.194	61.34	0.222	69.90	0.222	70.00

The kinetic parameters obtained for all the mechanisms for CoRu-MCM-41 are shown in Table 5. It can be inferred from data that the value of the rate constant (k) increases with increasing temperature with only one of the 6 mechanisms which is FT-3. Therefore, for CoRu-MCM-41, the model FT-3 is best fitted with the kinetic data.

Table 5. Kinetic parameters obtained from proposed mechanisms for CoRu-MCM-41

Model	Temperature					
	210 °C	R^2	240 °C	R^2	270 °C	R^2
FT-1	$k = 0.872 \pm 0.0037$ $K_1 = 21.19 \pm 1.79$	0.88	$k = 0.23 \pm 0.027$ $K_1 = 16.49 \pm 7.66$	0.98	$k = 0.38 \pm 0.118$ $K_1 = 3.92 \pm 2.46$	0.98
FT-2	$k = 0.87 \pm 0.0037$ $K_1 = 21.19 \pm 1.79$	0.88	$k = 0.74 \pm 0.0014$ $K_1 = 21.19 \pm 0.84$	0.72	$k = 0.69 \pm 0.0067$ $K_1 = 6.05 \pm 0.34$	0.93
FT-3	**$k = 1.32 \pm 0.000029$** **$K_1 = 5.04 \pm 0.000194$** **$K_2 = 2.877 \pm 0.000468$**	0.97	**$k = 2.21 \pm 0.017$** **$K_1 = 4.02 \pm 0.103$** **$K_2 = 0.51 \pm 0.0064$**	0.96	**$k = 3.22 \pm 0.068$** **$K_1 = 1.41 \pm 0.05$** **$K_2 = 0.5 \pm 0.019$**	0.97
FT-4	$k = 0.443 \pm 1.69$ $K_1 = 1.88 \pm 19.53$ $K_2 = 2.49 \pm 34.08$	0.98	$k = 0.85 \pm 0.015$ $K_1 = 11.09 \pm 0.686$ $K_2 = 0.365 \pm 0.0077$	0.98	$k = 2.01 \pm 0.037$ $K_1 = 3.13 \pm 0.11$ $K_2 = 0.228 \pm 0.0045$	0.98
FT-5	$k = 0.536 \pm 0.000355$ $K_1 = 21.19 \pm 0.403$ $K_2 = 2.42 \pm 0.0023$	0.97	$k = 0.48 \pm 0.000205$ $K_1 = 21.19 \pm 0.265$ $K_2 = 2.33 \pm 0.0015$	0.90	$k = 0.41 \pm 0.009$ $K_1 = 0.33 \pm 0.014$ $K_2 = 11.09 \pm 0.396$	0.97
FT-6	$k = 0.704 \pm 0.0022$ $K_1 = 0.931 \pm 0.0039$ $K_2 = 31.29 \pm 3.21$ $K_3 = 1.77 \pm 0.0074$	0.96	$k = 0.67 \pm 0.0012$ $K_1 = 0.897 \pm 0.0021$ $K_2 = 31.29 \pm 1.83$ $K_3 = 1.69 \pm 0.0041$	0.87	$k = 0.243 \pm 0.00018$ $K_1 = 3.14 \pm 0.0052$ $K_2 = 31.29 \pm 0.76$ $K_3 = 1.67 \pm 0.0027$	0.83

In order to determine the activation energy and the frequency factor from the Arrhenius equation, the logarithm of the rate constant was plotted against the inverse of reaction temperature as shown in Figure 8. The activation energy was determined to be 32.21 kJ/mol and the frequency factor was 4099 kmol/KgCat.hr. (atm)^2 for CoRu-MCM-41 catalyst. Table 6 shows that the rate constant increases with the increase of reaction temperature. However, the two adsorption equilibrium constants (K_1 and K_2) decrease with the increase of reaction temperature. Since adsorption is an exothermic process, the adsorption equilibrium constant decreases with rise in temperature.

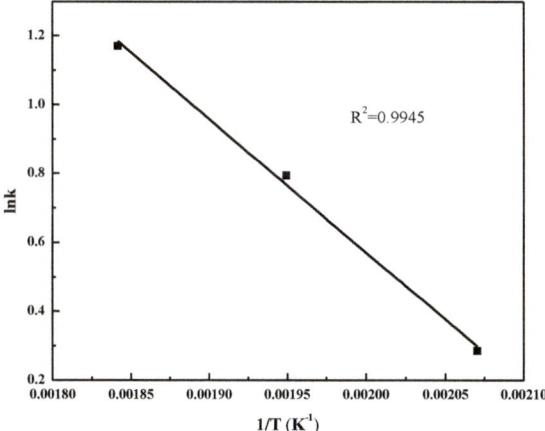

Figure 8. Arrhenius plot-based on Langmuir–Hinselwood (LH) model for FT synthesis over CoRu-MCM-41 catalyst.

Table 4 shows the experimental data of the kinetic runs for CoRu-SBA-15 at 210 °C, 240°C, and 270 °C. When the data are fit against all the kinetic models, FT-2 was the best-fitted model obtained for CoRu-SBA-15. The kinetic parameters for all of the proposed mechanisms are shown in Table 6.

Table 6. Kinetic parameters obtained from proposed mechanisms for CoRu-SBA-15.

Model	Temperature					
	210 °C	R^2	240 °C	R^2	270 °C	R^2
FT-1	k = 3.02 ± 0.092 K_1 = 0.46 ± 0.015	0.94	k = 2.01 ± 0.059 K_1 = 0.655 ± 0.021	0.94	k = 2.01 ± 0.026 K_1 = 0.727 ± 0.01	0.98
FT-2	**k = 9.08 ± 0.266** **K_1 = 0.166 ± 0.005**	0.94	**k = 11.09 ± 0.281** **K_1 = 0.118 ± 0.0033**	0.95	**k = 13.12 ± 0.005** **K_1 = 0.109 ± 0.000046**	0.99
FT-3	k = 11.09 ± 0.698 K_1 = 2.23 ± 0.145 K_2 = 20.62 ± 1.49	0.82	k = 2.01 ± 0.043 K_1 = 10.79 ± 0.285 K_2 = 17.51 ± 0.536	0.79	k = 11.09 ± 0.337 K_1 = 2.01 ± 0.063 K_2 = 18.10 ± 0.645	0.95
FT-4	k = 11.09 ± 0.379 K_1 = 0.161 ± 0.0057 K_2 = 0.761 ± 0.026	0.94	k = 2.01 ± 0.0026 K_1 = 0.283 ± 0.00042 K_2 = 2.31 ± 0.0032	0.93	k = 2.01 ± 0.037 K_1 = 3.13 ± 0.11 K_2 = 0.228 ± 0.0045	0.98
FT-5	k = 2.01 ± 0.026 K_1 = 0.189 ± 0.0031 K_2 = 4.41 ± 0.062	0.92	k = 2.01 ± 0.0237 K_1 = 0.193 ± 0.0029 K_2 = 3.92 ± 0.050	0.94	k = 2.01 ± 0.032 K_1 = 0.205 ± 0.0036 K_2 = 3.62 ± 0.063	0.98
FT-6	k = 2.01 ± 0.037 K_1 = 0.988 ± 0.021 K_2 = 0.894 ± 0.031 K_3 = 1.57 ± 0.032	0.90	k = 6.05 ± 0.253 K_1 = 0.273 ± 0.012 K_2 = 0.266 ± 0.014 K_3 = 3.56 ± 0.153	0.94	k = 2.01 ± 0.018 K_1 = 0.937 ± 0.0094 K_2 = 2.18 ± 0.057 K_3 = 1.09 ± 0.011	0.97

The activation energy and frequency factor of this catalyst were evaluated from the Arrhenius equation by plotting the logarithm of the rate constant to the inverse of reaction temperature as shown in Figure 9. The activation energy was determined to be 13.39 kJ/mol and the frequency factor was 254 kmol/KgCat.hr. atm. Table 6 shows that for FT-3 mechanism, the reaction rate constant increases with reaction temperature and adsorption equilibrium constant (K_1) decreases with the increase of reaction temperature.

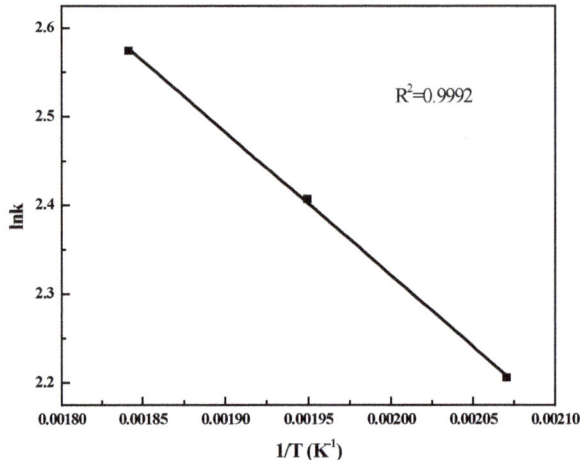

Figure 9. Arrhenius plot-based on LH model for FT synthesis over CoRu-SBA-15 catalyst.

Table 4 shows the experimental data for CoRu-KIT-6 catalyst. All the models were fitted with this experimental kinetic data and FT-6 was found to be the best fitted model for CoRu-KIT-6. The kinetic parameters for all the proposed mechanisms are shown in Table 7 and the activation energy from Figure 10 was determined to be 12.59 kJ/mol and the frequency factor was 39 kmol/KgCat.hr. atm.

Table 7. Kinetic parameters obtained from all the proposed mechanisms for CoRu-KIT-6.

Model	Temperature					
	210 °C	R^2	240 °C	R^2	270 °C	R^2
FT-1	k = 2.18 ± 8.89 K_1 = 0.73 ± 3.28	0.97	k = 0.79 ± 0.543 K_1 = 2.13 ± 1.91	0.98	k = 1.38 ± 1.42 K_1 = 1.19 ± 1.44	0.99
FT-2	k = 1.42 ± 0.00044 K_1 = 11.09±0.038	0.95	k = 1.29±0.022 K_1 = 11.09±2.09	0.98	k = 2.02±3.30 K_1 = 1.94±8.71	0.98
FT-3	k = 11.09±0.316 K_1 = 0.892±0.031 K_2 = 0.279±0.013	0.96	k = 10.09±0.166 K_1 = 0.882±0.018 K_2 = 0.279±0.0073	0.98	k = 11.09±0.088 K_1 = 1.94±0.016 K_2 = 15.64±0.15	0.97
FT-4	k = 2.01±0.052 K_1 = 0.122±0.0038 K_2 = 7.27±0.203	0.97	k = 2.01±0.088 K_1 = 1.195±0.067 K_2 = 0.694±0.033	0.98	k = 2.01±0.072 K_1 = 0.355±0.015 K_2 = 2.29±0.089	0.99
FT-5	k = 2.01±0.051 K_1 = 0.121±0.0038 K_2 = 7.27±0.203	0.97	k = 1.46±0.00069 K_1 = 10.08±0.055 K_2 = 0.91±0.00049	0.92	k = 2.01±0.028 K_1 = 0.842±0.022 K_2 = 1.62±0.024	0.99
FT-6	k = 1.71±0.02 K_1 = 1.39±0.02 K_2 = 0.828±0.018 K_3 = 1.61±0.021	0.95	k = 2.01±0.039 K_1 = 1.11±0.027 K_2 = 0.454±0.013 K_3 = 2.21±0.048	0.98	k = 2.41±0.040 K_1 = 0.91±0.017 K_2 = 0.315±0.0071 K_3 = 3.12±0.056	0.99

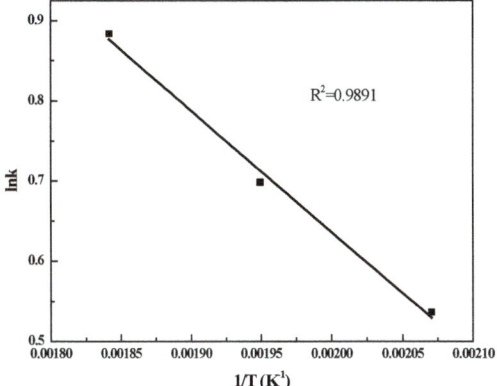

Figure 10. Arrhenius plot-based on LH model for FT synthesis over CoRu-KIT-6 catalyst.

Based on our experimental and kinetic model data, it can be concluded that only one of the six mechanisms for each catalyst is statistically relevant for fitting the model. The rate constant (k), for some of the other five mechanisms, does not show an increasing trend with the increase in temperature or remains constant, while the equilibrium constants, K_1 and K_2, did not show the decreasing trend with the increase in temperature. Thus, FT-3, FT-2, and FT-6 mechanisms were considered as kinetically relevant model equations for CoRu-MCM-41, CoRu-SBA-15, and CoRu-KIT-6, respectively.

The results from our studies in a microreactor are similar to those reported in literature. Mansouri et al., developed a similar mechanism to estimate kinetic parameters for FT synthesis using cobalt-based catalyst with silica support and found that the experimental data were best fitted with surface reaction mechanism proposed based on Langmuir-Hineshelwood model and the optimal activation for the proposed kinetic model was found to be 31.57 kJ/mol [88]. Very recently, Sonal et al., detailed mechanistic approach for FT synthesis based on Langmuir–Hinshelwood–Hougen–Watson (LHHW) and Eley–Rideal using Fe–Co based catalyst. They claimed that a mechanism based on the adsorption and desorption have a satisfactory fit to the experimental data with the activation energies for the formation of methane, paraffin and olefin to be around 70 kJ/mol, 113 kJ/mol, and 91 kJ/mol, respectively [89]. In order to improve the efficiency of FT synthesis significantly, detailed kinetic rate expressions were derived which is very similar to our work reported in literature for both fixed bed reactor as well as microreactor using iron or cobalt-based catalyst [80,81,90,91]. The elementary steps in the above studies were used to develop kinetic mechanisms considering FT reactions with and without water gas shift (WHS) reactions occurring on the surface of the catalysts forming intermediates with active sites. A similar approach was considered in this present study where CO*, COH$_2$* are assumed to form as intermediates, where * is an active site of the catalyst. From the activation energy calculations, shown in Figures 8–10, the FT activation energy is observed in the order, CoRu-MCM-41 > CoRu-SBA-15 > CoRu-KIT-6. Almost 20 kJ/mol less activation energy was obtained for SBA-15 supported catalyst than that of MCM-41 catalyst. The activation energy of KIT-6 supported catalyst is a bit less than that of SBA-15 supported catalyst. This reflects that activation energy depends on metal–support interactions in different mesoporous catalysts. The variation of activity in different mesoporous catalysts might arise due to experimental uncertainties and operating conditions [92]. In addition, the FT activation energy is sensitive to the reactor system. Sun et al., [93] reported that the activation energy in a microchannel microreactor is smaller than that observed in a fixed bed reactor (FBR).

Figure 11 shows the variation of the model predicted rate with the experimental rate for all catalysts. The best fitted models i.e., FT-3, FT-2, and FT-6 for CoRu-MCM-41, CoRu-SBA-15, CoRu-KIT-6, respectively, were chosen to plot the graph for predicted and experimental rates. The correlation coefficient (R^2) for all cases was above or equal to 0.95 with an error band of ±15% to ±30%. This indicates that the error between experimental and predicted values lies within the statistical permissible limits at all reaction temperatures for all the catalysts and consistent with the mechanistic models proposed in the literature. Moazami et al., conducted kinetic studies for FT synthesis in a fixed bed reactor with cobalt-based catalyst over silica support and found that 60% of the results were predicted with a relative error of less than 15%, while the rest of the proposed kinetic models has error less than 32% with confidence interval of 0.99 [94]. They also proposed a pseudo-homogenous one-dimensional model to evaluate the kinetic performance of the catalyst and achieved less than 8% error with the predicted data for kinetic experiments [95]. More recently, Marchese et al., performed kinetic studies with Co-Pt/γ-Al_2O_3 catalyst in a lab-scale tubular reactor and reported an error band around ± 25% with a confidence level of 0.95 stating it lies in the suitable acceptable limits with many mechanistic models proposed in the literature [89,96–98].

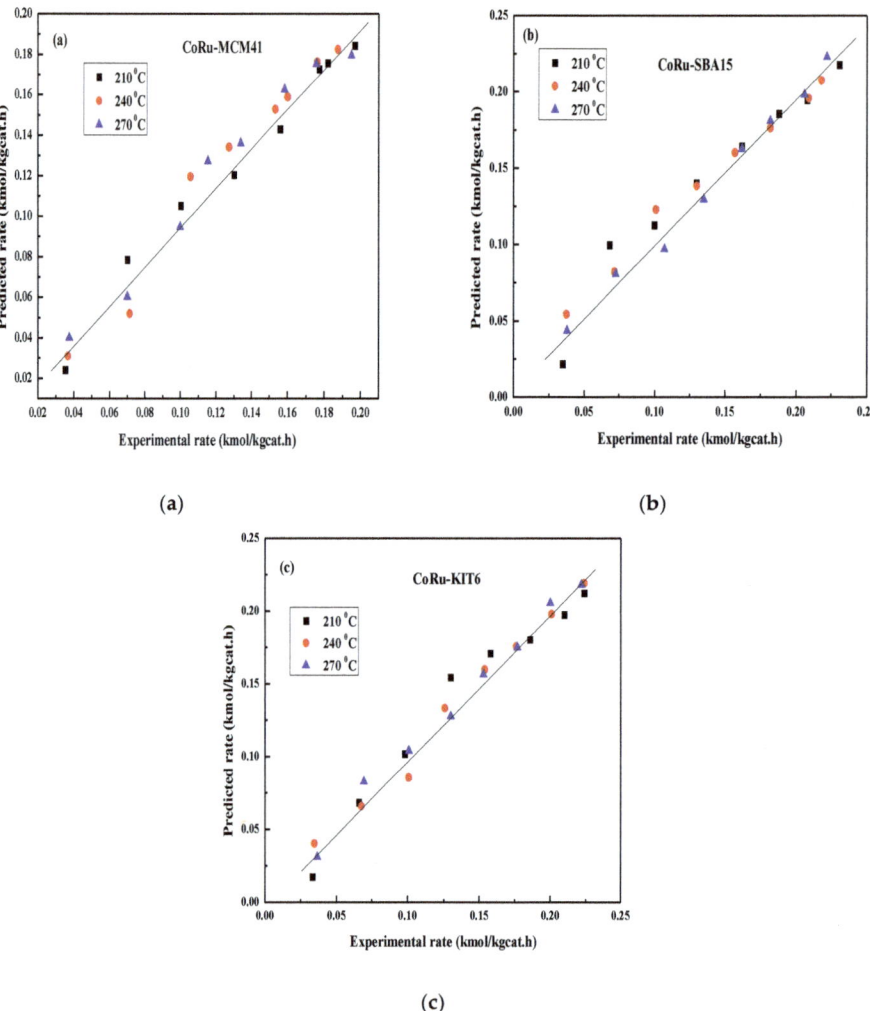

Figure 11. Experimental rate vs predicted rate for all temperature of three different catalysts: (**a**) CoRu-MCM-41; (**b**) CoRu-SBA-15; (**c**) CoRu-KIT-6.

2.5. Deactivation Studies

In order to further understand how the interaction of Co and Ru metals with different mesoporous silica supports affects the stability of the FT catalysts, deactivation studies were performed. Figure 12 shows the deactivation rates of the catalysts tested continuously for 60 h at 240 °C, 1 atm, and H_2:CO ratio of 2:1. All the catalysts maintained fairly consistent CO conversion with very little fluctuation during the first 10 h. More specifically, the catalysts maintained 65%–79% CO conversion in the first 10 h of the reaction with CoRu-KIT-6 exhibiting the highest conversion and CoRu-MCM-41, the lowest. The activity of all the catalysts dropped by 20% after 24 h and then started to decline further. At the end of 60 h, the activity of the CoRu-MCM-41 dropped by 70% whereas, in the case of CoRu-KIT-6 and CoRu-SBA-15, the CO conversion decreased by 64% and 84%, respectively. Our results suggest that in terms of stability, the support has a significant impact on FT performance. More significantly, MCM-41 and KIT-6 supports are more stable when compared to the FT stability studies with SBA-15.

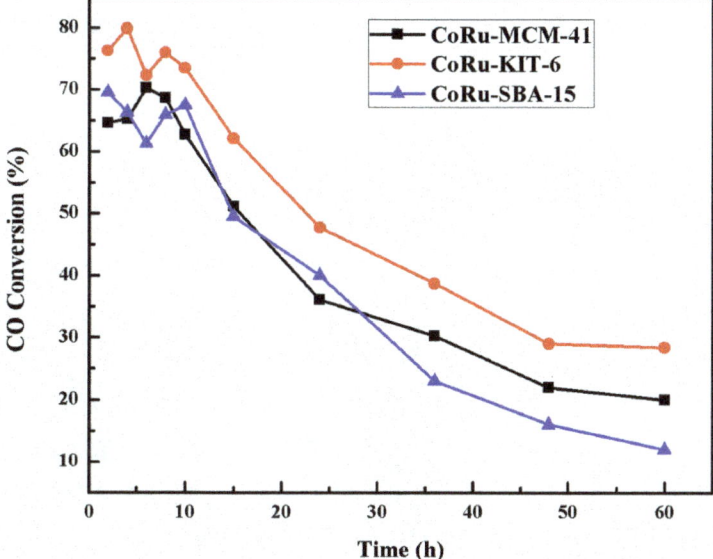

Figure 12. Deactivation studies of the catalysts at T = 240 °C, P = 1 atm and H_2:CO = 2:1.

Many deactivation mechanisms have been proposed for FT studies that include catalysts poisoning, sintering, oxidation, the effect of water, carbon deposition and surface reconstruction [99]. The syngas used in our studies is a mixture of ultrahigh pure 5.0 CO and H_2 gases; therefore, there is very little or no chance of catalyst deactivation due to poisoning by the gas feed at the inlet to the reactor. It was also observed in our previous studies that the support (SiO_2 vs TiO_2) can enhance the stability of the catalyst to resist deactivation [38,40]. Iglesia et al., noticed that silica supported materials are less stable when compared to the other supports like Al_2O_3 [100] in their FT studies with Co-based catalyst. They also reported that CoRu-TiO_2 and CoRu-SiO_2 were found to have almost the same activation energy upon the addition of Ru to the Co catalyst; however, there were strong differences in the deactivation rates of catalysts depending upon the support [101]. Based on our CO- conversion studies, the ability of catalysts to withstand or retard the FT deactivation rate was in the order of CoRu-KIT-6 > CoRu-MCM-41 > CoRu-SBA-15.

3. Materials and Methods

3.1. Materials

The reagents used for catalysis synthesis were of analytical grade with no further purification. Tetramethyl orthosilicate, 99% (TMOS) and ammonium hydroxide, Tetraethyl orthosilicate reagent grade, 98% (TEOS), Pluronic acid (P123), Hydrochloric acid (HCl), Cetyltrimethylammoniumbromide (CTAB), $Co(NO_3)_2·6H_2O$, $RuCl_3·xH_2O$ were purchased from Sigma Aldrich. Ethanol (anhydrous), Butanol and acetone, ACS grade, were obtained from Fischer Scientific, Branchburg, New Jersey, USA.

3.2. Fabrication of Microreactor

The microchannel microreactor and the respective cover channel were fabricated using 3D printing technology. Typically, the microreactor and its cover channel are designed using AutoCAD software which is schematically shown in Figure 13. The design is based on the split and recombination principle which has 11 microchannels of 500 μm × 500 μm × 2.4 cm as reaction zone in between them. This stainless-steel 3D printed microreactor is assembled in a custom-built heating block with an inlet and outlet system which facilitates the flow of syngas through the channels.

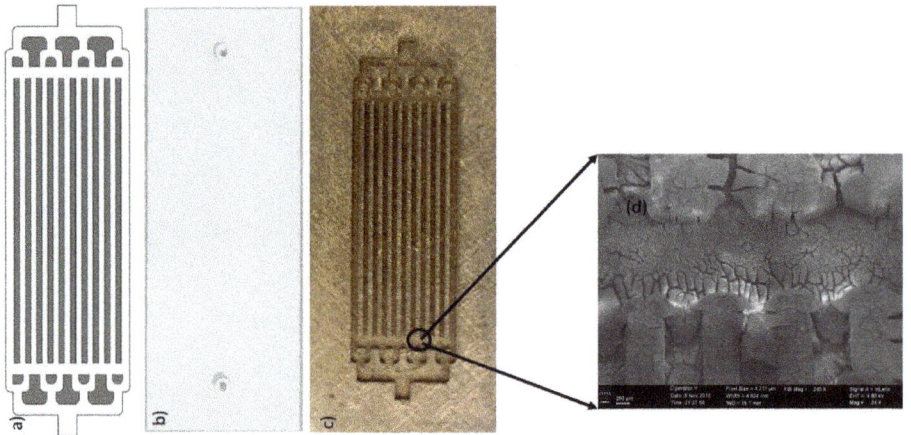

Figure 13. (**a**) and (**b**): AutoCAD design of the microreactor and cover channel (**c**) 3D printed reactor, (**d**) SEM image of the microchannels coated with catalyst prior to FT studies.

3.3. Catalyst Synthesis and Loading

Three types of Co-Ru based nanocatalysts supported by different mesoporous silica supports i.e., MCM-41, SBA-15 and KIT-6 supports are used in this study. A constant metal loading of 10%Co and ~5% Ru in weight was maintained in all preparations and this metal loading was also determined using the amount of the precursor. Three catalysts using different mesoporous support—10%Co5%Ru-MCM-41, 10%Co5%Ru-SBA-15 and 10%Co5%Ru-KIT-6—were synthesized using the one-pot hydrothermal procedure (as shown below) [48]. The catalysts were labeled as CoRu-MCM-41, CoRu-SBA-15, and CoRu-KIT-6 in this manuscript.

For the synthesis of CoRu-MCM-41, TMOS, CTAB, DI-water, and ethanol were used in a molar ratio of 1:0.13:130:20 as described elsewhere [48]. In short, CTAB was dissolved in DI-water at 30 °C to produce a clear solution. The metal precursors were dissolved in ethanol in a separate beaker. The precursor, TMOS, which is a limiting agent for this synthesis, was added dropwise to the mixture of the two solutions prepared previously. Ammonium hydroxide was added dropwise to precipitate metal hydroxides in the solution, till the final pH was ~10. The precipitate was stirred for 3 h, followed by 18 h of aging at 65 °C. The precipitate was then washed with DI water till the filtrate reached a pH of 7, finally washed with ethanol and filtered. The filtered material is air-dried for a day and then oven-dried at 110 °C for 24 h. The dried catalyst is calcined at 550 °C for 16 h with a ramp rate of 2 °C/min to remove the CTAB template.

For the synthesis of CoRu-SBA-15: TEOS, CTAB, water, ethanol, pluronic acid, and hydrochloric acid were mixed in molar ratios of 1:0.081:41:7.5:0.0168:5.981. In a typical synthesis procedure, P123 was dissolved in 2M HCl at 35 °C till a clear solution was obtained. Another solution was prepared by dissolving CTAB in DI water at 35 °C until a homogenous mixture was produced. These two solutions were mixed and stirred for 35 min. Ethanol containing metal precursors were added dropwise into the solution and stirred for 30 min. Afterwards, TEOS which was limiting reagent in this procedure was also added dropwise and stirred for 20 h at 35 °C. The aqueous solution was aged for 48 h at 98 °C followed by air drying for 24 h. The material was then oven-dried at 110 °C for 24 h. Finally, the dried material is then calcined in stepwise fashion with heating rate of 1 °C/min at 350 °C, 450 °C, and 550 °C for 8 h each, respectively to remove CTAB and pluronic acid.

In the case of CoRu-KIT-6, TEOS, P123, HCl, DI water, and butanol were mixed in a molar ratio of 1:0.017:1.83:195:1.31 [102]. For a typical procedure, P123 was added to HCl at 35 °C till a clear solution was obtained. A separate solution was prepared with butanol containing metal precursors and poured to the previous solution and stirred until a homogeneous solution was obtained. To this mixture, TEOS, which was the limiting reagent, was added dropwise and stirred at 500 rpm for 24 h. The final solution was aged for 24 h at 100 °C, followed by air drying for 24 h under the fume hood. The material is oven-dried at 110 °C for 24 h and then calcined at 550 °C for 4 h, to remove P123, the structure directing agent, SDA, with heating and cooling rates of 1 °C/min.

The catalyst is loaded into the microchannels of the microreactor using a PVA suspension containing the catalyst, DI water, binder PVA (polyvinyl alcohol 98%–99% hydrolyzed MW: 31000) and acetic acid of weight ratio 1:5:0.25:0.05. This suspension with well-dispersed catalyst was dip-coated and dried in air and then calcined in presence of air at 400 °C for 2 h with heating and cooling rates of 5 °C/min. Figure 13d shows the SEM image of the catalyst coated microreactor prior to the in-situ reduction.

3.4. Catalyst Characterization

Specific surface area, pore size, pore volume and TPR studies of the catalyst were carried out using Micromeritics, 3-Flex instrument. The Brunner–Emmett–Teller (BET) method was used to calculate the surface area of the catalyst where an equation was obtained from adsorption isotherm in the relative pressure range of 0.07–0.03. The surface area was calculated from adsorption isotherm in the relative pressure range of $P/P_0 = 0.07$–0.3 using the Brunner–Emmett–Teller (BET) equation. The total volume per gram of catalyst was determined from the amount of N_2 adsorbed at $P/P_0 = 1$. The N_2 desorption from the catalyst surface provides information about the pore size distribution using BJH (Barret–Joyner–Halenda) plots [103]. The H_2 temperature programmed reduction (TPR) analysis was also done with the same instrument which has a TCD detector to monitor the reduction signals of the catalyst. Around 50 mg of the catalyst was loaded into the quartz sample tube in which a stream of 10% H_2/Ar at flowrate 110 mL/min was passed through and the temperature is increased to 1000 °C with 10 °C/min ramp rate. The small and wide-angle powder x-ray diffraction (XRD) were recorded using D8 Discover X-ray and Rigaku SmartLab X-ray diffractometers, respectively, with Cu K-alpha radiation (wavelength = 0.15418 nm) radiation generated at 40 mA and 40 kV. The step size and time per step used in these measurements are 0.05° and 3 secs/step, respectively. The crystal sizes of the metal oxides were determined using the Scherrer equation. In the Scherrer equation below, τ stands for the crystal size, λ is the wavelength of the Cu Kα radiation, β is the full width half maximum and θ is the Bragg diffraction angle.

$$\tau = \frac{0.9\lambda}{\beta * \cos\theta} \tag{7}$$

The morphology and the size of the catalysts were analyzed using transmission electron (TEM Carl Zeiss Libra 120) at 120 KeV and scanning electron microscopy (Zeiss Auriga FIB/FESEM). The sample for TEM was prepared by dispersing a small quantity of catalyst in 3 mL of ethanol followed by vortex dispersion and sonication for a few minutes. Then the suspension was drop coated on a carbon-coated copper grid of 300 μm mesh size, followed by drying in an oven at 100 °C for 12 h.

The elemental composition and oxidation states of the metals were analyzed using Energy Dispersive X-ray spectrometry (Zeiss Auriga FIB/FESEM obtained from Carl Zeiss, Oberkochen, Germany) and oxidation states by X-ray photoelectron spectroscopy (XPS-Escalab Xi+-Thermo Scientific obtained from Thermo Scientific, West Sussex, UK), respectively.

3.5. Fischer–Tropsch Synthesis in Microreactor and Kinetic Data Collection

An in-house LabVIEW automated experimental setup was built to carry out the FT experiments for precise control over the operating conditions. The experimental setup is shown in Figure 14. The flowrates of the syngas mixture which is a mixture of hydrogen and carbon monoxide were controlled by precalibrated mass flow controllers obtained from cole parmer with flow rates ranging from 0–1 sccm. Nitrogen was used as a carrier gas into the system and was controlled by Aalborg mass flow controller with a maximum flow rate of 10 sccm. The upstream and downstream pressures were continuously monitored by pressure gauges obtained from Cole–Parmer and the data are fed to Aalborg solenoid valve from which the reaction pressure is controlled and kept constant throughout the reaction. All these controllers are operated by LabVIEW 2018 program. The product stream is directly fed to the GC-MS (Agilent Technologies 7890B GC and 5977 MSD). Prior to the start of the kinetic experiments, the microreactors were reduced ex-situ in Carbolite Gero tubular furnace with 10% H_2Ar. To compensate the losses while transferring the microreactor to the heating block the microreactor containing the catalyst was reduced again in-situ for 6 h at 350 °C before the start of FT reaction. The kinetic studies were performed by varying the weight hourly space velocity (WHSV = Wcat/$F_{CO,in}$, where W_{Cat} = weight of the catalyst and $F_{CO,in}$ = molar flow rate of CO in feed) in the range of ~25.2–3.15 kgcat.h/kmol. The reactions were performed with syngas having a feed molar ratio (H_2/CO) of 2:1 at 210 °C, 240 °C, and 270 °C while the reaction pressure was maintained at 1 atm. Based on our previous FT studies using this setup and preliminary runs, all reactions reached a steady state after an hour at each setpoint of WHSV. Deactivation studies were also performed for all three catalysts at 240 °C using syngas feed molar ration of 2:1. CO conversion was calculated based on following equation:

$$X_{CO}\% = \frac{F_{CO,in} - F_{CO,out}}{F_{CO,in}} \times 100 \qquad (8)$$

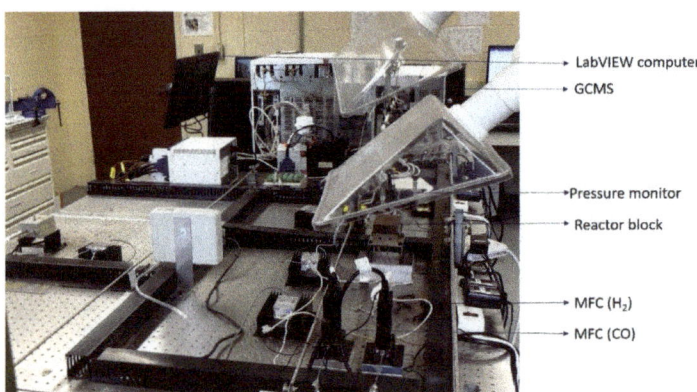

Figure 14. Experimental set-up for FT synthesis in a 3-D printed stainless steel microreactor.

4. Conclusions

Three different types of mesoporous silica supported Co-Ru based catalysts were synthesized using the one-pot hydrothermal method and performance for FT was evaluated. These catalysts resulted in high surface area with hexagonal ordered mesoporous structure as supported by BET, low angle XRD and TEM studies. The interaction between the metal and different types of support has a significant effect on the kinetic and stability studies of FT synthesis. Six mechanistic models were developed based on the Langmuir–Hinshelwood and Eley-Radiel mechanisms. The best fitted model for all catalysts was obtained on the basis of non-linear regression by comparing with the objective function which is equation-7 in this paper. The proposed model FT-3 was best fitted with the kinetic data for CoRu-MCM-41 catalyst. Whereas, FT-2 and FT-6 were well fitted with the kinetic data for CoRu-SBA-15 and CoRu-KIT-6, respectively. CoRu-KIT-6 was found to be more active than other catalysts with a low activation energy of 12.59 kJ/mol, whereas for CoRu-SBA-15 and CoRu-MCM-41 the activation energies are 13.39 and 32.21 kJ/mol, respectively. An average error of 5.65%, 1.76%, and 3.70% was obtained for catalysts CoRu-MCM-41, CoRu-SBA-15, CoRu-KIT-6, respectively considering the best fitted model explained above for FT synthesis. The predicted data provided by kinetic models were satisfactory with the experimental data. These results highlight the potential of the mechanistic FT models as well as reaction mechanisms to further improve the performance of FT synthesis. In addition, this information can help to design more active and selective catalysts for the optimized FT process. Furthermore, all catalysts exhibited significant resistance to the deactivation rate following the order CoRu-KIT-6> CoRu-MCM-41>CoRu-SBA-15. This study suggests that even if the support is of same type, the structure of the support plays a vital role in catalyst performance for FT synthesis.

Supplementary Materials: The following are available online at http://www.mdpi.com/2073-4344/9/10/872/s1, Figure S1: title, Table S1: title.

Author Contributions: Methodology, analysis and experimental investigation by N.M. and S.B.; project administration and supervision by S.A., and D.K.; all the authors contributed to the discussion of the experimental results as well as writing and editing of the manuscript.

Funding: This project received funding from NSF CREST (#260326) and UNC-ROI (#110092). This work is performed at North Carolina A&T State University and Joint School of Nanoscience and Nanoengineering, a member of the Southeastern Nanotechnology Infrastructure Corridor (SENIC) and National Nanotechnology Coordinated Infrastructure (NNCI), which is supported by the National Science Foundation (Grant ECCS-1542174).

Acknowledgments: The authors gratefully acknowledge Dr.Kyle Nowlin and Mr. Klinton Davis for TEM support, and Dr. Xin Li for XRD support.

Conflicts of Interest: The authors declare no conflict of interest

Appendix A

Derivation of Kinetic model:
FT-3 kinetic model
Dual-site adsorption of CO and H$_2$, the surface reaction is rate controlling step (RCS)

$$CO + * \underset{k_{-1}}{\overset{k_1}{\rightleftharpoons}} CO* \quad K_1 = \frac{k_1}{k_{-1}} \text{(Adsorption 1)} \tag{A1}$$

$$H_2 + * \underset{k_{-2}}{\overset{k_2}{\rightleftharpoons}} H_2* \quad K_2 = \frac{k_2}{k_{-2}} \text{(Adsorption 2)} \tag{A2}$$

$$CO* + H_2* \overset{k}{\rightarrow} C + D + 2* \text{ (Surface reaction) (RCS)} \tag{A3}$$

From step 1 (Adsorption 1 (rapid reaction)),
The rate of formation of CO* is,

$$r_{CO*} = k_1 p_{CO} C_* - k_{-1} C_{CO*} \tag{A4}$$

From step 2 (Adsorption 2 (rapid reaction [89])),
The rate of formation of H_2^* is,

$$r_{H_2*} = k_2 p_{H_2} C_* - k_{-2} C_{H_2*} \tag{A5}$$

From step 3 (Surface reaction 3),

$$-r_{CO} = k C_{CO*} C_{H_2*} \tag{A6}$$

According to pseudo steady-state hypothesis (PSSH), the rate of formation of the intermediate is zero.
So,

$$r_{CO*} = 0 \tag{A7}$$

$$r_{H_2*} = 0 \tag{A8}$$

So, putting the value of r_{CO*} from Equation (A4) in Equation (A7), we have,

$$\begin{aligned}
& k_1 p_{CO} C_* - k_{-1} C_{CO*} = 0 \\
& \Rightarrow k_1 p_{CO} C_* = k_{-1} C_{CO*} \\
& \Rightarrow C_{CO*} = \frac{k_1 p_{CO} C_*}{k_{-1}} \\
& \Rightarrow C_{CO*} = K_1 p_{CO} C_*
\end{aligned} \tag{A9}$$

So, putting the value of r_{H_2*} from Equation (A5) in Equation (A8), we have,

$$\begin{aligned}
& k_2 p_{H_2} C_* - k_{-2} C_{H_2*} = 0 \\
& \Rightarrow k_2 p_{H_2} C_* = k_{-2} C_{H_2*} \\
& \Rightarrow C_{H_2*} = \frac{k_2 p_{H_2} C_*}{k_{-2}} \\
& \Rightarrow C_{H_2*} = K_2 p_{H_2} C_*
\end{aligned} \tag{A10}$$

Taking the values of C_{CO*} and C_{H_2*} from Equations (A9) and (A10) and putting in Equation (A6), we have,

$$-r_{CO} = k K_1 K_2 p_{CO} p_{H_2} C_*^2 \tag{A11}$$

Making the catalyst active site balance.
Considering, the total site is,

$$\begin{aligned}
& C_T = 1 \\
& \Rightarrow (No.\, of\, vacant\, sites) + (No.\, of\, occupied\, sites) = 1 \\
& \Rightarrow (C_*) + (C_{CO*} + C_{H_2*}) = 1 \\
& \Rightarrow (C_*) + (K_1 p_{CO} C_* + K_2 p_{H_2} C_*) = 1 \\
& \Rightarrow C_*(1 + K_1 p_{CO} + K_2 p_{H_2}) = 1 \\
& \Rightarrow C_* = \frac{1}{(1 + K_1 p_{CO} + K_2 p_{H_2})}
\end{aligned} \tag{A12}$$

Putting the value of C_* from Equation (A12) in Equation (A11), we get,

$$\Rightarrow -r_{CO} = \frac{k K_1 K_2 p_{CO} p_{H_2}}{(1 + K_1 p_{CO} + K_2 p_{H_2})^2} \text{(FT - 3)}$$

[Taking values of C_{CO*} and C_{H_2*} from Equations (A9) and (A10)]

References

1. Fischer, F.; Tropsch, H. The synthesis of petroleum at atmospheric pressures from gasification products of coal. *Brennstoff-Chemie* **1926**, *7*, 97–104.
2. Steynberg, A.P. Chapter 1—Introduction to Fischer-Tropsch Technology. In *Studies in Surface Science and Catalysis*; Steynberg, A., Dry, M., Eds.; Elsevier: Amsterdam, The Netherlands, 2004; Volume 152, pp. 1–63.
3. Davis, B.H. Overview of reactors for liquid phase Fischer—Tropsch synthesis. *Catal. Today* **2002**, *71*, 249–300. [CrossRef]
4. Wood, D.A.; Nwaoha, C.; Towler, B.F. Gas-to-liquids (GTL): A review of an industry offering several routes for monetizing natural gas. *J. Nat. Gas Sci. Eng.* **2012**, *9*, 196–208. [CrossRef]
5. Pohar, A. Process Intensification through Microreactor Application. *Chem. Biochem. Eng. Q.* **2009**, *23*, 537–544.
6. Lerou, J.J.; Tonkovich, A.L.; Silva, L.; Perry, S.; McDaniel, J. Microchannel reactor architecture enables greener processes. *Chem. Eng. Sci.* **2010**, *65*, 380–385. [CrossRef]
7. Hessel, V.; Löb, P.; Holger, L. *Industrial and Real-Life Applications of Micro-Reactor Process Engineering for Fine and Functional Chemistry*; Elsevier: Amsterdam, The Netherlands, 2006; Volume 159, pp. 35–46.
8. Ouyang, X.; Besser, R.S. Development of a microreactor-based parallel catalyst analysis system for synthesis gas conversion. *Catal. Today* **2003**, *84*, 33–41. [CrossRef]
9. Gavriilidis, A.; Angeli, P.; Cao, E.; Yeong, K.K.; Wan, Y.S.S. Technology and Applications of Microengineered Reactors. *Chem. Eng. Res. Design* **2002**, *80*, 3–30. [CrossRef]
10. Myrstad, R.; Eri, S.; Pfeifer, P.; Rytter, E.; Holmen, A. Fischer-Tropsch synthesis in a microstructured reactor. *Catal. Today* **2009**, *147*, S301–S304. [CrossRef]
11. LeViness, S.; Deshmukh, S.R.; Richard, L.A.; Robota, H.J. Velocys Fischer—Tropsch synthesis technology—New advances on state-of-the-art. *Top. Catal.* **2014**, *57*, 518–525. [CrossRef]
12. Mazanec, T.; Perry, S.; Tonkovich, L.; Wang, Y. Microchannel gas-to-liquids conversion-thinking big by thinking small. In *Studies in Surface Science and Catalysis*; Elsevier: Amsterdam, The Netherlands, 2004; Volume 147, pp. 169–174.
13. Deshmukh, S.R.; Tonkovich, A.L.Y.; Jarosch, K.T.; Schrader, L.; Fitzgerald, S.P.; Kilanowski, D.R.; Lerou, J.J.; Mazanec, T.J. Scale-up of microchannel reactors for Fischer—Tropsch synthesis. *Ind. Eng. Chem. Res.* **2010**, *49*, 10883–10888. [CrossRef]
14. Almeida, L.C.; Sanz, O.; D'olhaberriague, J.; Yunes, S.; Montes, M. Microchannel reactor for Fischer—Tropsch synthesis: Adaptation of a commercial unit for testing microchannel blocks. *Fuel* **2013**, *110*, 171–177. [CrossRef]
15. Schulz, H. Short history and present trends of Fischer—Tropsch synthesis. *Appl. Catal. A Gen.* **1999**, *186*, 3–12. [CrossRef]
16. Borg, Ø.; Eri, S.; Blekkan, E.A.; Storsæter, S.; Wigum, H.; Rytter, E.; Holmen, A. Fischer—Tropsch synthesis over γ-alumina-supported cobalt catalysts: Effect of support variables. *J. Catal.* **2007**, *248*, 89–100. [CrossRef]
17. Chu, W.; Chernavskii, P.A.; Gengembre, L.; Pankina, G.A.; Fongarland, P.; Khodakov, A.Y. Cobalt species in promoted cobalt alumina-supported Fischer—Tropsch catalysts. *J. Catal.* **2007**, *252*, 215–230. [CrossRef]
18. Hilmen, A.; Schanke, D.; Hanssen, K.; Holmen, A. Study of the effect of water on alumina supported cobalt Fischer—Tropsch catalysts. *Appl. Catal. A Gen.* **1999**, *186*, 169–188. [CrossRef]
19. Storsæter, S.; Tøtdal, B.; Walmsley, J.C.; Tanem, B.S.; Holmen, A. Characterization of alumina-, silica-, and titania-supported cobalt Fischer—Tropsch catalysts. *J. Catal.* **2005**, *236*, 139–152. [CrossRef]
20. Hilmen, A.; Schanke, D.; Holmen, A. TPR study of the mechanism of rhenium promotion of alumina-supported cobalt Fischer-Tropsch catalysts. *Catal. Lett.* **1996**, *38*, 143–147. [CrossRef]
21. Morales, F.; De Groot, F.M.; Gijzeman, O.L.; Mens, A.; Stephan, O.; Weckhuysen, B.M. Mn promotion effects in Co/TiO_2 Fischer—Tropsch catalysts as investigated by XPS and STEM-EELS. *J. Catal.* **2005**, *230*, 301–308. [CrossRef]
22. Iglesia, E.; Soled, S.L.; Fiato, R.A. Fischer-Tropsch synthesis on cobalt and ruthenium. Metal dispersion and support effects on reaction rate and selectivity. *J. Catal.* **1992**, *137*, 212–224. [CrossRef]
23. Li, J.; Coville, N.J. The effect of boron on the catalyst reducibility and activity of Co/TiO_2 Fischer—Tropsch catalysts. *Appl. Catal. A Gen.* **1999**, *181*, 201–208. [CrossRef]

24. Feltes, T.E.; Espinosa-Alonso, L.; De Smit, E.; D'Souza, L.; Meyer, R.J.; Weckhuysen, B.M.; Regalbuto, J.R. Selective adsorption of manganese onto cobalt for optimized Mn/Co/TiO$_2$ Fischer—Tropsch catalysts. *J. Catal.* **2010**, *270*, 95–102. [CrossRef]
25. Li, J.; Jacobs, G.; Zhang, Y.; Das, T.; Davis, B.H. Fischer—Tropsch synthesis: Effect of small amounts of boron, ruthenium and rhenium on Co/TiO$_2$ catalysts. *Appl. Catal. A Gen.* **2002**, *223*, 195–203. [CrossRef]
26. Arai, H.; Mitsuishi, K.; Seiyama, T. TiO$_2$-supported fe–co, co–ni, and ni–fe alloy catalysts for fischer-tropsch synthesis. *Chem. Lett.* **1984**, *13*, 1291–1294. [CrossRef]
27. Duvenhage, D.; Coville, N. Fe: Co/TiO2 bimetallic catalysts for the Fischer—Tropsch reaction: Part 2. The effect of calcination and reduction temperature. *Appl. Catal. A Gen.* **2002**, *233*, 63–75. [CrossRef]
28. Ying, X.; Zhang, L.; Xu, H.; Ren, Y.-L.; Luo, Q.; Zhu, H.-W.; Qu, H.; Xuan, J. Efficient Fischer—Tropsch microreactor with innovative aluminizing pretreatment on stainless steel substrate for Co/Al$_2$O$_3$ catalyst coating. *Fuel Process. Technol.* **2016**, *143*, 51–59. [CrossRef]
29. Ernst, B.; Libs, S.; Chaumette, P.; Kiennemann, A. Preparation and characterization of Fischer—Tropsch active Co/SiO$_2$ catalysts. *Appl. Catal. A Gen.* **1999**, *186*, 145–168. [CrossRef]
30. Li, J.; Jacobs, G.; Das, T.; Zhang, Y.; Davis, B. Fischer—Tropsch synthesis: Effect of water on the catalytic properties of a Co/SiO$_2$ catalyst. *Appl. Catal. A Gen.* **2002**, *236*, 67–76. [CrossRef]
31. Barbier, A.; Tuel, A.; Arcon, I.; Kodre, A.; Martin, G.A. Characterization and catalytic behavior of Co/SiO$_2$ catalysts: Influence of dispersion in the Fischer—Tropsch reaction. *J. Catal.* **2001**, *200*, 106–116. [CrossRef]
32. Yang, Y.; Xiang, H.-W.; Tian, L.; Wang, H.; Zhang, C.-H.; Tao, Z.-C.; Xu, Y.-Y.; Zhong, B.; Li, Y.-W. Structure and Fischer—Tropsch performance of iron—Manganese catalyst incorporated with SiO$_2$. *Appl. Catal. A Gen.* **2005**, *284*, 105–122. [CrossRef]
33. Moradi, G.; Basir, M.; Taeb, A.; Kiennemann, A. Promotion of Co/SiO$_2$ Fischer—Tropsch catalysts with zirconium. *Catal. Commun.* **2003**, *4*, 27–32. [CrossRef]
34. Chen, W.; Fan, Z.; Pan, X.; Bao, X. Effect of confinement in carbon nanotubes on the activity of Fischer—Tropsch iron catalyst. *J. Am. Chem. Soc.* **2008**, *130*, 9414–9419. [CrossRef] [PubMed]
35. Bahome, M.C.; Jewell, L.L.; Hildebrandt, D.; Glasser, D.; Coville, N.J. Fischer—Tropsch synthesis over iron catalysts supported on carbon nanotubes. *Appl. Catal. A Gen.* **2005**, *287*, 60–67. [CrossRef]
36. Abbaslou, R.M.M.; Tavassoli, A.; Soltan, J.; Dalai, A.K. Iron catalysts supported on carbon nanotubes for Fischer—Tropsch synthesis: Effect of catalytic site position. *Appl. Catal. A Gen.* **2009**, *367*, 47–52. [CrossRef]
37. Nagineni, V.S.; Zhao, S.; Potluri, A.; Liang, Y.; Siriwardane, U.; Seetala, N.V.; Fang, J.; Palmer, J.; Kuila, D. Microreactors for Syngas Conversion to Higher Alkanes: Characterization of Sol–Gel-Encapsulated Nanoscale Fe–Co Catalysts in the Microchannels. *Ind. Eng. Chem. Res.* **2005**, *44*, 5602–5607. [CrossRef]
38. Mehta, S.; Deshmane, V.; Zhao, S.; Kuila, D. Comparative Studies of Silica-Encapsulated Iron, Cobalt, and Ruthenium Nanocatalysts for Fischer—Tropsch Synthesis in Silicon-Microchannel Microreactors. *Ind. Eng. Chem. Res.* **2014**, *53*, 16245–16253. [CrossRef]
39. Zhao, S.; Nagineni, V.S.; Seetala, N.V.; Kuila, D. Microreactors for Syngas Conversion to Higher Alkanes: Effect of Ruthenium on Silica-Supported Iron–Cobalt Nanocatalysts. *Ind. Eng. Chem. Res.* **2008**, *47*, 1684–1688. [CrossRef]
40. Abrokwah, R.Y.; Rahman, M.M.; Deshmane, V.G.; Kuila, D. Effect of titania support on Fischer-Tropsch synthesis using cobalt, iron, and ruthenium catalysts in silicon-microchannel microreactor. *Mol. Catal.* **2019**, *478*, 110566. [CrossRef]
41. Gutmann, B.; Koeckinger, M.; Glotz, G.; Ciaglia, T.; Slama, E.; Zadravec, M.; Pfanner, S.; Gruber-Woelfler, H.; Maier, M.; Kappe, C.O. Design and 3D Printing of a Stainless Steel Reactor for Continuous Difluoromethylations Using Fluoroform. *React. Chem. Eng.* **2017**, *2*. [CrossRef]
42. Monaghan, T.; Harding, M.J.; Harris, R.A.; Friel, R.J.; Christie, S.D. Customisable 3D printed microfluidics for integrated analysis and optimisation. *Lab Chip* **2016**, *16*, 3362–3373. [CrossRef]
43. Scotti, G.; Matilainen, V.; Kanninen, P.; Piili, H.; Salminen, A.; Kallio, T.; Franssila, S. Laser additive manufacturing of stainless steel micro fuel cells. *J. Power Sources* **2014**, *272*, 356–361. [CrossRef]
44. Scotti, G.; Kanninen, P.; Matilainen, V.-P.; Salminen, A.; Kallio, T. Stainless steel micro fuel cells with enclosed channels by laser additive manufacturing. *Energy* **2016**, *106*, 475–481. [CrossRef]
45. Capel, A.J.; Edmondson, S.; Christie, S.D.; Goodridge, R.D.; Bibb, R.J.; Thurstans, M. Design and additive manufacture for flow chemistry. *Lab Chip* **2013**, *13*, 4583–4590. [CrossRef] [PubMed]

46. Amin, R.; Knowlton, S.; Hart, A.; Yenilmez, B.; Ghaderinezhad, F.; Katebifar, S.; Messina, M.; Khademhosseini, A.; Tasoglu, S. 3D-printed microfluidic devices. *Biofabrication* **2016**, *8*, 022001. [CrossRef] [PubMed]
47. Reintjens, R.; Ager, D.J.; De Vries, A.H. Flow chemistry, how to bring it to industrial scale? *Chim. Oggi* **2015**, *33*, 21–24.
48. Abrokwah, R.Y.; Deshmane, V.G.; Kuila, D. Comparative performance of M-MCM-41 (M: Cu, Co, Ni, Pd, Zn and Sn) catalysts for steam reforming of methanol. *J. Mol. Catal. A Chem.* **2016**, *425*, 10–20. [CrossRef]
49. Deshmane, V.G.; Abrokwah, R.Y.; Kuila, D. Synthesis of stable Cu-MCM-41 nanocatalysts for H_2 production with high selectivity via steam reforming of methanol. *Int. J. Hydrogen Energy* **2015**, *40*, 10439–10452. [CrossRef]
50. Abrokwah, R.Y.; Deshmane, V.G.; Owen, S.L.; Kuila, D. Cu-Ni Nanocatalysts in Mesoporous MCM-41 and TiO_2 to Produce Hydrogen for Fuel Cells via Steam Reforming Reactions. *Adv. Mater. Res.* **2015**, *1096*, 161–168. [CrossRef]
51. Tatineni, B.; Basova, Y.; Rahman, A.; Islam, S.; Rahman, M.; Islam, A.; Perkins, J.; King, J.; Taylor, J.; Kumar, D.; et al. Development of Mesoporous Silica Encapsulated Pd-Ni Nanocatalyst for Hydrogen Production. In *Production and Purification of Ultraclean Transportation Fuels*; American Chemical Society: Washington, DC, USA, 2011; Volume 1088, pp. 177–190.
52. Taghizadeh, M.; Akhoundzadeh, H.; Rezayan, A.; Sadeghian, M. Excellent catalytic performance of 3D-mesoporous KIT-6 supported Cu and Ce nanoparticles in methanol steam reforming. *Int. J. Hydrogen Energy* **2018**, *43*, 10926–10937. [CrossRef]
53. Lanzafame, P.; Perathoner, S.; Centi, G.; Frusteri, F. Synthesis and characterization of Co-containing SBA-15 catalysts. *J. Porous Mater.* **2007**, *14*, 305–313. [CrossRef]
54. Sing, K.S. Reporting physisorption data for gas/solid systems with special reference to the determination of surface area and porosity (Recommendations 1984). *Pure Appl. Chem.* **1985**, *57*, 603–619. [CrossRef]
55. Biz, S.; Occelli, M.L. Synthesis and characterization of mesostructured materials. *Catal. Rev.* **1998**, *40*, 329–407. [CrossRef]
56. Zhao, D.; Feng, J.; Huo, Q.; Melosh, N.; Fredrickson, G.H.; Chmelka, B.F.; Stucky, G.D. Triblock copolymer syntheses of mesoporous silica with periodic 50 to 300 angstrom pores. *Science* **1998**, *279*, 548–552. [CrossRef] [PubMed]
57. Karthikeyan, G.; Pandurangan, A. Post synthesis alumination of KIT-6 materials with Ia3d symmetry and their catalytic efficiency towards multicomponent synthesis of 1H-pyrazolo[1,2-]phthalazine-5,10-dione carbonitriles and carboxylates. *J. Mol. Catal. A Chem.* **2012**, *361*, 58–67. [CrossRef]
58. Schwertmann, U.; Cambier, P.; Murad, E. Properties of Goethites of Varying Crystallinity. *Clays Clay Miner.* **1985**, *33*, 369–378. [CrossRef]
59. Li, G.; Zhang, C.; Wang, Z.; Huang, H.; Peng, H.; Li, X. Fabrication of mesoporous Co_3O_4 oxides by acid treatment and their catalytic performances for toluene oxidation. *Appl. Catal. A Gen.* **2018**, *550*, 67–76. [CrossRef]
60. Liu, X.; He, J.; Yang, L.; Wang, Y.; Zhang, S.; Wang, W.; Wang, J. Liquid-phase oxidation of cyclohexane to cyclohexanone over cobalt-doped SBA-3. *Catal. Commun.* **2010**, *11*, 710–714. [CrossRef]
61. Li, Z.; Miao, Z.; Wang, X.; Zhao, J.; Zhou, J.; Si, W.; Zhuo, S. One-pot synthesis of ZrMo-KIT-6 solid acid catalyst for solvent-free conversion of glycerol to solketal. *Fuel* **2018**, *233*, 377–387. [CrossRef]
62. Hess, C.; Tzolova-Müller, G.; Herbert, R. The Influence of Water on the Dispersion of Vanadia Supported on Silica SBA-15: A Combined XPS and Raman Study. *J. Phys. Chem. C* **2007**, *111*, 9471–9479. [CrossRef]
63. Bhoware, S.S.; Singh, A.P. Characterization and catalytic activity of cobalt containing MCM-41 prepared by direct hydrothermal, grafting and immobilization methods. *J. Mol. Catal. A Chem.* **2007**, *266*, 118–130. [CrossRef]
64. Yao, Q.; Lu, Z.-H.; Yang, K.; Chen, X.; Zhu, M. Ruthenium nanoparticles confined in SBA-15 as highly efficient catalyst for hydrolytic dehydrogenation of ammonia borane and hydrazine borane. *Sci. Rep.* **2015**, *5*, 15186. [CrossRef]
65. Panpranot, J.; Goodwin, J.G., Jr.; Sayari, A. Synthesis and characterics of MCM-41 supported CoRu catalysts. *Catal. Today* **2002**, *77*, 269–284. [CrossRef]

66. Martínez, A.; López, C.; Márquez, F.; Díaz, I. Fischer-Tropsch synthesis of hydrocarbons over mesoporous Co/SBA-15 catalysts: The influence of metal loading, cobalt precursor, and promoters. *J. Catal.* **2003**, *220*, 486–499. [CrossRef]
67. Lim, S.; Ciuparu, D.; Yang, Y.; Du, G.; Pfefferle, L.D.; Haller, G.L. Improved synthesis of highly ordered Co-MCM-41. *Microporous Mesoporous Mater.* **2007**, *101*, 200–206. [CrossRef]
68. Jiang, T.; Shen, W.; Zhao, Q.; Li, M.; Chu, J.; Yin, H. Characterization of CoMCM-41 mesoporous molecular sieves obtained by the microwave irradiation method. *J. Solid State Chem.* **2008**, *181*, 2298–2305. [CrossRef]
69. Haddad, G.J.; Goodwin, J.G., Jr. The impact of aqueous impregnation on the properties of prereduced vs. precalcined Co/SiO$_2$. *J. Catal.* **1995**, *157*, 25–34. [CrossRef]
70. Cai, Q.; Li, J. Catalytic properties of the Ru promoted Co/SBA-15 catalysts for Fischer—Tropsch synthesis. *Catal. Commun.* **2008**, *9*, 2003–2006. [CrossRef]
71. Van Der Laan, G.P.; Beenackers, A. Kinetics and selectivity of the Fischer—Tropsch synthesis: A literature review. *Catal. Rev.* **1999**, *41*, 255–318. [CrossRef]
72. Yates, I.C.; Satterfield, C.N. Intrinsic kinetics of the Fischer-Tropsch synthesis on a cobalt cvatalyst. *Energy Fuels* **1991**, *5*, 168–173. [CrossRef]
73. Iglesia, E.; Reyes, S.C.; Soled, S.L. Reaction-transport selectivity models and the design of Fischer-Tropsch catalysts. *Chem. Ind. N. Y. Marcel Dekker* **1993**, *51*, 199–257.
74. Zennaro, R.; Tagliabue, M.; Bartholomew, C.H. Kinetics of Fischer—Tropsch synthesis on titania-supported cobalt. *Catal. Today* **2000**, *58*, 309–319. [CrossRef]
75. Das, T.K.; Conner, W.A.; Li, J.; Jacobs, G.; Dry, M.E.; Davis, B.H. Fischer–Tropsch synthesis: Kinetics and effect of water for a Co/SiO$_2$ catalyst. *Energy Fuels* **2005**, *19*, 1430–1439. [CrossRef]
76. Almeida, L.; Sanz, O.; Merino, D.; Arzamendi, G.; Gandía, L.; Montes, M. Kinetic analysis and microstructured reactors modeling for the Fischer—Tropsch synthesis over a Co–Re/Al$_2$O$_3$ catalyst. *Catal. Today* **2013**, *215*, 103–111. [CrossRef]
77. Yang, C.-H.; Massoth, F.; Oblad, A. Kinetics of CO+ H$_2$ reaction over Co–Cu–Al$_2$O$_3$ catalyst. Hydrocarbon synthesis from carbon monoxide and hydrogen. *J. Am. Chem. Soc.* **1979**, *178*, 35–46.
78. Pannell, R.B.; Kibby, C.L.; Kobylinski, T.P. A Steady-State Study of Fischer-Tropsch Product Distributions Over Cobalt, Iron and Ruthenium. In *Studies in Surface Science and Catalysis*; Seivama, T., Tanabe, K., Eds.; Elsevier: Amsterdam, The Netherlands, 1981; Volume 7, pp. 447–459.
79. Outi, A.; Rautavuoma, I.; Van der Baan, H.S. Kinetics and mechanism of the fischer tropsch hydrocarbon synthesis on a cobalt on alumina catalyst. *Appl. Catal.* **1981**, *1*, 247–272. [CrossRef]
80. Sun, Y.; Jia, Z.; Yang, G.; Zhang, L.; Sun, Z. Fischer-Tropsch synthesis using iron based catalyst in a microchannel reactor: Performance evaluation and kinetic modeling. *Int. J. Hydrogen Energy* **2017**, *42*, 29222–29235. [CrossRef]
81. Mirzaei, A.A.; Kiai, R.M.; Atashi, H.; Arsalanfar, M.; Shahriari, S. Kinetic study of CO hydrogenation over co-precipitated iron–nickel catalyst. *J. Ind. Eng. Chem.* **2012**, *18*, 1242–1251. [CrossRef]
82. Sarup, B.; Wojciechowski, B.W. Studies of the fischer-tropsch synthesis on a cobalt catalyst II. Kinetics of carbon monoxide conversion to methane and to higher hydrocarbons. *Can. J. Chem. Eng.* **1989**, *67*, 62–74. [CrossRef]
83. Bepari, S.; Pradhan, N.C.; Dalai, A.K. Selective production of hydrogen by steam reforming of glycerol over Ni/Fly ash catalyst. *Catal. Today* **2017**, *291*, 36–46. [CrossRef]
84. Fogler, H.S. *Elements of Chemical Reaction Engineering*, 3rd ed.; Prentice Hall PTR: Upper Saddle River, NJ, USA, 1999.
85. Van Santen, R.A.; Ciobîcâ, I.M.; Van Steen, E.; Ghouri, M.M. Mechanistic Issues in Fischer-Tropsch Catalysis. In *Advances in Catalysis*; Academic Press: Cambridge, MA, USA, 2011; Volume 54, pp. 127–187.
86. Van Santen, R.A.; Ghouri, M.M.; Shetty, S.; Hensen, E.M.H. Structure sensitivity of the Fischer-Tropsch reaction; Molecular kinetics simulations. *Catal. Sci. Technol.* **2011**, *1*, 891–911. [CrossRef]
87. Markvoort, A.J.; Van Santen, R.A.; Hilbers, P.A.J.; Hensen, E.J.M. Kinetics of the Fischer-Tropsch reaction. *Angew. Chem.-Int. Ed.* **2012**, *51*, 9015–9019. [CrossRef]
88. Mansouri, M.; Atashi, H.; Mirzaei, A.A.; Jangi, R. Kinetics of the Fischer-Tropsch synthesis on silica-supported cobalt-cerium catalyst. *Int. J. Ind. Chem.* **2013**, *4*, 1. [CrossRef]
89. Pant, K.K.; Upadhyayula, S. Detailed kinetics of Fischer Tropsch synthesis over Fe-Co bimetallic catalyst considering chain length dependent olefin desorption. *Fuel* **2019**, *236*, 1263–1272. [CrossRef]

90. Mosayebi, A.; Abedini, R. Detailed kinetic study of Fischer—Tropsch synthesis for gasoline production over CoNi/HZSM-5 nano-structure catalyst. *Int. J. Hydrogen Energy* **2017**, *42*, 27013–27023. [CrossRef]
91. Abbasi, M.; Mirzaei, A.A.; Atashi, H. The mechanism and kinetics study of Fischer-Tropsch reaction over iron-nickel-cerium nano-structure catalyst. *Int. J. Hydrogen Energy* **2019**, *44*, 24667–24679. [CrossRef]
92. Sun, Y.; Wei, J.; Zhang, J.P.; Yang, G. Optimization using response surface methodology and kinetic study of Fischer—Tropsch synthesis using SiO_2 supported bimetallic Co–Ni catalyst. *J. Nat. Gas Sci. Eng.* **2016**, *28*, 173–183. [CrossRef]
93. Sun, Y.; Yang, G.; Zhang, L.; Sun, Z. Fischer-Tropsch synthesis in a microchannel reactor using mesoporous silica supported bimetallic Co-Ni catalyst: Process optimization and kinetic modeling. *Chem. Eng. Proc. Process Intensif.* **2017**, *119*, 44–61. [CrossRef]
94. Moazami, N.; Wyszynski, M.L.; Rahbar, K.; Tsolakis, A.; Mahmoudi, H. A comprehensive study of kinetics mechanism of Fischer-Tropsch synthesis over cobalt-based catalyst. *Chem. Eng. Sci.* **2017**, *171*, 32–60. [CrossRef]
95. Moazami, N.; Wyszynski, M.L.; Mahmoudi, H.; Tsolakis, A.; Zou, Z.; Panahifar, P.; Rahbar, K. Modelling of a fixed bed reactor for Fischer—Tropsch synthesis of simulated N_2-rich syngas over Co/SiO_2: Hydrocarbon production. *Fuel* **2015**, *154*, 140–151. [CrossRef]
96. Marchese, M.; Heikkinen, N.; Giglio, E.; Lanzini, A.; Lehtonen, J.; Reinikainen, M. Kinetic Study Based on the Carbide Mechanism of a Co-Pt/γ-Al_2O_3 Fischer—Tropsch Catalyst Tested in a Laboratory-Scale Tubular Reactor. *Catalysts* **2019**, *9*, 717. [CrossRef]
97. Bhatelia, T.; Li, C.e.; Sun, Y.; Hazewinkel, P.; Burke, N.; Sage, V. Chain length dependent olefin re-adsorption model for Fischer—Tropsch synthesis over Co-Al_2O_3 catalyst. *Fuel Process. Technol.* **2014**, *125*, 277–289. [CrossRef]
98. Todic, B.; Bhatelia, T.; Froment, G.F.; Ma, W.; Jacobs, G.; Davis, B.H.; Bukur, D.B. Kinetic Model of Fischer—Tropsch Synthesis in a Slurry Reactor on Co–Re/Al_2O_3 Catalyst. *Ind. Eng. Chem. Res.* **2013**, *52*, 669–679. [CrossRef]
99. Jahangiri, H.; Bennett, J.; Mahjoubi, P.; Wilson, K.; Gu, S. A review of advanced catalyst development for Fischer—Tropsch synthesis of hydrocarbons from biomass derived syn-gas. *Catal. Sci. Technol.* **2014**, *4*, 2210–2229. [CrossRef]
100. Iglesia, E. Design, synthesis, and use of cobalt-based Fischer-Tropsch synthesis catalysts. *Appl. Catal. A Gen.* **1997**, *161*, 59–78. [CrossRef]
101. Iglesia, E.; Soled, S.L.; Fiato, R.A.; Via, G.H. Bimetallic Synergy in Cobalt Ruthenium Fischer-Tropsch Synthesis Catalysts. *J. Catal.* **1993**, *143*, 345–368. [CrossRef]
102. Kleitz, F.; Choi, S.H.; Ryoo, R. Cubic Ia 3 d large mesoporous silica: Synthesis and replication to platinum nanowires, carbon nanorods and carbon nanotubes. *Chem. Commun.* **2003**, *17*, 2136–2137. [CrossRef]
103. Barrett, E.P.; Joyner, L.G.; Halenda, P.P. The Determination of Pore Volume and Area Distributions in Porous Substances. I. Computations from Nitrogen Isotherms. *J. Am. Chem. Soc.* **1951**, *73*, 373–380. [CrossRef]

© 2019 by the authors. Licensee MDPI, Basel, Switzerland. This article is an open access article distributed under the terms and conditions of the Creative Commons Attribution (CC BY) license (http://creativecommons.org/licenses/by/4.0/).

MDPI
St. Alban-Anlage 66
4052 Basel
Switzerland
Tel. +41 61 683 77 34
Fax +41 61 302 89 18
www.mdpi.com

Catalysts Editorial Office
E-mail: catalysts@mdpi.com
www.mdpi.com/journal/catalysts

www.ingramcontent.com/pod-product-compliance
Lightning Source LLC
LaVergne TN
LVHW071935080526
838202LV00064B/6612